T0399022

NEW STRATEGIES TARGETING CANCER METABOLISM

NEW STRATEGIES TARGETING CANCER METABOLISM

Anticancer Drugs, Synthetic Analogs, and Antitumor Agents

GALAL H. ELGEMEIE
Department of Chemistry, Helwan University, Cairo, Egypt

REHAM A. MOHAMED-EZZAT
Chemistry of Natural and Microbial Products Department, National Research Center, Cairo, Egypt

Elsevier
Radarweg 29, PO Box 211, 1000 AE Amsterdam, Netherlands
The Boulevard, Langford Lane, Kidlington, Oxford OX5 1GB, United Kingdom
50 Hampshire Street, 5th Floor, Cambridge, MA 02139, United States

For information on all Elsevier publications visit our website at https://www.elsevier.com/books-and-journals

ISBN: 978-0-12-821783-2

Publisher: Susan Dennis
Editorial Project Manager: Andrea R. Dulberger
Production Project Manager: Kumar Anbazhagan
Cover Designer: Greg Harris

Typeset by STRAIVE, India

Contents

CHAPTER 1

Medicinal chemistry of anticancer agents

1. Introduction

Pyrimidines, purines, nucleosides, and nucleotides are a biologically essential category of compounds in which several of its analogs constitute nucleic acid components. Pyrimidines and purines are azaheterocyclic bases. To the C-1 position of a sugar where the pyrimidine or a purine is attached, this corresponding structure is the nucleoside, whereas a heterocyclic base-sugar phosphoric acid unit is known as the nucleotide. A dinucleotide is composed of two nucleotide units joined via phosphate groups. These compounds are of main biochemical interest as the backbone of nucleic acid molecules (RNA, DNA) is a polynucleotide chain that is made up of various nucleotide units attached via phosphate groups. In other circumstances, specific names are utilized: cytidine is cytosine-β-D-riboside. Monophosphate, diphosphate, and triphosphate derivatives of nucleosides are nucleotides with one, two, or three phosphate groups. Many pyrimidine and purine derivatives, including fluoro-, mercapto-, and aza-substituents, are utilized in pharmaceuticals; thus, their metabolism has to be examined in living organisms [1].

Today, chemotherapeutic drugs target nearly every step of cell growth and division. The enzyme dihydrofolate reductase (DHFR) was the first enzyme to be targeted utilizing chemotherapy [2]. Dihydrofolate reductase (DHFR) inhibitors are an essential category of drugs, as manifested through their utility as anticancer, antimalarial, antibacterial, and antifungal agents. Progress in understanding the biochemical basis of mechanisms important for antiproliferative impacts and enzyme selectivity has revived the interest in antifolates for cancer chemotherapy and prompted the scientists to design and develop new and selective human DHFR inhibitors, accordingly resulting in developing novel generation of DHFR inhibitors [3–6].

In the production of thymine nucleotides, DHFR performs a supporting but crucial function [2]. Thymidylate synthase (TS) catalyzes the de novo synthesis of deoxythymidylate and is a rate-limiting enzyme of synthesizing DNA.

New Strategies Targeting Cancer Metabolism
https://doi.org/10.1016/B978-0-12-821783-2.00004-2

The essential site of action of the classic antifolate methotrexate is direct inhibition of DHFR, but it inhibits TS indirectly through reducing levels of the TS co-substrate 5,10-methylenetetrahydrofolate. Both the limitations and success of the developed drugs resulted in searching a novel and more efficacious TS inhibitor. There is newly discovered evidence that some new antifolate TS inhibitors are potent against a wide range of neoplasms, such as mesothelioma and lung carcinomas [7]. Thymidylate synthase inhibitors continue to reveal a promising potency in treating cancer [8–20].

The enzyme thymidylate synthase adds a methyl group to uracil to generate thymine. This may seem like a little difference, but the extra hydrophobic bulk of this methyl group is critical to distinguish thymine from the other bases via transcription factors, enhancers, repressors, and other DNA-binding proteins. Folate, the cofactor molecule, is an important vitamin for humans, which delivers the methyl group to thymidylate synthase. The folate molecule is oxidized as the methyl is transferred from folate to uracil during the process. DHFR restores the reduced form of folate, making it ready for the next round of thymine synthesis. If DHFR action is blocked, the cell dies [2].

Folate metabolism is the target of two main drug categories, the thymidylate synthase inhibitors (e.g., 5-fluorouracil) and the folate antagonists (e.g., methotrexate). These agents are broadly utilized in cancer chemotherapy and for treating autoimmune diseases such as rheumatoid arthritis [21,22]. Antimetabolites play a remarkable role in treating several malignant and non-malignant diseases, such as rheumatoid arthritis, parasitic and bacterial infections. Novel antimetabolites have become an area for anticancer drug expansion. More potent thymidylate synthase inhibitors have also been developed [23]. Various novel anticancer agents are undergoing a subsequent clinical development. A flood of rationally designed medications with new therapeutic impacts has appeared, ranging from direct cellular toxicity to the inhibition of blood vessel formation and of metastases. These drugs are the camptothecin analogs, taxanes (e.g., docetaxel), novel generation of thymidylate synthase inhibitors (e.g., raltitrexed), nucleoside analogs (e.g., gemcitabine), and oral alternatives to intravenous 5-fluorouracil [24].

2. Biochemistry of nucleotide metabolism

Nucleotides are biosynthesized via two metabolic pathways: de novo synthesis and salvage pathways. The de novo synthesis is assigned to the synthesis of pyrimidines and purines from precursor molecules; the salvage pathway is assigned to the transformation of preformed pyrimidines and purines derived from nucleic acid turnover and through adding the ribose-5-phosphate to the base. De novo synthesis of purines depends on metabolism of the one-carbon compounds [25].

Purine nucleotides and thymidine monophosphate (thymidylate) are synthesized using one-carbon moieties of various redox states. They are also utilized in the metabolism of various amino acids, e.g., homocysteine and serine, the initiation of protein biosynthesis in mitochondria and bacteria through formylation of methionine, and the methylation of several metabolites. These one-carbon reactions use coenzymes derived from folate or folic acid. Folate is a vitamin needed by humans due to their inability to be synthesized in their bodies. Folic acid (Fig. 1.1) is pteroylglutamic acid, a compound comprising glutamic acid, *p*-aminobenzoic acid (PABA), and heterobicylic pteridine (Fig. 1.2). The combination of the latter two yields pteroic acid [25].

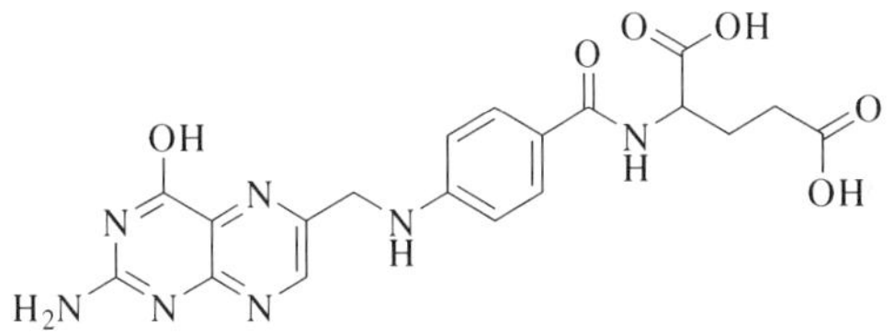

Fig. 1.1 Folic acid.

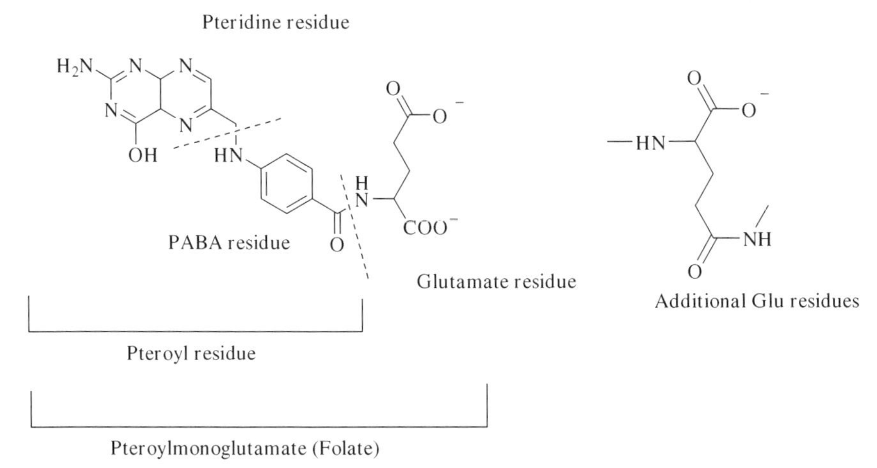

Fig. 1.2 Chemical structure of folate: Pteroylglutamic acid, p-aminobenzoic acid (PABA), and a heterobicylic pteridine.

Most antifolates used in oncology are similar in chemical structure to the naturally essential vitamin, folic acid, which play effective roles in one-carbon metabolism that produces substrates necessary for nucleotide synthesis [26].

As folates are essential for the cancer cell metabolism, antifolates are designed to inhibit either thymidylate synthase (TS), dihydrofolate reductase (DHFR), or the purine de novo pathway [27]. Hence, antifolates exert their antiproliferative activity [28]. Folates are present in various forms in mammalian cells. The principal structure for these forms is folic acid [29], *N*-[4 (2-amino-4-hydroxy-pteridin-6-ylmethylamino)-benzoyl]-L(+)-glutamic acid [30], which contains a pteridine ring, *para*-aminobenzoic acid (PABA) [31,32], and a glutamate residue [33].

Its importance as a coenzyme returns to its ability to be reduced in two cumulative stages by dihydrofolate reductase (DHFR) [34] to form 7,8-dihydrofolate reductase and then 5,6,7,8-tetrahydrofolate (Fig. 1.3), as it cannot perform its functions in its native oxidized state.

5,6,7,8-Tetrahydrofolate can transport single-carbon units into various metabolic processes [29,35]. The essential cofactor for DNA synthesis is tetrahydrofolate [35–37] that can transfer single-carbon units by being attached to the folate molecule at the *N*-5 or *N*-10 position or can be attached to both positions to form an extra ring. The carbon transfer process is catalyzed by thymidylate synthase. And during conversion of dUMP to TMP, oxidation of the folate ring occurs. Following this process, one dihydrofolate molecule is obtained per molecule of TMP produced. Consequently, dihydrofolate must be converted back to tetrahydrofolate in the cells by DHFR.

The DHFR inhibition is the principal biochemical role of methotrexate. And if this inhibition occurs, the tetrahydrofolate is consumed by TMP

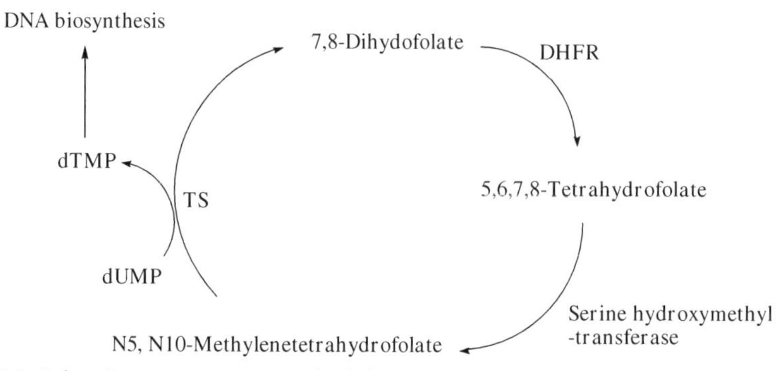

Fig. 1.3 Folate interconversions and inhibition.

synthesis, so de novo purine and thymidine nucleotide production will be ceased [29].

Humans do not have the enzymes capable of synthesizing PABA or of linking glutamate to pteroic acid. Folate is being reduced and transformed to N5-methyltetrahydrofolate in the intestines, and then it is secreted to the circulation. Tissue requirements for folate are met via uptake from plasma. Approximately two-thirds of the folate in the plasma is linked to proteins. With the synthesis of methionine, N5-methyltetrahydrofolate transfers its —CH_3 group to homocysteine within tissue cells [25].

Homocysteine methyltransferase, a vitamin B12 coenzyme-dependent enzyme, catalyzes this process, which is considered to be the main site of interdependence of these two vitamins. Tetrahydrofolate is transformed to polyglutamyl forms in tissues via an ATP-dependent synthetase. Pteroyl pentaglutamate is the most common type in the liver. Folate-dependent enzymes prefer reduced polyglutamyl forms as coenzymes, which are individually substituted with one of many one-carbon moieties. The conversion of folate **F** to tetrahydrofolate **FH4** is performed in two stages: **F** is reduced to 7,8-dihydrofolate **FH2**, which is then reduced to 5,6,7,8-tetrahydrofolate **FH4**. Dihydrofolate reductase, a single NADPH-linked enzyme, catalyzes both of these processes (Fig. 1.4), Methotrexate (Fig. 1.5), a structural analog of FH 2, is a potent inhibitor of dihydrofolate reductase and is utilized in chemotherapy of neoplastic disease [25].

Several one-carbon folate derivatives (with various redox states) operate as one-carbon carriers in various metabolic pathways. The one-carbon moiety is covalently linked to one or both of the nitrogen atoms at the 5- and 10-positions of the pteroic acid moiety of tetrahydrofolate in all of these processes. Fig. 1.6 depicts six different types of carriers.

Folinic acid (5-formyltetrahydrofolate), also termed citrovorum factor or leucovorin, is chemically stable and is utilized clinically to reverse or prevent the toxic impact of folate antimetabolites, such as pyrimethamine and methotrexate. The most common source of one-carbon fragments is serine, a non-essential amino acid. The synthesis of purine nucleotides (N10-formyl FH 4) is the first process that uses folate-mediated one-carbon transfer reactions;

As a crucial step in the biosynthesis of DNA, methylation of deoxyuridylate to thymidylate (N5,N10-methylene FH 4), impairment of this reaction is the cause of the clinical signs of folate deficiency; then the synthesis of formylmethionyl-tRNA (N10-formyl FH4) is needed for the initiation of the synthesis of protein in mitochondria and in prokaryotes; and the

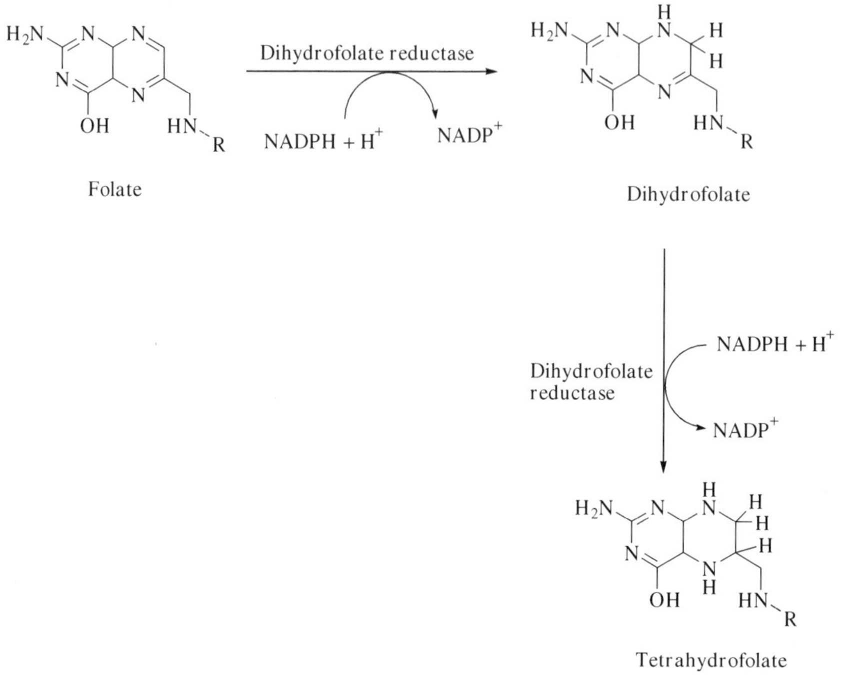

Fig. 1.4 Conversion of folate to tetrahydrofolate.

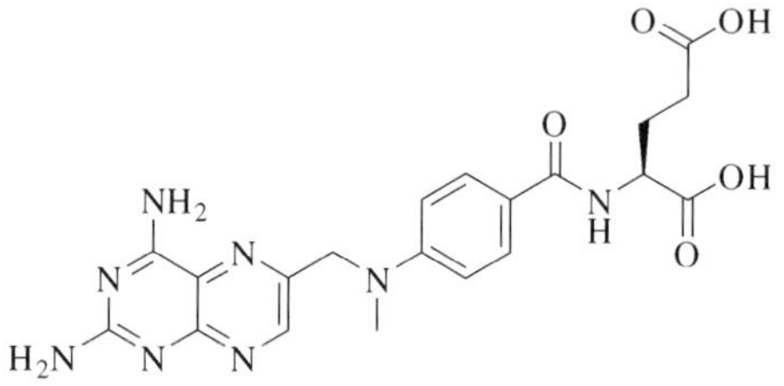

Fig. 1.5 Methotrexate.

transformation of homocysteine to methionine (N5-methyl FH4) occurs. The biosynthesis of purine nucleotides (N10-formyl FH 4) is the first pathway that utilizes folate-mediated one-carbon transfer reactions.

Defects in folate carrier (hereditary folate malabsorption), deficiency of N5,N10-methylene FH 4 reductase, or functional deficiency of N5-methyl FH4 methyltransferase as a result of the defects in vitamin B12 metabolism or formiminotransferase are all examples of inherited folate transport and metabolism disorders.

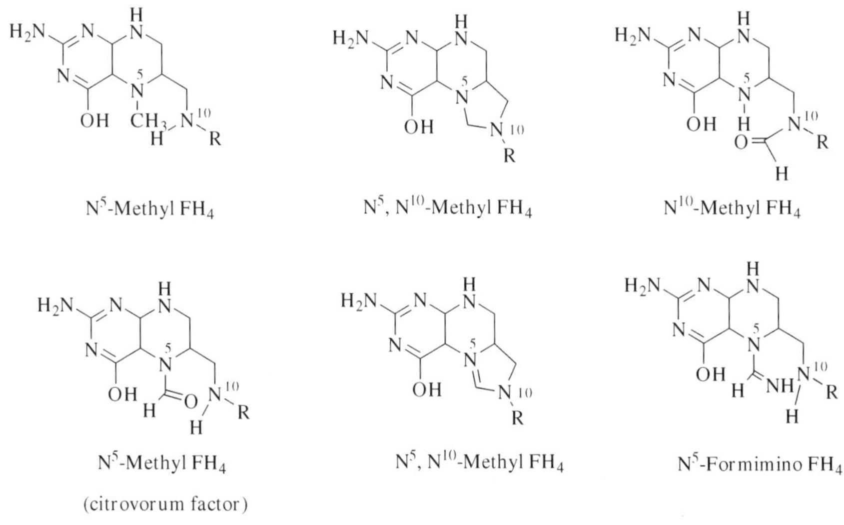

Fig. 1.6 Redox states of folate derivatives.

5-Phosphoribosyl-1-pyrophosphate (PRPP) is a main intermediate in nucleotide biosynthesis. It is essential for de novo synthesis of pyrimidine and purine nucleotides and the salvage pathways, in which purines are transformed to their respective nucleotides through the transfer of ribose 1-phosphate group from PRPP to the base (Fig. 1.7) [25].

The pyrophosphate group of ATP is transferred to ribose 5-phosphate in this process, yielding PRPP, a high-energy molecule. Inorganic phosphate

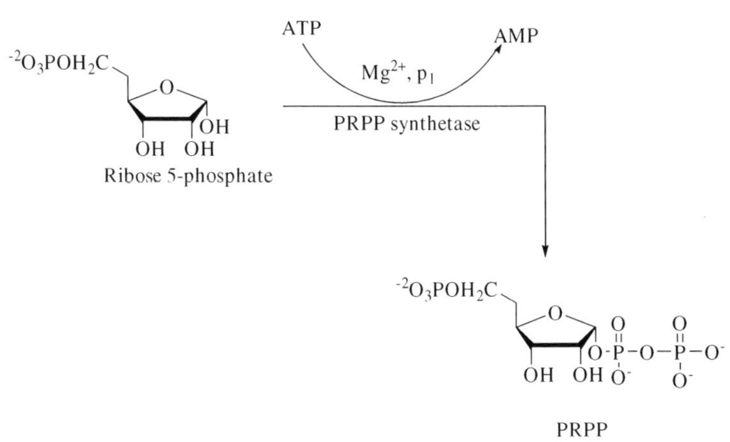

ATP
AMP
Mg^{2+}, p_1
PRPP synthetase
Ribose 5-phosphate
PRPP

Fig. 1.7 Route to 5-phosphoribosyl-1-pyrophosphate (PRPP).

(P I, which acts as an allosteric activator) is an absolute requirement for PRPP synthetase.

Many nucleotides, which are the final products of the pathway for which PRPP is an important substrate, block the enzyme. The PRPP synthetase variants with enhanced catalytic potency have been produced as a result of mutations in this gene, resulting in uric acid overproduction. The pentose phosphate pathway (PPP) is considered as the main source of ribose 5-phosphate.

-**Purine nucleotides** can be produced using two distinct routes. Free purine bases are used in the salvage route, which are converted to their corresponding ribonucleotides via the phosphoribosyltransferases. The de novo pathway uses glycine, 2,5-diamino-5-oxopentanoic acid (glutamine), N10-formyl FH4, aspartate, hydrogencarbonate, and PRPP in synthesizing inosinic acid (IMP), which is then transformed to GMP and AMP.

De novo synthesis: The atoms of the purine ring derived from array of sources (Fig. 1.8). C 2 and C 8 are derived from N 10-formyl FH 4. An entire incorporation of glycine gives C 4, C 5, and N 7. The aspartic acid's α-amino group provides N 1. N 3 and N 9 are provided by glutamine's amide group. Carbon dioxide (or HCO_3) is the source of C 6 [25].

For the sequential synthesis of IMP the purine biosynthesis has the following characteristics:

1. Upon the incorporation of the first —NH group of the purine ring, the glycosidic bond is formed. This process does not use free purines or purine nucleosides as intermediates. The pyrimidine ring is completely generated before the addition of ribose 5-phosphate in pyrimidine biosynthesis.

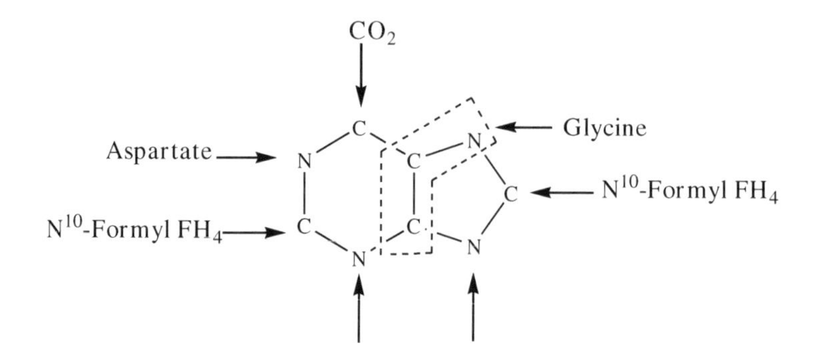

Fig. 1.8 Structure of the purine ring.

2. Several ATP-driven processes are involved in biosynthesis. Electrophilic carbonyl groups and nucleophilic amino groups are incorporated at the appropriate positions to achieve the cyclizations of the rings.
3. Inosinic acid (IMP) is the first purine nucleotide generated, and it serves as a precursor for the production of guanylic acid (GMP) and adenylic acid (AMP) in two separate routes (Fig. 1.9).
4. The cytosol contains all of the essential enzymes. This is also for the enzymes of salvage pathways, interconversion of nucleotide, and degradation.
5. The placenta and liver are highly active in de novo synthesis. Preformed purines are generated in the liver and transferred through red blood cells (RBCs) to non-hepatic tissues like bone marrow. They are highly efficient at salvaging purines and have little or no xanthine oxidase potency, which is responsible for oxidizing free purines.

Salvage pathways: Salvage routes are pathways that are constituted via the re-use of purine bases after they have been converted to their respective nucleotides. Extrahepatic tissues are particularly dependent on these pathways. Purines are formed during the intermediate metabolism of nucleotides and polynucleotide degradation. The phosphoribosyltransferase process is primarily responsible for salvage. Two phosphoribosyltransferases are found

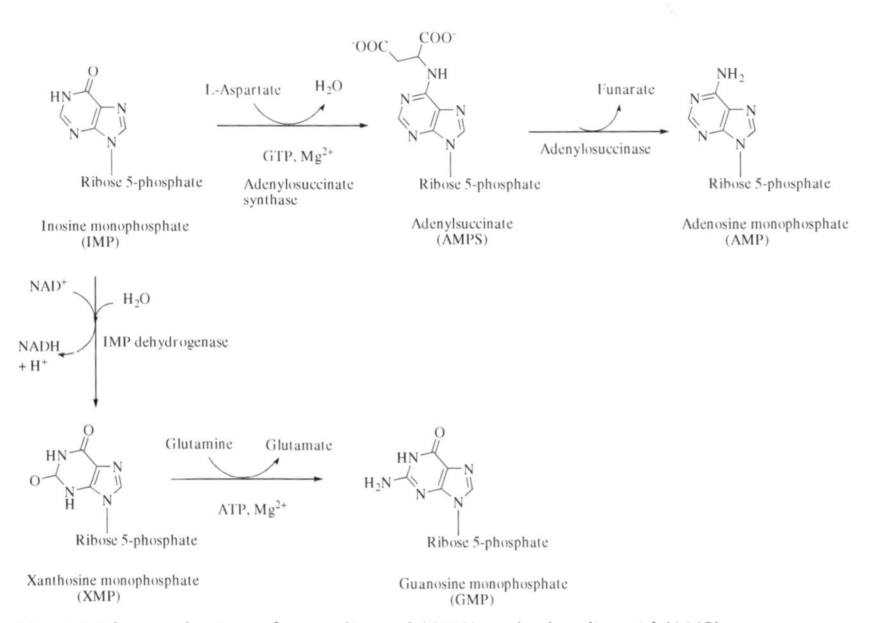

Fig. 1.9 The production of guanylic acid (GMP) and adenylic acid (AMP).

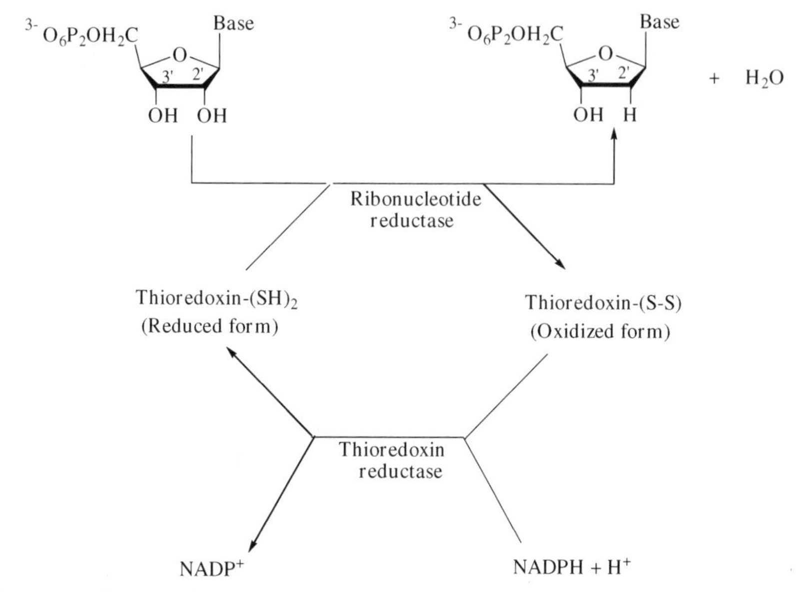

Fig. 1.10 Salvage pathways.

in human tissue (Figs. 1.10 and 1.11). The enzyme adenine phosphoribosyltransferase (APRT) catalyzes the conversion of adenine to AMP. HGPRT catalyzes the synthesis of hypoxanthine guanine phosphoribosyltransferase (IMP) and guanine phosphoribosyltransferase (GMP) from guanine and hypoxanthine. Other purines (8-azaguanine, 6-thiopurine, and allopurinol) are also converted to their respective ribonucleotides by HGPRT. Only two high-energy bonds are used during salvage, whereas at least six high-energy bonds are involved during de novo synthesis of GMP or AMP. Nucleoside kinase and purine nucleoside phosphorylase are used in a less significant salvage pathway. The synthesis of inosine, deoxyinosine, guanosine, or deoxyguanosine could be catalyzed via the phosphorylase.

Adenosine deaminase, on the other hand, can convert the last two nucleosides to inosine and deoxyinosine. The phosphorylase's primary purpose appears to be the generation of free hypoxanthine and guanine, which can then be converted to uric acid.

Deficiency of purine nucleoside phosphorylase or adenosine deaminase leads to immunodeficiency disease. In muscle, the purine nucleotide cycle, which is considered as unique nucleotide reutilization pathway, involves three enzymes: adenylosuccinate synthetase; myoadenylate deaminase, and adenylosuccinate lyase.

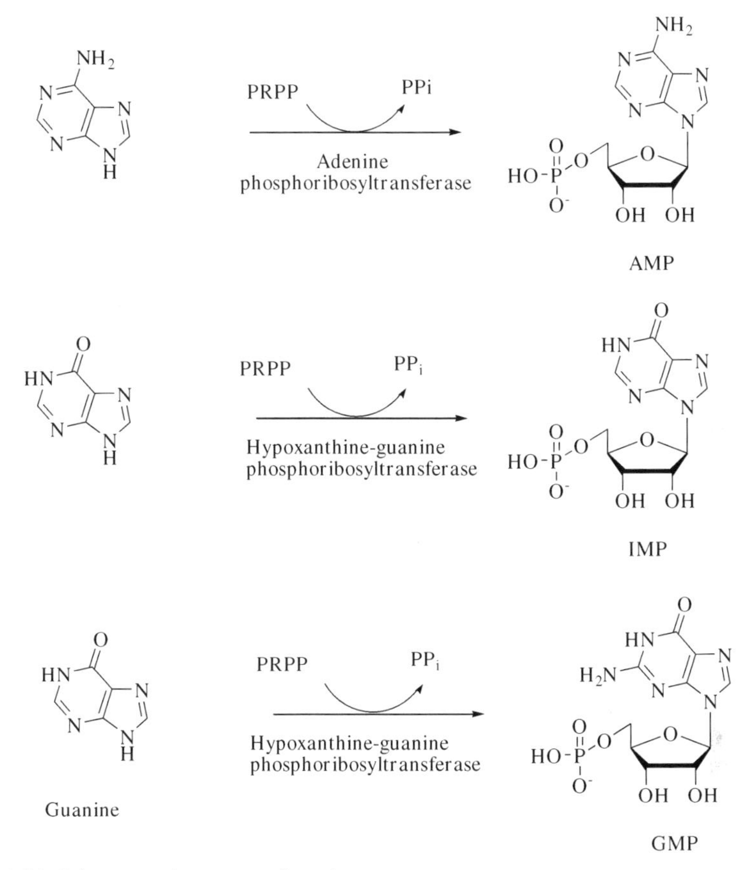

Fig. 1.11 Salvage pathways-continued.

AMP is transformed to IMP with the production of NH_3 in this cycle, and IMP is then retransformed to AMP. The deficiency of myoadenylate deaminase causes benign muscle disorder. In mammalian tissues, adenosine kinase is broadly distributed. Dietary purines obtained from food are mostly transformed to uric acid via intestinal xanthine oxidase and do not participate in the salvage pathways.

DNA polymerases and RNA polymerases use nucleoside and deoxynucleoside triphosphates as substrates. They are formed in two phases from monophosphates. Kinases that are base specific rather than sugar specific catalyze the transformation to diphosphates. Phosphate is usually obtained from ATP. Other triphosphates or dATP may, however, be utilized in specific instances. The diphosphates are transformed to the triphosphates via the enzyme nucleoside diphosphate kinase. The phosphate donor and acceptor both have a lack of base or sugar specificity. The transformation of ADP to ATP takes place via mitochondrial oxidative phosphorylation coupled to electron transport [25].

The transformation of ribonucleotides to the deoxy form takes place at the diphosphate level. The reaction is catalyzed by ribonucleoside diphosphate reductase. This enzyme is present in all tissues and species. The enzyme, which oxidizes two sulfhydryl groups to a disulfide, is the immediate source of reducing equivalents.

Regeneration of the ribonucleotide reductase is carried out via way of means of thioredoxin, a coenzyme of polypeptide dithiol (12,000 MW), which additionally performs a function in different protein disulfide reductase reactions. In thioredoxin, cysteine residues in the sequence – Cys – Gly – Pro – Cys – are transformed to cystine disulfide. Reduced thioredoxin is regenerated via thioredoxin reductase (TrxR), a flavoprotein enzyme that utilizes NADPH H. Ribonucleotide reductase is composed of two subunits, B 1 and B 2, each of which has no catalytic function. B 1 is a dimer wherein every monomer includes two allosteric effector binding sites and a substrate binding site.

The one sort of effector site confers substrate specificity, whereas the other is regulatory. A pair of sulfhydryl groups are present in B 1 and are necessary for catalytic activity. B 2 is a dimer with one non-heme Fe(III) and an organic free radical delocalized over the aromatic ring of a tyrosine residue in each of its polypeptide chains [25].

The catalytic site's formation is performed through the interaction of B 1 and B 2. A free radical mechanism implies the iron atom of B 2, tyrosyl residues, and sulfhydryl groups of B 1.

By inactivating the free radical, the hydroxyurea as an antineoplastic agent inhibits the ribonucleotide reductase. Ribonucleotide reductase is regulated to keep a balanced supply of deoxynucleotides required for the synthesis of DNA. Excess dATP, for example, causes a decrease in synthesizing all deoxyribonucleotides, while ATP promotes the generation of dCDP and dUDP. TTP binding stimulates the production of dGDP and thus dGTP.

The production of dADP and thus dATP is stimulated by the binding of dGTP. These nucleotide effectors are able to equalize the amounts of the four deoxyribonucleotides necessary for synthesizing DNA through binding to diverse regulatory sites.

Regulation of de novo purine biosynthesis is necessary as it consumes energy as well as glutamine, glycine, aspartate, and N 10-formyl FH 4. Regulation is carried out at the PRPP synthetase reaction, the amidophosphoribosyltransferase reaction, and the steps involved in forming GMP and AMP from IMP.

As an allosteric activator, PRPP synthetase needs inorganic phosphate. Its potency is determined by the intracellular concentrations of the end products of PRPP-dependent pathways. Purine and pyrimidine nucleotides are the final products. De novo purine biosynthesis is aided by increased levels of intracellular PRPP.

Amidophosphoribosyltransferase reaction is the rate-determining step of the de novo route. Amidophosphoribosyltransferase is an allosteric enzyme that requires a divalent cation to function. The enzyme is inhibited by GMP and AMP, which bind at distinct sites.

Pyrimidine nucleotides, in relatively high concentrations, also inhibit the enzyme. Inhibition by GMP and AMP is competitive with respect to PRPP.

The human placental enzyme is present in small and large forms with M.W. 133,000 and 270,000. The small form is catalytically active. Ribonucleotides act on the conversion of the active form to the large one, while PRPP performs the contradictory action. The regulatory actions of amidophosphoribosyltransferase and PRPP synthetase are co-ordinated. PRPP synthase is activated when the intracellular concentration of adenine ribonucleotides decreases. This leads to the increase in synthesis of PRPP, which transforms the inactive form of amidophosphoribosyltransferase to the active one and increases the purine nucleotide biosynthesis.

In the generation of GMP and AMP from IMP, ATP is needed for GTP synthesis, and GTP is required to yield ATP. Additionally, the inhibition of adenylosuccinate synthetase is performed via AMP, while the IMP dehydrogenase is inhibited via GMP.

Purines and their nucleotides are degraded during turnover of endogenous nucleic acids as well as the degradation of ingested nucleic acids (Fig. 1.12), during which most of the purines are transformed to uric acid.

Degradation of purine nucleoside phosphates (IMP, AMP, XMP, and GMP) starts with the hydrolysis via 5-nucleotidase, to generate inosine, adenosine, xanthosine, guanosine, and phosphate. Adenosine (Adenocor) is transformed to inosine via the adenosine deaminase. For inosine, xanthosine, and guanosine, the following stage is catalyzed via purine nucleoside phosphorylase and implies a phosphorylation and a cleavage to yield ribose-1-phosphate and 1,9-dihydro-6*H*-purin-6-one (hypoxanthine), xanthine, and 2-amino-1,9-dihydro-6*H*-purin-6-one (guanine). The pentose sugars are excreted or metabolized further. Purine nucleoside phosphorylase is involved in purine salvage, and its deficiency causes a

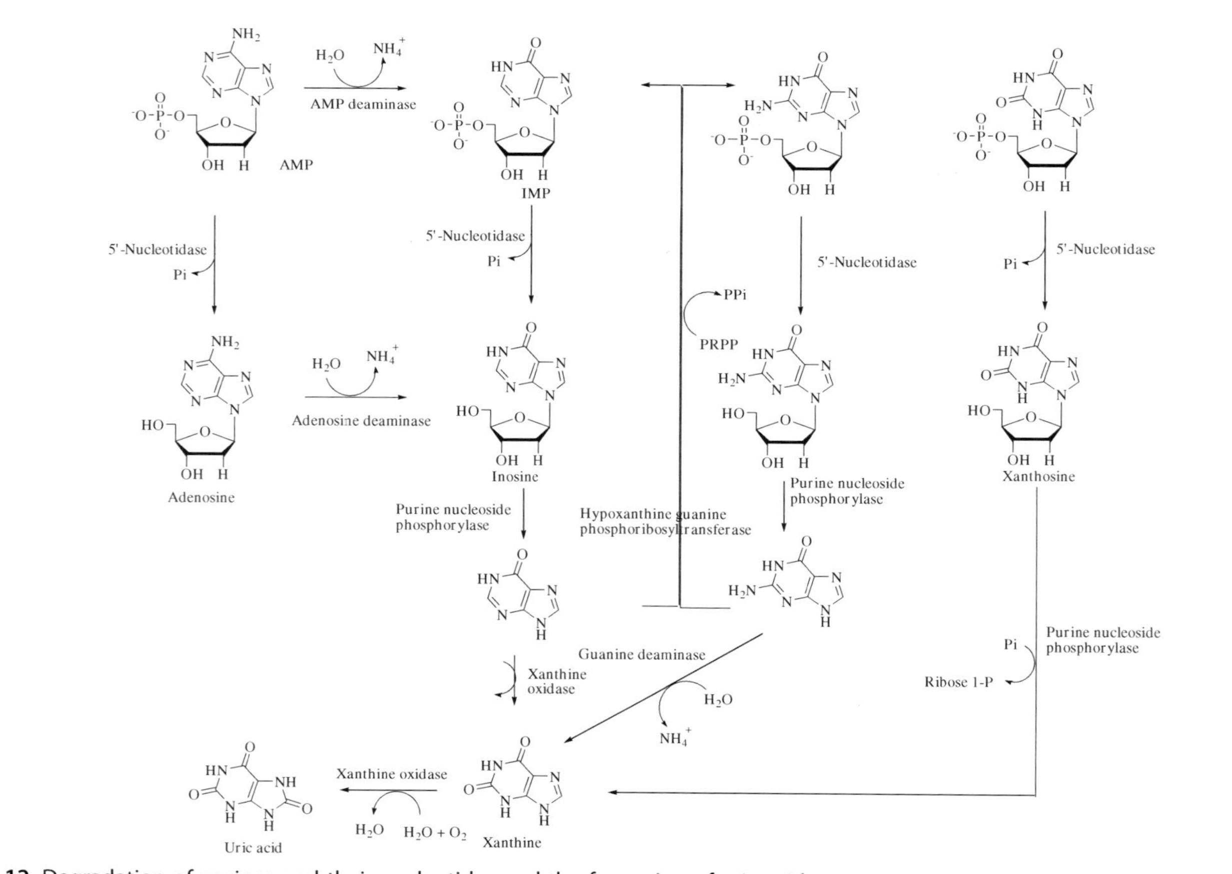

Fig. 1.12 Degradation of purines and their nucleotides and the formation of uric acid.

reduction in cell-mediated immunity. Guanine and hypoxanthine are transformed to xanthine via the guanine aminohydrolase and xanthine oxidase. Thus, all purine nucleosides generate 3,7-dihydro-1*H*-purine-2,6-dione (xanthine). Xanthine oxidase (XAO) is found in most tissues and is particularly active in the intestinal mucosa and liver. Hydrogen peroxide and uric acid are the end products of the xanthine oxidase reaction. Uric acid is excreted in the urine. Urate oxidase, a liver enzyme found in mammals other than primates, transforms uric acid to (2,5-dioxo-4-imidazolidinyl) urea (allantoin). In humans, carbonyl diamide (urea), uric acid, and 2-amino-1-methyl-5*H*-imidazol-4-one (creatinine) are the end products of nitrogen metabolism. Urea is considered as the primary end product. Uric acid production and excretion are considered as a balanced process in humans [25].

3. Targets for chemotherapeutic agents and inhibition of biosynthetic pathways

The key stages in which the antimetabolite drugs exert their action during the biosynthesis of DNA are outlined in Fig. 1.13.

3.1 Inhibitors of uridylic acid

The production of pyrimidine nucleotides begins with the formation of a heterocyclic system from aspartate carbamoylation, which is then cyclized to dihydroorotate. Its dehydrogenation produces orotate, which is then converted to orotidylate by reacting with phosphoribosyl pyrophosphate (PRPP). Finally, decarboxylation produces uridylic acid (uridine monophosphate, UMP) (Fig. 1.14). Uridine monophosphate is the precursor to other pyrimidine nucleotides, after its transformation to the corresponding nucleoside triphosphate (UTP). (2*S*)-2-[(2-Phosphonoacetyl)amino]butanedioate (*N*-phosphonoacetyl-L-aspartate; PALA), an inhibitor of aspartate transcarbamoylase that acts as a transition-state derivative (Fig. 1.15), is one of the compounds that inhibit reactions of this pathway. Also, it has been subjected to clinical trials [38,39].

3.2 Inhibitors of 2′-deoxyribonucleotides

The 2’-OH group is replaced by a hydrogen atom in the biosynthesis of deoxyribonucleotides (immediate precursors of DNA) (Fig. 1.16). The enzyme ribonucleotide reductase (RNR), (nucleoside diphosphate reductase), catalyzes this reaction on ribonucleoside-5′-diphosphates (NDPR).

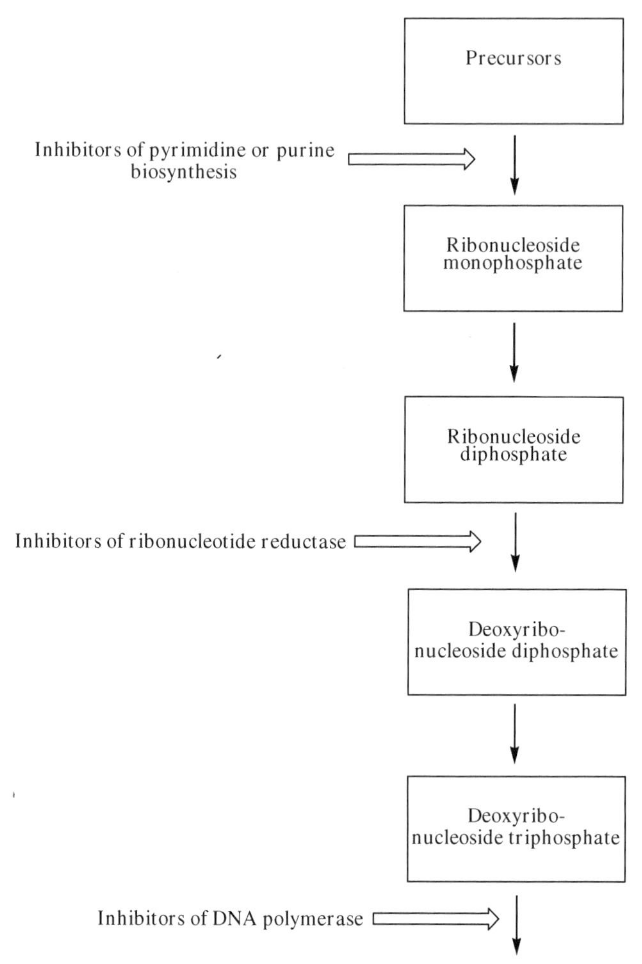

Fig. 1.13 The key stages in which the antimetabolite drugs exert their action during the biosynthesis of DNA.

3.3 Inhibitors of thymidylic acid

Thymidylate synthase (TS) catalyzes the transformation of [(2*R*,3*S*,5*R*)-5-(2,4-dioxo-3,4-dihydropyrimidin-1(2*H*)-yl)-3-hydroxyoxolan-2-yl]methyl dihydrogen phosphate (deoxyuridine monophosphate, dUMP) to thymidine monophosphate (thymidylate, TMP) in a reductive methylation that requires transferring a carbon atom from the 5,10-methylenetetrahydrofolate (cofactor) to the pyrimidine ring in its 5 position. This conversion, that is, the only de novo source of thymidylate, is a constituent of the thymidylate cycle

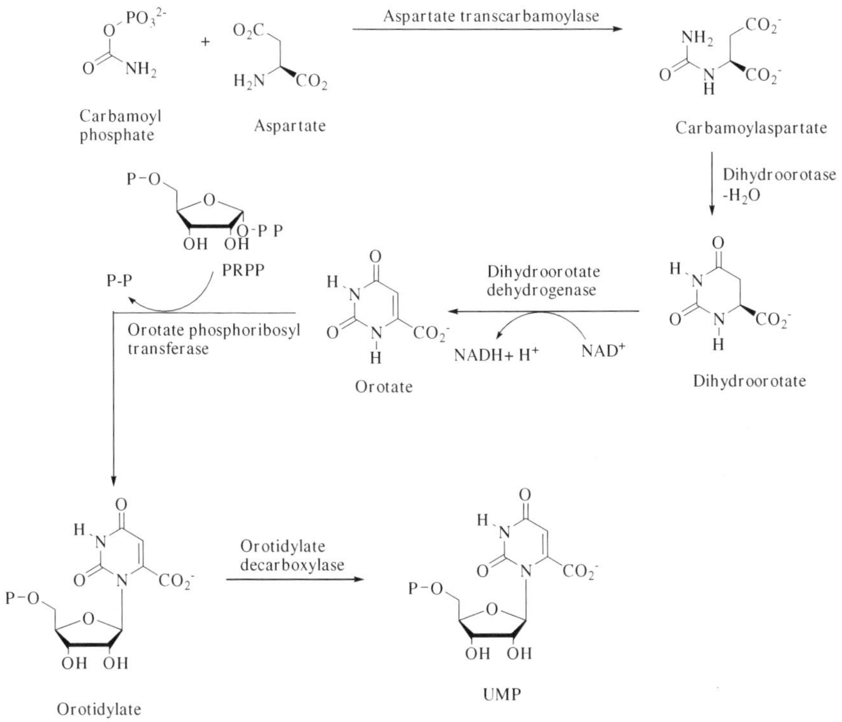

Fig. 1.14 Route to the uridylic acid (uridine monophosphate, UMP).

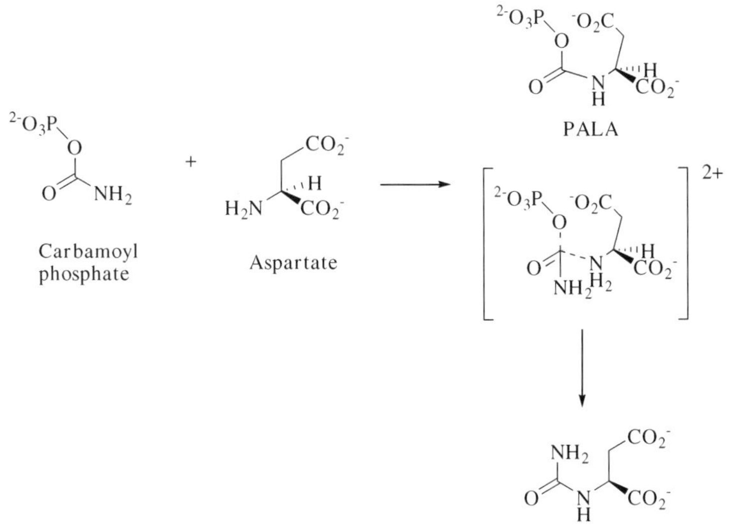

Fig. 1.15 (2S)-2-[(2-Phosphonoacetyl)amino]butanedioate (N-phosphonoacetyl-L-aspartate (PALA) as a transition-state derivative.

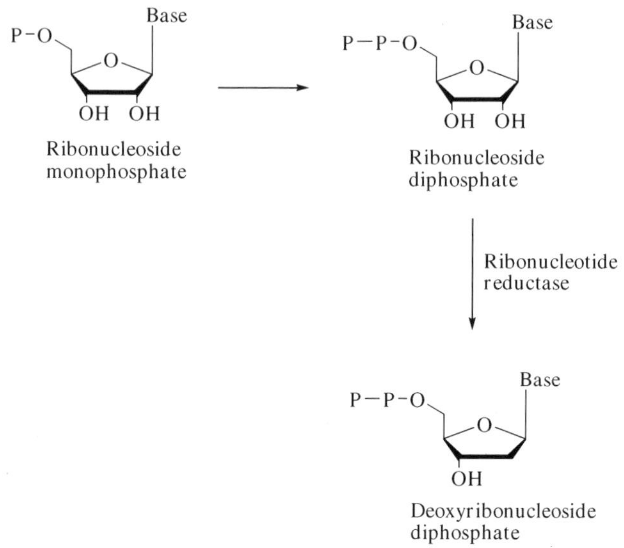

P-O Base O OH OH
Ribonucleoside monophosphate

P—P-O Base O OH OH
Ribonucleoside diphosphate

Ribonucleotide reductase

P—P-O Base O OH
Deoxyribonucleoside diphosphate

Fig. 1.16 The biosynthesis of deoxyribonucleotides.

(Fig. 1.17), where two other enzymes participate, known as dihydrofolate reductase (DHFR) and serine hydroxymethyl transferase (SHMT). SHMT catalyzes the generation of 5,10-methylenetetrahydrofolate from THF, coupled with the transformation of serine into glycine, with pyridoxal 5′-phosphate (P5P, PLP) as a cofactor. In the reaction catalyzed via TS, the 5,10-methylene-THF thus generated contributes its $CH_2 <$group to dUMP, being converted into dihydrofolate. DHFR completes the cycle via reducing DHF to THF. Although uracil methylation appears to be a minor structural modification, the increased lipophilicity and bulk associated with the $—CH_3$ group is critical for appropriate discrimination of thymine from the other bases existing in DNA chains via transcription factors, enhancers, repressors, and other DNA-binding proteins.

3.4 Inhibitors of dihydrofolate reductase (DHFR)

Folic acid and its metabolites (often referred to as folates) are coenzymes in a variety of metabolic reactions. They play a key role in the de novo synthesis of thymidine monophosphate (TMP) (thymidylic acid) and purine nucleotides, as well as the transfer of one-carbon unit. Folate-dependent enzymes are obvious targets for cancer treatment.

In mammals, folic acid is supplied from the diet and reduced to THF via the dihydro-folic reductase, utilizing NADPH (cofactor). Subsequent

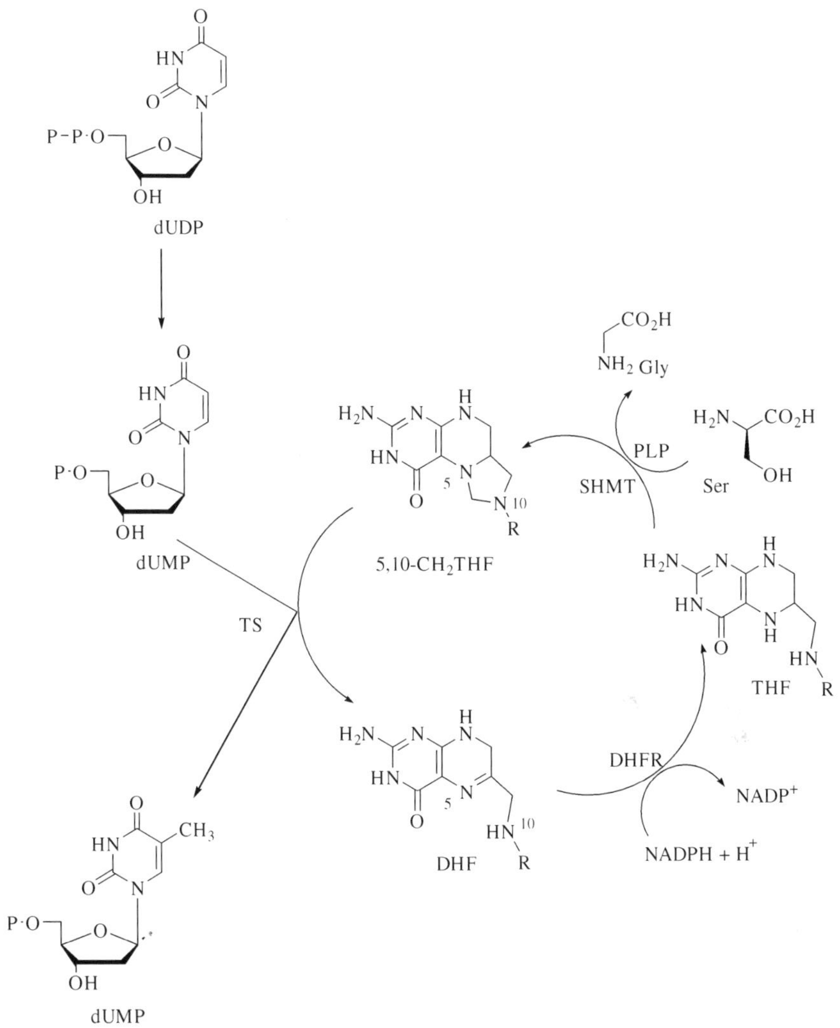

Fig. 1.17 The thymidylate cycle and the inhibitors of thymidylic acid.

conversions of THF afforded 5,10-methylene-THF, 5,10-methenyl-THF (5,10-CH=THF), 5-formyl-THF, and 10-formyl-THF, which are designated as folinic acids and are implied in transferring of one-carbon units.

Due to the significant involvement of folinic acids in synthesizing thymidylate and purine bases, DHFR inhibition results in the cell death. DHFR is a small protein with a large active site. DHF binds close to the cofactor (NADPH) in a pocket buried deep within the enzyme, catalyzing the transfer of the pro-R hydrogen of the C-4 position of the dihydropyridine ring in the cofactor onto the C=N double bonds of dihydrofolic acid and folic acid.

Folic acid analogs are considered as the most potent inhibitors of DHFR. They vary from the natural ligand in that they carry a 2,4-diaminopyrimidine unit, such as methotrexate (MTX, amethopterin) and aminopterin (AM). Classical antifolates are the inhibitors in which the side chain terminates in a glutamic acid residue, as in folic acid, while the inhibitors that contain lipophilic substituents are the non-classical antifolates.

3.4.1 Classical DHFR inhibitors

MTX and AM were designed by substituting the hydroxyl group at C-4 of the native substrate (DHF) by -NH group. The underlying assumption was that the two ligands would bind in the same way and that the 4-amino group of MTX would occupy the position normally occupied by the DHF oxygen in the binding site. However, when X-ray diffraction structures of DHFR with MTX and DHF were performed, various binding modes were found. Hydrogen bonds and other interactions with bridging water molecules bind both ligands to the DHFR active site. MTX is approximately 3 pKa units more basic than folic acid as it has an electron-releasing —NH group conjugated with the basic guanidine fragment in place of an electron-withdrawing carbonyl, thus so binds in a protonated form [39]. The electrostatic interaction and an extra H-bond implying the 4-amino group result in a 103-fold stronger binding affinity than folate [2]. The presence of the NADPH cofactor is required for MTX to bind to DHFR, and this is a model of a form of enzyme inhibition known as slow, tight-binding inhibition [40]. The differences in NADPH to NADP and NADH ratios in both kinds of cells appear to be the cause of MTX's selective toxicity in malignant cells compared to normal cells [41].

The toxicity of MTX in malignant cells is selective compared to normal cells. AM was the first antifolate to be used in cancer therapy, but it was indicated that MTX is less toxic and has better pharmacokinetic features, and this compound is now the only classical antifolate in clinical utility for treating non-Hodgkin's lymphoma, acute lymphocytic leukemia, and choriocarcinoma, as well as in several combination therapies.

Its side effects are the myelosuppression and the damage to liver, the gastrointestinal tract, and kidneys. MTX is frequently associated with the calcium salt of leucovorin (N5-formyltetrahydrofolic acid), one of the folinic

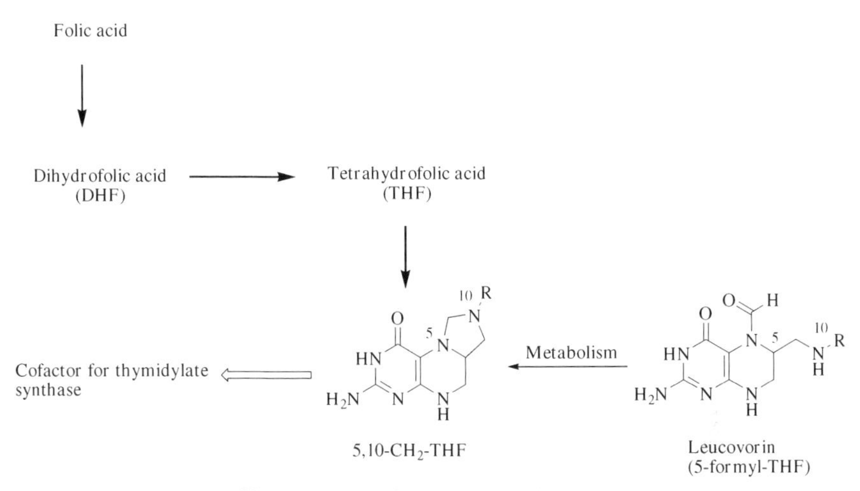

Fig. 1.18 Conversion of leucovorin to the 5,10-methylene-THF.

acids, in order to partially mitigate its bone marrow toxicity. By the RFC, folinic acid (leucovorin) can enter the cell, and then it is transformed to 5,10-methylene-THF without the involvement of DHFR, thus circumventing its inhibition (Fig. 1.18) [39].

The essential targets for antifolate drugs and their relationships with the biosynthesis of nucleic acid are accomplished in Fig. 1.19.

3.4.2 Non-classical (lipophilic) DHFR inhibitors

Compounds that are not substrates for active folate transport systems enter the cells via passive diffusion when the glutamic chain is suppressed. Their advantage is that they act on methotrexate-resistant cancer cells due to transport defects.

Contrastingly, the lack of the glutamic acid side chain prevents polyglutamation, and thus, these compounds are not maintained inside the cells, necessitating longer therapies. Trimetrexate and piritrexim are two of these compounds.

3.5 Inhibitors of the de novo purine biosynthesis pathway

Contrary to pyrimidine nucleotide biosynthesis, where a preformed heterocyclic moiety is linked to PRPP, purine nucleotides involve the gradual construction of the purine ring.

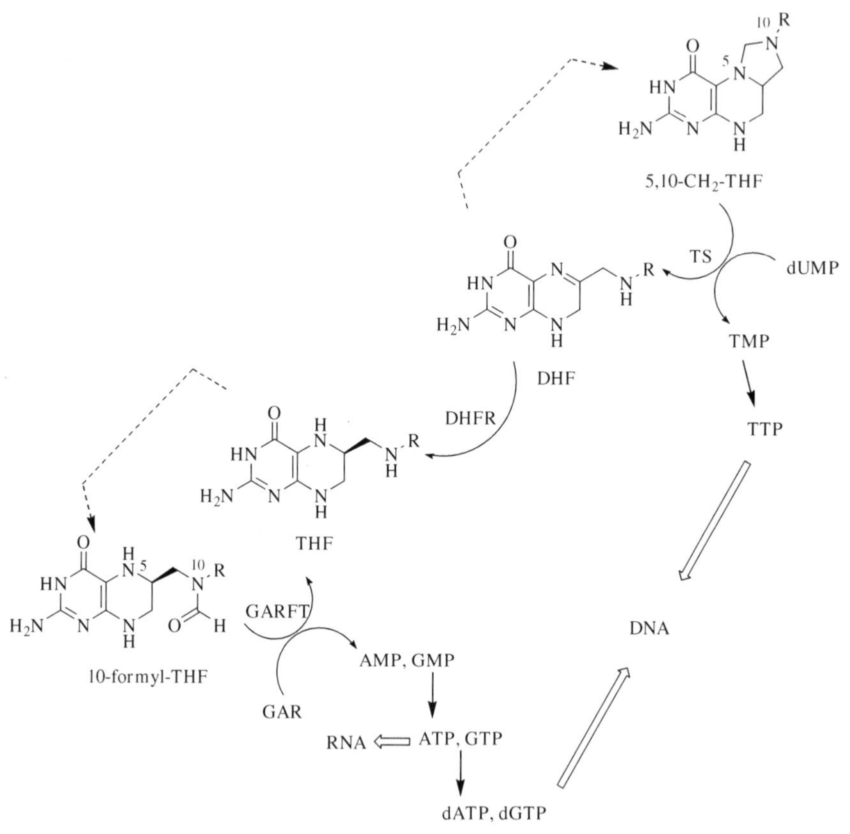

Fig. 1.19 Targets for antifolate drugs.

The generation of inosine monophosphate (IMP), the precursor of GTP, ATP, dATP, and dGTP essential for DNA and RNA synthesis, is resulted from the de novo pathway.

The pyrophosphate group of phosphoribosyl pyrophosphate (PRPP) is displaced via nucleophilic substitution by a molecule of NH_3 formed through the glutamine hydrolysis to glutamic acid, which is the first irreversible step in de novo purine biosynthesis.

Both reactions are catalyzed via **phosphoribosyl pyrophosphate** PRPP amidotransferase, whose main inhibitors are thiopurines such as 6-mercaptopurin, acting via feedback mechanisms. The inhibition of numerous enzymes involved in purine production and nucleic acids' misincorporation is a complex mechanism of action for these anticancer medicines. The third reaction in the de novo purine biosynthesis is the conversion of glycinamide ribonucleotide (GAR) into its formylderivative (FGAR) utilizing 10-formyl-THF (formyl donor). The enzyme that catalyzes this stage is called **glycinamide ribonucleotide formyltransferase (GARFT)**.

The first selective and active GARFT inhibitor was lometrexol (DDATH-B), which is a folate analog lacking the 5 and 10 nitrogen atoms; it is unable to involve in transferring single carbon units. Contrastively, lometrexol possesses a 2-aminopyrimidin-4-one subunit that is identical to that of the THF cofactor, making it distinct from the 2,4-diaminopyrimidine pattern found in most DHFR inhibitors. Its glutamate side chain facilitates its delivery into cells via the RFC and MFR transport systems, as well as its polyglutamation via folylpolyglutamate synthase (FPGS). Clinical trials on lometrexol were conducted, but unanticipated findings of postponed cumulative toxicity [39,42] led a quest for second-generation antimetabolites with a better profile.

Phosphoribosylformylglycinamidine synthetase catalyzes the conversion of formylglycinamide ribonucleotide to formylglycinamidine ribonucleotide using ammonia and glutamine. The enzyme activates the amide group adjacent to the ribose ring to nucleophilic attack via its conversion into iminoether.

The enzyme converts the amide group next to the ribose ring into iminoether, which makes it vulnerable to nucleophilic assault.

Additionally, another catalytic site of the enzyme hydrolyzes 2-amino-4-carbamoylbutanoic acid (glutamine) to ammonia and 2-aminopentanedioic acid (glutamic acid), which is then directed to the first site, and then after an addition/elimination mechanism, it yields the formylglycinamidine ribonucleotide. Some analogs of glutamine carrying a diazomethyl moiety possess antitumor potency as their ability to inhibit various reactions in which glutamine (cofactor) is a participant, particularly the one catalyzed by formylglycinamidine ribonucleotide synthetase.

Some antifolate drugs (e.g., MTX) inhibit **5-aminoimidazole-4-carboxamide ribonucleotide formyltransferase** enzyme, although it is not their primary target [39].

6-Mercaptopurine (MP) and 6-thioguanine (TG) are the most potent synthetic purine derivatives estimated as anticancer agents. These are some of the oldest cancer chemotherapy drugs in use; MP is utilized to treat myeloblastic and lymphoblastic leukemias, while TG is used to treat acute nonlymphocytic leukemia. 6-Mercaptopurine is converted into thioinosinic acid via intracellular metabolism by hypoxanthine guanine phosphoribosyl transferase (HGPRT). The thioinosinic acid indicates the cell cycle S-phase-specific cytotoxicity. Intracellular activation causes a number of enzymes in the de novo purine synthesis pathway to be inhibited, resulting in misincorporation into RNA and DNA.

Thus, thioinosine monophosphate is formed through incorporating a ribose phosphate unit to MP catalyzed by HGPRT, inhibits PRPP amidotransferase, the first enzyme in the de novo biosynthesis of purines, via a retroinhibition mechanism. Other enzymes involved, which are also inhibited leading to lower levels of GMP and AMP, are listed below:

a. HGPRT is itself a result of competition between MP and its natural substrate, hypoxanthine.

b. Inosinic dehydrogenase is an enzyme that converts inosinic acid (IMP) to xanthylic acid (XMP), which is a precursor to guanylic acid (GMP).

c. Adenylosuccinate synthetase catalyzes the first stage of the conversion of inosinic acid to adenylic acid (AMP).

Finally, thioinosine monophosphate is converted into thioguanylic acid, which is misincorporated into RNA and DNA. This results in the single-strand DNA breaks and DNA-protein cross-links via alterating DNA-repair mechanisms. Thioguanine, after being converted to thioguanylic acid by HGPRT, acts in a similar manner. *S*-Methylation by thiopurine methyltransferase (TPMT) and oxidation via xanthine oxidase to an 8-oxo analog and then to 6-thioxo-7,9-dihydro-1*H*-purine-2,8(3*H*,6*H*)-dione (6-thiouric acid, TUA) are the two main degradative routes of MP.

Allopurinol is a competitive inhibitor of xanthine oxidase and is a structural analog of hypoxanthine. It is also a substrate for xanthine oxidase, where it is transformed to alloxanthine in a slow manner, which causes the inhibition of the enzyme. As the *S*-methyl analog is not a substrate for purine phosphoribosyl transferases, *S*-methylation is another catabolic pathway of MP. Some analogs of thiopurines have been designed to sustain protection from the degradation reactions, such as the nitroimidazole derivatives thiamiprine (guaneran) and azathioprine (Imuran). These compounds act as prodrugs and are thought to be activated via an SNAr mechanism implying thiol nucleophilic attack on the 5-position of the 4-nitro imidazole ring, followed by thiopurine elimination as a leaving group. Although azathioprine is a remarkable immunosuppressant agent, utilized to treat autoimmune diseases, none of these prodrugs are more efficient as anticancer agents than the parent compounds [39,43].

3.6 Inhibitors of adenosine deaminase

Pentostatin (2′-deoxycoformycin) and coformycin are natural products (isolated from Streptomyces sp.) and are analogs of hypoxanthosine (inosine) and deoxyinosine, which contain a seven-membered ring (modified purine ring). They are active inhibitors of a denosine deaminase, the enzyme that degrades deoxyadenosine through its conversion into deoxyinosine. Pentostatin is combined with adenosine-derived antitumor drugs to enhance their half-life. Pentostatin is also an antitumor agent that has utility in treating some leukemia, as hairy cell leukemia. The mechanism of its antitumor potency is unclear as it comprises the following circumstances.

Pentostatin is misincorporated into DNA after being triphosphorylated. The inhibition of adenosine deaminase causes a buildup of adenosine, which causes the enzyme *S*-adenosylhomocysteine hydrolase to be retroinhibited. Most methyltransferases that utilize *S*-adenosylmethionine as a cofactor are competitively inhibited by the *S*-adenosylhomocysteine that then accumulates. As a consequence, the mechanisms relating to nucleic acid methylation are disrupted.

High levels of deoxyadenosine triphosphate, which is an inhibitor of RNR, the enzyme that removes the 2′-hydroxy group of the ribose ring during DNA biosynthesis, occur from deoxyadenosine accumulation.

3.7 Inhibitors of late stages in the syntheses of DNA

Anticancer prodrugs that are activated to their triphosphates via phosphorylation catalyzed by kinases comprise various ribonucleoside and deoxyribonucleoside analogs [39,44]. The triphosphates act via misincorporation into DNA after bioactivation leading to alterations in DNA repair, and the chain elongates in a slow manner. Inhibition of DNA polymerase is another mechanism of antitumor action of these drugs, as are other mechanisms (e.g., inhibition of PNP or RNR). A common issue is connected with these drugs as their cytotoxicity to lymphoid cells causes a remarkable immunosuppression.

3.7.1 *Pyrimidine nucleosides*

The synthesis of both DNA and RNA requires two purine and two pyrimidine nucleoside triphosphates, which are derived from either a newly synthesized (de novo pathway) or a salvage pathway, which is the recycling of partially degraded nucleotides (Fig. 1.20).

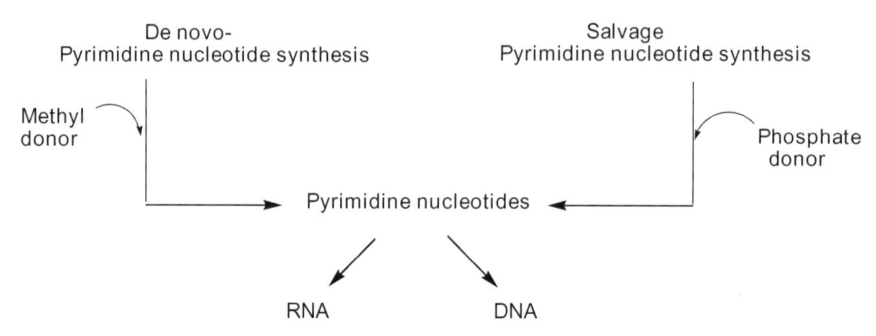

Fig. 1.20 Synthesis of DNA and RNA.

The de novo synthesis of the pyrimidine structure is demonstrated in single step through the condensation of two main precursor molecules via the enzyme aspartate carbamoyltransferase. The *N*-phosphonoacetyl-L-aspartate, an inhibitor of this enzyme, has been produced and evaluated as an anticancer drug. It has indicated some potency against tumors in vivo, although it may be more effective as a modulator of the action of other antimetabolites. The steps in the pyrimidine nucleotide synthesis that are most frequently relevant to the therapeutics, involving deoxyuridine and thymidine nucleotides, are outlined in Fig. 1.21 (dashed lines showed the sites of drug inhibition).

The unique existence of thymidine in DNA and the link between DNA synthesis and cellular proliferation affirm the importance of these steps. The main enzyme in this part of the pathway for synthesizing thymidine nucleotides is thymidylate synthase. The transformation of dUMP to TMP via thymidylate synthase is inhibited both directly and indirectly by many antimetabolites [29].

Cytarabine (Ara-C), fazarabine, gemcitabine (dFdC), and azacitidine are examples of anticancer drugs with a modified ribose ring. Among the arabinose-derived nucleosides, cytarabine (Ara-C) is used to treat a variety of leukemias, comprising AML and non-Hodgkin lymphoma. Cytarabine's incorporation into DNA after being activated to the matching triphosphate inhibits strand elongation; it is considered a prime example of an anticancer medication that acts exclusively in the S-phase of the cell cycle. Because of this S-phase specificity, cells must be exposed to cytotoxic doses for an extended period of time to acquire optimal cytotoxic activity. The fast deamination of cytarabine by cytosine deaminase to the physiologically active metabolite uracil arabinoside reduces its activity [45]. As a result,

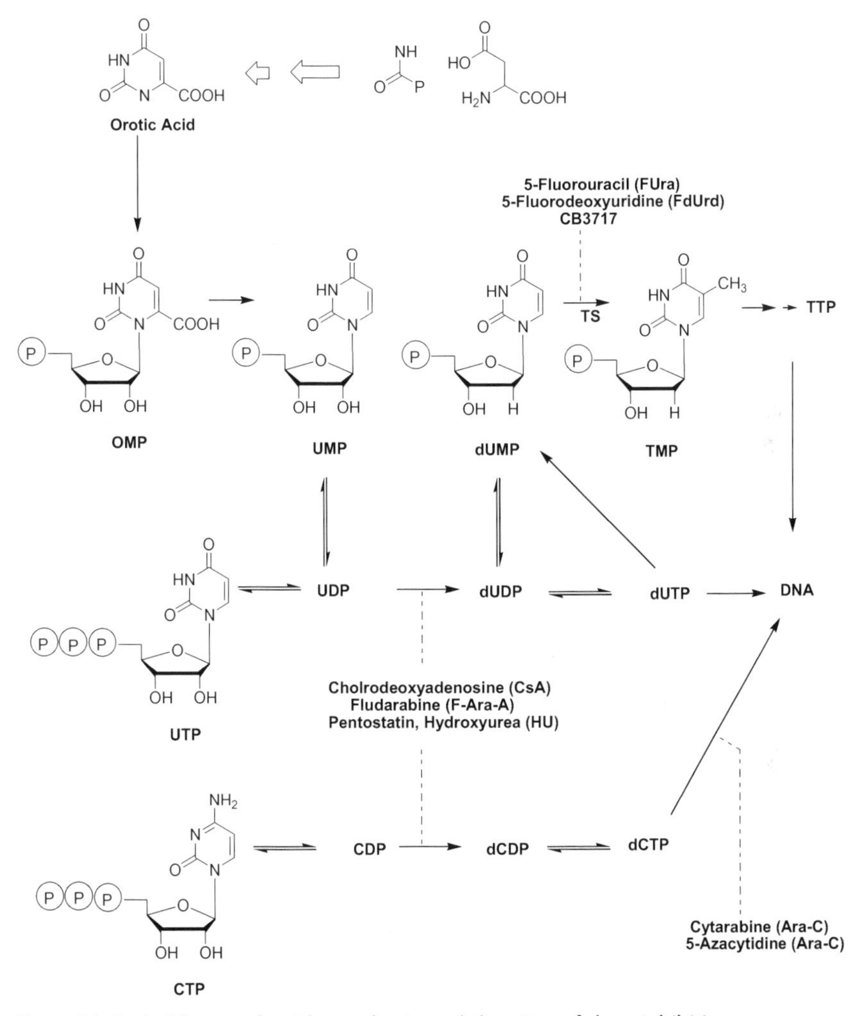

Fig. 1.21 Pyrimidine nucleotide synthesis and the sites of drug inhibition.

the hunt for safe and effective cytarabine formulations and derivatives has intensified [46]. Fazarabine is an aza analog that has shown to be particularly effective in animal models of solid tumors [47].

Gemcitabine, like cytarabine, stops the cell cycle at the S phase and is an RNR inhibitor in its diphosphate form. Gemcitabine is the most often used nucleoside analog in treating pancreatic cancer, non-small-cell lung cancer, breast cancer, and bladder cancer. Eventually, 5-azacitidine and its 2′-deoxy analog are misincorporated into nucleic acids after being triphosphorylated. Because cytarabine is rapidly deaminated to its uracil analog by cytidine deaminase, it has a short half-life in plasma.

3.7.2 Purine nucleosides

Cladribine and fludarabine are utilized in cancer therapy, particularly as the second-line treatment for B-cell chronic lymphocytic leukemia (CLL) patients who have not responded to alkylating drugs. Other purine nucleosides (such as nelarabine, clofarabine, and forodesine) have lately entered clinical studies [48]. These compounds enter cells via nucleoside-specific membrane transporters and must subsequently be transformed into active triphosphate forms.

4. Examples of antineoplastic agents

Antineoplastic agents are extensively utilized in cancer therapy as they can inhibit growth via disrupting cell division and through destroying the actively growing cells. The commonly utilized antineoplastic agents are as follows [49–52] (Table 1.1 and Fig. 1.22) [53–56].

- Alkylating agents that interfere with normal mitosis and cell division) e.g., cyclophosphamide (Cytoxan), chlorambucil (Leukeran), carmustine (BiCNU).
- Antimetabolites that interfere with folic acid, pyrimidine, and purine synthesis, e.g., methotrexate (Mexate), mercaptopurine (Purinethol), fluorouracil (Adrucil).
- natural products (antimitotic agents) that block mitosis and cause metaphase arrest, e.g., vincristine (Oncovin), vinblastine (Velban), paclitaxel (Taxol).
- Antibiotics that cause single- and double-strand DNA breaks, e.g., doxorubicin (Adriamycin), actinomycin D (Cosmegen), bleomycin (Bleo).
- Miscellaneous agents, e.g., hydroxyurea (Hydrea), which acts as an antimetabolite in S-phase, and estrogens, which interfere with proteins and hormone receptors in all phases of cell cycle.

Some of these antineoplastic agents are also being utilized in treating nonmalignant diseases, such as the use of methotrexate for rheumatoid arthritis [57], cyclophosphamide 2 for multiple sclerosis [58], and 5-fluorouracil for psoriasis [59].

Table 1.1 Examples of antineoplastic agents.

Antineoplastic agents	Examples
Alkylating agents	Altretamine, Myleran, Carmustine, Bendamustine, Cytoxan, Chlorambucil, Deticene, Mechlorethamine, Isofosfamide, Melphalan, Belustine, Procarbazin, Zanosar, THIO-TEPA, Yondelis, Temozolomide Platinum coordination complexes: Oxaliplatin, Cisplatin, Carboplatin
Antibiotics	Bleomicin, Cosmegen, Cerubidine, Doxorubicin, Epiadriamycin, Idarubicin, Mutamycin, Mitozantrone, Mithramycin, Valstar
Antimetabolites	Antifolates: Methotrexate, Pralatrexate, Pemetrexed, Trimetrexate Pyrimidine analogs: Capecitabine, Azacitidine, Gemcitabine, Cytarabine, Floxuridine, Decitabine, Fluorouracil, Trifluridine/Tipracil Purine analogs: Cladribine, Azathioprine, Fludarabine, Thioguanine, Mercaptopurine
Biologic response modifiers	Denileukin Diftitox, Aldesleukin, Interferon Gamma
Histone deacetylase inhibitors	Panobinostat, Belinostat (PXD101), Vorinostat, Antibiotic FR 901228
Hormonal agents	Antiandrogens: Abiraterone (CB-7598), Enzalutamide, Erleada, Casodex, Flutamide, Nilutamide, Ciproterona Antiestrogens (e.g., aromatase inhibitors): Aromasin, Faslodex, Anastrozole, Femara, Tamoxifen, Keoxifene, Farestone Gonadotropin releasing hormone analogs: Zoladex, Histrelinum, Degarelix, Triptoreline, Leuprolide Peptide hormones: Octreotide, Angiopeptin acetate, Signifor
Monoclonal antibodies	Blinatumomab, Atezolizumab, Brentuximab, Bevacizumab, Alemtuzumab, Cemiplimab, Daratumumab, Dinutuximab, Cetuximab, Durvalumab, Gemtuzumab, Inotuzumab Ozogamicin, Elotuzumab, Ipilimumab, Mogamulizumab, Necitumumab, Moxetumomab Pasudotox, Nivolumab, Avelumab, Pembrolizumab, Olaratumab, Ofatumumab, Panitumumab, PertuzumabRamucirumabRituximab TositumomabTrastuzumab

Continued

Table 1.1 Examples of antineoplastic agents—cont'd

Antineoplastic agents	Examples
Protein kinase inhibitors	Calquence, Tovok, Abemaciclib, Inlyta, Alectinib, Velcade, Binimetinib, Bosutinib, Cabozantinib, ALUNBRIG, Carfilzomib, Cobimetinib, ZYKADIA, Aliqopa, Tafinlar, Dacomitinib (PF299804, PF299), Crizotinib, Sprycel, Enasidenib, Encorafenib (LGX818), Duvelisib, Iressa, Erlotinib, Xospata, IMBRUVICA, Zydelig, Glasdegib, Imatinib, MLN-2238, Lenvatinib, Ivosidenib, Lapatinib, Midostaurin, LOXO-101, Nilotinib, Loratinib, Nerlynx, Olaparib, Niraparib, Mereletinib, Ponatinib, Pazopanib (GW-786034), Pexidartinib (PLX3397), Palbociclib, BAY 73–4506, Ruxolitinib, Selumetinib, Rucaparib, Erismodegib, Sunitinib, Sorafenib, (BMN-673), Zactima, Zelboraf, Trametinib, Brukinsa, Vismodegib
Taxanes	Paclitaxel, Docetaxel, Cabazitaxel
Topoisomerase inhibitors	Irinotecan, VePesid, Topotecan, Vumon
Vinca alkaloids	Vincristine, Vinblastin, Navelbine
Miscellaneous	Bexarotene, asparaginate, Everolimus, Halaven, Ixabepilone, Hydroxycarbamide, Omacetaxine, Revlimid, Pomalidomide, Lysodren, Telotristat, Thalidomide, Torisel, Tagraxofusp, Venetoclax

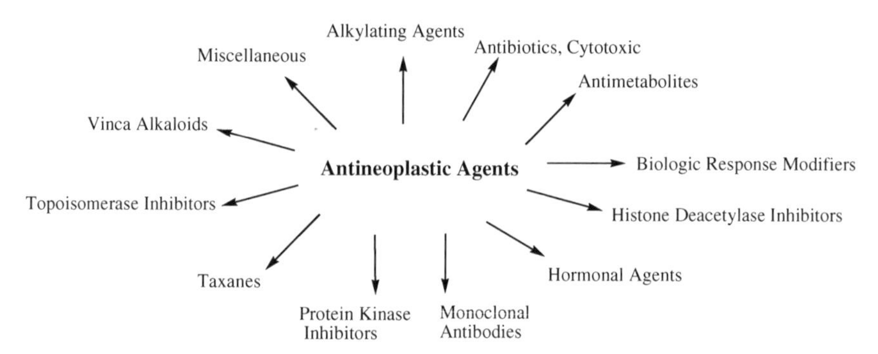

Fig. 1.22 Antineoplastic agents.

References

[1] Aaron JJ, Trajkovska FS. Purines, pyrimidines, and nucleotides. In: Worsfold P, Townshend A, Poole C, editors. Encyclopedia of analytical science. 2nd ed. Oxford: Elsevier; 2005. p. 393–402.
[2] Goodsell DS. The molecular perspective: methotrexate. Oncologist 1999;4(4):340–1.
[3] Raimondi MV, Randazzo O, La Franca M, et al. DHFR inhibitors: reading the past for discovering novel anticancer agents. Molecules 2019;24(6):1140.
[4] He J, Qiao W, An Q, Yang T, Luo Y. Dihydrofolate reductase inhibitors for use as antimicrobial agents. Eur J Med Chem 2020;195, 112268.
[5] McCormack JJ. Dihydrofolate reductase inhibitors as potential drugs. Med Res Rev 1981;1(3):303–31.
[6] Capasso C, Supuran CT. Sulfa and trimethoprim-like drugs—antimetabolites acting as carbonic anhydrase, dihydropteroate synthase and dihydrofolate reductase inhibitors. J Enzyme Inhib Med Chem 2014;29(3):379–87.
[7] Lehman NL. Future potential of thymidylate syntase inhibitors in cancer therapy. Expert Opin Investig Drugs 2002;11(12):1775–87.
[8] Walters CL, Arend RC, Armstrong DK, Naumann RW, Alvarez RD. Folate and folate receptor alpha antagonists mechanism of action in ovarian cancer. Gynecol Oncol 2013;131(2):493–8.
[9] Ackland SP, Peters GJ. Thymidine phosphorylase: its role in sensitivity and resistance to anticancer drugs. Drug Resist Updat 1999;2(4):205–14.
[10] Takemura Y, Jackman AL. Folate-based thymidylate synthase inhibitors in cancer chemotherapy. Anticancer Drugs 1997;8(1):3–16.
[11] Showler MS, Weiser BP. A possible link to uracil DNA glycosylase in the synergistic action of HDAC inhibitors and thymidylate synthase inhibitors. J Transl Med 2020;18(1):377.
[12] Ackland SP, Clarke SJ, Beale P, Peters GJ. Thymidylate synthase inhibitors. Update Cancer Ther 2006;1(4):403–27.
[13] Jackman AL, Judson IR. The new generation of thymidylate synthase inhibitors in clinical study. Expert Opin Investig Drugs 1996;5(6):719–36.
[14] Van Triest B, Pinedo HM, Giaccone G, Peters GJ. Downstream molecular determinants of response to 5-fluorouracil and antifolate thymidylate synthase inhibitors. Ann Oncol 2000;11(4):385–91.
[15] Touroutoglou N, Pazdur R. Thymidylate synthase inhibitors. Clin Cancer Res 1996;2(2):227–43.
[16] Jackman C. Folate-based thymidylate synthase inhibitors as anticancer drugs. Ann Oncol 1995;6(9):871–81.
[17] Galvani E, Peters GJ, Giovannetti E. Thymidylate synthase inhibitors for non-small cell lung cancer. Expert Opin Investig Drugs 2011;20(10):1343–56.
[18] Rustum YM, Harstrick A, Cao S, Vanhoefer U, Yin MB, Wilke H, Seeber S. Thymidylate synthase inhibitors in cancer therapy: direct and indirect inhibitors. J Clin Oncol 1997;15(1):389–400.
[19] Papamichael. The use of thymidylate synthase inhibitors in the treatment of advanced colorectal cancer: current status. Stem Cells 2000;18(3):166–75.
[20] Anon. Clark perspectives on new chemotherapeutic agents in the treatment of colo rectal cancer. Semin Oncol 1997;24(5):S18-19–24.
[21] Ulrich CM, Robien K, Sparks R. Pharmacogenetics and folate metabolism—a promising direction. Pharmacogenomics 2002;3(3):299–313.
[22] Robien K, Boynton A, Ulrich CM. Pharmacogenetics of folate-related drug targets in cancer treatment. Pharmacogenomics 2005;6(7):673–89.
[23] Scagliotti GV, Selvaggi G. Antimetabolites and cancer: emerging data with a focus on antifolates. Expert Opin Ther Pat 2006;16(2):189–200.

[24] Mani S, Ratain MJ. Promising new agents in oncologic treatment. Curr Opin Oncol 1996;8(6):525–34.
[25] Bhagavan NV, Ha C. Essentials of medical biochemistry with clinical cases. 2nd ed. Academic Press; 2015. p. 465–87.
[26] Damaraju VL, Cass CE, Sawyer MB. Renal conservation of folates role of folate transport proteins. Vitam Horm 2008;79:185–202.
[27] McGuire JJ. Anticancer antifolates: current status and future directions. Curr Pharm Des 2003;9(31):2593–613.
[28] Frouin I, Prosperi E, Denegri M, Negri C, Donzelli M, Rossi L, Riva F, Stefanini M, Scovassi AI. Different effects of methotrexate on DNA mismatch repair proficient and deficient cells. Eur J Cancer 2001;37(9):1173–80.
[29] William BP, Raymond WR, William DE, Jonathan M. The anticancer drugs antimetabolites. 2nd ed. Oxford University Press; 1994. p. 69.
[30] Wu Y, Kochat H. Process for synthesizing antifolates; 2005. US 20050020833 A1.
[31] Sahr T, Ravanel S, Basset G, Nichols BP, Hanson AD, Rebeille F. Folate synthesis in plants: purification, kinetic properties, and inhibition of aminodeoxychorismate synthase. Biochem J 2006;396(1):157–62.
[32] Anthony JC. The complete book of enzyme therapy. Avery; 1999. p. 57.
[33] Simon P, Michael JR. Quinazoline derivatives as metabolically inert antifolate compounds; 2006. WO 2006029385 A2.
[34] Wright DL, Anderson AC. Antifolate agents: a patent review (2006–2010). Expert Opin Ther Pat 2011;21(9):1293–308.
[35] Askari BS, Krajinovic M. Dihydrofolate reductase gene variations in susceptibility to disease and treatment outcomes. Curr Genomics 2010;11(8):578–83.
[36] Field MS, Anderson DD, Stover PJ. Mthfs is an essential gene in mice and a component of the purinosome. Front Genet 2011;2:36.
[37] Salcedo-Sora JE, Ochong E, Beveridge S, Johnson D, Nzila A, Biagini GA, Stocks PA, O'Neill PM, Krishna S, Bray PG, et al. The molecular basis of folate salvage in Plasmodium falciparum: characterization of two folate transporters. J Biol Chem 2011;286 (52):44659–68.
[38] Redei I, Green F, Hoffman JP, Weiner LM, Scher R, O'Dwyer PJ. Phase II trial of PALA and 6-methylmercaptopurine riboside (MMPR) in combination with 5-fluorouracil in advanced pancreatic cancer. Invest New Drugs 1994;12(4):319–21.
[39] Avendano C, Menendez JC. Chapter 2—Antimetabolites that interfere with nucleic acid biosynthesis. In: Medicinal chemistry of anticancer drugs. 2nd ed. Elsevier Science; 2015. p. 23–79.
[40] Blakley RL, Cocco L. Role of isomerization of initial complexes in the binding of inhibitors to dihydrofolate reductase. Biochemistry 1985;24(18):4772–7.
[41] Kamen BA, Whyte-Bauer W, Bertino JR. A mechanism of resistance to methotrexate. NADPH but not NADH stimulation of methotrexate binding to dihydrofolate reductase. Biochem Pharmacol 1983;32(12):1837–41.
[42] Sessa C, de Jong J, D'Incalci M, Hatty S, Pagani O, Cavalli F. Phase I study of the antipurine antifolate lometrexol (DDATHF) with folinic acid rescue. Clin Cancer Res 1996;2(7):1123–7.
[43] Aberra FN, Lichtenstein GR. Review article: monitoring of immunomodulators in inflammatory bowel disease. Aliment Pharmacol Ther 2005;21:307–19.
[44] Johnson SA. Clinical pharmacokinetics of nucleoside analogues: focus on haematological malignancies. Clin Pharmacokinet 2000;39(1):5–26.
[45] Ohta T, Hori H, Ogawa M, Miyahara M, Kawasaki H, Taniguchi N, Komada Y. Impact of cytidine deaminase activity on intrinsic resistance to cytarabine in carcinoma cells. Oncol Rep 2004;12(5):1115–20.
[46] Hamada A, Kawaguchi T, Nakano M. Clinical pharmacokinetics of cytarabine formulations. Clin Pharmacokinet 2002;41(10):705–18.

[47] Ben-Baruch N, Denicoff AM, Goldspiel BR, O'Shaughnessy JA, Cowan KH. Phase II study of fazarabine (NSC 281272) in patients with metastatic colon cancer. Invest New Drugs 1993;11(1):71–4.
[48] Robak T, Lech-Maranda E, Korycka A, Robak E. Purine nucleoside analogs as immunosuppressive and antineoplastic agents: mechanism of action and clinical activity. Curr Med Chem 2006;13(26):3165–89.
[49] Rogers B. Health hazards to personnel handling antineoplastic agents. In: Emmett EA, editor. Occupational medicine, health problems of health care workers, vol. 2 (3). Philadelphia, PA: Hanley & Belfus, Inc.; 1987. p. 513–24.
[50] Chabner BA, Allegra CJ, Curt GA, Calabresi P, Chapter 51. Antineoplastic agents. In: Hardman JG, Limbird LE, Molinoff PB, Ruddon RW, Gilman AG, editors. Goodman and Gilman's: the pharmacological basis of therapeutics. 9th ed. New York: McGraw-Hill; 1996. p. 1233–87.
[51] Jochimsen PR. Handling of cytotoxic drugs by health care workers: a review of the risks of exposure. Drug Saf 1992;7(5):374–80.
[52] McFarland HM, Almuete V, Brisby J, Delman B, Farmer D, Grant L, Hudock S, Lewis J, Malong M, Roche K, Topping D, Steinberg J, Wyrik J. 2001 guide for the administration and use of cancer chemotherapeutic agents. Oncology special edition 2001;4:96–105.
[53] Anon, https://www.ncbi.nlm.nih.gov/books/NBK548022/.
[54] Zimmerman HJ. Hepatotoxic effects of oncotherapeutic and immunosuppressive agents. In: Zimmerman HJ, editor. Hepatotoxicity: the adverse effects of drugs and other chemicals on the liver. 2nd ed. Philadelphia: Lippincott; 1999. p. 673–708.
[55] DeLeve LD. Cancer chemotherapy. In: Kaplowitz N, DeLeve LD, editors. Drug-induced liver disease. 3rd ed. Amsterdam: Elsevier; 2013. p. 541–67.
[56] Wellstein A. General principles in the pharmacotherapy of cancer. In: Brunton LL, Hilal-Dandan R, Knollman BC, editors. Goodman & Gilman's the pharmacological basis of therapeutics. 13th ed. New York: McGraw-Hill; 2018. p. 1161–6.
[57] Baker GL, Kahl LE, Zee BC, Stolzer BL, Agarwal AK, Medsger Jr TA. Malignancy following treatment of rheumatoid arthritis with cyclophosphamide. Am J Med 1987;83(1):1–9.
[58] Moody DJ, Kagan J, Liao D, Ellison GW, Myers LW. Administration of monthly-pulse cyclophosphamide in multiple sclerosis patients. J Neuroimmunol 1987;14:161–73.
[59] Abel EA. Immunosuppressant and cytotoxic drugs: unapproved uses or indications. Clin Dermatol 2000;18:95–101.

CHAPTER 2

Antifolate-based anticancer drugs

1. Introduction

Antimetabolites are potent chemotherapeutic agents for hematologic malignancies and many solid tumors. Folate antagonists are one of the three principal and main categories of antimetabolites. Antifolate-based drugs have striking anticancer activities and are effective in treating various types of malignancies. They were the first to reach clinics 65 years ago.

Antifolates disrupt cellular proliferation through blocking the folate-dependent one-carbon biosynthesis and methylation process.

Antifolates form active polyglutamate derivatives that are retained in tumor cells, thus leading to a sustained inhibition of their target enzymes. Antifolates are currently approved for treating cancer, such as pralatrexate and methotrexate, which are the inhibitors of dihydrofolate reductase, and pemetrexed, which targets thymidylate synthase and 5-aminoimidazole-4-carboxamide ribonucleotide (AICAR) transformylase, thus leading to the accumulation of polyglutamate form of folates. New folate analogs and conjugates, which are currently used in clinical trials, exploit membrane folate receptors for transport into tumor cells through endocytic mechanism [1].

Antifolate drug development is based on agents designed to overcome various aspects of the drug resistance. Currently, novel antifolate agents are in different stages of clinical development. Antifolate toxicity has been sporadic and difficult to anticipate clinically. Supplementing antifolate with vitamin B12 and folic acid has been revealed to reduce the toxicity of some drugs and increase their therapeutic index. Many antifolate-based anticancer drugs are divided into two types: classical antifolate drugs (e.g., methotrexate, pralatrexate, raltitrexed, and pemetrexed) and nonclassical antifolate drugs, (e.g., trimetrexate). Both the types have unique pharmacological characteristics and are in various stages of clinical development, which are described in this chapter.

New Strategies Targeting Cancer Metabolism
https://doi.org/10.1016/B978-0-12-821783-2.00007-8

2. Classical antifolate drugs (polyglutamate antifolate)

2.1 Methotrexate (MTX) drug

The DHFR inhibitor, MTX (Fig. 2.1), is a classical polyglutamate antifolate, as it bears a side chain consisting of *p*-aminobenzogylglutamic acid [2].

MTX is a chemotherapeutic drug that inhibits DNA synthesis and induces apoptosis [3–5]. MTX is employed in the therapy of solid tumors, leukemias [6–8], severe asthma [7], along with other pathological conditions [9] (such as cancer [10–17], melanoma—which is a highly aggressive form of cancer [18]—chronic inflammatory diseases [12], and osteosarcoma [19]). It can act as an immunosuppressive agent in the treatment of some autoimmune diseases [7,8,15]. Additionally, MTX in combination with technetium-99m-labeled folate act as a tumor imaging agent [20], and it can also form conjugates with other compounds [21].

In 1945 since its discovery, methotrexate has been a conventional therapy for treating a variety of disorders, comprising cancer, pulmonary diseases, and inflammatory diseases. The major physiological interactions of MTX comprise folate pathway, prostaglandins, adenosine, cytokines, and leukotrienes. Methotrexate is a second-line therapy for pulmonary sarcoidosis and is drug of choice for patients who are not candidates for corticosteroid therapy.

The use of methotrexate in rheumatoid arthritis and oncology patients was the focus of a number of studies. Many studies are conducted on the oral use of MTX and MTX in treating patients with chronic sarcoidosis. Polyglutamation of MTX changes its pharmacokinetic and pharmacodynamic characteristics and also increases its activity. Although the majority of methotrexate is excreted in the urine, bile excretion plays a significant role due to considerable enterohepatic recirculation. When corticosteroids are contraindicated in sarcoidosis patients, a better understanding of its pharmacokinetic characteristics is needed to optimize therapy [22]. MTX is the

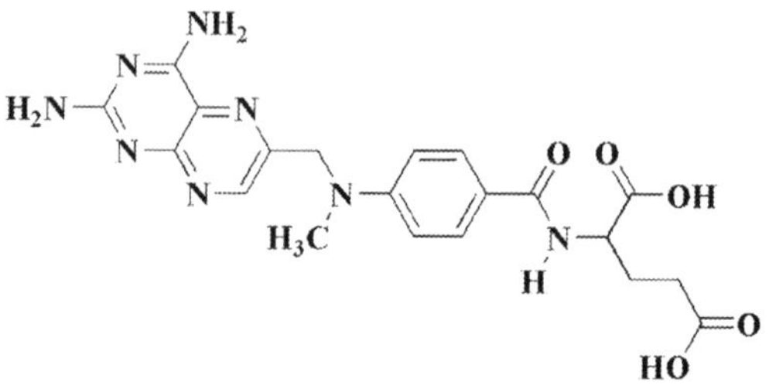

NH_2, H_2N, N, N, N, N, N, H_3C, O, O, OH, N, H, O, HO

Fig. 2.1 Methotrexate.

first-line disease-modifying antirheumatic drug (DMARD) in the treatment of rheumatoid arthritis and has also been shown to be effective in treating other autoimmune diseases. MTX has also been used in treating certain cancers. The most common side effects of MTX are gastrointestinal (such as nausea and vomiting), while some central nervous system (CNS)-related side effects (lethargy and fatigue, headache, and vertigo) are observed. Therefore, administering the drug subcutaneously rather than orally does not always alleviate its side effects [23]. A combination of methotrexate and hydroxyethyl starch can be also used in the treatment of cancer [24].

The mechanism of action is that cell cycle-specific antifolate analog becomes active in the S-phase of the cell cycle as it enters cells via specific transport systems mediated by the folate receptor protein and the reduced folate carrier. Then, inside the cells, folylpolyglutamate synthase (FPGS) converts MTX into its polyglutamates, which have a cytotoxic property. The MTX polyglutamates binds to dihydrofolate reductase (DHFR) and inhibits its activity, thus limiting the availability of one-carbon fragments, which ultimately result in the inhibition of de novo thymidylate and purine syntheses. Due to the inhibition of thymidylate synthesis, the misincorporation of dUTP into DNA occurs instead of dTTP, thus leading to the inhibition of DNA synthesis.

Its mechanisms of resistance are as follows: (1) decreased carrier-mediated transport of drug into the cell is via a decreased expression and/ or activity of folate receptor protein (FRP) or reduced folate carrier (RFC); (2) decreased formation of cytotoxic methotrexate polyglutamates occurs through either a decreased expression of FPGS or an increased expression of γ-glutamyl hydrolase (GGH); (3) increased expression of the target enzyme DHFR is via either gene amplification or increased transcription, translation, and/or posttranslational events; (4) there is a reduced binding affinity of DHFR for MTX; and (5) also, there is a decrease in the expression of mismatch repair enzymes that may play a role in drug resistance [25].

2.2 Pralatrexate drug

Pralatrexate (PTXT) (Fig. 2.2), a dihydrofolate reductase inhibitor used for treating various malignancies potentially [26,27], is a novel antineoplastic folate analog, and named as the (2*S*)-2-[[4-[(1*RS*)-1-[(2,4-diaminopteridin-6-yl)methyl]but-3-ynyl] benzoyl] amino] pentanedioic acid [28] (Folotyn; Allos Therapeutics Inc.). In 2009, the U.S. Food and Drug

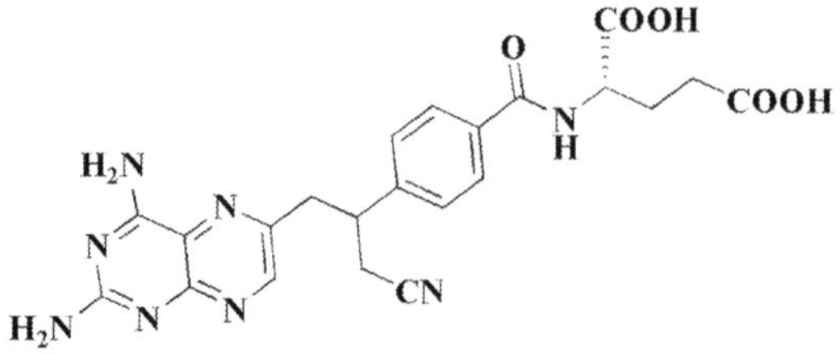

Fig. 2.2 Pralatrexate.

Administration granted accelerated approval for Folotyn that is used as a single agent for treating patients with relapsed or refractory peripheral T-cell lymphoma (PTCL) [29–38]. However, it is considered as the first drug to obtain FDA approval for treating patients with relapsed or refractory PTCL. The European Medicines Agency, on the other hand, has refused the marketing authorization. None of the current treatments have yielded adequate results. Pralatrexate seems to be one of the most promising agents in treating patients with peripheral T-cell lymphomas (PTCLs) [39], which are a heterogeneous group of T-cell neoplasms [40].

Pralatrexate (PDX) is a folate antagonist that resembles methotrexate (MTX) in structure [41]. The folate analog was developed to overcome the folate analog methotrexate's limitations. Compared with methotrexate in preclinical studies, pralatrexate revealed superior intracellular transport by the reduced folate carrier and increased accumulation within cells via enhanced polyglutamylation [26]. Pralatrexate (PDX) has higher affinity for tumor cells than methotrexate [42]. The enhanced activity of pralatrexate compared to MTX is attributed to its significantly faster rate of transport and polyglutamation, with the former being less essential when the carrier is saturated. The poor affinity of pralatrexate for PCFT predicts a lower level of enterohepatic circulation and higher fecal excretion when compared to MTX [43,44]. Pralatrexate was more potent than methotrexate in a panel of solid tumor lines [45]. Pralatrexate displays a potency and pattern of activity, which is distinct from its predecessors like methotrexate (MTX) [46]. It has a high affinity for the one-carbon reduced folate carrier, which results in the better cellular internalization of the drug, and it has a greater antitumor effect than methotrexate [47].

Pralatrexate is a dihydrofolate reductase competitive inhibitor. Polyglutamylation is also competitively inhibited by pralatrexate by its binding to folylpolyglutamyl synthetase. This inhibition results in the loss of thymidine and other biological molecules that require single-carbon transfer for their synthesis [48].

Pralatrexate was designed to have a higher affinity for folylpolyglutamate synthetase (FPGS) and reduced folate carrier (RFC). RFC is the primary transporter of folates and antifolates into cells, leading to extensive internalization and accumulation in tumor cells. Inside the cell, FPGS can efficiently form pralatrexate polyglutamates. Pralatrexate has shown different degrees of efficacy in peripheral T-cell lymphoma, with response rates varying between many subtypes of the disease. While phase III investigations are still ongoing, early clinical trials suggest that pralatrexate could be a viable novel therapeutic for PTCL [49,50]. The prognosis for peripheral T-cell lymphoma (PTCL) is dismal, especially in patients with relapsed/refractory (R/R) disease [51]. Pralatrexate has a significant single-agent activity in patients with relapsed/refractory T-cell lymphoma [52]. While pralatrexate (PDX) has been effectively developed for treating T-cell lymphoma, the mechanism behind its T-cell selectivity and acquired resistance is yet unknown [53]. In addition, pralatrexate may be a treatment option for patients with relapsed or refractory transformed mycosis fungoides (tMF), which is an aggressive disease with poor prognosis [54]. Pralatrexate is being investigated in the treatment of various malignancies. The cytotoxicity patterns of pralatrexate and methotrexate were similar, with pralatrexate being more potent. Pralatrexate potentiated the effects of EGFR inhibitors, platinum drugs, and antimetabolites. The cytotoxicity of pralatrexate was dose and time dependent, which was associated with high FPGS mRNA expression. Pralatrexate resistance was linked to lower RFC-1 expression, while methotrexate resistance was linked to higher DHFR expression, implying various mechanisms of acquired resistance. In a panel of solid tumor lines, pralatrexate was more effective than methotrexate [45]. In treating advanced-stage disseminated extranodal natural killer/T-cell lymphoma, pralatrexate was utilized as a useful bridge to the transplantation of the allogeneic hematopoietic stem cell [55]. The majority of patients with PTCL have a dismal prognosis and cannot be cured without stem cell transplantation. Pralatrexate was developed with the goal of impeding folate metabolism by inhibiting dihydrofolate reductase (DHFR) and being more efficiently internalized into tumor cells, [40] to improve cellular uptake and retention of the drug [56].

Pralatrexate has been tested in clinical studies—as a single agent and in combination with other drugs—for nonlymphoma, Hodgkin's disease, and solid malignancies. Pralatrexate has a significant activity in vitro, and responses were seen in patients with aggressive T-cell lymphomas in early phase I/II trials. Pralatrexate's activity was demonstrated across a spectrum of heavily pretreated patients with different aggressive T-cell lymphoma subtypes. Studies on pralatrexate have demonstrated that at various doses and schedules, it has shown an efficacy in treating cutaneous T-cell lymphoma. Mucositis, reversible thrombocytopenia, and fatigue were the most common side effects found in these studies [40,57]. Pralatrexate showed an activity in treating lymphoma in both preclinical and clinical investigations. Supplementation of pralatrexate with folate and vitamin B12 is recommended to reduce its toxic effects [58]. Mucositis has been reported as the most common dose-modifying adverse event [59], and this can be mitigated by the utility of leucovorin along with folic acid and cyanocobalamin [60,61].

Pralatrexate exhibits antineoplastic and immunosuppressive activities, which are similar to those of methotrexate. The survival rates for children with high-risk neuroblastoma have not been improved despite breakthroughs in multimodality treatment techniques. As a result, intensive research is still underway to find the best new agent or combination of chemotherapeutic medicines for treating high-risk neuroblastoma. Pralatrexate has been shown to effectively inhibit cell growth and viability. Pralatrexate's increased potency compared to methotrexate, a medication with high toxicity levels, implies that it could be a safer option to methotrexate as a chemotherapeutic agent in the treatment of high-risk neuroblastoma patients [62].

Many lymphoid malignancies, including chemotherapy-resistant T-cell lymphoma, respond well to pralatrexate. According to new combination trials, pralatrexate is extremely synergistic with gemcitabine and histone deacetylase inhibitors, such as romidepsin [63–65] and bortezomib [65]. It is a proteasome inhibitor that has indicated some potency in T-cell lymphoma patients. Bortezomib in combination with pralatrexate acts a new and potentially crucial platform for treating T-cell malignancies [66]. This combination was typically safe and showed the modest efficacy in treating relapsed or refractory multiple myeloma [67]. Also, the combination of pralatrexate and oxaliplatin is safe, is well tolerated, and has the modest efficacy in treating in advanced esophagogastric cancer (EGC). Pharmacogenomic analysis may be relevant to the use of pralatrexate in combination with platinum agents [68] and carboplatin [69]. Additionally, the combination of oral bexarotene and pralatrexate is active with high response rates and minimal toxicity for cutaneous T-cell lymphomas [70]. Other combination of pralatrexate and belinostat, wherein said therapeutically effective amount, leads to a synergistic antiproliferative effect on cancer cell growth [71].

Many inventions are provided that accomplish the syntheses of the salts of pralatrexate, their polymorphic forms, pharmaceutical compounds comprising these salts and at least one pharmaceutically acceptable excipient, and the use of pharmaceutical compound for treating conditions related to human tumors [72,73] and for treating relapsed or refractory peripheral T-cell lymphoma [74]. Other alpha polyglutamated pralatrexate and liposome-containing formulations have been synthesized to treat hyperproliferative disorders (e.g., cancer) and disorders of the immune system (e.g., an autoimmune disease such as rheumatoid arthritis [75]).

2.3 Tomudex drug (raltitrexed; ZD1649)

The chemical name of Tomudex™ (raltitrexed, ZDI694; Zeneca Pharmaceuticals, Macclesfield, UK) (ICI D1694) is *N*-(5-[*N*-(3,4-dihydro-2-methyl-4-oxoquinazolin-6-ylmethyl)-*N*-methylamino]-2-theonyl-L-glutamic acid (Fig. 2.3) [76,77]. It is a new thymidylate synthase inhibitor [78–80] and is considered as a potential anticancer drug [2]. It is active in several preclinical and clinical settings [81,82].

Raltitrexed, as a quinazoline folate analog, can selectively inhibit thymidylate synthase (TS) [83], which is a key enzyme in the de novo synthesis of thymidine triphosphate (TTP), a nucleotide needed for deoxyribonucleic acid (DNA) synthesis. DNA fragmentation and cell death occur when TS is inhibited [83]. Intracellular retention of polyglutamated forms of raltitrexed causes prolonged inhibitory effects [84]. This allows a convenient dosage schedule of a single IV injection once every 3 weeks [85]. Raltitrexed is a radiation sensitizing agent [86]. It induces cell cycle arrest in the S-phase [87].

Tomudex inhibits DNA replication and development in rapidly developing tissues by blocking critical steps in one-carbon metabolism [88]. Tomudex, TDX, is a second-generation folate analog that, when polyglutamated, is a potent inhibitor of thymidylate synthase (TS), the enzyme that converts dUMP to dTMP [89,90]. It is an alternative antifolate drug to 5-fluorouracil (5-FU) to inhibit thymidylate synthase (TS) by lowering

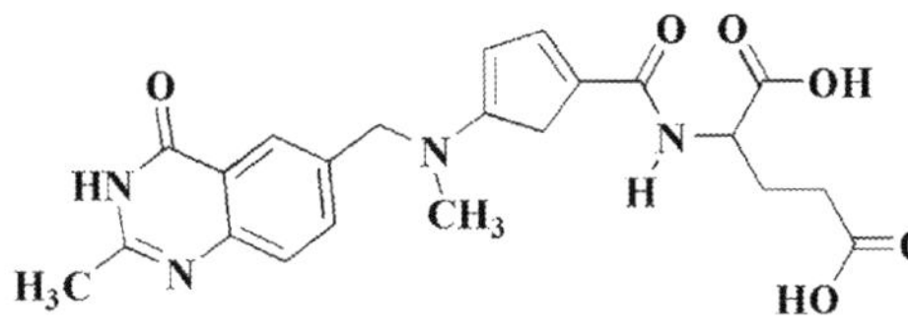

O
HN
H_3C
N
N
CH_3
O O
OH
N
H
O
HO

Fig. 2.3 Tomudex.

dihydrofolate reductase (DHFR) potency. The clinical trial demonstrates that combining raltitrexed with other anticancer drugs can improve the ability of raltitrexed to inhibit TS [91,92]. The overexpression of thymidylate synthase (TS) is widely accepted as a major molecular mechanism responsible for Tomudex (TDX) resistance [93].

Tomudex is a classical antifolate employing carrier-mediated cellular uptake mechanisms (RFC) and is an FPGS substrate. Because polyglutamates are more effective TS inhibitors than the original drug and are maintained inside the cells for longer periods of time, polyglutamation is a serious pharmacological concern. The latter property has lent itself to a convenient, infrequent clinical dosing regimen [94].

The Institute of Cancer Research and Zeneca Pharmaceuticals collaborated to design the specific folate-based quinazoline TS inhibitors.

Tomudex, the first of these medications to reach advanced clinical development, is currently undergoing phase III trials. Tomudex shows potency in a range of tumor types, including breast cancer (objective response rate of 26%) and advanced colorectal cancer, and it shows acceptable toxic effects: The most common WHO grade 3 and 4 adverse events were self-limiting reversible increases in liver transaminases, nausea and vomiting, transient leucopenia, diarrhea, and malaise or tiredness. Alopecia, stomatitis/mucositis, and skin toxicity were notable for their mild intensity and low incidence. Tomudex is a novel cytotoxic agent used for the treatment of colorectal cancer and marks the effective culmination of a rational drug design program. Its effect on other types of tumors will be studied in the future [95]. It is approved for clinical use in several European countries [96]. Tomudex is about to enter the Italian pharmaceutical market [97].

Tomudex is a drug recently introduced for the treatment of advanced colorectal cancer [98–102]. Raltitrexed received its first regulatory approval for the first-line treatment of advanced colorectal cancer [103].

It was the first new treatment for colorectal cancer in about 40 years (UK 1995) and the first new drug for nearly 35 years. Tomudex belongs to a class of drugs that utilize the reduced folate carrier (RFC) for its uptake into the cell and that are excellent substrates for folylpolyglutamate synthetase (FPGS) [104,105].

In addition, the antitumor activities of Tomudex and SN38, and their combination, were studied in human colon tumor cell lines [106].

Several medicines, including raltitrexed, target TS as an anticancer target in treating colorectal and other tumors. Drug resistance can be mediated by TS overexpression, which can be due to enhanced gene transcription and mRNA translation. Reduced cellular uptake and polyglutamylation of TS-targeting medications (e.g., raltitrexed), increased drug efflux, altered cytotoxic drug metabolism (e.g., 5-FU), and other processes can reduce TS-targeting therapies' efficiency [107].

Membrane transport is considered as a critical determinant of the Tomudex's antitumor potency utilized in cancer chemotherapy, and impaired uptake of antifolates is a frequent mode of drug resistance [108]. Increased thymidylate synthase activity and diminished intracellular drug absorption and polyglutamation are all mechanisms of Tomudex resistance. However, little is known about other mechanisms of resistance, such as a possible protection against Tomudex-induced apoptosis mediated by bcl-2 [109].

Tomudex as a drug acts as a topoisomerase I inhibitor and interferes with signal transduction pathway [110]. It is also considered as a drug with increased target specificity [111].

Raltitrexed monotherapy, a direct, specific thymidylate synthase (TS) inhibitor, is an effective option in the first-line treatment for advanced colorectal cancer (ACC) when compared with high-dose leucovorin (LV) plus fluorouracil (5-FU). Raltitrexed has been proven to be a viable first-line palliative treatment for ACC, with comparable efficacy and tolerability advantages (in terms of reduced incidence of stomatitis, leukopenia, and diarrhea) over 5-FU/LV [112].

The potential of Tomudex to interact with ionizing radiation was assessed in vitro and in vivo in comparison with that of 5-fluorouracil. The results achieved with Tomudex were equal to or better than those obtained with 5-fluorouracil in each assay. These findings revealed that clinical trial of Tomudex along with fractionated radiation therapy is warranted [113].

Tomudex is currently approved only in Europe and only used for treating colon cancer. While MTX is known to be unsuccessful in treating colon cancer, this is still a step forward for antifolates; thus, Tomudex expands the tumor range of antifolates [114].

It features a convenient dosing schedule and the possibility for decreased toxicity, both of which are significant advantages over current treatments. 5-Fluorouracil is being used to treat advanced colorectal cancer, usually in combination with other drugs like leucovorin. This results in complicated dosing schedules, increased activity, and the possibility of significant toxicity. Raltitrexed is a novel cytotoxic agent, rationally designed to inhibit a specific molecular target, thymidylate synthase. Unlike other existing agents, raltitrexed inhibits thymidylate synthase in a direct, specific, and noncompetitive manner, potentially improving the toxicity profile. It is retained within cells as a polyglutamate metabolite, thus allowing a more convenient dosing schedule than for 5-fluorouracil [115].

In contrast to MTX, Tomudex has low K_ms for folylpolyglutamate synthetase, and polyglutamate forms of the inhibitor are accumulated in both lymphoid and myeloid acute leukemia cells, paralleling the equivalent cytotoxicity found between lymphoid and myeloid leukemia cell lines. It was believed that a clinical trial of Tomudex in patients with acute myeloid leukemia is warranted [116]. In vitro investigations have indicated a schedule-dependent synergism between 5-fluorouracil (5-FU) and Tomudex. The combination of Tomudex, 5-FU, and LFA is well tolerated and active in head and neck cancer and colorectal cancer. Tomudex's actual mean dose intensity is higher than what is generally attained in monotherapy. Because there is not a clear pharmacokinetic interaction between 5-FU and Tomudex, it is possible that the synergism occurs at the cellular level [117]. As a single agent or in combination with 5-FU, Tomudex appears to be the most promising drug [118]. The different possible drug combinations with 5-FU currently confirm that current standards will be changed in the near future [119].

It is worthy to note that raltitrexed is ineffective as a second-line therapy for small-cell lung cancer [120]. Tomudex, a recent and expensive drug, seems to exhibit toxic effects [121].

2.4 Pemetrexed drug (Alimta)

L-Glutamic acid, *N*-[4-[2-(2-amino-4,7-dihydro-4-oxo-1*H*-pyrrolo [2,3-*d*]pyrimidin-5-yl)ethyl]benzoyl] [27] (Fig. 2.4), is a multitargeted antifolate that inhibits several folate-dependent enzymes (TS, DHFR, and GARFT),

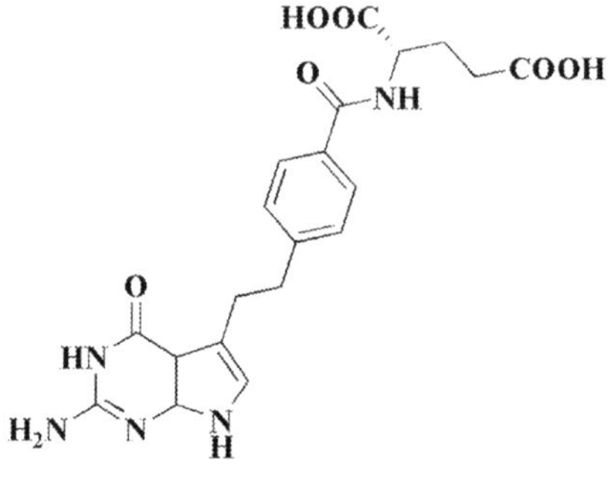

HOOC
COOH
O
NH
O
HN
H_2N
N
N
H

Fig. 2.4 Pemetrexed.

which play roles in the synthesis of purine and pyrimidine [122–150]. It is effective against malignant pleural mesothelioma [134,151–162], lung cancer [163] particularly, non-small-cell lung cancer [151–166], and many solid tumors [162–168].

Like raltitrexed and 5-fluorouracil, pemetrexed primarily inhibits thymidylate synthase (TS), leading to a decrease in the thymidine available for DNA synthesis. Pemetrexed also inhibits glycinamide ribonucleotide formyltransferase (GARFT) and dihydrofolate reductase (DHFR), which are essential for the de novo biosynthesis of thymidine and purine nucleotides. Pemetrexed is polyglutamated once it enters the cell via the reduced folate carrier. Glutamation boosts the cellular retention and intracellular half-life of pemetrexed, and it also produces polyglutamated metabolites, which are 60 times more potent in inhibiting TS. In the G1/S phase, pemetrexed induces the cell cycle arrest [169].

Pemetrexed (PEM) is an antimetabolite drug that interferes with enzymes involved in DNA synthesis and also with the folate-dependent metabolic processes necessary for homocysteine homeostasis and DNA replication [170]. Pemetrexed is considered as a first-line treatment for mesothelioma PEM, and it inhibits DNA and RNA syntheses. It suppresses many enzymes involved in the synthesis of pyrimidines and purines, and the increased activity of these enzymes in malignancies is frequently associated with PEM resistance [171].

Pemetrexed (PEME) (PMX) is a pharmaceutical used in treating tumors [172]. Pemetrexed (PMT) is utilized in treating non-small-cell lung cancer (NSCLC) patients [173,174]. It is one of the most commonly used NSCL medications, which causes patients to have a variety of therapeutic responses. These therapeutic responses have been shown to be highly individual specific, including adverse drug reactions (ADRs) and intended therapeutic outcomes [175]. Pemetrexed was authorized by the Food and Drug Administration (FDA) three times in the United States for treating locally progressed or metastatic nonsquamous NSCLC in 2004, 2008, and 2009. [176]. However, in clinical practice, the best use of pemetrexed and other medications includes a therapy schedule still a work in progress, especially for NSCLC with wild-type epidermal growth factor receptors (EGFR). Current investigations of a prospective therapeutic approach were made for NSCLC patients with wild-type EGFR who were treated with pemetrexed [177].

Pemetrexed disodium (PEM, Alimta), a small hydrophilic drug, is currently utilized in the treatment of patients suffering from lung cancer. PEM, on the other hand, has drawbacks such as low bioavailability, rapid elimination, and poor tumor cell selectivity and penetration [178].

Patients had good tolerability, moderate side effects, a high percentage of treatment completion, and a long, progression-free survival period when pemetrexed was used in the maintenance treatment of non-small-cell lung cancer. It is an ideal drug for improving the prognosis of patients [179], but with short half-life and toxic adverse effects [180].

Following its approval as a second-line monotherapy for treating locally advanced or metastatic nonsquamous NSCLC, pemetrexed has established itself as the first-line regimen in conjunction with cisplatin [181]. Pemetrexed (PEM) can also be used in combination with immune checkpoint blockade therapy to treat advanced non-small-cell lung cancer patients (NSCLC). PEM can efficiently sensitize human NSCLC cells to cytotoxic immune cells while modifying the expression of immunoregulatory molecules, according to these findings [182].

Pemetrexed targets various folate biosynthesis enzymes and is used to treat malignant pleural mesothelioma (MPM) and a variety of neoplasms [183,184], which are aggressive tumors, rare, treatment-resistant cancer characterized by poor prognosis. There is no effective treatment for MPM at this time. Pemetrexed (Pe) is one of the few chemotherapeutic drugs licensed for advanced-stage cancer, despite the drug's weak objective response [185]. So being a thymidylate synthase (TS) inhibitor, pemetrexed is used as a first-line chemotherapy regimen for MPM all over the world. However, there is little consensus for a second-line chemotherapy [186].

The overexpression of the thymidylate synthase (TS) gene limits its therapeutic efficacy in many clinical contexts [187]. Although it has been claimed that high levels of thymidylate synthase (TYMS) gene expression in malignant tumors are linked to a decreased sensitivity to the antifolate medication pemetrexed, no clear evidence for this has been found in routine clinical samples from patients treated with the drug [188].

Pemetrexed is eliminated by tubular secretion via human organic anion transporter 3 (hOAT3). Although proton pump inhibitors (PPIs) are commonly utilized in cancer patients, the drug interaction between PPIs and **pemetrexed** is unknown [189].

PMX-conjugated hyaluronan (HA-ADH-PMX) was synthesized for the first time to develop a new anticancer chemotherapeutic agent. Hyaluronan (HA) in blood is widely known as a disease marker of MPM [190]. There are inventions related to a pemetrexed-hyaluronic acid conjugate that is covalently bound to hyaluronic acid or a hyaluronic acid derivative, either directly or through a spacer (e.g., hydrazide spacer), which was developed as a pharmaceutical component for treating disseminated tumor [191].

Chemotherapeutic drugs do not target only cancer cells in treatment. During treatment, these drugs interact with both normal and abnormal cells.

Pemetrexed may have genotoxic potential on normal leukocytes [192]. Cutaneous reactions are associated with pemetrexed use (in approximately 20% of patients) [193]. To prevent cutaneous responses, pemetrexed prescribing information recommends oral dexamethasone before its infusion [194].

Although glucocorticoids were frequently used in conjunction with PEM to minimize toxicity during chemotherapy, it is unclear whether glucocorticoids could alter PEM efficacy in NSCLC [195]. So as pemetrexed is now a widely used chemotherapeutic agent, it is important to be aware of rare adverse events related to its administration, and it has often a limited efficacy due to drug resistance [196]. Chemoresistance may inhibit clinical efficacy after long-term use. Mechanisms responsible for chemoresistance to pemetrexed in NSCLC are plethoric but can be divided into two categories: tumor cells and their interactions with drugs [197]. There are also reports for the induction of fibrosing disorders [193].

Pemetrexed's acute renal toxicity has recently been characterized in the context of polychemotherapy, in which determining the individual liability of each agent is challenging. The optimal strategy with regard to renal complications in cancer patients is not clear. Acute or chronic loss in renal function generally results in a novel treatment line, possibly jeopardizing the overall success of the treatment [198].

It is worthy to note that one of the most common therapies for breast cancer is pemetrexed. Multidrug resistance (MDR) in breast cancer, on the other hand, severely limits the therapeutic efficiency of chemotherapies like pemetrexed. Nanomedicine is gaining traction as a viable alternative to technique for cancer MDR. Pemetrexed liposomal is a viable treatment strategy to overcome MDR breast cancer [199].

By screening an FDA-approved drug library, pemetrexed was identified as a potent anti-KSHV agent (IC_{50} of 90 nM). The etiological agent of Kaposi's sarcoma, primary effusion lymphoma, and multicentric Castleman's disease is Kaposi's sarcoma-associated herpesvirus (KSHV). Pemetrexed interferes with the lytic replication of viral DNA, leading to the reduction of infectious virions.

The antiviral effect of pemetrexed relies on the dTMP synthesis pathway that needs the folate-dependent enzymes. Pemetrexed also has a wide range of antiherpes virus potency. Pemetrexed inhibits the lytic replication of KSHV DNA by blocking dTMP synthesis. Pemetrexed has the potential to be used as an anti-KSHV agent [200].

Additionally, pemetrexed is a radiation-sensitizing agent, and it can act as a radiosensitizer. Pemetrexed is more efficacious in preclinical models when combined with the base excision repair inhibitor methoxyamine [201]. A phase I/II trial was conducted to determine the tolerability, feasibility, and anticancer activity of intrathecal pemetrexed combination with involved-field radiotherapy for leptomeningeal metastases from solid tumors. Myelosuppression, an elevation in hepatic aminotransferases, and radiculitis were among the most serious side effects [202].

Pemetrexed (PEM) has the potential for broader application in combination therapies [203]. Pemetrexed is now used to treat metastatic non-small cell lung cancer and malignant mesothelioma in conjunction with cisplatin [198,204]. Pemetrexed in combination with platinum was approved for the treatment of advanced lung adenocarcinoma due to its low toxicity and high efficacy. Pemetrexed plus cisplatin have the same clinical efficacy as docetaxel plus cisplatin in the treatment of advanced lung cancer, but the former may have a lower risk of side reactions [205]. The use of pemetrexed in combination with platinum medicines to treat advanced lung cancer in the elderly is studied to help patients achieve better treatment outcomes [206]. Platinum plus pemetrexed plus has become a standard of care in the first-line treatment for advanced nonsquamous non-small-cell lung cancer [207–210].

When used in combination with platinum-based medicines, pemetrexed (PEM) offers a superior efficacy and safety profile [211–213]. Pemetrexed in combination with cisplatin and gefitinib has shown to be effective in treating advanced lung cancer [214–216]. In East Asian patients with advanced non-squamous (NS) non-small-cell lung cancer (NSCLC) with activating EGFR mutations, pemetrexed was combined with gefitinib. When compared with the current standard of care, this combination may provide epidermal growth factor receptor (EGFR) mutation-positive patients alternative therapeutic options and improved clinical results [217].

It is also combined with bavituximab, which is an immunomodulatory chimeric monoclonal antibody that suppresses phosphatidylserine signaling, which enhances innate and adaptive immune responses [218,219]. In the treatment of NSCLC, the use of bevacizumab in combination with pemetrexed and cisplatin can result in a superior therapeutic impact with fewer adverse effects [220].

In combination with pemetrexed, abemaciclib, a dual cyclin-dependent kinase 4 and 6 inhibitor, has shown a preclinical potency in non-small-cell lung cancer (NSCLC) and shown an acceptable safety profile [221]. It is also combined with temsirolimus [222].

Gemcitabine and pemetrexed have been used as maintenance therapy for non-small-cell lung carcinoma (NSCLC) [223]. Preclinical rationale and phase II study results for prolonged pemetrexed infusion plus gemcitabine in refractory metastatic colorectal cancer indicated that the combination of pemetrexed and gemcitabine (experimental pemetrexed-gemcitabine) was ineffective and moderately toxic [224].

Docetaxel and pemetrexed are two regularly used drugs with the well-known effects in a single-phase cell culture. When compared to the standard single-phase experiments, sequential therapeutic effects can produce better results. Pem-Doc may be a more effective adenocarcinoma treatment [225].

Pemetrexed (PMX) comprises two carboxy functionalities that can bind metal ions in a coordinative manner. A multilayer PMX NP system was developed with each layer serving a different function. Coordinative interactions between zirconium (IV) ions and PMX molecules are used to build the metal-drug NP core, which is assembled in a bottom-up strategy [226]. Additionally, an oral delivery system of **pemetrexed** by improving its intestinal membrane permeability was designed. PMX was ionically complexed with *N*-deoxycholyl-L-lysyl-methylester and prepared; the partition coefficient and in vitro membrane permeability of PMX were significantly increased after complex formation, but it had identical cytotoxic and inhibitory effects on cancer cell proliferation/migration [227]. The invention's pemetrexed-disodium lipid complex, which has the benefits of good stability, powerful targeting, and low toxicity, can be mixed with a pharmaceutically acceptable carrier or any pharmacological formulation [228]. A pharmaceutical compound comprises a complex of **pemetrexed** and a co-former and one or more pharmaceutically acceptable excipients [229].

In addition, the electropolymerization of ortho-phenylenediammine (o-PD) was used to construct a novel molecularly imprinting polymer (MIP)-based sensor for the selective detection of pemetrexed (PMX) [230].

Fast and precise methods to detect the bound and free drug contained in the examined conjugate preparations were developed for research of putative macromolecular carriers for PMX [231].

Crystal forms of pemetrexed diacid are developed, which are readily produced for either laboratory scale or industrial scale [232]. Other oral pharmaceutical compound, which contains the water solution, anticancer drug pemetrexed, was invented, and a production method thereof, and more specifically, to an oral pharmaceutical compound was reported [233–236]. Other inventions are related to a ready-to-use type injection solution compound containing pemetrexed or its pharmaceutically acceptable salt [237–261]; the injection has an uniform particle size distribution, good stability, and low side effect, so that it can be used for treating cancer [262]. Polymorphs and amorphous form of pemetrexed or its pharmaceutical acceptable salts were also synthesized [263].

3. Nonclassical antifolate drugs (nonpolyglutamable antifolates)

They lack the glutamate moiety, e.g., trimetrexate and piritrexim.

3.1 Trimetrexate

Trimetrexate (TMTX; Neutrexin) (Fig. 2.5) is a quinazoline inhibitor of dihydrofolate reductase (DHFR) [251–253], and it is considered as an FDA-approved [264] cytotoxic drug [265]. It is a potent dihydrofolate reductase enzyme inhibitor (DHFR) [266,267], which is cytotoxic to various tumors resistant to methotrexate in vitro and in vivo. It has a higher lipophilicity than the parent antifolate and does not transport via the reduced folate carrier. In clinic, these characteristics exhibit a greater potency than those of methotrexate; its resistance to polyglutamylation may improve its therapeutic index [268].

As a nonclassical antifolate, trimetrexate is undergoing significant clinical trials. Trimetrexate's efficacy in preclinical studies was highly schedule dependent, with repeated dosing schedules being advantageous. On all regimens studied in humans, phase I investigations have revealed that myelosuppression is the most toxic impact of trimetrexate. In phase II trials, a 5-day schedule will be evaluated; trials in numerous tumor types and the role of schedule are currently underway [268].

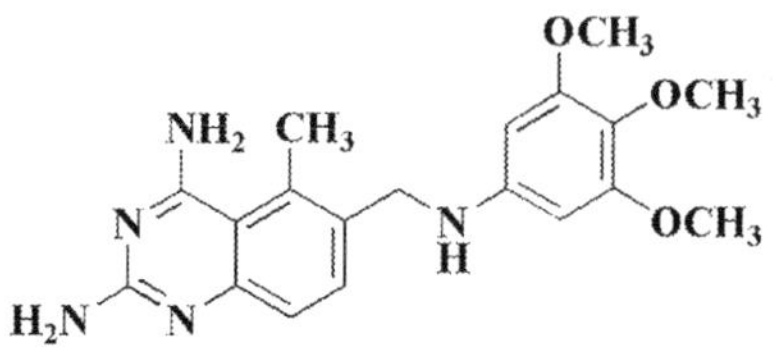

OCH_3 OCH_3 NH_2 CH_3 N N H OCH_3 H_2N N

Fig. 2.5 Trimetrexate.

It is also considered as a folate antagonist that selectively depletes proliferating alloreactive precursors in vitro in a dose- and time-dependent way [269]. Trimetrexate (TMQ) is a potent nonclassical inhibitor, which has no selectivity for the pathogen DHFR and must be utilized in conjunction with host rescue [270]. Trimetrexate has been estimated in clinical trials for the second-line treatment for recurrent high-grade osteosarcoma (HGOS) patients, unfortunately indicating modest potency [271]. The response rate for TMTX in conjunction with 5-fluorouracil and leucovorin (=TFL) is evaluated in treating advanced gastric cancer in phase II trial [272]. Additionally, trimetrexate inhibits chordoma cell growth [273]. Moreover, the joint effects of the combinations of the GARFT inhibitor AG2034 and trimetrexate (TMQ) to inhibit the growth of HCT-8 human ileocecal adenocarcinoma cells were studied [274].

It is worthy to note that the trimetrexate, a lipophilic antifolate licensed by the FDA to treat *Pneumocystis carinii* infection in AIDS patients, is a potent inhibitor of *Trigonoscuta cruzi* DHFR function. Trimetrexate acts as *P. carinii* dihydrofolatereductase inhibitors that are currently utilized in treating *P. carinii* pneumonia. *P. carinii* is a nonpathogenic fungus prevalent in healthy human's respiratory tract. It is a leading cause of death in AIDS patients. However, trimetrexate has various adverse effects [275,276].

In human bone marrow cells, raloxifene (RAL) in combination with trimetrexate/5-fluorouracil was evaluated to identify the optimal regimens and cellular mechanisms of action for mitigating trimetrexate cytotoxicity [277]. In addition, trimetrexate has a potent activity against malaria. It is, however, often regarded to be toxic and thus unsuitable for treating malaria [278].

References

[1] Visentin M, Zhao R, Goldman ID. The antifolates. Hematol Oncol Clin North Am 2012;26(3):629–48.

[2] Wu Y, Kochat H. Process for synthesizing antifolates; 2005. US 20050020833 A1.

[3] McGuire JJ, Coward JK. Folylpolyglutamate synthetase as a target for therapeutic intervention. Drugs Future 2003;28(10):967–74.

[4] Celtikci B, Lawrance AK, Wu Q, Rozen R. Methotrexate-induced apoptosis is enhanced by altered expression of methylenetetrahydrofolate reductase. Anticancer Drugs 2009;20(9):787–93.

[5] Galbiatti ALS, Caldas HC, Padovani Jr JA, Pavarino EC, Goloni-Bertollo EM. Sensitivity of human laryngeal squamous cell carcinoma Hep-2 to metrotexate chemotherapy. Exp Oncol 2012;34(4):367–9.

[6] Leclerc GJ, Mou C, Leclerc GM, Mian AM, Barredo JC. Histone deacetylase inhibitors induce FPGS mRNA expression and intracellular accumulation of long-chain

methotrexate polyglutamates in childhood acute lymphoblastic leukemia: implications for combination therapy. Leukemia 2010;24(3):552–62.
[7] Rubino FM. Separation methods for methotrexate, its structural analogues and metabolites. J Chromatogr B Biomed Sci Appl 2001;764(1–2):217–54.
[8] Kisliuk RL. Deaza analogs of folic acid as antitumor agents. Curr Pharm Des 2003;9(31):2615–25.
[9] Matherly LH, Hou Z. Structure and function of the reduced folate carrier a paradigm of a major facilitator superfamily mammalian nutrient transporter. Vitam Horm 2008;79145–84.
[10] Nurul M, Syarbani E, Howarth GS, Tran CD. Zinc supplementation alone is effective for partial amelioration of methotrexate-induced intestinal damage. Altern Ther Health Med 2015;21(Suppl):222–31.
[11] Matherly LH, Hou Z, Deng Y. Human reduced folate carrier: translation of basic biology to cancer etiology and therapy. Cancer Metastasis Rev 2007;26(1):111–28.
[12] Wohlrab J, Wohlrab J, Neubert RHH, Michael J, Neubert RHH, Naumann S. Methotrexate for topical application in an extemporaneous preparation. J Ger Soc Dermatol 2015;13(9):891–901.
[13] Matherly LH, Hou Z. Structure and function of the reduced folate carrier: a paradigm of a major facilitator superfamily mammalian nutrient transporter. In: Vitamins and hormones; 2008. p. 145–84 [(San Diego, CA, United States). 79(Folic Acid and Folates)].
[14] Chabner BA, Longo DL. Cancer chemotherapy and biotherapy: principles and practice. 5th. Wolters Kluwer: Lippincott Williams & Wilkins; 2011. p. 765.
[15] Truchuelo T, Alcantara J, Moreno C, Vano-Galvan S, Jaen P. Focal skin toxicity related to methotrexate sparing psoriatic plaques. Dermatol Online J 2010;16(6):16.
[16] Jackman AL, Forster M, Ng M. Targeting thymidylate synthase by antifolate drugs for the treatment of cancer. Cancer Drug Des Discovery 2008;198–226.
[17] Meirim MG, Neuse EW, N'da DD. Carrier-bound methotrexate. I. Water-soluble polyaspartamide-methotrexate conjugates with ester links in the polymer-drug spacer. J Appl Polym Sci 2001;82(8):1844–9.
[18] Nihal M, Wu J, Wood GS. Methotrexate inhibits the viability of human melanoma cell lines and enhances Fas/Fas-ligand expression, apoptosis and response to interferon-alpha: rationale for its use in combination therapy. Arch Biochem Biophys 2014;563:101–7.
[19] Bienemann K, Staege MS, Howe SJ, Sena-Esteves M, Hanenberg H, Kramm CM. Targeted expression of human folylpolyglutamate synthase for selective enhancement of methotrexate chemotherapy in osteosarcoma cells. Cancer Gene Ther 2013;20(9):514–20.
[20] Okarvi SM, Jammaz IA. Preparation and in vitro and in vivo evaluation of technetium-99m-labeled folate and methotrexate conjugates as tumor imaging agents. Cancer Biother Radiopharm 2006;21(1):49–60.
[21] Wei W, Fountain M, Magda D, Zhong W, Lecane P, Mesfin M, Miles D, Sessler JL. Gadolinium texaphyrin-methotrexate conjugates. Towards improved cancer chemotherapeutic agents. Org Biomol Chem 2005;3(18):3290–6.
[22] Maksimovic V, Pavlovic-Popovic Z, Vukmirovic S, Cvejic J, Mooranian A, Al-Salami H, Mikov M, Golocorbin-Kon S. Molecular mechanism of action and pharmacokinetic properties of methotrexate. Mol Biol Rep 2020;47(6):4699–708.
[23] L.D. Garcia, L. Gibson, W. Lambert, B.W. Li, L. Zhu. Sustained release formulation of methotrexate as a disease-modifying antirheumatic drug (DMARD) and an anti-cancer agent. US20110280932A1. 2011.
[24] Goszczyński T., Boratyński J., Wietrzyk J., Filip-Psurska B., Kempińska K. A conjugate of methotrexate and hydroxyethyl starch for use in the treatment cancer. WO2013127885A1. 2013.

[25] Chu E, DeVita Jr VT. Physicians' cancer chemotherapy drug manual 2019. Jones & Bartlett Learning; 2018. p. 610.
[26] Molina JR. Pralatrexate, a dihydrofolate reductase inhibitor for the potential treatment of several malignancies. IDrugs 2008;11(7):508–21.
[27] Kristan DR, Celeste L. Pemetrexed: a multitargeted antifolate. Clin Ther 2005;27(9): 1343–82.
[28] Sastry RVRP, Venkatesan CS, Sastry BS, Mahesh K. Identification and characterization of forced degradation products of pralatrexate injection by LC-PDA and LC-MS. J Pharm Biomed Anal 2016;131:400–9.
[29] Ho AL, Lipson BL, Sherman EJ, Xiao H, Fury MG, Apollo A, Seetharamu N, Sima CS, Haque S, Lyo JK, et al. A phase II study of pralatrexate with vitamin B12 and folic acid supplementation for previously treated recurrent and/or metastatic head and neck squamous cell cancer. Invest New Drugs 2014;32(3):549–54.
[30] Malik SM, Liu K, Qiang X, Sridhara R, Tang S, McGuinn Jr WD, Verbois SL, Marathe A, Williams GM, Bullock J, et al. Folotyn (pralatrexate injection) for the treatment of patients with relapsed or refractory peripheral T-cell lymphoma: U.S. Food and Drug Administration drug approval summary. Clin Cancer Res 2010;16 (20):4921–7.
[31] O'Connor OA, Amengual J, Colbourn D, Deng C, Sawas A. Pralatrexate: a comprehensive update on pharmacology, clinical activity and strategies to optimize use. Leuk Lymphoma 2017;58(11):2548–57.
[32] Advani RH, Ansell SM, Lechowicz MJ, Beaven AW, Loberiza F, Vose JM, Carson KR, Evens AM, Foss F, Horwitz S, et al. A phase II study of cyclophosphamide, etoposide, vincristine and prednisone (CEOP) alternating with pralatrexate (P) as front line therapy for patients with peripheral T-cell lymphoma (PTCL): final results from the T-cell consortium trial. Br J Haematol 2016;172(4):535–44.
[33] Maruyama D, Nagai H, Maeda Y, Nakane T, Shimoyama T, Nakazato T, Sakai R, Ishikawa T, Izutsu K, Ueda R, et al. Phase I/II study of pralatrexate in Japanese patients with relapsed or refractory peripheral T-cell lymphoma. Cancer Sci 2017;108 (10):2061–8.
[34] Makita S, Maeshima AM, Maruyama D, Izutsu K, Tobinai K. Forodesine in the treatment of relapsed/refractory peripheral T-cell lymphoma: an evidence-based review. Onco Targets Ther 2018;11:1–7.
[35] Koch E, Story SK, Geskin LJ. Preemptive leucovorin administration minimizes pralatrexate toxicity without sacrificing efficacy. Leuk Lymphoma 2013;54(11):2448–51.
[36] Reungwetwattana T, Jaramillo S, Molina JR. Current and emerging therapies in peripheral T-cell lymphoma: focus on pralatrexate. Clin Med Insights: Ther 2011;3:125–35.
[37] Howman RA, Prince HM. New drug therapies in peripheral T-cell lymphoma. Expert Rev Anticancer Ther 2011;11(3):457–72.
[38] Gisselbrecht C. Peripheral T-cell lymphoma: tailoring of new drugs to the histology? Onco Targets Ther 2011;13(9):565–70.
[39] Dondi A, Bari A, Pozzi S, Ferri P, Sacchi S. The potential of as a treatment of peripheral T-cell lymphoma. Expert Opin Investig Drugs 2014;23(5):711–8.
[40] Foss FM. Evaluation of the pharmacokinetics, preclinical and clinical efficacy of pralatrexate for the treatment of T-cell lymphoma. Expert Opin Drug Metab Toxicol 2011;7(9):1141–52.
[41] McPherson JP, Vrontikis A, Sedillo C, Halwani AS, Gilreath JA. Pralatrexate monitoring using a commercially available methotrexate assay to avoid potential drug interactions. Pharmacotherapy 2016;36(2):e8–e11.
[42] Fury MG, Krug LM, Azzoli CG, Sharma S, Kemeny N, Wu N, Kris MG, Rizvi NA. A phase I clinical pharmacologic study of pralatrexate in combination with probenecid

in adults with advanced solid tumors. Cancer Chemother Pharmacol 2006;57 (5):671–7.
[43] Visentin M, Unal ES, Zhao R, Goldman I. David. The membrane transport and polyglutamation of pralatrexate: a new-generation dihydrofolate reductase inhibitor. Cancer Chemother Pharmacol 2013;72(3):597–606.
[44] Visentin M, Unal ES, Zhao R, Goldman ID. The membrane transport and polyglutamation of pralatrexate: a new-generation dihydrofolate reductase inhibitor. Cancer Chemother Pharmacol 2013;72(3):597–606.
[45] Serova M, Bieche I, Sablin M-P, Pronk GJ, Vidaud M, Cvitkovic E, Faivre S, Raymond E. Single agent and combination studies of pralatrexate and molecular correlates of sensitivity. Br J Cancer 2011;104(2):272–80.
[46] Kinahan C, Mangone MA, Scotto L, Marchi E, O'Connor OA, Kinahan C, Mangone MA, Visentin M, Cho HJ. The anti-tumor activity of pralatrexate (PDX) correlates with the expression of RFC and DHFR mRNA in preclinical models of multiple myeloma. Oncotarget 2020;11(18):1576–89.
[47] Abramovits W, Oquendo M, Granowski P, Gupta A, Cather J. Pralatrexate (Folotyn). Skinmed 2012;10(4):244–6.
[48] Package Insert-FOLOTYN® (pralatrexate injection). Allos Therapeutics, Inc. Westminster, CO. 2012. https://www.pbm.va.gov/PBM/clinicalguidance/drugmonographs/PralatrexateMonograph.pdf.
[49] Rodd AL, Ververis K, Karagiannis TC. Safety and efficacy of pralatrexate in the management of relapsed or refractory peripheral T-cell lymphoma. Clin Med Insights: Oncol 2012;6:305–14.
[50] O'Connor OA, Pro B, Pinter-Brown L, Bartlett N, Popplewell L, Coiffier B, Lechowicz MJ, Savage KJ, Shustov AR, Gisselbrecht C, et al. Pralatrexate in patients with relapsed or refractory peripheral T-cell lymphoma: results from the pivotal PROPEL study. J Clin Oncol 2011;29(9):1182–9.
[51] Xiaonan H, Yuqin S, Jun Z, Huiqiang H, Bing B, Huilai Z, Xiaoyan K, Yuankai S, Guodong L, Chunxiao C, et al. Pralatrexate in Chinese patients with relapsed or refractory peripheral T-cell lymphoma: a single-arm, multicenter study. Target Oncol 2019;14(2):149–58.
[52] O'Connor OA, Horwitz S, Hamlin P, Portlock C, Moskowitz CH, Sarasohn D, Neylon E, Mastrella J, Hamelers R, MacGregor-Cortelli B, et al. Phase II-I-II study of two different doses and schedules of pralatrexate, a high-affinity substrate for the reduced folate carrier, in patients with relapsed or refractory lymphoma reveals marked activity in T-cell malignancies. J Clin Oncol 2009;27(26):4357–64.
[53] Scotto L, Kinahan C, Casadei B, Mangone M, Douglass E, Murty VV, Marchi E, Ma H, George C, Montanari F, et al. Generation of pralatrexate resistant T-cell lymphoma lines reveals two patterns of acquired drug resistance that is overcome with epigenetic modifiers. Genes Chromosomes Cancer 2020;59(11):639–51.
[54] Foss F, Horwitz SM, Coiffier B, Bartlett N, Popplewell L, Pro B, Pinter-Brown LC, Shustov A, Furman RR, Haioun C, et al. Pralatrexate is an effective treatment for relapsed or refractory transformed mycosis fungoides: a subgroup efficacy analysis from the PROPEL study. Clin Lymphoma Myeloma Leuk 2012;12(4):238–43.
[55] Yao-Chung L, Ting-An L, Hao-Yuan W, Po-Shen K, Chia-Jen L, Liang-Tsai H, Jyh-Pyng G, Yao-Chung L, Ting-An L, Hao-Yuan W, et al. Pralatrexate as a bridge to allogeneic hematopoietic stem cell transplantation in a patient with advanced-stage extranodal nasal-type natural killer/T cell lymphoma refractory to first-line chemotherapy: a case report. J Med Case Reports 2020;14(1):43.
[56] Grem JL, Evande RE, Schwarz JK, Kos ME, Meza JL. A phase 1 clinical trial of sequential pralatrexate followed by a 48-hour infusion of 5-fluorouracil given every other week in adult patients with solid tumors. Cancer 2015;121(21):3862–8.

[57] Hong JY, Yoon DH, Yoon SE, Kim SJ, Lee HS, Eom H, Lee HW, Shin D, Koh Y, Yoon S, et al. Pralatrexate in patients with recurrent or refractory peripheral T-cell lymphomas: a multicenter retrospective analysis. Sci Rep 2019;9(1):20302.
[58] Parker T, Barbarotta L, Foss F. Pralatrexate: treatment of T-cell non-Hodgkin's lymphoma. Future Oncol 2013;9(1):21–9.
[59] Foss FM, Parker TL, Girardi M, Li A. Effect of leucovorin administration on mucositis and skin reactions in patients with peripheral T-cell lymphoma or cutaneous T-cell lymphoma treated with pralatrexate*. Leuk Lymphoma 2019;60(12):2927–30.
[60] Zhao JC, Jaszczur SM, Afifi S, Foss F. Pralatrexate injection for the treatment of patients with relapsed or refractory peripheral T-cell lymphoma. Expert Rev Hematol 2020;13(6):577–83.
[61] Casanova M, Medina-Perez A, Moreno-Beltran M, Mata-Vazquez M, Rueda A. Critical appraisal of pralatrexate in the management of difficult-to-treat peripheral T cell lymphoma. Ther Clin Risk Manag 2011;7:401–8.
[62] Clark RA, Lee S, Qiao J, Chung DH, Clark RA, Lee S. Preclinical evaluation of the anti-tumor activity of pralatrexate in high-risk neuroblastoma cells. Oncotarget 2020;11(32):3069–77.
[63] Amengual JE, Lichtenstein R, Lue J, Sawas A, Deng C, Lichtenstein E, Khan K, Atkins L, Rada A, Kim HA, et al. A phase 1 study of romidepsin and pralatrexate reveals marked activity in relapsed and refractory T-cell lymphoma. Blood 2018;131(4):397–407.
[64] Jain S, Jirau-Serrano X, Zullo KM, Scotto L, Palermo CF, Cremers S, Sastra SA, Olive KP, Thomas T, Bhagat G, et al. Preclinical pharmacologic evaluation of pralatrexate and romidepsin confirms potent synergy of the combination in a murine model of human T-cell lymphoma. Clin Cancer Res 2015;21(9):2096–106.
[65] Marchi E, Mangone M, Zullo K, O'Connor OA. Pralatrexate pharmacology and clinical development. Clin Cancer Res 2013;19(24):6657–61.
[66] Marchi E, Paoluzzi L, Scotto L, Seshan VE, Zain JM, Zinzani PL, O'Connor OA. Pralatrexate is synergistic with the proteasome inhibitor bortezomib in in vitro and in vivo models of T-cell lymphoid malignancies. Clin Cancer Res 2010;16 (14):3648–58.
[67] Dunn TJ, Dinner S, Price E, Coutre SE, Gotlib J, Hao Y, Berube C, Medeiros BC, Liedtke M. A phase 1, open-label, dose-escalation study of pralatrexate in combination with bortezomib in patients with relapsed/refractory multiple myeloma. Br J Haematol 2016;173(2):253–9.
[68] Malhotra U, Mukherjee S, Fountzilas C, Boland P, Miller A, Patnaik S, Attwood K, Yendamuri S, Adjei A, Kannisto E, et al. Pralatrexate in combination with oxaliplatin in advanced esophagogastric cancer: a phase II trial with predictive molecular correlates. Mol Cancer Ther 2020;19(1):304–11.
[69] del Carmen MG, Supko JG, Horick NK, Rauh-Hain JA, Clark RM, Campos SM, Krasner CN, Atkinson T, Birrer MJ. Phase 1 and 2 study of carboplatin and pralatrexate in patients with recurrent, platinum-sensitive ovarian, fallopian tube, or primary peritoneal cancer. Cancer (Hoboken, NJ, USA) 2016;122(21):3297–306.
[70] Duvic M, Kim YH, Zinzani PL, Horwitz SM. Results from a phase I/II open-label, dose-finding study of pralatrexate and oral bexarotene in patients with relapsed/refractory cutaneous T-cell lymphoma. Clin Cancer Res 2017;23(14):3552–6.
[71] Reddy G. Combination therapy using belinostat and pralatrexate to treat lymphoma. PCT Int. Appl; 2016. WO 2016205203 A1 20161222.
[72] Lahiri S, Gupta N, Singh HK. Preparation of pralatrexate salts and compositions for the treatment of neoplasms. PCT Int. Appl; 2014. WO 2014020553 A1 20140206.
[73] Zhou Z, Zhang A, Zhang X, Hu Z, Zhu X, Yuan Z. Crystal form of pralatrexate and preparation method thereof. Faming Zhuanli Shenqing; 2015. CN 104628727 A 20150520.

[74] Tiseni PS, Biljan T. Crystalline forms of pralatrexate and dosage forms for treatment of T-cell lymphoma. PCT Int. Appl; 2012. WO 2012061469 A2 20120510.
[75] Niyikiza C, Moyo VM. Alpha polyglutamated pralatrexate and uses thereof. PCT Int. Appl; 2019. WO 2019157129 A1 20190815.
[76] Ferguson PJ, Collins O, Dean NM, DeMoor J, Sha-Li C, Vincent MD, Koropatnick J. Antisense down-regulation of thymidylate synthase to suppress growth and enhance cytotoxicity of 5-FUdR, 5-FU and tomudex in HeLa cells. Br J Pharmacol 1999;127(8):1777–86.
[77] Koropatnick DJ, Vincent MD, Dean NM. Antisense oligonucleotides against thymidylate synthase in cancer therapy. PCT Int. Appl; 1999. WO 9915648 A1 19990401.
[78] Samlowski WE, Lew D, Kuebler PJ, Kolodziej MA, Medina JE, Mangan KF, Moore DFJ, Schuller DE, Ensley JF. Evaluation of Tomudex in patients with recurrent or metastatic squamous cell carcinoma of the head and neck: a Southwest Oncology Group study. Invest New Drugs 1999;16(3):271–4.
[79] Labianca R, Pessi A, Facendola G, Pirovano M, Luporini G. Modulated 5-fluorouracil (5-FU) regimens in advanced colorectal cancer: a critical review of comparative studies. Eur J Cancer 1996;32A(Suppl. 5):S7–12.
[80] Peters GJ, Smitskamp-Wilms E, Smid K, Pinedo HM, Jansen G. Determinants of activity of the antifolate thymidylate synthase inhibitors tomudex (ZD1694) and GW1843U89 against mono- and multilayered colon cancer cell lines under folate-restricted conditions. Cancer Res 1999;59(21):5529–35.
[81] Rustum YM, Cao S. New drugs in therapy of colorectal cancer: preclinical studies. Semin Oncol 1999;26(6):612–20.
[82] Sotelo-Mundo RR, Ciesla J, Dzik JM, Rode W, Maley F, Maley GF, Hardy LW, Montfort WR. Crystal structures of rat thymidylate synthase inhibited by Tomudex, a potent anticancer drug. Biochemistry 1999;38(3):1087–94.
[83] Astra Zeneca Canada Inc. TOMUDEX® product monograph. Mississauga, Ontario; 6 September 2001. BCCA Cancer Drug Manual© 2001.
[84] Gunasekara NS, Faulds D. Raltitrexed. A review of its pharmacological properties and clinical efficacy in the management of advanced colorectal cancer. Drugs 1998;55 (3):423–35.
[85] Cunningham D, Zalcberg JR, Rath U, et al. 'Tomudex' (ZD1694): results of a randomised trial in advanced colorectal cancer demonstrate efficacy and reduced mucositis and leucopenia. The 'Tomudex' Colorectal Cancer Study Group. Eur J Cancer 1995;31A(12):1945–54.
[86] James R, Price P, Valentini V. Raltitrexed (Tomudex) concomitant with radiotherapy as adjuvant treatment for patients with rectal cancer: preliminary results of phase I studies. Eur J Cancer 1999;35(Suppl 1):S19–22.
[87] Matsui SI, Arredondo MA, Wrzosek C, et al. DNA damage and p53 induction do not cause ZD1694-induced cell cycle arrest in human colon carcinoma cells. Cancer Res 1996;56(20):4715–23.
[88] de Rotte MCFJ, den Boer E, Calasan MB, Heijstek MW, te Winkel ML, Heil SG, Lindemans J, Jansen G, Peters GJ, Kamphuis SSM, et al. Personalized medicine of methotrexate therapy. Ned Tijdschr Klin Chem Laboratoriumgeneeskd 2012;37 (1):50–3.
[89] Drake JC, Allegra CJ, Moran RG, Johnston PG. Resistance to tomudex (ZD1694): multifactorial in human breast and colon carcinoma cell lines. Biochem Pharmacol 1996;51(10):1349–55.
[90] Muggia FM, Blessing JA, Homesley HD, Sorosky J. Tomudex (ZD1694, NSC 639186) in platinum-pretreated recurrent epithelial ovarian cancer: a phase II study by the Gynecologic Oncology Group. Cancer Chemother Pharmacol 1998;42 (1):68–70.

[91] Sabri NH, Halim SNA, Zain SM, Lee VS. Modification of anti-cancer co-crystal for thymidylate synthase inhibition: molecular dynamics study. Chiang Mai J Sci 2018;45 (6):2361–73.
[92] Law AA, Collie-Duguid ESR, Smith TAD. Influence of resistance to 5-fluorouracil and tomudex on [18F]-FDG incorporation, glucose transport and hexokinase activity. Int J Oncol 2012;41(1):378–82.
[93] Wang W, McLeod HL, Cassidy J, Collie-Duguid ESR. Mechanisms of acquired chemoresistance to 5-fluorouracil and tomudex: thymidylate synthase dependent and independent networks. Cancer Chemother Pharmacol 2007;59(6):839–45.
[94] Hughes LR, Stephens TC, Boyle FT, Jackman AL. Raltitrexed (Tomudex), a highly polyglutamatable antifolate thymidylate synthase inhibitor: design and preclinical activity. Antifolate Drugs Cancer Ther 1999;147–65.
[95] Cunningham D, Zalcberg J, Smith I, Gore M, Pazdur R, Burris 3rd H, Meropol NJ, Kennealey G, Seymour L. 'Tomudex ' (ZD1694): a novel thymidylate synthase inhibitor with clinical antitumour activity in a range of solid tumours.' Tomudex 'International Study Group. Ann Oncol: Off J Eur Soc Med Oncol 1996;7(2):179–82.
[96] Phan J, Koli S, Minor W, Dunlap RB, Berger SH, Lebioda L. Human thymidylate synthase is in the closed conformation when complexed with dUMP and Raltitrexed, an antifolate drug. Biochemistry 2001;40(7):1897–902.
[97] Sobrero AF. The potential role of Tomudex in the treatment of advanced colorectal cancer. Tumori 1997;83(2):576–80.
[98] Ross P, Heron J, Cunningham D. Cost of treating advanced colorectal cancer: a retrospective comparison of treatment regimens. Eur J Cancer 1996;32A(Suppl 5):S13–7.
[99] Blackledge G. New developments in cancer treatment with the novel thymidylate synthase inhibitor raltitrexed (Tomudex). Br J Cancer 1998;77(Suppl. 2, What Are the Critical Issues When Investigating New Chemotherapies for Colorectal Cancer?):29–37.
[100] Haller DG. An overview of adjuvant therapy for colorectal cancer. Eur J Cancer 1995;31A(7–8):1255–63.
[101] Jackman AL, Calvert AH. Folate-based thymidylate synthase inhibitors as anticancer drugs. Ann Oncol: Off J Eur Soc Med Oncol 1995;6(9):871–81.
[102] Gorlick R, Bertino JR. Drug resistance in colon cancer. Semin Oncol 1999; 26(6):606–11.
[103] Summerhayes M. Raltitrexed and the treatment of advanced colorectal cancer. J Oncol Pharm Pract 1996;2(4):225–35.
[104] Jackman AL, Boyle FT, Harrap KR. Tomudex (ZD1694): from concept to care, a program in rational drug discovery. Invest New Drugs 1996;14(3):305–16.
[105] Smith I, Jones A, Spielmann M, Namer M, Green MD, Bonneterre J, Wander HE, Hatschek T, Wilking N, et al. A phase II study in advanced breast cancer: ZD1694 ("Tomudex") a novel direct and specific thymidylate synthase inhibitor. Br J Cancer 1996;74(3):479–81.
[106] Kimbell R, Jackman AL. In vitro studies with ZD1694 (Tomudex) and SN38 in human colon tumor cell lines. In: Pfleiderer W, Rokos H, editors. Chemistry and biology of pteridines and folates 1997, Proceedings of the international symposium on pteridines and folates, 11th, Berchtesgaden, Germany, June 15–20, 1997; 1997. p. 249–52.
[107] Berg RW, Ferguso PJ, DeMoor JM, Vincen MD, Koropatnick J. The means to an end of tumor cell resistance to chemotherapeutic drugs targeting thymidylate synthase: shoot the messenger. Curr Drug Targets 2002;3(4):297–309.
[108] Matherly LH. Molecular and cellular biology of the human reduced folate carrier. Prog Nucleic Acid Res Mol Biol 2001;67:131–62.

[109] Orlandi L, Bearzatto A, Abolafio G, De Marco C, Daidone MG, Zaffaroni N. Involvement of bcl-2 and p21waf1 proteins in response of human breast cancer cell clones to tomudex. Br J Cancer 1999;81(2):252–60.
[110] Verweij J. New promising anticancer agents in development. What comes next? Cancer Chemother Pharmacol 1996;38(Suppl):S3–S10.
[111] De Jonge MJA, Verweij J. Moving the frontiers of cancer chemotherapy for solid tumors by changing the scope of drug development. Hematology 1996;1(3):183–98.
[112] Cocconi G, Cunningham D, Van Cutsem E, Francois E, Gustavsson B, van Hazel G, Kerr D, Possinger K, Hietschold SM. Open, randomized, multicenter trial of raltitrexed versus fluorouracil plus high-dose leucovorin in patients with advanced colorectal cancer. Tomudex Colorectal Cancer Study Group. J Clin Oncol Off J Am Soc Clin Oncol 1998;16(9):2943–52.
[113] Teicher BA, Ara G, Chen Y, Recht A, Coleman C. Norman interaction of Tomudex with radiation in vitro and in vivo. Int J Oncol 1998;13(3):437–42.
[114] McGuire JJ. Anticancer antifolates: current status and future directions. Curr Pharm Des 2003;9(31):2593–613.
[115] Judson IR. 'Tomudex' (raltitrexed) development: preclinical, phase I and II studies. Anti-Cancer Drugs 1997;8(Suppl. 2):S5–9.
[116] Longo GSA, Gorlick R, Tong WP, Ercikan E, Bertino JR. Disparate affinities of antifolates for folylpolyglutamate synthetase from human leukemia cells. Blood 1997;90 (3):1241–5.
[117] Caponigro F, Avallone A, McLeod H, Carteni G, De Vita F, Casaretti R, Morsman J, Blackie R, Budillon A, De Lucia L, et al. Phase I and pharmacokinetic study of tomudex combined with 5-fluorouracil plus levofolinic acid in advanced head and neck cancer and colorectal cancer. Clin Cancer Res 1999;5(12):3948–55.
[118] Bokemeyer C, Hartmann JT, Kanz L. Current aspects of adjuvant and palliative chemotherapy in colorectal carcinoma. Praxis 1997;86(39):1510–6.
[119] Louvet C. Chemotherapy of advanced colorectal cancers. Ann Chir 1997; 51(4):361–7.
[120] Woll PJ, Basser R, Le Chevalier T, Drings P, Manga GP, Adenis A, Seymour L, Smith F, Thatcher N. Phase II trial of raltitrexed ("Tomudex") in advanced small-cell lung cancer. Br J Cancer 1997;76(2):264–5.
[121] Locher C, Auperin A, Boige V, Alzieu L, Pignon JP, Abbas M, Ducreux M. Assessment of the cost of first line chemotherapy in metastatic colorectal cancer. Preliminary results in the FFCD 9601 trial. Gastroenterol Clin Biol 2001;25 (8–9):749–54.
[122] Wang Y, Zhao R, Goldman ID. Biochem Pharmacol 2003;65(7):1163–70.
[123] Sigmond J, Backus HHJ, Wouters D, Temmink OH, Jansen G, Peters GJ. Induction of resistance to the multitargeted antifolate Pemetrexed (ALIMTA) in WiDr human colon cancer cells is associated with thymidylate synthase overexpression. Biochem Pharmacol 2003;66(3):431–8.
[124] Bunn PAJ. Incorporation of pemetrexed (Alimta) into the treatment of non-small cell lung cancer (thoracic tumors). Semin Oncol 2002;29(3 Suppl. 9):17–22.
[125] Tesei A, Ricotti L, De Paola F, Amadori D, Frassineti GL, Zoli W. In vitro schedule-dependent interactions between the multitargeted antifolate LY231514 and gemcitabine in human colon adenocarcinoma cell lines. Clin Cancer Res 2002;8(1):233–9.
[126] Guo Z, Wang H, Xin C, Chen L. Synthesis of antifolate alimta. Youji Huaxue 2006;26(4):546–50.
[127] Calvert H. Pemetrexed (Alimta): a promising new agent for the treatment of breast cancer. Semin Oncol 2003;30(2 Suppl 3):2–5.
[128] Socinski MA. Pemetrexed (Alimta) in small cell lung cancer. Semin Oncol 2005; 32(2 Suppl. 2):S1–4.

[129] Goldman ID, Zhao R. Molecular, biochemical, and cellular pharmacology of pemetrexed. Semin Oncol 2002;29(6 Suppl. 18):3–17.
[130] Fossella FV. Pemetrexed for treatment of advanced non-small cell lung cancer. Semin Oncol 2004;31(1 Suppl 1):100–5.
[131] Adjei AA. Current data with pemetrexed (Alimta) in non-small-cell lung cancer. Clin Lung Cancer 2003;4(Suppl 2S):64–7.
[132] Kut V, Patel JD, Argiris A. Pemetrexed: a novel antifolate agent enters clinical practice. Expert Rev Anticancer Ther 2004;4(4):511–22.
[133] Zhao R, Zhang S, Hanscom M, Chattopadhyay S, Goldman ID. Loss of reduced folate carrier function and folate depletion result in enhanced pemetrexed inhibition of purine synthesis. Clin Cancer Res 2005;11(3):1294–301.
[134] Puto K, Garey JS. Pemetrexed therapy for malignant pleural mesothelioma. Ann Pharmacother 2005;39(4):678–83.
[135] Adjei AA. Pharmacology and mechanism of action of pemetrexed. Clin Lung Cancer 2004;5(Suppl. 2):S51–5.
[136] Calvert AH. Biochemical pharmacology of pemetrexed. Oncology (Williston Park) 2004;18(13 Suppl 8):13–7.
[137] Pearce HL, Alice MM. The evolution of cancer research and drug discovery at Lilly research laboratories. Adv Enzyme Regul 2005;45229–55.
[138] Scagliotti G. An evaluation of pemetrexed in second-line treatment of non-small cell lung cancer. Expert Opin Pharmacother 2005;6(16):2855–66.
[139] Chattopadhyay S, Zhao R, Tsai E, Schramm VL, Goldman ID. The effect of a novel transition state inhibitor of methylthioadenosine phosphorylase on pemetrexed activity. Mol Cancer Ther 2006;5(10):2549–55.
[140] Adjei AA. Pemetrexed (Alimta): a novel multitargeted antifolate agent. Expert Rev Anticancer Ther 2003;3(2):145–56.
[141] Robinson DM, Keating GM, Wagstaff AJ. Pemetrexed: a review of its use in malignant pleural mesothelioma and non-small-cell lung cancer. Am J Cancer (Auckland, N Z) 2004;3(6):387–99.
[142] Zhao D, Tang C, Shi T, Cheng M. Synthesis of a multitargeted antifolate-pemetrexed. Zhongguo Yaowu Huaxue Zazhi 2008;18(6):445–8.
[143] Curtin NJ, Hughes AN. Pemetrexed disodium, a novel antifolate with multiple targets. Lancet Oncol 2001;2(5):298–306.
[144] Eismann U, Oberschmidt O, Ehnert M, Fleeth J, Luedtke FE, Struck S, Schulz L, Blatter J, Lahn MM, Ma D, et al. Pemetrexed: mRNA expression of the target genes TS, GARFT and DHFR correlates with the in vitro chemosensitivity of human solid tumors. Int J Clin Pharmacol Ther 2005;43(12):567–9.
[145] Ngeow J, Toh C. The role of pemetrexed combined with gemcitabine for non-small-cell lung cancer. Curr Drug Targets 2010;11(1):61–6.
[146] Min T, Ye H, Zhang P, Liu J, Zhang C, Shen W, Wang W, Shen L. Water-soluble poly(ethylene glycol) prodrug of pemetrexed: synthesis, characterization, and preliminary cytotoxicity. J Appl Polym Sci 2009;111(1):444–51.
[147] Bischof M, Weber K, Blatter J, Wannenmacher M, Latz D. Interaction of pemetrexed disodium (ALIMTA, multitargeted antifolate) and irradiation in vitro. Int J Radiat Oncol Biol Phys 2002;52(5):1381–8.
[148] Yang T, Chang G, Chen K, Hung H, Hsu K, Sheu G, Hsu S. Sustained activation of ERK and Cdk2/cyclin-A signaling pathway by pemetrexed leading to S-phase arrest and apoptosis in human non-small cell lung cancer A549 cells. Eur J Pharmacol 2011;663(1–3):17–26.
[149] Giovannetti E, Lemos C, Tekle C, Smid K, Nannizzi S, Rodriguez JA, Ricciardi S, Danesi R, Giaccone G, Peters GJ. Molecular mechanisms underlying the synergistic interaction of erlotinib, an epidermal growth factor receptor tyrosine kinase inhibitor,

with the multitargeted antifolate pemetrexed in non-small-cell lung cancer cells. Mol Pharmacol 2008;73(4):1290–300.
[150] Zhang Y, Trissel LA. Physical instability of frozen pemetrexed solutions in PVC bags. Ann Pharmacother 2006;40(7–8):1289–92.
[151] Zhang Y, Trissel LA. Physical and chemical stability of pemetrexed in infusion solutions. Ann Pharmacother 2006;40(6):1082–5.
[152] Stapleton SL, Reid JM, Thompson PA, Ames MM, McGovern RM, McGuffey L, Nuchtern J, Dauser R, Blaney SM. Plasma and cerebrospinal fluid pharmacokinetics of pemetrexed after intravenous administration in non-human primates. Cancer Chemother Pharmacol 2007;59(4):461–6.
[153] Manegold C. Pemetrexed: its promise in treating non-small-cell lung cancer. Oncology (Williston Park, NY) 2004;18(8 Suppl. 5):43–8.
[154] Nagai S, Takenaka K, Sonobe M, Wada H, Tanaka F. Schedule-dependent synergistic effect of pemetrexed combined with gemcitabine against malignant pleural mesothelioma and non-small cell lung cancer cell lines. Chemotherapy 2008;54(3):166–75.
[155] Andrea T. Malignant mesothelioma. Springer; 2011. p. 140.
[156] Lansiaux A, Lokiec F. Pemetrexed: from preclinic to clinic. Bull Cancer 2007;94 (Spec):S134–8.
[157] Richard RB, Maurie M, Marcus R. Principles and practice of gynecologic oncology. 5th ed. Philadelphia: Wolters Kluwer Health/Lippincott Williams & Wilkins; 2009. p. 441.
[158] Nagai H, Yasuda H, Hatachi Y, Xue D, Sasaki T, Yamaya M, Sakamori Y, Togashi Y, Masago K, Ito I, et al. Nitric oxide (NO) enhances pemetrexed cytotoxicity via NO-cGMP signaling in lung adenocarcinoma cells in vitro and in vivo. Int J Oncol 2012;41(1):24–30.
[159] Racanelli AC, Rothbart SB, Heyer CL, Moran RG. Therapeutics by cytotoxic metabolite accumulation: pemetrexed causes ZMP accumulation, AMPK activation, and mammalian target of rapamycin inhibition. Cancer Res 2009;69(13):5467–74.
[160] Harvey IP, David PC, David HJ, John DM, Giorgio VS, Andrew TT. Principles and Practice of Lung Cancer: The Official Reference Text of the International Association for the Study of Lung Cancer (IASLC). 4th ed. Lippincott Williams & Wilkins; 2010. p. 672.
[161] Buque A, Muhialdin JS, Munoz A, Calvo B, Carrera S, Aresti U, Sancho A, Rubio I, Lopez-Vivanco G. Molecular mechanism implicated in Pemetrexed-induced apoptosis in human melanoma cells. Mol Cancer 2012;1125.
[162] Hanauske AR, Chen V, Paoletti P, Niyikiza C. Pemetrexed disodium: a novel antifolate clinically active against multiple solid tumors. Oncologist 2001;6(4):363–73.
[163] Ozasa H, Oguri T, Uemura T, Miyazaki M, Maeno K, Sato S, Ueda R. Significance of thymidylate synthase for resistance to pemetrexed in lung cancer. Cancer Sci 2010;101 (1):161–6.
[164] Wu M, Hsiao Y, Huang C, Huang Y, Yang W, Chan H, Chang JT, Ko J. Genetic determinants of pemetrexed responsiveness and nonresponsiveness in non-small cell lung cancer cells. J Thorac Oncol 2010;5(8):1143–51.
[165] Hagner N, Joerger M. Cancer chemotherapy: targeting folic acid synthesis. Cancer Manag Res 2010;2293–301.
[166] Baldwin CM, Perry CM. Pemetrexed. A review of its use in the management of advanced non-squamous non-small cell lung cancer. Drugs 2009;69(16):2279–302.
[167] Kim JH, Lee K, Jung Y, Kim TY, Ham HS, Jong H, Jung KH, Im S, Kim T, Kim NK, et al. Cytotoxic effects of pemetrexed in gastric cancer cells. Cancer Sci 2005;96 (6):365–71.
[168] Chaigneau L, Villanueva C, Thierry-Vuillemin A, Legat-Fagnoni C, N'Guyen T, Maurina T, Lorgis V, Pivot X. Pemetrexed development in oncology. Bull Cancer 2007;94(spec):S142–8.

[169] Paz-Ares L, Bezares S, Tabernero J, et al. Review of a promising new agent—pemetrexed disodium. Cancer 2003;97(Suppl. 8):2056–63 [2. Eli Lilly Canada Inc. ALIMTA® Product Monograph. Toronto, Ontario; 21 May 2004].
[170] Matama G, Tokito T, Takeoka H, Hiraoka Y, Matsuo N, Nakamura M, Ishii H, Kinoshita T, Azuma K, Yamada K, et al. Aplastic anemia in a lung adenocarcinoma patient receiving Pemetrexed. Invest New Drugs 2017;35(5):662–4.
[171] Xu Y, Jiang X, Zhang L, Chen X, Huang Z, Lu J. Establishment and characterization of Pemetrexed-resistant NCI-H460/PMT cells. Anticancer Agents Med Chem 2019;19(6):731–9.
[172] Fu B, Ferronato C, Fine L, Meunier F, Luis Valverde J, Ferro Fernandez VR, Giroir-Fendler A, Chovelon J. Removal of Pemetrexed from aqueous phase using activated carbons in static mode. Chem Eng J (Amsterdam, Netherlands) 2021;405:127016.
[173] Amano Y, Ohta S, Sakura KL, Ito T. Pemetrexed-conjugated hyaluronan for the treatment of malignant pleural mesothelioma. Eur J Pharm Sci 2019;138:105008.
[174] Wei D, Hu R, Li H. Effect of Fuzhengkang compound adjuvant pemetrexed in treatment of advanced non-small cell lung cancer and its influence on serum CEA level. Zhongguo Zhongyiyao Keji 2016;23(5):588–9.
[175] Zhang X, Zhang D, Huang L, Li G, Chen L, Ma J, Li M, Wei M, Zhou W, Zhou C, et al. Discovery of novel biomarkers of therapeutic responses in Han Chinese pemetrexed -based treated advanced NSCLC patients. Front Pharmacol 2019;10:944.
[176] Minami S, Kijima T. Pemetrexed as indispensable drug for the treatment of non-squamous non-small cell lung cancer: the role in induction and maintenance therapy. Cancer Cell Microenviron 2014;1(2):e157/1–e157/14.
[177] Koyama N, Suzuki M. Clinical and biological significance of erlotinib therapy after pemetrexed in non-small cell lung cancer with wild-type EGFR. Neoplasma 2015;62(6):996–1004.
[178] Yang W, Yang L, Xia Y, Cheng L, Zhang J, Meng F, Yuan J, Zhong Z. Lung cancer specific and reduction-responsive chimaeric polymersomes for highly efficient loading of pemetrexed and targeted suppression of lung tumor in vivo. Acta Biomater 2018;70:177–85.
[179] Luo X, Dai Y, Tian W, Dou Y, Zhu J, Li S, Wang H. Efficacy in the maintenance treatment of non small cell lung cancer. Hebei Yixue 2016;22(3):433–6.
[180] He K, Liu J, Gao Y, Hao Y, Yang X, Huang G. Preparation and evaluation of Stearylamine-bearing pemetrexed disodium-loaded cationic liposomes in vitro and in vivo. AAPS PharmSciTech 2020;21(5):193.
[181] Minami S, Kijima T. Pemetrexed in maintenance treatment of advanced non-squamous non-small-cell lung cancer. Lung Cancer: Targets Ther 2015;6:13–25.
[182] Okimoto T, Kotani H, Iida Y, Koyanagi A, Tanino R, Tsubata Y, Isobe T, Harada M. Pemetrexed sensitizes human lung cancer cells to cytotoxic immune cells. Cancer Sci 2020;111(6):1910–20.
[183] van den Hombergh E, de Rouw N, van den Heuvel M, Croes S, Burger DM, Derijks J, van Erp NP, ter Heine R. Simple and rapid quantification of the multi-enzyme targeting antifolate pemetrexed in human plasma. Ther Drug Monit 2020;42(1):146–50.
[184] He X, Jiang Y, Wang L, Cui L. Comparative analysis of efficacy of docetaxel, gemcitabine, pemetrexed disodium and cisplatin in treatment of advanced non-small cell lung cancer. Xiandai Zhenduan Yu Zhiliao 2015;26(22):5148–50.
[185] Cova E, Pandolfi L, Colombo M, Frangipane V, Inghilleri S, Morosini M, Mrakic-Sposta S, Moretti S, Monti M, Pignochino Y, et al. Pemetrexed -loaded nanoparticles targeted to malignant pleural mesothelioma cells: an in vitro study. Int J Nanomedicine 2019;14:773–85.
[186] Hamamoto Y, Takeoka S, Mouri A, Fukusumi M, Wakuda K, Ibe T, Honma C, Arimoto Y, Yamada K, Wagatsuma M, et al. Orotate phosphoribosyltransferase is

overexpressed in malignant pleural mesothelioma: dramatically responds one case in high OPRT expression. Rare Dis 2016;4(1):e1165909/1–9.
[187] Abu Lila AS, Fukushima M, Huang C, Wada H, Ishida T. Systemically administered RNAi molecule sensitizes malignant pleural mesotheliomal cells to pemetrexed therapy. Mol Pharm 2016;13(11):3955–63.
[188] Chamizo C, Zazo S, Domine M, Cristobal I, Garcia-Foncillas J, Rojo F, Madoz-Gurpide J. Thymidylate synthase expression as a predictive biomarker of pemetrexed sensitivity in advanced non-small cell lung cancer. BMC Pulm Med 2015;15:132/1–7.
[189] Ikemura K, Hamada Y, Kaya C, Enokiya T, Muraki Y, Nakahara H, Fujimoto H, Kobayashi T, Iwamoto T, Okuda M. Lansoprazole exacerbates pemetrexed-mediated hematologic toxicity by competitive inhibition of renal basolateral human organic anion transporter 3. Drug Metab Dispos 2016;44(10):1543–9.
[190] Sakura K, Sasai M, Ito O, Ota S, Amano Y. Pemetrexed-hyaluronic acid conjugate and pharmaceutical composition containing the same for treatment of disseminated tumor. Jpn. Kokai Tokkyo Koho; 2019. JP 2019156771 A 20190919.
[191] Qin Y, Sekine I, Hanazono M, Morinaga T, Fan M, Takiguchi Y, Tada Y, Shingyoji M, Yamaguchi N, Tagawa M. AMPK activation induced in pemetrexed-treated cells is associated with development of drug resistance independently of target enzyme expression. Mol Oncol 2019;13(6):1419–32.
[192] Cobanoglu H, Coskun M, Coskun M, Cayi A. Evaluation of genotoxic potential of pemetrexed by using in vitro alkaline comet assay. Marmara Pharm J 2017;21 (3):544–50.
[193] Ishikawa K, Sakai T, Saito-Shono T, Miyawaki M, Osoegawa A, Sugio K, Ono A, Mori H, Nishida H, Yokoyama S, et al. Pemetrexed-induced scleroderma-like conditions in the lower legs of a patient with non-small cell lung carcinoma. J Dermatol 2016;43(9):1071–4.
[194] Clark SK, Anselmo LM. Incidence of cutaneous reactions with pemetrexed: comparison of patients who received three days of oral dexamethasone twice daily to patients who did not. J Oncol Pharm Pract 2019;25(7):1645–50.
[195] Ge H, Ke J, Xu N, Li H, Gong J, Li X, Song Y, Zhu H, Bai C. Dexamethasone alleviates pemetrexed-induced senescence in non-small-cell lung cancer. Food Chem Toxicol 2018;119:86–97.
[196] Gamazon ER, Trendowski MR, Wen Y, Wing C, Delaney SM, Huh W, Wong S, Cox NJ, Dolan ME. Gene and MicroRNA perturbations of cellular response to pemetrexed implicate biological networks and enable imputation of response in lung adenocarcinoma. Sci Rep 2018;8(1):1–13.
[197] Liang J, Lu T, Chen Z, Zhan C, Wang Q. Mechanisms of resistance to pemetrexed in non-small cell lung cancer. Transl Lung Cancer Res 2019;8(6):1107–18.
[198] Assayag M, Rouvier P, Gauthier M, Costel G, Cluzel P, Mercadal L, Deray G, Bagnis CI. Renal failure during chemotherapy: renal biopsy for assessing subacute nephrotoxicity of pemetrexed. BMC Cancer 2017;17:770/1–8.
[199] Bai F, Yin Y, Chen T, et al. Development of liposomal pemetrexed for enhanced therapy against multidrug resistance mediated by ABCC5 in breast cancer. Int J Nanomedicine 2018;13:1327–39.
[200] Chen J, Zhang H, Chen X. Pemetrexed inhibits Kaposi's sarcoma-associated herpesvirus replication through blocking dTMP synthesis. Antiviral Res 2020;180:104825.
[201] Oleinick NL, Biswas T, Patel R, Tao M, Patel R, Weeks L, Sharma N, Dowlati A, Gerson SL, Fu P, et al. Radiosensitization of non-small-cell lung cancer cells and xenografts by the interactive effects of pemetrexed and methoxyamine. Radiother Oncol 2016;121(2):335–41.
[202] Pan Z, Yang G, He H, Cui J, Li W, Yuan T, Chen K, Jiang T, Gao P, Sun Y, et al. Intrathecal pemetrexed combined with involved-field radiotherapy as a first-line intra-

CSF therapy for leptomeningeal metastases from solid tumors: a phase I/II study. Ther Adv Med Oncol 2020;12, 1758835920937953.
[203] Rosch JG, DuRoss AN, Landry MR, Sun C. Development of a pemetrexed /folic acid nanoformulation: synthesis, characterization, and efficacy in a murine colorectal cancer model. ACS Omega 2020;5(25):15424–32.
[204] Li B, Zhang J, Su Y, Yi H, Wang Z, Zhao L, Sun S, Fu H. Overexpression of PTEN may increase the effect of pemetrexed on A549 cells via inhibition of the PI3K/AKT/mTOR pathway and carbohydrate metabolism. Mol Med Rep 2019;20(4):3793–801.
[205] Zhou J, Ben S. Clinical effects of pemetrexed or docetaxel combined with cisplatin as the first line treatment for advanced lung adenocareinoma. Zhongguo Aizheng Zazhi 2015;25(9):671–6.
[206] Ba X, Han J, Zhao G, Bian L. Study on the treatment of advanced lung adenocarcinoma in the elderly with pemetrexed combined with platinum drugs. Pak J Pharm Sci 2019;32(5, Suppl):2415–8.
[207] Lu S, Cheng Y, Zhou C, Wang J, Yang JC, Zhang P, Zhang X, Wang X, Orlando M, Wu Y. Meta-analysis of first-line pemetrexed plus platinum treatment in compared to other platinum-based doublet regimens in elderly east Asian patients with advanced nonsquamous non-small-cell lung cancer. Clin Lung Cancer 2016;17(5):e103–12.
[208] Nash A, Francis P. Osimertinib for use in the treatment of non-small cell lung cancer in EGFR kinase inhibitor-naive subjects. PCT Int. Appl; 2020. WO 2020201097 A1 20201008.
[209] Pangeni R, Choi JU, Panthi VK, Byun Y, Park JW. Enhanced oral absorption of pemetrexed by ion-pairing complex formation with deoxycholic acid derivative and multiple nanoemulsion formulations: preparation, characterization, and in vivo oral bioavailability and anticancer effect. Int J Nanomedicine 2018;13:3329–51.
[210] Noronha V, Patil VM, Joshi A, Menon N, Chougule A, Mahajan A, Janu A, Purandare N, Kumar R, More S, et al. Gefitinib versus gefitinib plus pemetrexed and carboplatin chemotherapy in EGFR-mutated lung cancer. J Clin Oncol 2020; 38(2):124–37.
[211] Ando Y, Hayashi T, Ujita M, Murai S, Ohta H, Ito K, Yamaguchi T, Funatsu M, Ikeda Y, Imaizumi K, et al. Effect of renal function on pemetrexed-induced haematotoxicity. Cancer Chemother Pharmacol 2016;78(1):183–9.
[212] Sato Y, Matsuda S, Maruyama A, Nakayama J, Miyashita T, Udagawa H, Umemura S, Yanagihara K, Ochiai A, Tomita M, et al. Metabolic characterization of antifolate responsiveness and non-responsiveness in malignant pleural mesothelioma cells. Front Pharmacol 2018;9:1129.
[213] Pu X, Li W, Lu B, Wang Z, Yang M, Fan W, Meng L, Lv Z, Xie Y, Wang J. Single pemetrexed is noninferior to platinum-based pemetrexed doublet as first-line treatment on elderly Chinese patients with advanced nonsquamous nonsmall cell lung cancer. Medicine (Philadelphia, PA, USA) 2017;96(11):e6002.
[214] Zhang Y. Clinical efficacy and side effects of pemetrexed combined with cisplatin and gefitinib in treatment of advanced lung adenocarcinoma. Shiyong Linchuang Yiyao Zazhi 2016;20(17):145–6.
[215] Kenmotsu H, Yamamoto N, Yamanaka T, Yoshiya K, Takahashi T, Ueno T, Goto K, Daga H, Ikeda N, Sugio K, et al. Randomized phase III study of pemetrexed plus cisplatin versus vinorelbine plus cisplatin for completely resected stage II to IIIA, nonsquamous non-small-cell lung cancer. J Clin Oncol 2020;38(19):2187–96.
[216] Gerogianni I, Pitaraki E, Jagirdar RM, Kouliou O, Giannakou L, Giannopoulos S, Papazoglou E, Hatzoglou C, Gourgoulianis KI, Zarogiannis SG. 2-Deoxy-glucose enhances the effect of cisplatin and pemetrexed in reducing malignant pleural mesothelioma cell proliferation but not spheroid growth. Anticancer Res 2019; 39(7):3809–14.

[217] Cheng Y, Murakami H, Yang P, He J, Nakagawa K, Kang JH, Kim J, Wang X, Enatsu S, Puri T, et al. Randomized phase II trial of gefitinib with and without pemetrexed as first-line therapy in patients with advanced nonsquamous non-small-cell lung cancer with activating epidermal growth factor receptor mutations. J Clin Oncol 2016;34 (27):3258–66.
[218] Grilley-Olson JE, Weiss J, Ivanova A, Villaruz LC, Moore DT, Stinchcombe TE, Lee C, Shan JS, Socinski MA. Phase Ib study of Bavituximab with carboplatin and pemetrexed in chemotherapy-naive advanced nonsquamous non-small-cell lung cancer. Clin Lung Cancer 2018;19(4):e481–7.
[219] Kozuki T, Nogami N, Hataji O, Tsunezuka Y, Seki N, Harada T, Fujimoto N, Bessho A, Takamura K, Takahashi K, et al. Open-label, multicenter, randomized phase II study on docetaxel plus bevacizumab or pemetrexed plus bevacizumab for treatment of elderly (aged ≥75 years) patients with previously untreated advanced non-squamous non-small cell lung cancer: TORG1323. Transl Lung Cancer Res 2020;9(3):459–70.
[220] Yang H, Jin R, Su W, Hong L. Clinical analysis of bevacizumab combined with pemetrexed and cisplatin in treatment of non-small cell lung cancer. Linchuang Junyi Zazhi 2015;43(10):1053–5.
[221] Kim ES, Kelly K, Paz-Ares LG, Garrido P, Jalal S, Mahadevan D, Gutierrez M, Provencio M, Schaefer E, Shaheen M, et al. Abemaciclib in combination with single-agent options in patients with stage IV non-small cell lung cancer: a phase Ib study. Clin Cancer Res 2018;24(22):5543–51.
[222] Waqar SN, Baggstrom MQ, Morgensztern D, Williams K, Rigden C, Govindan R. A phase I trial of temsirolimus and pemetrexed in patients with advanced non-small cell lung cancer. Chemotherapy (Basel, Switzerland) 2016;61(3):144–7.
[223] Hu X, Pu K, Feng X, Wen S, Fu X, Guo C, He W. Role of gemcitabine and pemetrexed as maintenance therapy in advanced NSCLC: a systematic review and meta-analysis of randomized controlled trials. PLoS ONE 2016;11(3):e0149247/1–e0149247/18.
[224] Passardi A, Fanini F, Turci L, Foca F, Rosetti P, Ruscelli S, Casadei Gardini A, Valgiusti M, Dazzi C, Marangolo M. Prolonged pemetrexed infusion plus gemcitabine in refractory metastatic colorectal cancer: preclinical rationale and phase II study results. Oncologist 2017;22(8):886–e79.
[225] Wang L, Wang Y, Guan Q, Liu Y, He T, Wang J. Establishment of a first-line second-line treatment model for human pulmonary adenocarcinoma. Oncol Lett 2016;12 (6):4461–6.
[226] Steinborn B, Hirschle P, Hoehn M, Bauer T, Barz M, Wuttke S, Wagner E, Laechelt U. Core-shell functionalized zirconium-pemetrexed coordination nanoparticles as carriers with a high drug content. Adv Ther (Weinheim, Germany) 2019;2 (11):1900120.
[227] Kimura M, Imamura F, Inoue T, Nishino K, Uchida J, Kumagai T, Okami J, Higasiyama M, Kamoshida S. Protein and mRNA expression of folic acid-associated enzymes as biomarkers for the cytotoxicity of the thymidylate synthase-targeted drugs, pemetrexed and S-1, in non-small cell lung cancer. Mol Clin Oncol 2017;7(1):15–23.
[228] Liang C, Liu H. Pemetrexed disodium-lipid complex and preparation method thereof. Faming Zhuanli Shenqing; 2018. CN 109077997 A 20181225.
[229] Purandare SM, Malhotra G, Rao DR, Kankan RN, Pathi SL, Puppla R. Pemetrexed complexes and pharmaceutical compositions containing pemetrexed complexes. U.S. Pat. Appl. Publ; 2015. US 20150359898 A1 20151217.
[230] Ozcelikay G, Karadas-Bakirhan N, Taskin-Tok T, Ozkan SA. A selective and molecular imaging approach for anticancer drug: pemetrexed by nanoparticle accelerated molecularly imprinting polymer. Electrochim Acta 2020;354:136665.

[231] Ciekot J, Goszczynski TM, Boratynski J. Methods for preliminary determination of pemetrexed in macromolecular drug-carrier systems. Acta Pharm (Warsaw, Poland) 2016;66(1):147–53.
[232] Lin Y, Wang K, Lo W, Lin W, Cheng W. Crystalline forms of pemetrexed diacid and manufacturing processes therefor. U.S. Pat. Appl. Publ; 2016. US 20160122355 A1 20160505.
[233] Hill J, Vargo C, Smith M, Streeter J, Carbone DP. Safety of dose-reduced pemetrexed in patients with renal insufficiency. J Oncol Pharm Pract 2019;25(5):1125–9.
[234] Cheng Q, Zhong J, He L, Yang X, Xue J, Wang Q. Preparation of **pemetrexed** related substances and its pharmaceutical salts. Faming Zhuanli Shenqing; 2020. CN 111592547 A 20200828.
[235] Shah DM, Wader GR. Process for preparation of stable salts of pemetrexed. PCT Int. Appl; 2017. WO 2017168442 A1 20171005.
[236] Wang N, Song X, Wang J, Tong Y, Yan Y. Preparation method of pemetrexed disodium hydrate for treating malignant pleural mesothelioma. Faming Zhuanli Shenqing; 2020. CN 111333658 A 20200626.
[237] Son MW, Jang SU, Won DH, Kim YM, Hwang HD, Min DH. Stable pharmaceutical composition comprising pemetrexed or pharmaceutically acceptable salt thereof. Repub Korean Kongkae Taeho Kongbo; 2016. KR 2016058564 A 20160525.
[238] Sohn M, Jang S, Won D, Kim Y, Hwang H, Min D. Stable pharmaceutical composition comprising pemetrexed or pharmaceutically acceptable salt thereof. PCT Int. Appl; 2016. WO 2016080687 A1 20160526.
[239] Kumar S, Kane P, Bhowmick SB, Pandit UK, Mistry NB, Varu RK. Stable injectable solution of pemetrexed. PCT Int. Appl; 2018. WO 2018002956 A1 20180104.
[240] Mailatur Sivaraman M, Patel H, Patel BV, Kannekanti R, Sanka M, Mohammad R, Chakali P. Ready to infuse pemetrexed liquid compositions. PCT Int. Appl; 2018. WO 2018127743 A1 20180712.
[241] Gao Z, Wang X, Zhao Y, Liu G. Injection for skin koji culture plug magnetic self-assembling nano composite particle preparation method [Machine Translation]. Faming Zhuanli Shenqing; 2020. CN 111840547 A 20201030.
[242] Imai Y., Hide T. Solution preparation for injection containing pemetrexed with high storage stability. Jpn. Kokai Tokkyo Koho 2017; JP 2017014153 A 20170119.
[243] Hu H, Ding Z, Liu Y. Pemetrexed disodium lyophilized powder injection and preparation method thereof. Faming Zhuanli Shenqing; 2018. CN 108992413 A 20181214.
[244] Imai Y, Hayashi T, Matsuzaki S. Pemetrexed injectable solution preparations comprising pharmaceutically acceptable salt. . Jpn. Kokai Tokkyo Koho; 2019. JP 2019123684 A 20190725.
[245] Araki R, Keira T, Masuda Y, Tanaka T, Yamada H, Hamamoto T. Effects of proton pump inhibitors on severe haematotoxicity induced after first course of pemetrexed/carboplatin combination chemotherapy. J Clin Pharm Ther 2019;44(2):276–84.
[246] Khattar D, Khana R, Yadav M, Burman K. Pharmaceutical compositions of pemetrexed. U.S. Pat. Appl. Publ; 2015. US 20150111905 A1 20150423.
[247] Li X, Ren J, Qian Y. Pemetrexed disodium pharmaceutical composition and preparation method therefor. PCT Int. Appl; 2018. WO 2018049997 A1 20180322.
[248] Khattar D, Khanna R, Yadav A, Bhandari V, Modi SR, Hemlata SR. Stable liquid compositions of pemetrexed for use in treatment of malignant pleural mesothelioma and non-small cell lung cancer. Indian Pat. Appl; 2019. IN 201711030501 A 20190719.
[249] Kumar S, Varu RK, Patel N, Kane P, Bhowmick SB. Intravenous infusion dosage form for pemetrexed. U.S. Pat. Appl. Publ; 2019. US 20190105262 A1 20190411.

[250] Avci R. Lyophilized pharmaceutical composition of pemetrexe. Turk. Pat. Appl; 2014. TR 2013000019 A2 20140721.
[251] Cho JU, Kim GH, Seo MH, Lee SW. Stabilized pharmaceutical compositions containing pemetrexed or its salts. Repub. Korean Kongkae Taeho Kongbo; 2016. KR 2016139514 A 20161207.
[252] Prasad V. Pharmaceutical formulations of pemetrexed. Indian Pat. Appl; 2017. IN 2015CH04324 A 20170224.
[253] Park YJ, Shin MJ, Jin HC, Choi HY, Choi NH. Stabilized pemetrexed formulation. U.S; 2016. US 9265832 B2 20160223.
[254] Birsan E, Maria R, Stancu C. Pharmaceutical composition comprising pemetrexed. PCT Int. Appl; 2016. WO 2016207443 A1 20161229.
[255] Cho J, Kim G, Seo M, Lee S. Stabilized pharmaceutical compositions containing pemetrexed or its salts. PCT Int. Appl; 2016. WO 2016190712 A2 20161201.
[256] Singh S, Alagarswamy A, Malladi M, Gite S. Stable pharmaceutical formulations of pemetrexed for treatment of lung cancer. U.S. Pat. Appl. Publ; 2020. US 20200155556 A1 20200521.
[257] Nayani A, Bothra M, George AK. A stable, ready to use aqueous pharmaceutical composition of pemetrexed. Indian Pat. Appl; 2020. IN 201921017365 A 20201106.
[258] Matsuyama K, Otori T, Maegawa T, Kimachi T, Sato M. Medoxomil, hemiacetal ester of pemetrexed, production method therefor, and pharmaceutical composition containing ester prodrug. PCT Int. Appl; 2018. WO 2018096853 A1 20180531.
[259] Chandrashekhar K, Nagaraju B. Stable liquid formulations of pemetrexed. PCT Int. Appl; 2016. WO 2016199053 A1 20161215.
[260] Li X, Ren J, Qian Y. A kind of pemetrexed disodium pharmaceutical composition and preparation method thereof. Faming Zhuanli Shenqing; 2018. CN 107837237 A 20180327.
[261] Chen F, Krill SL. Pemetrexed formulations. U.S. Pat. Appl. Publ; 2017. US 20170239250 A1 20170824.
[262] Lv H, Li W, Li Q, Jiang T, Feng R. Injection containing pemetrexed and preparation method thereof. Faming Zhuanli Shenqing; 2016. CN 106176608 A 20161207.
[263] Palle RV, Nariyam SM, Patel VB, Vinjamuri RRS, Devarakonda SN, Yarraguntla SR, Mudapaka VK, Nalivela V. Solid forms of pemetrexed. U.S; 2013. US 8507508 B2 20130813.
[264] Zhang Q, Nguyen T, McMichael M, Velu SE, Zou J, Zhou X, Wu H. New small-molecule inhibitors of dihydrofolate reductase inhibit *Streptococcus mutans*. Int J Antimicrob Agents 2015;46(2):174–82.
[265] Krause K, Suter D, Salmon P, Myburgh R, Pepper M. Transgenic cell selection. U.S. Pat. Appl. Publ; 2016. US 20160010094 A1 20160114.
[266] Gibson MW, Dewar S, Ong HB, Sienkiewicz N, Fairlamb AH. Trypanosoma brucei DHFR-TS revisited: characterisation of a bifunctional and highly unstable recombinant dihydrofolate reductase-thymidylate synthase. PLoS Negl Trop Dis 2016;10(5): e0004714/1–e0004714/20.
[267] Castellano G, Manuel J. Boosting the effects of methotrexate through the combined use with lipophilic statins. PCT Int. Appl; 2015. WO 2015015039 A1 20150205.
[268] O'Dwyer PJ, DeLap RJ, King SA, Grillo-Lopez AJ, Hoth DF, Leyland-Jones B. Trimetrexate: clinical development of a nonclassical antifolate. NCI Monogr 1987;5:105–9.
[269] Szabolcs P, Park K, Marti L. deOliveria D, Lee Y, Colvin MO, Kurzberg J. Superior depletion of alloreactive T cells from peripheral blood stem cell and umbilical cord blood grafts by the combined use of trimetrexate and interleukin-2 immunotoxin. Biol Blood Marrow Transplant 2004;10(11):772–83.

[270] Gangjee A, Kurup S, Namjoshi O. Dihydrofolate reductase as a target for chemotherapy in parasites. Curr Pharm Des 2007;13(6):609–39.
[271] Hattinger CM, Tavanti E, Fanelli M, Vella S, Picci P, Serra M. Pharmacogenomics of genes involved in antifolate drug response and toxicity in osteosarcoma. Expert Opin Drug Metab Toxicol 2017;13(3):245–57.
[272] Blanke CD, Chansky K, Christman KL, Hundahl SA, Issell BF, Van Veldhuizen PJJ, Budd GT, Abbruzzese JL, MacDonald JS. S9511: a southwest oncology group phase II study of trimetrexate, 5-fluorouracil, and leucovorin in unresectable or metastatic adenocarcinoma of the stomach. Am J Clin Oncol 2010;33(2):117–20.
[273] Xia M, Huang R, Sakamuru S, Alcorta D, Cho M, Lee D, Park DM, Kelley MJ, Sommer J, Austin CP. Identification of repurposed small molecule drugs for chordoma therapy. Cancer Biol Ther 2013;14(7):638–47.
[274] Fang H, Yu T, Tan M. Efficient experiment design and nonparametric modeling of drug interaction. Front Biosci (Elite Ed) 2010;E2(1):258–65.
[275] Zare-Shahabadi V. Quantitative structure-activity relationships of dihydrofolatereductase inhibitors. Med Chem Res 2016;25(12):2787–97.
[276] Senkovich O, Schormann N, Chattopadhyay D. Structures of dihydrofolate reductase-thymidylate synthase of *Trypanosoma cruzi* in the folate-free state and in complex with two antifolate drugs, trimetrexate and methotrexate. Acta Crystallogr D Biol Crystallogr 2009;65(7):704–16.
[277] Das JR, Fryar EB, Green S, Southerland WM, Bowen D. The protection against trimetrexate cytotoxicity in human bone marrow by sequence-dependent administration of raloxifene, 5-fluorouracil/trimetrexate. Anticancer Res 2006;26(6B):4279–86.
[278] Nzila A, Okombo J, Becker RP, Chilengi R, Lang T, Niehues T. Anticancer agents against malaria: time to revisit? Trends Parasitol 2010;26(3):125–9.

CHAPTER 3

Purine-based anticancer drugs

1. Introduction

Purine-based drugs have striking anticancer potencies and can act as an efficacious therapy. They represent a category of cytotoxic agents with high anticancer potencies. They have a similar chemical structure and mechanism of action. Their cytotoxicity could be caused by a variety of mechanisms.

Synergistic interactions between purine-based drugs and other cytotoxic agents, e.g., anthracycline, alkylating agents, cytarabine, monoclonal antibodies, and antitumor antibiotics, have been accomplished in both preclinical and clinical investigations. Clinical trials have confirmed their efficacy as a single agent or in combination protocols in treating various syndromes. Their toxicity profiles are similar for all drugs, which have primarily dose-limiting side effects. Many purine-based anticancer drugs have been depicted in this chapter.

2. Mercaptopurine drug

Mercaptopurine (6-MP) is a chemotherapeutic agent of the antimetabolite class [1] and one from the thiopurine drug therapy [2]; it is considered as a clinical important antitumor drug [3–5]. 6-Thioguanine (6TG) and 6-mercaptopurine are analogs of the natural purines: guanine and hypoxanthine. Both mercaptopurine and thioguanine are substrates for hypoxanthineguanine phosphoribosyltransferase and are transformed into 6-thioinosine monophosphate (T-IMP) and ribonucleotide 6-thioguanosine monophosphate (6-thioGMP). The accumulation of these monophosphates inhibits various key metabolic processes. These thiopurine bases remain effective agents in initiating and maintaining remissions in patients with acute lymphocytic and myelocytic leukemia. Despite their clinical relevance, 6TG and 6MP have some therapeutic drawbacks that have fueled the search for purine analogs that improve therapeutic efficacy. Other novel mercaptopurine and thioguanine analogs, as well as their nucleosides, have been designed and synthesized to enhance the antitumor efficacy [6]. 6-MP interferes with the synthesis of nucleic acids, consequently halting cancer growth

New Strategies Targeting Cancer Metabolism
https://doi.org/10.1016/B978-0-12-821783-2.00005-4

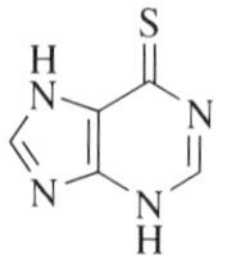

Fig. 3.1 Mercaptopurine.

and causing cell death [7,8]. 6-MP has been utilized because of its lifespan of more than 35 years in treating several types of cancers [9–24].

6-Mercaptopurine drug (Fig. 3.1) has shown effectiveness in treating autoimmune diseases, as well as in organ transplantation and in treating leukemia [25–28], e.g., acute lymphoblastic leukemia (ALL) [29–42] commonly in childhood and adult leukemia [43], acute myeloid leukemia (AML) [44], and chronic myelocytic leukemia [45]. It has also been discovered as a promising candidate for the central nervous system (CNS) active cancer drug development that can cross the blood-brain barrier [46].

6-MP is a well-established immunosuppressive drug [47–55]. It is utilized in the treatment of pediatric non-Hodgkin's lymphoma, psoriatic arthritis, polycythemia vera, and inflammatory bowel disease (IBD), such as ulcerative colitis [56–63], and Crohn's disease [64,65], and colorectal cancer. The drug indicates irritation on gastric exposure. Its bioavailability is low, with a half-life of less than 2 h. Targeted delivery of 6-MP will be both efficacious and safe therapy for colon diseases [66].

6-MP is a commonly used immunosuppressive medication. 6-MP is commonly utilized immunosuppressive drug that is effective in various inflammatory disorders. Although 6-MP has been utilized in the treatment of asthma, in airway epithelial cells, its role and mechanism of action are unknown [67]. 6-MP is considered as a well-established immunosuppressant to normalize cerebrospinal fluid (CSF) lymphocyte frequencies in opsoclonus-myoclonus syndrome (OMS) and function as a steroid sparer [68]. 6-MP has been indicated to have beneficial effects in a cell-specific fashion on smooth muscle cells (SMC), endothelial cells, and macrophages [69].

6-MP is a known antiviral agent and anticancer drug that inhibits the synthesis of DNA, RNA, and proteins [70], and induces immunosuppression and cytotoxicity [71]. A study that used chemotherapy drug 6-mercaptopurine (6MP) as anti-CoV-2 showed that it has little to no stable intermolecular interaction with papain-like protease (PLpro) of the SARS-CoV-2 (CoV2), which is involved in viral replication and immune response evasion. It is quickly dissociated or remained highly mobile [72].

Mercaptopurine is primarily utilized in treating cancer, autoimmune disorders, and organ transplant. Brain cancer (most commonly astrocytomas) has been connected to chronic thiopurine therapy, and Tpmt status has been associated with this risk [73]. 6-MP is an essential medicine utilized in the treatment of leukemia in the clinics. 6-MP suffers, however, from low oral bioavailability, poor water solubility, and remarkable side effects [74,75]. Indefinite use of these drugs is tempered by long-term risks [76]; dose-limiting toxicities can restrict its effectiveness [77,78].

Three cytosolic enzymes metabolize mercaptopurine. One of these, thiopurine *S*-methyltransferase (TPMT), catalyzes the methylation of mercaptopurine to 6-methylmercaptopurine, which renders the drug inactive [79]. 6-MP is transformed to an active metabolite 6-thioguanine nucleotide (6-TGN) via enzymic reaction, including thiopurine methyltransferase (TPMT) [80], which is a cytoplasmic transmethylase present in both eukaryotes and prokaryotes. In humans, it indicates its existence in almost all of the tissues, predominantly in the kidney and liver. TPMT is one of the phase II metabolic pathway's key metabolic enzymes, catalyzing the methylation of thiopurine medicines, which are utilized to treat patients with neoplasia, autoimmune disease, and transplant recipients [81].

TPMT is a cytosolic enzyme involved in the metabolism of thiopurine drugs such as 6-MP and azathioprine [82–84]. Various mutations in the TPMT gene correlate with low enzyme potency and subsequent adverse effects, mainly myelotoxicity [85]. TPMT activity displays an interindividual variability, which is primarily due to genetic polymorphism. Patients with intermediate or deficient TPMT activity are at risk for toxicity after receiving standard doses of thiopurine drugs [86,87].

Various inventions and researches are conducted on 6-mercaptopurine, and its analogs have been reported for antitumor drug preparations and many other applications [88–129].

3. 6-Thioguanine

6-Thioguanine (6-TG) (Fig. 3.2) is a nucleobase guanine antineoplastic purine analog drug and is considered as antimetabolic agent that belongs

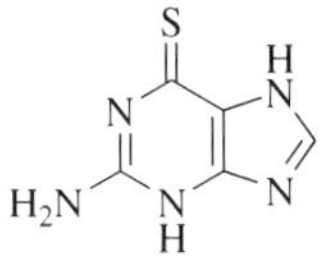

Fig. 3.2 Thioguanine.

to thiopurine drug [130], and is commonly used chemotherapeutic drug [131]. 6-TG is widely utilized as cytotoxic [132] anticancer agent, immunosuppressant [54,133,134], and antiviral drug [135]. 60 years ago, 6-TG was introduced into the clinic [136]. It is a cytotoxic drug used in treating myelogenous leukemia [137], acute leukemia [138], inflammatory bowel disease [139,140], Crohn's disease, acquired immune deficiency syndrome, and childhood acute lymphoblastic leukemia [139]. It is considered as a clinical drug for acute myeloid leukemia [141–143].

As mentioned before, thiopurine S-methyltransferase (TPMT) is responsible for catalyzing the *S*-methylation of 6-thioguanine, resulting in its inactivation [144]. 6-TG is inactive prodrug that needs intracellular metabolism for activation to cytotoxic metabolites. Thiopurine methyltransferase (TPMT) is one of the most essential enzymes in this process, metabolizing both 6-TG and 6-MP to various methylated metabolites comprising methylthioguanosine monophosphate (meTGMP) and methylthioinosine monophosphate (meTIMP), with various suggested cytotoxic and pharmacological features. While meTIMP is a potent inhibitor of de novo purine synthesis (DNPS) and contributes significantly to the cytotoxic effects of meTGMP, 6-MP does not add much to the effects of 6-TG, and the cytotoxicity of 6-TG appears to be more dependent on the incorporation of thioguanine) nucleotides (TGNs) into DNA rather than the inhibition of the DNPS [145].

6-Thioguanine is considered as an anticancer drug with narrow safe dose range and severe side effects. [146]. It has high adverse effects in human body [147]. 6-TG therapy in childhood acute lymphoblastic leukemia leads to portal hypertension and chronic hepatotoxicity [148]. It has high systemic toxicity as a result of the lack of target specificity. Thus, enhancing target selectivity should improve and enhance the drug safety [149].

Invention is reported relates to a method of eliminating toxicity associated with 6-TG drug treatment for leukemia. And this relates to the utility of new organometallic complexes of transition metals with 6-thioguanine or its analogs for treating and/or in the prevention of cancer [150]. In addition, thioguanine is utilized in the synthesis of a drug used in the treatment of ADSL deficiency [151]. Also, various aryl and acyclic analogs of 6-TG were synthesized [152]. Many oral sustained-release preparations comprise thioguanine as an antitumor active drug [153]. Other biomedical applications and studies associated with 6-TG have been performed [154–162].

4. Azathioprine

Azathioprine (AZA) (Fig. 3.3) is an imidazolyl analog of 6-mercaptopurine. Azathioprine is transformed into 6-mercaptopurine in the body where it blocks purine metabolism and then blocks the DNA synthesis [163]. AZA is considered as common immunosuppressive agent [164,165]. So, FDA approved this drug as an immunosuppressive and antiarrhythmic drug [166]. It is used effectively for treating ileitis, jejunoileitis, and gastroduodenal Crohn's disease [167]. Azathioprine is a first-line immunomodulatory drug that is widely utilized for treating patients with neuromyelitis optica spectrum disorders (NMOSDs), which are demyelinating autoimmune diseases in CNS [168]. In 1961, azathioprine was first utilized as an immunosuppressive agent in kidney transplant patients, and then it was widely utilized as a noncorticosteroid drug for treating autoimmune diseases [169], rheumatoid arthritis [170] systemic lupus erythematosus (SLE) [171], and myasthenia gravis (MG) [172]. Azathioprine is an immunosuppressive active pharmaceutical ingredient that can be taken orally or intravenously to prevent organ transplant rejection [173]. It is considered as immunomodulatory drug used for the maintenance of remission in Crohn's disease (CD). It is to achieve long-term clinical remission, allowing children to grow and develop normally [174,175]. Many autoimmune diseases especially following solid organ transplantation are treated with 6-mercaptopurine (6-MP) and its prodrug azathioprine. Their properties are mediated via 6-methylmercaptopurine (6-MMP) and 6-thioguanine nucleotides (6-TGN) and active metabolites. The relationships between concentrations of azathioprine metabolites and hepatotoxicity or myelotoxicity have not been confirmed [176].

The thioguanine derivative, azathioprine, is a prodrug of 6-mercaptopurine that is further metabolized via different enzymes that exist in the gut and liver and have shown an efficacy in the treatment of

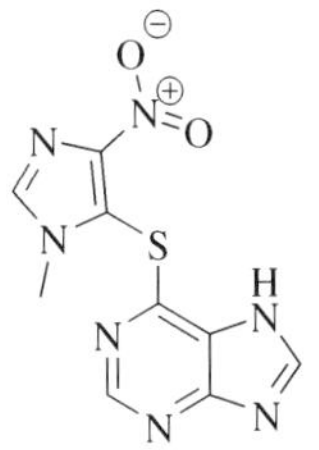

Fig. 3.3 Azathioprine.

inflammatory bowel disease (ulcerative colitis) [177]. It is still the most often used immunomodulator in treating inflammatory bowel disease (IBD) [178–180], provides long-term disease remission in IBDs, but has remarkable side effects, most of which are associated with a single nucleotide polymorphism in the gene for thiopurine methyltransferase (TPMT), which ensures the degradation and efficacy of AZA [181]. It is effective in treating disorders of immune regulation and acute lymphoblastic leukemia [182].

The pancreatitis risk of azathioprine and also its immunosuppressive effect are influenced by the potency of thiopurine methyltransferase (TPMT) and by some genetic mutations [183]. Genetic polymorphisms and potency of thiopurine methyltransferase (TPMT) have been linked to azathioprine toxicity in various populations and its efficacy [184]. The thiopurine S-methyltransferase (TPMT)/azathioprine (AZA) gene-drug pair is one of the most remarkable pharmacogenetic markers [185].

Although azathioprine is an essential drug mainly utilized in the therapy of the autoimmune system disorders, it induces various hazard effects that limit and restrict its use [186]. The long-term usage of azathioprine has been associated with a higher risk of certain cancers; hence, its effective detection is necessary [187].

5. Pentostatin

Pentostatin (2′-deoxycoformycin, dCF) (Nipent; SuperGen, San Ramon, CA) (Fig. 3.4) is a purine nucleoside analog, and it is produced due to the fermentation of Streptomyces antibiotics. It is considered as a tight binding inhibitor of adenosine deaminase (ADA), a key enzyme in the cellular metabolism of purines. Children with congenital absence of ADA suffer from severe combined immune deficiency (SCID) syndrome and atrophy of lymphoid tissues. The fact that pentostatin is lymphocytotoxic was predicted, and it turned out to be correct; this discovery stimulated more

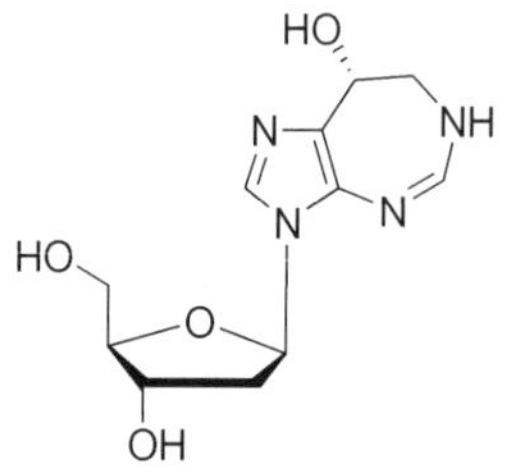

HO
N
NH
N
N
HO
O
OH

Fig. 3.4 Pentostatin.

research into lymphoid neoplasms [188,189]. Pentostatin drug can be effectively used to treat T-cell lymphomas [190,191] and B-cell chronic lymphocytic leukemia (B-CLL) [192,193]. Pentostatin was approved by Food and Drug Administration (FDA) for treating hematologic malignancies [194].

A number of hematologic malignancies are currently treated with pentostatin. In addition to its anticancer characteristics, it has been shown to have significant immunosuppressive qualities. Pentostatin affects the immune system by reducing the quantity and function of lymphocytes. Its immunosuppressive properties make it a promising therapy option for steroid-refractory aGVHD (acute graft-versus-host disease) [195–198]. It is considered as a drug of choice in treating hairy cell leukemia (HCL). It has drastically changed the natural history of the disease [199–205].

Nipent is a profound inhibitor of the enzyme adenosine deaminase [206], leading to the accumulation of metabolites that inhibit ribonucleotide reductase, which in turn causes the inhibition of the DNA synthesis. Pentostatin was the first of the purine analogs to undergo considerable research as an anticancer drug and the first to receive US Food and Drug Administration approval. In hairy cell leukemia, it is a highly successful first-line monotherapy. As a single agent, pentostatin is effective against chronic lymphocytic leukemia, but it looks to be even more effective when combined with the alkylating drug cyclophosphamide or chlorambucil [206]. Pentostatin exerts potent immunomodulatory effects [207]. In numerous lymphoid neoplasms, pentostatin has a remarkable anticancer effect. Pentostatin is beneficial in the treatment of indolent lymphoid malignancies and may be effective in lymphocyte research that is not related to cancer [208]. Pentostatin was expected to be the most effective in neoplasms with high intracellular concentrations of ADA, such as acute lymphocytic leukemia (ALL), particularly T-cell ALL [205]. Although pentostatin was found to be effective in ALL, it needed large doses and had significant side effects that overshadowed the therapeutic impacts. Pentostatin, on the other hand, was found to be particularly effective in hairy cell leukemia (HCL), a B-cell neoplasm with low intracellular concentrations of ADA. Pentostatin has since been revealed to have potency in prolymphocytic leukemia, cutaneous T-cell lymphomas, chronic lymphocytic leukemia, low-grade non-Hodgkin's lymphomas, and adult T-cell lymphoma leukemia. It increases the effectiveness of vidarabine against viruses and acute myeloid leukemia cells.

Pentostatin is inactive in renal carcinoma and melanoma, but has not been adequately estimated in other solid tumors [205]. Pentostatin is a well-tolerated and safe medication but, like all chemotherapeutic agents, it may be associated with some toxicities [209,210]. The toxic effects of pentostatin include CNS depression, renal failure, immunosuppression, opportunistic infections, and keratoconjunctivitis. In the absence of preexisting bone marrow compromise, pentostatin furnishes only mild myelosuppression [205]. Apart from its usage as an anticancer medicine, pentostatin has the potential to be used as an immunosuppressive drug, as an antimalarial compound, as an antiviral agent, and in the protection of cells of the central nervous system (CNS) from damage induced by anoxia and ischemia [205]. A combination regimen comprising pentostatin and rituximab (Rituxan) may have a useful impact in patients with resistant/relapsing CLL [211]. Both Rituximab and Nipent have single-agent potency in B-cell malignancies, comprising indolent and intermediate-grade non-Hodgkin's lymphoma (NHL). Combining rituximab and pentostatin is considered as an efficient treatment option as both drugs possess a limited toxicity profile [212,213]. Additionally, pentostatin is combined with many other drugs [214–218].

Novel precursors of pentostatin, pentostatin analogs, and their derivatives have been manufactured. Many aspects of the inventions and clinical trials have been studied [219–225].

6. Cladribine

Cladribine (2-chloro-2′-deoxyadenosine) (Fig. 3.5) [226–228] is a novel chemotherapeutic [229] antileukemic drug. Cladribine is a cytotoxic drug; in addition to cytotoxicity, the mode of action may include immunomodulatory mechanisms [230]. Cladribine from the therapies has been named pulsed immune reconstitution therapies or pulsed immunosuppressive therapies [231]. Cladribine was first developed in the 1970s to treat a variety of blood malignancies. Due to the ability of the molecule to suppress T and B lymphocytes preferentially, it has been developed into an oral formulation

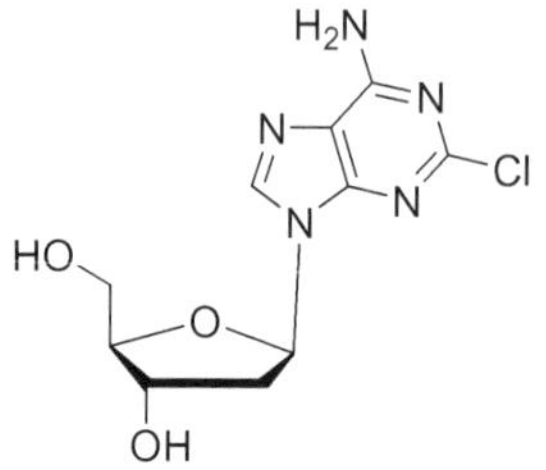

H_2N N N Cl N N HO O OH

Fig. 3.5 Cladribine.

for treating multiple sclerosis (MS) [232–234]. Multiple sclerosis (MS) is one of the most common neurological disorders in young adults. It is a chronic neurodegenerative and inflammatory disease when autoreactive B and T cells have downstream consequences that lead to neuronal loss and demyelination [235]. Cladribine depletes lymphocytes and reduces T and B cells in a dose-dependent manner. It was approved for the therapy of highly active relapsing/remitting multiple sclerosis [236–240].

CdA is a nucleoside analog potent in B-cell chronic lymphocytic leukemia (B-CLL) [241–244], and it is used as an effective treatment for lymphoproliferative disorders [245].

Cladribine as a single agent or in combination with therapeutic antibodies showed effectiveness in the treatment of patients with non-Hodgkin's lymphoma and chronic lymphocytic leukemia [246]. It can be combined with many agents such as simvastatin, which is a commonly used cholesterol-lowering medicine, was evaluated in terms of its effect on apoptosis and DNA damage in CLL cells as a single agent and in combination with cladribine and fludarabine [247,248].

Cladribine has been revealed to possess hypomethylating features and has potency as a single agent or in combination with other therapies for mantle cell lymphoma [249]. Cladribine, unlike many other drugs, is toxic to both indolent and dividing lymphoid malignancies [250–252]. Cladribine is currently in use as a chemotherapeutic agent in pediatric acute myelogenous leukemia and chronic lymphoid malignancies [253]. It is also considered as the drug of choice in the first-line setting due to the very high complete remission rate and prolonged duration of response following a single seven-day infusion [254]. Cladribine is an adenosine deaminase-resistant deoxyadenosine analog with a unique mechanism of action. It is directly toxic toward both proliferating and resting human monocytes and lymphocytes. Additionally, it is demonstrated as the first-line treatment for hairy cell leukemia HCL [255–257]. Cladribine is a safe and effective treatment for HCL patients, and it is the first-line treatment for symptomatic HCL [258]. Cladribine is cytotoxic to both proliferating and quiescent cells, and its mechanism of action could take various different forms. The FDA approved it for the treatment of HCL, as well as for the treatment of chronic lymphocytic leukemia in various European nations; however, it indicates high efficacy in both myeloid and lymphoid malignancies [259]. 2-CdA is usually administered as continuous or intermittent i.v. infusion [228,251,255,260–262]. It was approved by Food and Drug Administration (FDA) for the treatment of hematological malignancies [263]. Cladribine appears to interfere or prevent lymphocyte proliferation through the inhibition of the DNA repair. Cladribine's pharmacokinetics best fit a 2-compartment, first-order elimination model. Of the conditions that have been treated with cladribine, hairy cell leukemia (HCL) has revealed the most significant response. Overall response rates in clinical investigations have ranged from 80% to 100%, with a large majority of these being complete remissions; median durations of responses (from 9 to 16 months). Other conditions that have responded to cladribine are acute leukemia, chronic lymphocytic leukemia (CLL), low-grade lymphomas, chronic myeloid leukemia, cutaneous T-cell lymphoma, and Waldenstrom macroglobulinemia. The drug is inactive against solid tumors. The main, dose-limiting adverse effects of cladribine are fever, immunosuppression, bone marrow suppression, local skin reactions, neurologic and renal effects [264,265], and cardiac toxicities [266].

7. Fludarabine phosphate

Fludarabine phosphate, 9-β-D-arabinosyl-2-fluoroadenine (2-fluoro-ara-AMP) (F-ara-A), Fludara (Fig. 3.6), is approved adenosine analog [267]. It is considered as the 2-fluoro, 5′-monophosphate analog of 9-β-D-arabinofuranosyl adenine (Vidarabine, ara-A) with the advantages of resistance to de-amination via adenosine deaminase (ADA) and enhanced solubility [268].

It is an adenine nucleoside analog resistant to adenosine deaminase that indicates a promising therapeutic potency in clinical treating of lymphocytic hematologic malignancies [269]. It is approved for the clinical treatment of hematological malignancies [270–272]. Its mechanism of action is via the inhibition of DNA synthesis and the cytolytic impacts via inducing endonuclease independent apoptosis [273].

Fludara entered clinical trials in 1982. Fludara has a marked potency in indolent lymphoproliferative disorders, but it shows an inactivity in solid

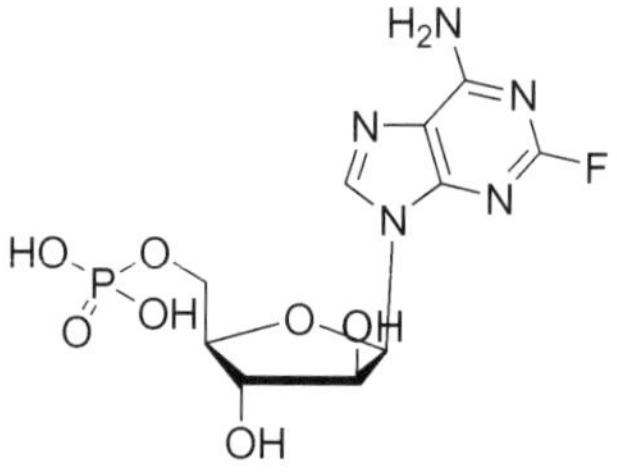

Fig. 3.6 Fludarabine.

tumors. Fludara's exact mechanism of action is uncertain. Fludara has been considered as the most potent single agent in chronic lymphocytic leukemia (CLL) in single arm and clinical trials that are compared [274–276]. Although fludarabine is the standard first-line therapy in treating chronic lymphocytic leukemia (CLL), only few data are available about the resistance of B cells to this purine nucleoside analog in vivo [277].

It is useful in treating Hodgkin disease, lymphoma, and leukemia [278,279], and relapsed indolent B-cell non-Hodgkin's lymphoma [280]. Fludarabine is an effective nucleoside analog for treating leukemias when utilized as a single agent or together [281]. Other strategies have utilized fludarabine to minimize immunological function, thus facilitating nonmyeloablative stem cell transplants. Fludarabine is a prodrug that is transformed to the free nucleoside 9-β-D-arabinosyl-2-fluoroadenine (F-ara-A), which enters the cells and accumulates mainly as the 5′-triphosphate, F-ara-ATP [282].

F-ara-A is transported into cells, where it is transformed to its triphosphate (F-ara-ATP), the essential active metabolite. Deoxycytidine kinase is the essential enzyme responsible for the activation and metabolism. The differential transport and phosphorylation of F-ara-A and accumulation of F-ara-ATP by cancer and normal cells may comprise the metabolic basis of its positive therapeutic index.

The main action of fludarabine is the inhibition of DNA synthesis. Through the DNA polymerases, F-ara-ATP competes with deoxyadenosine triphosphate for its incorporation into the A sites of the elongating DNA strand, terminating DNA synthesis at the incorporation sites. That action is potentiated through the decrease of cellular dATP that results from the inhibition of ribonucleotide reductase via F-ara-ATP. In vitro experiments exhibited that DNA polymerase δ is able to excise the incorporated F-ara-AMP residues from DNA with its 3′–5′ exonuclease potency. The terminal incorporation of F-ara-AMP into DNA leads to the deletion of genetic material. That mechanism may be essential for the observed Fludara's mutagenicity, and eventually, its cytotoxic effects [269].

This antimetabolite has also indicated "in vitro" antiproliferative potency against experimental models of solid mammary tumor [283]. Other investigations reported the combined effect of radiation and fludarabine on human squamous carcinoma cell lines in vitro, providing data for future decisions on head and neck chemoradiotherapy regimen [284]. Additionally, fludarabine is considered an effective repair inhibitor of radiation-induced chromosome breaks, and indomethacin, an inhibitor of prostaglandin synthesis [285].

8. Clofarabine

Clofarabine, 2-chloro-2′-fluoro-2′-deoxyarabinosyladenine, (Cl-F-ara-ATP) [ClF] (Fig. 3.7) [286,287], is a halogenated purine nucleoside analog [288], is a cytotoxic drug, and is one of the potential novel medicines [289]; it is considered as an analog of 2′-deoxyadenosine connecting biochemical potencies of its prototypes: fludarabine (2-fluoro-arabinosyladenine) and cladribine (2-chloro-2′-deoxyadenosine). It is an antileukemic drug that targets either of the two subunits of RNR, RNR-α, which affects noncanonical RNR-α functions [290]. Clofarabine, a more toxic nucleoside analog drug, is characterized by significantly various molecular interactions with poly(propyleneimine) dendrimers than fludarabine, resulting in dramatically various cellular outcomes [291]. Clofarabine is a ribonucleotide reductase inhibitor approved by the FDA that has demonstrated a significant antiretroviral potency in transformed cell lines [292]. In both activated CD4+ T cells and macrophages, clofarabine is an effective HIV-1 inhibitor. Clofarabine is a dual-action HIV-1 replication inhibitor that inhibits DNA polymerase activity of HIV-1 reverse transcriptase while also limiting dNTP substrates for viral DNA synthesis [293]. In comparison with its congeners, clofarabine is a novel anticancer medication that is more efficacious (at low dosages) and has a higher oral bioavailability [294].

Clofarabine was synthesized in 1992 and 2004 and was then approved for treating pediatric patients with refractory/relapsed acute lymphoblastic leukemia (ALL) [294–298] in the third line of treatment. It has shown a promise in both pediatric and adult patients who are candidates for allogenic hematopoietic stem cell transplantation (HSCT) [299]. Clofarabine for injection is a second-generation nucleoside analog approved in Europe (Evoltra) and the United States (Clolar) for treating pediatric relapsed or refractory acute lymphoblastic leukemia [300]. Clofarabine is the most-developed compound, and single-agent experience as well as combinations with other

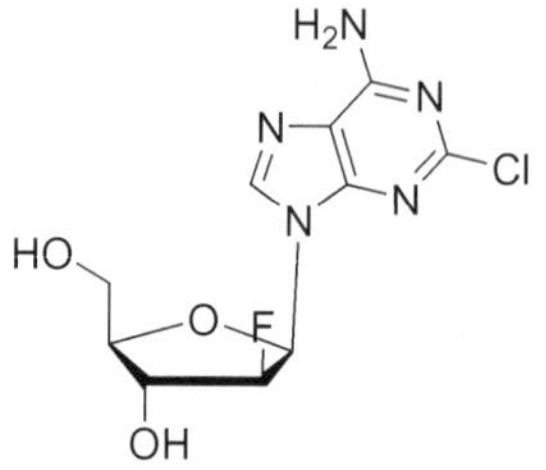

Fig. 3.7 Clofarabine.

active medicines is being investigated in acute myeloid leukemia [301–304]. It acts by inhibiting DNA synthesis, the enzyme ribonucleotide reductase, and repair and activation of mitochondrial repair processes [296]. Clofarabine is designed to combine the most favorable pharmacokinetic features of cladribine and fludarabine. Clofarabine acts through inhibiting DNA polymerases and ribonucleotide reductase, and it induces apoptosis in both cycling and noncycling cells. Phase I and II clinical investigations showed its efficacy in hematological malignancies, and in 2004, clofarabine was approved by the United States Food and Drug Administration for treating pediatric relapsed/refractory acute lymphoblastic leukemia after at least 2 prior chemotherapy regimens [305,306]. Clofarabine would be efficient in the treatment of indolent leukemias [302,307], while clofarabine in conjunction would be efficient in the treatment of hematological malignancies, lymphoma, and leukemia [308,309]. Clofarabine (CAFdA) has been synthesized and introduced into clinical trials. CAFdA is phosphorylated primarily through deoxynucleoside salvage mechanisms [310]. It is undergoing phase I-II clinical studies for treating hematopoietic malignancies [311]. Hepatotoxicity is considered as nonhematologic dose-limiting toxicity for clofarabine [312]. Also, the utilization of the drug is restricted by potentially severe neurotoxicity [299]. Clofarabine has been associated with neurological and cardiac toxicities [266].

9. Nelarabine

Nelarabine (Arranon) (Fig. 3.8) is a chemotherapeutic [313,314], an antileukemia drug, and was synthesized from guanine arabinoside [315]. Modern pharmacology has begun to be influenced by marine-based medications, and several anticancer drugs derived from marine compounds have been licensed for clinical utility, as nelarabine, a bioactive molecule, is derived from the marine environment with anticancer potency [316]. Nelarabine is the

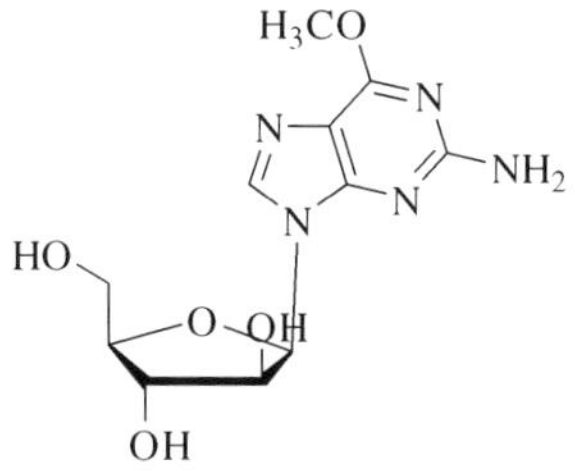

H_3CO N N NH_2 N N HO O OH OH

Fig. 3.8 Nelarabine.

prodrug of 9-β-arabinofuranosylguanine (ara-G) and is therapeutically categorized as a purine nucleoside analog. Ara-G is an active compound of nelarabine that is intracellularly phosphorylated to a triphosphate form, which causes inhibition of the DNA synthesis [317]. Nelarabine, a water-soluble prodrug of ara-G, is a T-cell-specific purine nucleoside analog [318]. It is currently being estimated in clinical trials [319,320]. It has promising antileukemic potency [321]. It can be utilized for preparing pharmaceuticals in the treatment of leukemia [322].

Nelarabine (Glaxo-SmithKline) or GW506U78 has shown potency as a single agent in patients with T-cell malignancies that have relapsed or are refractory to other therapy [323]. It is active as a single agent in recurrent T-cell leukemia [324]. Nelarabine is the only drug specifically approved for (R/R T-ALL/T-LBL) relapsed or refractory (R/R) T-cell acute lymphoblastic leukemia/lymphoma (T-ALL/T-LBL), but its usage is based on limited available data [325,326]. The US Food and Drug Administration (FDA) approved it in 2005 for treating T-cell acute lymphoblastic leukemia and T-cell lymphoblastic lymphoma that have not responded to or have relapsed following treatment with at least two chemotherapy regimens [303,327–336].

Nelarabine has displayed an impressive single-agent clinical potency in T-cell acute lymphoblastic leukemia (T-ALL) [337]. This novel purine nucleoside analog has recently become available for both children and adults with refractory or relapsed T-cell acute lymphoblastic leukemia (T-ALL) and T-lymphoblastic lymphoma (T-LBL) [338–340]. Nelarabine is an effective salvage therapy in T-cell acute lymphoblastic leukemia relapsing following transplantation [341]. Nelarabine has recently received an accelerated approval by the US Food and Drug Administration for the treatment of refractory or relapsed pediatric acute lymphoblastic leukemia as well as in refractory-relapsed T-cell acute lymphoblastic leukemia or T-cell lymphoblastic lymphoma. Nelarabine was developed based on preclinical and clinical data with other nucleoside analogs or normal deoxynucleotides like dGTP. Triphosphate is considered as the active metabolite for both of these purine nucleoside analogs, according to cell line studies. During clinical studies, pharmacodynamic and pharmacokinetic researches have confirmed the importance of triphosphate levels in achieving clinical responses. Clofarabine's use in acute leukemias and nelarabine's use in T-cell disorders have been suggested by a number of phase I and II clinical studies [312]. More than 40 years has passed since nelarabine was first developed as a medicinal drug to treat various types of hematologic malignancies. Studies have revealed an efficacy in the treatment of T-cell lymphoma and leukemia, and further work is ongoing as novel dosing schedules of nelarabine and combinations with other chemotherapeutic agents are examined [342]. Etoposide and fludarabine might have a synergistic effect with nelarabine through the inhibition of the ribonucleotide reductase and via preparing cell cycle for G1/S phase [343]. Additionally, the efficient CSF penetration of nelarabine and ara-G encourages more research into nelarabine's role in the prevention and treatment of CNS leukemia [344].

Nelarabine is transformed to ara-G via adenosine deaminase and transported into cells via a nucleoside transporter. Ara-G is consequently phosphorylated to form ara-G triphosphate (ara-GTP), which triggers the therapeutic effect through inhibiting synthesis of DNA. The pharmacokinetics of nelarabine has been thoroughly examined, and the results show that ara-GTP accumulates preferentially in malignant T cells. Clinical responses to nelarabine have been seen in a variety T-cell malignancies and appear to be linked to a higher intracellular concentration of ara-GTP than nonresponders.

As a result, this unique drug property of nelarabine accounts for its clinical use in the treatment of pediatric and adult patients with refractory or relapsed T-cell lymphoblastic lymphoma or T-cell acute lymphoblastic leukemia. The most common side effect of nelarabine is neuropathy, which occurs at a rate that is proportional to the dose administered. There has been evidence of myelosuppression, with neutropenia and thrombocytopenia being the most common hematologic consequences [345–350]. Nelarabine's neurotoxicity is a dose-dependent side effect [351]. It is considered as a life-threatening condition induced by nelarabine [352–354]. Neurotoxicity has been identified as the most common and severe toxicity in previous investigations; also, severe liver injury might be induced by nelarabine [355]. So, the utility of the drug is restricted by potentially severe neurotoxicity [356,357] and myelosuppression [358].

For the first time, an efficient and pure β-anomer synthesis of the clinical medication nelarabine from easily available vidarabine has been produced [359]. Many inventions included nelarabine as antitumor drug in the injection components and consider it from its pharmaceutically active ingredients, and belongs to the pharmacy's technical field [360–368]. Many studies and applications on nelarabine have been performed [369–378], and the drug nelarabine demonstrated the good future of industrial application [379].

References

[1] Supandi HY, Harmita AR. Simultaneous analysis of 6-mercaptopurine, 6-methylmercaptopurine, and 6-thioguanosine-5′-monophosphate in dried blood spot using ultra performance liquid chromatography tandem mass spectrometry. Indones J Chem 2018;18(3):544–51.

[2] Balbir S, Tirath K, Aditya G. Impact of thiopurine S-methyl transferase gene (TPMT) polymorphism on thiopurine drug metabolism: a way to individualized therapy. J Pharm Res 2013;2(5):1–7.

[3] Wang J, Yu X, Zhou C, Lin Y, Chen C, Pan G, Mei X. Improving the dissolution and bioavailability of 6-mercaptopurine via co-crystallization with isonicotinamide. Bioorg Med Chem Lett 2015;25(5):1036–9.

[4] Yang Y, Zhou S, Ouyang R, Yang Y, Tao H, Feng K, Zhang X, Xiong F, Guo N, Zong T, et al. Improvement in the anticancer activity of 6-mercaptopurine via combination with bismuth(III). Chem Pharm Bull 2016;64(11):1539–45.

[5] Suresh Kumar S, Athimoolam S, Sridhar B. Hydrogen bonding motifs, spectral characterization, theoretical computations and anticancer studies on chloride salt of 6-mercaptopurine: an assembly of corrugated lamina shows enhanced solubility. J Mol Struct 2015;1098:332–41.

[6] Elgemeie GH. Thioguanine, mercaptopurine: their analogs and nucleosides as antimetabolites. Curr Pharm Des 2003;9(31):2627–42.

[7] Abou-Egla MH, Abou-El-Naga AM, Ramadan MM, El-Etriby AH. Molecular and cytogenetic studies on the effects of the anticancer drug (6-mercaptopurine) with or without zinc protection on pregnant mothers and their newborns, of albino mice. J Environ Sci (Mansoura, Egypt) 2009;38:179–99.

[8] Oancea I, Movva R, Das I, Aguirre de Carcer D, Schreiber V, Yang Y, Purdon A, Harrington B, Proctor M, Wang R, et al. Colonic microbiota can promote rapid local improvement of murine colitis by thioguanine independently of T lymphocytes and host metabolism. Gut 2017;66(1):59–69.

[9] Raza A, Barakzai Q. Pharmacogenomic related toxicity of 6-mercaptopurine in childhood acute lymphoblastic leukemia. Pak J Pharmacol 2009;26(1):55–60.

[10] Schmiegelow K, Nersting J, Nielsen SN, Heyman M, Wesenberg F, Kristinsson J, Vettenranta K, Schroeder H, Weinshilboum R, Jensen KL, et al. Maintenance therapy of childhood acute lymphoblastic leukemia revisited—should drug doses be adjusted by white blood cell, neutrophil, or lymphocyte counts? Pediatr Blood Cancer 2016;63(12):2104–11.

[11] Choughule KV, Barnaba C, Joswig-Jones CA, Jones JP. In vitro oxidative metabolism of 6-mercaptopurine in human liver: insights into the role of the molybdoflavoenzymes aldehyde oxidase, xanthine oxidase and xanthine dehydrogenase. Drug Metab Dispos 2014;42(8):1334–40.

[12] Filhon B, Dumesnil C, Van Dreden P, Schneider P, Vasse M, Vannier JP. Isolated decrease in factor V in children treated by 6-mercaptopurine for acute lymphoblastic leukemia. Thromb Res 2014;134(5):1164–6.

[13] Vang SI, Schmiegelow K, Frandsen T, Rosthoej S, Nersting J. Mercaptopurine metabolite levels are predictors of bone marrow toxicity following high-dose methotrexate therapy of childhood acute lymphoblastic leukaemia. Cancer Chemother Pharmacol 2015;75(5):1089–93.

[14] Farfan MJ, Salas C, Canales C, Silva F, Villarroel M, Kopp K, Torres JP, Santolaya ME, Morales J. Prevalence of TPMT and ITPA gene polymorphisms and effect on mercaptopurine dosage in Chilean children with acute lymphoblastic leukemia. BMC Cancer 2014;14:299/1–5.

[15] Wennerstrand P, Maartensson L, Soederhaell S, Zimdahl A, Appell ML. Methotrexate binds to recombinant thiopurine S-methyltransferase and inhibits enzyme activity after high-dose infusions in childhood leukaemia. Eur J Clin Pharmacol 2013;69(9):1641–9.
[16] Tanaka Y, Manabe A, Nakadate H, Kondoh K, Nakamura K, Koh K, Utano T, Kikuchi A, Komiyama T. The activity of the inosine triphosphate pyrophosphatase affects toxicity of 6-mercaptopurine during maintenance therapy for acute lymphoblastic leukemia in Japanese children. Leuk Res 2012;36(5):560–4.
[17] Stanczyk M, Sliwinski T, Trelinska J, Cuchra M, Markiewicz L, Dziki L, Bieniek A, Bielecka-Kowalska A, Kowalski M, Pastorczak A, et al. Role of base-excision repair in the treatment of childhood acute lymphoblastic leukaemia with 6-mercaptopurine and high doses of methotrexate. Mutat Res Genet Toxicol Environ Mutagen 2012;741(1–2):13–21.
[18] Li F, Fridley BL, Matimba A, Kalari KR, Pelleymounter L, Moon I, Ji Y, Jenkins GD, Batzler A, Wang L, et al. Ecto-5′-nucleotidase and thiopurine cellular circulation: association with cytotoxicity. Drug Metab Dispos 2010;38(12):2329–38.
[19] Lbayrak M, Konyssova U, Kaya Z, Gursel T, Guntekin S, Percin EF, Kocak U. Thiopurine methyltransferase polymorphisms and mercaptopurine tolerance in Turkish children with acute lymphoblastic leukemia. Cancer Chemother Pharmacol 2011;68(5):1155–9.
[20] Sorouraddin M, Khani M, Amini K, Naseri A, Asgari D, Rashidi M. Simultaneous determination of 6-mercaptopurine and its oxidative metabolites in synthetic solutions and human plasma using spectrophotometric multivariate calibration methods. Bioimpacts 2011;1(1):53–62.
[21] de Beaumais TA, Fakhoury M, Medard Y, Azougagh S, Zhang D, Yakouben K, Jacqz-Aigrain E. Determinants of mercaptopurine toxicity in paediatric acute lymphoblastic leukemia maintenance therapy. Br J Clin Pharmacol 2011;71(4):575–84.
[22] Adam de Beaumais T, Jacqz-Aigrain E. Pharmacogenetic determinants of mercaptopurine disposition in children with acute lymphoblastic leukemia. Eur J Clin Pharmacol 2012;68(9):1233–42.
[23] Stork LC, Matloub Y, Broxson E, La M, Yanofsky R, Sather H, Hutchinson R, Heerema NA, Sorrell AD, Masterson M, et al. Oral 6-mercaptopurine versus oral 6-thioguanine and veno-occlusive disease in children with standard-risk acute lymphoblastic leukemia: report of the Children's Oncology Group CCG-1952 clinical trial. Blood 2010;115(14):2740–8.
[24] Larsen Rikke H, Hjalgrim Lisa L, Grell K, Kristensen K, Pedersen LG, Brunner ED, Als-Nielsen B, Schmiegelow K, Nersting J. Pharmacokinetics of tablet and liquid formulations of oral 6-mercaptopurine in children with acute lymphoblastic leukemia. Cancer Chemother Pharmacol 2020;86(1):25–32.
[25] Guijarro LG, Roman ID, Fernandez-Moreno M, Gisbert JP, Hermandez-Breijo B. Is the autophagy induced by thiopurines beneficial or deleterious? Curr Drug Metab 2012;13(9):1267–76.
[26] Diamai S, Warjri W, Saha D, Negi DPS. Sensitive determination of 6-mercaptopurine based on the aggregation of phenylalanine-capped gold nanoparticles. Colloids Surf A Physicochem Eng Asp 2018;538:593–9.
[27] Coulthard SA, Hogarth LA, Little M, Matheson EC, Redfern CPF, Minto L, Hall AG. The effect of thiopurine methyltransferase expression on sensitivity to thiopurine drugs. Mol Pharmacol 2002;62(1):102–9.
[28] Coulthard SA, Hogarth LA. Old drugs-current perspectives. Curr Pharmacogenomics 2004;2(2):163–73.

[29] Gulbanul A, Xie Y, Shen S. Research progress on relationship between NUDT15 gene polymorphism and mercaptopurine drug tolerance. Zhongguo Xiandai Yixue Zazhi 2020;30(18):58–62.
[30] Han J, Mei S, Xu J, Zhang D, Jin S, Zhao Z, Zhao L. Simultaneous UPLC-MS/MS determination of 6-mercaptopurine, 6-methylmercaptopurine and 6-thioguanine in plasma: application to the pharmacokinetic evaluation of novel dosage forms in beagle dogs. Curr Pharm Des 2020;26(46):6013–20.
[31] Liu C, Janke LJ, Yang JJ, Evans WE, Schuetz JD, Relling MV. Differential effects of thiopurine methyltransferase (tpmt) and multidrug resistance-associated protein gene 4 (mrp4) on mercaptopurine toxicity. Cancer Chemother Pharmacol 2017;80(2): 287–93.
[32] Li Z, Wang H, Sun H. Research advance of mercaptopurine drugs in individualized treatment of children with acute lymphoblastic leukemia. Shanghai Yiyao 2015;36(19): 12–5. 64.
[33] Greene BHC, Alhatab DS, Pye CC, Brosseau CL. Electrochemical-surface enhanced raman spectroscopic (EC-SERS) study of 6-thiouric acid: a metabolite of the chemotherapy drug azathioprine. J Phys Chem C 2017; [Ahead of Print].
[34] Tanaka Y, Kato M, Hasegawa D, Urayama KY, Nakadate H, Kondoh K, Nakamura K, Koh K, Komiyama T, Manabe A. Susceptibility to 6-MP toxicity conferred by a NUDT15 variant in Japanese children with acute lymphoblastic leukaemia. Br J Haematol 2015;171(1):109–15.
[35] Tiphaine A, Hjalgrim LL, Nersting J, Breitkreutz J, Nelken B, Schrappe M, Stanulla M, Thomas C, Bertrand Y, Leverger G, et al. Evaluation of a pediatric liquid formulation to improve 6-mercaptopurine therapy in children. Eur J Pharm Sci 2016;83:1–7.
[36] Hanff LM, Mathot RAA, Smeets O, Postma DJ, Ramnarain S, Vermes A, Pieters R, Zwaan CM. A novel 6-mercaptopurine oral liquid formulation for pediatric acute lymphoblastic leukemia patients – results of a randomized clinical trial. Int J Clin Pharmacol Ther 2014;52(8):653–62.
[37] Zaboli M, Raissi H, Zaboli M. Investigation of nanotubes as the smart carriers for targeted delivery of mercaptopurine anticancer drug. J Biomol Struct Dyn 2021; [Ahead of Print].
[38] Yang JJ, Landier W, Yang W, Liu C, Hageman L, Cheng C, Pei D, Chen Y, Crews KR, Kornegay N, et al. Inherited NUDT15 variant is a genetic determinant of mercaptopurine intolerance in children with acute lymphoblastic leukemia. J Clin Oncol 2015;33(11):1235–42.
[39] Choe YH, Greenwald RB, Conover CD, Zhao H, Longley CB, Guan S, Zhao Q, Xia J. PEG prodrugs of 6-mercaptopurine for parenteral administration using benzyl elimination of thiols. Oncol Res 2004;14(9):455–68.
[40] White P. Oral suspension for treatment of acute lymphoblastic leukemia comprising 6-mercaptopurine. PCT Int Appl; 2013. WO 2013034931 A1 20130314.
[41] Chen LJ, Nightingale G, Baer MR. Mercaptopurine-induced fever: hypersensitivity reaction in a patient with acute lymphoblastic leukemia. Pharmacotherapy 2010;30(1): e65–7.
[42] Azimi F, Esmaeilzadeh A, Ramazani A. Clinical significance of ITPA rs67002563 polymorphism in patients with acute lymphoblastic leukemia treated with 6-mercaptopurine. Pharmacol Res 2015;102:61–2.
[43] Kouwenberg TW, van den Bosch BJC, Bierau J, te Loo DMWM, Coenen MJH, Hagleitner MM. Dosage of 6-mercaptopurine in relation to genetic TPMT and ITPA variants: toward individualized pediatric acute lymphoblastic leukemia maintenance treatment. J Pediatr Hematol Oncol 2020;42(2):e94–7.

[44] Daenen SMGJ, De Wolf JTM, Vellenga E, Van Imhoff GW, Smit JW, Van Den Berg-De RE. 6-Mercaptopurine, still valuable for the palliative treatment of acute myeloid leukaemia. Hematology (Reading, U K) 2001;6(4):231–40.
[45] Ordentlich P, Yan Y, Zhou S, Heyman RA. Identification of the antineoplastic agent 6-mercaptopurine as an activator of the orphan nuclear hormone receptor Nurr1. J Biol Chem 2003;278(27):24791–9.
[46] Banh TN, Kode NR, Phadtare S. CNS cancer cell line cytotoxicity profiles of some 2, 6, 9-substituted purines: a comparative five-dose testing study. Lett Drug Des Discovery 2012;9(5):500–4.
[47] Huang H, Chang H, Tsai M, Chen J, Wang M. 6-Mercaptopurine attenuates tumor necrosis factor-α production in microglia through Nur77-mediated transrepression and PI3K/Akt/mTOR signaling-mediated translational regulation. J Neuroinflammation 2016;13:78/1–78/20.
[48] Marinkovic G, Kroon J, Hoogenboezem M, Hoeben KA, Ruiter MS, Kurakula K, Otermin Rubio I, Vos M, de Vries CJM, van Buul JD, et al. Inhibition of GTPase Rac1 in endothelium by 6-mercaptopurine results in immunosuppression in nonimmune cells: new target for an old drug. J Immunol 2014;192(9):4370–8.
[49] Ogungbenro K, Aarons L. Physiologically based pharmacokinetic modelling of methotrexate and 6-mercaptopurine in adults and children. Part 2: 6-mercaptopurine and its interaction with methotrexate. J Pharmacokinet Pharmacodyn 2014;41(2):173–85.
[50] Kumar GP, Sanganal JS, Phani AR, Manohara C, Tripathi SM, Raghavendra HL, Janardhana PB, Amaresha S, Swamy KB, Prasad RGSV. Anti-cancerous efficacy and pharmacokinetics of 6-mercaptopurine loaded chitosan nanoparticles. Pharmacol Res 2015;100:47–57.
[51] Nakamura T. Azathiopurine: a new "old" drug. Gekkan Yakuji 2012;54(6):957–60.
[52] Sochacka J, Pawelczak B, Sobczak A. Application of molecular docking to study 6-mercaptopurine-binding to human serum albumin. Ann Acad Med Silesiensis 2011;65(3):41–8.
[53] Trombino S, Cassano R, Cilea A, Ferrarelli T, Muzzalupo R, Picci N. Synthesis of L-lysine-based prodrugs of 5-aminosalicylic acid and 6-mercaptopurine for colon specific release. Int J Pharm 2011;420(2):290–6.
[54] Ren X, XuY KP. Photo-oxidation of 6-thioguanine by UVA: the formation of addition products with low molecular weight thiol compounds. Photochem Photobiol 2010;86(5):1038–45.
[55] Reichardt C, Guo C, Crespo-Hernandez CE. Excited-state dynamics in 6-Thioguanosine from the femtosecond to microsecond time scale. J Phys Chem B 2011;115(12):3263–70.
[56] Anon. Mercaptopurine 75-mg capsules. Int J Pharm Compd 2013;17(3):234.
[57] Bermejo F, Gisbert JP. Usefulness of salicylate and thiopurine coprescription in steroid-dependent ulcerative colitis and withdrawal strategies. Ther Adv Chronic Dis 2010;1(3):107–14.
[58] Kakuta Y, Kinouchi Y, Shimosegawa T. Pharmacogenetics of thiopurines for inflammatory bowel disease in East Asia: prospects for clinical application of NUDT15 genotyping. J Gastroenterol 2018;53(2):172–80.
[59] Sandborn WJ. Azathioprine/6-mercaptopurine: mechanisms of action, pharmacology and toxicology. Falk Symp 2001;119(Immunosuppression in Inflammatory Bowel Diseases):91–100.
[60] Conklin LS, Cuffari C, Okazaki T, Miao Y, Saatian B, Chen T, Tse M, Brant SR, Li X. 6-Mercaptopurine transport in human lymphocytes: correlation with drug-induced cytotoxicity. J Dig Dis 2012;13(2):82–93.

[61] Oancea I, Png CW, Das I, Lourie R, Winkler IG, Eri R, Subramaniam N, Jinnah HA, McWhinney BC, Levesque J, et al. A novel mouse model of veno-occlusive disease provides strategies to prevent thioguanine-induced hepatic toxicity. Gut 2013;62 (4):594–605.
[62] Sanderson JD, Marinaki AM, Smith MA, Blaker PA. Non-thiopurine methyltransferase related effects in 6-mercaptopurine therapy. U.S. Pat. Appl. Publ; 2013. US 20130022971 A1 20130124.
[63] Shih DQ, Nguyen M, Zheng L, Ibanez P, Mei L, Kwan LY, Bradford K, Ting C, Targan SR, Vasiliauskas EA. Split-dose administration of thiopurine drugs: a novel and effective strategy for managing preferential 6-MMP metabolism. Aliment Pharmacol Ther 2012;36(5):449–58.
[64] Marakova K, Piestansky J, Mikus P. Determination of drugs for Crohn's disease treatment in pharmaceuticals by capillary electrophoresis hyphenated with tandem mass spectrometry. Chromatographia 2017;80(4):537–46.
[65] Kolatch B, Hotovely-Salomon A. Treatment of Crohn's disease with delayed-release 6-mercaptopurine. PCT Int. Appl; 2015. WO 2015168448 A1 20151105.
[66] Jagdale SC, Hude RU. Guar gum and HPMC coated colon targeted delivery of 6-mercapto-purine. Nat Prod J 2017;7(1):18–29.
[67] Kurakula K, Hamers AA, van Loenen P, de Vries CJM. 6-mercaptopurine reduces cytokine and Muc5ac expression involving inhibition of NFκB activation in airway epithelial cells. Respir Res 2015;16:1–9.
[68] Pranzatelli MR, Tate ED, Allison TJ. 6-Mercaptopurine modifies cerebrospinal fluid T cell abnormalities in paediatric opsoclonus-myoclonus as steroid sparer. Clin Exp Immunol 2017;190(2):217–25.
[69] Ruiter MS, van Tiel CM, Doornbos A, Marinkovic G, Strang AC, Attevelt NJM, de Waard V, de Winter RJ, Steendam R, de Vries CJM. Stents eluting 6-mercaptopurine reduce neointima formation and inflammation while enhancing strut coverage in rabbits. PLoS ONE 2015;10(9):e0138459/1–e0138459/16.
[70] Swain BC, Mukherjee SK, Rout J, Sakshi MPP, Mukherjee M, Tripathy U. A spectroscopic and computational intervention of interaction of lysozyme with 6-mercaptopurine. Anal Bioanal Chem 2020;412(11):2565–77.
[71] Pacheco NM, Alves ANL, Fortini AS, Burattini MN, Sumita NM, Srougi M, Chocair PR. Therapeutic drug monitoring of azathioprine: a review. J Bras Patol Med Lab 2008;44(3):161–7.
[72] Bosken YK, Cholko T, Lou Y, Wu K, Chang CA. Insights into dynamics of inhibitor and ubiquitin-like protein binding in SARS-CoV-2 papain-like protease. Front Mol Biosci 2020;7:174.
[73] Hosni-Ahmed A, Barnes JD, Wan J, Jones TS. Thiopurine methyltransferase predicts the extent of cytotoxicity and DNA damage in astroglial cells after thioguanine exposure. PLoS ONE 2011;6(12):e29163.
[74] Qiu J, Cheng R, Zhang J, Sun H, Deng C, Meng F, Zhong Z. Glutathione-sensitive hyaluronic acid-mercaptopurine prodrug linked via carbonyl vinyl sulfide: a robust and CD44-targeted nanomedicine for leukemia. Biomacromolecules 2017;18 (10):3207–14.
[75] Xu L, Chen J, Yan Y, Lu T. Improving the solubility of 6-mercaptopurine via cocrystals and salts. Cryst Growth Des 2012;12(12):6004–11.
[76] Kennedy NA, Kalla R, Warner B, Gambles CJ, Musy R, Reynolds S, Dattani R, Nayee H, Felwick R, Harris R, et al. Thiopurine withdrawal during sustained clinical remission in inflammatory bowel disease: relapse and recapture rates, with predictive factors in 237 patients. Aliment Pharmacol Ther 2014;40(11–12):1313–23.
[77] Huynh T, Murray J, Flemming CL, Kamili A, Hofmann U, Cheung L, Roundhill EA, Yu DMT, Webber HT, Schwab M, et al. CCI52 sensitizes tumors to

6-mercaptopurine and inhibits MYCN-amplified tumor growth. Biochem Pharmacol (Amsterdam, Netherlands) 2020;172:113770.
[78] Rebelo A, Oliveira J, Sousa C. Severe mercaptopurine-induced hypoglycemia in acute lymphoblastic leukemia. Pediatr Hematol Oncol 2020;37(3):245–7.
[79] Saiz-Rodriguez M, Ochoa D, Belmonte C, Roman M, Martinez-Ingelmo C, Ortega-Ruiz L, Sarmiento-Iglesias C, Herrador C, Abad-Santos F. Influence of thiopurine S-methyltransferase polymorphisms in mercaptopurine pharmacokinetics in healthy volunteers. Basic Clin Pharmacol Toxicol 2019;124(4):449–55.
[80] Jayachandran D, Lainez-Aguirre J, Rundell A, Vik T, Hannemann R, Reklaitis G, Ramkrishna D. Model-based individualized treatment of chemotherapeutics: Bayesian population modeling and dose optimization. PLoS ONE 2015;10(7):e0133244/1–e0133244/24.
[81] Katara P, Kuntal H. TPMT polymorphism: when shield becomes weakness. Interdiscip Sci: Comput Life Sci 2016;8(2):150–5.
[82] Corominas H, Baiget M. Clinical utility of thiopurine S-methyltransferase genotyping. Am J PharmacoGenomics 2004;4(1):1–8.
[83] Wang L, Sullivan W, Toft D, Weinshilboum R. Thiopurine S-methyltransferase pharmacogenetics: chaperone protein association and allozyme degradation. Pharmacogenetics 2003;13(9):555–64.
[84] Roberts RL, Wallace MC, Drake JM, Stamp LK. Identification of a novel thiopurine S-methyltransferase allele (TPMT*37). Pharmacogenet Genomics 2014;24(6):320–3.
[85] Cabaleiro T, Roman M, Gisbert JP, Abad-Santos F. Utility of assessing thiopurine S-methyltransferase polymorphisms before azathioprine therapy. Curr Drug Metab 2012;13(9):1277–93.
[86] Chrzanowska M, Kuehn M, Januszkiewicz-Lewandowska D, Kurzawski M, Drozdzik M. Thiopurine S-methyltransferase phenotype-genotype correlation in children with acute lymphoblastic leukemia. Acta Pol Pharm 2012;69(3):405–10.
[87] Roman M, Cabaleiro T, Ochoa D, Novalbos J, Chaparro M, Gisbert JP, Abad-Santos F. Validation of a genotyping method for analysis of TPMT polymorphisms. Clin Ther 2012;34(4):878–84.
[88] Yuan J, Mu Y, Zhao J, Xu W, Chen J. Preparation of 6-mercaptopurine copolymer with good antitumor activity. Faming Zhuanli Shenqing; 2014. CN 103936922 A 20140723.
[89] Miron T, Wilchek M, Shvidel L, Berrebi A, Arditti FD. S-allyl derivatives of 6-mercaptopurine are highly potent drugs against human B-CLL through synergism between 6-mercaptopurine and allicin. Leuk Res 2012;36(12):1536–40.
[90] Kirchherr H, Shipkova M, von Ahsen N. Improved method for therapeutic drug monitoring of 6-thioguanine nucleotides and 6-methylmercaptopurine in whole-blood by LC/MSMS using isotope-labeled internal standards. Ther Drug Monit 2013;35(3):313–21.
[91] Mulla H, Leary A, White P, Pandya HC. A step toward more accurate dosing for mercaptopurine in childhood acute lymphoblastic leukemia. J Clin Pharmacol 2012;52 (10):1610–3.
[92] Hassanpour A, Ebrahimiasl S, Youseftabar-Miri L, Ebadi A, Ahmadi S, Eslami M. A DFT study on the electronic detection of mercaptopurine drug by boron carbide nanosheets. Comput Theor Chem 2021;1198, 113166.
[93] Pannico M, Musto P. SERS spectroscopy for the therapeutic drug monitoring of the anticancer drug 6-mercaptopurine: molecular and kinetic studies. Appl Surf Sci 2021;539:148225.
[94] Yang Y, Ostadhosseini N. A theoretical investigation on the mercaptopurine drug interaction with boron nitride nanocage: solvent and density functional effect. Phys E (Amsterdam, Netherlands) 2021;125:114337.

[95] Suman P, Rao TS, Reddy KVSRK. Development and validation of stability indicating RP-HPLC method for determination of mercaptopurine-an anti cancer drug in pharmaceutical formulations. Eur J Biomed Pharm Sci 2018;5(5):1–8.
[96] Zhang F, Liu H, Liu Q, Su X. An enzymatic ratiometric fluorescence assay for 6-mercaptopurine by using MoS2 quantum dots. Microsc Acta 2018;185(12):1–8.
[97] Parulekar GT. Determination and assay of some antineoplastic agents in pure form and in their pharmaceutical preparations using titrimetric method. World J Pharm Res 2017;6(14):1018–23.
[98] Gomar M, Yeganegi S. Adsorption of 5-fluorouracil, hydroxyurea and mercaptopurine drugs on zeolitic imidazolate frameworks (ZIF-7, ZIF-8 and ZIF-9). Microporous Mesoporous Mater 2017;252:167–72.
[99] Radhika R, Shankar R, Vijayakumar S, Kolandaivel P. Role of 6-mercaptopurine in the potential therapeutic targets DNA base pairs and G-quadruplex DNA: insights from quantum chemical and molecular dynamics simulations. J Biomol Struct Dyn 2018;36(6):1369–401.
[100] Fernandez-Banares F, Piqueras M, Guagnozzi D, Robles V, Ruiz-Cerulla A, Casanova MJ, Gisbert JP, Busquets D, Arguedas Y, Perez-Aisa A, et al. Collagenous colitis: requirement for high-dose budesonide as maintenance treatment. Dig Liver Dis 2017;49(9):973–7.
[101] Mulla H, Buck H, Price L, Parry A, Bell G, Skinner R. Acceptability' of a new oral suspension formulation of mercaptopurine in children with acute lymphoblastic leukemia. J Oncol Pharm Pract 2016;22(3):387–95.
[102] Ghoshal S, Kushwaha SKS, Tiwari P, Srivastava M. Comparative loading and release of 6-mercaptopurine from functionalized multiwalled carbon nanotubes using various methods. Int J Pharm Pharm Res 2015;4(1):25–38.
[103] Tan M, Zheng H, Zhang X, Yu Z, Ye Z, Shi L, Ding K. Inclusion-interaction assembly strategy for constructing pH/redox responsive micelles for controlled release of 6-mercaptopurine. Mater Sci Appl 2015;6(7):605–16.
[104] La Duke KE, Ehling S, Cullen JM, Baumer W. Effects of azathioprine, 6-mercaptopurine, and 6-thioguanine on canine primary hepatocytes. Am J Vet Res 2015;76(7):649–55.
[105] Kostereli Z, Severin K. Array-based sensing of purine derivatives with fluorescent dyes. Org Biomol Chem 2015;13(35):9231–5.
[106] Tanaka Y, Manabe A, Fukushima H, Suzuki R, Nakadate H, Kondoh K, Nakamura K, Koh K, Fukushima T, Tsuchida M, et al. Multidrug resistance protein 4 (MRP4) polymorphisms impact the 6-mercaptopurine dose tolerance during maintenance therapy in Japanese childhood acute lymphoblastic leukemia. Pharmacogenomics J 2015;15(4):380–4.
[107] Zhao W, Bai J, Li H, Chen T, Tang Y. Tubulin structure-based drug design for the development of novel 4β-sulfur-substituted podophyllum tubulin inhibitors with antitumor activity. Sci Rep 2015;5:10172.
[108] Ogungbenro K, Aarons L. Physiologically based pharmacokinetic model for 6-mercpatopurine: exploring the role of genetic polymorphism in TPMT enzyme activity. Br J Clin Pharmacol 2015;80(1):86–100.
[109] Yu D, Shi J, Wang J, Hao J, Wu H. Mercaptopurine composition lyophilized tablet and production method thereof. Faming Zhuanli Shenqing; 2015. CN 104546766 A 20150429.
[110] Tueluemen T, Ayata A, Oezen M, Suetcue R, Canatan D. The protective effect of capparis ovata on 6-mercaptopurine-induced hepatotoxicity and oxidative stress in rats. J Pediatr Hematol Oncol 2015;37(4):290–4.
[111] Gao W, Tang J. Application of 6-mercaptopurine in manufacture of medicine for inhibiting tumor cell metastasis and propagation. Faming Zhuanli Shenqing; 2015. CN 104490890 A 20150408.

[112] Liu X, Xia T, Zhu Y, Guo X, Xie Z. Thiopurine methyltransferase polymorphisms and thiopurine toxicity in treatment of acute lymphoblastic leukemia. Nanjing Yike Daxue Xuebao, Ziran Kexueban 2014;34(9):1279–83.
[113] Han G, Liu R, Han M, Jiang C, Wang J, Du S, Liu B, Zhang Z. Label-free surface-enhanced Raman scattering imaging to monitor the metabolism of antitumor drug 6-mercaptopurine in living cells. Anal Chem (Washington, DC, U S) 2014;86 (23):11503–7.
[114] Zhao Q, Wang C, Liu Y, Wang J, Gao Y, Zhang X, Jiang T, Wang S. PEGylated mesoporous silica as a redox-responsive drug delivery system for loading thiol-containing drugs. Int J Pharm (Amsterdam, Netherlands) 2014;477 (1–2):613–22.
[115] Ghoshal S, Kushwaha SKS, Srivastava M, Tiwari P. Drug loading and release from functionalized multiwalled carbon nanotubes loaded with 6-mercaptopurine using incipient wetness impregnation method. Am J Adv Drug Deliv 2014;2 (2):213–23.
[116] D'souza PF, Shenoy A, Moses SR, Shabaraya AR. Anti tumor activity of mercaptopurine in combination with Trikatu and Gomutra on 20-methylcholantrene induced carcinogenesis. J Appl Pharm Sci 2013;3(8):20–4.
[117] Chowdhary R, Pai RS, Singh G. Development and evaluation of 6-mercaptopurine and metoclopramide polypill formulation for oral administration: in-vitro and ex vivo studies. Int J Pharm Invest 2013;3(4):217–24.
[118] Tang W, Zhang M, Zeng X. Establishment of dsDNA/GNs/chit/GCE biosensor and electrochemical study on interaction between 6-mercaptopurine and DNA. Biomed Mater Eng 2014;24(1):1071–7.
[119] Kevadiya BD, Chettiar SS, Rajkumar S, Bajaj HC, Gosai KA, Brahmbhatt H. Evaluation of clay/poly (L-lactide) microcomposites as anticancer drug, 6-mercaptopurine reservoir through in vitro cytotoxicity, oxidative stress markers and in vivo pharmacokinetics. Colloids Surf B Biointerfaces 2013;112:400–7.
[120] Kaur G, Hearn MTW, Bell TDM, Saito K. Release kinetics of 6-mercaptopurine and 6-thioguanine from bioinspired core-crosslinked thymine functionalised polymeric micelles. Aust J Chem 2013;66(8):952–8.
[121] Wang W, Fang C, Wang X, Chen Y, Wang Y, Feng W, Yan C, Zhao M, Peng S. Modifying mesoporous silica nanoparticles to avoid the metabolic deactivation of 6-mercaptopurine and methotrexate in combinatorial chemotherapy. Nanoscale 2013;5(14):6249–53.
[122] Duijvestein M, Molendijk I, Roelofs H, Vos ACW, Verhaar AP, Reinders MEJ, Fibbe WE, Verspaget HW, van den Brink GR, Wildenberg ME, et al. Mesenchymal stromal cell function is not affected by drugs used in the treatment of inflammatory bowel disease. Cytotherapy 2011;13(9):1066–73.
[123] Huang Y, Zhao S, Shi M, Liang H. A microchip electrophoresis strategy with online labeling and chemiluminescence detection for simultaneous quantification of thiol drugs. J Pharm Biomed Anal 2011;55(5):889–94.
[124] Ali I, Rahis-Uddin SK, Rather MA, Wani WA, Haque A. Advances in nano drugs for cancer chemotherapy. Curr Cancer Drug Targets 2011;11(2):135–46.
[125] Senthil V, Kumar RS, Nagaraju CVV, Jawahar N, Ganesh GNK, Gowthamarajan K. Design and development of hydrogel nanoparticles for mercaptopurine. J Adv Pharm Technol Res 2010;1(3):334–7.
[126] Zheng H, Rao Y, Shu K, Yin Y. Mercaptopurine drugs based novel amphiphilic macromolecular prodrug, and preparation and application thereof. Faming Zhuanli Shenqing; 2010. CN 101829338 A 20100915.
[127] Yamauchi T, Ueda T. Anticancer drugs. Ketsueki, Shuyoka 2004;49(Suppl. 4):362–7.
[128] Neuberger J. Immunosuppression after liver transplantation. Graft (Thousand Oaks, CA, USA) 2003;6(2):110–9.

[129] Dervieux T, Hancock ML, Pui C, Rivera GK, Sandlund JT, Ribeiro RC, Boyett J, Evans WE, Relling MV. Antagonism by methotrexate on mercaptopurine disposition in lymphoblasts during up-front treatment of acute lymphoblastic leukemia. Clin Pharmacol Ther (St Louis, MO, USA) 2003;73(6):506–16.
[130] Ishtikhar M, Khan A, Chang C, Lin LT, Wang SSS, Khan R. Hasan effect of guanidine hydrochloride and urea on the interaction of 6-thioguanine with human serum albumin: a spectroscopic and molecular dynamics based study. J Biomol Struct Dyn 2016;34(7):1409–20.
[131] Fimognari C, Lenzi M, Ferruzzi L, Turrini E, Scartezzini P, Poli F, Gotti R, Guerrini A, Carulli G, Ottaviano V, et al. Mitochondrial pathway mediates the antileukemic effects of Hemidesmus indicus, a promising botanical drug. PLoS ONE 2011;6(6), e21544.
[132] Lennard L. Assay of 6-thioinosinic acid and 6-thioguanine nucleotides, active metabolites of 6-mercaptopurine, in human red blood cells. J Chromatogr Biomed Appl 1987;423:169–78.
[133] Pelin M, De Iudicibus S, Fusco L, Taboga E, Pellizzari G, Lagatolla C, Martelossi S, Ventura A, Decorti G, Stocco G. Role of oxidative stress mediated by glutathione-S-transferase in thiopurines' toxic effects. Chem Res Toxicol 2015;28(6):1186–95.
[134] Mishra P, Mahajan RP, Sharma A, Bhandari G, Mahajan SC. The use of azathioprine in inflammatory bowel diseases—a review. Int J Invent Pharm Sci 2013;1(4):300–14.
[135] Mikawa T, Oya J, Takashima J. Manufacture of 6-thioguanine and 6-thioguanosine with Actinoplanes. Jpn. Kokai Tokkyo Koho; 1990. JP 02265496 A 19901030.
[136] Munshi PN, Lubin M, Bertino JR. 6-Thioguanine: a drug with unrealized potential for cancer therapy. Oncologist 2014;19(7):760–5.
[137] Jeanbart L, Kourtis IC, van der Vlies AJ, Swartz MA, Hubbell JA. 6-Thioguanine-loaded polymeric micelles deplete myeloid-derived suppressor cells and enhance the efficacy of T cell immunotherapy in tumor-bearing mice. Cancer Immunol Immunother 2015;64(8):1033–46.
[138] Elion GB. The purine path to chemotherapy. Science (Washington, DC, U S) 1989;244(4900):41–7.
[139] Shen Y, Shen Y, Zhang L, Shi C, Eslami M. A computational study on the thioguanine drug interaction with silicon carbide graphyne-like nanosheets. Monatsh Chem 2020;151(12):1797–804.
[140] Bayoumy AB, van Liere ELSA, Simsek M, Warner B, Loganayagam A, Sanderson JD, Anderson S, Nolan J, de Boer NK, Mulder CJJ, et al. Efficacy, safety and drug survival of thioguanine as maintenance treatment for inflammatory bowel disease: a retrospective multi-centre study in the United Kingdom. BMC Gastroenterol 2020;20(1):296.
[141] Lin H, Kuan Y, Chu H, Cheng S, Pan H, Chen W, Sun C, Lin T. Disulfiram and 6-thioguanine synergistically inhibit the enzymatic activities of USP2 and USP21. Int J Biol Macromol 2021;176:490–7.
[142] Grabowska-Jadach I, Drozd M, Kulpinska D, Komendacka K, Pietrzak M. Modification of fluorescent nanocrystals with 6-thioguanine: monitoring of drug delivery. Appl Nanosci 2020;10(1):83–93.
[143] Li H, An X, Zhang D, Li Q, Zhang N, Yu H, Li Z. Transcriptomics analysis of the tumor-inhibitory pathways of 6-thioguanine in MCF-7 cells via silencing DNMT1 activity. Onco Targets Ther 2020;13:1211–23.
[144] Moreno-Guerrero SS, Ramirez-Pacheco A, Dorantes-Acosta EM, Medina-Sanson A. Analysis of genetic polymorphisms of thiopurine S-methyltransferase (TPMT) in Mexican pediatric patients with cancer. Rev Investig Clin 2013;65(2):156–64.
[145] Karim H, Ghalali A, Lafolie P, Vitols S, Fotoohi AK. Differential role of thiopurine methyltransferase in the cytotoxic effects of 6-mercaptopurine and 6-thioguanine on human leukemia cells. Biochem Biophys Res Commun 2013;437(2):280–6.

[146] Zhang W, Wang Y, Wang Y, Xu Z. Highly reproducible and fast detection of 6-thioguanine in human serum using a droplet-based microfluidic SERS system. Sens Actuators B 2019;283:532–7.

[147] Karimi-Maleh H, Shojaei AF, Tabatabaeian K, Karimi F, Shakeri S, Moradi R. Simultaneous determination of 6-mercaptopruine, 6-thioguanine and dasatinib as three important anticancer drugs using nanostructure voltammetric sensor employing Pt/MWCNTs and 1-butyl-3-methylimidazolium hexafluoro phosphate. Biosens Bioelectron 2016;86:879–84.

[148] Rawat D, Gillett PM, Devadason D, Wilson DC, McKiernan PJ. Long-term follow-up of children with 6-thioguanine-related chronic hepatoxicity following treatment for acute lymphoblastic leukaemia. J Pediatr Gastroenterol Nutr 2011;53(5):478–9.

[149] Zhang X, Elfarra AA. Toxicity mechanism-based prodrugs: glutathione-dependent bioactivation as a strategy for anticancer prodrug design. Expert Opin Drug Discovery 2018;13(9):815–24.

[150] Mitra R, Samuelson AG. Anticancer transition metal-thioguanine complexes. Indian Pat. Appl; 2016. IN 2013CH06111 A 20160624.

[151] Zhu W, Pan W. Use of thioguanine in preparation of drug for treating adsl deficiency. PCT Int. Appl; 2020. WO 2020140730 A1 20200709.

[152] Banh TN, Kode NR, Phadtare S. Aryl and acyclic unsaturated derivatives of thioguanine and 6-mercaptopurine: synthesis and cytotoxic activity. Lett Drug Des Discovery 2011;8(8):709–16.

[153] Zheng X, Liu X, Shao Q, Deng Q, Hu Q, Tan J. Oral sustained-release preparation of antitumor agent and preparation method thereof. Faming Zhuanli Shenqing; 2019. CN 109999009 A 20190712.

[154] Wang R, Yue L, Yu Y, Zou X, Song D, Liu K, Liu Y, Su H. Gold nanoparticles modify the photophysical and photochemical properties of 6-thioguanine: preventing DNA oxidative damage. J Phys Chem C 2016;120(26):14410–5.

[155] Huynh T, Wojnarowicz A, Sosnowska M, Srebnik S, Benincori T, Sannicolo F, D'Souza F, Kutner W. Cytosine derivatized bis(2,2′-bithienyl)methane molecularly imprinted polymer for selective recognition of 6-thioguanine, an antitumor drug. Biosens Bioelectron 2015;70:153–60.

[156] Ferguson LR, Hill CL, Morecombe P. Induction of resistance to 6-thioguanine and cytarabine by a range of anticancer drugs in Chinese Hamster AA8 cells. Eur J Cancer, Part A 1992;28A(4–5):736–42.

[157] Choi J, Lee Y, Jee J. Thiopurine drugs repositioned as tyrosinase inhibitors. Int J Mol Sci 2018;19(1):77/1–77/15.

[158] Qin B, McClarty G. Effect of 6-thioguanine on *Chlamydia trachomatis* growth in wild-type and hypoxanthine-guanine phosphoribosyltransferase-deficient cells. J Bacteriol 1992;174(9):2865–73.

[159] Popovic M, Banic B, Dakovic-Svajcer K, Nerudova J. Biochemical interaction of rifamycin and purine drugs. Zb Rad Prir-Mat Fak Univ Novom Sadu Ser Hem 1990;18:25–32.

[160] Testorelli C, Frigerio S, Lauletta M. Reversal of drug-induced immunogenicity in L1210 murine leukemia by "in vitro" treatment with 6-thioguanine. J Exp Clin Cancer Res 1991;10(3):149–55.

[161] Barco S, Gennai I, Bonifazio P, Maffia A, Barabino A, Arrigo S, Tripodi G, Cangemi G. A rapid and robust HPLC-DAD method for the monitoring of Thiopurine metabolites in whole blood: application to paediatric patients with inflammatory bowel disease. Curr Pharm Anal 2015;11(2):80–5.

[162] Kakar S, Singh R. 6-Thioguanine loaded magnetic microspheres as a new drug delivery system to cancer patients. Afr J Pharm Pharmacol 2014;8(31):786–92.

[163] Eswar KA, Sharada N, Bhogini T. Formulation development and in vitro evaluation of azathioprine tablets for colon drug delivery system. World J Pharm Pharm Sci 2016;5 (8):1067–90.
[164] Sodeifian G, Razmimanesh F, Saadati Ardestani N, Sajadian SA. Experimental data and thermodynamic modeling of solubility of azathioprine, as an immunosuppressive and anti-cancer drug, in supercritical carbon dioxide. J Mol Liq 2020;299, 112179.
[165] Wu S, He C, Tang T, Li Y. A review on co-existent Epstein-Barr virus-induced complications in inflammatory bowel disease. Eur J Gastroenterol Hepatol 2019;31 (9):1085–91.
[166] Klangjorhor J, Chaiyawat P, Teeyakasem P, Sirikaew N, Phanphaisarn A, Settakorn J, Lirdprapamongkol K, Yama S, Svasti J, Pruksakorn D. Mycophenolic acid is a drug with the potential to be repurposed for suppressing tumor growth and metastasis in osteosarcoma treatment. Int J Cancer 2020;146(12):3397–409.
[167] Helmy AM, Ibrahim EA. Development and in vitro evaluation of polysaccharide-based system for intestinal delivery of azathioprine for treatment of gastroduodenal Crohn's disease, jejunoileitis and ileitis. World J Pharm Pharm Sci 2017;6(6):314–27.
[168] Li X, Mei S, Gong X, Zhou H, Yang L, Zhou A, Liu Y, Li X, Zhao Z, Zhang X. Relationship between azathioprine metabolites and therapeutic efficacy in chinese patients with neuromyelitis optica spectrum disorders'. BMC Neurol 2017;17:130/1–8.
[169] Kong R, Gao Y, Gao J, Dai S, Xu X, Zhao D. A case report of azathioprine in the treatment of drug fever caused by arteritis and literature review. Fujian Yiyao Zazhi 2016;38(3):100–1.
[170] Tsuchiya A, Aomori T, Sakamoto M, Takeuchi A, Suzuki S, Jibiki A, Otsuka N, Ishioka E, Kaneko Y, Takeuchi T, et al. Effect of genetic polymorphisms of azathioprine-metabolizing enzymes on response to rheumatoid arthritis treatment. Pharmazie 2017;72(1):22–8.
[171] Hu J, Wang M, Xiao X, Zhang B, Xie Q, Xu X, Li S, Zheng Z, Wei D, Zhang X. A novel long-acting azathioprine polyhydroxyalkanoate nanoparticle enhances treatment efficacy for systemic lupus erythematosus with reduced side effects. Nanoscale 2020;12(19):10799–808.
[172] Lorenzoni PJ, Kay CSK, Zanlorenzi MF, Ducci RD, Werneck LC, Scola RH. Myasthenia gravis and azathioprine treatment: adverse events related to thiopurine S-methyl-transferase (TPMT) polymorphisms. J Neurol Sci 2020;412:116734.
[173] Kulkarni NM. Detection of heavy metals in azathiorpine API drugs. Int J Chem Phys Sci 2018;7(3):88–91.
[174] Djuric Z, Saranac L, Budic I, Pavlovic V, Djordjevic J. Therapeutic role of methotrexate in pediatric Crohn's disease. Bosn J Basic Med Sci 2018;18(3):211–6.
[175] Liu Q, Wang Y, Mei Q, Han W, Hu J, Hu N. Measurement of red blood cell 6-thioguanine nucleotide is beneficial in azathioprine maintenance therapy of Chinese Crohn's disease patients. Scand J Gastroenterol 2016;51(9):1093–9.
[176] Zochowska D, Zegarska J, Hryniewiecka E, Samborowska E, Jazwiec R, Tszyrsznic W, Borowiec A, Dadlez M, Paczek L. Determination of concentrations of azathioprine metabolites 6-thioguanine and 6-methylmercaptopurine in whole blood with the use of liquid chromatography combined with mass spectrometry. Transplant Proc 2016;48(5):1836–9.
[177] Priya MR, Ajay S, Govind B, Mahajan SC, Pooja M. Formulation and evaluation of microbial triggered colon specific delivery of azathioprine. World J Pharm Res 2015;4 (9):1597–615.
[178] Heerasing NM, Ng JF, Dowling D. Does lymphopenia or macrocytosis reflect 6-thioguanine levels in patients with inflammatory bowel disease treated with azathioprine or 6-mercaptopurine? Intern Med J 2016;46(4):465–9.

[179] Ferraro S, Leonardi L, Convertino I, Blandizzi C, Tuccori M. Is there a risk of lymphoma associated with anti-tumor necrosis factor drugs in patients with inflammatory bowel disease? A systematic review of observational studies. Front Pharmacol 2019;10:247.
[180] Lim SZ, Chua EW. Revisiting the role of thiopurines in inflammatory bowel disease through pharmacogenomics and use of novel methods for therapeutic drug monitoring. Front Pharmacol 2018;9:1107.
[181] Harmand P, Solassol J. Thiopurine drugs in the treatment of ulcerative colitis: identification of a novel deleterious mutation in TPMT. Genes 2020;11(10):1212.
[182] Marinaki AM, Arenas-Hernandez M. Reducing risk in thiopurine therapy. Xenobiotica 2020;50(1):101–9.
[183] Taha N, Hosein K, Grant-Orser A, Lin-Shaw A, Mura M. TPMT and HLA-DQA1-HLA-DRB genetic profiling to guide the use of azathioprine in the treatment of interstitial lung disease: first experience. Pulm Pharmacol Ther 2021;66, 101988.
[184] Silva CRG, Fialho SL, Barbosa J, Araujo BCR, Carneiro G, Sebastiao RCO, Mussel WN, Yoshida MI, de Freitas-Marques MB. Compatibility by a nonisothermal kinetic study of azathioprine associated with usual excipients in the product quality review process. J Braz Chem Soc 2021;32(3):638–51.
[185] Rucci F, Cigoli MS, Marini V, Fucile C, Mattioli F, Robbiano L, Cavallari U, Scaglione F, Perno CF, Penco S, et al. Combined evaluation of genotype and phenotype of thiopurine S-methyltransferase (TPMT) in the clinical management of patients in chronic therapy with azathioprine. Drug Metab Pers Ther 2019;34(1):1–10.
[186] El-Ashmawy IM, Bayad AE. Folic acid and grape seed extract prevent azathioprine-induced fetal malformations and renal toxicity in rats. Phytother Res 2016;30 (12):2027–35.
[187] Selvi SV, Nataraj N, Chen S, Prasannan A. An electrochemical platform for the selective detection of azathioprine utilizing a screen-printed carbon electrode modified with manganese oxide/reduced graphene oxide. New J Chem 2021;45(7):3640–51.
[188] Spiers ASD. Deoxycoformycin (pentostatin): clinical pharmacology, role in the chemotherapy of cancer, and use in other diseases. Haematologia 1996;27(2):55–84.
[189] Spiers ASD. Pentostatin (2prime prime or minute-deoxycoformycin): clinical pharmacology, role in cancer chemotherapy, and future prospects. Am J Ther 1995;2 (3):196–216.
[190] Dearden CE. Role of single-agent purine analogues in therapy of peripheral T-cell lymphomas. Semin Hematol 2006;43(2 Suppl 2):S22–6.
[191] Tsimberidou A, Giles F, Duvic M, Fayad L, Kurzrock R. Phase II study of pentostatin in advanced T-cell lymphoid malignancies: update of an M. D. Anderson cancer center series. Cancer (New York, NY, USA) 2004;100(2):342–9.
[192] Johnson SA, Catovsky D, Child JA, Newland AC, Milligan DW, Janmohamed R. Phase I/II evaluation of pentostatin (2′-deoxycoformycin) in a five day schedule for the treatment of relapsed/refractory b-cell chronic lymphocytic leukemia. Invest New Drugs 1998;16(2):155–60.
[193] Sauter C, Lamanna N, Weiss MA. Pentostatin in chronic lymphocytic leukemia. Expert Opin Drug Metab Toxicol 2008;4(9):1217–22.
[194] Robak T, Korycka A, Lech-Maranda E, Robak P. Current status of older and new purine nucleoside analogues in the treatment of lymphoproliferative diseases. Molecules 2009;14(3):1183–226.
[195] Margolis J, Vogelsang G. An old drug for a new disease: pentostatin (Nipent) in acute graft-versus-host disease. Semin Oncol 2000;27(2, Suppl. 5):72–7.
[196] Schmitt T, Luft T, Hegenbart U, Tran TH, Ho AD, Dreger P. Pentostatin for treatment of steroid-refractory acute GVHD: a retrospective single-center analysis. Bone Marrow Transplant 2011;46(4):580–5.

[197] Klein SA, Bug G, Mousset S, Hofmann W, Hoelzer D, Martin H. Long term outcome of patients with steroid-refractory acute intestinal graft host disease after treatment with pentostatin. Br J Haematol 2011;154(1):143–6.

[198] Watanabe M, Nakamura C, Fujii S, Kaneko H, Hirata H, Tsudo M. Donor lymphocyte infusion followed by pentostatin, cyclophosphamide and rituximab therapy is effective for relapsed chronic lymphocytic leukemia after allogeneic stem cell transplant. Leuk Lymphoma 2013;54(12):2750–2.

[199] Janus A, Robak T. Moxetumomab pasudotox for the treatment of hairy cell leukemia. Expert Opin Biol Ther 2019;19(6):501–8.

[200] Kolesar JM, Morris AK, Kuhn JG. Purine nucleoside analogs: fludarabine, pentostatin, and cladribine. Part 2. Pentostatin. J Oncol Pharm Pract 1996;2(4):211–24.

[201] Tadmor T. Purine analog toxicity in patients with hairy cell leukemia. Leuk Lymphoma 2011;52(Suppl. 2):38–42.

[202] Kraut EH, Neff JC, Bouroncle BA, Gochnour D, Grever MR. Immunosuppressive effects of pentostatin. J Clin Oncol 1990;8(5):848–55.

[203] Ng JP, Nolan B, Chan-Lam D, Coup AJ, McKenna D. Successful treatment of aplastic variant of hairy-cell leukaemia with deoxycoformycin. Hematology (Amsterdam, Netherlands) 2002;7(4):259–62.

[204] Grever MR. Pentostatin: impact on outcome in hairy cell leukemia. Hematol Oncol Clin North Am 2006;20(5):1099–108.

[205] Al-Razzak LA, Benedetti AE, Waugh WN, Stella VJ. Chemical stability of pentostatin (NSC-218321), a cytotoxic and immunosuppressant agent. Pharm Res 1990;7(5):452–60.

[206] Dillman RO. Pentostatin (Nipent) in the treatment of chronic lymphocyte leukemia and hairy cell leukemia. Expert Rev Anticancer Ther 2004;4(1):27–36.

[207] Bethge WA, Kerbauy FR, Santos EB, Gooley T, Storb R, Sandmaier BM. Extracorporeal photopheresis combined with pentostatin in the conditioning regimen for canine hematopoietic cell transplantation does not prevent GVHD. Bone Marrow Transplant 2014;49(9):1198–204.

[208] O'Dwyer PJ, Wagner B, Leyland-Jones B, Wittes RE, Cheson BD, Hoth DF. 2′-Deoxycoformycin (pentostatin) for lymphoid malignancies. Rational development of an active new drug. Ann Intern Med 1988;108(5):733–43.

[209] Margolis J, Grever MR. Pentostatin (Nipent): a review of potential toxicity and its management. Semin Oncol 2000;27(2, Suppl. 5):9–14.

[210] Kraut EH. Phase II trials of pentostatin (Nipent) in hairy cell leukemia. Semin Oncol 2000;27(2, Suppl. 5):27–31.

[211] Tsiara SN, Kapsali HD, Chaidos A, Christou L, Bourantas KL. Treatment of resistant/relapsing chronic lymphocytic leukemia with a combination regimen containing deoxycoformycin and rituximab. Acta Haematol 2004;111(4):185–8.

[212] Drapkin R. Pentostatin and rituximab in the treatment of patients with B-cell malignancies. Oncology (Williston Park, NY) 2000;14(6 Suppl 2):25–9.

[213] Kay NE, Wu W, Kabat B, La Plant B, Lin TS, Byrd JC, Jelinek DF, Grever MR, Zent CS, Call TG, et al. Pentostatin and rituximab therapy for previously untreated patients with B-cell chronic lymphocytic leukemia. Cancer (Hoboken, NJ, USA) 2010;116(9):2180–7.

[214] Bosanquet AG, Richards SM, Wade R, Else M, Matutes E, Dyer MJS, Rassam SMB, Durant J, Scadding SM, Raper SL, et al. Drug cross-resistance and therapy-induced resistance in chronic lymphocytic leukaemia by an enhanced method of individualised tumour response testing. Br J Haematol 2009;146(4):384–95.

[215] Willis CR, Goodrich A, Park K, Waselenko JK, Lucas M, Reese A, Diehl LF, Grever MR, Byrd JC, Flinn IW. A phase I/II study examining pentostatin, chlorambucil, and theophylline in patients with relapsed chronic lymphocytic leukemia and non-Hodgkin's lymphoma. Ann Hematol 2006;85(5):301–7.

[216] do Carmo GM, de Sa MF, Gressler LT, Baldissera MD, Monteiro SG, Grando TH, Henker LC, Mendes RE, Stefani LM, Da Silva AS. Cordycepin (3′-deoxyadenosine) and pentostatin (deoxycoformycin) against *Trypanosoma cruzi*. Exp Parasitol 2019;19947–51.
[217] Tedeschi A, Rossi D, Motta M, Quaresmini G, Rossi M, Coscia M, Anastasia A, Rossini F, Cortelezzi A, Nador G, et al. A phase II multi-center trial of pentostatin plus cyclophosphamide with ofatumumab in older previously untreated chronic lymphocytic leukemia patients. Haematologica 2015;100(12):e501–4.
[218] Morabito F, Callea I, Console G, Stelitano C, Sculli G, Filangeri M, Oliva B, Musolino C, Iacopino P, Brugiatelli M. The in vitro cytotoxic effect of mitoxantrone in combination with fludarabine or pentostatin in B-cell chronic lymphocytic leukemia. Haematologica 1997;82(5):560–5.
[219] Phiasivongsa P, Redkar S. Synthesis and manufacture of pentostatin and its precursors, analogs and derivatives. PCT Int. Appl; 2005. WO 2005027838 A2 20050331.
[220] Byrd JC, Grever MR, Flinn IW, Waselenko JK. Combination therapy for lymphoproliferative diseases with pentostatin and methylxanthines. PCT Int. Appl; 2000. WO 2000050082 A1 20000831.
[221] Li N, Zhang L, Chen L, Ying X, Di S. Crystal form of pentostatin and preparation method and use thereof. Faming Zhuanli Shenqing; 2018. CN 108586556 A 20180928.
[222] Kraut EH, Grever MR. Past and present role of pentostatin in hairy cell leukemia. Adv Blood Disord 2000;5(Hairy Cell Leukemia):151–61.
[223] Sands H, Redkar S, Ravivarapu H. Polymer-based pharmaceutical formulations for targeting specific regions of the gastrointestinal tract. PCT Int. Appl; 2004. WO 2004041195 A2 20040521.
[224] Kramer I. Viability of microorganisms in novel antineoplastic and antiviral drug solutions. J Oncol Pharm Pract 1998;4(1):32–7.
[225] Calderon Cabrera C, de la Cruz VF, Marin-Niebla A, Carrillo Cruz E, Rios Herranz E, Espigado Tocino I, Prats Martin C, Falantes JF, Martino Galiana ML, Perez-Simon JA. Pentostatin plus cyclophosphamide and bexarotene is an effective and safe combination in patients with mycosis fungoides/sezary syndrome. Br J Haematol 2013;162(1):130–2.
[226] Hentosh P, Peffley DM. The cladribine conundrum: deciphering the drug's mechanism of action. Expert Opin Drug Metab Toxicol 2010;6(1):75–81.
[227] Albertioni F, Lindemalm S, Eriksson S, Juliusson G, Liliemark J. Relationship between cladribine (CdA) plasma, intracellular CdA-5′-triphosphate (CdATP) concentration, deoxycytidine kinase (dCK), and chemotherapeutic activity in chronic lymphocytic leukemia (CLL). Adv Exp Med Biol 1998;431693–7.
[228] Goodman GR, Beutler E, Saven A. Cladribine in the treatment of hairy-cell leukaemia. Best Pract Res Clin Haematol 2003;16(1):101–16.
[229] Cottam HB, Carson DA. 2-chlorodeoxyadenosine (cladribine): rational development of a novel chemotherapeutic agent edited by: Huang, Ziwei. Drug Discovery Res 2007;393–407.
[230] Korsen M, Alonso SB, Peix L, Broeker BM, Dressel A. Cladribine exposure results in a sustained modulation of the cytokine response in human peripheral blood mononuclear cells. PLoS ONE 2015;10(6):e0129182/1–e0129182/14.
[231] Sorensen PS, Sellebjerg F. Pulsed immune reconstitution therapy in multiple sclerosis. Ther Adv Neurol Disord 2019;12, 1756286419836913.
[232] Rammohan K, Coyle PK, Sylvester E, Galazka A, Dangond F, Grosso M, Leist TP. The development of Cladribine tablets for the treatment of multiple sclerosis: a comprehensive review. Drugs 2020;80(18):1901–28.
[233] Spurgeon S, Yu M, Phillips JD, Epner EM. Cladribine: not just another purine analogue? Expert Opin Investig Drugs 2009;18(8):1169–81.

[234] Faissner S, Gold R. Oral therapies for multiple sclerosis. Cold Spring Harb Perspect Med 2019;9(1), a032011.
[235] Boyko AN, Boyko OV. Cladribine tablets' potential role as a key example of selective immune reconstitution therapy in multiple sclerosis. Degener Neurol Neuromuscular Dis 2018;8:35–44.
[236] Lebrun-Frenay C, Berestjuk I, Tartare-Deckert S, Cohen M, Cohen M. Effects on melanoma cell lines suggest no significant risk of melanoma under cladribine treatment. Neurol Ther 2020;9(2):599–604.
[237] Laugel B, Borlat F, Galibert L, Vicari A, Weissert R, Chvatchko Y, Bruniquel D. Cladribine inhibits cytokine secretion by T cells independently of deoxycytidine kinase activity. J Neuroimmunol 2011;240–241:52–7.
[238] Schreiner TL, Miravalle A. Current and emerging therapies for the treatment of multiple sclerosis: focus on cladribine. J Cent Nerv Syst Dis 2012;4:1–14.
[239] Sipe JC. Cladribine tablets: a potential new short-course annual treatment for relapsing multiple sclerosis. Expert Rev Neurother 2010;10(3):365–75.
[240] Yavuz S, Cetin A, Akdemir A, Doyduk D, Disli A, Celik Turgut G, Sen A, Yildirir Y. Synthesis and functional investigations of computer designed novel cladribine-like compounds for the treatment of multiple sclerosis. Arch Pharm (Weinheim, Germany) 2017;350(11).
[241] Smal C, Lisart S, Ferrant A, Bontemps F, Van Den Neste E. Inhibition of the ERK pathway promotes apoptosis induced by 2-chloro-2′-deoxyadenosine in the B-cell leukemia cell line EHEB. Nucleosides Nucleotides Nucleic Acids 2006;25 (9–11):1009–12.
[242] Cardoen S, Van Den NE, Smal C, Rosier JF, Delacauw A, Ferrant A, Van den Berghe G, Bontemps F. Resistance to 2-chloro-2′-deoxyadenosine of the human B-cell leukemia cell line EHEB. Clin Cancer Res 2001;7(11):3559–66.
[243] Robak T. Therapy of chronic lymphocytic leukaemia with purine nucleoside analogues: facts and controversies. Drugs Aging 2005;22(12):983–1012.
[244] Robak T. Therapy of chronic lymphocytic leukemia with purine analogs and monoclonal antibodies. Transfus Apher Sci 2005;32(1):33–44.
[245] Galmarini CM, Mackey JR, Dumontet C. Nucleoside analogues: mechanisms of drug resistance and reversal strategies. Leukemia 2001;15(6):875–90.
[246] Kohnke PL, Mactier S, Almazi JG, Crossett B, Christopherson RI. Fludarabine and cladribine induce changes in surface proteins on human B-lymphoid cell lines involved with apoptosis, cell survival, and antitumor immunity. J Proteome Res 2012;11 (9):4436–48.
[247] Podhorecka M, Halicka D, Klimek P, Kowal M, Chocholska S, Dmoszynska A. Simvastatin and purine analogs have a synergic effect on apoptosis of chronic lymphocytic leukemia cells. Ann Hematol 2010;89(11):1115–24.
[248] Podhorecka M, Halicka D, Klimek P, Kowal M, Chocholska S, Dmoszynska A. Resveratrol increases rate of apoptosis caused by purine analogues in malignant lymphocytes of chronic lymphocytic leukemia. Ann Hematol 2011;90(2):173–83.
[249] Ghai V, Sharma K, Abbi Kamal KS, Shimko S, Epner Elliot M. Current approaches to epigenetic therapy for the treatment of mantle cell lymphoma. Adv Exp Med Biol 2013;779257–66.
[250] Lotfi K, Juliusson G, Albertioni F. Pharmacological basis for cladribine resistance. Leuk Lymphoma 2003;44(10):1705–12.
[251] Saven A, Lee T, Kosty M, Piro L. Cladribine and mitoxantrone dose escalation in indolent non-Hodgkin's lymphoma. J Clin Oncol 1996;14(7):2139–44.
[252] Robak T. Purine nucleoside analogues in the treatment of myleoid leukemias. Leuk Lymphoma 2003;44(3):391–409.

[253] Bellezza I, Tucci A, Minelli A. 2-Chloroadenosine and human prostate cancer cells. Anticancer Agents Med Chem 2008;8(7):783–9.
[254] Fanta PT, Saven A. Hairy cell leukemia. Cancer Treat Res 2008;142193–209.
[255] Greyz N, Saven A. Cladribine: from the bench to the bedside—focus on hairy cell leukemia. Expert Rev Anticancer Ther 2004;4(5):745–57.
[256] Robak T, Korycka A, Robak E. Older and new formulations of cladribine: pharmacology and clinical efficacy in hematological malignancies. Front Anti-Cancer Drug Discovery 2010;1:497–524.
[257] Claussen MC, Korn T. Immune mechanisms of new therapeutic strategies in MS – Teriflunomide. Clin Immunol (Amsterdam, Netherlands) 2012;142(1):49–56.
[258] Ogura M. Hairy cell leukemia therapy by cladribine. Nihon Rinsho Jpn J Clin Med 2004;62(7):1337–42.
[259] Robak T, Korycka-Wolowiec A, Robak E. Pharmacology and clinical efficacy of cladribine in hematological malignancies—older and newer formulations. Top Anti-Cancer Res 2012;291–333.
[260] Robak T, Korycka A, Robak E. Older and new formulations of cladribine. Pharmacology and clinical efficacy in hematological malignancies. Recent Pat Anti-Cancer Drug Discovery 2006;1(1):23–38.
[261] Johnston JB. Mechanism of action of pentostatin and cladribine in hairy cell leukemia. Leuk Lymphoma 2011;52(Suppl. 2):43–5.
[262] Arner ES. On the phosphorylation of 2-chlorodeoxyadenosine (CdA) and its correlation with clinical response in leukemia treatment. Leuk Lymphoma 1996;21 (3–4):225–31.
[263] Robak T, Robak P. Purine nucleoside analogs in the treatment of rarer chronic lymphoid leukemias. Curr Pharm Des 2012;18(23):3373–88.
[264] Baltz JK, Montello MJ. Cladribine for the treatment of hematologic malignancies. Clin Pharm 1993;12(11):805–13 [quiz 860-2].
[265] Vilpo J, Koski T, Vilpo L. Calcium antagonists potentiate P-glycoprotein-independent anticancer drugs in chronic lymphocytic leukemia cells in vitro. Haematologica 2000;85(8):806–13.
[266] Jensen K, Johnson L, Jacobson PA, Kachler S, Kirstein MN, Lamba J, Klotz K. Cytotoxic purine nucleoside analogues bind to A1, A2A, and A3 adenosine receptors. Naunyn Schmiedebergs Arch Pharmacol 2012;385(5):519–25.
[267] Kolesar JM, Morris AK, Kuhn JG. Purine nucleoside analogs: fludarabine, pentostatin, and cladribine. Part 1. Fludarabine. J Oncol Pharm Pract 1996;2(3):160–81.
[268] Chun HG, Leyland-Jones B, Cheson BD. Fludarabine phosphate: a synthetic purine antimetabolite with significant activity against lymphoid malignancies. J Clin Oncol 1991;9(1):175–88.
[269] Plunkett W, Huang P, Gandhi V. Metabolism and action of fludarabine phosphate. Semin Oncol 1990;17(5 Suppl. 8):3–17.
[270] https://go.drugbank.com/drugs/DB01073.
[271] Laurent D, Pradier O, Schmidberger H, Rave-Frank M, Frankenberg D, Hess CF. Radiation rendered more cytotoxic by fludarabine monophosphate in a human oropharynx carcinoma cell-line than in fetal lung fibroblasts. J Cancer Res Clin Oncol 1998;124(9):485–92.
[272] Hong JS, Waud WR, Levasseur DN, Townes TM, Wen H, McPherson SA, Moore BA, Bebok Z, Allan PW, Secrist III JA, et al. Excellent in vivo bystander activity of fludarabine phosphate against human glioma xenografts that express the *Escherichia coli* purine nucleoside phosphorylase gene. Cancer Res 2004;64(18):6610–5.
[273] Stoica GS, Greenberg HE, Rossoff LJ. Corticosteroid responsive fludarabine pulmonary toxicity. Am J Clin Oncol 2002;25(4):340–1.

[274] Keating MJ, O'Brien S, Mclaughlin P, Dimopoulos M, Gandhi V, Plunkett W, Lerner S, Kantarjian H, Estey E. Clinical experience with fludarabine in hemato-oncology. Hematol Cell Ther 1996;38(Suppl. 2):S83–91.
[275] Wilson PK, Mulligan SP, Christopherson RI. Metabolic response patterns of nucleotides in B-cell chronic lymphocytic leukaemias to cladribine, fludarabine and deoxycoformycin. Leuk Res 2004;28(7):725–31.
[276] Petersen AJ, Brown RD, Gibson J, Pope B, Luo X, Schutz L, Wiley JS, Joshua DE. Nucleoside transporters, bcl-2 and apoptosis in CLL cells exposed to nucleoside analogs in vitro. Eur J Haematol 1996;56(4):213–20.
[277] Moussay E, Palissot V, Vallar L, Poirel HA, Wenner T, El Khoury V, Aouali N, Van Moer K, Leners B, Bernardin F, et al. Determination of genes and microRNAs involved in the resistance to fludarabine in vivo in chronic lymphocytic leukemia. Mol Cancer 2010;9.
[278] Robak T, Drzewoski J. Clinical pharmacology of fludarabine. Pol Tyg Lek 1991;46 (48–49):942–4.
[279] Hayakawa M. Current and future development of fludarabine phosphate (Fludara). BIO Clin 2005;20(13):1188–93.
[280] Ogawa Y. Phase I/II and pharmacokinetic study of oral fludarabine phosphate in relapsed indolent B-cell non-Hodgkin's lymphoma. Ketsueki Shuyoka 2008;56 (1):32–8.
[281] Gandhi V, Huang P, Chapman AJ, Chen F, Plunkett W. Incorporation of fludarabine and 1-β-D-arabinofuranosylcytosine 5′-triphosphates by DNA polymerase α: affinity, interaction, and consequences. Clin Cancer Res 1997;3(8):1347–55.
[282] Gandhi V, Plunkett W. Cellular and clinical pharmacology of fludarabine. Clin Pharmacokinet 2002;41(2):93–103.
[283] Pierige F, De Marco C, Orlotti N, Dominici S, Biagiotti S, Serafini S, Zaffaroni N, Magnanil M, Rossi L. Cytotoxic activity of 2-Fluoro-ara-AMP and 2-Fluoro-ara-AMP-loaded erythrocytes against human breast carcinoma cell lines. Int J Oncol 2010;37(1):133–42.
[284] Nitsche M, Christiansen H, Hermann RM, Lucke E, Peters K, Rave-Frank M, Schmidberger H, Pradier O. The combined effect of fludarabine monophosphate and radiation as well as gemcitabine and radiation on squamous carcinoma tumor cell lines in vitro. Int J Radiat Biol 2008;84(8):643–57.
[285] Gregoire V, Hunter NR, Brock WA, Hittleman WN, Plunkett W, Milas L. Improvement in therapeutic ratio of radiotherapy for a murine sarcoma by indomethacin plus fludarabine. Radiat Res 1996;146(5):548–53.
[286] Chen LS, Plunkett W, Gandhi V. Polyadenylation inhibition by the triphosphates of deoxyadenosine analogues. Leuk Res 2008;32(10):1573–81.
[287] Lubecka K, Kaufman-Szymczyk A, Fabianowska-Majewska K. Inhibition of breast cancer cell growth by the combination of Clofarabine and sulforaphane involves epigenetically mediated CDKN2A upregulation. Nucleosides Nucleotides Nucleic Acids 2018;37(5):280–9.
[288] Steensma DP. Novel therapies for myelodysplastic syndromes. Hematol Oncol Clin North Am 2010;24(2):423–41.
[289] Montalban-Bravo G, Garcia-Manero G. Novel drugs for older patients with acute myeloid leukemia. Leukemia 2015;29(4):760–9.
[290] Long MJC, Zhao Y, Aye Y. Clofarabine commandeers the RNR-a-ZRANB3 nuclear signaling Axis. Cell Chem Biol 2020;27(1):122–133.e5.
[291] Gorzkiewicz M, Deriu MA, Studzian M, Janaszewska A, Grasso G, Pulaski L, Appelhans D, Danani A, Klajnert-Maculewicz B. Fludarabine-specific molecular interactions with maltose-modified poly(propyleneimine) dendrimer enable effective cell

entry of the active drug form: comparison with clofarabine. Biomacromolecules 2019;20(3):1429–42.
[292] Daly MB, Kim B, Roth ME, Maldonado JO, Clouser CL, Mansky LM, Roth ME, Maldonado JO, Mansky LM, Roth ME, et al. Dual anti-HIV mechanism of clofarabine. Retrovirology 2016;1320.
[293] Daly MB, Roth ME, Bonnac L, Maldonado JO, Xie J, Clouser CL, Patterson SE, Kim B, Mansky LM. Dual anti-HIV mechanism of clofarabine. Retrovirology 2016;13:20/1–20/12.
[294] Majda K, Lubecka K, Kaufman-Szymczyk A, Fabianowska-Majewska K. Clofarabine (2-chloro-2′-fluoro-2′-deoxyarabinosyladenine)- -biochemical aspects of anticancer activity. Acta Pol Pharm 2011;68(4):459–66.
[295] Lindemalm S, Liliemark J, Gruber A, Eriksson S, Karlsson MO, Wang Y, Albertioni F. Comparison of cytotoxicity of 2-chloro-2′-arabino-fluoro-2′-deoxyadenosine (Clofarabine) with cladribine in mononuclear cells from patients with acute myeloid and chronic lymphocytic leukemia. Haematologica 2003;88(3):324–32.
[296] Jhaveri KD, Chidella S, Fishbane S, Allen SL. Clofarabine-induced kidney toxicity. J Oncol Pharm Pract 2014;20(4):305–8.
[297] Robak P, Robak T. Older and new purine nucleoside analogs for patients with acute leukemias. Cancer Treat Rev 2013;39(8):851–61.
[298] Robak T. New purine nucleoside analogs for acute lymphoblastic leukemia. Clin Cancer Drugs 2014;1(1):2–10.
[299] Korycka A, Lech-Maranda E, Robak T. Novel purine nucleoside analogues for hematological malignancies. Recent Pat Anticancer Drug Discov 2008;3(2):123–36.
[300] Bonate PL, Cunningham CC, Gaynon P, Jeha S, Kadota R, Lam GN, Razzouk B, Rytting M, Steinherz P, Weitman S. Population pharmacokinetics of clofarabine and its metabolite 6-ketoclofarabine in adult and pediatric patients with cancer. Cancer Chemother Pharmacol 2011;67(4):875–90.
[301] Faderl S, Gandhi V, Kantarjian HM. Potential role of novel nucleoside analogs in the treatment of acute myeloid leukemia. Curr Opin Hematol 2008;15(2):101–7.
[302] Thudium KE, Ghoshal S, Fetterly GJ, Haese JPD, Karpf AR, Wetzler M. Synergism between clofarabine and decitabine through p53R2: a pharmacodynamic drug-drug interaction modeling. Leuk Res 2012;36(11):1410–6.
[303] Larson RA. Three new drugs for acute lymphoblastic leukemia: nelarabine, clofarabine, and forodesine. Semin Oncol 2007;34(6 Suppl 5):S13–20.
[304] Gulbis AM, Culotta KS, Jones RB, Andersson BS. Busulfan and metronidazole: an often forgotten but significant drug interaction. Ann Pharmacother 2011;45(7–8), e39.
[305] Lech-Maranda E, Korycka A, Robak T. Clofarabine as a novel nucleoside analogue approved to treat patients with haematological malignancies: mechanism of action and clinical activity. Mini Rev Med Chem 2009;9(7):805–12.
[306] Wang G, Liu M, Huang Y. Clofarabine: a new therapeutic agent for acute lymphocytic leukemia. Zhongguo Xinyao Yu Linchuang Zazhi 2008;27(7):538–42.
[307] Gandhi V, Plunkett W, Bonate PL, Du M, Nowak B, Lerner S, Keating MJ. Clinical and pharmacokinetic study of clofarabine in chronic lymphocytic leukemia: strategy for treatment. Clin Cancer Res 2006;12(13):4011–7.
[308] Robak T. New nucleoside analogs for patients with hematological malignancies. Expert Opin Investig Drugs 2011;20(3):343–59.
[309] Lu J, Huang X. Application of clofarabine on hematological malignancy. Zhonghua Xueyexue Zazhi 2010;31(1):66–8.
[310] Robak T, Lech-Maranda E, Korycka A, Robak E. Purine nucleoside analogs as immunosuppressive and antineoplastic agents: mechanism of action and clinical activity. Curr Med Chem 2006;13(26):3165–89.

[311] Lech-Maranda E, Korycka A, Robak T. Pharmacological and clinical studies on purine nucleoside analogs—new anticancer agents. Mini Rev Med Chem 2006;6(5):575–81.
[312] Gandhi V, Plunkett W. Clofarabine and nelarabine: two new purine nucleoside analogs. Curr Opin Oncol 2006;18(6):584–90.
[313] Jayaraman A, Jamil K. Drug targets for cell cycle dysregulators in leukemogenesis: in silico docking studies. PloS ONE 2014;9(1):e86310.
[314] Burke MP, Borland KM, Litosh VA. Base-modified nucleosides as chemotherapeutic agents: past and future. Curr Top Med Chem (Sharjah, United Arab Emirates) 2016;16 (11):1231–41.
[315] Xia R, Sun L, Yang X, Qu G. Synthesis of nelarabine. Zhongguo Yiyao Gongye Zazhi 2015;46(12):1278–80.
[316] Barreca M, Spano V, Montalbano A, Cueto M, Marrero ARD, Deniz I, Erdogan A, Bilela LL, Moulin C, Taffin-De-Givenchy E, et al. Marine anticancer agents: an overview with a particular focus on their chemical classes. Mar Drugs 2020;18(12):619.
[317] Yamauchi T, Uzui K, Nishi R, Tasaki T, Ueda T. A nelarabine-resistant T-lymphoblastic leukemia CCRF-CEM variant cell line is cross-resistant to the purine nucleoside phosphorylase inhibitor forodesine. Anticancer Res 2014;34 (9):4885–92.
[318] Abaza Y, Kantarjian HM, Faderl S, Jabbour E, Jain N, Thomas D, Kadia T, Borthakur G, Khoury JD, Burger J, et al. Hyper-CVAD plus nelarabine in newly diagnosed adult T-cell acute lymphoblastic leukemia and T-lymphoblastic lymphoma. Am J Hematol 2018;93(1):91–9.
[319] Ju Hee Y, Hong CR, Shin HY. Advancements in the treatment of pediatric acute leukemia and brain tumor—continuous efforts for 100% cure. Korean J Pediatr 2014;57 (10):434–9.
[320] Marks DI, Paietta EM, Moorman AV, Richards SM, Buck G, DeWald G, Ferrando A, Fielding AK, Goldstone AH, Ketterling RP, et al. T-cell acute lymphoblastic leukemia in adults: clinical features, immunophenotype, cytogenetics, and outcome from the large randomized prospective trial (UKALL XII/ECOG 2993). Blood 2009;114 (25):5136–45.
[321] Homminga I, Zwaan CM, Manz CY, Parker C, Bantia S, Smits WK, Higginbotham F, Pieters R, Meijerink JPP. In vitro efficacy of forodesine and nelarabine (ara-G) in pediatric leukemia. Blood 2011;118(8):2184–90.
[322] Gu Q, Sun X, Xu C. Method for synthesizing nelarabine from 6-chloroguanosine Faming Zhuanli Shenqing; 2007. CN 101092441 A 20071226.
[323] Curbo S, Karlsson A. Nelarabine: a new purine analog in the treatment of hematologic malignancies. Rev Recent Clin Trials 2006;1(3):185–92.
[324] Berg SL, Blaney SM, Devidas M, Lampkin TA, Murgo A, Bernstein M, Billett A, Kurtzberg J, Reaman G, Gaynon P, et al. Phase II study of nelarabine (compound 506U78) in children and young adults with refractory T-cell malignancies: a report from the Children's Oncology Group. J Clin Oncol 2005;23(15):3376–82.
[325] Candoni A, Lazzarotto D, Fanin R, Ferrara F, Curti A, Papayannidis C, Lussana F, Del Principe MI, Bonifacio M, Pizzolo G, et al. Nelarabine as salvage therapy and bridge to allogeneic stem cell transplant in 118 adult patients with relapsed/refractory T-cell acute lymphoblastic leukemia/lymphoma. A CAMPUS ALL study. Am J Hematol 2020;95(12):1466–72.
[326] Dunsmore KP, Winter SS, Devidas M, Wood BL, Esiashvili N, Chen Z, Eisenberg N, Briegel N, Hayashi RJ, Gastier-Foster JM, et al. Children's oncology group AALL0434: a phase III randomized clinical trial testing nelarabine in newly diagnosed T-cell acute lymphoblastic leukemia. J Clin Oncol 2020;38(28):3282–93.
[327] Dua SG, Jhaveri MD. MR imaging in nelarabine-induced myelopathy. J Clin Neurosci 2016;29:205–6.

[328] Eryilmaz E. Multi-targeted anti-leukemic drug design with the incorporation of silicon into Nelarabine: how silicon increases bioactivity. Eur J Pharm Sci 2019;134:266–73.
[329] Fullmer A, O'Brien S, Kantarjian H, Jabbour E. Novel therapies for relapsed acute lymphoblastic leukemia. Curr Hematol Malig Rep 2009;4(3):148–56.
[330] Goekbuget N, Basara N, Baurmann H, Beck J, Brueggemann M, Diedrich H, Gueldenzoph B, Hartung G, Horst H, Huettmann A, et al. High single-drug activity of nelarabine in relapsed T-lymphoblastic leukemia/lymphoma offers curative option with subsequent stem cell transplantation. Blood 2011;118(13):3504–11.
[331] Ichikawa M, Horibe K. Nelarabine in the treatment of acute lymphoblastic leukemia. Ketsueki Naika 2014;68(2):183–8.
[332] Beesley AH, Palmer M, Ford J, Weller RE, Cummings AJ, Freitas JR, Firth MJ, Perera KU, de Klerk NH, Kees UR. In vitro cytotoxicity of nelarabine, clofarabine and flavopiridol in paediatric acute lymphoblastic leukaemia. Br J Haematol 2007;137 (2):109–16.
[333] Styczynski J, Kolodziej B, Rafinska B. Differential activity of nelarabine and clofarabine in leukemia and lymphoma cell lines. Wspolczesna Onkol 2009;13(6):281–6.
[334] Suzuki I, Kikkawa T, Watanabe S, Yonekura A. Development of novel therapeutic agent nelarabine for relapsed · refractory T cell acute lymphocytic leukemia/T cell lymphoblastic lymphoma. BIO Clin 2007;22(13):1195–200.
[335] Fan H. Nelarabine freeze-dried powder injection and preparation method thereof. Faming Zhuanli Shenqing; 2010. CN 101664391 A 20100310.
[336] Shang L. Intravenous injection of nelarabine and its preparation method. Faming Zhuanli Shenqing; 2010. CN 101926816 A 20101229.
[337] Winter SS, Dunsmore KP, Devidas M, Eisenberg N, Asselin BL, Wood BL, Leonard RN, Marcia S, Murphy J, Gastier-Foster JM, Carroll AJ, et al. Safe integration of nelarabine into intensive chemotherapy in newly diagnosed T-cell acute lymphoblastic leukemia: Children's Oncology Group Study AALL0434. Pediatr Blood Cancer 2015;62 (7):1176–83.
[338] Kanazawa T, Shiba N, Aizawa A, Okuno H, Tamura K, Tsukada S, Kumamoto T, Yamada S, Kobayashi Y, Arakawa H. Nelarabine resistance of childhood T-cell lymphoblastic leukemia/lymphoma cells. Kitakanto Med J 2011;61(2):119–26.
[339] Dai J, Ye D, Wang H, Dai Y. Injection containing nelarabine and ethylenediaminetetraacetic acid or ethylenediaminetetraacetate for treating T cell acute lymphoblastic leukemia and T cell lymphoblastic lymphoma, and its preparation method. Faming Zhuanli Shenqing; 2009. CN 101401786 A 20090408.
[340] Czarnik AW. Deuterium-enriched nelarabine. U.S. Pat. Appl. Publ; 2009. US 20090075930 A1 20090319.
[341] Forcade E, Leguay T, Vey N, Baruchel A, Delaunay J, Robin M, Socie G, Dombret H, Peffault de Latour R, Raffoux E. Nelarabine for T cell acute lymphoblastic leukemia relapsing after allogeneic hematopoietic stem cell transplantation: an opportunity to improve survival. Biol Blood Marrow Transplant 2013;19(7):1124–6.
[342] Kisor DF. Collaboration to meet a therapeutic need: the development of nelarabine. Clin Med: Ther 2009;1:1317–20.
[343] Kumamoto T, Goto H, Ogawa C, Hori T, Deguchi T, Araki T, Saito AM, Manabe A, Horibe K, Toyoda H. FLEND (nelarabine, fludarabine, and etoposide) for relapsed T-cell acute lymphoblastic leukemia in children: a report from Japan Children's Cancer Group. Int J Hematol 2020;112(5):720–4.
[344] Berg SL, Brueckner C, Nuchtern JG, Dauser R, McGuffey L, Blaney SM. Plasma and cerebrospinal fluid pharmacokinetics of nelarabine in nonhuman primates. Cancer Chemother Pharmacol 2007;59(6):743–7.
[345] Reilly KM, Kisor DF. Profile of nelarabine: use in the treatment of T-cell acute lymphoblastic leukemia. Onco Targets Ther 2009;2:219–28.

[346] Jain P, Kantarjian H, Ravandi F, Thomas D, O'Brien S, Kadia T, Burger J, Borthakur G, Daver N, Jabbour E, et al. The combination of hyper-CVAD plus nelarabine as frontline therapy in adult T-cell acute lymphoblastic leukemia and T-lymphoblastic lymphoma: MD Anderson Cancer Center experience. Leukemia 2014;28(4):973–5.
[347] Yamauchi T, Uzui K, Nishi R, Shigemi H, Ueda T. Reduced drug incorporation into DNA and antiapoptosis as the crucial mechanisms of resistance in a novel nelarabine-resistant cell line. BMC Cancer 2014;14:547/1–9.
[348] Cooper TM. Role of nelarabine in the treatment of T-cell acute lymphoblastic leukemia and T-cell lymphoblastic lymphoma. Ther Clin Risk Manag 2007;3 (6):1135–41.
[349] Czuczman MS, Porcu P, Johnson J, Niedzwiecki D, Kelly M, Hsi ED, Cook JR, Canellos G, Cheson BD. Results of a phase II study of 506U78 in cutaneous T-cell lymphoma and peripheral T-cell lymphoma: CALGB 59901. Leuk Lymphoma 2007;48(1):97–103.
[350] Kurtzberg J, Ernst TJ, Keating MJ, Gandhi V, Hodge JP, Kisor DF, Lager JJ, Stephens C, Levin J, Krenitsky T, et al. Phase I study of 506U78 administered on a consecutive 5-day schedule in children and adults with refractory hematologic malignancies. J Clin Oncol 2005;23(15):3396–403.
[351] Amer-Salas N, Cladera-Serra A, Gonzalez-Morcillo G, Rodriguez-Camacho JM. Nelarabine-associated myelopathy in a patient with acute lymphoblastic leukaemia: case report. J Oncol Pharm Pract 2021;27(1):244–9.
[352] Lonetti A, Cappellini A, Bertaina A, Locatelli F, Pession A, Buontempo F, Evangelisti C, Evangelisti C, Orsini E, Zambonin L, et al. Improving nelarabine efficacy in T cell acute lymphoblastic leukemia by targeting aberrant PI3K/AKT/mTOR signaling pathway. J Hematol Oncol 2016;9:114/1–114/16.
[353] Ngo D, Patel S, Kim EJ, Brar R, Koontz MZ. Nelarabine neurotoxicity with concurrent intrathecal chemotherapy: case report and review of literature. J Oncol Pharm Pract 2015;21(4):296–300.
[354] Lonetti A, Buontempo F, Evangelisti C, Orsini E, Martelli AM, Cappellini A, Bertaina A, Locatelli F, Pession A, Evangelisti C, et al. Improving nelarabine efficacy in T cell acute lymphoblastic leukemia by targeting aberrant PI3K/AKT/mTOR signaling pathway. J Hematol Oncol 2016;9(1):114.
[355] Iino M. Severe liver injury following nelarabine chemotherapy for T-cell lymphoblastic lymphoma. [Rinsho Ketsueki] Jpn J Clin Hematol 2009;50(1):49–51.
[356] Kawakami M, Taniguchi K, Yoshihara S, Ishii S, Kaida K, Ikegame K, Okada M, Watanabe S, Nishina T, Hamada H, et al. Irreversible neurological defects in the lower extremities after haploidentical stem cell transplantation: possible association with nelarabine. Am J Hematol 2013;88(10):853–7.
[357] Korycka A, Lech-Maranda E, Robak T. Novel purine nucleoside analogues for hematological malignancies. Front Anti-Cancer Drug Discovery 2010;1:219–40.
[358] Roecker AM, Stockert A, Kisor DF. Nelarabine in the treatment of refractory T-cell malignancies. Clin Med Insights: Oncol 2010;4:133–41.
[359] Xia R, Sun L-P, Qu G. Synthesis of nelarabine with pure β-anomer through late-stage C-H nitration/nitro-reduction. Heterocycles 2015;91(12):2386–93.
[360] Shang S, Li X, Zheng J, Xiao F. Nelarabine injection composition and its preparation method. Faming Zhuanli Shenqing; 2013. CN 103191051 A 20130710.
[361] Li M, Han F, Li Z, Wang X. Nelarabine crystalline compound and its preparation method thereof. Faming Zhuanli Shenqing; 2013. CN 103172687 A 20130626.
[362] Yang X. Antitumor drug nelarabine powder injection composition. Faming Zhuanli Shenqing; 2017. CN 106727332 A 20170531.
[363] Yang X. Method for preparing nelarabine compound for antitumor nelarabine powder injection composition. Faming Zhuanli Shenqing; 2017. CN 106831918 A 20170613.

[364] Yang X. Nelarabine compound for antitumor drug nelarabine composition. Faming Zhuanli Shenqing; 2017. CN 106749459 A 20170531.
[365] Yang X. Antitumor drug nelzarabine compound. Faming Zhuanli Shenqing; 2017. CN 106632555 A 20170510.
[366] Yang X. Method for preparation of antitumor drug nelzarabine. Faming Zhuanli Shenqing; 2017. CN 106589028 A 20170426.
[367] Yang X. Method of preparing nelarabine compound for antitumor drug nelarabine composition. Faming Zhuanli Shenqing; 2017. CN 106632557 A 20170510.
[368] Li X, Xia J, Hou J. Compound for treating leukemia and preparation method thereof. Faming Zhuanli Shenqing; 2017. CN 106397517 A 20170215.
[369] Yang X. Anti-tumor drug nelzarabine composition. Faming Zhuanli Shenqing; 2017. CN 106619690 A 20170510.
[370] Cheng G, Tang Q, Cao B. High drug loading system to co-deliver anticancer drugs and nucleic acids for cancer therapy. U.S. Pat. Appl. Publ; 2015. US 20150099005 A1 20150409.
[371] Rodriguez COJ, Plunkett W, Paff MT, Du M, Nowak B, Ramakrishna P, Keating MJ, Gandhi V. High-performance liquid chromatography method for the determination and quantitation of arabinosylguanine triphosphate and fludarabine triphosphate in human cells. J Chromatogr B Biomed Sci Appl 2000;745(2):421–30.
[372] Liu X, Xiang Z, Ma R. Crystal-form compound of antitumor drug nelarabine and preparation method thereof. Faming Zhuanli Shenqing; 2017. CN 106317150 A 20170111.
[373] Vidyadhara S, Sasidhar RLC, Rao BV, Saibabu T, Harika DL. A stability-indicating high performance liquid chromatographic method for the determination of Nelarabine. Orient J Chem 2016;32(1):601–7.
[374] Murakami H. New drugs of the world: 2005 (2). Fain Kemikaru 2006;35(11):82–92.
[375] Matutes E. Novel and emerging drugs for rarer chronic lymphoid leukaemias. Curr Cancer Drug Targets 2012;12(5):484–504.
[376] Corrigan DK, Salton NA, Preston C, Piletsky S. Towards the development of a rapid, portable, surface enhanced Raman spectroscopy based cleaning verification system for the drug nelarabine. J Pharm Pharmacol 2010;62(9):1195–200.
[377] Gu Q, Jin Z, Mi C, Jiang W, Li Z. Nelarabine injection and the preparation method thereof. Faming Zhuanli Shenqing; 2010. CN 101683321 A 20100331.
[378] Vigneron J, Astier A, Trittler R, Hecq JD, Daouphars M, Larsson I, Pourroy B, Pinguet F. SFPO and ESOP recommendations for the practical stability of anticancer drugs: an update. Ann Pharm Fr 2013;71(6):376–89.
[379] Xia R, Sun L. Efficient and green synthesis of purine arabinosides via CuO catalyzed dehydrazination in tap water. ARKIVOC (Gainesville, FL, USA) 2015;7:284–92.

CHAPTER 4

Pyrimidine-based anticancer drugs

1. Introduction

The pyrimidine analogs are a diverse group of agents with structural similarities but also slight differences in potencies and mechanisms of action. These antimetabolites interfere or compete with nucleoside triphosphates in the synthesis of RNA or DNA or both [1]. Many synthetic pyrimidine nucleoside analogs have been developed [2–7]. The main anticancer pyrimidine-based drugs, including the analogs of cytosine and uracil, are depicted in this chapter.

2. Cytosine analogs

2.1 Cytarabine (Ara-C, Cytosar U)

Cytarabine (Fig. 4.1) is considered as arabinose-derived nucleoside, 1-β-D-arabinofuranosyl cytosine, the 2′-epimer of cytidine. It is effective in many leukemias [8], such as acute myelogenous leukemia [9–12] and non-Hodgkin lymphoma.

The antitumor mechanisms of cytarabine are outlined in Fig. 4.2. Ara-C gets into the cell by nucleoside transport proteins, and once in the cytoplasm, demands activation for its cytotoxicity. Ara-C is transformed to Ara-cytidine monophosphate, through the deoxycytidine kinase, which is the rate-limiting step for the anabolism of Ara-C. Other steps transform the drug to aracytidine triphosphate, which inhibits DNA polymerase, delta DNA-polymerase, and beta DNA-polymerase. Cytarabine interferes with DNA elongation by various mechanisms. Ara-C is more potent in the S phase of the cell cycle [8].

Cytarabine is considered the conventional drug for acute myeloid leukemia (AML) treatment that targets the cell cycle or nuclear DNA [13–15]. Synergistic interactions between cytotoxic agents cytarabine and PNAs have been demonstrated in both preclinical and clinical investigations [16]. Many cytarabine derivatives have been synthesized, such as decitabine

New Strategies Targeting Cancer Metabolism
https://doi.org/10.1016/B978-0-12-821783-2.00006-6

Fig. 4.1 Cytarabine.

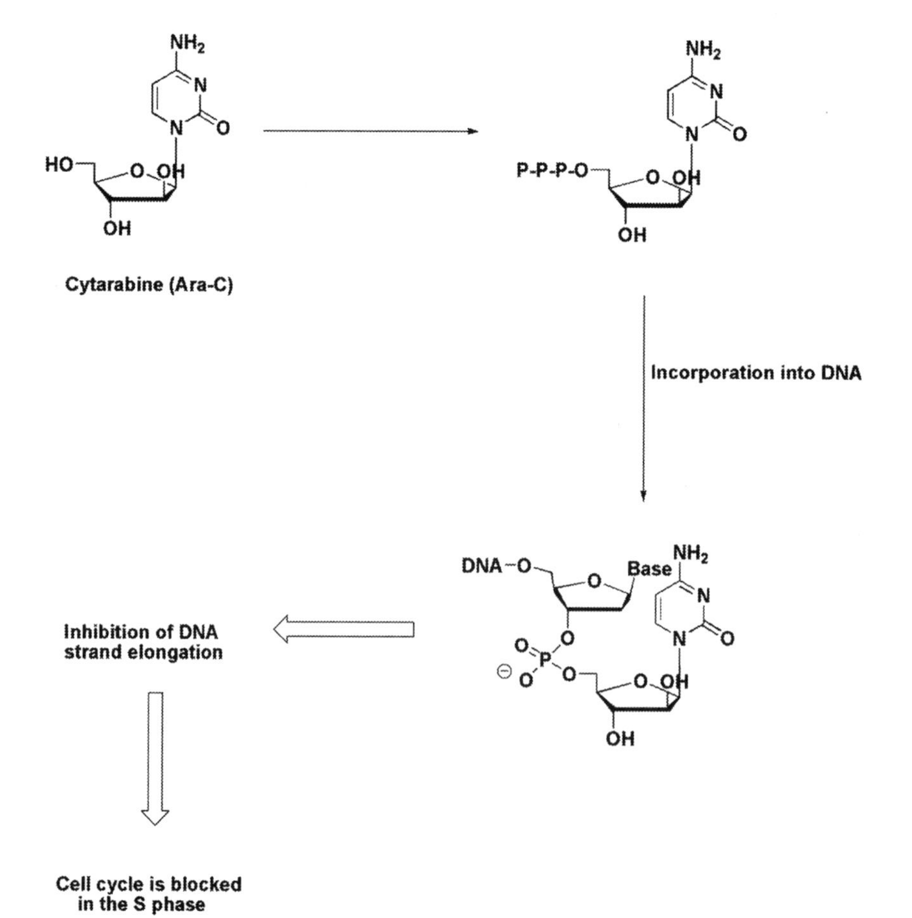

Fig. 4.2 The antitumor mechanisms of cytarabine.

[NSC 127716, dezocitidine, DAC, 2′-deoxy-5-azacytidine, Aza dC, Dacogen (trademark)], which is a cytarabine and deoxycytidine derivative with antileukemic potency. In vitro gene activation and cellular differentiation are induced via decitabine through a mechanism in which DNA hypomethylation is involved. Decitabine has been investigated in phase II trials for various types of leukemia and solid tumors. The drug has been revealed to possess very limited efficacy against solid tumors. However, it displays higher potency for treating hematological malignancies [17]. Additionally, cytarabine ocfosfate (trade name: Starasid) is cytarabine derivative, and it is an oral prodrug having stearyl group joined to phosphoric acid at 5′ position of arabinose moiety of Ara-C. The mode of action is the inhibition of DNA synthesis after transformation to Ara-CTP as in cytosine arabinoside (Ara-C). In the liver, the drug is metabolized, yielding the intermediate metabolite, C-C3PCA, which is transformed to Ara-C gradually [18]. Also, N^4-behenoyl ara-C (BHAC) [55726-47-1] has an activity against leukemia, and its mechanism of action in leukemic cells was studied [19].

Cytarabine is used in combination with other drugs such as idarubicin, which is at least as effective as daunorubicin for acute nonlymphocytic leukemia, and two studies show greater potency and longer survival with a cytarabine-idarubicin regimen than with a cytarabine-daunorubicin regimen. Other data revealed potency in breast cancer, acute lymphocytic leukemia, and lymphomas [20].

2.2 Elacytarabine

Elacytarabine (Fig. 4.3) is a new cytotoxic nucleoside analog. Its mechanism of action is similar to that of cytarabine [21]. It also shows a single-agent potency in patients suffering from advanced acute myeloid leukemia.

Elacytarabine (CP-4055) is a novel cytotoxic prodrug of cytarabine [22,23], ara-C-5′elaidic-acid-ester [24], and it is lipophilic fatty acid derivative of the nucleoside analog araC [25,26] and is considered as nucleoside

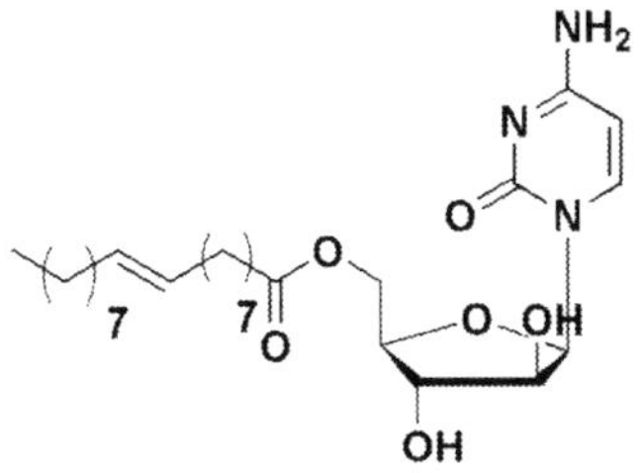

Fig. 4.3 Elacytarabine.

analog of cytosine arabinoside. Elacytarabine has the potential to possess a significant role in treating multiple malignancies in the future [27,28].

It is rationally designed to circumvent cytarabine resistance [29,30] associated with the decrease in cellular uptake, given the ability of the lipophilic drug moiety for the cell entry, and this could be done independently without requiring specialized nucleoside transporters or nuclear transport proteins, including the human equilibrative nucleoside transporter 1 [hENT1] [31–33]. Elacytarabine delivers higher intracellular levels of the active cytarabine metabolite in cells resistant to the parent drug [34]. It is active in cells with deficient nucleoside membrane transport and altered mismatch repair [25].

Elacytarabine was granted orphan drug designation status from the US FDA and the European Commission, with a fast-track approval designation from the FDA in 2010 [32]. It is currently utilized in the clinical trials in solid and hematological tumors [25,26].

Elacytarabine has monotherapy activity in patients with advanced acute myeloid leukemia (AML) [31,35]. Preclinical data were encouraging although the consequent clinical studies have not demonstrated success to indicate superiority of elacytarabine compared with standard of care as monotherapy in patients with AML. Clinical trials investigating anthracyclines in combination with elacytarabine are ongoing [36]. Elacytarabine is one of seven commonly utilized AML salvage regimens, including multiagent chemotherapy, high-dose cytarabine, supportive care hydroxyurea, and hypomethylating agents [37]. Recently, it was suggested that elacytarabine may show potency in patients with leukemia for whom cytarabine is ineffective [38]. Elacytarabine could be potent in some cases of non-small-cell lung cancer, and the combination of tyrosine kinase inhibitors and elacytarabine may show significant possible or additive synergistic effects [39]; it is also comprised in pharmaceutical components for cancer treatment [40].

2.3 Gemcitabine (dFdC, Gemfar)

Gemcitabine (Fig. 4.4), difluorodeoxycytidine [41], is a pyrimidine nucleoside analog of cytosine, which inhibits DNA synthesis and repair [42]. Gemcitabine is a novel antineoplastic agent [43,44]. It is considered as the leading marketed nucleoside analog and is utilized for treating pancreatic cancer, non-small-cell lung cancer [45–49], breast cancer, bladder cancer, and other carcinomas. The drug has an activity in non-Hodgkin and Hodgkin lymphomas and has been used in combination with other drugs

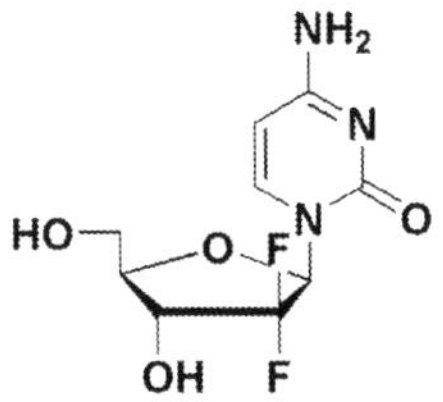

Fig. 4.4 Gemcitabine.

in salvage setting. Similar to cytarabine, gemcitabine blocks the cell cycle at the S phase. In addition, it is an inhibitor of ribonucleotide reductase in its diphosphate form.

Gemcitabine, 2′,2′-difluoro 2′-deoxycytidine (dFdC), is a deoxycytidine (dCyd) analog and the most remarkable analog developed since Ara-C. The evidence of its antitumor potency in the in vitro and in vivo tumor models has been confirmed clinically. Despite pharmacological and structural similarities to Ara-C, gemcitabine exhibits distinctive features of cellular pharmacology, mechanism of action, and metabolism [50]; gemcitabine is able to autoactivate its own mechanism of action [51].

Following influx via the cell membrane through nucleoside transporters, **gemcitabine** undergoes complex intracellular transformation to the nucleotides **gemcitabine** triphosphate (dFdCTP) and diphosphate (dFdCDP), which cause their cytotoxic actions. The cytotoxic potency of gemcitabine may be due to the various actions on DNA synthesis. The dFdCTP competes with the deoxycytidine triphosphate (dCTP) as an inhibitor of the DNA polymerase. dFdCDP is a potent inhibitor of ribonucleoside reductase, leading the deoxyribonucleotide pools to be depleted, which are essential for the synthesis of DNA and cause the potential effects of the dFdCTP. dFdCTP is incorporated into DNA, and after the incorporation of one more nucleotide, it results in DNA strand termination. This extra nucleotide may be essential in hiding dFdCTP from the DNA repair enzymes, as the incorporation of the dFdCTP into DNA appears to be resistant to normal mechanisms of the DNA repair [50]. Gemcitabine can be mainly inactivated through the action of deoxycytidine deaminase to the difluorodeoxyuridine [50,52]. Additionally, 5′-nucleotidase opposes the action of the nucleoside kinases via catalyzing transformation of the nucleotides to nucleosides [50].

In clinical trials, gemcitabine reveals an efficacy like the most potent drugs utilized in treating nonmicrocytic lung cancer. It exhibits an acceptable toxicity profile, in which the hematological toxicity is a significant side effect [43]. Gemcitabine has revealed potency in metastatic breast cancer not only as a single agent, but also in different combination regimens. It has a unique mechanism of action, which comprises masked DNA chain termination, cytotoxic self-potentiation, and potent inhibition of DNA repair [53]. Additionally, it has a remarkable single-agent activity in non-small-cell lung cancer (NSCLC). Gemcitabine and paclitaxel is an active and well-tolerated combination in advanced NSCLC. As both are cell cycle and phase specific in their mechanism of action, frequent exposure should optimize potency [54]. Gemcitabine has potency in solid tumors and leukemia, which needs phosphorylation via deoxycytidine kinase (dCK). Decreased membrane transport is a mechanism of resistance to gemcitabine. For facilitating the gemcitabine uptake and prolong retention in the cell, a lipophilic prodrug (CP-4126) was prepared, which is considered as a promising novel anticancer drug; it is a membrane transporter independent and has an elaidic fatty acid esterified at the 5′-position. Intraperitoneally administered CP-4126 was as active as gemcitabine in various xenografts [55].

Recently, RX-3117 is developed and it is an oral, small-molecule cytidine analog with an enhanced pharmacological profile relative to gemcitabine and other nucleoside analogs. It has an excellent potency against several cancer cell lines and xenografts, including gemcitabine-resistant variants, and it has an excellent oral bioavailability; it is not a substrate for the degradation enzyme cytidine deaminase. RX-3117 has revealed a clinical potency in pancreatic cancer and refractory bladder cancer [56].

2.4 Sapacitabine (CYC-682, CS-682)

Sapacitabin (1-[2-C-cyano-2-deoxy-β-D-arabino-pentfuranosyl]*N*4-palmitoyl cytosine; CYC682 or CS-682) (Fig. 4.5) is a novel 2′-deoxycytidine derivative [57]. It is an oral deoxycytidine nucleoside analog [58,59] and is considered as *N*4-palmitoyl derivative of an oral prodrug [8]. It is activated by intestinal and plasma amidases. This compound treats both acute myeloid

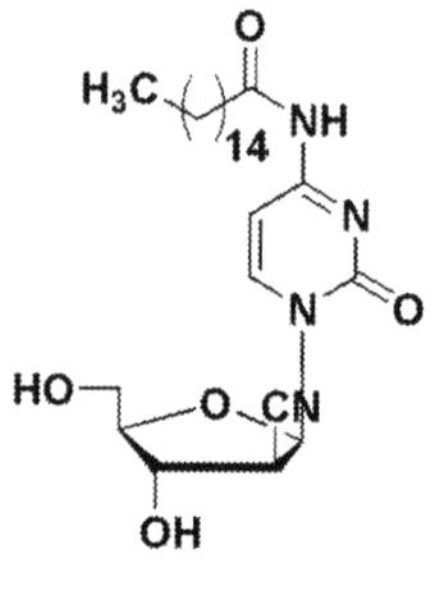

O
H_3C
14
NH
N
N
O
HO
O
CN
OH

Fig. 4.5 Sapacitabine.

leukemia and MDS. The application and clinical potency of sapacitabine differ from those of gemcitabine or cytarabine [8].

Sapacitabine has recently appeared as an effective agent due to its easy oral administration, acceptable toxicity profile, and good efficacy. Preclinical and clinical results have been very promising [60–62]. It has a unique mechanism of action that differ from that of cytarabine. Sapacitabine has potency against leukemia and (myelodysplastic syndromes) MDS. Phase II studies are ongoing [58,63]. Sapacitabine is utilized in acute myeloid leukemia (AML) maintenance therapy [64–74]. It is also used in the preparation of a medicament in treating a proliferative disorder [73,74] and in the preparation of a medicament in the treating cutaneous T-cell lymphoma (CTCL) [75]. Sapacitabine is still in single-agent phases of its development, and clinical experience is quickly accumulated [76]. Sapacitabine exhibited an antiproliferative potency across different concentrations in various cell lines, comprising those indicated to be resistant to various anticancer drugs. Sapacitabine is biotransformed via plasma, gut, and liver amidases into CNDAC and leads to cell cycle arrest predominantly in G_2/M phase. The correlation between sensitivity to sapacitabine and the expression of critical factors involved in resistance to nucleoside analogs as human equilibrative nucleoside transporter 1, cytosolic 5′-nucleotidase, DNA polymerase-α, and deoxycytidine kinase (dCK) was not observed. However, sapacitabine revealed a cytotoxic potency against dCK-deficient L1210 cells, revealing that in some cells, a dCK-independent mechanism of action may be involved. Sapacitabine indicated a synergistic effect when combined with gemcitabine and a sequence-specific synergy with oxaliplatin and doxorubicin [77].

Sapacitabine is an orally bioavailable prodrug of the nucleoside analog 2’-C-cyano-2′-deoxy-1-β-D-arabino-pentofuranosylcytosine (CNDAC) [the active metabolite of the anticancer drug, sapacitabine] [78]. Both the active metabolite and the prodrug are used in clinical trials for solid tumors and hematological malignancies. CNDAC has a unique mechanism of action after being incorporated into DNA; it induces single-strand breaks (SSBs) that are transformed into the double-strand breaks (DSBs) when the cells go through the second S phase [79,80]. Its unique action distinguishes CNDAC from other deoxycytidine analogs [81,82]. This lesion needs homologous recombination (HR) for repair [83]. Sapacitabine indicates a marked reduction in the water solubility as the result of the fatty acid side chain on the N4 group of the cytosine moiety. Poor water solubility is one of the remarkable reasons by which it does not exert the maximum antitumor potency [84].

2.5 Azacitidine (Vidaza)

Azacitidine (Fig. 4.6) and its 2′-deoxy analog are triphosphorylated. They could be misincorporated into the nucleic acids.

5′-Azacitidine is considered as an analog of cytosine, varying from the natural nucleoside, as it has a nitrogen in lieu of carbon in the 5 position of the pyrimidine. Although it is being developed from 40 years ago approximately, it has only recently been utilized at low doses with success in treating the myelodysplastic syndromes (MDS). It has a hypomethylating potency; this drug exerts its action via the reinduction of the expression of the genes silenced through hypermethylation of the CpG islands in their promoters. It was the first agent employed to delay acute myeloblastic leukemia transformation and to prolong survival for patients with higher-risk MDS, and it was approved in 2004 by the US FDA for treating all MDS risk categories. Azacitidine allows transfusion independence in more than 40% of the treated MDS patients, and has opened an era in treating MDS and the utility of epigenetic drugs [85]. 5AZA-CdR is a very active antileukemic agent in experimental animals, which is more potent than the related antileukemic drug, cytosine arabinoside [86].

2.6 Fazarabine (Ara-AC)

Fazarabine (Fig. 4.7) is an aza analog. It shows observed potency in animal models, including solid tumors; thus, it has been submitted to clinical trials [8].

Fazarabine is a nucleoside analog comprising the pyrimidine base of 5-azacytidine and the arabinoside ring of 1-β-D-arabinofuranosylcytosine [87]. It combines the triazine base of 5-azacytidine with the arabinose sugar of cytosine arabinoside. It revealed potency against a variety of human solid tumor xenografts, including lung, colon, and breast cancers [88]. A phase I trial of fazarabine (ara-AC, 1-β-d-arabinofuranosyl-5-azacytosine, NSC

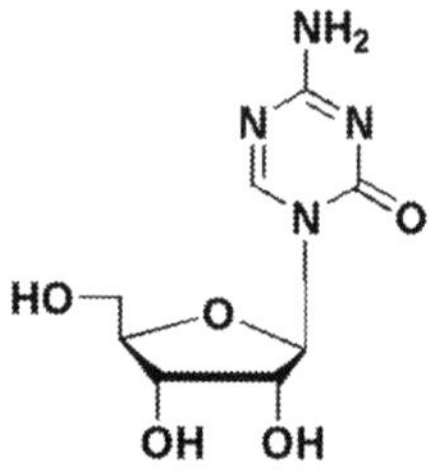

Fig. 4.6 Azacitidine.

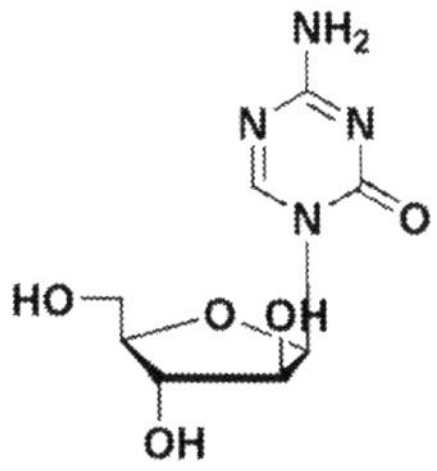

Fig. 4.7 Fazarabine.

281272) was performed in adults with solid tumor malignancies [89]. It is broadly active in preclinical tumor screening evaluations [90]. Fazarabine has revealed potency in the panel of 60-cultured human tumor lines of the National Cancer Institute [91]. Fazarabine has shown a broad spectrum of antitumor potency in experimental models [92,93]. A phase I trial of fazarabine was performed in children with refractory malignancies [94] and in adult patients with refractory cancer [95]. The clinical pharmacology of fazarabine was assessed in patients with several malignancies during a phase I trial [96]. Phase II trial of fazarabine was conducted in patients with advanced pancreatic adenocarcinoma [97]. Fazarabine has a broad-spectrum preclinical activity and was chosen for study in patients with incurable non-small-cell carcinoma of the lung. Some patients with metastatic non-small-cell lung cancer were treated with fazarabine. Fazarabine has no demonstrable potency on the metastatic non-small-cell carcinoma of the lung [98]. Phase II trial of fazarabine was carried out on metastatic breast cancer [99], advanced colorectal carcinoma [100], and advanced head and neck cancer [101].

3. Uracil analogs

3.1 5-Fluorouracil

5-Fluorouracil (5-FU) (Fig. 4.8) is a well-known primary chemotherapy drug from fluoropyrimidines (FPs). It is FDA-approved clinical cytostatic

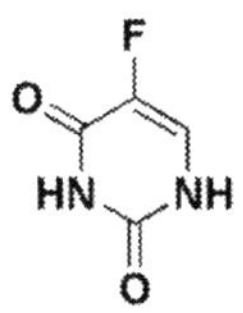

Fig. 4.8 5-Fluorouracil.

and cytotoxic drug, and it is considered as a typical chemotherapy drug that remains the backbone of treating several solid tumors [102,103]. It is a major component of various chemotherapy regimens [104]. It inhibits cell division by affecting DNA synthesis. It is one of the most promising anticancer drugs applied against breast cancer, stomach cancer, and many other ones [105]. 5-FU has showed a potential effect to treat premalignant and malignant lesions [106]. It is proposed that 5-FU has anticancer features via the interference with nucleotide synthesis and incorporation into DNA [107]; 5-fluorouracil inhibits thymidylate synthase (TS) and causes DNA damage through the misincorporation of dUTP and FdUTP into DNA under the conditions of dTTP depletion [108,109]. 5FU activity is strongly reduced against a subset of cancer cells termed cancer stem cells (CSCs), which are believed to be essential for chemoresistance and tumor recurrence [110], so drug resistance has greatly affected the clinical utility of 5-FU [111]. The multidrug resistance caused by long-term use and the nonspecific cytotoxicity caused the limitation of the clinical efficacy of 5-FU [112–116]. It may cause severe or life threatening side effects, and the treatment-related mortality rate is 0.2%–1.0% [117,118].

It has been utilized for treating colon cancer. 5-FU significantly reduces tumor burden, but the progression of metastasis with the phenotype of drug resistance is a main obstacle in successful anticancer therapy [119–125]; myelosuppression is one of the most frequent adverse effects [126].

5-Fluorouracil is considered as the second-line chemotherapy of gallbladder cancer (GBC) [127]. Also, it has been widely utilized in treating laryngeal squamous cell carcinoma (LSCC), the second most common head and neck squamous cell carcinoma [128], locally advanced rectal cancer (LARC) [129], breast cancer [130], gastric cancer (GC) [131–140], and locally advanced gastric cancer (LAGC) [141,142]. A fraction of GC patients acquire 5-Fu chemoresistance [143].

5-Fluorouracil-based chemotherapy is the first-line antineoplastic agent chemotherapy and the backbone of chemotherapy regimens approved for treating colorectal cancer (CRC) [130,144–150]. Many studies presented were to observe the therapeutic effect of fluorouracil combined with other drugs on CRC to explore its mechanism of action [151–153]. Resistance to chemotherapy is a main reason of treatment failure of CRC, and it is a well-known fact that cancer stem cells play a remarkable role in the acquisition of drug resistance [154–160]. So despite its effectiveness in CRC therapy, its clinical applications are restricted as 5-FU has various limitations, including its low bioavailability, high cytotoxicity, and short half-life [161]. In order to overcome the drawbacks of 5-FU and improve its therapeutic efficiency, many scientists have focused on designing a novel delivery system to deliver 5-FU to tumor sites. Innovative approaches and investigations were studied on FU for skin [162,163], brain [164,165], hyperkeratotic actinic keratosis [166], locoregional advanced nasopharyngeal carcinoma (NPC) [167], esophageal [168–171], ovarian [172], liver [173,174], hepatoblastoma (HB) [175], and pancreatic cancer (PC) [176,177]. The synergistic antitumor effect of 5-FU and a natural flavonoid, kaempferol, has been studied. Kaempferol-fluorouracil cocrystal has the potential to be developed as an efficient oral formulation of drug combination, which will overcome the weaknesses of each parent drug [178].

3.2 Capecitabine

Capecitabine (Fig. 4.9) is a tumor-selective fluoropyrimidine carbamate with antineoplastic properties being used in treating gastrointestinal and breast cancers. It is prodrug of 5-deoxy-5-fluorouridine, which is enzymatically transformed to 5-FU [8].

Capecitabine from oral fluoropyrimidines is used in treating solid tumors in Europe [179,180]. Its mode of action has been demonstrated to include TS inhibition [181]. Capecitabine has largely replaced 5-FU in different indications, comprising gastric cancer [182]. It is transformed to its only potent metabolite, FU, via thymidine phosphorylase. Capecitabine has indicated varying degrees of efficacy with acceptable tolerability in various cancers, including ovarian, prostate, pancreatic, and renal cells, with the largest amount of evidence in colorectal and metastatic breast cancer [183].

Capecitabine was developed as a prodrug of FU, with the goal of enhancing tolerability and intratumor drug concentrations via tumor-specific transformation to the active drug. Capecitabine is currently

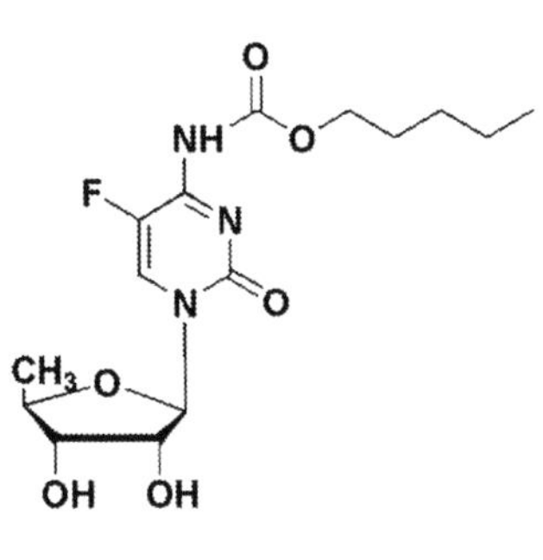

O
NH O
F N
N O
CH_3 O
OH OH

Fig. 4.9 Capecitabine.

approved by the FDA as first-line therapy for metastatic colorectal cancer. The drug has also been licensed for utility as a single agent in metastatic breast cancer patients who are resistant to both paclitaxel- and anthracycline-based regimens or in whom further anthracycline treatment is contraindicated and in combination with docetaxel after failure of prior anthracycline-based chemotherapy. Combination regimens and single agent have also indicated benefits in patients with pancreatic, prostate, ovarian, and renal cell cancers [183].

3.3 Tegafur

Tegafur (Fig. 4.10), an anticancer 5-fluorouracil prodrug [184], is an oral slow-release prodrug formulation of fluorouracil, which is readily absorbed through the gastrointestinal tract [185].

S-1 (Teysuno), an oral formulation containing the tegafur and the two enzyme modulators oteracil and gimeracil, has not been available in non-Asian countries until recently. In Japan, S-1 combined with cisplatin is the first-line treatment recommended in patients with gastric cancer [186]. The second-generation oral anticancer agent UFT, a combination of tegafur (TGF) and uracil, leads to a higher 5-FU concentration in the tumor tissues than that resulted via TGF or comparable doses of intravenous 5-FU. UFT has been studied clinically on gastric carcinoma. UFT is one of the first second-generation oral 5-FU prodrugs under research in Europe and North America. UFT is well tolerated and has cellular pharmacokinetic superiority over the first-generation 5-FU prodrug TGF. UFT has a more favorable toxicity profile than intravenous 5-FU [187].

3.4 Doxifluridine

Doxifluridine (Fig. 4.11), prodrug of fluorouracil, is composed of sugar-like moiety attached to the molecule of fluorouracil. It possesses a higher therapeutic index for various types of cancer.

Doxifluridine (5′-DFUR) is from the fluoropyrimidine drugs (FPs) and is located among the most widely used anticancer agents in the treatment of solid tumors [188,189]. Various methods for the preparation of doxifluridine

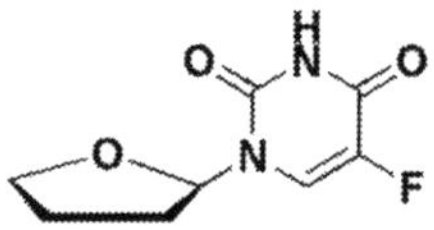

O H N O N O F

Fig. 4.10 Tegafur.

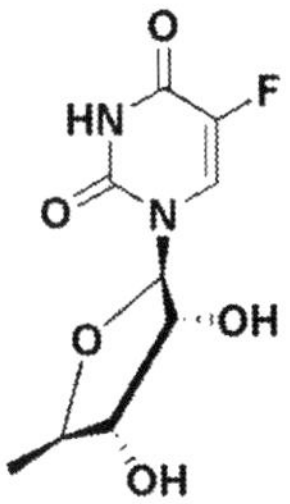

Fig. 4.11 Doxifluridine.

were invented [190–192], and other doxifluridine pharmaceutical compounds contain doxifluridine and a natural product; when the **doxifluridine** and this natural product are doing their function independently, it has an effect of treating liver damage resulted from the leukemic liver infiltration; when the two types of substance are combined, they have a stronger effect, and they can be developed into the drug utilized in the treatment of liver damage; compared with the present technology, it has an outstanding substantive characteristic and remarkable progression [193].

Additionally, doxifluridine has many applications in the treatment of cholangiocarcinoma [194], hepatocellular carcinoma [195], and gastric cancer [196].

3.5 Floxuridine (5-FUdR)

Floxuridine (Fig. 4.12) is a deoxynucleoside analog of 5-fluorouracil [197]. 5-Fluoro-2′-deoxyuridine is a chemotherapeutic nucleoside analog employed for treating colorectal cancer [198,199].

Various potent drugs in market contain pyrimidine nucleus like the anticancer agent floxuridine (FUdR, 5-fluoro-2-deoxyuridine) (antineoplastic). [200–202]. It is a very effective drug with high potency in treating numerous tumors but its usage is limited by its low cellular uptake efficiency [203]. It is considered as a thymidylate synthase (TS) inhibitor that is utilized against solid tumors [204]. Floxuridine acts as an inhibitor of DNA replication through binding to thymidylate synthase and is widely utilized in treating

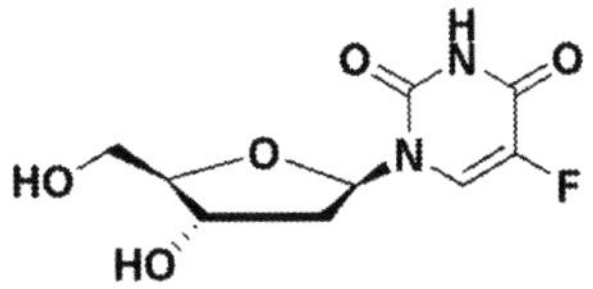

Fig. 4.12 Floxuridine.

colorectal cancer [205]. It was widely utilized in patients with tumor. But the severe side effects and poor activity have been estimated clinically, which resulted from the increased degradation cleavage of the FUdR to 5-FU via thymidine phosphorylase and reduced transporter-mediated entry into cells [206]. Because of the resistance and toxicity of anticancer drugs, different methods have been estimated to enhance their effects in treating cancer. Since boron nitride nanotubes (BNNTs) are nontoxic and biocompatible, they can act as fine drug delivery vehicles of these anticancer drugs to deliver them directly to the target cells. Floxuridine encapsulated into the boron nitride nanotube is estimated [207,208]. Based on the similarity in structure between the natural nucleoside thymidine (T) and the nucleoside analog floxuridine (F), F can be incorporated into the nucleic acid strands through either enzyme-mediated transcription or solid-phase synthesis [209,210]. Floxuridine combined with other drugs was considered as most efficient combination chemotherapy in cancer [211–214].

3.6 Carmofur

Carmofur (HCFU, Mifurol) 1-hexylcarbamoyl-5-fluorouracil (Fig. 4.13), oral fluoropyrimidine, is an antineoplastic 5-fluorouracil analog [215–219] and is widely utilized clinically either alone or in combination therapy [220]. Carmofur is an effective chemotherapeutic agent applied to various cancers [221]. It is clinically utilized as a chemotherapeutic agent and as a dual NAAA and FAAH inhibitor [222]. It has been widely utilized in Japan as a postoperative adjuvant chemotherapy agent for breast and colorectal cancer [223]. It was chosen as a practical antitumor drug [224]. The effects of carmofur were evaluated in preventing postoperative recurrence of the bladder cancer [225]. **Carmofur** is clinically used in colorectal cancers [226–236] and is considered as a promising drug to undergo subsequent animal studies and further clinical trials for treating pediatric patients suffering from brain tumors [226]. Carmofur is used in treating colon cancer [236,237], either in combination with other chemotherapeutic agents or

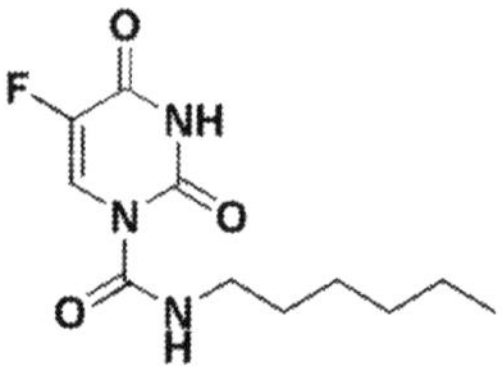

Fig. 4.13 Carmofur.

as a single agent [238]. Adjuvant utility of carmofur is useful for early breast cancer [239], and it was found that it has an antitumor effect on human gastric cancer [240,241]. Additionally, carmofur is clinically beneficial in thermo-chemotherapy [242]. Recently, it has been reported to possess a serious and infrequent association with leukoencephalopathy [243–246]. Recently, carmofur is revealed to inhibit the SARS-CoV-2 main protease (M(pro)). Carmofur inhibits viral replication in cells and is a promising lead compound to develop a novel antiviral therapy for COVID-19 [247]. Many other studies and applications have been performed on carmofur [248–266].

3.7 Trifluridine

Trifluridine (FTD) (Fig. 4.14), a tri-fluorinated thymidine-based analog, a thymidine-based chemotherapeutic, is a key component of the new orally administered nucleoside analog-type chemotherapeutic antitumor drug TAS-102 (TFTD), which consists of a thymidine phosphorylase inhibitor (TPI) and FTD [267–276] that enhances the bioavailability of FTD [277]. Trifluridine has a limited bioavailability after clinical administration as it is rapidly degraded via thymidine phosphorylase [278]. TAS-102 is effective in the therapy of some solid tumors. Trifluridine induces DNA fragmentation as it is incorporated into DNA molecules. Its anticancer potency takes place even after oral administration. Tipiracil is a nucleoside analog with a modified pyrrolidine system present instead of the sugar moiety. Tipiracil prevents trifluridine transformation to an inactive metabolite. The mechanism of this trifluridine protection is based on the inhibition of thymidine phosphorylase and, consequently, increasing trifluridine bioavailability. Tipiracil also acts as a platelet-derived endothelial cell growth factor and provides an indirect antiangiogenic impact that also has positive effects in cancer therapy of solid tumors [279]. After tumor progression with first- and second-line treatment, trifluridine (FTD) and tipiracil (TPI) have been revealed to be an option for treatment [280,281]. Trifluridine (FTD) targets thymidylate synthase (TS) that inhibits its activity and their nucleotides are

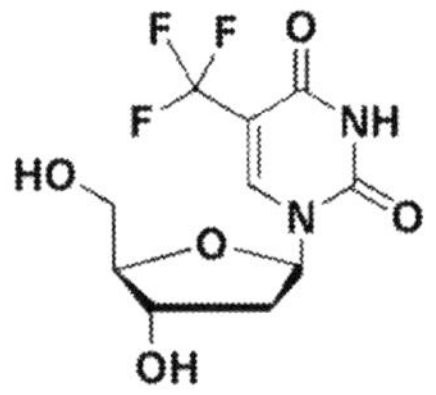

Fig. 4.14 Trifluridine.

incorporated into DNA [282]. FTD/TPI is a novel oral cytotoxic chemotherapy comprising trifluridine and a thymidine phosphorylase inhibitor, tipiracil [283,284]. Trifluridine/tipiracil (FTD/TPI) acts through affecting the DNA of tumor cells [285]. The nucleoside FTD/TPI or TAS-102 (Taiho Oncol., Lonsurf, Princeton, NJ, USA) is a combination tablet of tipiracil and trifluridine, a thymidine phosphorylase inhibitor, in a 1:0.5 M ratio [286,287]. Trifluridine (TFT) has a potent antiherpes simplex virus activity [288]. This drug was first approved for use in metastatic colorectal cancer (mCRC) patients [289–317]. Trifluridine or a trifluridine tipiracil compound is used in the preparation of a medicament used in the treatment of blood diseases. It is specifically involved in the preparation of a medicament in the treatment of blood diseases caused by gene mutation or β-Hb gene defect [318]. Trifluridine or trifluridine-tipiracil compound has applications in the treatment of blood diseases [319]. FTD/TPI is approved and well established worldwide for treating patients with metastatic gastrointestinal cancer [320,321] and advanced gastric cancer [322–324]; it also indicates a promising antitumor potency in heavily pretreated patients with advanced biliary tract carcinoma [325]. Trifluridine displays anticancer potencies after its oral administration despite its hydrophilic nature [326]. The synergistic effects of in vitro FTD/TPI and 5-fluoruracil (5-FU) are studied [327]. Sequential combinations of 5-fluorouracil and trifluridine /tipiracil (TAS-102) [328], and other combinations of capecitabine and trifluridine/tipiracil, which are synergistic in colon cancer [329], were studied.

References

[1] LiverTox. Clinical and research information on drug-induced liver injury [Internet]. Bethesda (MD): National Institute of Diabetes and Digestive and Kidney Diseases; 2012. Pyrimidine Analogues. [Updated 16 April 2017].

[2] Elgemeie GH, Salah AM, Abbas NS, Hussein HA, Mohamed RA. Nucleic acid components and their analogs: design and synthesis of novel cytosine thioglycoside analogs. Nucleosides Nucleotides 2017;36:139–50.

[3] Elgemeie GH, Alkhursani SA, Mohamed RA. New synthetic strategies for acyclic and cyclic pyrimidinethione nucleosides and their analogues. Nucleosides Nucleotides Nucleic Acids 2019;38(1):1–76.

[4] Elgemeie GH, Mohamed RA. Microwave chemistry: synthesis of purine and pyrimidine nucleosides using microwave radiation. J Carbohydr Chem 2019;38(1):1–47.

[5] Elgemeie GH, Mohamed RA, Hussein HA, Jones PG. Crystal structure of N-(2-amino-5-cyano-4-methylsulfanyl-6-oxo-1,6-dihydropyrimidin-1-yl)-4-bromobenzenesulfonamide dimethylformamide monosolvate. Acta Crystallogr E Crystallogr Commun 2015;71:1322–4.

[6] Elgemeie GH, Salah AM, Mohamed RA, Jones PG. Crystal structure of (E)-2-amino-4-methylsulfanyl-6-oxo-1-{[(thiophen-2-yl)methylidene]amino}-1,6-dihydropyrimidine-5-carbonitrile. Acta Crystallogr E Crystallogr Commun 2015; 71:1319–21.
[7] Elgemeie GH, Salah AM, Abbas NS, Hussein HA, Mohamed RA. Pyrimidine non-nucleoside analogs: a direct synthesis of a novel class of N-substituted amino and N-sulfonamide derivatives of pyrimidines. Nucleosides Nucleotides Nucleic Acids 2017;36(3):213–23.
[8] Avendano C, Menendez JC. Medicinal chemistry of anticancer drugs. 2nd ed. Elsevier Science; 2015. p. 72.
[9] Österroos A, Kashif M, Haglund C, Blom K, Höglund M, Andersson C, Gustafsson MG, Eriksson A, Larsson R. Combination screening in vitro identifies synergistically acting KP372-1 and cytarabine against acute myeloid leukemia. Biochem Pharmacol 2016;118:40–9.
[10] Wiernik PH, Goldman JM, Dutcher JP, Robert A, Kyle RA. Neoplastic diseases of the blood. 5th ed. Springer; 2013. p. 381.
[11] Handin RI, Lux SE, Stossel TP. Blood: principles and practice of hematology. 2nd ed. vol. 1. Philadelphia, PA: Lippincott Williams & Wilkins; 2003. p. 508.
[12] Skeel RT. Handbook of cancer chemotherapy. 7th ed; 2007. p. 460.
[13] Wei-ping Y, Juan L. Novel agents inhibit human leukemic cells. Acta Pharmacol Sin 2012;33(2):210–1.
[14] Freeman C, Keane N, Swords R, Giles F. Vosaroxin: a new valuable tool with the potential to replace anthracyclines in the treatment of AML? Expert Opin Pharmacother 2013;14(10):1417–27.
[15] Liang A, Zhang H, Fang Y, Li Y, Zhu D, Chen F, Lu H, Fu J, Bo L. Sequential arabinosylcytosin with or without fludarabine in paracmastic patients with acute myeloid leukemia. Pharmazie 2012;67(7):635–8.
[16] Robak T, Korycka A, Kasznicki M, Wrzesien-Kus A, Smolewski P. Purine nucleoside analogues for the treatment of hematological malignancies: pharmacology and clinical applications. Curr Cancer Drug Targets 2005;5(6):421–44.
[17] Anonymous. Decitabine: 2′-deoxy-5-azacytidine, Aza dC, DAC, dezocitidine, NSC 127716. Drugs R&D 2003;4(6):352–8.
[18] Tsukagoshi S. A new antileukemic drug, cytarabine ocfosfate. Gan to kagaku ryoho. Cancer Chemother 1993;20(12):1877–83.
[19] Nakamura T, Ueda T. Pharmacokinetics and action mechanism of acyl ara-C in leukemia chemotherapy. In: Ishigami J, editor. Recent Adv. Chemother., Proc. Int. Congr. Chemother., 14th, Anticancer Sect, vol. 1; 1985. p. 109–12.
[20] Cersosimo RJ. Idarubicin: an anthracycline antineoplastic agent. Clin Pharm 1992;11 (2):152–67.
[21] Ashton Acton Q. Acute myeloid leukemia: new insights for the healthcare professional. ScholarlyEditions; 2013. p. 148.
[22] O'Brien S, Rizzieri DA, Vey N, Ravandi F, Krug UO, Sekeres MA, Dennis M, Venditti A, Berry DA, Jacobsen TF, et al. Elacytarabine has single-agent activity in patients with advanced acute myeloid leukaemia. Br J Haematol 2012;158(5):581–8.
[23] Peters GJ. Novel developments in the use of antimetabolites. Nucleosides Nucleotides Nucleic Acids 2014;33(4–6):358–74.
[24] Adema AD, Smid K, Losekoot N, Honeywell RJ, Verheul HM, Myhren F, Sandvold ML, Peters GJ. Metabolism and accumulation of the lipophilic deoxynucleoside analogs elacytarabine and CP-4126. Invest New Drugs 2012;30(5):1908–16.
[25] Sandvold ML, Galmarini C, Myhren F, Peters G. The activity of the lipophilic nucleoside derivatives elacytarabine and CP-4126 in a panel of tumor cell lines resistant to nucleoside analogues. Nucleosides Nucleotides Nucleic Acids 2010;29(4–6):386–93.

[26] Giles FJ, Vey N, Rizzieri D, Ravandi F, Prebet T, Borthakur G, Jacobsen TF, Hagen S, Nilsson B, O'Brien S. Phase I and pharmacokinetic study of elacytarabine, a novel 5′-elaidic acid derivative of cytarabine, in adults with refractory hematological malignancies. Leukemia 2012;26(7):1686–9.

[27] Burke AC, Giles FJ. Elacytarabine—lipid vector technology overcoming drug resistance in acute myeloid leukemia. Expert Opin Investig Drugs 2011;20(12):1707–15.

[28] Pignata S, Amant F, Scambia G, Sorio R, Breda E, Rasch W, Hernes K, Pisano C, Leunen K, Lorusso D, et al. A phase I-II study of elacytarabine (CP-4055) in the treatment of patients with ovarian cancer resistant or refractory to platinum therapy. Cancer Chemother Pharmacol 2011;68(5):1347–53.

[29] Keane N, Freeman C, Swords R, Giles FJ. Elacytarabine: lipid vector technology under investigation in acute myeloid leukemia. Expert Rev Hematol 2013;6(1):9–24.

[30] Adema AD, Losekoot N, Smid K, Kathmann I, Myhren F, Sandvold ML, Peters GJ. Induction of resistance to the lipophilic cytarabine prodrug elacytarabine (CP-4055) in CEM leukemic cells. Nucleosides Nucleotides Nucleic Acids 2010;29(4–6):394–9.

[31] Knapper S, Chevassut T, Duarte R, Bergua JM, Salamero O, Johansen M, Jacobsen TF, Hals P, Rasch W, Gianella-Borradori A, et al. Elacytarabine in relapsed/refractory acute myeloid leukaemia: an evaluation of clinical efficacy, pharmacokinetics, cardiac safety and effects on lipid profile. Leuk Res 2014;38(3):346–51.

[32] DiNardo CD, O'Brien S, Gandhi VV, Ravandi F. Elacytarabine (CP-4055) in the treatment of acute myeloid leukemia. Future Oncol 2013;9(8):1073–82.

[33] Rizzieri D, Vey N, Thomas X, Huguet-Rigal F, Schlenk RF, Krauter J, Kindler T, Gjertsen BT, Blau IW, Jacobsen TF, et al. A phase II study of elacytarabine in combination with idarubicin and of human equilibrative nucleoside transporter 1 expression in patients with acute myeloid leukemia and persistent blasts after the first induction course. Leuk Lymphoma 2014;55(9):2114–9.

[34] Burke A, Swords R, Kelly K, Giles FJ. Elacytarabine: antimetobolite oncolytic. Drugs Future 2009;34(12):941–9.

[35] DiNardo CD, O'Brien S, Gandhi VV, Ravandi F. Elacytarabine (CP-4055) in the treatment of acute myeloid leukemia [Erratum to document cited in CA159:508176]. Future Oncol 2015;11(10):1584.

[36] Rein LAM, Rizzieri DA. Clinical potential of elacytarabine in patients with acute myeloid leukemia. Ther Adv Hematol 2014;5(6):211–20.

[37] Roboz GJ, Rosenblat T, Arellano M, Gobbi M, Altman JK, Montesinos P, O'Connell C, Solomon SR, Pigneux A, Vey N, et al. International randomized phase III study of elacytarabine versus investigator choice in patients with relapsed/refractory acute myeloid leukemia. J Clin Oncol Off J Am Soc Clin Oncol 2014;32(18):1919–26.

[38] Giles F, Rizzieri D, Ravandi F, Swords R, Jacobsen TF, O'Brien S. Elacytarabine, a novel 5′-elaidic acid derivative of cytarabine, and idarubicin combination is active in refractory acute myeloid leukemia. Leuk Res 2012;36(4):e71–3.

[39] Bruheim S, Sandvold ML, Malandsmo GM, Fodstad O. Antitumor activity of elacytarabine combined with bevacizumab, cetuximab and trastuzumab in human NSCLC xenografts. Anticancer Res 2013;33(9):3615–21.

[40] Ahrabi S, Myhren F, Eriksen OH. Parenteral formulations of cytarabine derivatives for cancer treatment. PCT Int Appl 2012. WO 2012008845 A1 20120119.

[41] Bergman AM, Pinedo HM, Peters GJ. Determinants of resistance to 2′,2′-difluorodeoxycytidine (gemcitabine). Drug Resist Updat 2002;5:19–33.

[42] Thomas AL, Cox G, Sharma RA, Steward WP, Shields F, Jeyapalan K, Muller S, O'Byrne KJ. Gemcitabine and paclitaxel associated pneumonitis in non-small cell lung cancer: report of a phase I/II dose-escalating study. Eur J Cancer 2000;36:2329–34.

[43] Vazquez Polo A, Najera Perez MD, Diaz Carrasco MS, San Miguel Zamora MT, Is RG. Gemcitabine and its role in nonmicrocytic lung cancer. Farm Clin 1997;14 (1):36–42.
[44] Shimizu T, Fukuoka M. Gemcitabine. Igaku no Ayumi 2005;215(5):342–6.
[45] Von Hoff DD, Evans DB, Hruban RH. Pancreatic cancer. Sudbury, MA: Jones and Bartlett; 2005. p. 453.
[46] Chang AE, Ganz PA, Hayes DF, Kinsella T, Pass HI, Schiller JH, Strecher VJ. Oncology: an evidence-based approach. Internal Medicine; 2006. p. 775.
[47] Kaufman HL, Wadler S, Antman K. Molecular targeting in oncology. Humana Press; 2008. p. 211.
[48] Wolpin BM. Pancreatic cancer: hematology/oncology clinics of North America. The Clinics: Internal Medicine; 2015. p. 763.
[49] Cavalli F, Kaye' SB, Hansen HH, Armitage JO, Piccart-Gebhart M. Textbook of medical oncology. 4th ed; 2009. p. 165.
[50] Balboni B, El Hassouni B, Honeywell RJ, Sarkisjan D, Giovannetti E, Poore J, Heaton C, Peterson C, Benaim E, Lee YB, et al. RX-3117 (fluorocyclopentenyl cytosine): a novel specific antimetabolite for selective cancer treatment. Expert Opin Investig Drugs 2019;28(4):311–22.
[51] Sugiyama E, Kaniwa N, Kim SR, Kikura-Hanajiri R, Hasegawa R, Maekawa K, Saito Y, Ozawa S, Sawada J, Kamatani N, Furuse J, Ishii H, Yoshida T, Ueno H, Okusaka T, Saijo N. Pharmacokinetics of gemcitabine in Japanese cancer patients: the impact of a cytidine deaminase polymorphism. J Clin Oncol 2007;25(1):32–42.
[52] Bergman AM, Adema AD, Balzarini J, Bruheim S, Fichtner I, Noordhuis P, Fodstad O, Myhren F, Sandvold ML, Hendriks HR, et al. Antiproliferative activity, mechanism of action and oral antitumor activity of CP-4126, a fatty acid derivative of gemcitabine, in in vitro and in vivo tumor models. Invest New Drugs 2011;29(3):456–66.
[53] Mini E, Nobili S, Caciagli B, Landini I, Mazzei T. Cellular pharmacology of gemcitabine. Ann Oncol 2006;17(Suppl. 5):v7–12.
[54] Heinemann V. Gemcitabine in metastatic breast cancer. Expert Rev Anticancer Ther 2005;5(3):429–43.
[55] Gillenwater HH, Stinchcombe TE, Qaqish BF, Tyann M, Hensing TA, Socinski MA. A phase II trial of weekly paclitaxel and gemctiabine infused at a constant rate in patients with advanced non-small cell lung cancer. Lung Cancer 2005;47(3):413–9.
[56] Milano G, Chamorey A, Thyss A. Clinical pharmacology of nucleoside analogues bulletin du cancer; 2002. 89 Spec. No. S71–5.
[57] Serova M, Galmarini CM, Ghoul A, Benhadji K, Green SR, Chiao J, Faivre S, Cvitkovic E, Le Tourneau C, Calvo F, et al. Antiproliferative effects of sapacitabine (CYC682), a novel 2′-deoxycytidine-derivative, in human cancer cells. Br J Cancer 2007;97(5):628–36.
[58] Kantarjian H, Garcia-Manero G, O'Brien S, Faderl S, Ravandi F, Westwood R, Green SR, Chiao JH, Boone PA, Cortes J, et al. Phase I clinical and pharmacokinetic study of oral sapacitabine in patients with acute leukemia and myelodysplastic syndrome. J Clin Oncol 2010;28(2):285–91.
[59] Muggia F, Diaz I, Peters GJ. Nucleoside and nucleobase analogs in cancer treatment: not only sapacitabine, but also gemcitabine. Expert Opin Investig Drugs 2012;21 (4):403–8.
[60] Lim MY, Jamieson K. Profile of sapacitabine: potential for the treatment of newly diagnosed acute myeloid leukemia in elderly patients. Clin Interv Aging 2014;9:753–62.
[61] Norkin M, Richards AI. Sapacitabine in the treatment of acute myeloid leukemia. Expert Rev Anticancer Ther 2015;15(11):1261–6.

[62] Galmarini CM, Popowycz F, Joseph B. Cytotoxic nucleoside analogues: different strategies to improve their clinical efficacy. Curr Med Chem 2008;15(11):1072–82.
[63] Gianella-Borradori A, Chiao J. Methods of treatment using sapacitabine. U.S. Pat. Appl. Publ.; 2009. US 20090118315 A1 20090507.
[64] Chiao JH. Dosing regimen of acute myeloid leukemia comprising administering decitabine and sapacitabine or metabolite thereof. U.S. Pat. Appl. Publ.; 2020. US 20200129509 A1 20200430.
[65] Chiao J. Dosing regimen. PCT Int. Appl.; 2019. WO 2019081951 A1 20190502.
[66] Robak T, Szmigielska-Kaplon A, Pluta A, Grzybowska-Izydorczyk O, Wolska A, Czemerska M, Wierzbowska A. Novel and emerging drugs for acute myeloid leukemia: pharmacology and therapeutic activity. Curr Med Chem 2011;18(5):638–66.
[67] Czemerska M, Robak T, Wierzbowska A. The efficacy of sapacitabine in treating patients with acute myeloid leukemia. Expert Opin Pharmacother 2018;19(16):1835–9.
[68] Mccloskey J, Koprivnikar J, Faderl S. Sapacitabine in acute myelogenous leukemia expert opinion on orphan. Drugs 2015;3(12):1461–8.
[69] Chiao JH. Dosage regimen for sapacitabine and decitabine in combination for treating acute myeloid leukemia. U.S. Pat. Appl. Publ.; 2014. US 20140094428 A1 20140403.
[70] Green SR, Mackay R, Fleming IN. Combinations of sapacitabine or cndac with DNA methyltransferase inhibitors such as decitabine and procaine; 2013. US 8530445 B2 20130910.
[71] Chiao J. Dosage regimen for sapacitabine and decitabine in combination for treating acute myeloid leukemia. PCT Int. Appl; 2012. WO 2012140436 A1 20121018.
[72] Green SR, Choudhary AK, Fleming IN. Combination of sapacitabine and HDAC inhibitors stimulates cell death in AML and other tumour types. Br J Cancer 2010;103(9):1391–9.
[73] Burnett AK, Russell N, Hills RK, Panoskaltsis N, Khwaja A, Hemmaway C, Cahalin P, Clark RE, Milligan D. A randomised comparison of the novel nucleoside analogue sapacitabine with low-dose cytarabine in older patients with acute myeloid leukaemia. Leukemia 2015;29(6):1312–9.
[74] Chiao JH, Blake D, Zheleva D, Davis S, Green S, Shapiro G. Dosage regimen for sapacitabine and seliciclib. U.S. Pat. Appl. Publ; 2015. US 20150164933 A1 20150618.
[75] Chiao J. Use of sapacitabine to treat proliferative disease. PCT Int. Appl.; 2008. WO 2008132443 A1 20081106.
[76] Faderl S, Gandhi V, Kantarjian HM. Potential role of novel nucleoside analogs in the treatment of acute myeloid leukemia. Curr Opin Hematol 2008;15(2):101–7.
[77] Galmarini CM. Drug evaluation: sapacitabine—an orally available antimetabolite in the treatment of cancer. Curr Opin Investig Drugs (Thomson Sci) 2006;7(6):565–73.
[78] Al Abo M, Sasanuma H, Liu X, Rajapakse VN, Huang S, Kiselev E, Takeda S, Plunkett W, Pommier Y. TDP1 is critical for the repair of DNA breaks induced by sapacitabine, a nucleoside also targeting ATM- and BRCA-deficient tumors. Mol Cancer Ther 2017;16(11):2543–51.
[79] Liu X, Nowak B, Wang Y, Plunkett W. Sapacitabine, prodrug of CNDAC, is a nucleoside analog with a unique action mechanism of inducing DNA strand breaks. Chin J Cancer 2012;31(8):373–80.
[80] Liu X, Matsuda A, Plunkett W. Ataxia-telangiectasia and Rad3-related and DNA-dependent protein kinase cooperate in G2 checkpoint activation by the DNA strand-breaking nucleoside analogue 2'-C-cyano-2′-deoxy-1-β-D-arabino-pentofuranosylcytosine. Mol Cancer Ther 2008;7(1):133–42.
[81] Liu X, Jiang Y, Nowak B, Hargis S, Plunkett W. Mechanism-based drug combinations with the DNA strand-breaking nucleoside analog CNDAC. Mol Cancer Ther 2016;15(10):2302–13.

[82] Lai T, Ewald B, Zecevic A, Liu C, Sulda M, Papaioannou D, Garzon R, Blachly JS, Plunkett W, Sampath D. HDAC inhibition induces microRNA-182, which targets RAD51 and impairs HR repair to sensitize cells to sapacitabine in acute myelogenous leukemia. Clin Cancer Res 2016;22(14):3537–49.
[83] Liu X, Jiang Y, Nowak B, Qiang B, Cheng N, Chen Y, Plunkett W. Targeting BRCA1/2 deficient ovarian cancer with CNDAC-based drug combinations. Cancer Chemother Pharmacol 2018;81(2):255–67.
[84] Obata T, Suzuki Y, Ogawa N, Kurimoto I, Yamamoto H, Furuno T, Sasaki T, Tanaka M. Improvement of the antitumor activity of poorly soluble sapacitabine (CS-682) by using soluplus as a surfactant. Biol Pharm Bull 2014;37(5):802–7.
[85] Santini V. Azacitidine: activity and efficacy as an epigenetic treatment of myelodysplastic syndromes. Expert Rev Hematol 2009;2(2):121–7.
[86] Momparler RL. Pharmacology of 5-aza-2′-deoxycytidine (decitabine). Semin Hematol 2005;42(3 Suppl. 2):S9–16.
[87] Goldberg RM, Reid JM, Ames MM, Sloan JA, Rubin J, Erlichman C, Kuffel MJ, Fitch TR. Phase I and pharmacological trial of fazarabine (Ara-AC) with granulocyte colony-stimulating factor. Clin Cancer Res 1997;3(12, Pt. 1):2363–70.
[88] Ben-Baruch N, Denicoff AM, Goldspiel BR, O'Shaughnessy JA, Cowan KH. Phase II study of fazarabine (NSC 281272) in patients with metastatic colon cancer. Invest New Drugs 1993;11(1):71–4.
[89] Bailey H, Tutsch KD, Arzoomanian RZ, Tombes MB, Alberti D, Bruggink J, Wilding G. Phase I clinical trial of fazarabine as a twenty-four-hour continuous infusion. Cancer Res 1991;51(4):1105–8.
[90] Heideman RL, McCully C, Balis FM, Poplack DG. Cerebrospinal fluid pharmacokinetics and toxicology of intraventricular and intrathecal arabinosyl-5-azacytosine (fazarabine, NSC 281272) in the nonhuman primate. Invest New Drugs 1993;11(2–3):135–40.
[91] Barchi JJJ, Cooney DA, Ahluwalia GS, Gharehbaghi K, Covey JM, Hochman I, Paull KD, Jayaram HN. Studies on the mechanism of action of 1-β-D-arabinofuranosyl-5-azacytosine (fazarabine) in mammalian lymphoblasts. J Exp Ther Oncol 1996;1(3):191–203.
[92] Wilhelm M, O'brien S, Rios MB, Estey E, Keating MJ, Plunkett W, Sorenson M, Kantarjian HM. Phase I study of arabinosyl-5-azacytidine (fazarabine) in adult acute leukemia and chronic myelogenous leukemia in blastic phase. Leuk Lymphoma 1999;34(5/6):511–8.
[93] Amato R, Ho D, Schmidt S, Krakoff IH, Raber M. Phase I trial of a 72-h continuous-infusion schedule of fazarabine. Cancer Chemother Pharmacol 1992;30(4):321–4.
[94] Heideman RL, Gillespie A, Ford H, Reaman GH, Balis FM, Tan C, Sato J, Ettinger LJ, Packer RJ, Poplack DG. Phase I trial and pharmacokinetic evaluation of fazarabine in children. Cancer Res 1989;49(18):5213–6.
[95] Surbone A, Ford HJ, Kelley JA, Ben-Baruch N, Thomas RV, Fine R, Cowan KH. Phase I and pharmacokinetic study of arabinofuranosyl-5-azacytosine (fazarabine, NSC 281272). Cancer Res 1990;50(4):1220–5.
[96] Ho DH, Brown N, Lin JR, Covington W, Newman RA, Raber M, Amato R, Schmidt S, Krakoff IH. Clinical pharmacology of 1-beta-D-arabinofuranosyl-5-azacytosine (fazarabine) following 72-hour infusion. Drug Metab Dispos 1991;19(3):643–7.
[97] Casper ES, Schwartz GK, Kelsen DP. Phase II trial of fazarabine (arabinofuranosyl-5-azacytidine) in patients with advanced pancreatic adenocarcinoma. Invest New Drugs 1992;10(3):205–9.
[98] Williamson SK, Crowley JJ, Livingston RB, Panella TJ, Goodwin JW. Phase II trial and cost analysis of fazarabine in advanced non-small cell carcinoma of the lung: a Southwest Oncology Group study. Invest New Drugs 1995;13(1):67–71.

[99] Walters RS, Theriault RL, Holmes FA, Hortobagyi GN, Esparza L. Phase II trial of fazarabine (ARA-AC, arabinosyl-5-azacytosine) in metastatic breast cancer. Invest New Drugs 1992;10(1):43–4.
[100] Hubbard KP, Daugherty K, Ajani JA, Pazdur R, Levin B, Abbruzzese JL. Phase II trial of fazarabine in advanced colorectal carcinoma. Invest New Drugs 1992;10(1):39–42.
[101] Kuebler JP, Metch B, Schuller DE, Keppen M, Hynes HE. Phase II study of fazarabine in advanced head and neck cancer. A Southwest Oncology Group study. Invest New Drugs 1991;9(4):373–4.
[102] Ramakrishna P, Harish BK. Preparation and invitro evaluation of succinyl chitosan blended methyl cellulose microspheres for controlled release of 5-fluorouracil. World J Pharm Pharm Sci 2020;9(6):1454–67.
[103] Biagioni A, Staderini F, Peri S, Versienti G, Schiavone N, Cianchi F, Papucci L, Magnelli L. 5-Fluorouracil conversion pathway mutations in gastric cancer. Biology (Basel, Switzerland) 2020;9(9):265.
[104] Divsalar A, Ghobadi R. The presence of deep eutectic solvents of reline and glyceline on interaction and side effect of anti-cancer drug of 5-fluorouracil: Bovine liver catalase as a target. J Mol Liq 2021;323, 114588.
[105] Ewert de Oliveira B, Junqueira Amorim OH, Lima LL, Rezende RA, Mestnik NC, Bagatin E, Leonardi GR. Five fluorouracil innovative drug delivery systems to enhance bioavailability for topical use. J Drug Delivery Sci Technol 2021;61, 102155.
[106] Christensen S, Van der Roest B, Besselink N, Janssen R, Boymans S, Martens J, Yaspo M, Priestley P, Kuijk E, Cuppen E, et al. 5-Fluorouracil treatment induces characteristic T>G mutations in human cancer. bioRxiv 2019;1–41.
[107] Yokogawa T, Yano W, Tsukioka S, Osada A, Wakasa T, Ueno H, Hoshino T, Yamamura K, Fujioka A, Fukuoka M, et al. dUTPase inhibition confers susceptibility to a thymidylate synthase inhibitor in DNA-repair-defective human cancer cells. Cancer Sci 2021;112(1):422–32.
[108] Lin E, Huang C. Crystal structure of the single-stranded DNA-binding protein SsbB in complex with the anticancer drug 5-fluorouracil: extension of the 5-fluorouracil interactome to include the oligonucleotide/oligosaccharide-binding fold protein. Biochem Biophys Res Commun 2021;534:41–6.
[109] Moracci L, Crotti S, Traldi P, Agostini M. Mass spectrometry in the study of molecular complexes between 5-fluorouracil and catechins. J Mass Spectrom 2021;56(1), e4682.
[110] Sethy C, Kundu CN. 5-Fluorouracil (5-FU) resistance and the new strategy to enhance the sensitivity against cancer: implication of DNA repair inhibition. Biomed Pharmacother 2021;137, 111285.
[111] Xu T, Guo P, Pi C, He Y, Yang H, Hou Y, Feng X, Jiang Q, Wei Y, Zhao L. Synergistic effects of curcumin and 5-fluorouracil on the hepatocellular carcinoma in vivo and vitro through regulating the expression of COX-2 and NF-κB. J Cancer (Wyoming, Australia) 2020;11(13):3955–64.
[112] Yao X, Tu Y, Xu Y, Guo Y, Yao F, Zhang X. Endoplasmic reticulum stress confers 5-fluorouracil resistance in breast cancer cell via the GRP78/OCT4/lncRNA MIAT/AKT pathway. Am J Cancer Res 2020;10(3):838–55.
[113] Wang Y, Li G, Li Z, Zhong S. Comparison of efficacy between docetaxel + fluorouracil + cisplatin regimen and docetaxel + fluorouracil + lobaplatin regimen in the treatment of gastric cancer. Xiandai Zhongliu Yixue 2020;28(10):104–6.
[114] Alnuqaydan AM, Rah B, Almutary AG, Chauhan SS. Synergistic antitumor effect of 5-fluorouracil and withaferin-A inducesendoplasmic reticulum stress-mediated autophagy and apoptosis in colorectal cancer cells. Am J Cancer Res 2020;10(3):799–815.
[115] Zhang G, Luo X, Zhang W, Chen E, Xu J, Wang F, Cao G, Ju Z, Jin D, Huang X, et al. CXCL-1 3 regulates resistance to 5-fluorouracil in colorectal cancer. Cancer Res Treat 2020;52(2):622–33.

[116] Woermann B, Bokemeyer C, Burmeister T, Koehne C, Schwab M, Arnold D, Blohmer J, Borner M, Brucker S, Cascorbi I, et al. Dihydropyrimidine dehydrogenase testing prior to treatment with 5-fluorouracil, capecitabine, and Tegafur: a consensus paper. Oncol ResTreat 2020;43(11):628–36.
[117] Naren G, Wang L, Zhang X, Cheng L, Yang S, Yang J, Guo J, Nashun B. The reversible reproductive toxicity of 5-fluorouracil in mice. Reprod Toxicol 2021;101:1–8.
[118] Waghela BN, Vaidya FU, Dave G, Pathak C. Inhibition of NADPH oxidase activity augments 5-fluorouracil mediated cell death in human colon carcinoma HCT-116 cells. Int J Adv Res 2020;8(7):865–74.
[119] Wang B, Ma N, Zheng X, Li X, Ma X, Hu J, et al. GDF15 repression contributes to 5- fluorouracil resistance in human colon cancer by regulating epithelial-mesenchymal transition and apoptosis. Biomed Res Int 2020;2020, 2826010.
[120] Genovese S, Epifano F, Preziuso F, Slater J, Nangia-Makker P, Majumdar APN, Fiorito S. Gercumin synergizes the action of 5-fluorouracil and oxaliplatin against chemoresistant human cancer colon cells. Biochem Biophys Res Commun 2020;522(1):95–9.
[121] Wu X, Huang X, Lin X, Li Y, Chen Y, Huang Z. Data mining the prognostic biomarkers: miRNAs targeting the chemotherapeutic agent 5-fluorouracil in colon cancer. Clin Lab(Mainz, Germany) 2020;66(3):375–81.
[122] Ouyang M, Chen X, Zhu D, Zhang W, Luo Z, Liu C. Role of twist gene in resistance to oxaliplatin and fluorouracil of colon cancer Lovo cells. Guangdong Yixue 2016;37 (22):3354–8.
[123] Feng X, Wu J. Assessment of long-term recurrence, metastasis and blood indexes after sustained-released fluorouracil implantation in radical operation for colon cancer. Hainan Yixueyuan Xuebao 2016;22(19):2298–301.
[124] Othman MH, Zayed GM, Ali UF, Abdellatif AAH. Colon-specific tablets containing 5-fluorouracil microsponges for colon cancer targeting. Drug Dev Ind Pharm 2020;46 (12):2081–8.
[125] Ishibashi M, Ishii M, Yamamoto S, Mori Y, Shimizu S. Possible involvement of TRPM2 activation in 5-fluorouracil-induced myelosuppression in mice. Eur J Pharmacol 2021;891, 173671.
[126] Lingyue L, Xueli B, Tingbo L. Advances in systematic treatment of gallbladder cancer. Zhongguo Zhongliu Linchuang 2020;47(21):14–9.
[127] Ding S, Tang Z, Jiang Y, Luo P, Qing B, Wei Y, Zhang S, Tang R. HDAC1 regulates the chemosensitivity of laryngeal carcinoma cells via modulation of interleukin-8 expression. Eur J Pharmacol 2021;896, 173923.
[128] Cristobal I, Rubio J, Santos A, Torrejon B, Carames C, Imedio L, Mariblanca S, Luque M, Sanz-Alvarez M, Zazo S, et al. MicroRNA-199b downregulation confers resistance to 5-fluorouracil treatment and predicts poor outcome and response to neoadjuvant chemoradiotherapy in locally advanced rectal cancer patients. Cancers 2020;12(6):1655.
[129] Saif MW. Alternative treatment options in patients with colorectal cancer who encounter fluoropyrimidine-induced cardiotoxicity. Onco Targets Ther 2020;13:10197–206.
[130] Kim DS, Min K, Lee SK. Cell cycle dysregulation is associated with 5-fluorouracil resistance in gastric cancer cells. Anticancer Res 2020;40(6):3247–54.
[131] Shan T, Sun J, Li W. Clinical study on Shenlian capsules combined with ECF chemotherapy regimen in treatment of advanced gastric cancer. Xiandai Yaowu Yu Linchuang 2020;35(9):148–51.
[132] Zhou W, Chen Q, Luo H. Efficacy and safety of XELOX and FOLFOX4 regimens in treatment of patients with advanced gastric carcinoma. Shiyong Linchuang Yiyao Zazhi 2016;20(9):64–6.

[133] Li W, Zhao P. Effects comparison of docetaxel in combination with oxaliplatin + S-1 and docetaxel in combination with cisplatin + fluorouracil on advanced gastric cancer. Hainan Yixueyuan Xuebao 2016;22(17):2015–8.
[134] Li H, Lv J, Guo J, Wang S, Liu S, Ma Y, Liang Z, Wang Y, Qi W, Qiu W. 5-Fluorouracil enhances the chemosensitivity of gastric cancer to TRAIL via inhibition of the MAPK pathway. Biochem Biophys Res Commun 2021;540:108–15.
[135] Monti M, Morgagni P, Nanni O, Framarini M, Saragoni L, Marrelli D, Roviello F, Petrioli R, Romario UF, Rimassa L, et al. Preoperative or perioperative docetaxel, oxaliplatin, and capecitabine (GASTRODOC regimen) in patients with locally-advanced resectable gastric cancer: a randomized phase-II trial. Cancer 2020;12 (10):2790.
[136] Sah BK, Zhang B, Zhang H, Li J, Yuan F, Ma T, Shi M, Xu W, Zhu Z, Liu W, et al. Neoadjuvant FLOT versus SOX phase II randomized clinical trial for patients with locally advanced gastric cancer. Nat Commun 2020;11(1):6093.
[137] Li M, Chen H, He J, Xie J, Xia J, Liu H, Shi Y, Guo Z, Yan H. A qualitative classification signature for post-surgery 5-fluorouracil-based adjuvant chemoradiotherapy in gastric cancer. Radiother Oncol 2021;155:65–72.
[138] Sugisawa N, Nishino H, Higuchi T, Park JH, Yamamoto J, Tashiro Y, Kawaguchi K, Bouvet M, Unno M, Hoffman RM. A gemcitabine plus 5-fluorouracil combination inhibits gastric-cancer liver metastasis in a PDOX model: a novel treatment strategy. Anticancer Res 2020;40(10):5393–7.
[139] Zhan Z, Wang X, Yu J, Zheng J, Huang Y, Guo Z. Efficacy and safety of fluorouracil/ leucovorin-paclitaxel-oxaliplatin regimen (POF) in first-line treatment of advanced gastric signet ring cell carcinoma. Xiandai Zhongliu Yixue 2020;28(12):114–8.
[140] Kulkarni V, Thungappa SC, Patil S, Sarathy V, Krishnamurthy KP, Kumar R, Naik R. Safety and efficacy of neoadjuvant DOF [Docetaxel, Oxaliplatin, 5-fluorouracil] chemotherapy regimen in patients with locally advanced gastric and gastro-esophageal junction cancers: a single center experience from India. J Cancer Ther 2020;11 (5):237–50.
[141] Liu W, Zhang H, Ma J, Cheng Y, Ran J, Li Y. Observation of hypofractionated three dimensional conformal intensity modulated radiation therapy for locally advanced pancreatic carcinoma. Linchuang Zhongliuxue Zazhi 2016;21(9):826–30.
[142] Gao S, Song D, Liu Y, Yan H, Chen X. Helicobacter pylori CagA protein attenuates 5-Fu sensitivity of gastric cancer cells through upregulating cellular glucose metabolism. Onco Targets Ther 2020;13:6339–49.
[143] Smith T, Affram K, Nottingham EL, Han B, Amissah F, Krishnan S, Trevino J, Agyare E. Application of smart solid lipid nanoparticles to enhance the efficacy of 5-fluorouracil in the treatment of colorectal cancer. Sci Rep 2020;10(1):16989.
[144] Qiao G, Yang B, Zhang Q, Sun Y, Wu Y, Mao L. Effect of selective regional infusion chemotherapy combined with regional implantation of fluorouracil sustained-release agent on postoperative prognosis of patients with advanced colorectal cancer. Guangdong Yixue 2016;37(23):3590–3.
[145] Qin S, Deng Y, Bi F, Liu T, Liu Y, Zhang S, Xu J, Shu Y, Xu N, Wu C, et al. Efficacy and safety of bevacizumab in combination with fluoropyrimidine-based chemotherapy for the treatment of advanced metastatic colorectal cancer: a prospective, non-intervention and post-marketing multicenter clinical study (REACT). Linchuang Zhongliuxue Zazhi 2016;21(10):865–73.
[146] Zhao Y, Feng X, Chen Y, Selfridge JE, Gorityala S, Du Z, Wang JM, Hao Y, Cioffi G, Conlon RA, et al. Fluorouracil enhances the antitumor activity of the glutaminase inhibitor CB-839 against PIK3CA-mutant colorectal cancers. Cancer Res 2020;80 (21):4815–27.

[147] Entezar-Almahdi E, Mohammadi-Samani S, Tayebi L, Farjadian F. Recent advances in designing 5-fluorouracil delivery systems: a stepping stone in the safe treatment of colorectal cancer. Int J Nanomedicine 2020;15:5445–58.
[148] Shahi S, Ang C, Mathivanan S. A high-resolution mass spectrometry-based quantitative metabolomic workflow highlights defects in 5-fluorouracil metabolism in cancer cells with acquired chemoresistance. Biology (Basel, Switzerland) 2020;9(5):96.
[149] Lin P, Lee M, Huang C, Tai T, Chen J, Chen C, Su Y. Synergistic antiproliferative effect of Ribociclib (LEEO11) and 5-fluorouracil on human colorectal cancer. Anticancer Res 2020;40(11):6265–71.
[150] Kadowaki S, Masuishi T, Ura T, Sugiyama K, Mitani S, Narita Y, Taniguchi H, Muro K. A triplet combination of FOLFOXIRI plus cetuximab as first-line treatment in RAS wild-type, metastatic colorectal cancer: a dose-escalation phase Ib study. Int J Clin Oncol 2021; [Ahead of Print].
[151] Zhang H, Chen M, Liu Y, Dong X, Zhang C, Jiang H, Chen X. Paroxetine combined with fluorouracil plays a therapeutic role in mouse models of colorectal cancer with depression through inhibiting IL-22 expression to regulate the MAPK signaling pathway. Exp Ther Med 2020;20(6):240.
[152] Wei L, Chen J, Wen J, Wu D, Ma X, Chen Z, Huang J. Efficacy of oxaliplatin/5-fluorouracil/capecitabine-cetuximab combination therapy and its effects on K-Ras mutations in advanced colorectal cancer. Med Sci Monit 2020;26, e919031.
[153] Raigorodskaya MP, Turchinovich A, Tsypina IM, Zgoda VG, Nikulin SV, Maltseva DV. Laminin 521 modulates the cytotoxic effect of 5-fluorouracil on HT29 colorectal cancer cells. Appl Biochem Microbiol 2020;56(8):870–4.
[154] Ukai S, Sakamoto N, Taniyama D, Harada K, Honma R, Maruyama R, Naka K, Hinoi T, Takakura Y, Shimizu W, et al. KHDRBS3 promotes multi-drug resistance and anchorage-independent growth in colorectal cancer. Cancer Sci 2021. https://doi.org/10.1111/cas [Ahead of Print].
[155] Chen K, Chen Y, Ueng S, Hwang T, Kuo L, Hsieh P. Neutrophil elastase inhibitor (MPH-966) improves intestinal mucosal damage and gut microbiota in a mouse model of 5-fluorouracil-induced intestinal mucositis. Biomed Pharmacother 2021;134, 111152.
[156] Li S, Zheng S. Down-regulation of circ_0032833 sensitizes colorectal cancer to 5-fluorouracil and oxaliplatin partly depending on the regulation of miR-125-5p and MSI1. Cancer Manag Res 2020;12:11257–69.
[157] Xian Z, Hu B, Wang T, Zeng J, Cai J, Zou Q, Zhu P. lncRNA UCA1 contributes to 5-fluorouracil resistance of colorectal cancer cells through miR-23b-3p/ZNF281 axis. Onco Targets Ther 2020;13:7571–83.
[158] Cho Y, Ro EJ, Yoon J, Mizutani T, Kang D, Park J, Il Kim T, Clevers H, Choi K. 5-FU promotes stemness of colorectal cancer via p53-mediated WNT/β-catenin pathway activation. Nat Commun 2020;11(1):5321.
[159] Kishore C, Bhadra P. Current advancements and future perspectives of immunotherapy in colorectal cancer research. Eur J Pharmacol 2021;893, 173819.
[160] Wang L, Wang R, Wei G, Zhang R, Zhu Y, Wang Z, Wang S, Du G. Cryptotanshinone alleviates chemotherapy-induced colitis in mice with colon cancer via regulating fecal-bacteria-related lipid metabolism. Pharmacol Res 2021;163, 105232.
[161] Osman AM, Al-Johani HS, Kamel FO, Abdel-Hakim AAO, Huwait EA, Sayed-Ahmed MM. 5-Fluorouracil and simvastatin loaded solid lipid nanoparticles for effective treatment of colorectal cancer cells. Int J Pharm 2020;16(3):205–13.
[162] Patel G, Yadav BKN. Study of 5-fluorouracil loaded chitosan nanoparticles for treatment of skin cancer. Recent Pat Nanotechnol 2020;14(3):210–24.

[163] Rata DM, Cadinoiu AN, Atanase LI, Popa M, Mihai C, Solcan C, Ochiuz L, Vochita G. Topical formulations containing aptamer-functionalized nanocapsules loaded with 5-fluorouracil—an innovative concept for the skin cancer therapy. Korean J Couns Psychother 2021;119, 111591.

[164] Shinde G, Shiyani S, Shelke S, Chouthe R, Kulkarni D, Marvaniya K. Enhanced brain targeting efficiency using 5-FU (fluorouracil) lipid-drug conjugated nanoparticles in brain cancer therapy. Prog Biomater 2020;9(4):259–75.

[165] Zheng R, Ma D, Li Z, Zhang H. MiR-145 regulates the chemoresistance of hepatic carcinoma cells against 5-fluorouracil by targeting toll-like receptor 4. Cancer Manag Res 2020;12:6165–75.

[166] Moore AY, Nguyen M, Moore S. Cyclic calcipotriene 0.005% foam and 1% 5-fluorouracil cream after cryotherapy in treatment of hyperkeratotic actinic keratosis: a retrospective study. J Am Acad Dermatol 2020. https://doi.org/10.1016/j.jaad.2020.07.010 [Ahead of Print].

[167] Zhan Z, Tao H, Qiu W, Liu Z, Zhang R, Liao K, Li G, Yuan Y, Yuan T, Zheng R. Clinical value of nedaplatin-based chemotherapy combined with radiotherapy for locoregional advanced nasopharyngeal carcinoma: a retrospective, propensity score-matched analysis. J Cancer (Wyoming, Australia) 2020;11(23):6782–9.

[168] Takahashi K, Osaka Y, Ota Y, Watanabe T, Iwasaki K, Tachibana S, Nagakawa Y, Katsumata K, Tsuchida A. Phase II study of docetaxel, cisplatin, and 5-fluorouracil chemoradiotherapy for unresectable esophageal cancer. Anticancer Res 2020;40 (5):2827–32.

[169] Huang Y, Wang L, Liu Z, Liu L. Aspirin promotes apoptosis of esophageal cancer cells induced by 5-FU through inhibiting β-catenin/EMT signaling pathway. Zhongguo Yaolixue Tongbao 2020;36(5):73–7.

[170] Tanishima Y, Nishikawa K, Arakawa Y, Matsumoto A, Yuda M, Tanaka Y, Mitsumori N, Yanaga K. Five-year outcomes of chemotherapy with docetaxel, cisplatin, and 5-fluorouracil followed by oesophagectomy in oesophageal cancer. Anticancer Res 2020;40(10):5829–35.

[171] Jiang DM, Sim H, Espin-Garcia O, Chan BA, Natori A, Lim CH, Moignard S, Chen EX, Liu G, Darling G, et al. Chemoradiotherapy using carboplatin plus paclitaxel versus cisplatin plus fluorouracil for esophageal or gastroesophageal junction cancer. Oncology 2021;99(1):49–56.

[172] Zhen S, Liang G. Carboplatin and 5-fluorouracil in treating stage III-IV ovarian cancer with peritoneal hyperthermic perfusion chemotherapy. Zhongguo Yaoye 2020;29 (2):67–9.

[173] Chen Y. Mechanism of astragaloside II combined with 5-fluorouracil in inhibiting liver cell proliferation. Hainan Yixueyuan Xuebao 2016;22(17):2035–8.

[174] Zhi L. 5-Fluorouracil compounds. U.S. Pat. Appl. Publ; 2020. US 20200399227 A1 20201224.

[175] Almeida JZ, Lima LF, Vieira LA, Maside C, Ferreira ACA, Araujo VR, Duarte ABG, Raposo RS, Bao SN, Campello CC, et al. 5-Fluorouracil disrupts ovarian preantral follicles in young C57BL6J mice. Cancer Chemother Pharmacol 2021. https://doi.org/10.1007/s00280-020-04217-7 [Ahead of Print].

[176] El-Mahdy HA, El-Husseiny AA, Kandil YI, Gamal El-Din AM. Diltiazem potentiates the cytotoxicity of gemcitabine and 5-fluorouracil in PANC-1 human pancreatic cancer cells through inhibition of P-glycoprotein. Life Sci 2020;262, 118518.

[177] Pishvaian MJ, Wang H, He AR, Hwang JJ, Smaglo BG, Kim SS, Weinberg BA, Weiner LM, Marshall JL, Brody JR. A phase I/II study of veliparib (ABT-888) in combination with 5-fluorouracil and oxaliplatin in patients with metastatic pancreatic cancer. Clin Cancer Res 2020;26(19):5092–101.

[178] Lv W, Liu X, Dai X, Long X, Chen J. A 5-fluorouracil-kaempferol drug-drug cocrystal: a ternary phase diagram, characterization and property evaluation. CrstEngComm 2020;22(46):8127–35.

[179] Uemura M, Yamamoto H, Haraguchi N, Takemasa I, Mizushima T, Ikeda M, Ishii H, Sekimoto M, Doki Y, Mori M. Chemotherapy and molecular targeting therapeutics for colorectal cancer. Sogo Rinsho 2009;58(9):1965–71.

[180] Aiba K. 5-fluorouracil and oral fluoropyrimidines. Igaku no Ayumi 2005;215 (5):333–41.

[181] Kenny LM, Contractor KB, Stebbing J, Al-Nahhas A, Palmieri C, Shousha S, Coombes RC, Aboagye EO. Altered tissue 3′-deoxy-3′-[18F]fluorothymidine pharmacokinetics in human breast cancer following capecitabine treatment detected by positron emission tomography. Clin Cancer Res 2009;15(21):6649–57.

[182] Schellens JHM. Capecitabine. Oncologist 2007;12(2):152–5.

[183] Walko CM, Lindley C. Capecitabine: a review. Clin Ther 2005;27(1):23–44.

[184] dos Santos PM, Hall AJ, Manesiotis P. Stoichiometric molecularly imprinted polymers for the recognition of anti-cancer pro-drug tegafur. J Chromatogr B 2016;1021: 197–203.

[185] Kohne CH, Peters GJ. UFT: mechanism of drug action. Oncology (Williston Park, NY) 2000;14(10 Suppl 9):13–8.

[186] Mahlberg R, Lorenzen S, Thuss-Patience P, Heinemann V, Pfeiffer P, Mohler M. New perspectives in the treatment of advanced gastric cancer: S-1 as a novel oral 5-FU therapy in combination with cisplatin. Chemotherapy 2017;62(1):62–70.

[187] Takiuchi H, Ajani JA. Uracil-tegafur in gastric carcinoma: a comprehensive review. J Clin Oncol 1998;16(8):2877–85.

[188] Hishinuma E, Rico EG, Hiratsuka M. In vitro assessment of fluoropyrimidine-metabolizing enzymes: dihydropyrimidine dehydrogenase, dihydropyrimidinase, and β-ureidopropionase. J Clin Med 2020;9(8):2342.

[189] Takada M, Toi M. Role of oral fluoropyrimidine anticancer agents. Igaku no Ayumi 2017;261(5):514–20.

[190] Xiao G, He H, Li P, Chen Z. Uridine derivatives and method for the preparation of doxifluridine. Faming Zhuanli Shenqing; 2020. CN 111072734 A 20200428.

[191] Zhou J, Wang Y, Yang Z. A doxifluridine capsule for treating cancer and its making method. Faming Zhuanli Shenqing; 2016. CN 105343029 A 20160224.

[192] Zhang Q, Wang D, Zheng X, Wang C, Xie J. Synthetic method of doxifluridine. Faming Zhuanli Shenqing; 2018. CN 108117574 A 20180605.

[193] Xue J. Doxifluridine pharmaceutical composition and its medical use. Faming Zhuanli Shenqing; 2016. CN 105796585 A 20160727.

[194] Kang MH, Lee WS, Go S, Kim MJ, Lee US, Choi HJ, Kim DC, Lee J, Kim H, Bae KS, et al. Can thymidine phosphorylase be a predictive marker for gemcitabine and doxifluridine combination chemotherapy in cholangiocarcinoma?: case series. Medicine (Philadelphia, PA, United States) 2014;93(28), e305.

[195] Xue F, Lin X, Cai Z, Liu X, Ma Y, Wu M. Doxifluridine -based pharmacosomes delivering miR-122 as tumor microenvironments-activated nanoplatforms for synergistic treatment of hepatocellular carcinoma. Colloids Surf B Biointerfaces 2021;197, 111367.

[196] Kang YK, Chang HM, Yook JH, Ryu MH, Park I, Min YJ, Zang DY, Kim GY, Yang DH, Jang SJ, et al. Adjuvant chemotherapy for gastric cancer: a randomised phase 3 trial of mitomycin-C plus either short-term doxifluridine or long-term doxifluridine plus cisplatin after curative D2 gastrectomy (AMC0201). Br J Cancer 2013;108 (6):1245–51.

[197] Nenajdenko V. Fluorine in heterocyclic chemistry vol 2: 6-membered heterocycles. Springer International Publishing; 2014, 588.

[198] Vivian D, Polli JE. Synthesis and in vitro evaluation of bile acid prodrugs of floxuridine to target the liver. Int J Pharm 2014;475:597–604.
[199] Wei N, Zhang B, Wang Y, He XH, Xu LC, Li GD, Wang YH, Wang GZ, Huang HZ, Li WT. Transarterial chemoembolization with raltitrexed-based or floxuridine - based chemotherapy for unresectable colorectal cancer liver metastasis. Clin Transl Oncol 2019;21(4):443–50.
[200] Ajila E, Roy RAK, Sandhya SM, Prasobh GR, Devi MSP, Dhanya S. A review on anticancer activity of pyrimidine derivatives by leuckart reaction. World J Pharm Pharm Sci 2019;8(12):501–7.
[201] Sato A, Yamamoto A, Shimotsuma A, Ogino Y, Funayama N, Takahashi Y, Hiramoto A, Wataya Y, Kim H. Intracellular microRNA expression patterns influence cell death fates for both necrosis and apoptosis. FEBS Open Bio 2020. https://doi.org/10.1002/2211-5463.12995.
[202] Sato A, Hiramoto A, Kim H, Wataya Y. Anticancer strategy targeting cell death regulators: switching the mechanism of anticancer floxuridine-induced cell death from necrosis to apoptosis. Int J Mol Sci 2020;21(16):5876.
[203] Chirio D, Peira E, Battaglia L, Ferrara B, Barge A, Sapino S, Giordano S, Dianzani C, Gallarate M. Lipophilic prodrug of floxuridine loaded into solid lipid nanoparticles: in vitro cytotoxicity studies on different human cancer cell lines. J Nanosci Nanotechnol 2018;18(1):556–63.
[204] Yan Y, Qing Y, Pink JJ, Gerson SL. Loss of uracil DNA glycosylase selectively resensitizes p53-mutant and -deficient cells to 5-FdU. Mol Cancer Res 2018;16(2):212–21.
[205] Wang H, Zhao Y, Zhang Z. Age-dependent effects of floxuridine (FUdR) on senescent pathology and mortality in the nematode Caenorhabditis elegans. Biochem Biophys Res Commun 2019;509(3):694–9.
[206] Sun Y, Ke Y, Li C, Wang J, Tu L, Hu L, Jin Y, Chen H, Gong J, Yu Z. Bifunctional and unusual amino acid β- or γ-ester prodrugs of nucleoside analogues for improved affinity to ATB0,+ and enhanced metabolic stability: an application to floxuridine. J Med Chem 2020;63(19):10816–28.
[207] Ghahremani S, Samadizadeh M, Khaleghian M, Zabarjad SN. Theoretical study of encapsulation of floxuridine anticancer drug into BN (9,9-7) nanotube for medical application. Phosphorus Sulfur Silicon Relat Elem 2020;195(4):293–306.
[208] Shurpik DN, Sevastyanov DA, Zelenikhin PV, Subakaeva EV, Evtugyn VG, Osin YN, Cragg PJ, Stoikov II. Hydrazides of glycine-containing decasubstituted pillar [5]arenes: synthesis and encapsulation of floxuridine. Tetrahedron Lett 2018;59 (50):4410–5.
[209] Ma Y, Liu H, Mou Q, Yan D, Zhu X, Zhang C. Floxuridine -containing nucleic acid nanogels for anticancer drug delivery. Nanoscale 2018;10(18):8367–71.
[210] Mou Q, Ma Y, Pan G, Xue B, Yan D, Zhang C, Zhu X. DNA Trojan horses: self-assembled floxuridine-containing DNA polyhedra for cancer therapy. Angew Chem Int Ed 2017;56(41):12528–32.
[211] Liang X, Gao C, Cui L, Wang S, Wang J, Dai Z. Self-assembly of an amphiphilic janus camptothecin-floxuridine conjugate into liposome-like nanocapsules for more efficacious combination chemotherapy in cancer. Adv Mater(Weinheim, Germany) 2017;29(40). n/a.
[212] Qiang W, Shi H, Wu J, Ji M, Wu C. Hepatic arterial infusion combined with systemic chemotherapy for patients with extensive liver metastases from gastric cancer. Cancer Manag Res 2020;12:2911–6.
[213] Tran BT, Kim J, Ahn D. Systemic delivery of aptamer-drug conjugates for cancer therapy using enzymatically generated self-assembled DNA nanoparticles. Nanoscale 2020;12(45):22945–51.

[214] Huang P, Wang G, Wang Z, Zhang C, Wang F, Cui X, Guo S, Huang W, Zhang R, Yan D. Floxuridine-chlorambucil conjugate nanodrugs for ovarian cancer combination chemotherapy. Colloids Surf B Biointerfaces 2020;194, 111164.
[215] Takeda Y, Yoshizaki I, Nonaka Y, Yanagie H, Matsuzawa A, Eriguchi M. Docetaxel alone or orally combined with 5-fluorouracil and its derivatives: effects on mouse mammary tumor cell line MM2 in vitro and in vivo. Anticancer Drugs 2001;12 (8):691–8.
[216] Hu Y, Liu Y, Shen X, Fang X, Qu S. Studies on the interaction between 1-hexylcarbamoyl-5-fluorouracil and bovine serum albumin. J Mol Struct 2005;738 (1–3):143–7.
[217] Yasutomi M, Takahashi T, Kodaira S, Hojo K, Kato T, Ogawa M, Ichihashi H, Tominaga T, Tamada R, Kunii Y, et al. Prospective controlled study on the usefulness of carmofur as a postoperative adjuvant chemotherapy for colorectal cancer. Gan to kagaku ryoho Cancer Chemother 1997;24(13):1953–60.
[218] Tokuda C, Miyama Y, Takeda H, Fukui H. Antitumor activity of 1-hexylcarbamoyluracil which is carmofur substituted fluorine for hydrogen. Gan To Kagaku Ryoho 2000;27(7):1065–7.
[219] Dementiev A, Joachimiak A, Nguyen H, Nguyen H, Gorelik A, Illes K, Nagar B, Shabani S, Gelsomino M, Ahn Eun-Young E, et al. Molecular mechanism of inhibition of acid ceramidase by carmofur. J Med Chem 2019;62(2):987–92.
[220] Sato A, Matsukawa M, Kurihara M. Oral chemotherapeutic agents for gastric and colorectal cancer. Biotherapy (Tokyo, Japan) 2001;15(4):439–46.
[221] Liu P, Ma S, Liu H, Han H, Wang S. HCFU inhibits cervical cancer cells growth and metastasis by inactivating Wnt/β-catenin pathway. J Cell Biochem 2017;120(7). Online ahead of print.
[222] Wu K, Xiu Y, Zhou P, Qiu Y, Zhou P, Qiu Y, et al. A new use for an old drug: carmofur attenuates lipopolysaccharide (LPS)-induced acute lung injury via inhibition of FAAH and NAAA activities. Front Pharmacol 2019;10, 10818.
[223] Fujikawa A, Tsuchiya K, Katase S, Kurosaki Y, Hachiya J. Diffusion-weighted MR imaging of carmofur -induced leukoencephalopathy. Eur Radiol 2001;11 (12):2602–6.
[224] Imasaka K, Ueda H, Azuma T, Nagai T. Application of PHYCON 6600 to achieve sustained release of an antitumor drug (carmofur). Drug Des Deliv 1989;5(2):159–65.
[225] Nishio S, Kishimoto T, Maekawa M, Kawakita J, Morikawa Y, Funai K, Hayahara N, Yuki K, Nishijima T, Yasumoto R. Study on effectiveness of carmofur (Mifurol) in urogenital carcinoma, especially bladder cancer, as a post-operative adjuvant chemotherapeutic agent. Hinyokika kiyo Acta Urol Jap 1987;33(2):295–303.
[226] Doan NB, Mirza SP, Doan NB, Nguyen HS, Montoure A, Mueller WM, Kurpad S, Al-Gizawiy MM, Rand SD, Schmainda KM, et al. Acid ceramidase is a novel drug target for pediatric brain tumors. Oncotarget 2017;8(15):24753–61.
[227] Kameyama M, Nakamori S, Imaoka S, Utsunomiya J, Oshima A, Kikkawa N, Hioki K, Fukuda I, Mori T, Yasutomi M. Some problems of TS measurement after administration of fluoropyrimidines in colorectal cancer. Kinki Cooperative Study Group of Chemotherapy for Colorectal Carcinoma. Gan to kagaku ryoho Cancer Chemother 1993;20(14):2195–9.
[228] Ito K, Yamaguchi A, Miura K, Kato T, Baba S, Matsumoto S, Ishii M, Takagi H. Oral adjuvant chemotherapy with carmofur (HCFU) for colorectal cancer: five-year follow-up. Tokai HCFU Study Group—third study on colorectal cancer. J Surg Oncol 1996;63(2):107–11.
[229] Nakamura T, Ohno M, Tabuchi Y, Kamigaki T, Fujii H, Yamagishi H, Kuroda Y. Optimal duration of oral adjuvant chemotherapy with carmofur in the colorectal

cancer patients: the Kansai carmofur study group trial III. Int J Oncol 2001;19 (2):291–8.
[230] Osterlund P, Elomaa I, Virkkunen P, Joensuu H. A phase I study of raltitrexed (Tomudex) combined with carmofur in metastatic colorectal cancer. Oncology 2001;61(2):113–9.
[231] Kotake K, Koyama Y, Shida S, Tajima Y, Ishikawa H, Kanazawa K, Miyata M, Nagamachi Y, Iwasaki Y, Omoto R, et al. Neo-adjuvant chemotherapy with carmofur for colorectal cancer—a multi-institutional randomized controlled study. Gan to kagaku ryoho Cancer Chemother 2002;29(11):1917–24.
[232] Ito K, Yamaguchi A, Miura K, Kato T, Koike A, Takagi H. Prospective adjuvant therapy with mitomycin C and carmofur (HCFU) for colorectal cancer, 10-year follow-up: Tokai HCFU Study Group, the first study for colorectal cancer. J Surg Oncol 1996;62(1):4–9.
[233] Sakamoto J, Kodaira S, Hamada C, Ito K, Maehara Y, Takagi H, Sugimachi K, Nakazato H, Ohashi Y. An individual patient data meta-analysis of long supported adjuvant chemotherapy with oral carmofur in patients with curatively resected colorectal cancer. Oncol Rep 2001;8(3):697–703.
[234] Oesterlund P, Orpana A, Elomaa I, Repo H, Joensuu H. Raltitrexed treatment promotes systemic inflammatory reaction in patients with colorectal carcinoma. Br J Cancer 2002;87(6):591–9.
[235] Watanabe M, Nishida O, Kunii Y, Kodaira S, Takahashi T, Tominaga T, Hojyo K, Kato T, Niimoto M, Kunitomo K, et al. Randomized controlled trial of the efficacy of adjuvant immunochemotherapy and adjuvant chemotherapy for colorectal cancer, using different combinations of the intracutaneous streptococcal preparation OK-432 and the oral pyrimidines 1-hexylcarbamoyl-5-fluorouracil and uracil/tegafur. Int J Clin Oncol 2004;9(2):98–106.
[236] Zhang Z, Williams GR, Wells CJR, King AM, Davies G, Bear JC. pH-responsive nanocomposite fibres allowing MRI monitoring of drug release. J Mater Chem B 2020;8(32):7264–74.
[237] Sato S, Ueyama T, Fukui H, Miyazaki K, Kuwano M. Anti-tumor effects of carmofur on human 5-FU resistant cells. Gan To Kagaku Ryoho 1999;26(11):1613–6.
[238] Sakamoto J, Hamada C, Rahman M, Kodaira S, Ito K, Nakazato H, Ohashi Y, Yasutomi M. An individual patient data meta-analysis of adjuvant therapy with carmofur in patients with curatively resected colon cancer. Jpn J Clin Oncol 2005;35(9):536–44.
[239] Morimoto K, Koh M. Postoperative adjuvant use of carmofur for early breast cancer. Osaka City Med J 2003;49(2):77–83.
[240] Nakatani K, Watanabe A, Nishiwada T, Sawada H, Okumura T, Yamada Y, Yano T, Shino Y, Nakano H. Effect of concomitant use of anticancer drugs and a Ca2+ antagonist, on human gastric cancer transplanted into nude mice. Nihon Gan Chiryo Gakkai shi 1990;25(1):98–102.
[241] Grohn P, Heinonen E, Kumpulainen E, Lansimies H, Lantto A, Salmi R, Pyrhonen S, Numminen S. Oral carmofur in advanced gastrointestinal cancer. Am J Clin Oncol 1990;13(6):477–9.
[242] Liu Y, Wang C, Zheng C, Wang Z, Wu H, Qu S. Microcalorimetric study on the enhanced antitumor effects of 1-hexylcarbamoyl-5-fluorouracil by combination with hyperthermia on K-562 cell line. Thermochim Acta 2001;369(1–2):51–7.
[243] Matsumoto S, Nishizawa S, Murakami M, Noma S, Sano A, Kuroda Y. Carmofur -induced leukoencephalopathy: MRI. Neuroradiology 1995;37(8):649–52.
[244] Akiba T, Okeda R, Tajima T. Metabolites of 5-fluorouracil, α-fluoro-β-alanine and fluoroacetic acid, directly injure myelinated fibers in tissue culture. Acta Neuropathol 1996;92(1):8–13.

[245] Suzuki T, Koizumi J, Uchida K, Shiraishi H, Hori M. Carmofur-induced organic mental disorders. Jpn J Psychiatry Neurol 1990;44(4):723–7.
[246] Kuzuhara S, Ohkoshi N, Kanemaru K, Hashimoto H, Nakanishi T, Toyokura Y. Subacute leucoencephalopathy induced by carmofur, a 5-fluorouracil derivative. J Neurol 1987;234(6):365–70.
[247] Jin Z, Zhao Y, Zhang B, Wang H, Zhu Y, Zhu C, Hu T, Du X, Duan Y, Yu J, et al. Structural basis for the inhibition of SARS-CoV-2 main protease by antineoplastic drug carmofur. Nat Struct Mol Biol 2020;27(6):529–32.
[248] Wang Y, Cao X, Zhang C, Zhu Y. Improved method for synthesis of anticancer drug carmofur. Jingxi Yu Zhuanyong Huaxuepin 2005;13(10):11–3.
[249] Meng S, Zhao X, Jiang C, Liu Y. Study on the anticancer activity in vitro of carmofur immunoliposome specifically against human colorectal carcinoma. Shenyang Yaoke Daxue Xuebao 2003;20(5):370–2.
[250] Liu Y, Wang C, Zheng C, Wu H, Wang Z, Qu S. Microcalorimetric study of the metabolism of U-937 cells undergoing apoptosis induced by the combined treatment of hyperthermia and chemotherapy. J Therm Biol 2002;27(2):129–35.
[251] Ono T, Nagasue N, Kohno H, Hayashi T, Uchida M, Yukaya H, Yamanoi A. Adjuvant chemotherapy with epirubicin and carmofur after radical resection of hepatocellular carcinoma: a prospective randomized study. Semin Oncol 1997;24(2, Suppl. 6):S6.18–25.
[252] Yukawa H. Concentration of 5-FU in tumor tissues and flow cytometric DNA ploidy analysis after administration of antimetabolic anticancer agents in colorectal cancers. Wakayama Igaku 1991;42(4):741–58.
[253] Clifford RE, Govindarajah N, Bowden D, Sutton P, Glenn M, Parsons JL, Vimalachandran D, Darvish-Damavandi M, Buczacki S, Darvish-Damavandi M, et al. Targeting acid ceramidase to improve the radiosensitivity of rectal cancer. Cell 2020;9(12):2693.
[254] Ma C, Hu Y, Wang J, Townsend JA, Marty MT, Lagarias PI, Kolocouris A. Ebselen, disulfiram, carmofur, PX-12, Tideglusib, and Shikonin are nonspecific promiscuous SARS-CoV-2 main protease inhibitors. ACS Pharmacol Transl Sci 2020;3 (6):1265–77.
[255] Comlekci E, Kutlu HM, Vejselova SC. Toward stimulating apoptosis in human lung adenocarcinoma cells by novel nano-carmofur compound treatment. Anticancer Drugs 2021;32(6):657–63.
[256] Domracheva I, Muhamadejev R, Petrova M, Liepinsh E, Gulbe A, Shestakova I, et al. 1,2-Dimyristoyl-sn-glycero-3-phosphocholine (DMPC) increases carmofur stability and in vitro antiproliferative effect. Toxicol Rep 2015;2:377–83.
[257] Sanchez-Vazquez B, Amaral Aderito JR, Pasparakis G, Williams GR, Yu D. Electrosprayed Janus particles for combined photo-chemotherapy. AAPS PharmSciTech 2017;18(5):1460–8.
[258] Ito K, Yamaguchi A, Miura K, Kato T, Baba S, Matsumoto S, Ishii M, Takagi H. Effect of oral adjuvant therapy with carmofur (HCFU) for distant metastasis of colorectal cancer. Int J Clin Oncol 2000;5(1):29–35.
[259] Kageyama T, Toizumi A, Tamura Y. Inhibition of HCFU absorption after resection for gastric cancer—application of hydroxyaluminium gel. Gan to kagaku ryoho Cancer Chemother 2001;28(6):803–7.
[260] Tadaoka N, Takahashi N, Yoshinaga K, Yamada T, Inoue K, Mizutani H, Inomata Y, Sakuyama T, Kuroda T. A case of recurrent gastric cancer with liver metastasis responding to combination chemotherapy with cisplatin and carmofur. Gan to kagaku ryoho Cancer Chemother 1997;24(6):729–32.
[261] Osawa S, Shiroto H, Kondo Y, Nakanishi Y, Fujisawa J, Miyakawa K, Oku T, Nishimura A, Uchino J. Randomized controlled study on adjuvant immunochemotherapy

with carmofur (HCFU) for noncuratively resected and unresected gastric cancer. Gan to kagaku ryoho Cancer Chemother 1996;23(3):327–31.

[262] Akaishi S, Kobari M, Yusa T, Yamanami H, Furukawa T, Matsuno S. Transfer of carmofur (HCFU) to the serum, bile, pancreatic juice and pancreatic tumor tissue in the cases of peri-pancreatic head cancer. Gan to kagaku ryoho Cancer Chemother 1995;22(5):659–64.

[263] Katsumata K, Yamashita S, Hoshino S, Tani C, Yamamoto K, Kimura K. Evaluation of neoadjuvant combination chemotherapy with pirarubicin, carmofur and tamoxifen citrate on outpatients with advanced and recurrence breast cancer. Gan to kagaku ryoho Cancer Chemother 1994;21(2):273–5.

[264] Kajanti MJ, Pyrhonen SO. Phase II trial of oral carmofur in advanced pancreatic carcinoma. Ann Oncol 1991;2(10):765–6.

[265] Sipila P, Kivinen S, Grohn P, Vesala J, Heinonen E. Phase II evaluation of peroral carmofur, cyclophosphamide, and hexamethylmelamine as a second-line therapy in advanced epithelial ovarian carcinoma. Gynecol Oncol 1989;34(1):27–9.

[266] Zhu L, Nie H. Carmofur suppository as adjuvant agent for treating cervical carcinoma and its preparation. Faming Zhuanli Shenqing; 2006. CN 1857267 A 20061108.

[267] Kataoka Y, Iimori M, Niimi S, Tsukihara H, Wakasa T, Saeki H, Oki E, Maehara Y, Kitao H. Cytotoxicity of trifluridine correlates with the thymidine kinase 1 expression level. Sci Rep 2019;9(1):1–8.

[268] Suzuki N, Nakagawa F, Takechi T. Trifluridine /tipiracil increases survival rates in peritoneal dissemination mouse models of human colorectal and gastric cancer. Oncol Lett 2017;14(1):639–46.

[269] Kasi PM, Kotani D, Cecchini M, Shitara K, Ohtsu A, Ramanathan RK, Hochster HS, Grothey A, Yoshino T. Chemotherapy induced neutropenia at 1-month mark is a predictor of overall survival in patients receiving TAS-102 for refractory metastatic colorectal cancer: a cohort study. BMC Cancer 2016;16:467/1–7.

[270] Kitao H, Morodomi Y, Niimi S, Kiniwa M, Shigeno K, Matsuoka K, Kataoka Y, Iimori M, Tokunaga E, Saeki H, et al. The antibodies against 5-bromo-2′-deoxyuridine specifically recognize trifluridine incorporated into DNA. Sci Rep 2016;6:25286.

[271] Yamashita F, Komoto I, Oka H, Kuwata K, Takeuchi M, Nakagawa F, Yoshisue K, Chiba M. Exposure-dependent incorporation of trifluridine into DNA of tumors and white blood cells in tumor-bearing mouse. Cancer Chemother Pharmacol 2015;76(2):325–33.

[272] Matsuoka K, Iimori M, Niimi S, Tsukihara H, Watanabe S, Kiyonari S, Kiniwa M, Ando K, Tokunaga E, Saeki H, et al. Trifluridine induces p53-dependent sustained G2 phase arrest with its massive misincorporation into DNA and few DNA strand breaks. Mol Cancer Ther 2015;14(4):1004–13.

[273] van der Velden DL, Opdam FL, Voest EE. TAS-102 for treatment of advanced colorectal cancers that are no longer responding to other therapies. Clin Cancer Res 2016;22(12):2835–9.

[274] Ohashi S, Kikuchi O, Nakai Y, Ida T, Saito T, Kondo Y, Yamamoto Y, Mitani Y, Vu THN, Fukuyama K, et al. Synthetic lethality with trifluridine/tipiracil and checkpoint kinase 1 inhibitor for esophageal squamous cell carcinoma. Mol Cancer Ther 2020;19(6):1363–72.

[275] Edahiro K, Iimori M, Kobunai T, Morikawa-Ichinose T, Miura D, Kataoka Y, Niimi S, Wakasa T, Saeki H, Oki E, et al. Thymidine kinase 1 loss confers trifluridine resistance without affecting 5-fluorouracil metabolism and cytotoxicity. Mol Cancer Res 2018;16(10):1483–90.

[276] Kobunai T, Matsuoka K, Takechi T. ChIP-seq analysis to explore DNA replication profile in trifluridine -treated human colorectal cancer cells in vitro. Anticancer Res 2019;39(7):3565–70.

[277] Lenz H, Stintzing S, Loupakis F. TAS-102, a novel antitumor agent: a review of the mechanism of action. Cancer Treat Rev 2015;41(9):777–83.
[278] Cleary JM, Rosen LS, Yoshida K, Rasco D, Shapiro GI, Sun W. A phase 1 study of the pharmacokinetics of nucleoside analog trifluridine and thymidine phosphorylase inhibitor tipiracil (components of TAS-102) vs trifluridine alone. Invest New Drugs 2017;35(2):189–97.
[279] Novotny L, Al-Hasawi NA. Two-nucleoside combination in solid-tumors: TAS-102 provides clear benefits for the cancer patients. Res J Pharm, Biol Chem Sci 2020;11 (2):43–9.
[280] Siebenhuener A, De Dosso S, Meisel A, Wagner AD, Borner M. Metastatic colorectal carcinoma after second progression and the role of Trifluridine-Tipiracil (TAS-102) in Switzerland. Oncol ResTreat 2020;43(5):237–44.
[281] Kimura M, Usami E, Iwai M, Teramachi H, Yoshimura T. Severe neutropenia: a prognosticator in patients with advanced/recurrent colorectal cancer under oral trifluridine-tipiracil (TAS-102) chemotherapy. Pharmazie 2017;72(1):49–52.
[282] Sakamoto K, Yokogawa T, Ueno H, Oguchi K, Kazuno H, Ishida K, Tanaka N, Osada A, Yamada Y, Okabe H, et al. Crucial roles of thymidine kinase 1 and deoxyUTPase in incorporating the antineoplastic nucleosides trifluridine and 2′-deoxy-5-fluorouridine into DNA. Int J Oncol 2015;46(6):2327–34.
[283] Kawazoe A, Shitara K. Trifluridine/tipiracil for the treatment of metastatic gastric cancer. Expert Rev Gastroenterol Hepatol 2020;14(2):65–70.
[284] Kataoka Y, Iimori M, Fujisawa R, Morikawa-Ichinose T, Niimi S, Wakasa T, Saeki H, Oki E, Miura D, Tsurimoto T, et al. DNA replication stress induced by trifluridine determines tumor cell fate according to p53 status. Mol Cancer Res 2020;18 (9):1354–66.
[285] Andre T, Saunders M, Kanehisa A, Gandossi E, Fougeray R, Amellal NC, Falcone A. First-line trifluridine/tipiracil plus bevacizumab for unresectable metastatic colorectal cancer: SOLSTICE study design. Future Oncol 2020;16(4):21–9.
[286] Wheelden M, Yee NS. Clinical evaluation of the safety and efficacy oftrifluridine/ tipiracil in the treatment of advancedgastric/gastroesophageal junctionadenocarcinoma: evidence to date. Onco Targets Ther 2020;13:7459–65.
[287] Yoshino T, Kojima T, Bando H, Yamazaki T, Naito Y, Mukai H, Fuse N, Goto K, Ito Y, Doi T, et al. Effect of food on the pharmacokinetics of TAS-102 and its efficacy and safety in patients with advanced solid tumors. Cancer Sci 2016;107(5):659–65.
[288] Li J, Liu J, Wang R, Chen H, Li C, Zhao M, He F, Wang Y, Liu P. Trifluridine selectively inhibits cell growth and induces cell apoptosis of triple-negative breast cancer. Am J Cancer Res 2020;10(2):507–23.
[289] Shibutani M, Nagahara H, Fukuoka T, Iseki Y, Wang E, Okazaki Y, Kashiwagi S, Maeda K, Hirakawa K, Ohira M. Combining bevacizumab with trifluridine/thymidine phosphorylase inhibitor improves the survival outcomes regardless of the usage history of bevacizumab in front-line treatment of patients with metastatic colorectal cancer. Anticancer Res 2020;40(7):1–7.
[290] Suenaga M, Schirripa M, Cao S, Zhang W, Yang D, Dadduzio V, Salvatore L, Borelli B, Pietrantonio F, Ning Y, et al. Potential role of polymorphisms in the transporter genes ENT1 and MATE1/OCT2 in predicting TAS-102 efficacy and toxicity in patients with refractory metastatic colorectal cancer. Eur J Cancer 2017;86:197–206.
[291] Becerra CR, Yoshida K, Mizuguchi H, Patel M, Von Hoff D. A phase 1, open-label, randomized, crossover study evaluating the bioavailability of TAS-102 (Trifluridine/ Tipiracil) tablets relative to an oral solution containing equivalent amounts of Trifluridine and Tipiracil. J Clin Pharmacol 2017;57(6):751–9.
[292] Wang X, Zhou J, Li Y, Ge Y, Zhou Y, Bai C, Shen L. Pharmacokinetics, safety, and preliminary efficacy of oral trifle uridine/tipiracil in chinese patients with solid tumors: a phase 1b, open-label study. Clin Pharmacol: Adv Appl 2020;12:21–33.

[293] Masuishi T, Tsuji A, Kotaka M, Nakamura M, Kochi M, Takagane A, Shimada K, Denda T, Segawa Y, Tanioka H, et al. Phase 2 study of irinotecan plus cetuximab rechallenge as third-line treatment in KRAS wild-type metastatic colorectal cancer: JACCRO CC-08. Br J Cancer 2020;123(10):1490–5.
[294] Giuliani J, Bonetti A. Trifluridine /Tipiracil in heavily pretreated metastatic gastric cancer. A perspective based on pharmacological costs. Eur J Cancer 2020;138:77–9.
[295] Nose Y, Kagawa Y, Hata T, Mori R, Kawai K, Naito A, Sakamoto T, Murakami K, Katsura Y, Ohmura Y, et al. Neutropenia is an indicator of outcomes in metastatic colorectal cancer patients treated with FTD/TPI plus bevacizumab: a retrospective study. Cancer Chemother Pharmacol 2020;86(3):427–33.
[296] Giuliani J, Bonetti A. The onset of grade ≥ 3 neutropenia is associated with longer overall survival in metastatic colorectal cancer patients treated with trifluridine/tipiracil. Anticancer Res 2019;39(7):3967–9.
[297] Unseld M, Fischoeder S, Jachs M, Drimmel M, Siebenhuener A, Bianconi D, Kieler M, Puhr H, Minichsdorfer C, Winder T, et al. Different toxicity profiles predict third line treatment efficacy in metastatic colorectal cancer patients. J Clin Med 2020;9 (6):1772.
[298] Ogata M, Kotaka M, Ogata T, Hatachi Y, Yasui H, Kato T, Tsuji A, Satake H. Regorafenib vs trifluridine/tipiracil for metastatic colorectal cancer refractory to standard chemotherapies: a multicenter retrospective comparison study in Japan. PLoS ONE 2020;15(6), e0234314.
[299] Kashiwa M, Matsushita R. Comparative cost-utility analysis of Regorafenib and trifluridine/tipiracil in the treatment of metastatic colorectal cancer in Japan. Clin Ther 2020;42(7):1376–87.
[300] Kotani D, Kuboki Y, Horasawa S, Kaneko A, Nakamura Y, Kawazoe A, Bando H, Taniguchi H, Shitara K, Kojima T, et al. Retrospective cohort study of trifluridine/ tipiracil (TAS-102) plus bevacizumab versus trifluridine /tipiracil monotherapy for metastatic colorectal cancer. BMC Cancer 2019;19(1):1253.
[301] Carriles C, Jimenez-Fonseca P, Sanchez-Canovas M, Pimentel P, Carmona-Bayonas A, Garcia T, Carbajales-Alvarez M, Lozano-Blazquez A. Trifluridine /Tipiracil (TAS-102) for refractory metastatic colorectal cancer in clinical practice: a feasible alternative for patients with good performance status. Clin Transl Oncol 2019;21(12):1781–5.
[302] Suzuki N, Nakagawa F, Matsuoka K, Takechi T. Effect of a novel oral chemotherapeutic agent containing a combination of trifluridine, tipiracil and the novel triple angiokinase inhibitor nintedanib, on human colorectal cancer xenografts. Oncol Rep 2016;36(6):3123–30.
[303] Falcone A, Ohtsu A, Van Cutsem E, Mayer RJ, Buscaglia M, Bendell JC, Kopetz S, Bebeau P, Yoshino T. Integrated safety summary for trifluridine/tipiracil (TAS-102). Anticancer Drugs 2018;29(1):89–96.
[304] Kwakman JJM, Vink G, Vestjens JH, Beerepoot LV, de Groot JW, Jansen RL, Opdam FL, Boot H, Creemers GJ, van Rooijen JM, et al. Feasibility and effectiveness of trifluridine /tipiracil in metastatic colorectal cancer: real-life data from The Netherlands. Int J Clin Oncol 2018;23(3):482–9.
[305] Baba T, Kokuryo T, Yamaguchi J, Yokoyama Y, Uehara K, Ebata T, Nagino M. Pre-exposure to fluorouracil increased trifluridine incorporation and enhanced its antitumor effect for colorectal cancer. Anticancer Res 2018;38(3):1427–34.
[306] Matsuoka K, Takechi T. Combined efficacy and mechanism of trifluridine and SN-38 in a 5-FU-resistant human colorectal cancer cell lines. Am J Cancer Res 2017;7 (12):2577–86.
[307] Nakanishi R, Kitao H, Kiniwa M, Morodomi Y, Iimori M, Kurashige J, Sugiyama M, Nakashima Y, Saeki H, Oki E, et al. Monitoring trifluridine incorporation in the

peripheral blood mononuclear cells of colorectal cancer patients under trifluridine/tipiracil medication. Sci Rep 2017;7(1):1–8.
[308] Tanaka A, Sadahiro S, Suzuki T, Okada K, Saito G, Miyakita H. Retrospective study of regorafenib and trifluridine/tipiracil efficacy as a third-line or later chemotherapy regimen for refractory metastatic colorectal cancer. Oncol Lett 2018;16(5):6589–97.
[309] Pfeiffer P, Yilmaz M, Moller S, Zitnjak D, Krogh M, Petersen LN, Poulsen LO, Winther SB, Thomsen KG, Qvortrup C. TAS-102 with or without bevacizumab in patients with chemorefractory metastatic colorectal cancer: an investigator-initiated, open-label, randomised, phase 2 trial. Lancet Oncol 2020;21(3):412–20.
[310] Suzuki N, Tsukihara H, Nakagawa F, Kobunai T, Takechi T. Synergistic anticancer activity of a novel oral chemotherapeutic agent containing trifluridine and tipiracil in combination with anti-PD-1 blockade in microsatellite stable-type murine colorectal cancer cells. Am J Cancer Res 2017;7(10):2032–40. S1.
[311] Yasue F, Kimura M, Usam E, Go MM, Kawachi S, Mitsuoka M, Ikeda Y, Yoshimura T. Risk factors contributing to the development of neutropenia in patients receiving oral trifluridine -tipiracil (TAS-102) chemotherapy for advanced/recurrent colorectal cancer. Pharmazie 2018;73(3):178–81.
[312] Baba Y, Tamura T, Satoh Y, Gotou M, Sawada H, Ebara S, Shibuya K, Soeda J, Nakamura K. Panitumumab interaction with TAS-102 leads to combinational anticancer effects via blocking of EGFR-mediated tumor response to trifluridine. Mol Oncol 2017;11(8):1065–77.
[313] Fernandez Montes A, Vazquez Rivera F, Martinez Lago N, Covela Rua M, Cousillas Castineiras A, Gonzalez Villarroel P, de la Camara GJ, Mendez Mendez JC, Salgado Fernandez M, Candamio Folgar S, et al. Efficacy and safety of trifluridine/tipiracil in third-line and beyond for the treatment of patients with metastatic colorectal cancer in routine clinical practice: patterns of use and prognostic nomogram. Clin Transl Oncol 2020;22(3):351–9.
[314] Van Cutsem E, Mayer RJ, Laurent S, Winkler R, Gravalos C, Benavides M, Longo-Munoz F, Portales F, Ciardiello F, Siena S, et al. The subgroups of the phase III RECOURSE trial of trifluridine/tipiracil (TAS-102) versus placebo with best supportive care in patients with metastatic colorectal cancer. Eur J Cancer 2018;90:63–72.
[315] Patel AK, Barghout V, Yenikomshian MA, Germain G, Jacques P, Laliberte F, Duh MS. Real-world adherence in patients with metastatic colorectal cancer treated with Trifluridine plus Tipiracil or Regorafenib. Oncologist 2020;25(1):e75–84.
[316] Matsuoka K, Kobunai T, Nukatsuka M, Takechi T. Improved chemoradiation treatment using trifluridine in human colorectal cancer cells in vitro. Biochem Biophys Res Commun 2017;494(1–2):249–55.
[317] Rothkamm K, Christiansen S, Rieckmann T, Horn M, Frenzel T, Brinker A, Schumacher U, Stein A, Petersen C, Burdak-Rothkamm S. Radiosensitisation and enhanced tumour growth delay of colorectal cancer cells by sustained treatment with trifluridine/tipiracil and X-rays. Cancer Lett (N Y, NY, U S) 2020;493:179–88.
[318] Zhang R, Zhou D. Medical use of trifluridine or trifluridine tipiracil composition. PCT Int. Appl; 2020. WO 2020215999 A1 20201029.
[319] Zhou D, Zhang R. Application of trifluridine or trifluridine-tipiracil composition in the treatment of blood diseases. Faming Zhuanli Shenqing; 2019. CN 109999053 A 20190712.
[320] Fujimoto Y, Nakanishi R, Nukatsuka M, Matsuoka K, Ando K, Wakasa T, Kitao H, Oki E, Maehara Y, Mori M. Detection of trifluridine in tumors of patients with metastatic colorectal cancer treated with trifluridine/tipiracil. Cancer Chemother Pharmacol 2020;85(6):1029–38.
[321] Desbuissons G, Ngango L, Brocheriou I, Fournier P. IgA nephropathy associated with trifluridine/tipiracil: a case report. Nephron 2020;144(10):506–8.

[322] Peeters M, Cervantes A, Moreno Vera S, Taieb J. Trifluridine/tipiracil: an emerging strategy for the management of gastrointestinal cancers. Future Oncol 2018;14 (16):1629–45.

[323] Zhou K, Zhou J, Zhang M, Liao W, Li Q. Cost-effectiveness of trifluridine/tipiracil (TAS102) for heavily pretreated metastatic gastric cancer. Clin Transl Oncol 2020;22 (3):337–43.

[324] Shitara K, Doi T, Dvorkin M, Mansoor W, Arkenau H, Prokharau A, Alsina M, Ghidini M, Faustino C, Gorbunova V, et al. Trifluridine/tipiracil versus placebo in patients with heavily pretreated metastatic gastric cancer (TAGS): a randomised, double-blind, placebo-controlled, phase 3 trial. Lancet Oncol 2018;19(11):1437–48.

[325] Chakrabarti S, Zemla TJ, Ahn DH, Ou F, Fruth B, Borad MJ, Hartgers Mindy L, Wessling J, Walkes RL, Alberts SR, et al. Phase II trial of trifluridine/tipiracil in patients with advanced, refractory biliary tract carcinoma. Oncologist 2020;25 (5):380–e763.

[326] Takahashi K, Yoshisue K, Chiba M, Nakanishi T, Tamai I. Contribution of equilibrative nucleoside transporter(s) to intestinal basolateral and apical transports of anticancer trifluridine. Biopharm Drug Dispos 2018;39(1):38–46.

[327] Saif MW, Rosen L, Rudek MA, Sun W, Shepard DR, Becerra C, Yamashita F, Bebeau P, Winkler R. Open-label study to evaluate trifluridine /tipiracil safety, tolerability and pharmacokinetics in patients with advanced solid tumours and hepatic impairment. Br J Clin Pharmacol 2019;85(6):1239–46.

[328] Orlandi P, Gentile D, Banchi M, Cucchiara F, Di Desidero T, Cremolini C, Moretto R, Falcone A, Bocci G. Pharmacological effects of the simultaneous and sequential combinations of trifluridine/tipiracil (TAS-102) and 5-fluorouracil in fluoropyrimidine-sensitive colon cancer cells. Invest New Drugs 2020;38(1):92–8.

[329] Kim S, Jung JH, Lee HJ, Soh H, Lee SJ, Oh SJ, Chae SY, Lee JH, Lee SJ, Hong YS, et al. [18F]fluorothymidine PET informs the synergistic efficacy of capecitabine and trifluridine/tipiracil in Colon cancer. Cancer Res 2017;77(24):7120–30.

CHAPTER 5

Synthetic strategies for anticancer antifolates

1. Introduction

Folate antagonists are significant compounds and are universally accepted as one of the principal categories of antimetabolites [1,2]. Consequently, they are considered as main compounds in the pharmaceutical industry.

Up to date, synthetic strategies of anticancer antifolates have been exploited and highlighted throughout this chapter. Antifolates provide versatile and novel building blocks for pharmacological targets. Various strategies for the synthesis of many antifolates have been exhibited throughout this chapter to access novel and potent pharmacological agents. Various antifolates are regarded to be prospective chemotherapeutic agents that offer a major impact on the current medicinal research.

The paramount importance of antifolates triggered their broader utility in treating critical diseases. Sequentially, the novel insights of a broad array of new compounds have been established. Thus, well-recognized synthetic approaches have been set up for further drug discovery. The biological evaluation of the newly synthesized compounds was emphasized, and their synthetic procedures were also elaborated.

Crucial routes for the syntheses and development of new drugs are displayed through targeting the metabolic pathways. Following the new developments in the synthetic strategies of distinct antimetabolites, we surveyed the latest and up-to-date synthetic approaches of them that reflect their important significance in the distinctive fields.

Antifolates are used in the treatment of solid tumors [3], and they are used in a wide series of malignancies such as hematologic malignancies, malignant mesothelioma [4,5], and melanoma [6]. Besides their importance as anticancer agents [7–13] such as non-small-cell lung cancer [14,15] and other lung cancers [16,17], they are used in the treatment of other diseases such as rheumatoid arthritis [18], pneumocystis pneumonia infection [19], toxoplasmosis [20], relapsed or refractory cutaneous T-cell lymphoma [21], and cryptosporidiosis, which is a gastrointestinal disease [22]. They also have great

New Strategies Targeting Cancer Metabolism
https://doi.org/10.1016/B978-0-12-821783-2.00001-7

potential as immunomodulatory [23], antiparasitic [24,25], antiopportunistic agents [26–28], antiprotozoal [29], and antimicrobial, including antibacterial and antifungal agents [30–32]. Additionally, antifolate drugs have a remarkable importance in the treatment of malaria, which is one of the tropical parasitic diseases [33–36]. The synthesis of many antimalarial agents has been reported [37,38].

Antifolates have an affirmed record as clinically utilized oncolytic agents through the inhibition of the key enzymes thymidylate synthase (TS), which is an antifolate-based chemotherapeutic targets [39–47], also dihydrofolate reductase (DHFR), and other enzymes, which enable antifolates to find clinical usage.

Most antifolates used in oncology are similar in chemical structure to the naturally essential vitamin and folic acid (Fig. 5.1), which play effective roles in one-carbon metabolism that produces substrates necessary for nucleotide synthesis [48]. As folates are essential for the cancer cell metabolism, antifolates are designed to inhibit either thymidylate synthase (TS), dihydrofolate reductase (DHFR), or the purine de novo pathway. And hence, antifolates exert their antiproliferative activity [49]. Folates present in various forms in mammalian cells. The principal structure for these forms is folic acid [50], *N*-[4(2-amino-4-hydroxy-pteridin-6-ylmethylamino)-benzoyl]-L(+)-glutamic acid [51], which contains a pteridine ring, para-aminobenzoic acid (PABA) [52,53], and a glutamate residue [54].

Many structures were designed, synthesized [55–57], and are proved to be effective inhibitors such as DHFR inhibitors (e.g., compound **1–9**) [58–68], thymidylate synthase inhibitors (e.g., compound **10–13** [69–72]), inhibitors of folate receptors (FRs) (e.g., compound **14–16** [73]), inhibitors of de novo purine synthesis (GARFT) (e.g., compound **17**, **18** [74,75]), and inhibitors of the folylpolyglutamate synthetase (FPGS) (e.g., compound **19**, **20** [76,77]) (Figs. 5.2–5.6).

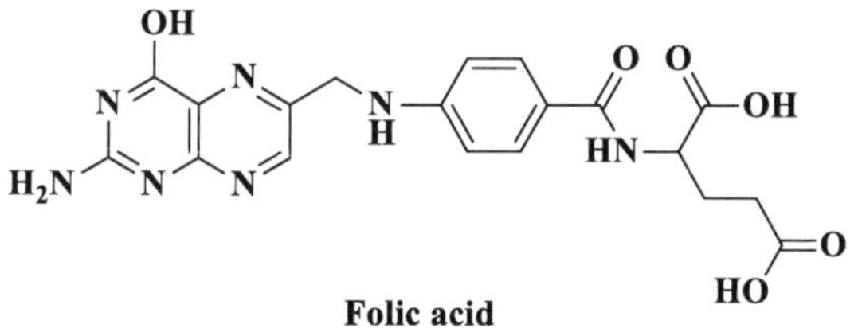

Folic acid

Fig. 5.1 Folic acid.

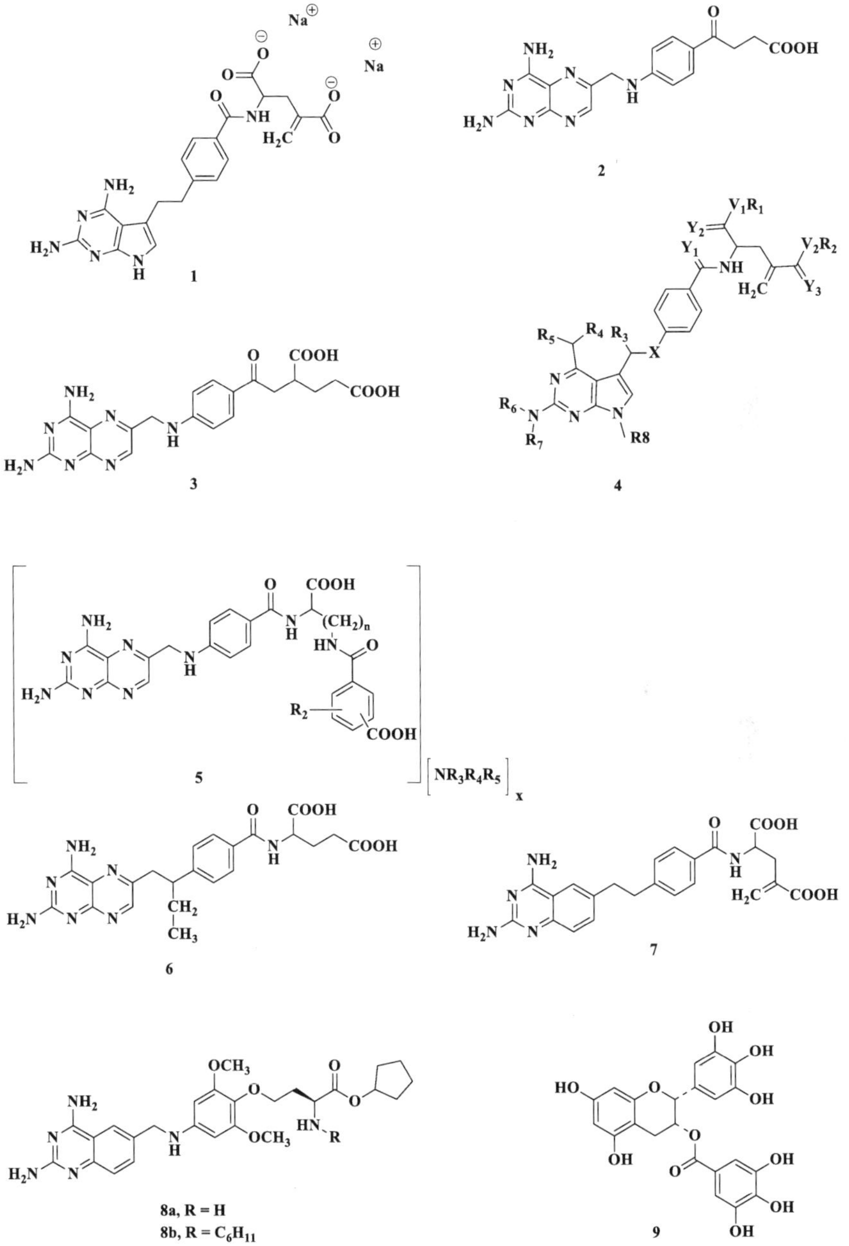

1

2

3

4

5

6

7

8a, R = H
8b, R = C_6H_{11}

9

Fig. 5.2 Inhibitors of the DHFR.

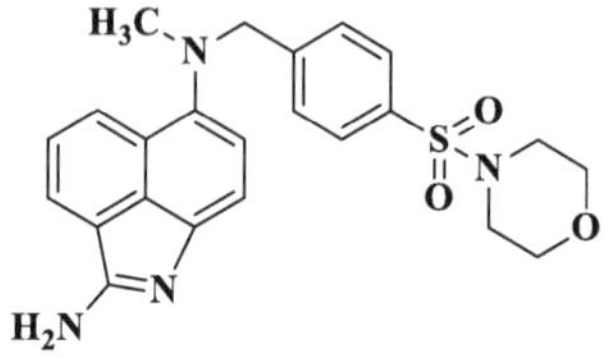

AG331, 10

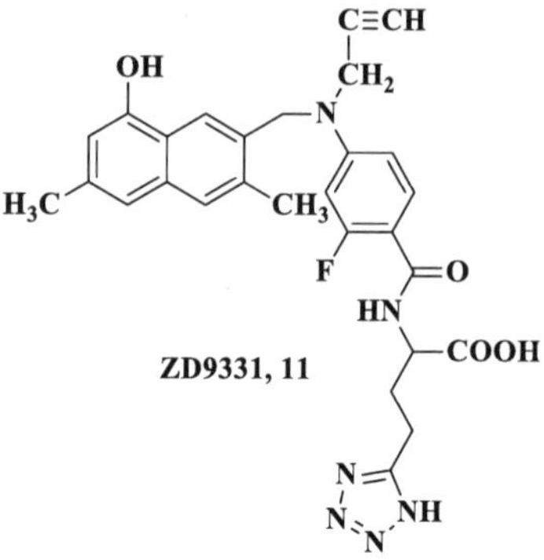

ZD9331, 11

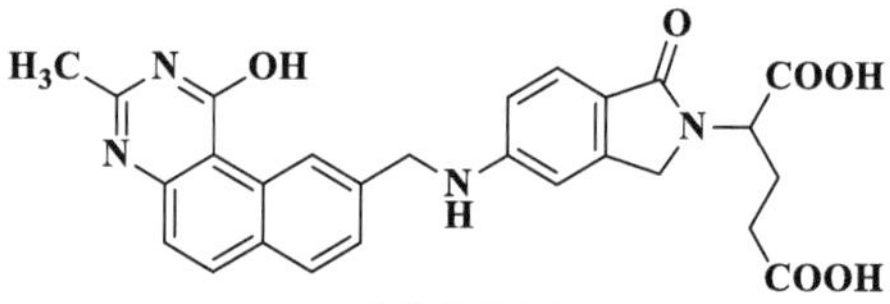

BW1843U89, 12

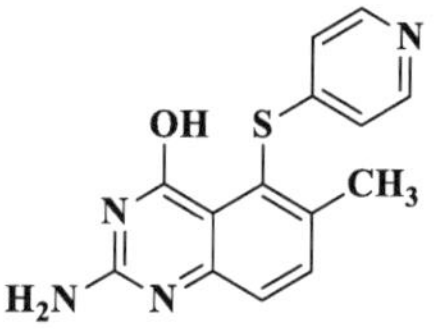

AG337, 13

Fig. 5.3 Inhibitors of thymidylate synthase.

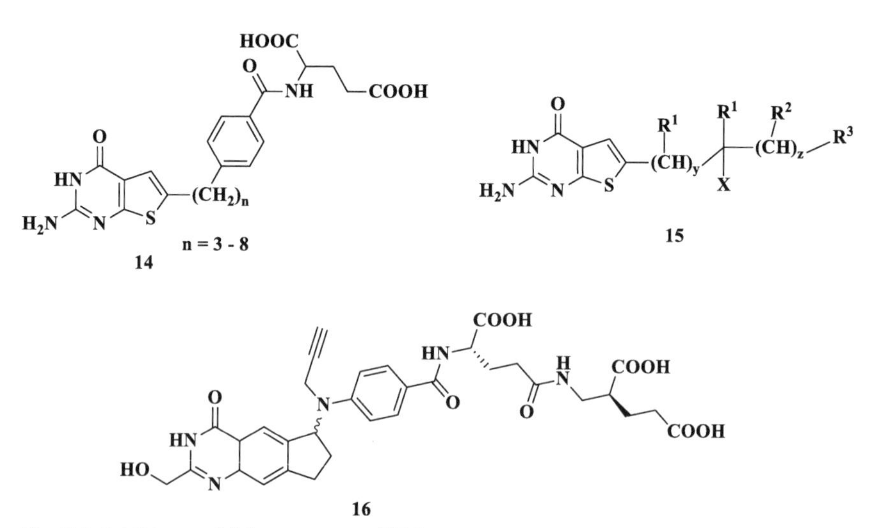

Fig. 5.4 Inhibitors of folate receptor (FRs).

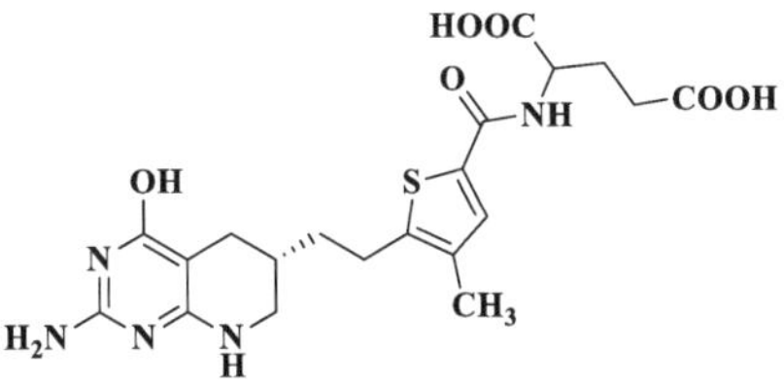

AG2037, 17

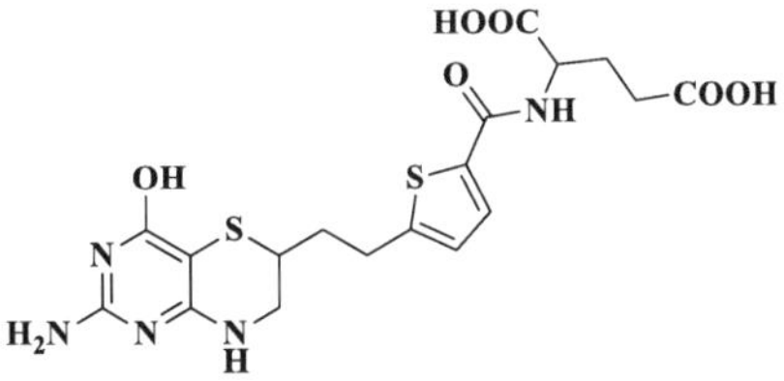

AG2034, 18

Fig. 5.5 Inhibitors of de novo purine synthesis (GARFT).

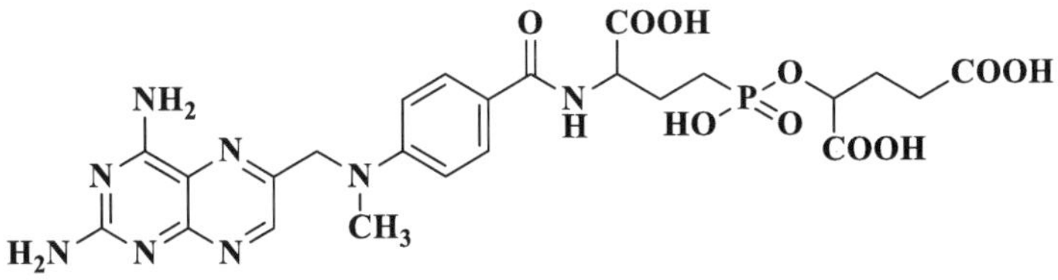

MTX-phosphonte, 19

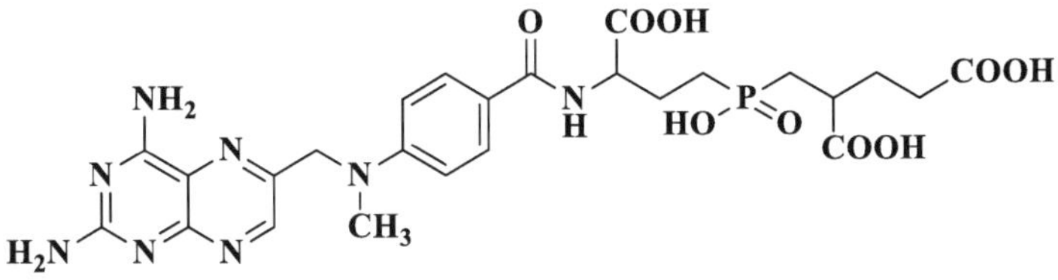

MTX-phosphinate, 20

Fig. 5.6 Inhibitors of folylpolyglutamate synthetase (FPGS).

These perspective highlights validate the significant importance of antifolates in the medicinal chemistry. Furthermore, many strategies for the syntheses of anticancer antifolates, including classical and nonclassical analogs, were addressed.

2. Classical antifolates: Syntheses and biological evaluation

2.1 Inhibitors of dihydrofolate reductase (DHFR)

The reaction of *N*-methyl benzoylglutamate **22** with **21** yielded compound **23**. The synthesis of the compound is accomplished in Scheme 5.1.

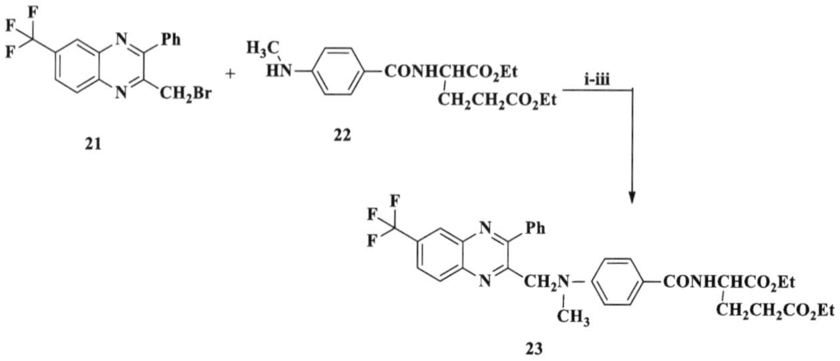

Scheme 5.1 Reagents and conditions: (i) DMA/RT/72h; (ii) DMA/50°C, 24h; (iii) DMA/TEA/100°C/10h.

The in vitro anticancer potency of compound **23** was estimated, and it was proved to be active against hDHFR (inhibition constant (K_i) = 200 nM) with a specificity of about 1000-folds as compared to hTS [78].

2.2 Inhibitors of thymidylate synthase (TS)

A 6- to 5-ring-fused analog **37** was synthesized as TS inhibitor and antitumor agent. The analogs **37** and **31** were accessed through the oxidative addition of 4-mercaptopyridine or the sodium salt of ethyl 4-mercaptobenzoate to the pyrrolopyrimidine derivative **28**. Compound **37** was synthesized via the deprotection of the ester followed by coupling with the L-glutamic acid diethyl ester and finally saponification (Schemes 5.2 and 5.3).

Compound **37** was a potent inhibitor of bacterial and human TS (IC_{50} = 21 and 42 nM). Compound **31** was 10 times less active than compound **37** against human TS but more than 4700-times less active than **37** against *Lactobacillus casei* TS. Compound **37** was neither an inhibitor nor a substrate of human FPGS derived from the CCRF-CEM cells. Compound **37** was cytotoxic to FaDu and CCRF-CEM tumor cell lines besides an FPGS-deficient subline of CCRF-CEM. It was demonstrated by the thymidine protection data that TS was the essential target of compound **37** [79].

Pyrazolopyrimidine analogs of the antitumor agent, LY231514, **38** (Fig. 5.7), have been synthesized (Schemes 5.4 and 5.5).

A palladium-catalyzed C—C coupling of the pyrazolopyrimidines **38–41** with the substituted glummate **42** was proved to be a principal synthetic step. Diethyl 4-formylbenzoyl-L-glutamale **(56)** reacted with the substituted pyrazolo[3,4-*d*]pyrimidine (**55**) to afford compound **57**.

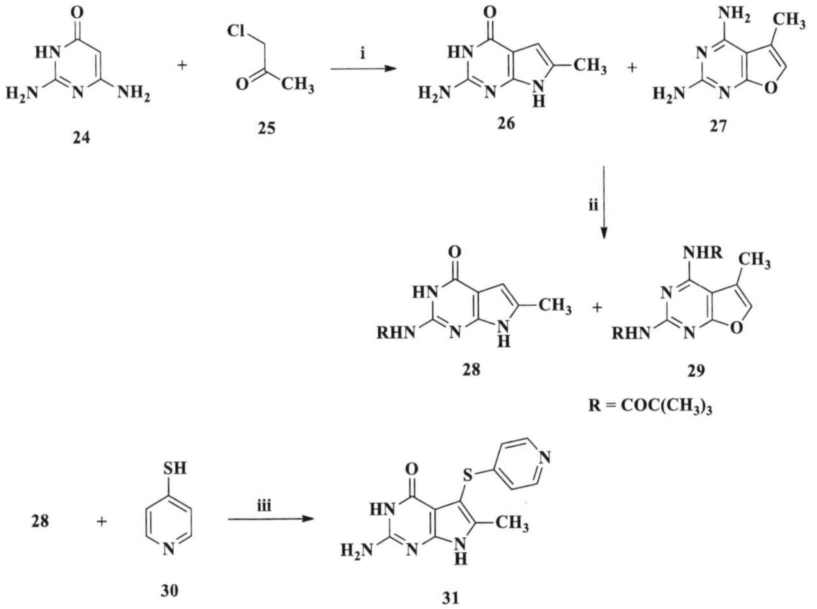

Scheme 5.2 Reagents and conditions: (i) DMF/60°C; (ii) $(CH_3)_3CCOCl/K_2CO_3$; (iii) I_2, $EtOH/H_2O$/reflux/16 h.

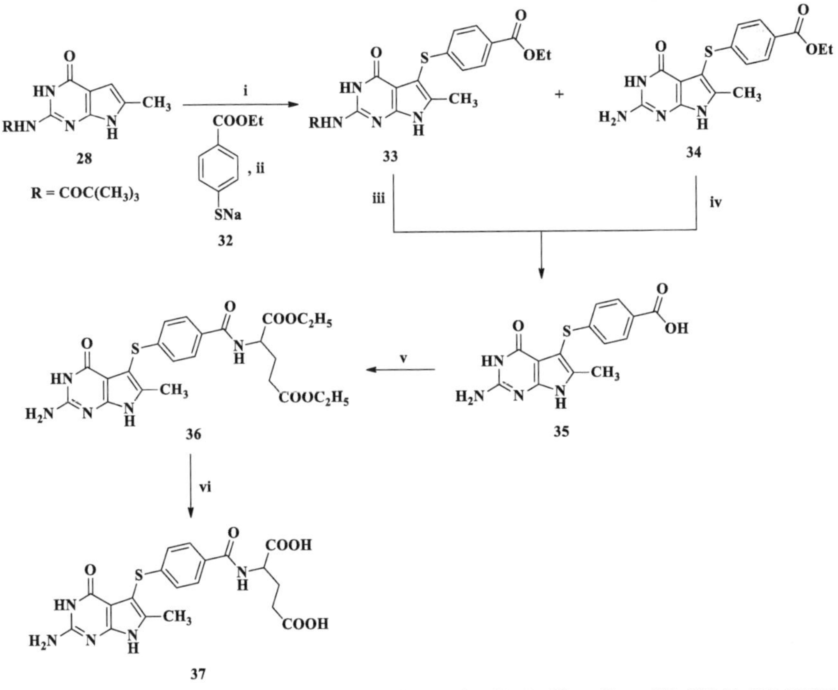

Scheme 5.3 Reagents and conditions: (i) $I_2/EtOH/H_2O$; (ii) reflux; (iii) 1 N NaOH/55°C; (iv) 1 N NaOH, RT; (v) *i*BuOCOCl/TEA/L-glutamic acid diethyl ester; (vi) 1 N NaOH/ $EtOH/H_2O$.

38

Fig. 5.7 LY231514.

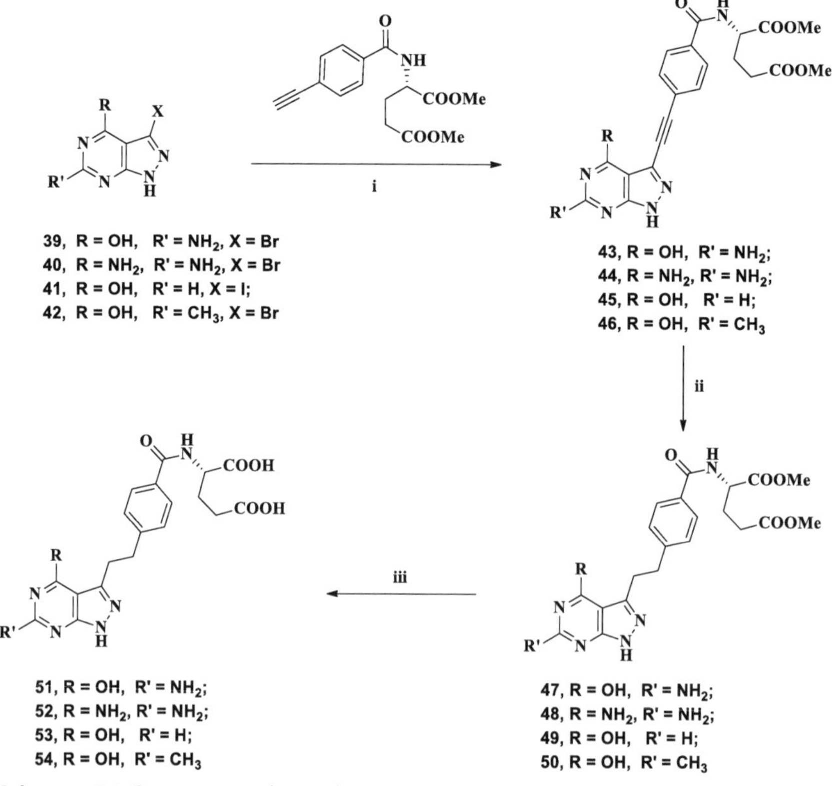

Scheme 5.4 Reagents and conditions: (i) $Pd(PPh_3)_4$/CuI/NEt_3/DMF; (ii) Pd-C/H_2; (iii) 1 N NaOH.

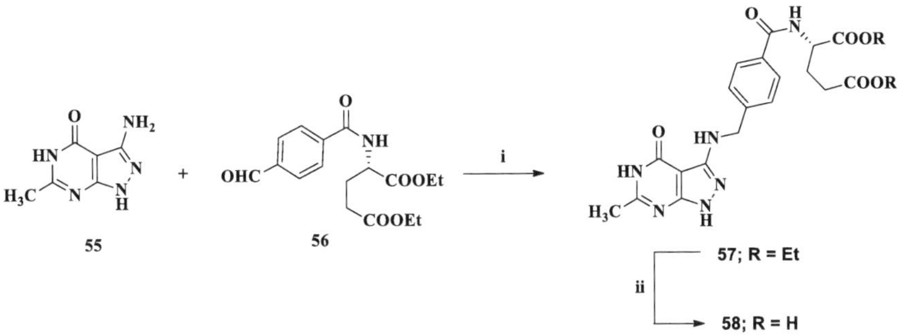

Scheme 5.5 Reagents and conditions: (i) Raney Nickel/70% AcOH/H_2; (ii) 1 N NaOH.

Compound **52** proved to possess in vitro cell growth inhibitory potency. The studies showed that compound **52** displayed a remarkable cytotoxic potency (IC_{50} = 0.018 pg/mL) [80].

The dideazafolic acid **66** was prepared from 2-amino-5-methylbenide as accomplished in Scheme 5.6.

Compound **66** is considered as an inhibitor of recombinant mouse thymidylate synthase, and inhibition was competitive with 5,10-methylenetetmhydrofolate as variable substrate (K_i = 2.3 pM). It was a substitute for murine folylpolyglutamate synthetase with kinetic properties (K_i = 28 pM) in respect to those of aminopterin, and the growth of L1210 cells was inhibited in culture (IC_{50} = 0.52 pM) [81].

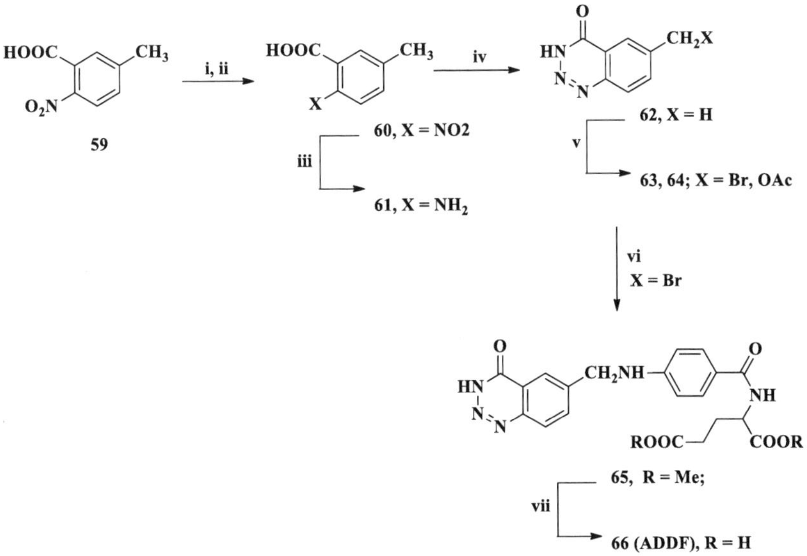

Scheme 5.6 Reagents and conditions: (i) $SOCl_2$; (ii) NH_2; (iii) H,/Pd/C; (iv) $NaNO_2$/HCl; (v) NBS/AcOH/hv; (vi) 4-$H_2NC_6H_4$CONHCH-(COOMe)CH_2CH_2COOMe; (vii) NaOH.

2.3 Inhibitors folate receptor (FR)

Pyrrolo[2,3-*d*]pyrimidine derivatives **73a–d** were prepared to access desired FR cellular uptake specificity of the antitumor agents. They inhibit the glycinamide ribonucleotide formyltransferase (GARFTase). Compounds **67a–d** were transformed to the acid chlorides and were allowed to react with azomethylene, and then the reaction with hydrogen bromide was followed to furnish the α-bromomethylketones **69a–d**. Condensation of the hydroxypyrimidine **70** with compound **69a–d** yielded compounds **71a–d**. Hydrolysis and then coupling and saponification generated compounds **73a–d** (Scheme 5.7).

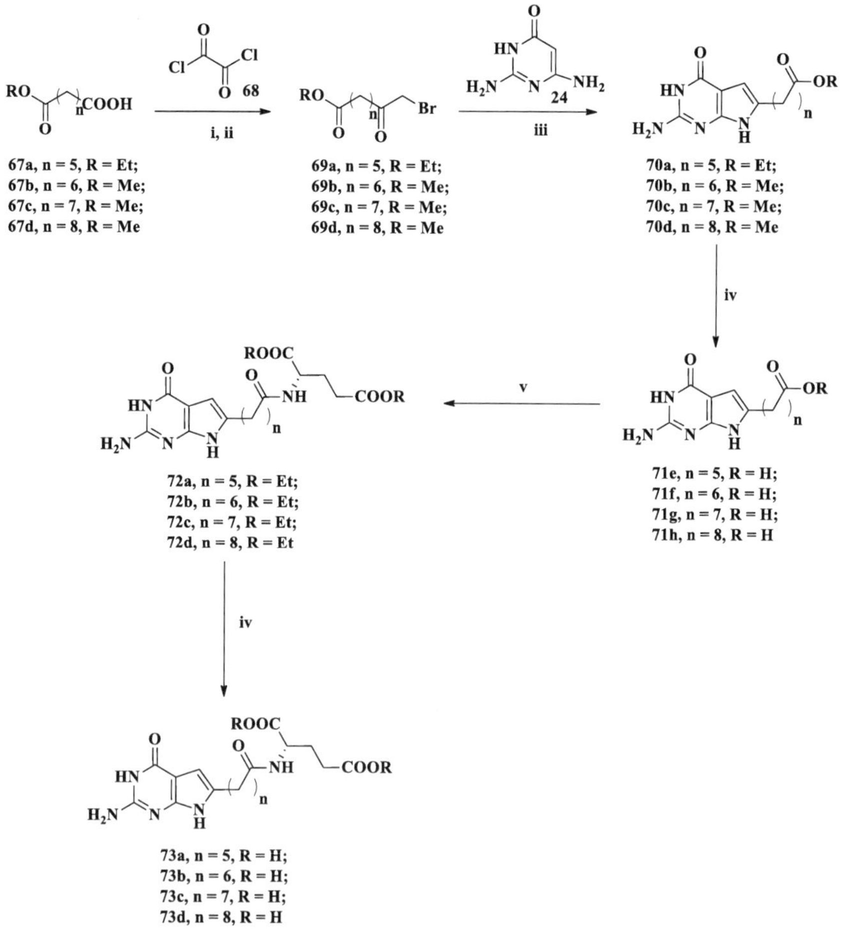

67a, n = 5, R = Et; 67b, n = 6, R = Me; 67c, n = 7, R = Me; 67d, n = 8, R = Me

69a, n = 5, R = Et; 69b, n = 6, R = Me; 69c, n = 7, R = Me; 69d, n = 8, R = Me

70a, n = 5, R = Et; 70b, n = 6, R = Me; 70c, n = 7, R = Me; 70d, n = 8, R = Me

71e, n = 5, R = H; 71f, n = 6, R = H; 71g, n = 7, R = H; 71h, n = 8, R = H

72a, n = 5, R = Et; 72b, n = 6, R = Et; 72c, n = 7, R = Et; 72d, n = 8, R = Et

73a, n = 5, R = H; 73b, n = 6, R = H; 73c, n = 7, R = H; 73d, n = 8, R = H

Scheme 5.7 Reagents and conditions: (i) CH_2N_2/0–5°C; (ii) 48% HBr/reflux; (iii) DMF/RT/3 days; (iv) 1 N NaOH/RT; (v) NMM/2,4-dimethoxy-6-chloro-triazine Diethyl L-glutamate/RT.

Selective cellular uptake of compound **73b–d** was revealed through FR-α and -β, associated with high-affinity binding and inhibition of de novo purine nucleotide biosynthesis through GARFTase; an active inhibition was resulted against FR-expressing Chinese hamster cells and human KB tumor cells in culture. It was established that a side-chain benzoyl group is not necessary for tumor-selective drug uptake via FR-α [82].

Substituted pyrrolopyrimidine with a thienoyl side chain **84–86** was prepared. Transformation of hydroxyl acetylen-thiophene carboxylic esters to thiophenyl-α-bromomethylketones and condensation with compound **24** yielded the pyrrolopyrimidine derivatives. Coupling and then saponification reactions generated compounds **84–86** (Scheme 5.8).

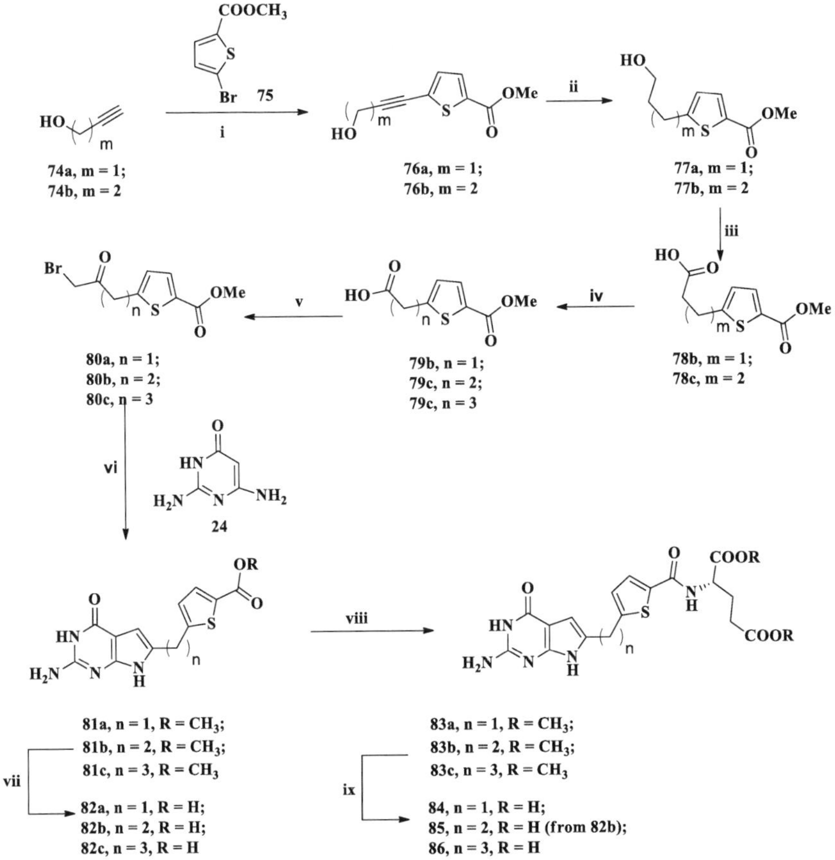

Scheme 5.8 Reagents and conditions: (i) $PdCl_2/CuI/Et_3N/PPh_3/CH_3CN$/MW/100°C/10 min; (ii) 10% $Pd/C/H_2$/55 psi/CH_3OH/4 h; (iii) CrO_3/H_2SO_4/0°C ~ RT; (iv) (i) $CH_2Cl_2/(COCl)_2$/reflux/1 h; (ii) azomethylene/Et_2O/RT/1 h; (iii) HBr/70–80°C/2 h; (v) DMF/RT/3 days; (vi) (i) 1 N NaOH/RT/12 h; (ii) 1 N HCl; (vii) 2-chloro-4,6-dimethoxy-1,3,5-triazine/*N*-methylmorpholine/diethyl L-glutamate/DMF/RT/12 h.

Selective inhibition of the proliferation of cells expressing FRs, or the proton-coupled folate transporter (PCFT), comprising IGROV1 and KB human tumor cells, was performed via compound **86**. The latter compound has an inhibitory activity toward β GARFTase. Compound **86** depleted cellular ATP pools. Compound **86** was effective in SCID mice with IGROV1 tumors. The results indicate an antitumor potency for **86** in vitro and in vivo, related to its selective membrane transport via PCFT and FRs over RFC and inhibition of GARFTase, demonstrating the 3-atom bridge as superior to other bridge lengths for the potency of this series [83].

Compounds **92** and **97** were active inhibitors of IGROV1 and KB human tumor cells with total selectivity for FR-α and PCFT over reduced folate carrier RFC.

The synthesis of compounds **92** and **97** was accomplished from intermediate **87** [84]. A Sonogashira coupling of **87** with different thiophene carboxylate derivatives **88 and 93** yielded compounds **89** and **94**. Further hydrogenation then saponification of compounds **89 and 94** gave compounds **92 and 97**. Compounds **89** and **94** were hydrolyzed to access the conformationally restricted analogs **90** and **95**. Compounds **88 and 93** were prepared via a peptide coupling of the L-glutamate diethyl ester hydrochloride with the bromo-substituted thiophene carboxylic acids (Scheme 5.9).

Tumor-targeted specificities of 6-substituted pyrrolopyrimidine analogs of **110** (Fig. 5.8), where the phenyl side chain is replaced by 3′,6′ (**107**, **101**), 2′,6′ (**109**, **103**), and 2′,5′ (**108**, **102**) pyridyls, were studied (Scheme 5.10).

Proliferation inhibition of isogenic Chinese hamster ovary (CHO) cells expressing FRs were ordered, **108** > **102** > **107** > **109** > **101**, with **103** indicating no activity, and **108** > **102** > **107** > **101**, with **103** and **109** being inactive. In competitive binding with [3H] folic acid, the antiproliferative effects toward FR-expressing cells were reflected. Compound **108** was potent against PCFT-expressing CHO cells (four times more active than compound **110**) and inhibited [^{3}H] methotrexate uptake via PCFT. In IGROV1 and KB tumor cells, **108** revealed $IC_{50} < 1$ nM, ~2–3 times more active than compound **110**. Compound **108** inhibited GARFTase in de novo purine biosynthesis and indicated in vivo potency toward subcutaneous IGROV1 tumor xenografts in SCID mice [86].

The pyrrolopyrimidine thienoyl antifolates in which the terminal L-glutamate of compound **120** was replaced by natural or unnatural amino acids were synthesized. Compounds **120–124** have a selective inhibitory effect toward FR-α-expressing Chinese hamster ovary cells. Antiproliferative effects of compounds **120** and **119, 121–124** toward FR-expressing

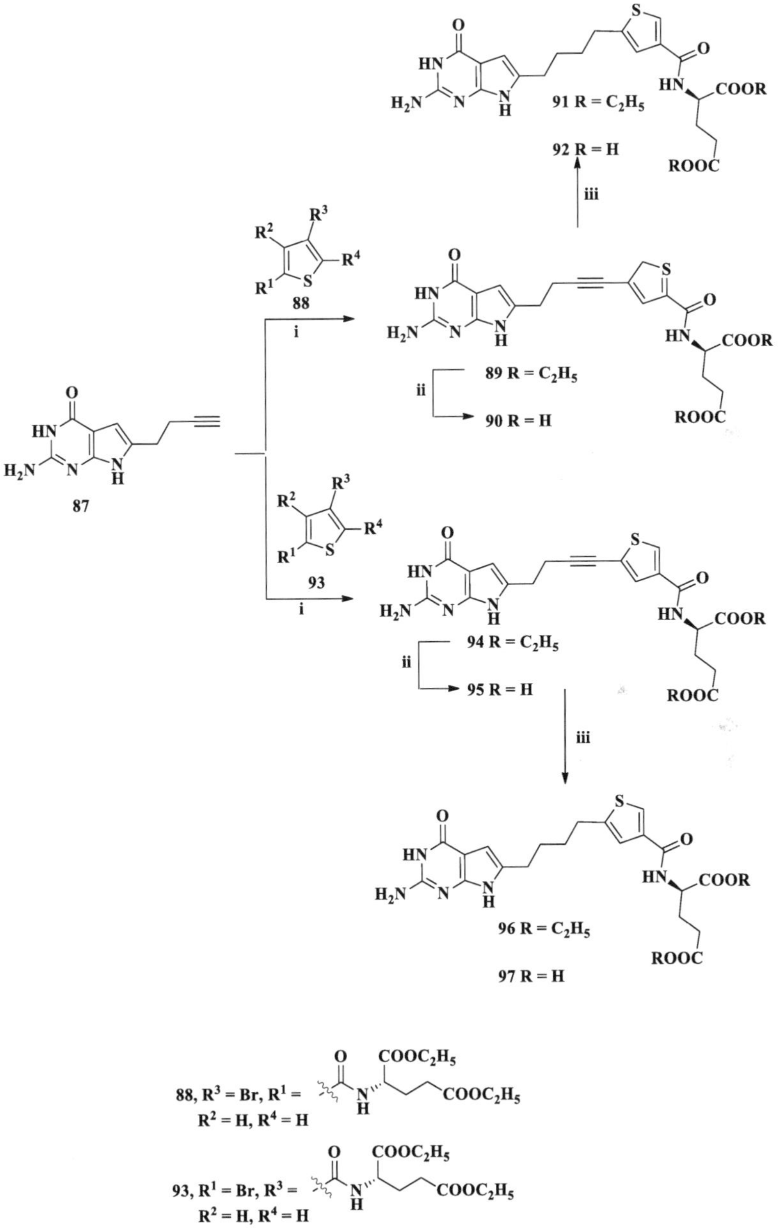

O
HN
H_2N N N H
S
O
HN COOR
ROOC
91 R = C_2H_5
92 R = H
iii

R^3 R^2 R^4 R^1 S
88
i
O
HN
H_2N N N H
S
O
HN COOR
89 R = C_2H_5
ii
90 R = H
ROOC

O
HN
H_2N N N H
87

R^3 R^2 R^4 R^1 S
93
i
O
HN
H_2N N N H
S
O
HN COOR
94 R = C_2H_5
ii
95 R = H
ROOC
iii

O
HN
H_2N N N H
S
O
HN COOR
96 R = C_2H_5
ROOC
97 R = H

88, R^3 = Br, R^1 = O $COOC_2H_5$ N H $COOC_2H_5$
R^2 = H, R^4 = H

93, R^1 = Br, R^3 = O $COOC_2H_5$ N H $COOC_2H_5$
R^2 = H, R^4 = H

Scheme 5.9 Reagents and conditions: (i) $Pd(0)(PPh_3)_4$/CuI/TEA/DMF/RT/12 h; (ii) (i) 1 N NaOH/$CHCl_3$/CH_3OH/RT/6 h; (ii) 1 N HCl; (iii) 10% Pd/C/H_2/55 psi/4 h [85].

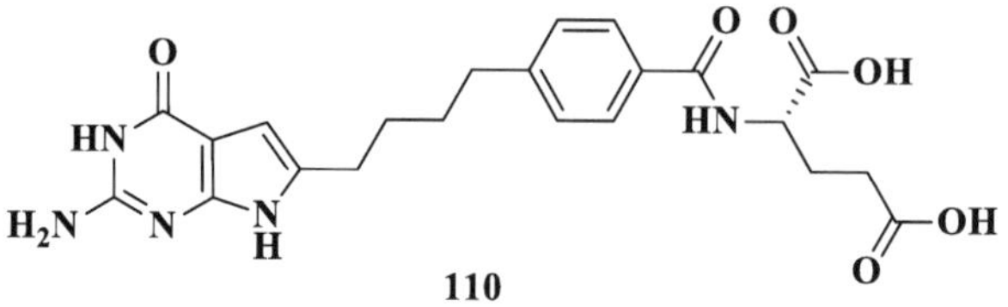

Fig. 5.8 Pyrrolopyrimidine antifolate derivative.

CHO cells were only partly reflected in binding affinities to FRs or in the docking scores with molecular models of FRs. In de novo purine biosynthesis in KB human tumor cells, compounds **120** and **122** (Fig. 5.9) were considered as active inhibitors of GARFTase. These investigations demonstrate the significance of the α- and γ-carboxylic acid groups, the conformation of the side chain for transporter binding, and the length of the amino acid, and biological potency of the substituted thienoyl antifolates.

Compounds **119**, **120**, and **121–124** were furnished from compound **113** (Scheme 5.11). Oxidation of compound **113** using periodic acid and pyridinium chlorochromate yielded the intermediate **114**. The latter was transformed to the acid chloride **115** and subsequently reacted with azomethylene, followed by reaction with hydrogen bromide to furnish the desired α-bromomethyl ketone **117**. 2,6-Diamino-3*H*-pyrimidin-4-one was condensed with compound **117** to yield the substituted pyrrolo[2,3-*d*]pyrimidine **118**. The latter was hydrolyzed to afford the corresponding free acid **119**. Further coupling with dimethyl (*S*)-2-aminohexanedioate or dimethyl L-aspartate or L-glutamate diethyl ester, methyl (*S*)-2-aminopentanoate, or methyl 4-aminobutanoate, utilizing the activating agent 2-chloro 4,6-dimethoxy-1,3,5-triazine to yield the targeted esters. Further saponification of the latter afforded the targeted compounds, **120** and **121–124** [87].

6-Substituted pyrrolopyrimidine antifolates with carbon bridge between the benzoyl-L-glutamate and the heterocycle **133–135** were prepared starting from methyl 4-formylbenzoate and a Wittig reaction with triphenylphosphonium bromide, then reduction and further transformation to the R-bromomethylketones. 2,4-Diamino-4-oxopyrimidine was condensed with the R-bromoketones. Further coupling and saponification yielded compounds **133-135** (Scheme 5.12).

Compounds **133–135** had negligible substrate potency for RFC but indicated selective inhibitory potencies toward Chinese hamster ovary cells that expressed FRs and toward FRR-expressing IGROV1 and KB human

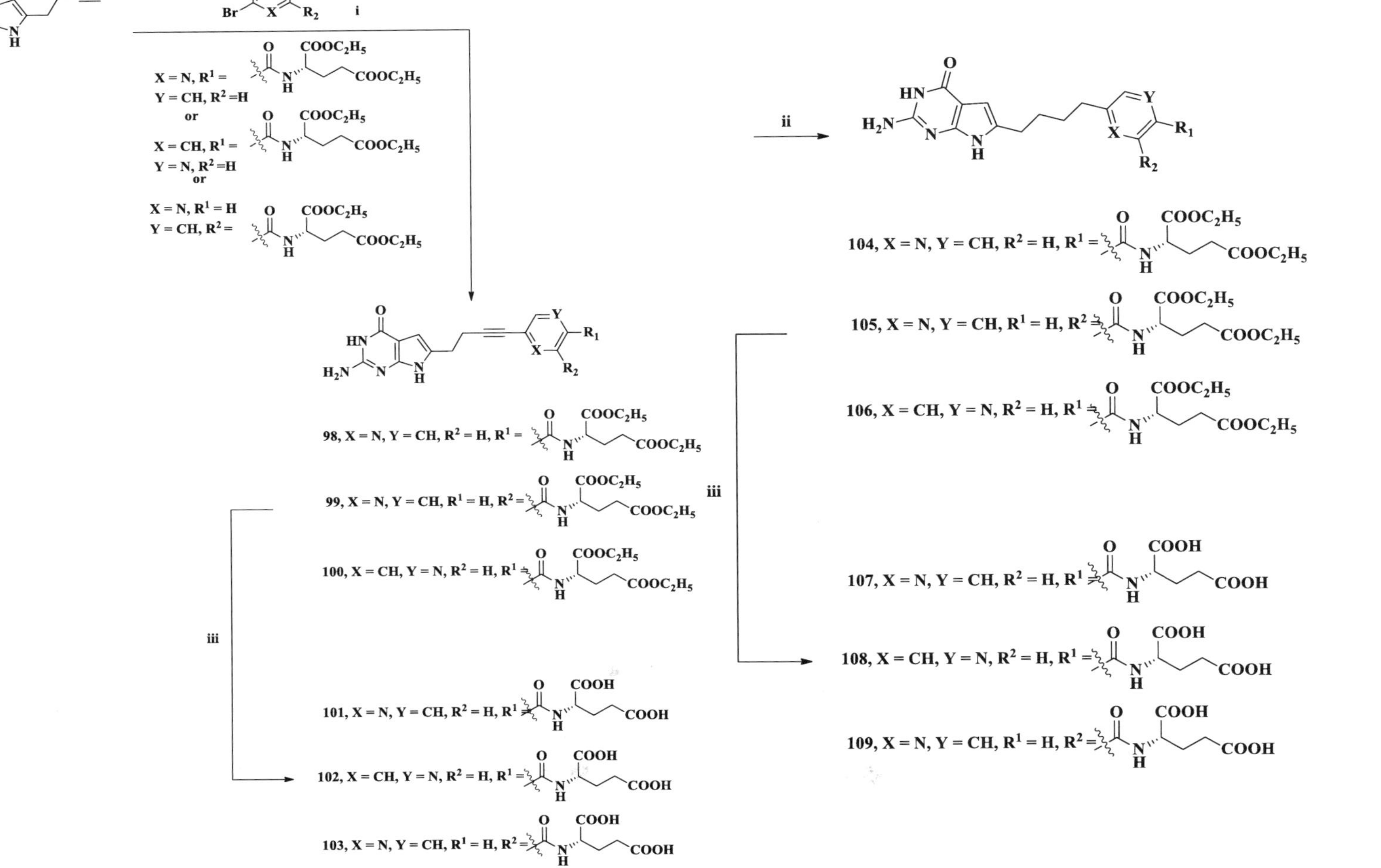

Scheme 5.10 Reagents and conditions: (i) $Pd(0)(PPh_3)_4$/CuI/TEA/DMF/RT/12 h; (ii) 5% Pd/C/H_2/55 psi/2 h; (iii) (i) 1 N NaOH/RT/6 h; (ii) 1 N HCl.

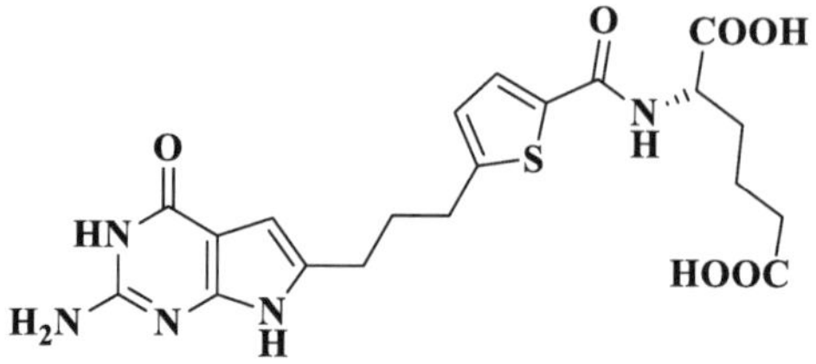

Fig. 5.9 Pyrrolopyrimidine thienoyl antifolate derivative.

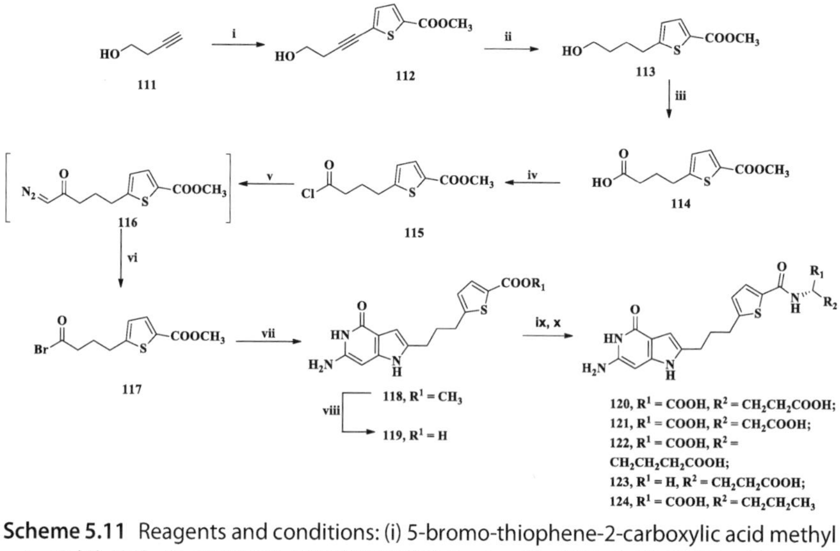

Scheme 5.11 Reagents and conditions: (i) 5-bromo-thiophene-2-carboxylic acid methyl ester/$PdCl_2$/PPh_3/CuI/TEA/CH_3CN/MW/100°C/10 min; (ii) 10% Pd/C/H_2/55 psi/CH_3OH/2 h; (iii) H_5IO_6/PCC/0°C—RT/3 h; (iv) CH_2Cl_2/$(COCl)_2$/reflux/1 h; (v) azomethylene/Et_2O/RT/1 h; (vi) HBr/70–80°C/2 h; (vii) 2,4-diamino-6-hydroxypyrimidine/DMF/RT/3 days; (viii) (i) 1 N NaOH/RT/12 h; (ii) 1 N HCl; (ix) 2-chloro-4,6-dimethoxy-1,3,5-triazine/*N*-methylmorpholine/L-glutamic acid diethyl ester hydrochloride or dimethyl (*S*)-2-aminohexanedioate or dimethyl L-aspartate/DMF/RT/12 h; (x) (i) 1 N NaOH/RT/12 h; (ii) 1 N HCl.

tumor cells. Inhibition of KB cell colony development was additionally observed. GARFTase was determined as the essential intracellular target of the pyrrolopyrimidines. The selective FR targeting, lack of RFC transport, and GARFTase inhibition leading to the active antitumor potency [88].

Compounds **141–145** were accessed using a modified method [89,90] utilizing α-bromo aldehydes **138a–e**, then the latter was condensed with

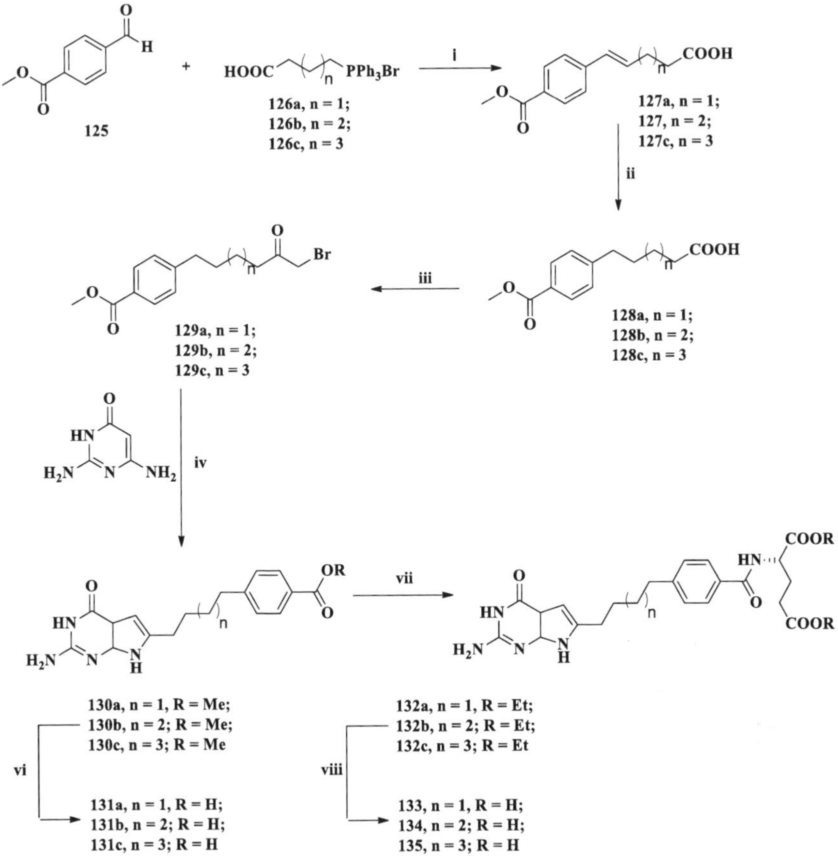

Scheme 5.12 Reagents and conditions: (i) NaH/DMSO/THF; (ii) 10% Pd/C/55Psi H_2; (iii) CH_2N_2/0–5°C; (iv) 48% HBr/reflux; (v) DMF/RT/3 days; (vi) 3 N NaOH/40–50°C; (vii) NMM/2,4-dimethoxy-6-chlorotriazine L-glutamate/RT; (viii) 1 N NaOH/RT.

the oxo-pyrimidine **24** as accomplished in Scheme 5.13. The aldehydes **136a–e** were α-brominated with cycloisopropylidene dibromomalonate **137** to furnish the corresponding α-bromo aldehydes **138a–e** [91]. The 5-substituted pyrrolopyrimidines **139a–e** were prepared via condensation of compound **138a–e** with 2,4-diamino-6-hydroxypyrimidine **24**. Further hydrolysis, coupling with L-glutamic acid diethyl ester, was followed utilizing *N*-methyl morpholine and substituted triazine as the activating agents, yielded the diesters **140a–e**. Subsequent saponification of the diesters afforded the 5-substituted pyrrolo[2,3-*d*]pyrimidines **141–145** [92].

5-Substituted thiophenyl pyrrolo[β,γ-d]pyrimidines **141–145** with varying chain lengths were prepared as hybrids of pemetrexed (PMX)

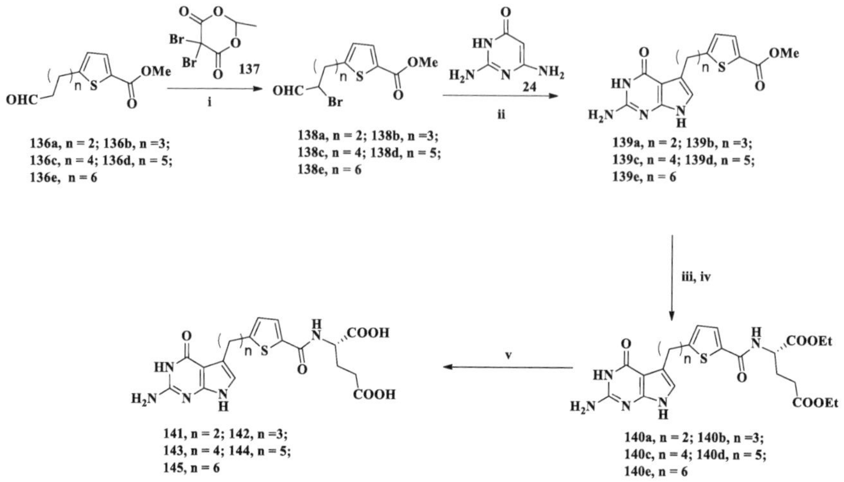

Scheme 5.13 Reagents and conditions: (i) HCl/RT; (ii) MeOH/45°C/CH_3COONa; (iii) 3 N NaOH; (iv) NMM/RT/2,4-dimethoxy-6-chlorotriazine/L-glutamic acid diethyl ester; (v) 1 N NaOH/RT.

and the 6-substituted thiophenyl pyrrolo[β,γ-d]pyrimidines with FR and PCFT uptake specificity over RFC and inhibition of de novo purine nucleotide biosynthesis at GARFTase. Analogs **141–145** inhibited KB human tumor cells. Compounds **142–144** were transported by PCFT, FRI, and RFC and, dissimilar to PMX, inhibited de novo purine nucleotide rather than thymidylate biosynthesis. The antiproliferative impacts of **142** and **143** are due to dual inhibitions of both AICARFTase and GARFTase [92].

Fluorinated pyrrolo[2,3-*d*]pyrimidines **150–153**, **161** were prepared (Schemes 5.14 and 5.15) and evaluated for selective cellular uptake by FRs or PCFT and for antitumor adequacy. Compounds **151**, **160**, **153**, and **161** indicated high in vitro antiproliferative potencies toward HeLa and engineered Chinese hamster ovary cells expressing PCFT or FRs. Compounds **151**, **160**, **153**, and **161** additionally inhibited the proliferation of A2780 epithelial ovarian cancer cells and IGROV1 cells; in IGROV1 cells with knockdown of FRα, compounds **160**, **153**, and **161** indicated a sustained inhibition related to uptake via PCFT. All compounds inhibited GARFTase. Molecular modeling studies approved in vitro cell-based results. In vivo potency of compound **153** was demonstrated with IGROV1 xenografts in extreme compromised immunodeficient mice [93].

Substituted thieno[2,3-*d*]pyrimidines with different bridge lengths were generated as selective FR substrates and as antitumor agents. The syntheses

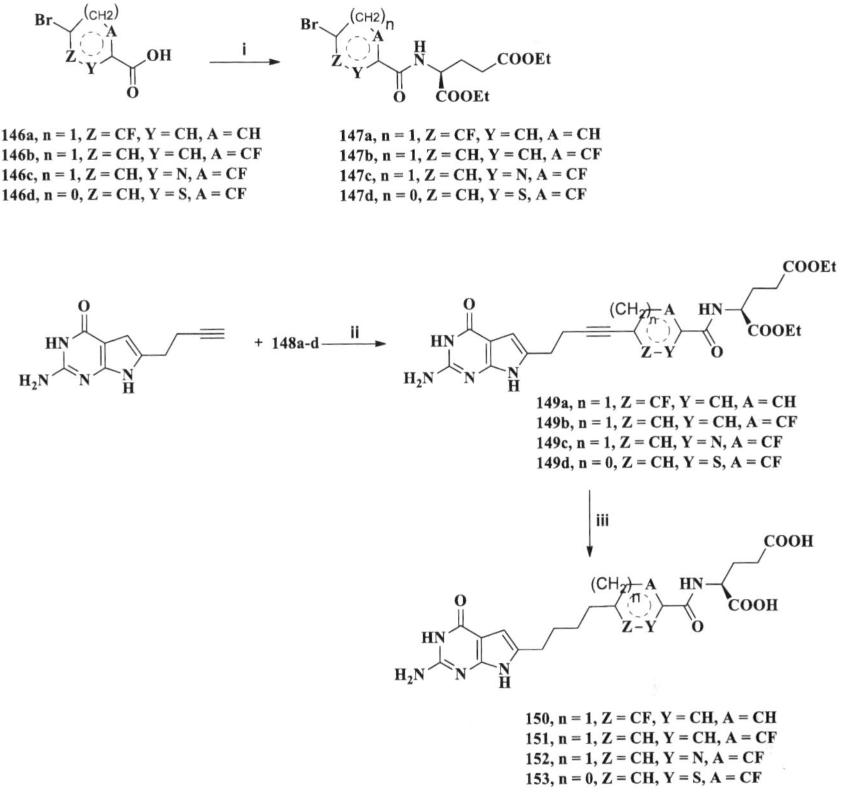

146a, n = 1, Z = CF, Y = CH, A = CH
146b, n = 1, Z = CH, Y = CH, A = CF
146c, n = 1, Z = CH, Y = N, A = CF
146d, n = 0, Z = CH, Y = S, A = CF

147a, n = 1, Z = CF, Y = CH, A = CH
147b, n = 1, Z = CH, Y = CH, A = CF
147c, n = 1, Z = CH, Y = N, A = CF
147d, n = 0, Z = CH, Y = S, A = CF

149a, n = 1, Z = CF, Y = CH, A = CH
149b, n = 1, Z = CH, Y = CH, A = CF
149c, n = 1, Z = CH, Y = N, A = CF
149d, n = 0, Z = CH, Y = S, A = CF

150, n = 1, Z = CF, Y = CH, A = CH
151, n = 1, Z = CH, Y = CH, A = CF
152, n = 1, Z = CH, Y = N, A = CF
153, n = 0, Z = CH, Y = S, A = CF

Scheme 5.14 Reagents and conditions: (i) 2,4-dimethoxy-6-chloro-triazine/NMM/L-glutamic acid diethyl ester/DMF/RT/12h; (ii) CuI/Pd(0), TEA/DMF/70°C/μW/12h; (iii) (i) 10% Pd/C, H_2 12 h; (ii) 1 N NaOH/rt./1 h.

were furnished from 4-iodobenzoate and allyl alcohols to yield the aldehydes, which were transformed to the substituted thiophenes **176–182**. The latter was cyclized with chloroformamidine to yield the thieno[2,3-*d*]pyrimidines **183–189**. Hydrolysis of the latter compounds, then coupling with diethyl-L-glutamate, followed by saponification, furnished the desired compounds **204–210** (Scheme 5.16).

Compounds **205–208** were active growth inhibitors (IC_{50} = 4.7–334 nM) of human tumor cells (IGROV1 and KB) that express FRs. Additionally, the growth of CHO cells that expressed FRs and not PCFT or RFC was inhibited by compounds **205–208**. The compounds were inactive toward CHO cells that lacked FRs; however, they contained either the PCFT or RFC. Along with in vitro and in situ enzyme activity assays, the mechanism of antitumor potency was determined as the dual inhibition

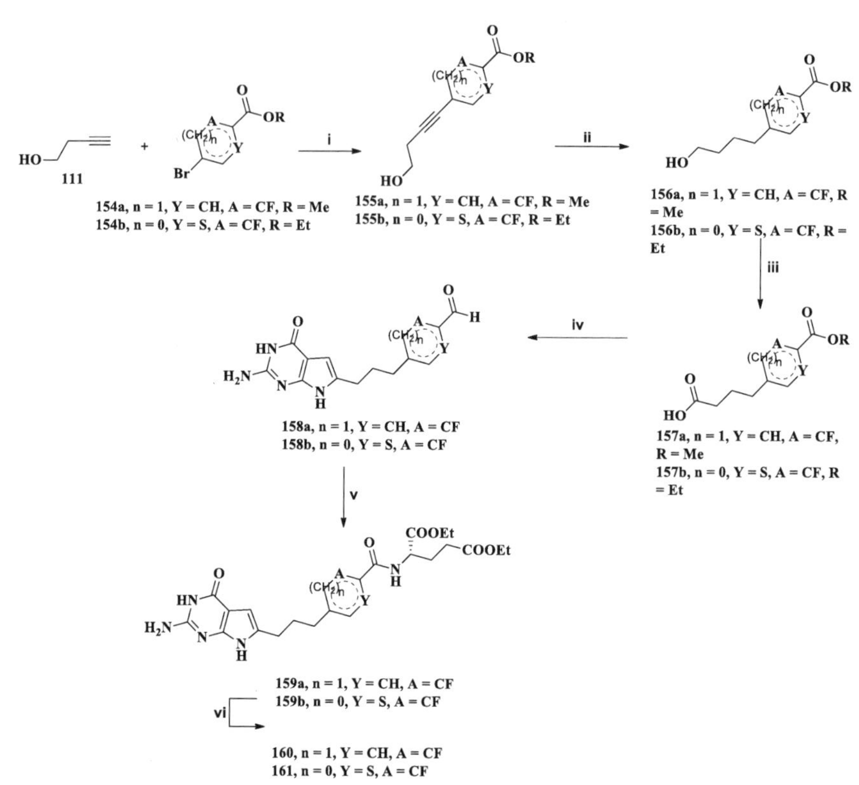

Scheme 5.15 Reagents and conditions: (i) $Ph_3P/PdCl_2$/TEA/ACN/CuI/μW/100°C/30 min; (ii) H_2, Pd/C; (iii) H_5IO_6/ACN/PCC/0°C to RT/3–12 h; (iv) (i) DCM/$(COCl)_2$/reflux/1 h; (ii) $(Et)_2O/CH_2N_2$/0°C—RT/1 h; (iii) $(Et)_2O$/48% HBr/50°C/2 h; (iv) compound **24**/DMF/rt., 3 days; (v) 1 N NaOH/RT/12 h; (v) CDMT/NMM/L-glutamic acid diethyl ester/DMF/RT/12 h; (vi) 1 N NaOH/RT/1 h.

of AICA ribonucleotide formyltransferase and GARFTase through nucleoside and AICA protection estimations. The dual inhibitory potency of the thienopyrimidine antifolates and the FR specificity show unique mechanistic characteristics for these compounds. The potent inhibitory impacts of compounds **205–208** toward cells expressing FRs and not PCFT indicate that the cellular uptake of this series of compounds via FRs does not rely upon the existence of PCFT and argues that direct coupling between these transporters is not obligatory [94].

6-Substituted pyrrolopyrimidine analogs **237** and **238** were generated as antitumor agents (Scheme 5.17). These compounds inhibited the

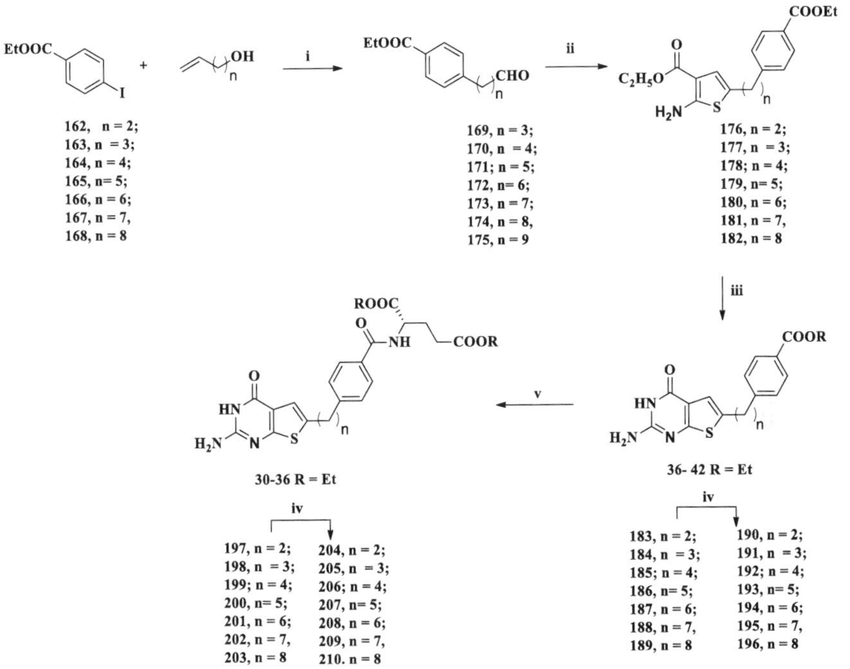

Scheme 5.16 Reagents and conditions: (i) Bu_4NCl/10%$Pd(AcO)_2$/LiOAc/LiCl/DMF/80°C/12h; (ii) CN-CH_2-COOEt/S/morpholine/RT/24h; (iii) DMSO/aminomethanecarbonimidoyl chloride hydrochloride/140°C/4h; (iv) 1N NaOH; (v) 2-Cl-4,6-dimethoxy-1,3,5-triazine/4-methylmorpholine/L-glutamic acid diethyl ester hydrochloride/DMF/10h.

proliferation of Chinese hamster ovary (CHO) sublines expressing FRs (IC_{50}s < 1 nM) or PCFT (IC_{50} < 7 nM). Compounds **237** and **238** inhibited IGROV1, KB, and SKOV3 human tumor cells, reflecting both PCFT and FR-α uptake. AGF152 (**233**), AGF163 (**234**), and 2,4-diamino-5-substituted-furo[2,3-*d*]pyrimidine thiophene regioisomers also inhibited the growth of FR-expressing KB and CHO cells. The analogs inhibited GARFTase. Crystal structures of human GARFTase complexed with **237** and **238** were studied. Compound **238** was effective in severe combined immunodeficient mice with SKOV3 tumors. A paradigm for selective tumor targeting is due to the selectivity of these compounds for FRs and PCFT over the ubiquitously expressed RFC.

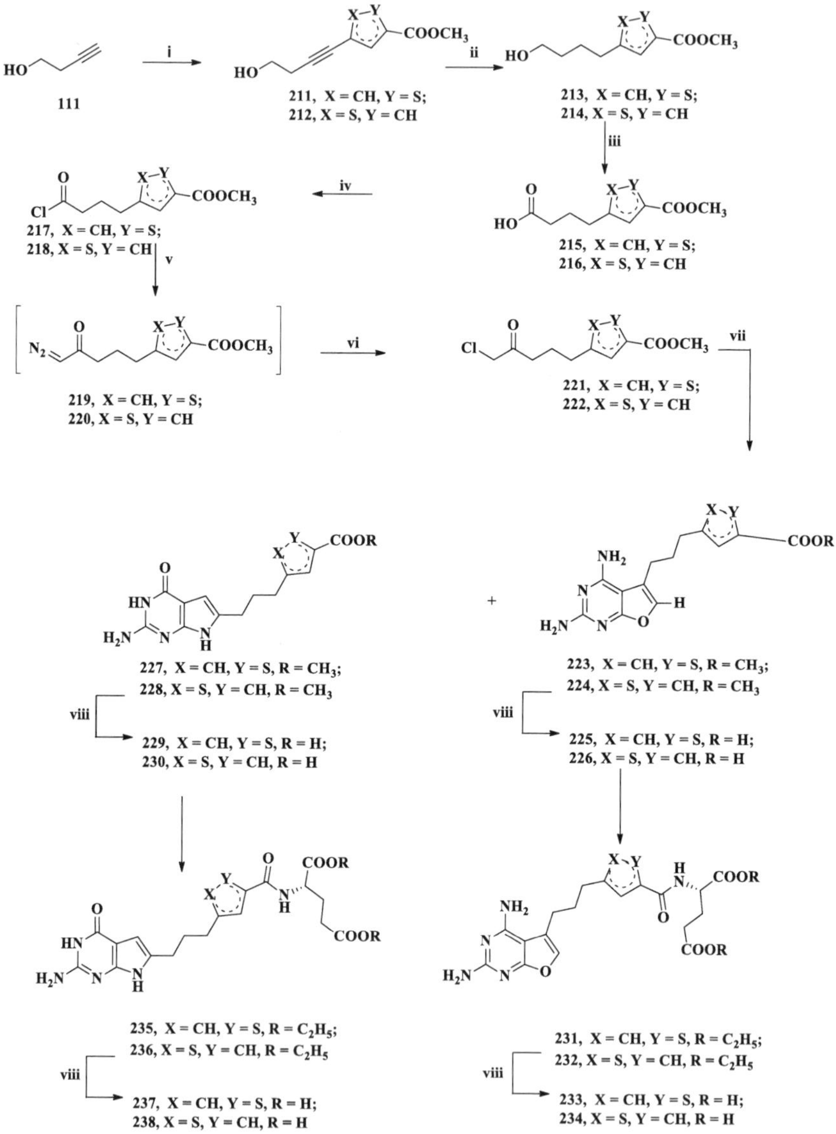

Scheme 5.17 Reagents and conditions: (i) methyl 5-bromothiophene-3-carboxylate or methyl 4-bromothiophene-2-carboxylate/$PdCl_2$/CuI/TEA/PPh_3/CH_3CN/100°C/6 h; (ii) 10% Pd/C/H2/55 psi/CH_3OH/4 h; (iii) 2.2 equiv. H_5IO_6/2 mol% PCC/CH_3CN/0°C/1 h; (iv) CH_2Cl_2/$(COCl)_2$/reflux/1 h; (v) azomethylene/Et_2O/RT/1 h; (vi) HCl/reflux/1.5 h; (vii) compound **24**/DMF/60°C/3 days; (viii) (i) 1 N NaOH; (ii) 1 N HCl; (i) 2-chloro-4,6-dimethoxy-1,3,5-triazine/*N*-methylmorpholine/diethyl L-glutamate hydrochloride/DMF/RT/12 h.

2.4 Antifolates targeting purine biosynthesis

5-Substituted pyrrolopyrimidine antifolates **244–249** with a benzoyl ring in the side chain and bridge carbons acted as antitumor agents.

From a modified method of the PMX synthesis [89], compounds **244–249** were accessed utilizing an α-bromo aldehyde condensation with the oxopyrimidine **24** (Scheme 5.18). Hence, a Heck coupling reaction [95] of aryl iodide **239** with compounds **111a–f** gives the coupled, unsaturated alcohols that rearrange to the vinyl alcohols and tautomerize to yield aldehydes **240a–f** [95]. Compounds **240a–f** were α-brominated with the

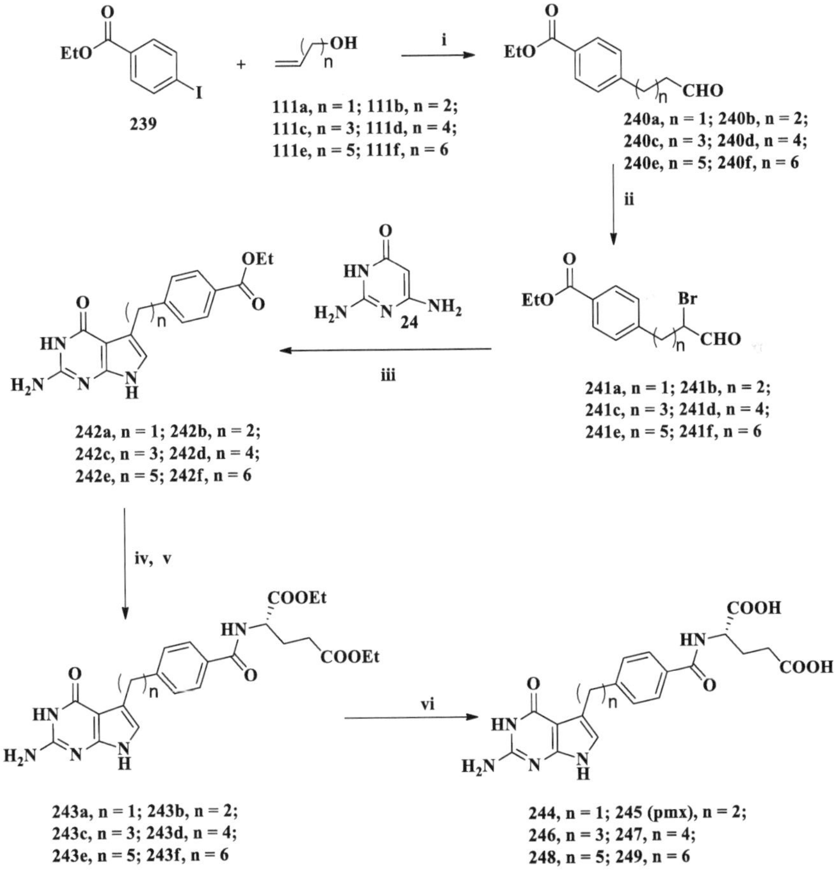

239

111a, n = 1; **111b**, n = 2; **111c**, n = 3; **111d**, n = 4; **111e**, n = 5; **111f**, n = 6

240a, n = 1; **240b**, n = 2; **240c**, n = 3; **240d**, n = 4; **240e**, n = 5; **240f**, n = 6

241a, n = 1; **241b**, n = 2; **241c**, n = 3; **241d**, n = 4; **241e**, n = 5; **241f**, n = 6

242a, n = 1; **242b**, n = 2; **242c**, n = 3; **242d**, n = 4; **242e**, n = 5; **242f**, n = 6

243a, n = 1; **243b**, n = 2; **243c**, n = 3; **243d**, n = 4; **243e**, n = 5; **243f**, n = 6

244, n = 1; **245 (pmx)**, n = 2; **246**, n = 3; **247**, n = 4; **248**, n = 5; **249**, n = 6

Scheme 5.18 Reagents and conditions: (i) $Pd(OAc)_2/LiAc/Bu_4NBr/70°C$; (ii) cycloisopropylidene dibromomalonate/HCl/RT. (iii) $CH_3OH/45°C/CH_3COONa$, (iv) 3 N NaOH; (v) NMM/2,4-dimethoxy-6-chlorotriazine/L-glutamic acid diethyl ester/rt.; (vi) 1 N NaOH/RT.

dioxodioxane [91] and generated the corresponding α-bromo aldehydes **241a–f**. Condensation of the latter compounds with compound **24** furnished compounds **242a–f**. Further hydrolysis followed by coupling with L-glutamic acid diethyl ester utilizing the chlorotriazine derivative and *N*-methyl morpholine as the activating agents yielded diesters **243a–f**. Subsequent saponification of the diesters generated compounds **244–249**.

The four-carbon-bridged compound **247** was the most potent derivative that inhibited the proliferation of FR-α-expressing Chinese hamster ovary and KB human tumor cells. Growth inhibition was totally or partially reversed in the presence of excess folic acid, revealing that FR-α is elaborated in cellular uptake, leading to S-phase accumulation and apoptosis. Antiproliferative impacts of compound **247** toward KB cells were protected via a surplus of adenosine (not thymidine), demonstrating de novo purine nucleotide biosynthesis as the desired pathway. However, the incomplete AICA protection indicated the inhibition of both glycinamide ribonucleotide formyltransferase (GARFTase) and AICA ribonucleotide formyltransferase (AICARFTase). Inhibition of AICARFTase and GARFTase by compound **247** was established via cellular metabolic assays leading to ATP pool depletion. The antifolate **247** is considered a dual inhibitor of AICARFTase and GARFTase because of essential mechanism of action [96].

The synthesis of the substituted glutamic acid **254** was described, which can be considered as a ring-contracted analog of the dideaza-tetrahydrofolic acid (DDATHF, **255**) (Fig. 5.10). As an inhibitor of the growth of human (CCRF-CEM) lymphoblastic leukemic cells in vitro, this compound displays a remarkable potency. It apparently acts through blocking de novo purine biosynthesis via the inhibition of GAR FTase.

Compound **249** underwent monobromination to yield compound **250**. Standard Pd-catalyzed coupling of compound **250** with dimethyl *N*-(4-ethynylbenzoyl)-L-glutamate (**251**) [97] was performed to generate

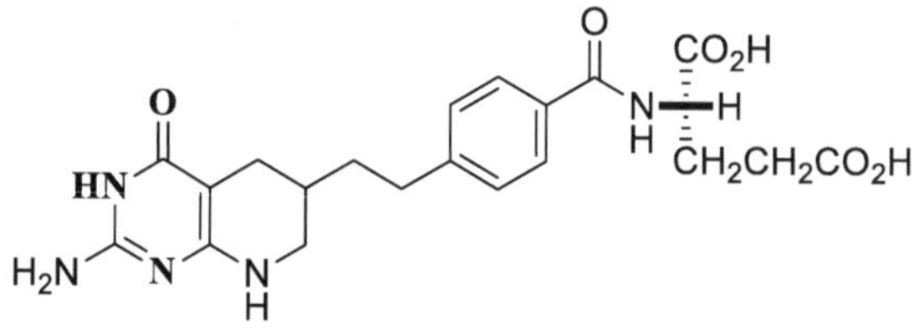

Fig. 5.10 Dideaza-tetrahydrofolic acid (DDATHF).

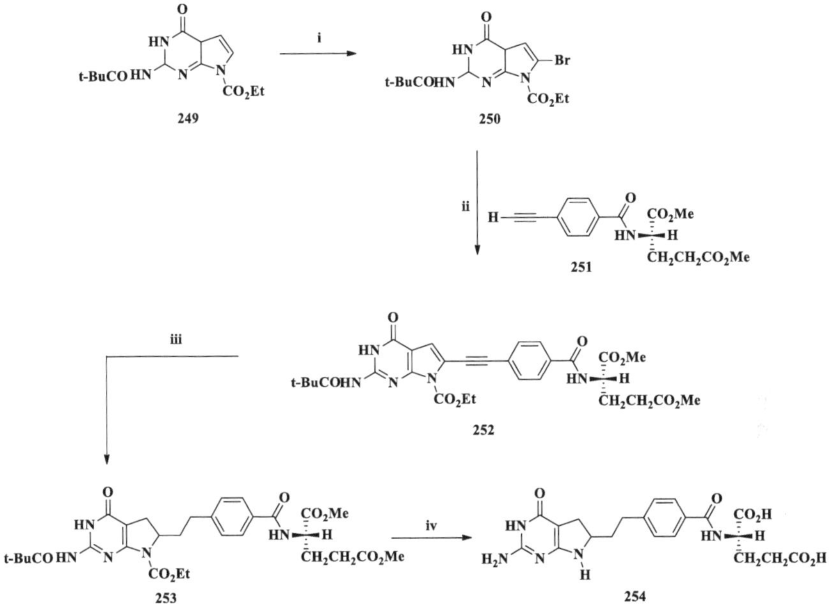

Scheme 5.19 Reagents and conditions: (i) NBS/CH_2Cl_2, (ii) $Pd(PPh_3)_4$/TEA/CuI; (iii) H_2/Pd/C; (iv) 1 N NaOH/HCl.

compound **252**; catalytic reduction of the pyrrole and acetylenic systems afforded compound **253**, the deprotection was followed to yield the desired compound **254** (Scheme 5.19).

Biological estimation indicated that **254** was a potent inhibitor of the growth of human (CCRF-CEM) lymphoblastic leukemic cells in vitro (IC_{50} = 0.071 μg/mL). This study shows that the potency of compound **254** is due to the inhibition of GAR FTase. So, compound **254** is considered as an inhibitor of purine biosynthesis [98].

2.5 Inhibitors of folylpolyglutamate synthetase (FPGS)

DL-3,3-Difluoroglutamic acid **256** was transformed to its di *tert*-butyl ester, compound **257**, via acid-catalyzed reaction with the 2-methylpropene. The EDC/HOBt-mediated coupling of compounds **257** and **258** followed by TFA and piperidine deprotection generated DL-β,β-difluorofolic acid **259** (Scheme 5.20).

Compound **259** is considered as a superior substrate for human CCRF-CEM folylpoly-γ-glutamate synthetase than folic acid ((*V*/*K*) c. seven times more prominent). Along these lines, the replacement of the

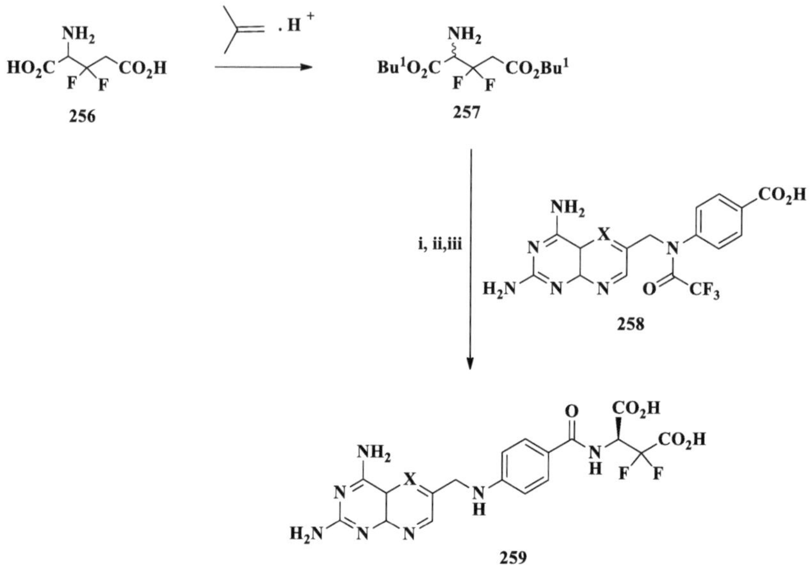

Scheme 5.20 Reagents and conditions: (i) EDC/HOBt; (ii) TFA; (iii) 0.1 M pip.

glutamate moiety of folic acid and methotrexate with 3,3-difluoroglutamic acid and 4-fluoroglutamic acid leads to folates and antifolates with varied polyglutamylation potency [99].

Synthesis of DL-γ,γ-F2MTX (**266**) was demonstrated via a previously reported method for the synthesis of MTX [100]. DL-4,4-F_2Glu (**260**) was first transformed to the corresponding di-*tert*-butyl ester, **261**, through acid-catalyzed reaction with the 2-methylpropene. Compound **261** was condensed with the *N*-Cbz-protected derivative of *p*-(*N*-methylamino) benzoic acid to furnish the difluoroglutamate (**263**), which was deprotected via catalytic hydrogenation to afford substituted difluoroglutamic acid di-*tert*-butyl ester (**264**). Coupling of compound **264** and the pteridinediamine hydrobromide in the presence of dimethylacetamide, and then further treating with trifluoroacetic acid yielded the DL-γ,γ-F_2MTX (**266**). Further DEAE-cellulose anion-exchange chromatography accessed pure DL-γ,γ-F_2MTX (**266**) as the triethylammonium salt from compound **264** (Scheme 5.21).

DL-4,4-Difluoroglutamic acid (DL-4,4-F_2Glu) and its methotrexate analog, DL-γ,γ difluoromethotrexate (DL-γ,γ-F2MTX), were prepared and estimated as alternate substrates or inhibitors of folate-dependent enzymes. The nitroaldol reaction of a difluorinated aldehyde ethyl hemiacetal with ethyl nitroacetate furnished DL-4,4-F_2Glu. The ligation of methotrexate to

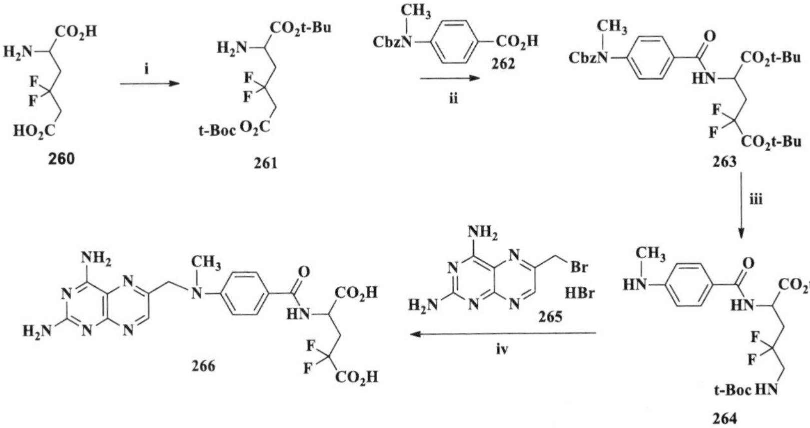

Scheme 5.21 Reagents and conditions: (i) 2-methylpropene/H+/CH_2Cl_2; (ii) DCC/HOBt/DMF; (iii) $H_2/Pd(OH)_2$/EtOH; (iv) TEA.

DL-4,4-F_2Glu, catalyzed by human FPGS, showed that DL-4,4-F_2Glu is a poor alternate substrate. DL-γ,γ F2MTX was prepared using *N*-[4-(methylamino)benzoyl]-4,4-difluoroglutamic acid di-*tert*-butyl ester; then the alkylation with the substituted pteridinediamine hydrobromide was followed. It was found that the DL-γ,γ-F2MTX was neither a substrate nor an inhibitor of the human FPGS. However, the fluorinated analog of MTX inhibits DHFR and cell growth with the same activity as MTX [101].

The difluoroornithine (AMPte-DL-4,4-F2Orn, **270**) was prepared and estimated as an inhibitor of DHFR, FPGS, and cell growth. Synthesis of compound **270** involved the utility of a protected form of DL-4,4-difluoroornithine **267**, which was derived from the difluoroglutamic acid. A pteroyl moiety was incorporated into compound **267** as previously reported by Piper et al. [18,19]. Compound **267** was condensed with the *N*-Cbz-protected derivative of *N*-methyl-4-aminobenzoic acid to yield compound **261**, which was then deprotected via catalytic hydrogenation to afford compound **263** (Scheme 5.22). Coupling of compound **263** to the pteridine diamine hydrobromide in dimethylacetamide followed by hydrolysis afforded compound **270**.

Biological potencies of **270** were compared directly to the corresponding nonfluorinated compound AMPte-L-Orn **271** (Fig. 5.11).

The fluorinated compound is a potent inhibitor of DHFR, but it is considered as a poor inhibitor of FPGS. Though, the analog is transported across the cell membrane and inhibits cell growth, as a result of the inhibition of

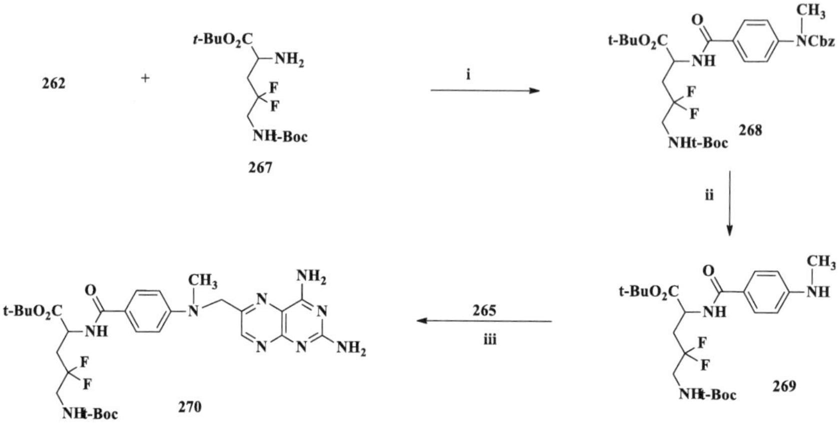

Scheme 5.22 Reagents and conditions: (i) DCC/HOBt/DMF; (ii) $H_2/Pd(OH)_2/EtOH$; (iii) TFA.

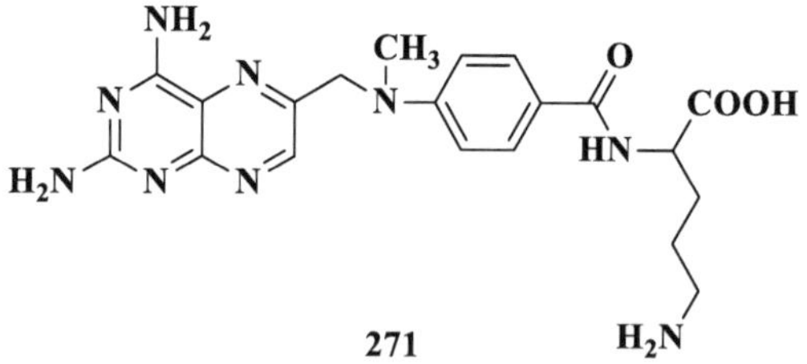

Fig. 5.11 AMPte-L-Orn.

DHFR. The data of the fluorinated compound oppose those of the corresponding nonfluorinated analog **271**, which is an active inhibitor of both DHFR and FPGS, but indicates very low cytotoxicity because of the poor transport.

The novel MTX derivative **270** was first tested as an inhibitor of human FPGS. Compound **270** was observed to be without inhibitory potency at concentrations as high as 150 íM. Conversely, the nonfluorinated compound **271** was affirmed as a strong inhibitor of FPGS. Despite the fact that the lack of FPGS inhibition by compound **270** was frustrating, other aspects of antifolate potency (i.e., inhibition of cell growth and DHFR) were pursued in order to evaluate the capability of compound **271** to cross cell membranes and to compare the cytotoxic impacts of compound **270** versus compound **271**. Compound **271** is an active inhibitor of DHFR and in respect to both the ornithine analog **271** and MTX. Given the potent inhibition of DHFR by compound **270**, this novel MTX derivative was estimated for its impact on the development of CCRF-CEM human leukemia cells. The study indicates that compound **270** is more active than is compound **271**. The growth and DHFR inhibitory potency study demonstrate that transport of the fluorinated ornithine analog **270** into cells is extensively improved over the nonfluorinated species **271** [102].

10-Deaza modifications of antifolate compounds having the disubstituted naphthalene ring instead of the benzene ring were synthesized and evaluated for antitumor potency. Substituted naphthalenes **274a–c** of 5-methyl-5,10-dideazaaminopterin, 5-ethyl-5,10-dideazaaminopterin, and 10-deazaaminopterin were synthesized through *C*-alkylations of the anion derived from 4-carboxy-1-naphthalene acetic acid dimethyl ester (**268**) by the diaminopteridine (**268a**) and 6-(bromomethyl)-2,4-diamino-5-methyl- and -5-ethyl-5-deazapteridines (**268a** and **268b**) followed by ester hydrolysis and then decarboxylation to afford naphthalene derivatives (**272a–c**) of the deazapteroic acid and the dideazapteroic acids. L-Glutamic acid dialkyl ester coupled with compound **272a–c** followed by mild ester hydrolysis yielded the desired compounds **274a–c** (Scheme 5.23).

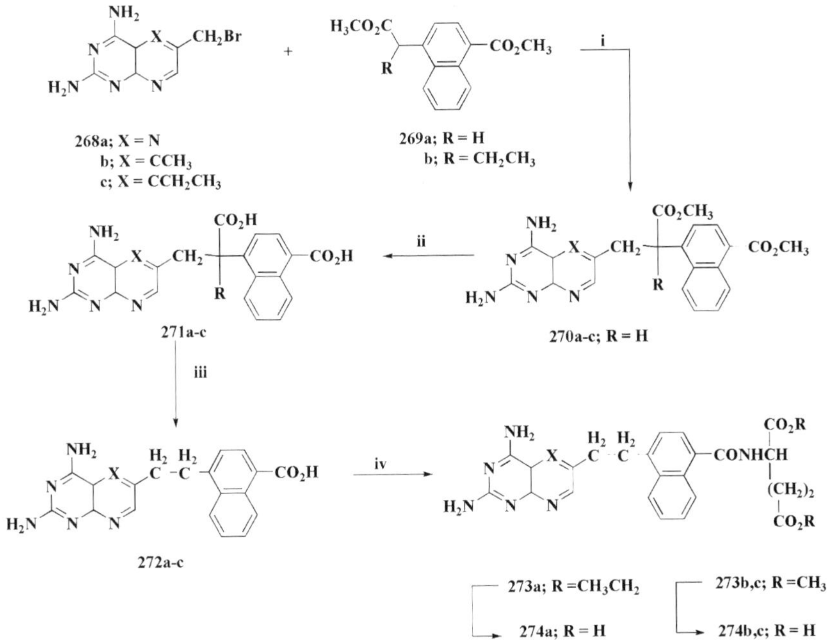

Scheme 5.23 Reagents and conditions: (i) NaH/DMF; (ii) ester hydrolysis; (iii) DMSO/reflux; (iv) peptide coupling.

Growth inhibition evaluations against three tumor cell lines (S180, L1210, and HL60) indicated the 10-DAM naphthalene analog **274a** to be four to six times more inhibitory than MTX but not as inhibitory as the deazaaminopterin; **274b** and **274c** were no more inhibitory than MTX. In tests against the E0771 mammary adenocarcinoma in mice, compound **274a** was less potent than MTX [103].

Compounds **284** and **290**, the thiophene analogs of the dideazafolic acid, were prepared. The 2-amino group was pivaloyled, followed by bromination, condensation with compound **282**, and then deprotection generated compound **284**. Compound **275** was treated with acetic anhydride to yield substituted benzoxazinone **285**, which on reaction with ammonia, sodium hydroxide was transformed to the dihydroquinazolin-4-one (**287**). Compound **287** was brominated, followed by condensation with compound **282** and ester cleavage, afforded compound **290** (Schemes 5.24 and 5.25).

As inhibitors of tumor cell growth in culture, the compounds were evaluated. Compounds **284** and **290** have IC_{50} values of 1.8 ± 0.1 and 2.1 ± 0.8 mM against CCRF-CEM human leukemic lymphoblasts [104].

Novel tetrahydropyridopteridine derivative **293** with various benzoyl and a benzyl substitution at the N7 position was prepared, as classical and nonclassical, linear tricyclic 5-deaza antifolates. Hydrolysis of compound **291** yielded the corresponding acid **292**. Coupling of the latter with

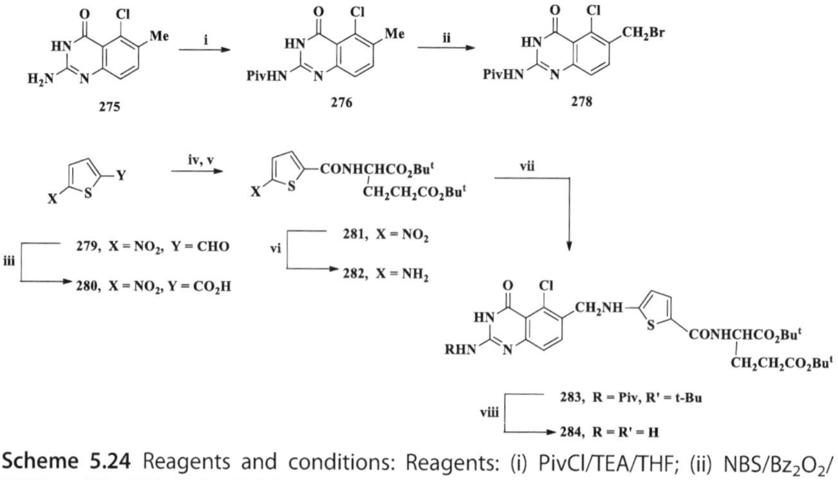

Scheme 5.24 Reagents and conditions: Reagents: (i) PivCl/TEA/THF; (ii) NBS/Bz_2O_2/$CHCl_3$; (iii) $KMnO_4$/NaH_2PO_4/aqMe_2CO; (iv) $SOCl_2$; (v) bis(2-methyl-2-propanyl) L-glutamate/TEA/CH_2Cl_2; (vi) iron powder, $FeSO_4$, aq CH_3OH; (vii) **278**, $NaHCO_3$/DMF; (viii) 1:3 TFA–CH_2Cl_2/5°C.

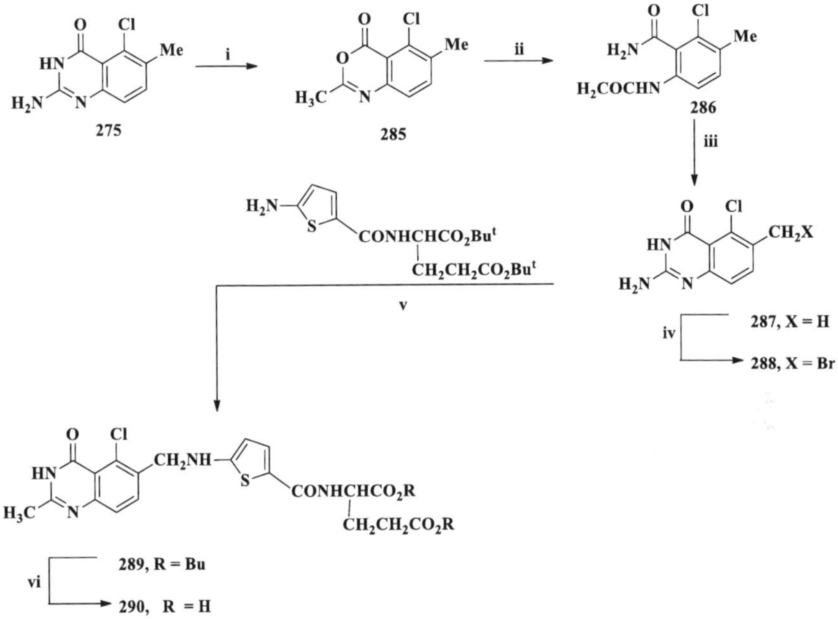

O Cl HN Me H_2N N **275** → (i) → O Cl O Me H_3C N **285** → (ii) → O Cl H_2N Me H_3COCHN **286** ↓ (iii)

O Cl HN CH_2X H_2N N

iv: **287, X = H** → **288, X = Br**

H_2N– S $CONHCHCO_2Bu^t$ $CH_2CH_2CO_2Bu^t$ (v)

O Cl HN CH_2NH– S $CONHCHCO_2R$ $CH_2CH_2CO_2R$ H_3C N

vi: **289, R = Bu** → **290, R = H**

Scheme 5.25 Reagents and conditions: (i) Ac_2O/reflux; (ii) NH_3/-33°C; (iii) 1 N NaOH/reflux; (iv) NBS/Bz_2O_2/$CHCl_3$; (v) $NaHCO_3$/DMF; (vi) 1:3 TFA–CH_2Cl_2/5°C.

[105] diethyl-L-glutamate followed by saponification [105] afforded the desired compound **293** (Scheme 5.26).

Biological estimation showed that this partial conformational modification for the classical analog **293** was detrimental to DHFR inhibitory

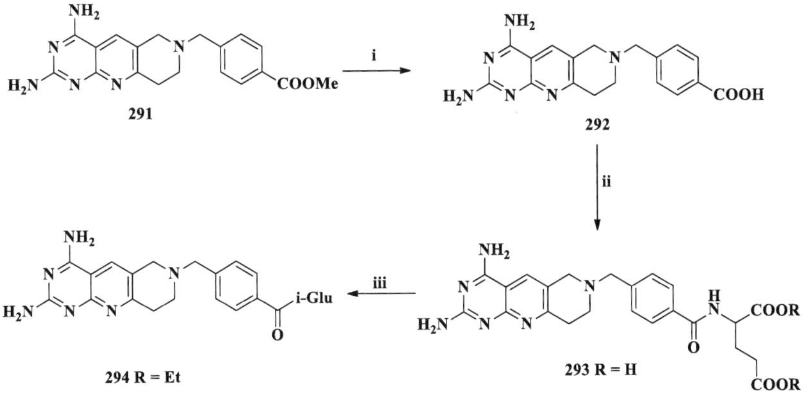

NH_2 N N H_2N N N COOMe **291** → (i) → NH_2 N N H_2N N N COOH **292** ↓ (ii)

NH_2 N N H_2N N N H N COOR O COOR **293 R = H** → (iii) → NH_2 N N H_2N N N i-Glu O **294 R = Et**

Scheme 5.26 Reagents and conditions: (i) 1 N NaOH/1 N HCl; (ii) *i*BuOCOCl/Et_3N/L-glutamic acid diethyl ester; (iii) 1 N NaOH/1 N HCl.

potency besides the antitumor potency comparable to methotrexate or 5-deaza MTX. However, the classical analog **293** was a superior substrate for FPGS than for MTX. These data indicate that a classical 5-deaza folate partially restricted through a bridge between the C7 and N10 positions retains FPGS substrate potency and that the antitumor potency of classical tricyclic analogs such as compound **293** would be affected by levels of FPGS in tumor systems [106].

Substituted furo- or pyrrolo[2,3-*d*]pyrimidines were synthesized as outlined in Scheme 5.27. The syntheses were accessed through the condensation of α-chloroketone with 2,6-diaminopyrimidin-4(3H)-one to yield the intermediates **294a** and **295a**, followed by hydrolysis, coupling, and saponification of the diethyl ester to generate compound **296b**.

With a single-carbon atom bridge, compounds **296b** are FPGS substrates. Compound **269b** is a highly effective FPGS substrate, revealing that the substituted furopyrimidines are significant lead structures for the design of antifolates with FPGS substrate potency. Compound **296b** are poor inhibitors of the growth of CCRF-CEM human leukemia cells in culture, demonstrating that single-carbon-bridged compounds in these series though conducive to FPGS substrate potency were not active inhibitors [107].

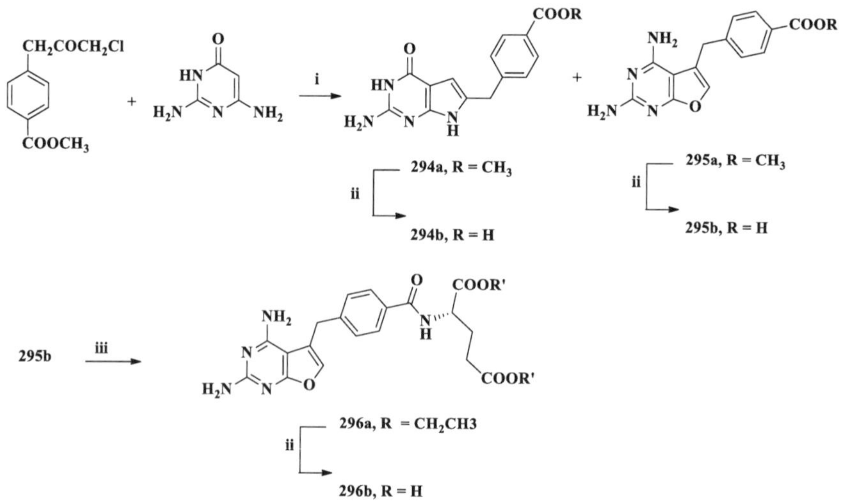

CH_2COCH_2Cl

$COOCH_3$

H_2N N NH_2 HN O

i

294a, R = CH_3

294b, R = H

295a, R = CH_3

295b, R = H

ii

295b iii

COOR'

296a, R = CH_2CH3

296b, R = H

Scheme 5.27 Reagents and conditions: (i) DMF/2 days/50–60°C; (ii) 1 N NaOH/RT/2 h/3 N HCl/0°C; (iii) *i*-BuOCOCl/Et_3N/L-glutamic acid diethyl ester/0°C—RT; (iv) 1 N NaOH/0°C/3–4 h/3 N HCl/0°C.

2.6 Dual inhibitors

6-Substituted pyrrolopyrimidine derivatives (**298**, **300a**–**300c**, and **300h**) have been furnished as accomplished in Scheme 5.28. The antiproliferative potencies of these compounds against A549, HL60, Hela, H1299, HT29, and HCT116 tumor cells were estimated. Most of the compounds displayed micromolar antiproliferative activities. Compound **300h**, the most active one, has GI50 value of 8.92, 1.72, and 0.73 μM against HL60, H1299, and A549 cells. Compound **300h** could elevate the G2/M-phase cell accumulations and that was exhibited through the cell cycle distribution assay. Compound **300h** indicated low activity in the induction of apoptosis. However, the formation of A549 cell colony was inhibited. These revealed that the tumor cell death depended on the irreversible impact of compound **300h** on clonogenicity and cell proliferation. It was implied via the determination of targeted pathway of compound **300h** that the antiproliferative activities of compound **300h** act via the dual inhibition of TS-DHFR [108].

The ethyl 4-aminobenzoate and but-3-en-2-one were exposed to Michael addition to generate the β-keto amine **301**. Selective α-bromination of compound **301** afforded compound **302**. Protection of

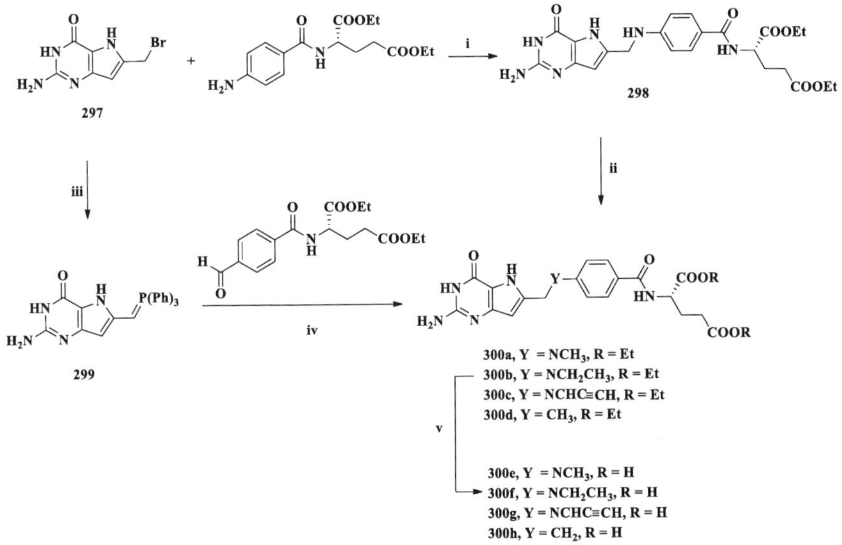

Scheme 5.28 Reagents and conditions: (i) DMAC/80°C; (ii) CH_3I/CH_3CH_2Br or 3-bromo-1-propyne/K_2CO_3/DMF/RT; (iii) $(C_6H_5)_3P$/anhydrous s DMF/argon/60°C/2 h/NaOMe/RT/1 h; (iv) (i) DMF/60°C/24 h; (ii) H_2, Pd/C; (v) 1 mol/LNaOH/RT.

the amine in **302** and further condensation with 2,6-diamino-3*H*-pyrimidin-4-one yielded the pyrrolopyrimidine derivative **303a**. Compound **303a** was hydrolyzed to give the corresponding free acid **303b**. Further coupling with (*S*)-di-tert-butyl 2-aminopentanedioate hydrochloride using the activating agent 1,3,5-triazine furnished the diester **304a**. The di-tert-butyl ester was deprotected to yield the corresponding acid **304b**. The reaction of the latter compound with formic acid and acetic anhydride, or trifluoroacetic anhydride, yielded compounds **305a–c** (Scheme 5.29).

Desired antifolates were generated and evaluated for selective cellular uptake via FRs or the PCF transporter. Toward engineered Chinese hamster

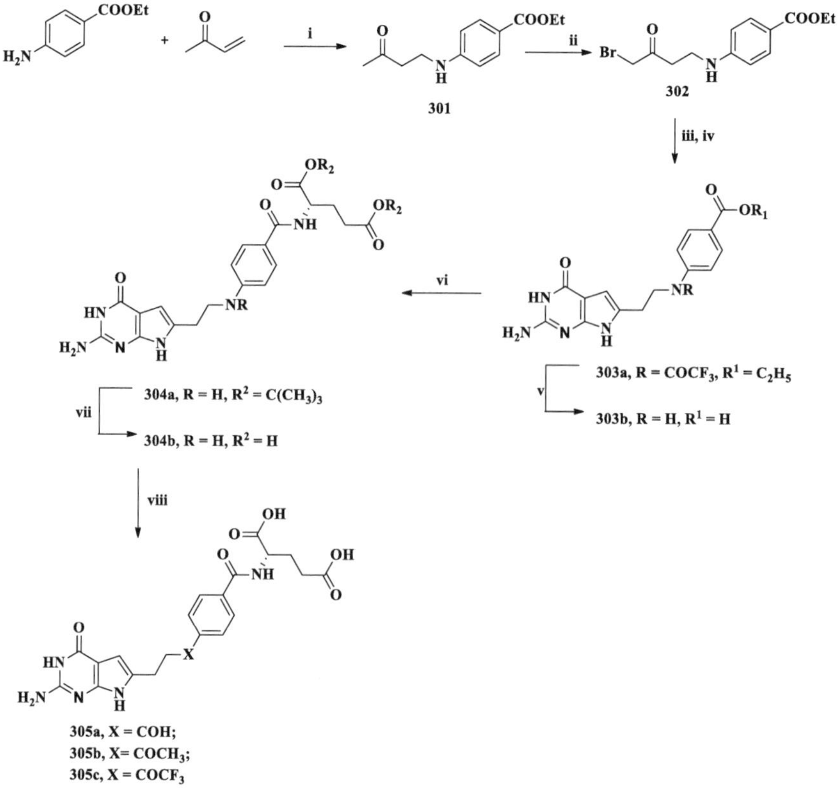

301

302

303a, R = $COCF_3$, $R^1 = C_2H_5$

303b, R = H, R^1 = H

304a, R = H, $R^2 = C(CH_3)_3$

304b, R = H, R^2 = H

305a, X = COH;
305b, X= $COCH_3$;
305c, X = $COCF_3$

Scheme 5.29 Reagents and conditions: (i) EtOH/reflux/5 h, 59%; (ii) CH_3COOH/33% HBr/Br_2/RT/2.5 h; (iii) $(CF_3CO)_2O$/RT/overnight; (iv) compound **24**/DMF/RT/3 days; (v) 1 N NaOH/RT/10 h; (vi) CDMT/$O(CH_2CH_2)_2NCH_3$/(*S*)-di-tert-butyl 2-aminopentanedioate hydrochloride/DMF/12 h; (vii) CH_2Cl_2/CF_3COOH/RT/2 h; (viii) $(CF_3CO)_2O$/RT/4 h or $(CH_3CO)_2O$/RT/12 h or $HCOOH/(CH_3CO)_2O$/1 h/reflux.

ovary cells expressing FRs, high in vitro antiproliferative potency was demonstrated by compounds **304b–305a–c**. Compounds **304b–305a–c** inhibited glycinamide ribonucleotide formyltransferase (GARFTase) and de novo purine biosynthesis. The bound conformations of compound **302** enquired that flexibility for attachment to both GARFTase and FR-α was indicated via the X-ray crystal structures for compound **304b** with GARFTase and FR-α. In mice bearing IGROV1 ovarian tumor xenografts, compound **304b** was effective. It was established that heteroatom substitutions in the 3-atom bridge region of the 6-substituted pyrrolopyrimidines provide desired antifolates that warrant further estimation as anticancer agents [109].

The substituted L-glutamic acid **318** was generated as a dual inhibitor of dihydrofolate reductase (DHFR) and thymidylate synthase (TS) and as an antitumor agent. The preparation of compound **318** was furnished through a multistep sequence, which involved the synthesis of the pyrrolo-[2,3-*d*] pyrimidine **311** from alpha-acetyl-gamma-butyrolactone. Protection of the —NH group of compound **312** and the 5-position regioselective iodination, then palladium-catalyzed coupling, was followed to yield intermediate **315**, which was transformed to compound **318** through reduction and saponification (Scheme 5.30).

The pyrrolopyrimidine ring in a 2,4-diamino binding mode in which the pyrrole nitrogen mimics the 4-amino moiety of the diaminopyrimidines was indicated via the determination of the X-ray crystal structure of the ternary complex of compound **318**, DHFR, and NADPH. This classical antifolate has an alternate mode of binding to DHFR. Compound **318** was more inhibitory than LY231514 against TS from *Escherichia coli* and *Lactobacillus casei*. Compound **318** was also more inhibitory against DHFR from human, *Pneumocystis carinii, and Toxoplasma gondii*. Estimation of **318** against MTX-resistant cell lines with identified mechanisms demonstrates that cross-resistance of **318** was much lower than that of MTX. Folylpoly-γ-glutamate synthetase studies and metabolite protection estimations propose that the antitumor potency of **318** against the growth of tumor cells in culture is due to dual inhibition of DHFR and TS. The analog **318** inhibited the growth of FaDu and CCRF-CEM cells in culture (ED_{50} = 12.5 and 7.0 nM) and was more potent against FaDu cells than MTX. Analog **318** was preclinically tested in the National Cancer Institute in vitro antitumor screening program and yielded IG_{50} values in the nanomolar range against a variety of tumor cell lines [110].

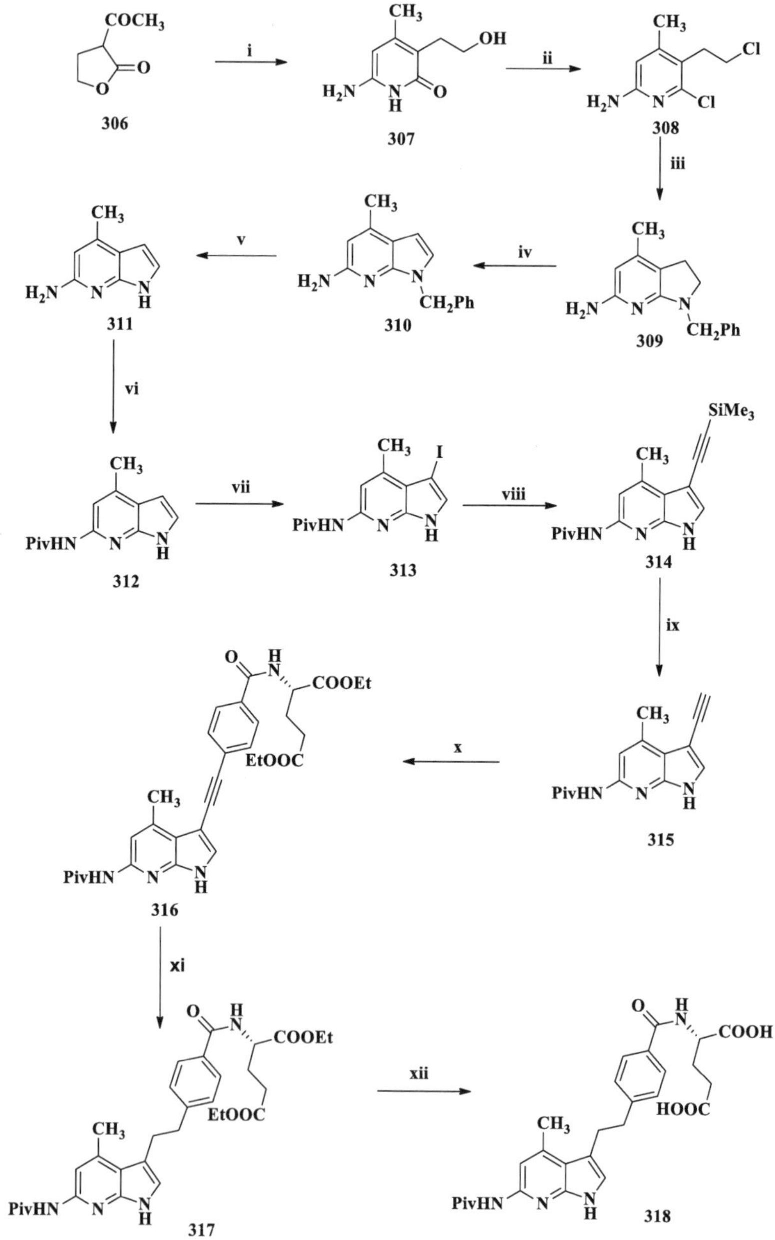

Scheme 5.30 Reagents and conditions: (i) EtOH/guanidine carbonate/TEA/reflux; (ii) $POCl_3$/reflux; (iii) $C_6H_5CH_2NH_2$/TEA/*n*-BuOH/90°C; (iv) MnO_2/1,4-dioxane/reflux; (v) Na/liq NH_3/−78°C. (vi) PivCl/pyridine/reflux; (vii) NIS/dark, RT; (viii) CuI/$HC_2Si(CH_3)_3$/Pd[P$(C_6H_5)_3]_4$/TEA/THF/RT; (ix) *n*-Bu4NF/THF/RT; (x) CuI/N-4-iodobenzoyl-L-glutamic acid diethyl ester/Pd[P$(C_6H_5)_3]_4$/TEA/THF/RT; (xii) 5% Pd/C/H_2/50 psi; (xiii) 1 N NaOH/CH_3OH/50°C.

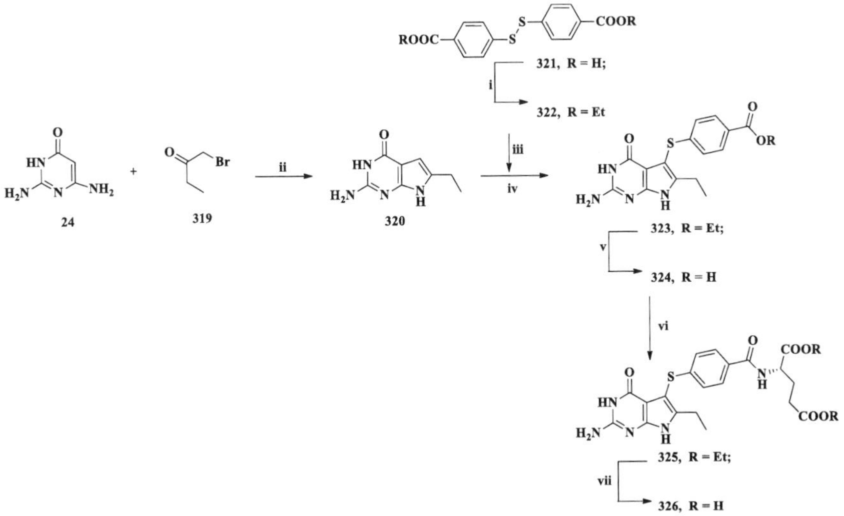

Scheme 5.31 Reagents and conditions: (i) $NaHCO_3$/EtI/RT/120 h; (ii) DMF/40–50°C/72 h; (iii) EtOH $NaBH_4$/RT, 30 min; (iv) EtOH/I_2/H_2O (2:1)/100–110°C; (v) 1 N NaOH/80°C/24 h; (vi) TEA/IBuOCOCl/L-glutamic acid diethyl ester hydrochloride/0°C—RT; (vii) 1 N NaOH/0°C/4 h, – RT/20 h.

A classical analog **326** as potential dual DHFR/TS inhibitors and as antitumor agents was synthesized (Scheme 5.31). The key intermediate in their preparation was the dihydropyrrolopyrimidine, **320**, to which different aryl thiols were 5-position attached through an oxidative addition reaction. For compound **326**, the ester furnished from the reaction was deprotected and coupled with L-glutamic acid diethyl ester and then saponification. Compound **326** was an active dual inhibitor of human DHFR (IC_{50}) 420 nM) and human TS (IC_{50}) 90 nM). Compound **326** was not a substrate for human FPGS. Metabolite protection estimations demonstrated TS as its essential target [111].

The antifolate **335** was synthesized and estimated as an antitumor agent. Compound **335** was furnished from the pyrrolopyrimidine **331**. Compound **335**, 2,4-diamino classical antifolate, has a potent inhibitory potency against both human thymidylate synthase (TS) and human dihydrofolate reductase (DHFR). Subsequent estimation of the mechanism of action of compound **335** implicated DHFR as its essential intracellular target. Compound **335** was FPGS substrate. The analog **335** also inhibited the growth of different human tumor cell lines in culture ($GI_{50} < 10^{-8}$ M). This evaluation demonstrates that the pyrrolopyrimidine scaffold is conducive to dual TS-DHFR and tumor inhibitory potency, and the activity is identified by the 4-position substituent [112].

The synthesis of compound **335** started from hydroxyacetone **327** (Scheme 5.32), which was first converted into the substituted furan **329** with malononitrile. Further condensation with guanidine furnished compound **331** [113]. 4-Sulfanylbenzoic acid was oxidatively added to the 6-position of compound **331** to afford compound **333** following a modification of a previous method [114]. The acid **333** coupled with diethyl L-glutamate utilizing the mixed anhydride method yielded the diester **334**. Hydrolysis of the coupled product **334**, followed by acidification, yielded the desired compound **335** (Scheme 5.32) [112].

The substituted glutamic acid **347** and its N-9 methyl analog **349** were furnished as significant dual inhibitors of dihydrofolate reductase (DHFR) and thymidylate synthase (TS) and as antitumor agents. The nonclassical antifolates, substituted furo[2,3-*d*]pyrimidines **339–342** with 3,4,5-trichloro, 3,4-dichloro, 3,4,5-trimethoxy, and 2,5-dimethoxy substituent, in the phenyl ring, were also prepared as potential inhibitors of DHFRs comprising those from *Toxoplasma gondii* and *Pneumocystis carinii*, which are organisms responsible for opportunistic infections in patients suffering from AIDS. The analogs were furnished through nucleophilic displacements of 2,4-diamino-5-(chloromethyl)furo[2,3-*d*]pyrimidine with substituted aniline or (*p*-aminobenzoy1)-L-glutamate. The key intermediate

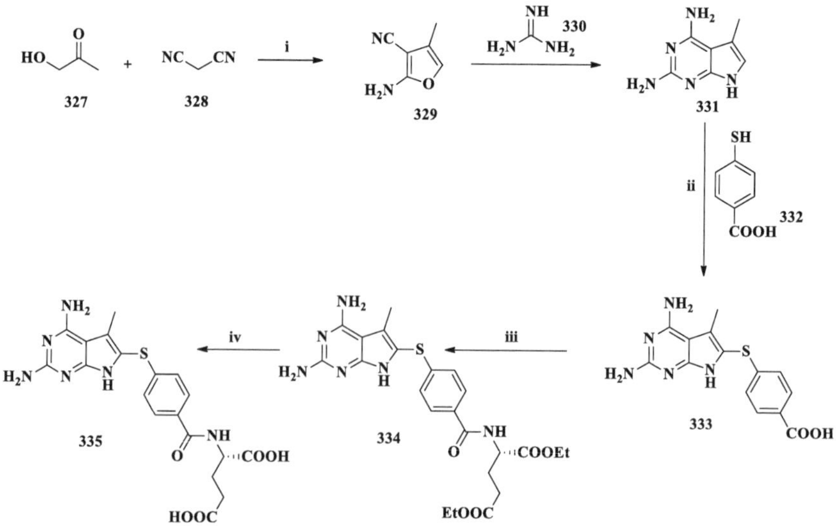

Scheme 5.32 Reagents and conditions: (i) NEt_3; (ii) $I_2/EtOH/H_2O$; (iii) *i*-BuOCOCl/TEA/L-glutamic acid diethyl ester; (iv) 1 N $NaOH/EtOH/H_2O$.

was in turn prepared from 1,3-dichloroacetone and 2,4-diamino-6-hydroxypyrimidine (Schemes 5.33 and 5.34)[115].

The compounds were in vitro evaluated against rat liver, (recombinant) human, *T. gondii*, *P. carinii*, and *Lactobacillus casei* DHFRs. The classical analogs revealed moderate to good DHFR inhibitory potency ($IC_{50}=10^{-6}$–10^{-8} M) with the N-CH_3 analog **349** about twice as active as **347**. The nonclassical analogs were inactive with IC_{50}s > 3 × 10–5 M. The classical analogs were also tested as inhibitors of TS (recombinant) human, human CCRF-CEM, and *L. casei*), GARFTase, and 5-aminoimidazole-4-carboxamide ribonucleotide formyltransferase and were observed to be inactive against these enzymes. Compound **349** was remarkably cytotoxic toward different tumor cell lines in culture. The nonclassical analogs were marginally potent. Both classical compounds were good substrates for human FPGS. Subsequent estimation of the cytotoxicity of compounds **347** and **349** in CCRF-CEM cells and its sublines, having defined mechanisms of methotrexate resistance, indicated that the analogs utilize the reduced folate/MTX-transport system and primarily

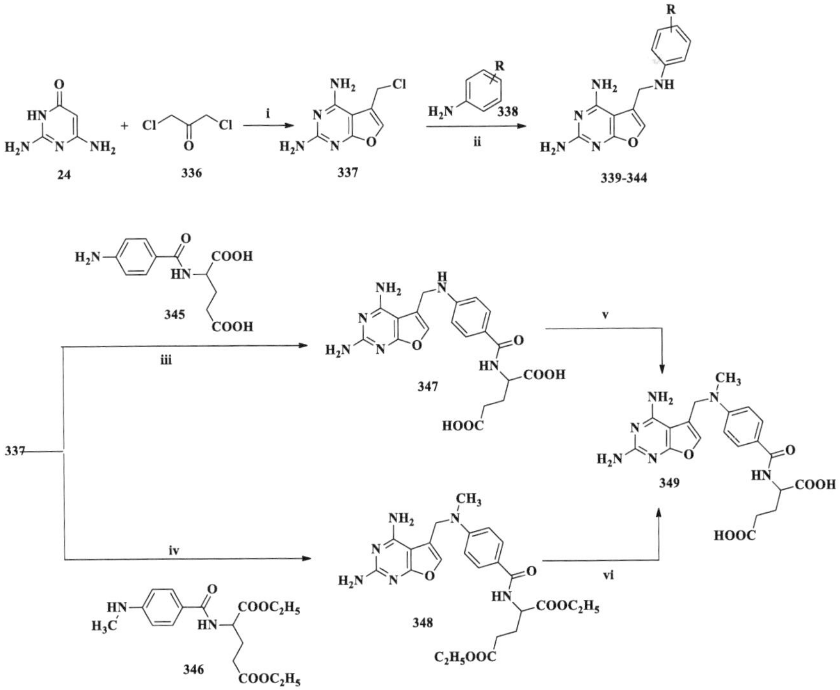

Scheme 5.33 Reagents and conditions: (i) DMF/RT.; (ii) K_2CO_3/DMSO; (iii) DMSO/45°C; (iv) DMSO/40°C; (v) HCHO/$NaCNBH_3$; (vi) 1 N NaOH/MeOH/H_2O.

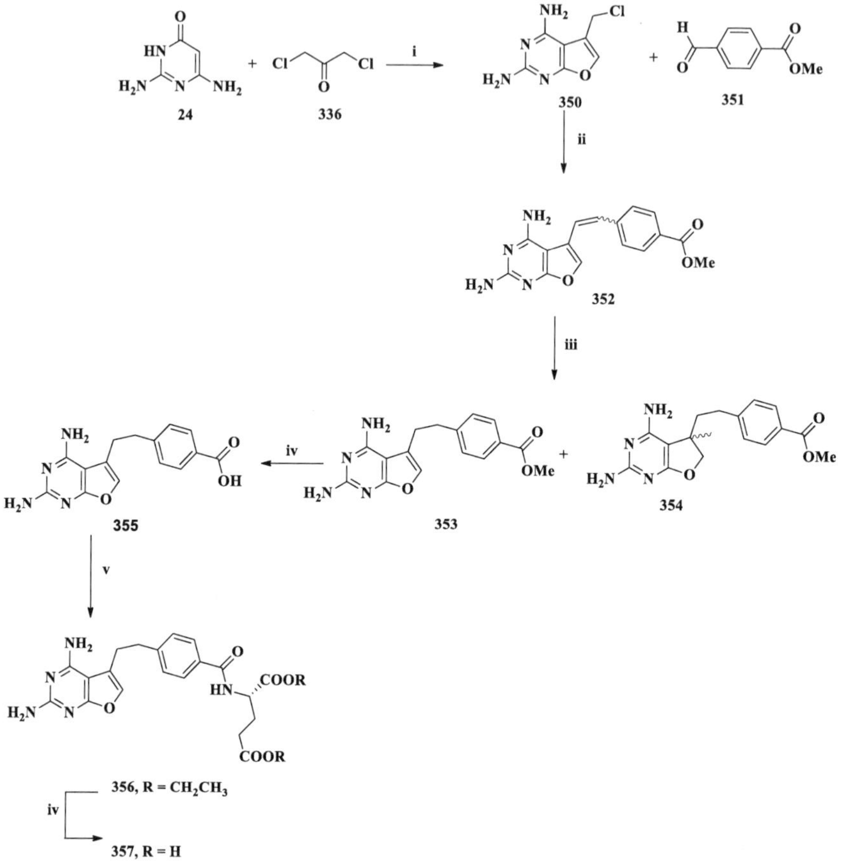

Scheme 5.34 Reagents and conditions: (i) DMF; (ii) Bu_3P/NaH/DMSO; (iii) H_2/25pai/MeOH/DMF; (iv) 1 N NaOH; (v) L-glutamic acid diethyl ester/*i*BuOCOCl/Et_3N.

inhibit DHFR and that poly-y-glutamylation was crucial to their mechanism of action. Protection estimations in the FaDu squamous cell carcinoma cell line showed that inhibition was totally reversed via leucovorin or the combination of hypoxanthine and thymidine. Besides, compounds **347** and **349** in differentiate to MTX, the FaDu cells were superior protected by thymidine alone than hypoxanthine alone, proposing a transcendently antithymidylate effect [115].

Classical 2,4-diaminofuro[2,3-*d*]pyrimidines are estimated as inhibitors of DHFR and TS and as antitumor agents. Particularly, —CH_2NHCH_2— and —CH_2CH_2— bridged analogs **363** and **357** were prepared (Scheme 5.35).

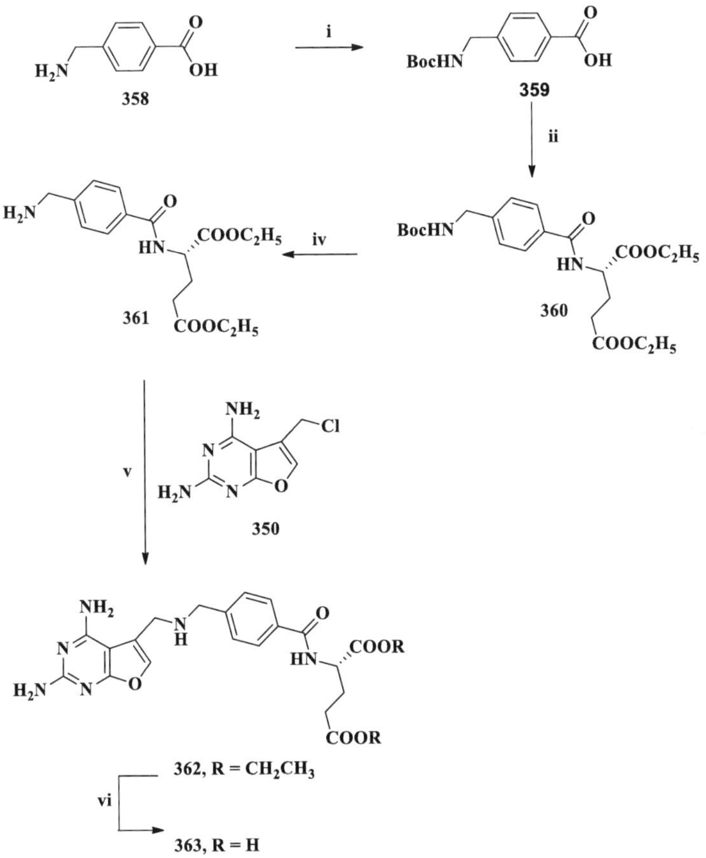

Scheme 5.35 Reagents and conditions: (i) Boc_2, NaOH, dioxane, H_2O; (ii) L-glutamic acid diethyl ester, *i*BuOCOCl, N-Me-morpholine; (iii) TFA, CH_2Cl_2; (iv) DMSO, K_2CO_3; (v) 1 N NaOH.

The in vitro estimation of the analogs as inhibitors of DHFRs from *Lactobacillus casei*, (recombinant) human, and human CCRF-CEM cells was studied. A moderate potency (IC_{50} 10^{-6}–10^{-7} M) was indicated by compound **357**. Compound **363** was inactive (IC_{50} M, CCRF-CEM). The compounds were also tested against TS from *L. casei* and (recombinant) human and were of low potency (IC_{50} M). The 3-atom-bridged analog **363** was more inhibitory to human TS than MTX.

Compound **363** revealed a low level of growth inhibitory potency. The inhibition of the growth of leukemia CCRF-CEM cells via both compounds parallels their inhibition of CCRF-CEM DHFR. Compound **357** was a good substrate for human FPGS derived from CCRF-CEM cells

(K_i 8.5 pM). Subsequent estimation of the growth inhibitory potency of compound **357** against the MTX-resistant subline of CCRF-CEM cells (RSOdm) with decreased FPGS showed that poly-γ-glutamylation was essential for its action. Protection evaluations with compound **357** in the FaDu squamous cell carcinoma cell line showed that inhibition was totally inverted by leucovorin or by a combination of hypoxanthine and thymidine, proposing an antifolate effect directed at DHFR [116].

The potencies of the 4-carbon atom-bridged antifolates on DHFR, TS, and FPGS as well as antitumor potency were reported. Two 4-carbon-bridged antifolates **370** and **372** (Scheme 5.36) have improved

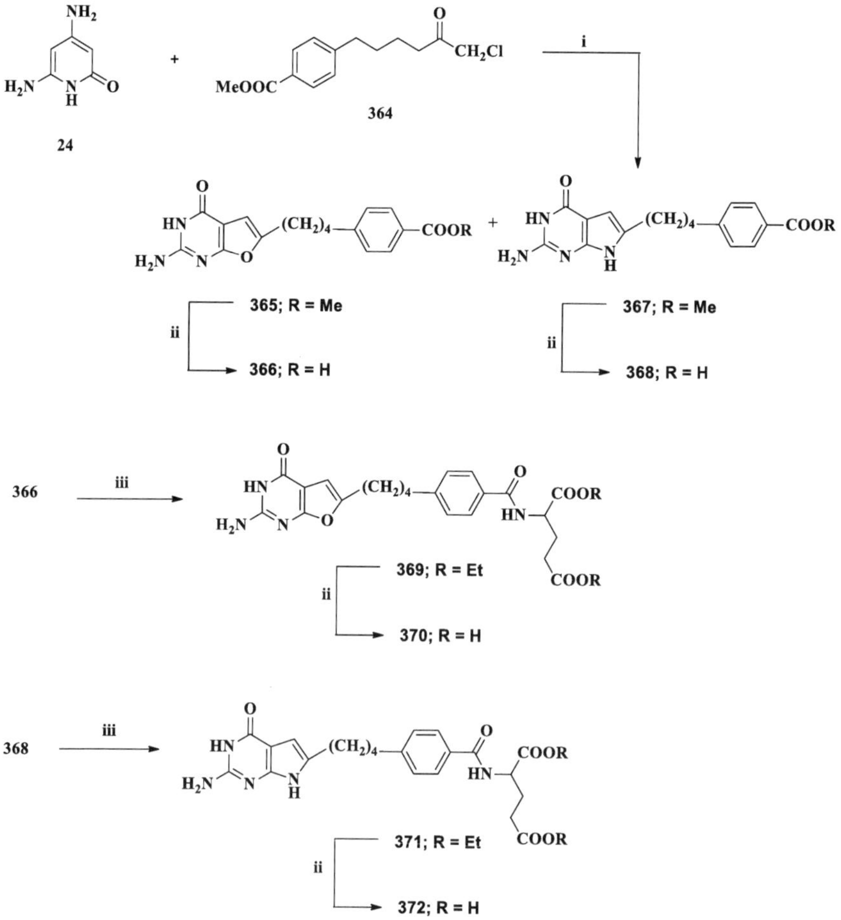

Scheme 5.36 Reagents and conditions: (i) DMF/N_2; (ii) NaOH; (iii) *i*BuOCOCl/TEA/DMF/N_2/L-glutamic acid diethyl ester.

FPGS substrate activity and inhibitory potency against tumor cells in culture ($EC_{50} \leq 10^{-7}$ M) in respect to the 2-carbon bridged compounds. These data affirm the hypothesis that the distance and orientation of the side chain *p*-aminobenzoyl-L-glutamate moiety related to the pyrimidine ring are a crucial determinant of the potency. Additionally, this evaluation shows that, for classical antifolates that are substrates for FPGS, poor inhibitory potency against isolated target enzymes is not essentially a predictor of a lack of antitumor potency [117].

The synthetic strategy for compounds **379m** is accomplished. The iodothienopyrimidin-4(3*H*)-one **376** exposed to microwave-assisted palladium-catalyzed coupling reactions with the aryl thiols yields compounds **379a–379m** and intermediate **377** for the preparation of the classical analog **379**. The thiophene-3-carboxylate derivative **374** was prepared from butyraldehyde **373** with sulfur, ethyl cyanoacetate, and triethylamine [118] It was [119] previously reported that ethyl 2-amino-5-methylthiophene-3-carboxylate reacted with the chloroformamidine hydrochloride to afford 2-amino-6-methylthieno[2,3-*d*]pyrimidin-4(3*H*)-one. Hence, heating a mixture of chloroformamidine hydrochloride and compound **374** afforded compound **375**. Under microwave irradiation, C5-bromination for the 2-amino-4-oxo-6-methylthieno[2,3-*d*]pyrimidine template was reported [120].

However, a mercuration methodology had reported [121] that could be adopted for the synthesis of the iodothienopyrimidin-4(3*H*)-one **376**. Extending this methodology to the synthesis of compound **376** required the 5-chloro-mercury derivative, which was accessed via mercuration of compound **375** with mercurate acetate in the presence of glacial acetic acid, followed by treatment with sodium chloride solution. Treatment of the mercury derivative with iodine gave compound **376**. The latter intermediate **376** was transformed to the desired compounds **379a–m** and **377**. Palladium-catalyzed cross-coupling reactions [122] to form carbon-sulfur bonds with arylthiols and arylbromides were required for the preparation of compounds **379a–m** and **377**. A mixture of compound **376**, *i*-Pr_2NEt, and arylthiols were heated using microwave irradiation to yield compounds **379a–m**. For the synthesis of the classical compound **379**, the dihydrothienopyrimidin-5-yl)thio]benzoate, **377**, was synthesized as shown for compound **379a–m**. Methyl 4-mercaptobenzoate was utilized to give compound **377**. Ester hydrolysis of compound **377** with sodium hydroxide yielded the corresponding free acid **378**. The acid **378** was coupled with 1,3,5-triazine and L-glutamic acid diethyl ester hydrochloride followed by purification to generate compound **380**. Hydrolysis of **380**, followed by acidification with hydrochloric acid, yielded compound **379** (Scheme 5.37) [123].

2.7 Multitargeted antifolate (MTA)

Synthesis of the pyrrolopyrimidine antitumor agent MTA (LY231514) was described. Manganese triacetate dihydrate-induced radical cyclization of the malonamide **380d** furnished the pyrrolidinone **381d**. Thiation of the latter

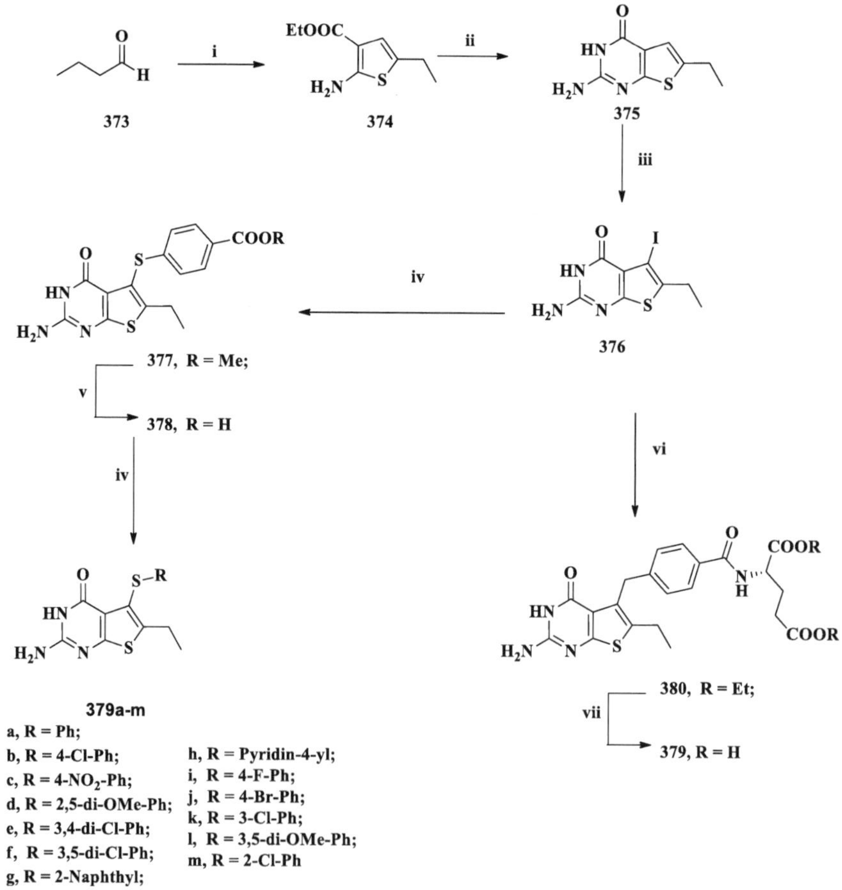

379a-m

a, R = Ph;
b, R = 4-Cl-Ph;
c, R = 4-NO_2-Ph;
d, R = 2,5-di-OMe-Ph;
e, R = 3,4-di-Cl-Ph;
f, R = 3,5-di-Cl-Ph;
g, R = 2-Naphthyl;
h, R = Pyridin-4-yl;
i, R = 4-F-Ph;
j, R = 4-Br-Ph;
k, R = 3-Cl-Ph;
l, R = 3,5-di-OMe-Ph;
m, R = 2-Cl-Ph

377, R = Me; 378, R = H

380, R = Et; 379, R = H

Scheme 5.37 Reagents and conditions: (i) ethyl cyanoacetate/S/TEA/DMF/55°C/3 h; (ii) aminomethanecarbonimidoyl chloride hydrochloride/DMSO2/120°C/1 h; (iii) (i) AcOH/$Hg(AcO)_2$/100°C, 3 h; (ii) CH_2Cl_2/I_2/RT/5 h; (iv) Thiols/Xantphos/$Pd_2(dba)_3$/i-Pr_2NEt/DMF/MW/190°C/30 min; (v) 1NNaOH/CH_3OH; (vi) (*S*)-2-Aminopentanedioic acid hydrochloride/*N*-methylmorpholine/CDMT/DMF/RT/5 h; (vii) 1 N NaOH/EtOH.

afforded the thiolactam **382d**. Cyclization with guanidine furnished the substituted dihydropyrrolo[2,3-*d*]pyrimidine **383d**. Pd-catalyzed coupling afforded the diethyl ester **384d**. Deprotection followed by saponification then yielded MTA (Scheme 5.38).

Various 7-substituted MTA derivatives were synthesized by the use of this methodology. These 7-substituted derivatives were affirmed to be devoid of any remarkable cell growth inhibitory potency in contradiction to a published patent. It was observed in a previous SAR evaluations on MTA that N-substitution at position 7 eliminated cell growth inhibitory potency. They were claimed that different 7-substituted MTA (e.g., **385a** and **385c**) were potent antitumor agents. However, these compounds are found to be inactive (IC_{50} > 20 pg/mL) as cell growth inhibitors [124].

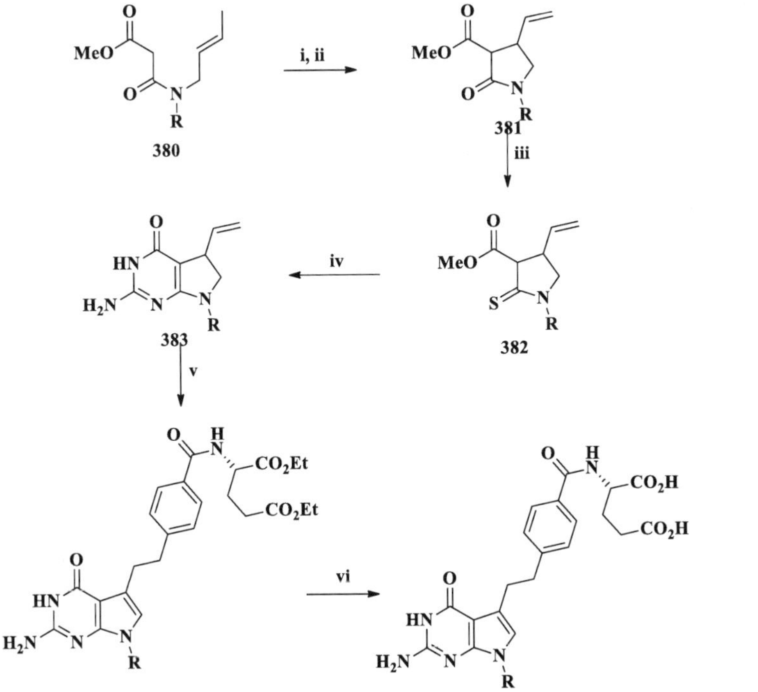

385 a, R = $CH(Me)C_6H_5$;
b, R = $CH_2C_6H_5$;
c, R = $CH_2C_6H_4OMe(\text{-}4')$
d, R = $CH_2C_6H_3(OMe)_2(\text{-}3',4')$

Scheme 5.38 Reagents and conditions: (i) $Mn(OAc)_3.2H_2O$; (ii) $Cu(OAc)_2.H_2O$; (iii) P_2S_5; (iv) guanidine; (v) diethyl-4-iodobenzoyl-L-glutamine/$Pd(OAc)_2$; (vi) NaOH.

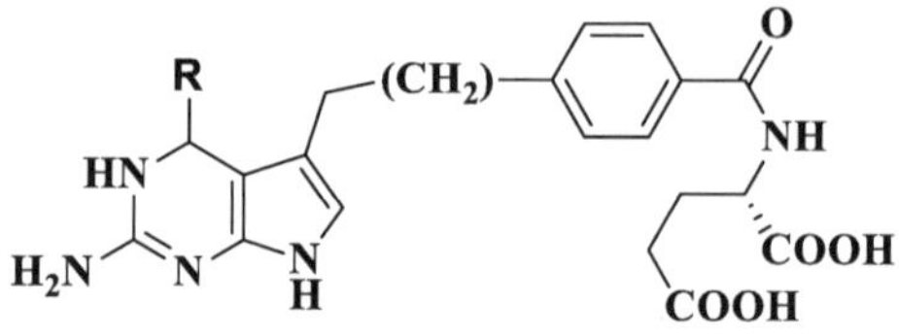

390, ALIMTA (disodium salt), R = OH, n = 1
395, TNP-351, R = NH_2, n = 2

Fig. 5.12 TNP-351, Alimta.

TNP-351, Alimta (Fig. 5.12), and homo-Alimta, a nonbridged analog of Alimta, have been synthesized via Michael addition of 1-nitroalkene with compound **24**, then the Nef reaction of the primary nitro Michael adduct was followed. Subsequent intramolecular cyclization of the pyrimidine 6-amino group with the resulting aldehyde afforded the corresponding pyrrolopyrimidine. A series of previously unknown 5-aryl pyrrolopyrimidines was synthesized through the same method from nitrostyrenes and pyrimidines. In a single step, the intermediate primary nitro Michael adduct can be synthesized via sonication of a mixture of the 6-aminopyrimidine, nitromethane, and arylaldehyde in the presence of ammonium acetate in acetic acid (Scheme 5.39) [125].

3. Nonclassical antifolates: Syntheses and biological evaluation

3.1 Inhibitors of dihydrofolate reductase (DHFR)

The synthesis of the substituted bicyclo[2.2.2]octane-1-carbonyl]-L-glutamic acid **404b** was accomplished in Scheme 5.40. Using the pteridinediamine **403**, the side chain precursor **402** was alkylated to afford compound **404a**. Further ester hydrolysis yielded compound **404b** (Scheme 5.40).

Antitumor and antifolate estimation of **404b** versus three tumor cell lines (S180, L1210, and HL60) and L1210 dihydrofolate reductase (DHFR) revealed it to be ineffective. Although compound **404b** is similar in structure to aminopterin, the bicyclooctane ring system instead of the phenyl ring in the *p*-aminobenzoate moiety makes ineffective stoichiometric binding [126].

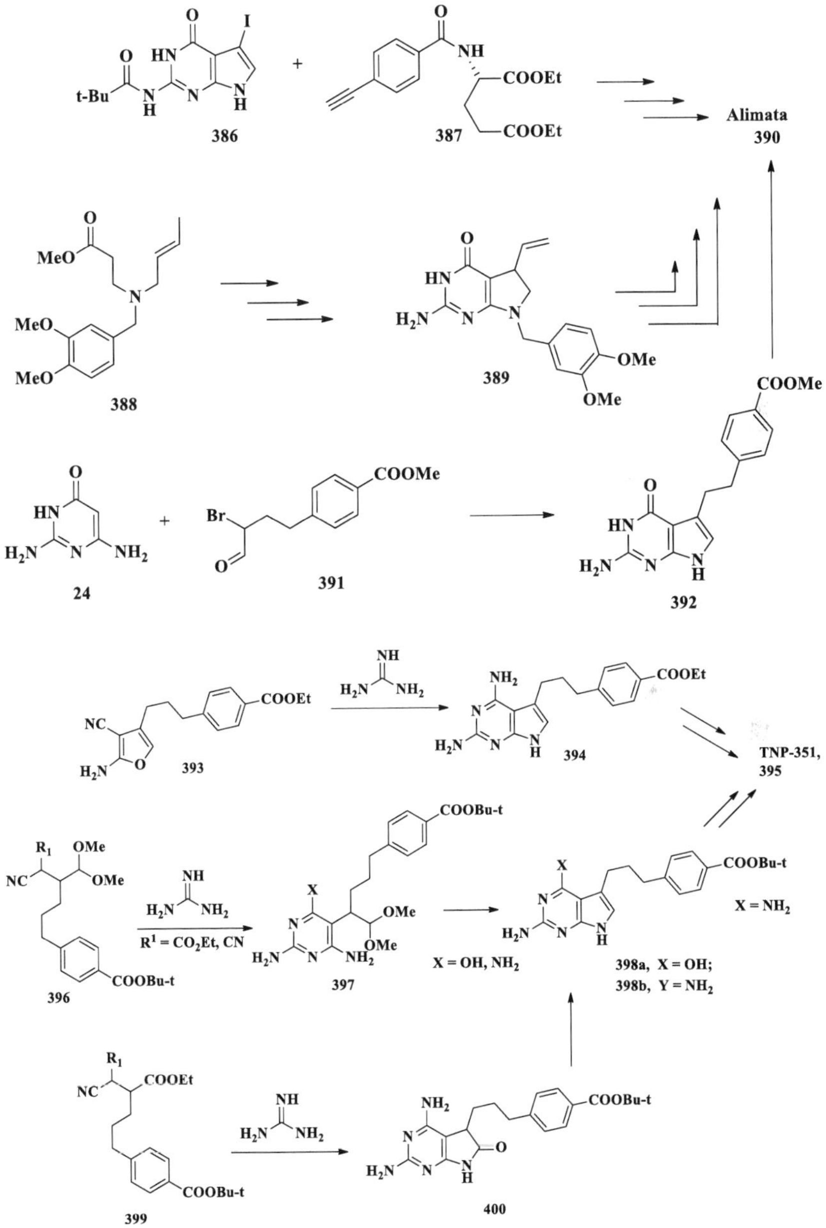
t-Bu
386
387
COOEt
Alimata
390
MeO
MeO
388
389
OMe
COOMe
24
Br
391
392
COOEt
NC
393
394
TNP-351,
395
COOBu-t
R₁
OMe
396
R¹ = CO₂Et, CN
397
X = OH, NH₂
398a, X = OH;
398b, Y = NH₂
X = NH₂
COOEt
399
400

Scheme 5.39

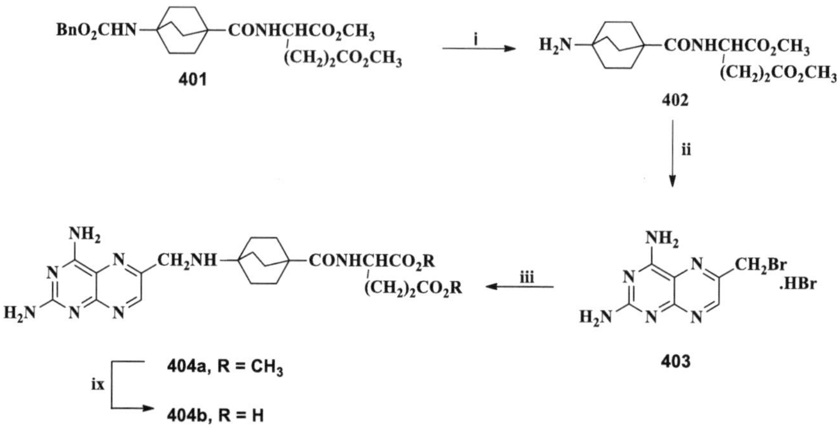

Scheme 5.40 Reagents and conditions: (i) 10% Pd/C/H_2/MeOH/1 atm/RT, 6 h; (ii) Me_2NAc/rt./48 h; (iii) 10% aq. NaOH/RT/20 h/aq. HCl at 0°C.

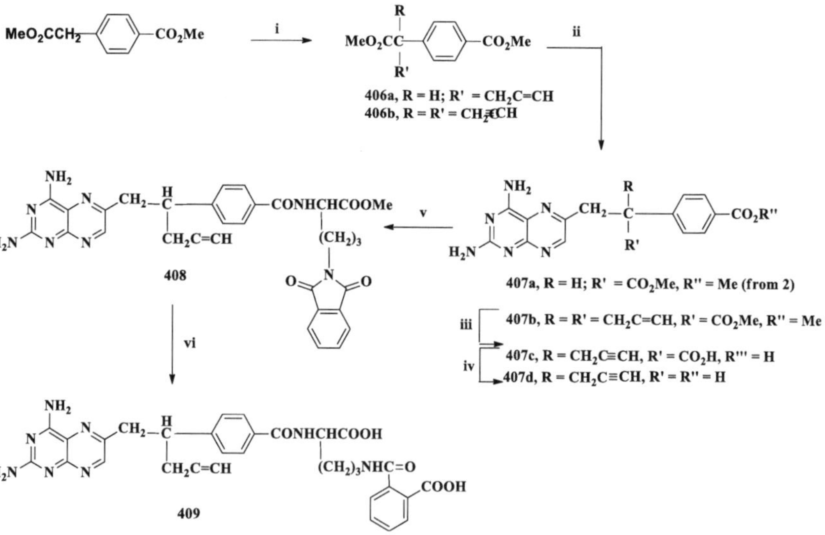

Scheme 5.41 Reagents and conditions: (i) $BrCH_2CtCH$/NaH/THF; (ii) 2,4-diamino-6-bromomethylpteridine hydrobromide/NaH/DMF; (iii) $MeOCH_2CH_2OH$/NaOH; (iv) DMSO/120°C; (v) methyl L-2-amino-5-phthalimido pentanoate/*i*-BuOCOCl/DMF; (vi) $Ba(OH)_2/CH_3OH$.

The synthesis of the hemiphthaloyl-L-ornithine **409** is shown in Scheme 5.41; this compound is considered as an analog of *N*R-(4-amino-4-deoxypteroyl)-*N*δ-hemiphthaloyl-L-ornithine (PT523, **405**, Fig. 5.13), a nonpolyglutamatable antifolate currently used in advanced preclinical development.

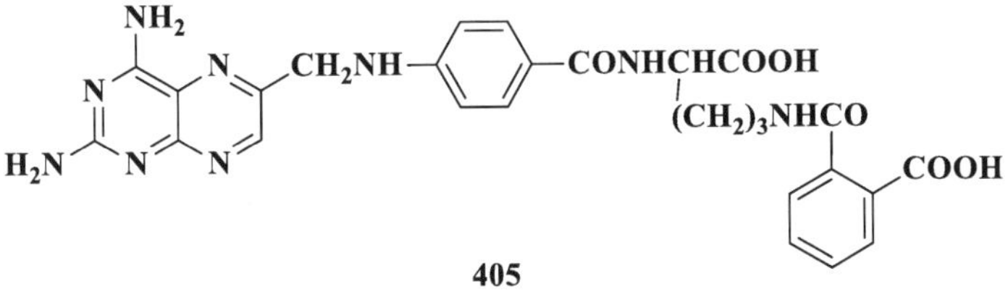

405

Fig. 5.13 *N* R-(4-amino-4-deoxypteroyl)-*Nä*-hemiphthaloyl-L-ornithine (PT523).

The dimethyl homoterephthalate reacted with propargyl bromide to afford the targeted monopropargyl derivative **406a**. Hydrolysis of the propargyl derivative **407b** using sodium hydroxide in ethylglycol monomethyl ether yielded the diacid **407c**. Subsequently, the latter was transformed to the key intermediate **407d** through heating in dimethylsulfoxide. Compound **407d** was condensed with the phthalimidopentanoate [127]. Treatment of the resulting phthalimide **408** with barium hydroxide yielded the dehydrate **409**.

The IC_{50} of compound **409** was 1.3 ± 0.35 nM, in respect to previously reported values 1.5 ± 0.39 nM for PT523 and 4.4 ± 0.1 nM for aminopterin (AMT) in a 72-h growth inhibition assay against cultures of CCRF-CEM human leukemic lymphoblasts. In a spectrophotometric assay of dihydrofolate reductase inhibition utilizing NADPH and dihydrofolate as the cosubstrates, the previously unreported mixed 10*R* and 10*S* diastereomers of compound **409** had *Ki* values of 0.60 ± 0.02 pM, in respect to the previously reported values of 0.33 ± 0.04 pM for PT523 and 3.70 ± 0.35 nM for aminopterin (AMT). As comparable to compound **405** in their capability to bind to DHFR and inhibit the growth of CCRFCEM cells, the mixed diastereomers of compound **409** were more potent than AMT although they were unable to form γ-polyglutamylated metabolites of the type formed in cells from AMT and other classical antifolates [128].

Tetrahydropyridopteridine derivatives with various benzoyl and benzyl substitutions at the N7 position were furnished (Scheme 5.42), as classical and nonclassical, partially restricted, linear tricyclic 5-deaza antifolates. The impact of conformational restriction of the C9-N10 (Ù2) and C6-C9 (Ù1) bonds through an ethyl bridge from the C7 to the N10 position of 5-deaza MTX on the inhibitory activity against DHFR was investigated from various sources and on antitumor potency.

The methodology for most of the desired compounds was constructive to build the tricyclic nucleus, tetrahydropyridopteridine **417**, then the regioselective alkylation of the N7 nitrogen.

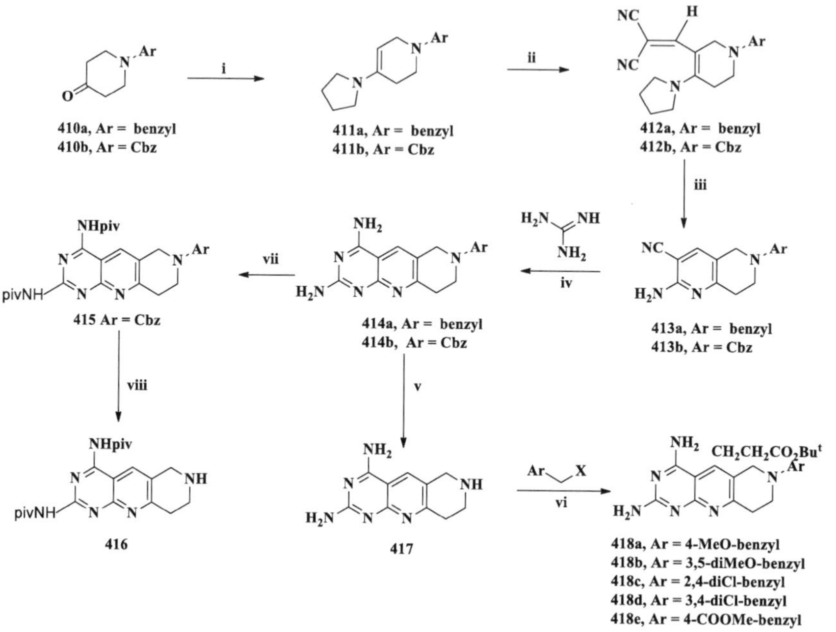

Scheme 5.42 Reagents and conditions: (i) *p*-TSOH/benzene/reflux; (ii) THF/−20°C/2 h; (iii) NH_3/CH_3OH/overnight; (iv) N_2/DMF/100°C; (v) 30% KOH/CH_3OH/N_2/reflux; (vi) DMF/TEA/N_2/DMAP/60–80°C/3–4 h; (vii) piv_2O/DMF/TEA/N_2/DMAP/60–80°C/1.5 h; (viii) CH_3Cl/Me_3SiI.

The partial conformational modification for the classical analog **419** was detrimental to DHFR inhibitory potency besides the antitumor activity compared to 5-deaza MTX or MTX. The classical analog **419** (Fig. 5.14), however, was a better FPGS substrate than MTX.

These data indicate that aclassical 5-deaza folate partially restricted through a bridge between the C7 and N10 positions retains FPGS substrate potency and that the antitumor potency of classical tricyclic analogs as compound **419** would be impacted via FPGS levels in tumor systems. The nonclassical analogs **414a,b** and **418a–d** indicated moderate to good selectivity against DHFR from pathogenic microbes as compared to rhDHFR. These data demonstrate that the 5-methyl group removal of piritrexim along with restriction of t_1 and t_2 can translate into selectivity for DHFR from pathogens [106].

The N9-H analogs were accessed through regiospecific reductive amination of the triaminoquinazoline with benzaldehydes, which was prepared from the substituted quinazoline in turn. The N9-CH_3 analogs were

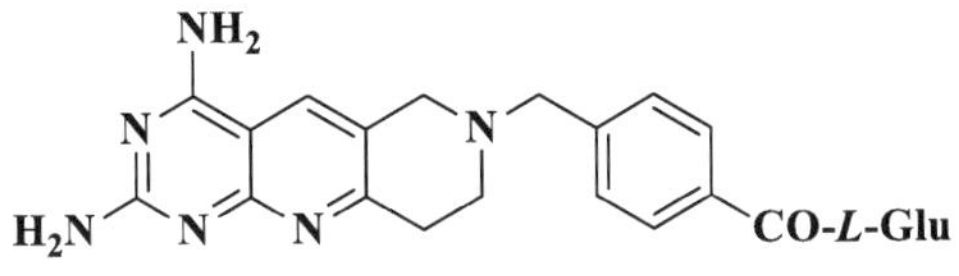

419

Fig. 5.14 Classical antifolate analog.

generated through a regiospecific reductive methylation of the respective N9-H precursors.

Compounds **422–435** were prepared from 2,4-diamino-6-nitro-quinazoline **420**, which was reduced to give the triaminoquinazoline **421**. The naphthaldehyde or benzaldehyde was then reductively aminated with **421**, to yield the desired compounds **422–430**. Reductive methylation of **423–425**, **427**, and **430**, using sodium cyanoborohydride and formaldehyde, generated the corresponding N9-methyl analogs **431–435** (Scheme 5.43).

6-Substituted 2,4-diaminoquinazolines were synthesized to enhance the cell penetration of a previously reported series of substituted pyridopyrimidines, which had indicated a remarkable activity and selectivity for *T. gondii* DHFR, but had much lower inhibitory effects on the growth of *T. gondii*

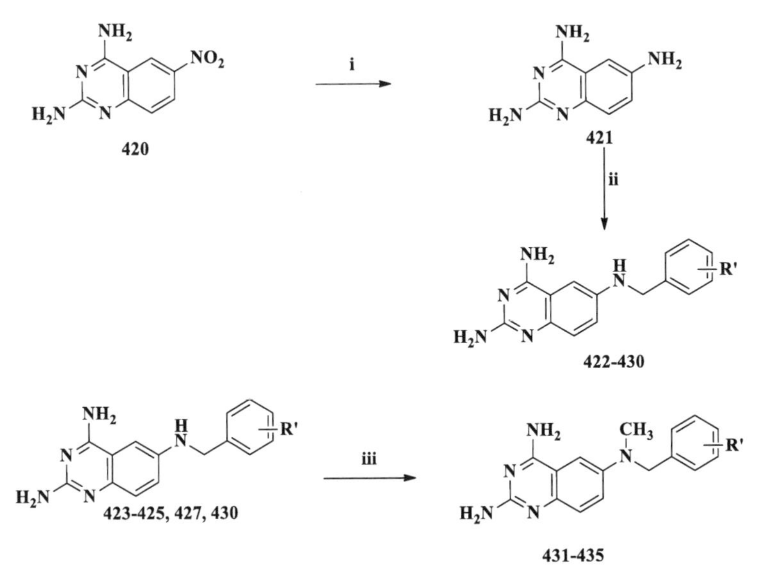

Scheme 5.43 Reagents and conditions: (i) Raney Ni/H_2 30 psi/DMF; (ii) Raney Ni/DMF/ArCHO/AcOH/H_2 30 psi; (iii) HCHO/HCl/$NaCNBH_3$.

cells in culture. The compounds were estimated as inhibitors of DHFR from human, *T. gondii*, *P. carinii*, rat liver, *Escherichia coli*, and *Lactobacillus casei*, and selected analogs were tested as inhibitors of the growth of tumor cells in culture. These compounds showed potent *T. gondii* DHFR inhibition besides the inhibition of the growth of *T. gondii* cells in culture. In addition, selected analogs were active inhibitors of tumor cell growth in culture in the National Cancer Institute's GI_{50}s in vitro screening program [129].

The pyridopyrimidine derivatives were furnished and estimated as inhibitors of *Toxoplasma gondii* (tg), *Pneumocystis carinii* (pc), and rat liver (rl) DHFR and as inhibitors of the growth of tumor cell lines in culture. The analogs **437–446** were designed to test the effects of the 5-position's substitutions, in the 2-atom bridge, and in the side chain phenyl ring on structure-activity and selectivity relationships of 2,4-diaminopyridopyrimidines against various DHFRs. Reductive amination of the carbonitrile **436** with different anilines yielded compounds **437–444**. Reductive methylation or nucleophilic substitution furnished the N10-methyl compounds **445–446** and **450** (Scheme 5.44).

The analogs **437–446** were designed to investigate the possibility of a greater diminishing activity against rlDHFR versus tgDHFR and/or pcDHFR due to the substitution of a hydrogen atom instead of a methyl group at the 5-position to generate greater selectivity for tgDHFR and/or pcDHFR versus rlDHFR. Compound **441** displayed selectivity ratios (IC_{50} of 23 and 15.7) for tgDHFR and pcDHFR. The 2′,5′-dichloro analog **441** displayed greater selectivity ratios for pcDHFR versus rlDHFR than trimethoprim (TMP), which is clinically used as therapeutic agent. Compound **441** was more active against tgDHFR and pcDHFR than trimethoprim (TMP). Compound **441** inhibited the growth of *P. carinii and T. gondii* cells in culture. In a *P. carinii* cell culture investigation, it displayed 88% cell growth inhibition at a concentration of 10 μM and exhibited marginal potency in an in vivo *T. gondii* mouse model (up to 7 days). Also, the growth of tumor cells was inhibited at micromolar to submicromolar concentrations, but was not efficacious in an in vivo evaluation [130].

5,7-Dinitro-3-phenyl-2-chloroquinoxaline **451** reacted with the anilines **452** in refluxing 1-propanol to afford the substituted quinoxalines **453**. Reduction of the latter yielded the targeted 5,7-diamino-3-phenylquinoxalines **454** (Scheme 5.45).

Compounds were evaluated as antitumor agents. Compound **454a** has an anticancer activity and showed a high degree of selectivity against leukemia (CCRF-CEM) and better than that of MTX.

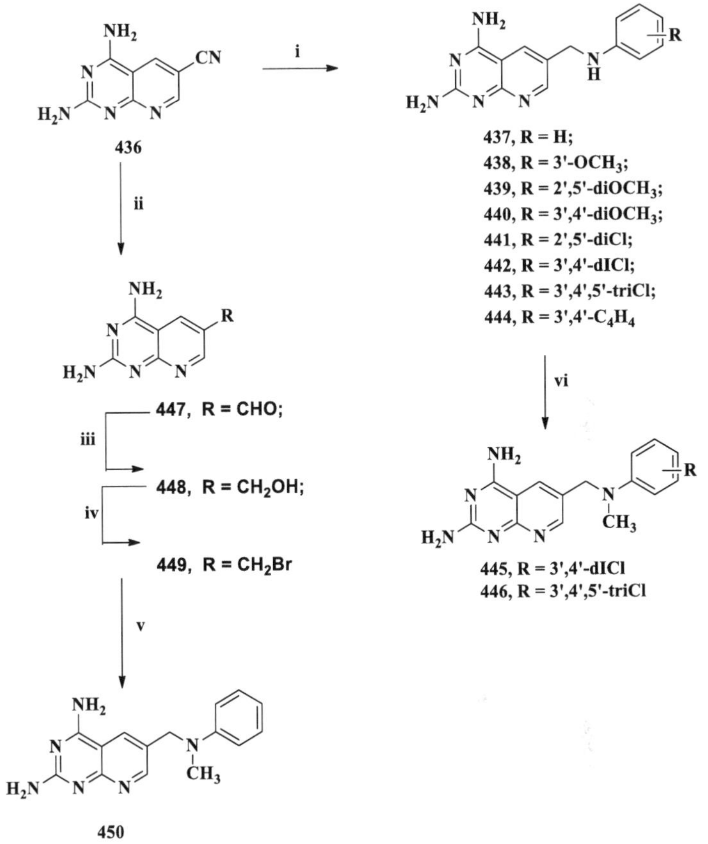

NH_2 N CN H_2N N N

436

i

NH_2 N H_2N N N N H R

437, R = H;
438, R = 3'-OCH_3;
439, R = 2',5'-diOCH_3;
440, R = 3',4'-diOCH_3;
441, R = 2',5'-diCl;
442, R = 3',4'-dICl;
443, R = 3',4',5'-triCl;
444, R = 3',4'-C_4H_4

ii

NH_2 N R H_2N N N

447, R = CHO;
iii
448, R = CH_2OH;
iv
449, R = CH_2Br

vi

NH_2 N H_2N N N N CH_3 R

445, R = 3',4'-dICl
446, R = 3',4',5'-triCl

v

NH_2 N H_2N N N N CH_3

450

Scheme 5.44 Reagents and conditions: (i) Raney Ni/H_2, atm press./$ArNH_2$/80% HOAc/; (ii) 97% HCOOH/Raney Ni; (v) $NaBH_4$/CH_3OH; (iv) 30% HBr/AcOH/AcOH; (vi) *N*-methylaniline/NaH/DMF; (vii) 37% HCHO/1 N HCl/$NaCNBH_3$/CH_3CN.

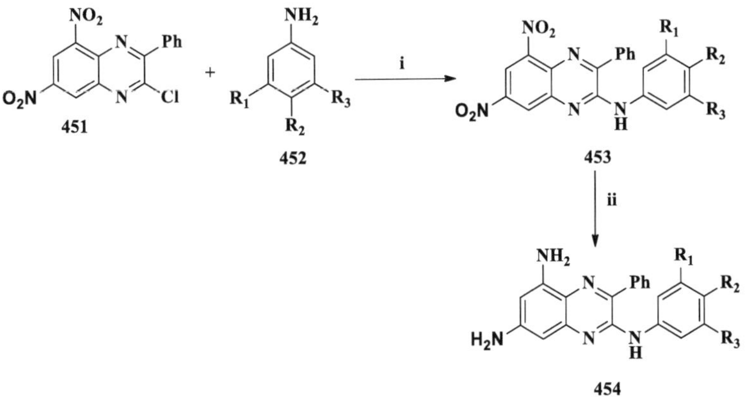

NO_2 N Ph O_2N N Cl

451

+

NH_2 R_1 R_3 R_2

452

i

NO_2 N Ph R_1 R_2 O_2N N N H R_3

453

ii

NH_2 N Ph R_1 R_2 H_2N N N H R_3

454

Scheme 5.45 Reagents and conditions: (i) 1-Propanol/reflux/1 h; (ii) $NH_2.NH_2$/H_2O/Pd/C/EtOH.

Identifications of the binding mode of the molecules with respect to MTX are indicated by the docking calculations of the complexes of the most potent compounds with hDHFR [131].

Refluxing 2-amino-5-methylbenzoic acid with thioacetamide yielded compound **457**, which was transformed to the (3*H*)-quinazolinone derivative **459** after bromination. And according to the method for synthesizing dithiocarbamates, compounds **461** were furnished by the reaction of the substituted quinazolinone with carbon disulfide and amines **460** (Scheme 5.46).

With dithiocarbamate side chains, the substituted quinazolinones were evaluated for their in vitro antitumor potency against human myelogenous leukemia K562 cells. The synthesized compound **461** showed an effective inhibitory activity against K562 cells ($IC_{50}=0.5$ lM) [132].

The chloropyrrolo[1,2-*a*]quinoxalines **462a,b** reacted with the corresponding anilines in refluxing 1-propanol to yield the anilinopyrroloquinoxalines **463** (Scheme 5.47).

Nucleophilic displacement of chloromethyl derivatives **462c,d** with the respective phenols **465** afforded compounds **466** (Scheme 5.48).

The synthesis of **471** was obtained as a result of the intramolecular ring closure of the new amide **470**. The latter yielded from the intermediate **469** in turn was synthesized through reduction of **468b** (Scheme 5.49).

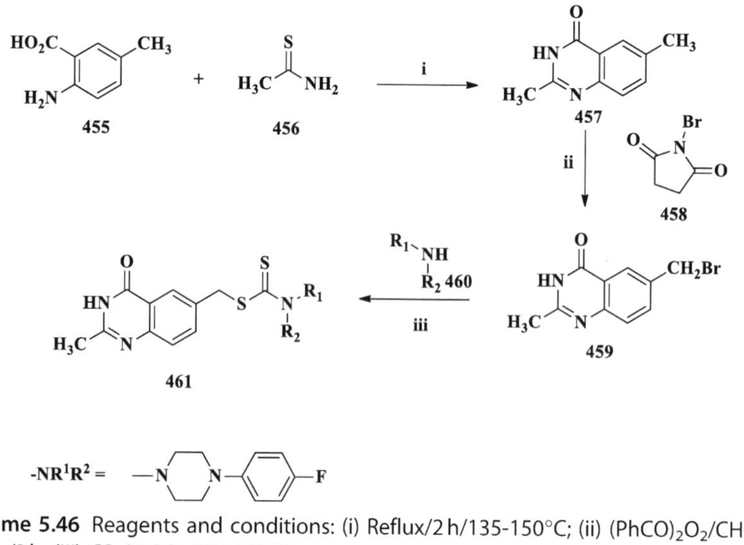

Scheme 5.46 Reagents and conditions: (i) Reflux/2 h/135-150°C; (ii) $(PhCO)_2O_2/CHCl_3$/reflux/3 h; (iii) CS_2/K_3PO_4/DMF/RT/2 h.

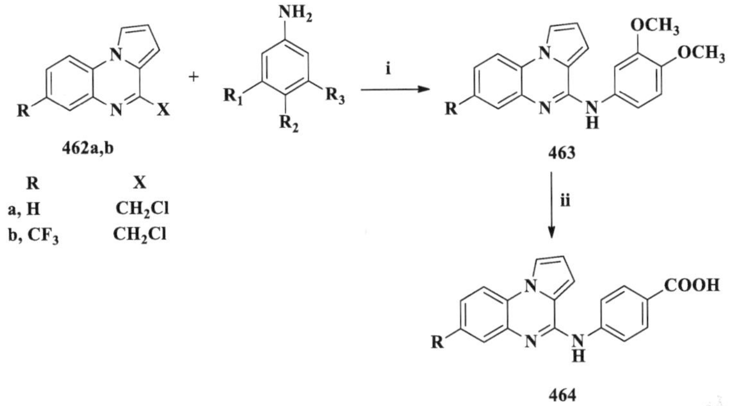

NH_2 OCH_3 OCH_3 N + i R R_1 R_3 R N N X N N H R_2

462a,b 463

R	X
a, H	CH_2Cl
b, CF_3	CH_2Cl

ii

N COOH R N N H

464

Scheme 5.47 Reagents and conditions: (i) PropOH/reflux/2–10 h; (ii) EtOH/1 M NaOH/reflux/4 h.

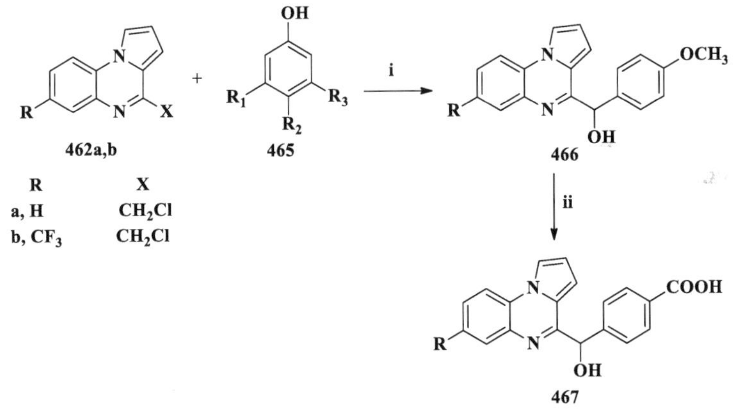

OH N OCH_3 + i R R_1 R_3 R N X N R_2 OH

462a,b 465 466

R	X
a, H	CH_2Cl
b, CF_3	CH_2Cl

ii

N COOH R N OH

467

Scheme 5.48 Reagents and conditions: (i) Cs_2Co_3/DMF/70°C/6 h/$CsHCO_3$/2 h; (ii) EtOH/1 M NaOH, 70 °C.

Effects of compounds against solid and hematological tumor-derived cell lines were shown [133].

6-Substituted pyrrolopyrimidines as potential antitumor agents were synthesized as accomplished. Compounds **478a–c** were furnished through condensation reactions from the substituted dihydropyrrolopyrimidin-4-one. All compounds displayed inhibitory effects against KB tumor cells in a submicromolar to nanomolar concentrations, while compounds **478a–c** also displayed nanomolar antiproliferative potencies toward A549 and SW620 cells in preliminary antiproliferation assay. Particularly, compounds

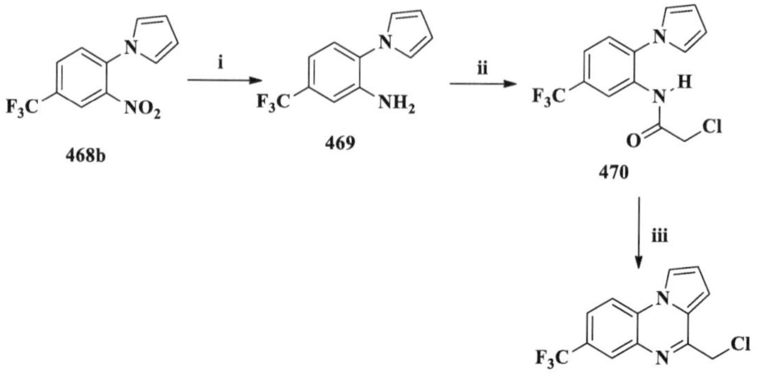

Scheme 5.49 Reagents and conditions: (i) $NH_2NH_2.H_2O$/Pd/C, 10%/EtOH/reflux/1.5 h; (ii) $ClCOCH_2Cl$, dioxane/C_5H_5N/reflux/4 h. (iii) $POCl_3$/reflux/4 h.

478a–c were remarkably more active than pemetrexed (PMX) and methotrexate (MTX) to A549 cells. The growth inhibition induced cell cycle arrest at G1-phase with S-phase suppression. Along with the data of nucleoside protection assays, inhibition assays of DHFR indicated that the DHFR was the intracellular target of the designed compounds. The estimations of the molecular modeling demonstrated two binding modes of the desired compounds with DHFR.

Target compounds **478a–c** were generated as depicted in Scheme 5.50. The key intermediate, substituted dihydropyrrolopyrimidin-6-yl acetate, compound **473**, was acquired by condensing of ethyl-4-chloroacetoacetate, **472**, and the 2,6-diaminopyrimidin-4(3H)-one, in the presence of sodium acetate-water. The ester **473** was first reduced to alcohol **473a** via treatment with lithium triethylborohydride, and the latter compound was then transformed to mesylate **473b** by methanesulfonyl chloride.

Reaction of sodium azide with compound **473b** and then hydrogenation were followed to afford the amine **476**. Condensation of the latter compound, then saponification reaction were followed to generate carboxylic acids **477b**. Subsequent condensation with different pyridinylmethylamines furnished desired compounds **478a–c** [134].

4-Fluoroaniline-substituted triazine-benzimidazole has been generated (Scheme 5.51). These analogs were further substituted with various amines. The novel compounds were estimated against 60 human tumor cell lines at one and five dose concentration levels. The most potent antitumor agents ($GI_{50}=2.87\,\mu M$) were found to be compounds **483**. The compounds were then estimated for their inhibitory potency to mammalian DHFR.

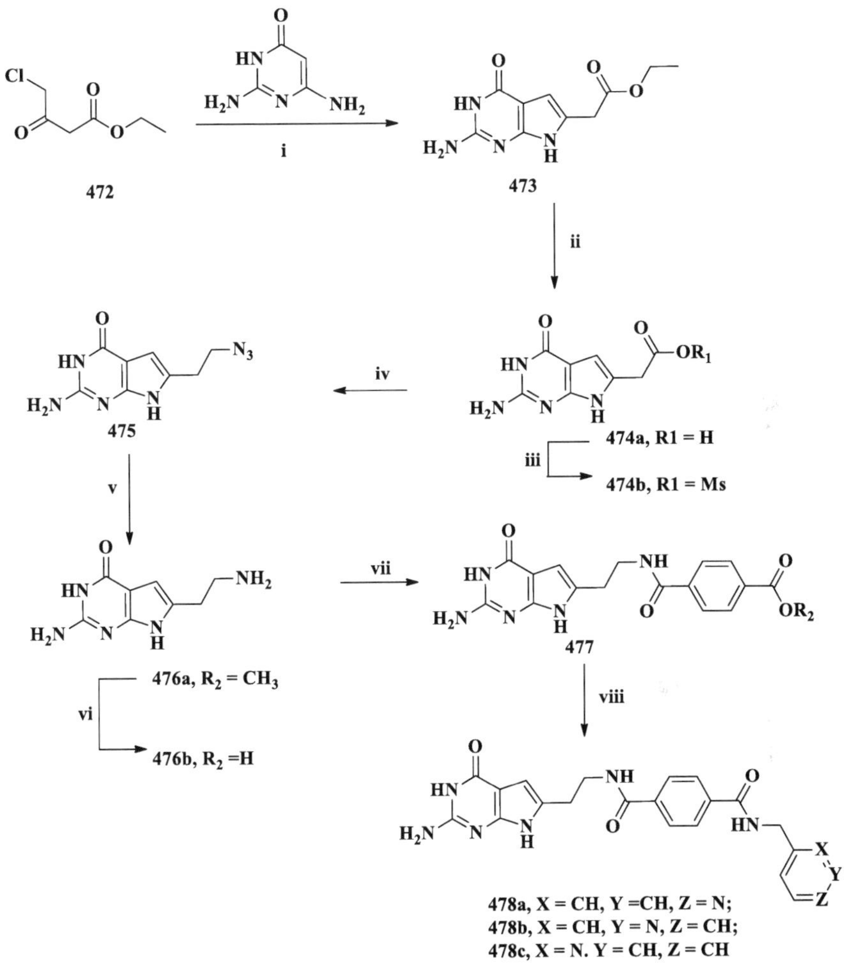

Scheme 5.50 Reagents and conditions: (i) NaOAc/H_2O/reflux/18 h; (ii) $LiEt_3BH$/THF/RT/3 h; (iii) CH_3SO_2Cl/TEA/DMF/0°C/0.5 h; (iv) NaN_3/DMF/70°C/8 h; (v) Pd/C, H_2, DMF, RT, 10 h; (vi) 4-(methoxycarbonyl) benzoic acid, EDCI, HOBt/TEA/DMF/RT/20 h; (vii) (i) 3 N NaOH/RT/1.5 h; (ii) 3 N HCl; (viii) HOBt/EDCI/TEA/DMF/RT/12 h.

Compound **483** is considered as the most potent compound in this series for the inhibition of DHFR ($IC_{50} = 2.0$ nM). Strong interacting features of the synthesized triazines toward calf thymus-DNA were also shown in DNA binding evaluations [135].

Dihydro-1,3,5-triazines with a heteroatom spiroring were furnished as shown in Scheme 5.52. All compounds displayed hDHFR inhibitory potency and antiproliferative activity against tumor cell lines (A549,

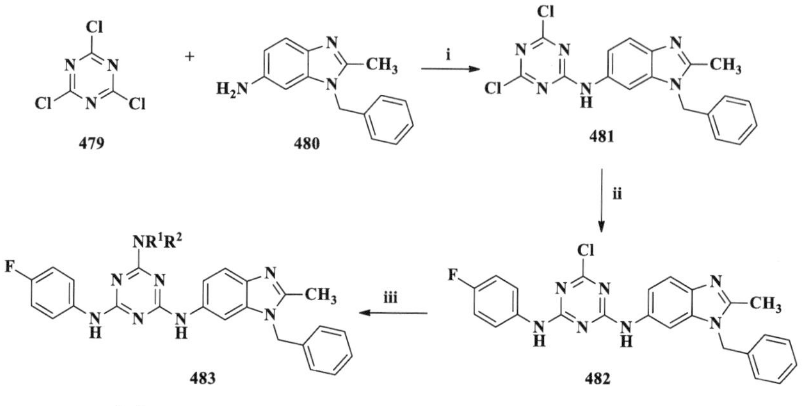

NR^1R^2 = 2-aminoethylmorpholine

Scheme 5.51 Reagents and conditions: (i) 10% $NaHCO_3$/THF/0–5°C; (ii) 1-amino-4-fluorobenzene/10% $NaHCO_3$/THF/RT; (iii) HNR^1R^2/1,4-dioxane/K_2CO_3/110°C.

HCT116, HepG2, HL-60, and MDA-MB-231). Compounds **488a, 488b**, **495a**, and **495b** indicated high hDHFR inhibitory potency with 7.46, 3.72, 6.46, and 4.08 nM IC_{50} values, in respect to MTX [136].

Compounds exhibited in vitro antiproliferative potency toward various tumor cell lines (IC_{50} values varying from 0.79 to 0.001 μM), which is better than MTX. The subsequent in vivo antitumor evaluation in a nude mouse A549 model indicated that compound **488a** could inhibit tumor growth. The results indicated that the insertion of sulfur and oxygen atoms into the spiroring could increase or maintain the hDHFR inhibition [137].

3.2 Inhibitors of thymidylate synthase (TS)

The synthesis of quinazolinone antifolates as potent thymidylate synthase (TS) inhibitor was accomplished, which involved coupling of the nitrobenzene derivative **503** with the substituted dihydro-4-oxoquinazoline **501** or **502** (Scheme 5.53).

With most of these new antifolates, good TS ($IC_{50} < 1$ μM) and growth inhibition ($IC_{50} = 0.1$–1 μM) were found. TS inhibitors in this series are not polyglutamated and penetrate the cell membrane through passive diffusion; they do not enquire the RFC for the cell entry. O, N, S, Cl, and CN besides the large —NH_2 and –SH substituents were favored by the enzyme. The concurrent incorporation of 7-methyl and 2′-fluoro substituents notably

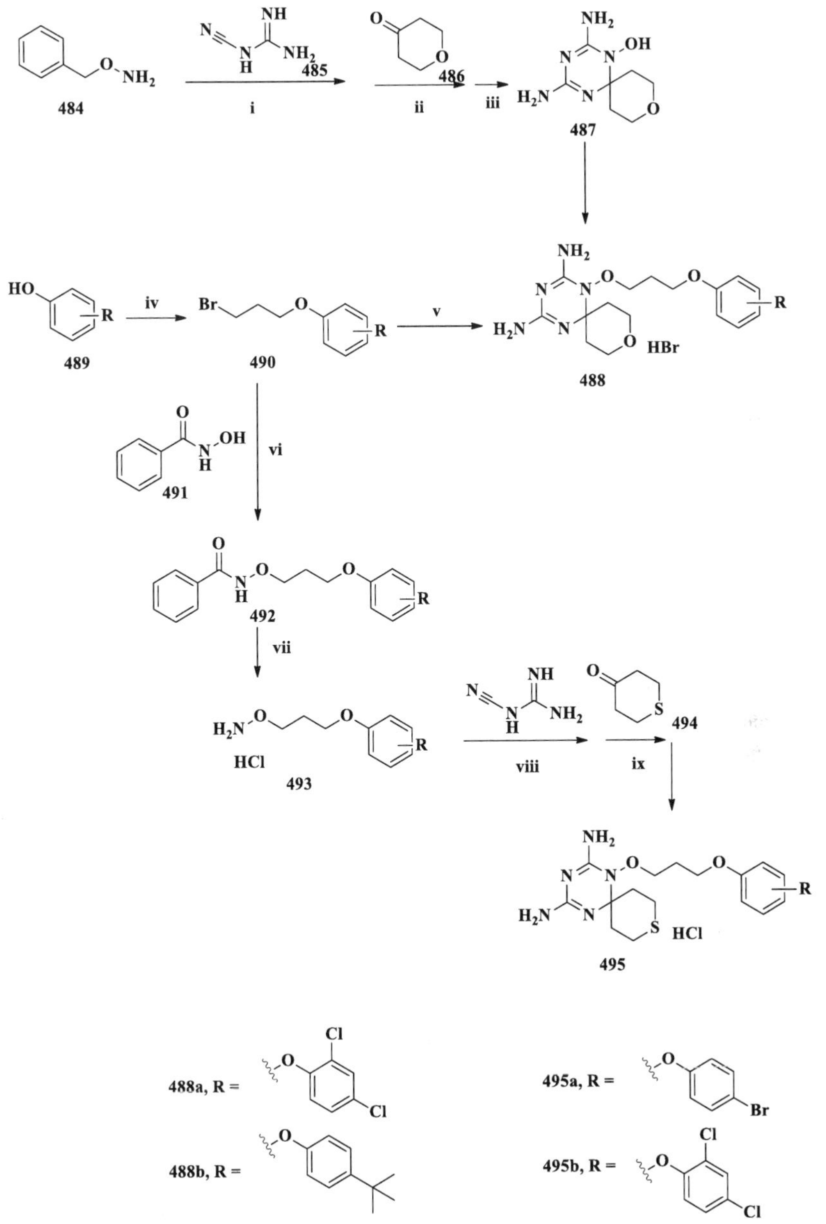

Scheme 5.52 Reagents and conditions: (i) dibromoalkane/CH_3CN/K_2CO_3/reflux/8 h; (ii) NaOH/DMF/8–24 h; (iii) NaOH/EtOH/reflux/10 h; (iv) HCl/EtOH/reflux/5 h; (v) EtOH/reflux/5-10 h; (vi) concentrated HCl/EtOH/RT; (vii) 10% Pb/C/90% EtOH/1 MPa/RT.

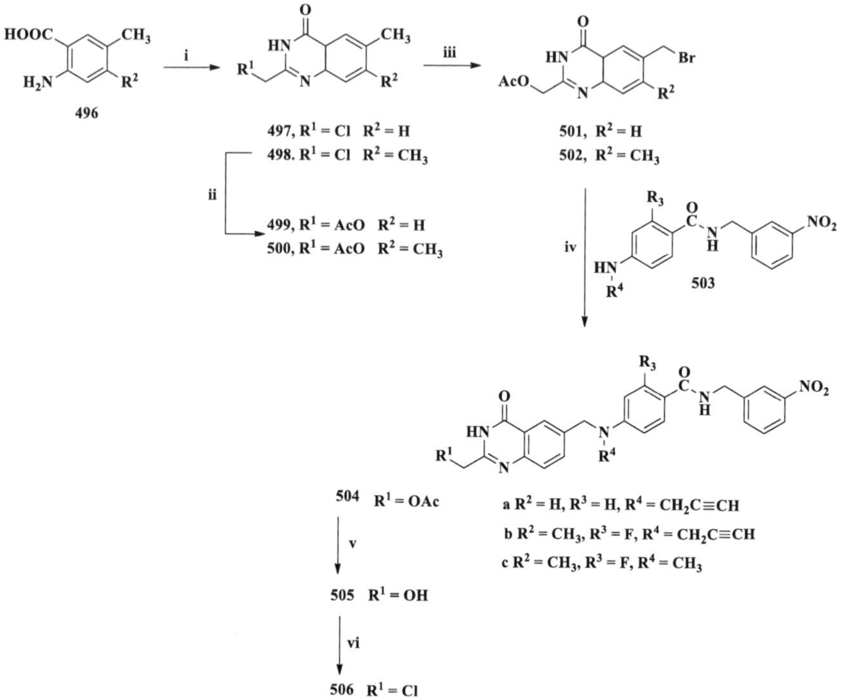

Scheme 5.53 Reagents and conditions: (i) $ClCH_2CN/MeONa/CH_3OH$; (ii) CH_3COONa/DMF; (iii) $NBS/CHCl_3/(PhCOO)_2$; (iv) $DMF/(CH_3)_2C_5H_3N$/80°C; (v) 2N $NaOH/CH_3OH$; (vi) $SOCl_2/CH_2Cl_2$.

yielded a series of highly active agents inhibiting cell growth (concentration < 1 μM) (**505**) [138].

The dideazafolic acid derivatives were prepared as inhibitors of thymidylate synthase and antitumor agents. This was demonstrated through routes to compound **511** and its regioisomer **509**, followed by DEPC coupling to the fluorobenzoic acid **512**, and then deprotection (Scheme 5.54). With the modified glutamate side chains, the resulting quinazoline-based antifolates, and particularly, the substituted tetrazole **514** indicated active TS and LI210 cell growth inhibitory potencies (e.g., compound **514**: TS $IC_{50} = 2.4$ nM, LI210 $IC_{50} = 1.3$ μM) [139].

3.3 Antifolates targeting purine biosynthesis

Pyrimido[4,5-*b*]azepine-based analog of DDATHF was synthesized as accomplished in Scheme 5.55. In vitro characterization of compound **528**

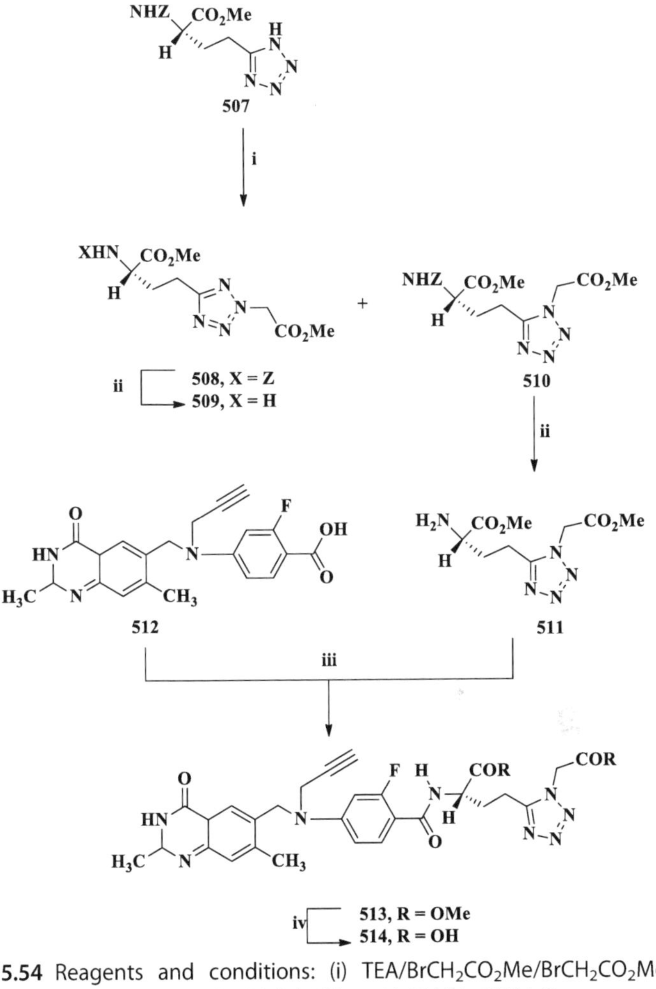

NHZ CO2Me H N N N N 507

i

XHN CO2Me H N N N N CO2Me + NHZ CO2Me CO2Me H N N N N 510

ii 508, X = Z 509, X = H

ii

O HN H3C N CH3 N F OH O 512

H2N CO2Me CO2Me H N N N N 511

iii

O HN H3C N CH3 N F H N H O COR COR N N N N

iv 513, R = OMe 514, R = OH

Scheme 5.54 Reagents and conditions: (i) TEA/$BrCH_2CO_2Me$/$BrCH_2CO_2Me$/CH_2Cl_2; (ii) H_2/10% Pd/C; (iii) DEPC/TEA/DMF; (iv) 1 N aq. NaOH/MeOH/H_2O.

against human T-cell-derived lymphoblastic leukemia (CCRF-CEM) cells afforded a 47 nM IC_{50} value. Reversal of the cytotoxicity of compound **528** could be effected via the addition of AICA and hypoxanthine, and not thymidine, showing that the locus of potency of compound **528** resides in the purine de novo biosynthetic pathway. The affinity measurements of compound **528** for murine trifunctional GARFT afforded a K_i of 147 nm similar to the result of compound **515** (Fig. 5.15) against the enzyme obtained from L1210 cells. In the lack of data on the transport features of compound **528**, its three times reduced cytotoxicity can be largely attributed to reduced FPGS

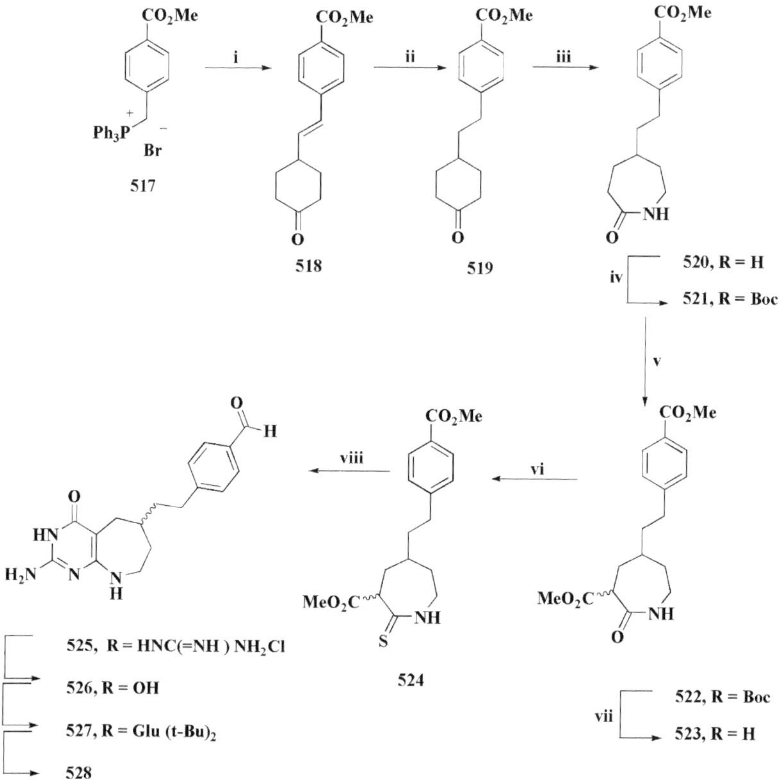

Scheme 5.55 Reagents and conditions: (i) DBU/THF/8h; (ii) EtOAc/H_2/Pd(C)/6h; (iii) H_3NO_4S/HCOOH/100°C/8h; (iv) $(Boc)_2O$/DMAP/TEA/CH_2Cl_2/24h; (v) LHMDS/TI-IF/ $CH_3OC(O)CN$/−78°C/8h; (vi) 0.20% TFA/CH_2C1_2/1h; (vii) P_4S_{10}/THF/60°C, 0.5h; (viii) Guanidinium chloride/NaOMe/90°C/10Torr/1h/6M HCl/H_2O; (ix) 1N NaOH, 60°C, 24h; (x) NMM/2-chloro-4,6-dimethoxy-l,3,5-triazine/(*S*)-di-tert-butyl 2-aminopentanedioate hydrochloride/DMF; (xi) 20% TFA/CH_2C1_2/8h/then 1N NaOH/6M HCl.

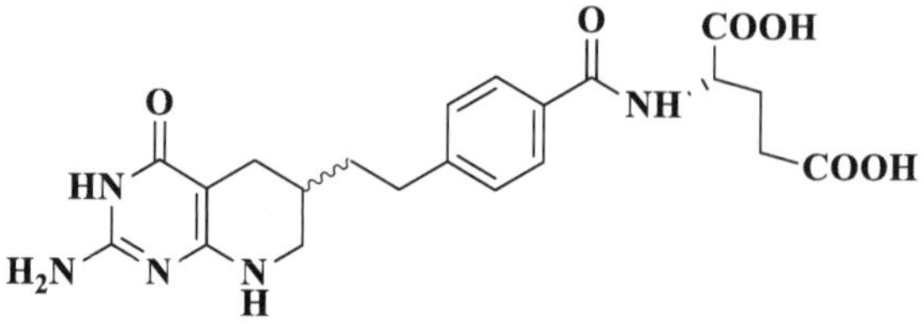

515, (DDATHF, (6RS))
516, (Lometrexol, (6R))

Fig. 5.15 DDATHF and lometrexol.

affinity. With a maximum velocity (V_{max}) of 797 nmol h^{-1} mg^{-1}, the K_m value for transformation of compound **528** to its diglutamate by hog liver FPGS was identified as 39 μM. Comparison of the first-order rate constants (K^1 values, defined as V_{max}/K_m) of compounds **528** and **516** (V_{max} = 977 nmol h^{-1} mg^{-1} and K_m = 16 μM) reveals that compound **528** (k^1 = 20) is threefold less active than compound **516** ($k1$ = 59) [140].

LSN 3213128 (**535a**) was synthesized (Schemes 5.56 and 5.57), which is considered as a new, nonclassical, selective, orally bioavailable antifolate with specific inhibitory potency for AICARFT. Inhibition of AICARFT with compound **535a** led to an elevation of 5-aminoimidazole

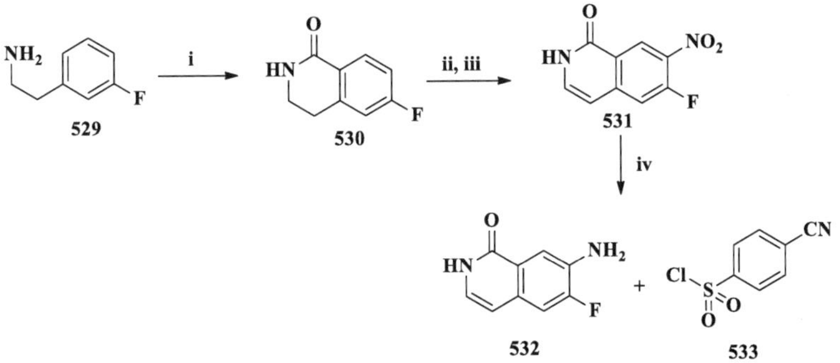

Scheme 5.56 Reagents and conditions: (i) Triphosgene/CH_2Cl_2/TEA/0°C/$AlCl_3$/0°C—rt.; (ii) KNO_3/H_2SO_4/0°C; (iii) MnO_2/dichloroethane/120°C; (iv) Pd/C/H_2/CH_3OH.

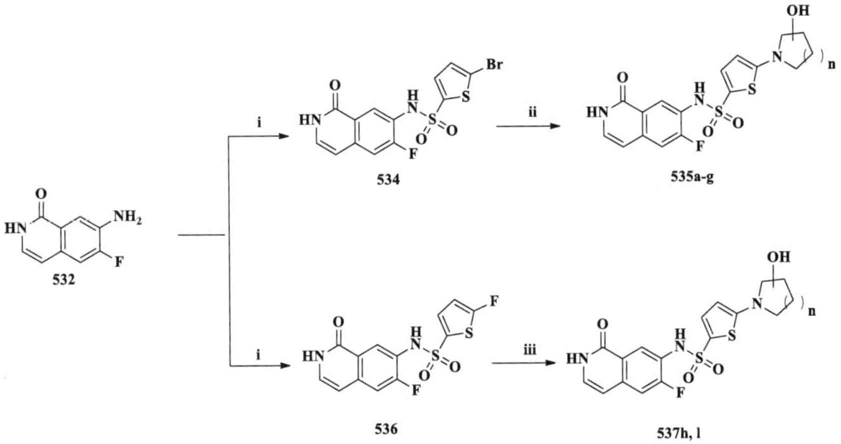

Scheme 5.57 Reagents and conditions: (i) sulfonyl chloride/pyridine; (ii) amine/CuBr/K_2CO_3 or $CsCO_3$/DMSO/100°C/16 h; (iii) amine/pyridine/110°C/40 h.

4-carboxamide ribonucleotide (ZMP) and growth inhibition in MDA-MB-231met2 and NCI-H460 cancer cell lines. Through in vivo investigations and treatment with this inhibitor, which is a freely permeable and does not need active uptake in a murine-based xenograft model of triple-negative breast cancer, TNBC, cell line led to tumor growth inhibition (TGI).

The antineoplastic agent **535a** displayed good drug-like physical properties and pharmacokinetic characteristics. It led to complete tumor growth inhibition when administered as a single agent dosed, while MTX failed to afford statistically remarkable TGI in the same evaluation [141].

3.4 Inhibitors of folylpolyglutamate synthetase (FPGS)

The discovery of a potent, nonpolyglutamatable, and selective inhibitor of GAR Tfase is shown in Scheme 5.58. The inhibitors were synthesized from the carboxylic acid **540** [142]. Coupling of the common precursor **540** with each side chain **545′a** and **545′b** [143,144] yielded compounds **538**, **542**, and **543**. Deprotection of compounds **538** and **543** furnished compounds **539** and **544**.

Compound **544**, which has a tetrazole in place of the γ-carboxylic acid in the L-glutamate subunit of the active GAR Tfase inhibitor **545** (Fig. 5.16), was potent in cellular-based functional assays displaying purine-sensitive cytotoxic potency (CCRF-CEM $IC_{50}=40\,nM$) and selectively inhibiting the rhGAR Tfase ($(Ki)=130\,nM$). Compound **544** was only 2.5 times less efficiently than **545** in cellular assays and fourfold less active against rhGAR Tfase. As **545** behaves, the functional potency of **544** in the cell-based assay benefits form and enquires transport into the cell via the RFC; however, it is independent of FPGS expression levels and polyglutamation unlike the 10-trifluoroacetyl-DDACTHF **545,** which was considered as potent GAR Tfase inhibitor [145] ((IC_{50}) 16 nM, CCRF-CEM; *K*i) 15 nM, rhGARTfase) [146].

Analogs of the antithrombotic and cardiovascular agent dipyridamole DP (**619)** (Fig. 5.17) were prepared. The structures of the compounds estimated as nucleoside transport inhibitors are reported. Pyrimidopyrimidines having identical groups at the 4,8- and 2,6-positions were prepared from compound **546**, synthesized from the disodium salt of dihydropyrimidopyrimidine-2,4,6,8-(3*H*,7*H*)-tetrone [147]. Compound **546** treated with surplus of the appropriate amine afforded the disubstituted dichloropyrimidopyrimidines (Scheme 5.59).

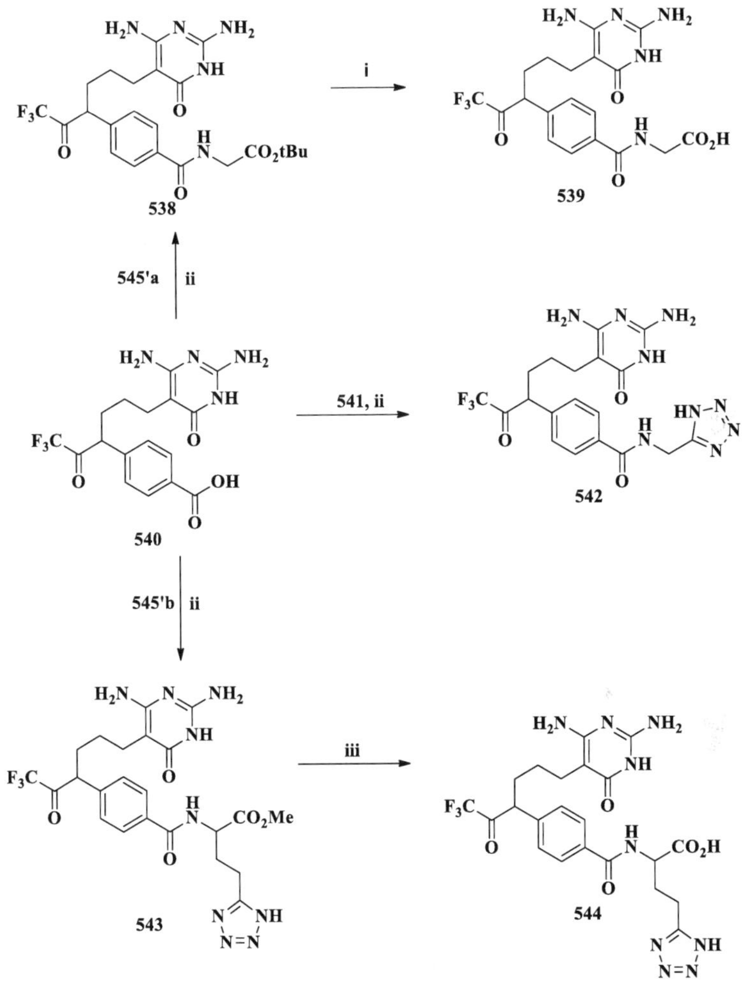

Scheme 5.58 Reagents and conditions: (i) TFA/$CHCl_3$; (ii) EDCI/$NaHCO_3$; (iii) NaOH/MeOH.

In the presence and absence of AGP acid glycoprotein, the analogs were estimated as inhibitors of 3*H*-thymidine uptake into the L1210 leukemia cells. Compounds with activity similar to that of DP were determined where the piperidino substituents at the 4,8-positions were replaced by piperonylamino, 3′,4′-dimethoxybenzylamino, or 4′-methoxybenzylamino groups [145,147].

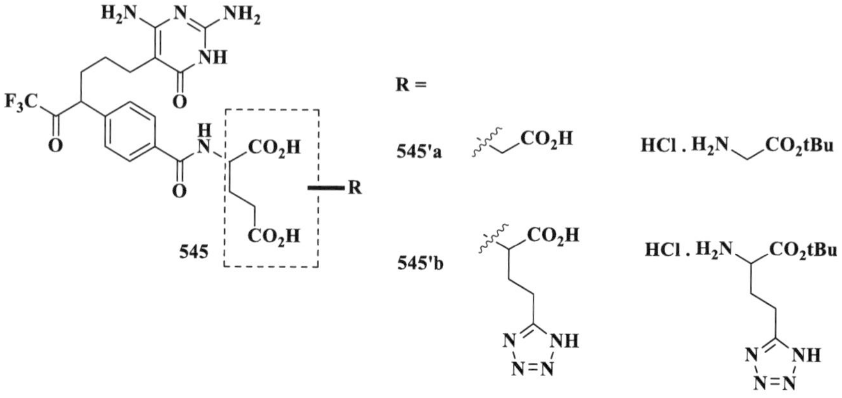

Fig. 5.16 10-Trifluoroacetyl-DDACTHF.

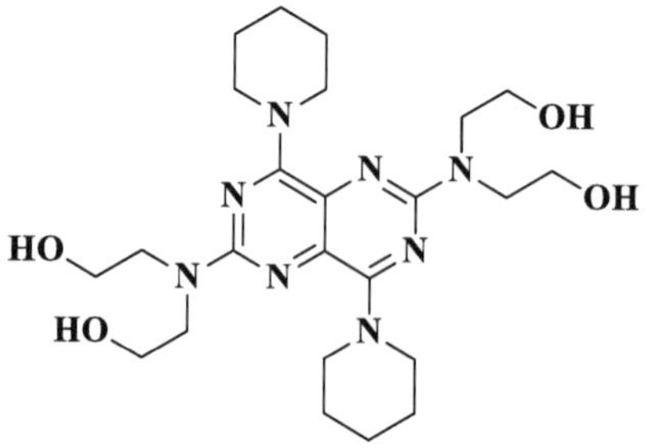

Fig. 5.17 Dipyridamole (DP).

3.5 Multitargeted antifolate (MTA)

Substituted pyrrolopyrimidines were generated as effective antifolates targeting both thymidylate and purine nucleotide biosyntheses. Compound **24** was condensed with ethyl-4-chloroacetoacetate; then the further hydrolysis yielded the substituted pyrrolopyrimidine. The latter compound was condensed with the appropriate amino acid methyl esters followed by

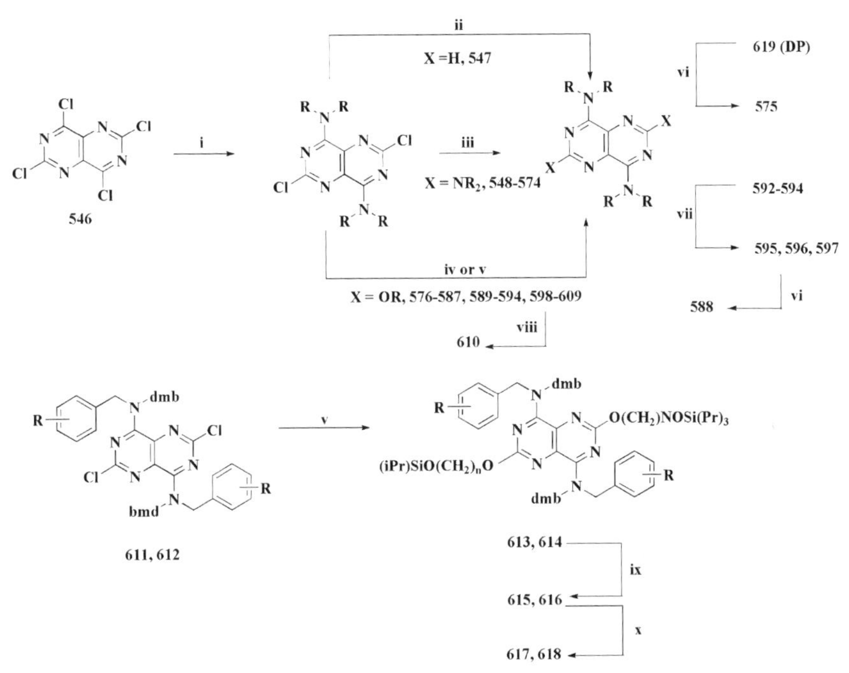

Scheme 5.59 Reagents and conditions:- (i) R_2NH/THF/25°C; (ii) Pd/C/H_2/THF/25°C; (iii) R_2NH/100–150°C; (iv) ROH/Na/reflux; (v) NaH/THF/ROH/reflux; (vi) CH_3I/NaH/DMF/25°C; (vii) 1 M HCl/THF/25°C; (viii) AcCl,/CH_3OH/THF/Pd/C/H_2/25°C; (ix) THF/TBAF/25°C; (x) TFA/25°C.

saponification and then the condensation with pyridin-3-ylmethanamine accessed compounds **623a**,**b** (Scheme 5.60).

The novel compounds displayed antiproliferative activities against a broad spectrum of tumor cell lines counting A549, HepG2, and KB. Toward KB cells, the growth inhibition of compound **623b** leads to cytotoxicity and G1/G2-phase accumulation, and was partially protected via surplus adenosine and thymidine, but was totally inverted in the combination of adenosine and thymidine, revealing both thymidylate and de novo purine nucleotide synthesis as the intended pathway.

Although the incomplete AICA protection, inhibition of both AICARFTase and GARFTase is proposed. The docking studies indicate that compound **623b** could bind and inhibit both TS and GARFTase

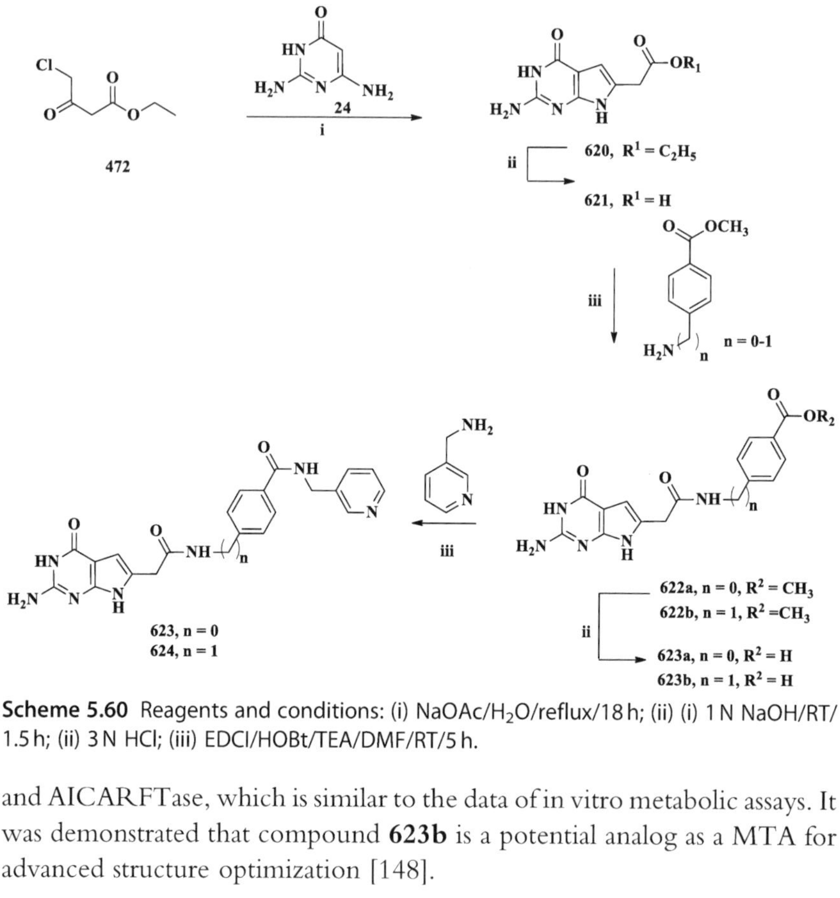

Scheme 5.60 Reagents and conditions: (i) NaOAc/H_2O/reflux/18 h; (ii) (i) 1 N NaOH/RT/1.5 h; (ii) 3 N HCl; (iii) EDCI/HOBt/TEA/DMF/RT/5 h.

and AICARFTase, which is similar to the data of in vitro metabolic assays. It was demonstrated that compound **623b** is a potential analog as a MTA for advanced structure optimization [148].

References

[1] Purcell WT, Ettinger DS. Novel antifolate drugs. Curr Oncol Rep 2003;5(2):114–25.

[2] Costi MP, Ferrari S. Update on antifolate drugs targets. Curr Drug Targets 2001;2 (2):135–66.

[3] Scagliotti GV, Selvaggi G. New data integrating multitargeted antifolates into treatment of first-line and relapsed non-small-cell lung cancer. Clin Lung Cancer 2008;9(Suppl. 3S):122–8.

[4] Sonvico F, Barbieri S, Colombo P, Barocelli E, Mucchino C, Cantoni AM, Petronini PG, Rusca M, Carbognani P, Ampollini L. Combined hyaluronate-based films loaded with pemetrexed and cisplatin for the treatment of malignant pleural mesothelioma: preliminary evaluation in an orthotopic tumor recurrence model. Eur J Pharm Sci 2018;123:89–97.

[5] Sato Y, Matsuda S, Maruyama A, Nakayama J, Miyashita T, Udagawa H, Umemura S, Yanagihara K, Ochiai A, Tomita M, Soga T, Tsuchihara K, Makinoshima H. Metabolic characterization of antifolate responsiveness and non-responsiveness in malignant pleural mesothelioma cells. Front Pharmacol 2018;9:1129.

[6] Giudice S, Benassi L, Bertazzoni G, Costi MP, Gelain A, Venturelli A, Bernardi C, Gualdi G, Coppi A, Rossi T, et al. New thymidylate synthase inhibitors induce apoptosis in melanoma cell lines. Toxicol In Vitro 2007;21(2):240–8.
[7] McGuire JJ. Anticancer antifolates: current status and future directions. Curr Pharm Des 2003;9(31):2593–613.
[8] Peters GJ, Hooijberg JH, Kaspers GJL, Jansen G. Folates and antifolates in the treatment of cancer; role of folic acid supplementation on efficacy of folate and nonfolate drugs. Trends Food Sci Technol 2005;16(6–7):289–97.
[9] Jie JL, Corey E. Chapter 8: Anticancer agents. In: Drug discovery: practices, processes, and perspectives. Wiley; 2013. p. 291.
[10] Sun X, Cross JA, Bognar AL, Baker EN, Smith CA. Folate-binding triggers the activation of folylpolyglutamate synthetase. J Mol Biol 2001;310(5):1067–78.
[11] Zhoua H, Rana G, Massonb J, Wanga C, Zhaoa Y, Songa Q. Novel tungsten phosphide embedded nitrogen-doped carbon nanotubes: a portable and renewable monitoring platform for anticancer drug in whole blood. Biosens Bioelectron 2018;105:226–35.
[12] Gelbert LM, Li S, Mullaney BP. Use of antifolates in patients with detectable levels of TTF-1 for the cancer treatment. PCT Int. Appl.; 2010. WO 2010021843 A1 20100225.
[13] Alexa-Stratulata T, Pešićc M, Gašparovićd AC, Trougakose IP, Rigantif C. What sustains the multidrug resistance phenotype beyond ABC efflux transporters? Looking beyond the tip of the iceberg. Drug Resist Updat 2019;46, 100643.
[14] Ge H, Ke J, Xu N, Li H, Gong J, Li X, Song Y, Zhu H, Bai C. Dexamethasone alleviates pemetrexed-induced senescence in non-small-cell lung cancer. Food Chem Toxicol 2018;119:86–97.
[15] Jiao H, Zhao X, Liu J, Ma T, Zhang Z, Zhang J, Wang J. In vivo imaging characterization and anticancer efficacy of a novel HER2 affibody and pemetrexed conjugate in lung cancer model. Nucl Med Biol 2019;68–69:31–9.
[16] Yang W, Yang L, Xia Y, Cheng L, Zhang J, Meng F, Yuan J, Zhong Z. Lung cancer specific and reduction-responsive chimaeric polymersomes for highly efficient loading of pemetrexed and targeted suppression of lung tumor in vivo. Acta Biomater 2018;70:177–85.
[17] Mahmud F, Jeon OC, Alam F, Maharjan R, Choi JU, Park J, Lee S, Park JW, Lee DS, Byun Y. Oral pemetrexed facilitates low-dose metronomic therapy and enhances antitumor efficacy in lung cancer. J Control Release 2018;284:160–70.
[18] McGuire JJ, Haile WH. Metabolism-blocked antifolates as potential anti-rheumatoid arthritis agents: 4-amino-4-deoxy-5,8,10-trideazapteroyl-,-4′-methyleneglutamic acid (CH-1504) and its analogs. Biochem Pharmacol 2009;77(7):1161–72.
[19] Shah K, Queener S, Cody V, Pace J, Gangjee A. Development of substituted pyrido [3,2-d]pyrimidines as potent and selective dihydrofolate reductase inhibitors for pneumocystis pneumonia infection. Bioorg Med Chem Lett 2019;29(15):1874–80.
[20] Hopper AT, Brockman A, Wise A, Gould J, Barks J, Radke JB, Sibley LD, Zou Y, Thomas S. Discovery of selective *Toxoplasma gondii* dihydrofolate reductase inhibitors for the treatment of toxoplasmosis. J Med Chem 2019;62(3):1562–76.
[21] Hui J, Przespo E, Elefante A. Pralatrexate: a novel synthetic antifolate for relapsed or refractory peripheral T-cell lymphoma and other potential uses. J Oncol Pharm Pract 2012;18(2):275–83.
[22] Kumar VP, Frey KM, Wang Y, Jain HK, Gangjee A, Anderson KS. Substituted pyrrolo[2,3-*d*]pyrimidines as *Cryptosporidium hominis* thymidylate synthase inhibitors. Bioorg Med Chem Lett 2013;23(19):5426–8.
[23] Jasmine Z, Owen O. Pralatrexate: basic understanding and clinical development. Expert Opin Pharmacother 2010;11(10):1705–14.
[24] Gangjee A, Kurup S, Namjoshi O. Dihydrofolate reductase as a target for chemotherapy in parasites. Curr Pharm Des 2007;13(6):609–39.

[25] Nduati E, Hunt S, Kamau EM, Nzila A. 2,4-diaminopteridine-based compounds as precursors for de novo synthesis of antifolates: a novel class of antimalarials. Antimicrob Agents Chemother 2005;49(9):3652–7.
[26] Gangjee A, Jain HD, Kurup S. Recent advances in classical and non-classical antifolates as antitumor and antiopportunistic infection agents: part I. Anticancer Agents Med Chem 2007;7(5):524–42.
[27] Gangjee A, Jain HD, Kurup S. Recent advances in classical and non-classical antifolates as antitumor and antiopportunistic infection agents: part II. Anticancer Agents Med Chem 2008;8(2):205–31.
[28] Gangjee A, Vidwans A, Elzein E, McGuire JJ, Queener SF, Kisliuk RL. Synthesis, antifolate, and antitumor activities of classical and nonclassical 2-amino-4-oxo-5-substituted-pyrrolo[2,3-*d*]pyrimidines. J Med Chem 2001;44(12):1993–2003.
[29] Ogwang S, Nguyen HT, Sherman M, Bajaksouzian S, Jacobs MR, Boom WH, Zhang G, Nguyen L. Bacterial conversion of folinic acid is required for antifolate resistance. J Biol Chem 2011;286(17):15377–90.
[30] Thiede JM, Kordus SL, Turman BJ, Buonomo JA, Aldrich CC, Minato Y, Baughn AD. Targeting intracellular *p*-aminobenzoic acid production potentiates the anti-tubercular action of antifolates. Sci Rep 2016;6:38083.
[31] Benkovic SJ, Fierke CA, Naylor AM. Insights into enzyme function from studies on mutants of dihydrofolate reductase. Science 1988;239(4844):1105–10.
[32] Hajian B, Scocchera E, Shoen C, Krucinska J, Viswanathan K, G-Dayanandan N, Erlandsen H, Estrada A, Mikušová K, Korduláková J, Cynamon M, Wright D. Drugging the folate pathway in *Mycobacterium tuberculosis*: the role of multi-targeting agents. Cell Chem Biol 2019;26(6):781–91.
[33] Patela TS, Bhatt JD, Dixit RB, Chudasama CJ, Patele BD, Dixit BC. Green synthesis, biological evaluation, molecular docking studies and 3DQSAR analysis of novel phenylalanine linked quinazoline-4(3*H*)-onesulphonamide hybrid entities distorting the malarial reductase activity in folate pathway. Bioorg Med Chem 2019;27:3574–86.
[34] Muller IB, Hyde JE. Folate metabolism in human malaria parasites—75 years on. Mol Biochem Parasitol 2013;188(1):63–77.
[35] Tarnchompoo B, Chitnumsub P, Jaruwat A, Shaw PJ, Vanichtanankul J, Poen S, Rattanajak R, Wongsombat C, Tonsomboon A, Decharuangsilp S, Anukunwithaya T, Arwon U, Kamchonwongpaisan S, Yuthavong Y. Hybrid inhibitors of malarial dihydrofolate reductase with dual binding modes that can forestall resistance. ACS Med Chem Lett 2018;9(12):1235–40.
[36] Mital A. Recent advances in antimalarial compounds and their patents. Curr Med Chem 2007;14(7):759–73.
[37] Patel TS, Vanparia SF, Patel UH, Dixit RB, Chudasama CJ, Patel BD, Dixit BC. Novel 2,3-disubstituted quinazoline-4(3H)-one molecules derived from amino acid linked sulphonamide as a potent malarial antifolates for DHFR inhibition. Eur J Med Chem 2017;129:251–65.
[38] Kamchonwongpaisan S, Quarrell R, Charoensetakul N, Ponsinet R, Vilaivan T, Vanichtanankul J, Tarnchompoo B, Sirawaraporn W, Lowe G, Yuthavong Y. Inhibitors of multiple mutants of *Plasmodium falciparum* dihydrofolate reductase and their antimalarial activities. J Med Chem 2004;47(3):673–80.
[39] Ferrari S, Calo S, Leone R, Luciani R, Costantino L, Sammak S, Di Pisa F, Pozzi C, Mangani S, Costi MP. 2'-Deoxyuridine 5′-monophosphate substrate displacement in thymidylate synthase through 6-hydroxy-2*H*-naphtho[1,8-*bc*]furan-2-one derivatives. J Med Chem 2013;56(22):9356–60.
[40] Gmeiner WH. Novel chemical strategies for thymidylate synthase inhibition. Curr Med Chem 2005;12(2):191–202.

[41] Costi MP, Tondi D, Rinaldi M, Barlocco D, Pecorari P, Soragni F, Venturelli A, Stroud RM. Structure-based studies on species-specific inhibition of thymidylate synthase. Biochim Biophys Acta 2002;1587(2–3):206–14.
[42] Wilson PM, Danenberg PV, Johnston PG, Lenz H, Ladner RD. Standing the test of time: targeting thymidylate biosynthesis in cancer therapy. Nat Rev Clin Oncol 2014;11:282–98.
[43] Chu E, Callender MA, Farrell Michael P, Schmitz JC. Thymidylate synthase inhibitors as anticancer agents: from bench to bedside. Cancer Chemother Pharmacol 2003;52 (Suppl:1):S80–90.
[44] Lehman NL. Future potential of thymidylate synthase inhibitors in cancer therapy. Expert Opin Investig Drugs 2002;11(12):1775–87.
[45] Rose MG, Farrell MP, Schmitz JC. Thymidylate synthase: a critical target for cancer chemotherapy. Clin Colorectal Cancer 2002;1(4):220–9.
[46] Ceppi P, Rapa I, Lo Iacono M, Righi L, Giorcelli J, Pautasso M, Bille A, Ardissone F, Papotti M, Scagliotti GV. Expression and pharmacological inhibition of thymidylate synthase and Src kinase in nonsmall cell lung cancer. Int J Cancer 2012;130 (8):1777–86.
[47] Matherly LH. Molecular and cellular biology of the human reduced folate carrier. Prog Nucleic Acid Res Mol Biol 2001;67:131–62.
[48] Damaraju VL, Cass CE, Sawyer MB. Renal conservation of folates role of folate transport proteins. Vitam Horm 2008;79:185–202.
[49] Frouin I, Prosperi E, Denegri M, Negri C, Donzelli M, Rossi L, Riva F, Stefanini M, Scovassi AI. Different effects of methotrexate on DNA mismatch repair proficient and deficient cells. Eur J Cancer 2001;37(9):1173–80.
[50] William BP, Raymond WR, William DE, Jonathan M. The anticancer drugs antimetabolites. 2nd ed. Oxford University Press; 1994. p. 69.
[51] Wu Y, Kochat H. Process for synthesizing antifolates. US 20050020833 A1; 2005.
[52] Sahr T, Ravanel S, Basset G, Nichols BP, Hanson AD, Rebeille F. Folate synthesis in plants: purification, kinetic properties, and inhibition of aminodeoxychorismate synthase. Biochem J 2006;396(1):157–62.
[53] Anthony JC. The complete book of enzyme therapy. Avery; 1999. p. 57.
[54] Simon P, Michael JR. Quinazoline derivatives as metabolically inert antifolate compounds. WO 2006029385 A2; 2006.
[55] Wang M, Yang J, Yuan M, Xue L, Li H, Tian C, Wang X, Liu J, Zhang Z. Synthesis and antiproliferative activity of a series of novel 6-substituted pyrido[3,2-d] pyrimidines as potential nonclassical lipophilic antifolates targeting dihydrofolate reductase. Eur J Med Chem 2017;128:88–97.
[56] Li H, Fang F, Liu Y, Xue L, Wang M, Guo Y, Wang X, Tian C, Liu J, Zhang Z. Inhibitors of dihydrofolate reductase as antitumor agents: design, synthesis and biological evaluation of a series of novel nonclassical 6-substituted pyrido[3,2-d] pyrimidines with a three- to five-carbon bridge. Bioorg Med Chem 2018;26:2674–85.
[57] Marques SM, Enyedy EA, Supuran CT, Krupenko NI, Krupenko SA, Santos MA. Pteridine-sulfonamide conjugates as dual inhibitors of carbonic anhydrases and dihydrofolate reductase with potential antitumor activity. Bioorg Med Chem 2010;18 (14):5081–9.
[58] Chelsea Therapeutics. New classical antifolates. US20080214585; 2008.
[59] Abraham A, McGuire JJ, Galivan J, Nimec Z, Kisliuk RL, Gaumont Y, Nair MG. Folate analogs. 34. Synthesis and antitumor activity of non-polyglutamylatable inhibitors of dihydrofolate reductase. J Med Chem 1991;34(1):222–7.
[60] Stoicescu D. Novel inhibitors of folic acid-dependent enzymes. US20100249141; 2010.

[61] Stoicescu D. Inhibitors of folic acid-dependent enzymes. US7718660; 2010.
[62] Dana-Farber Cancer Institute. Pharmaceutically active ornithine derivatives, ammonium salts thereof and methods of making same. US6989386; 2006.
[63] Grant SC, Kris MG, Young CW, Sirotnak FM. Edatrexate, an antifolate with antitumor activity: a review. Cancer Invest 1993;11(1):36–45.
[64] Takimoto CH. Antifolates in clinical development. Semin Oncol 1997;24(5 Suppl. 18):S18-40–51.
[65] Nair MG, Fayard ML, Lariccia JM, Amato AE, McGuire JJ, Galivan JH, et al. Metabolism blocked classical folate analog inhibitors of dihydrofolate reductase-1: synthesis and biological evaluation of mobiletrex. Med Chem Res 1999;9:176.
[66] Chroma Therapeutics. Pyrimidine derivatives useful as DHFR inhibitors. WO07129020; 2007.
[67] Chroma Therapeutics. DHFR enzyme inhibitors. WO07132146; 2007.
[68] Rodriguez-Lopez JN, Navarro-Peran EM, Cabezas-Herrera J. Dihydrofolate reductase inhibition by epigallocatechin gallate compounds. US20070249545; 2007.
[69] Varney MD, Marzoni GP, Palmer CL, Deal JG, Webber S, Welsch KM, Bacquet RJ, Barlett CA, Morse CA. Crystal-structure-based design and synthesis of benz[cd] indole-containing inhibitors of thymidylate synthase. J Med Chem 1992;35:663–76.
[70] Marsham PR, Wardleworth JM, Boyle FT, Hennequin LF, Kimbell R, Brown M, Jackman AL. Design and synthesis of potent non-polyglutamatable quinazoline antifolate thymidylate synthase inhibitors. J Med Chem 1999;42:3809–20.
[71] Duch DS, Banks S, Dev IK, Dickerson SH, Ferone R, Heath LS, Humphreys J, Knick V, Pendergast W, Singer S, Smith GK, Waters K, Wilson HR. Biochemical and cellular pharmacology of 1843U89, a novel benzoquinazoline inhibitor of thymidylate synthase. Cancer Res 1993;53(4):810–8.
[72] Webber SE, Bleckman TM, Attard J, Deal JG, Kathardekar V, Welsh KM, Webber S, Janson CA, Matthews DA. Design of thymidylate synthase inhibitors using protein crystal structures: the synthesis and biological evaluation of a novel class of 5-substituted quinazolinones. J Med Chem 1993;36:733–46.
[73] Duquesne University of the Holy Spirit. Thieno pyrimidine compounds. US20090326224; 2009.
[74] Varney MD, Palmer CL, Romines 3rd WH, Boritzki T, Margosiak SA, Almassy R, Janson CA, Barlett C, Howland EJ, Ferre R. Protein structure-based design, synthesis, and biological evaluation of 5-thia-2,6-diamino-4(3*H*)-oxopyrimidines: potent inhibitors of glycinamide ribonucleotide transformylase with potent cell growth inhibition. J Med Chem 1997;40:2502–24.
[75] Bartlett CA, Boritzki TJ, Dagostino EF, Margosiak SA, Palmer CL, Romines WH, et al. Biological properties of AG2037: a new inhibitor of glycinamide ribonucleotide formyltransferase with low affinity for the folate receptor. Proc Am Assoc Cancer Res 1999;40:291.
[76] Tsukamoto T, Haile WH, McGuire JJ, Coward JK. Mechanism-based inhibition of human folylpolyglutamate synthetase: design, synthesis, and biochemical characterization of a phosphapeptide mimic of the tetrahedral intermediate. Arch Biochem Biophys 1998;355:109.
[77] Valiaeva N, Bartley D, Konno T, Coward JK. Phosphinic acid pseudopeptides analogous to glutamyl-gamma-glutamate: synthesis and coupling to pteroyl azides leads to potent inhibitors of folylpoly-gamma-glutamate synthetase. J Org Chem 2001;66:5146.
[78] Loriga M, Piras S, Paglietti G, Costi MP, Venturelli A. Quinoxaline chemistry. Part 15. 4-[2-Quinoxalylmethylenimino]benzoylglutamates and -benzoates, 4-[2-quinoxalylmethyl-*N*-methylamino]benzoylglutamates as analogs of classical

antifolate agents. Synthesis, elucidation of structures and in vitro evaluation of antifolate and anticancer activities. Farmaco 2003;58(1):51–61.

[79] Gangjee A, Devraj R, McGuire JJ, Kisliuk RL. 5-Arylthio-substituted 2-amino-4-oxo-6-methylpyrrolo[2,3-*d*]pyrimidine antifolates as thymidylate synthase inhibitors and antitumor agents. J Med Chem 1995;38:4495–502.

[80] Taylor EC, Patel HH. Synthesis of pyrazolo[3,4-d]pyrimidine analogues of the potent antitumor agent N-{4-[2-(2-amino-4(3*H*)-oxo-7*H*-pyrrolo[2,3-*d*]pyrimidin-5-yl)ethyl]benzoyl}-L-glutamic acid (LY231514). Tetrahedron 1992;48 (37):8089–100.

[81] Rosowsky A, Forsch RA, Moran RG. Side chain modified 5-deazatolate and 5-deazatetrahydrotolate analogs as mammalian folylpolyglutamate synthetase and glycinamide ribonucleotide formyltransferase inhibitors: synthesis and in vitro biological evaluation. 47. N-[4-[[(3,4-Dihydro-4-oxo-1,2,3-benzotriazin-6-yl)methyl]amino]] benzoyl-L-glutamic acid, a novel a-ring analog of 2-desamino-5,8-dideazafolic acid. J Med Chem 1992;35(14):2626–30.

[82] Wang Y, Cherian C, Orr S, Mitchell-Ryan S, Hou Z, Raghavan S, Matherly LH, Gangjee A. Tumor-targeting with novel non-benzoyl 6-substituted straight chain pyrrolo[2,3-*d*]pyrimidine antifolates via cellular uptake by folate receptor α and inhibition of de novo purine nucleotide biosynthesis. J Med Chem 2013;56(21):8684–95.

[83] Wang L, Desmoulin SK, Cherian C, Polin L, White K, Kushner J, Fulterer A, Chang MH, Mitchell-Ryan S, Stout M, Romero MF, Hou Z, Matherly LH, Gangjee A. Synthesis, biological, and antitumor activity of a highly potent6-substituted Pyrrolo[2,3-*d*]pyrimidine thienoyl antifolate inhibitor with proton-coupled folate transporter and folate receptor selectivity over the reduced folate carrier that inhibits β-glycinamide ribonucleotide formyltransferase. J Med Chem 2011;54:7150–64.

[84] Wang L, Cherian C, Kugel Desmoulin S, Polin L, Deng Y, Wu J, Hou Z, White K, Kushner J, Matherly LH, Gangjee A. Synthesis and antitumor activity of a novel series of 6-substituted pyrrolo[2,3-d]pyrimidine thienoyl antifolate inhibitors of PurineBiosynthesis with selectivity for high affinity folate receptors andthe proton-coupled folate transporter over the reduced folate carrier for cellular entry. J Med Chem 2010;53:1306–18.

[85] Wang L, Cherian C, Kugel Desmoulin S, Mitchell-Ryan S, Hou Z, Matherly LH, Gangjee A. Synthesis and biological activity of 6-substituted pyrrolo[2,3-*d*]pyrimidine thienoyl regioisomers as inhibitors of de novo purine biosynthesis with selectivity for cellular uptake by high affinity folate receptors and the proton-coupled folate transporter over the reduced folate carrier. J Med Chem 2012;55:1758–70.

[86] Ravindra M, Wallace-Povirk A, Karim MA, Wilson MR, O'Connor C, White K, Kushner J, Polin L, George C, Hou Z, Matherly LH, Gangjee A. Tumor targeting with novel pyridyl 6-substituted pyrrolo[2,3-*d*]pyrimidine antifolates via cellular uptake by folate receptor α and the proton-coupled folate transporter and inhibition of de novo purine nucleotide biosynthesis. J Med Chem 2018;61(5):2027–40.

[87] Golani LK, George C, Zhao S, Raghavan S, Orr S, Wallace A, Wilson MR, Hou Z, Matherly LH, Gangjee A. Structure-activity profiles of novel 6-substituted pyrrolo [2,3-d]pyrimidine thienoyl antifolates with modified amino acids for cellular uptake by folate receptors α and β and the proton-coupled folate transporter. J Med Chem 2014;57(19):8152–66.

[88] Deng Y, Wang Y, Cherian C, Hou Z, Buck SA, Matherly LH, Gangjee A. Synthesis and discovery of high affinity folate receptor-specific glycinamide ribonucleotide formyltransferase inhibitors with antitumor activity. J Med Chem 2008;51(16):5052–63.

[89] Aso K, Imai Y, Yukishige K, Ootsu K, Akimoto H. Pyrrolo[β,γ-d]pyrimidine thymidylate synthase inhibitors: design and synthesis of one-carbon bridge derivatives. Chem Pharm Bull 2001;49:1280–7.

[90] Barnett CJ, Wilson TM, Kobierski ME. A practical synthesis of multitargeted antifolate LYβγ1514. Org Proc Res Dev 1999;3:184–8.
[91] Bloch R. 5,5-Dibromo-*β*,*β*-dimethyl-4,6-dioxo-1,3-dioxane—new brominating agent for saturated and alpha,beta-unsaturated carbonyl-compounds. Synthesis-Stuttgart 1978;1978(2):140–2. -1,γ-Dioxane" to be "-1,3-dioxane".
[92] Wang Y, Mitchell-Ryan S, Raghavan S, George C, Orr S, Hou Z, Matherly LH, Gangjee A. Novel 5-substituted pyrrolo[2,3-d]pyrimidines as dual inhibitors of glycinamide ribonucleotide formyltransferase and 5-aminoimidazole-4-carboxamide ribonucleotide formyltransferase and as potential antitumor agents. J Med Chem 2015;58 (3):1479–93.
[93] Ravindra M, Wilson MR, Tong N, O'Connor C, Karim M, Polin L, Wallace-Povirk A, White K, Kushner J, Hou Z, Matherly LH, Gangjee A. Fluorine-substituted pyrrolo[2,3-*d*]pyrimidine analogues with tumor targeting via cellular uptake by folate receptor α and the proton-coupled folate transporter and inhibition of de novo purine nucleotide biosynthesis. J Med Chem 2018;61(9):4228–48.
[94] Deng Y, Zhou X, Kugel Desmoulin S, Wu J, Cherian C, Hou Z, Matherly LH, Gangjee A. Synthesis and biological activity of a novel series of 6-substituted thieno[2,3-*d*] pyrimidine antifolate inhibitors of purine biosynthesis with selectivity for high affinity folate receptors over the reduced folate carrier and proton-coupled folate transporter for cellular entry. J Med Chem 2009;52(9):2940–51.
[95] Larock RC, Leung WY, Stolzdunn S. Synthesis of aryl-substituted aldehydes and ketones via palladium-catalyzed coupling of aryl halides and non-allylic unsaturated alcohols. Tetrahedron Lett 1989;30:6629–32.
[96] Mitchell-Ryan S, Wang Y, Raghavan S, Ravindra MP, Hales E, Orr S, Cherian C, Hou Z, Matherly LH, Gangjee A. Discovery of 5-substituted pyrrolo[2,3-*d*]pyrimidine antifolates as dual-acting inhibitors of glycinamide ribonucleotide formyltransferase and 5-aminoimidazole-4-carboxamide ribonucleotide formyltransferase in de novo purine nucleotide biosynthesis: implications of inhibiting 5-aminoimidazole-4-carboxamide ribonucleotide formyltransferase to ampk activation and antitumor activity. J Med Chem 2013;56(24):10016–32.
[97] Taylor EC, Wong GSK. Convergent and efficient palladium-effected synthesis of 5, 10-dideaza-5,6,7,8-tetrahydrofolic acid (DDATHF). J Org Chem 1989;54:3618–24.
[98] Taylor EC, Young WB, Spanka C. Synthesis of *N*-{4-[2-(2-Amino-5,6-dihydro-4 (3*H*)-oxo-7*H*-pyrrolo[2,3-*d*]pyrimidin-6-yl)- ethyl]benzoyl}-l-glutamic acid: a ring-contracted analogue of 5,10-Dideaza-5,6,7,8-tetrahydrofolic acid. J Org Chem 1996;61:1261–6.
[99] Hart BP, Haile WH, Licato NJ, Bolanowska WE, McGuire JJ, Coward JK. Synthesis and biological activity of folic acid and methotrexate analogues containing L-threo-(2S,4S)-4-fluoroglutamic acid and DL-3,3-difluoroglutamic acid. J Med Chem 1996;39(1):56–65.
[100] Piper JR, Montgomery JA. Preparation of 6-(bromomethyl)-2,4-pteridinediamine hydrobromide and its use in improved syntheses of methotrexate and related compounds. J Org Chem 1977;42:208–11.
[101] Tsukamoto T, Kitazume T, McGuire JJ, Coward JK. Synthesis and biological evaluation of DL-4,4-difluoroglutamic acid and DL-ç,ç-difluoromethotrexate. J Med Chem 1996;39:66–72.
[102] Tsukamoto T, Haile WH, McGuire JJ, Coward JK. Synthesis and biological evaluation of N^{α}-(4-Amino-4-deoxy-10-methylpteroyl)-DL-4,4-difluoroornithine. J Med Chem 1996;39:2536–40.
[103] Piper JR, Ramamurthy B, Johnson CA, Otter GM, Sirotnak FM. Analogues of 10-deazaaminopterin and 5-Alkyl-5,10-dideazaaminopterin with the 4-substituted 1-naphthoyl group in the place of 4-substituted benzoyl. J Med Chem 1996;39:614–8.

[104] Forsch RA, Wright JE, Rosowsky A. Synthesis and in vitro antitumor activity of thiophene analogues of 5-chloro-5,8-dideazafolic acid and 2-methyl-2-desamino-5-chloro-5,8-dideazafolic acid. Bioorg Med Chem 2002;10(6):2067–76.
[105] Gangjee A, Zeng Y, McGuire JJ, Kisliuk RL. Effect of C9-methyl substitution and C8-C9 conformational restriction on antifolate and antitumor activity of classical 5-substituted 2,4-diaminofuro[2,3-*d*]pyrimidines. J Med Chem 2000;43:3125–33.
[106] Gangjee A, Zeng Y, McGuire JJ, Kisliuk RL. Synthesis of classical and nonclassical, partially restricted, linear, tricyclic 5-deaza antifolates1. J Med Chem 2002;45:5173–81.
[107] Gangjee A, Yang J, McGuireb JJ, Kisliuk RL. Synthesis and evaluation of a classical 2,4-diamino-5-substitutedfuro[2,3-d]pyrimidine and a 2-amino-4-oxo-6-substitutedpyrrolo[2,3-*d*]pyrimidine as antifolates. Bioorg Med Chem 2006;14:8590–8.
[108] Tian C, Wang M, Han Z, Fang F, Zhang Z, Wang X, Liu J. Design, synthesis and biological evaluation of novel 6-substituted pyrrolo [3,2-*d*] pyrimidine analogues as antifolate antitumor agents. Eur J Med Chem 2017;138:630–43.
[109] Golani LK, Wallace-Povirk A, Deis SM, Wong J, Ke J, Gu X, Raghavan S, Wilson MR, Li X, Polin L, de Waal PW, White K, Kushner J, O'Connor C, Hou Z, Xu HE, Melcher K, Dann 3rd CE, Matherly LH, Gangjee A. Tumor targeting with novel 6-substituted pyrrolo [2,3-d] pyrimidine antifolates with heteroatom bridge substitutions via cellular uptake by Folate receptor α and the proton-coupled Ffolate transporter and inhibition of de novo purine nucleotide biosynthesis. J Med Chem 2016;59(17):7856–76.
[110] Gangjee A, Yu J, McGuire JJ, Cody V, Galitsky N, Kisliuk RL, Queener SF. Design, synthesis, and X-ray crystal structure of a potent dual inhibitor of thymidylate synthase and dihydrofolate reductase as an antitumor agent. J Med Chem 2000;43:3837–51.
[111] Gangjee A, Jain HD, Phan J, Lin X, Song X, McGuire JJ, Kisliuk RL. Dual inhibitors of thymidylate synthase and dihydrofolate reductase as antitumor agents: design, synthesis, and biological evaluation of classical and nonclassical pyrrolo[2,3-*d*]pyrimidine antifolates. J Med Chem 2006;49:1055–65.
[112] Gangjee A, Lin X, Kisliuk RL, McGuire JJ. Synthesis of *N*-{4-[(2,4-diamino-5--methyl-4,7-dihydro-3*H*- pyrrolo[2,3-*d*]pyrimidin-6-yl)thio]benzoyl}-l-glutamic acid and *N*-{4-[(2-Amino-4-oxo-5-methyl-4,7-dihydro-3*H*-pyrrolo[2,3-*d*]pyrimidin- 6-yl)thio]benzoyl}-l-glutamic acid as dual inhibitors of dihydrofolate reductase and thymidylate synthase and as potential antitumor agents. J Med Chem 2005;48:7215–22.
[113] Taylor EC, Patel HH, Jun J-G. A one-step ring transformation/ring annulation approach to pyrrolo[2,3-*d*]pyrimidines. A new synthesis of the potent dihydrofolate reductase inhibitor TNP-351. J Org Chem 1995;60:6684–7.
[114] Gangjee A, Lin X, Queener SF. Design, synthesis, and biological evaluation of 2,4-diamino-5-methyl-6-substituted pyrrolo[2,3-*d*]pyrimidines as dihydrofolate reductase inhibitors. J Med Chem 2004;47:3689–92.
[115] Gangjee A, Devraj R, McGuire JJ, Kisliuk RL, Queener SF, Barrows LR. Classical and nonclassical furo[2,3-d]pyrimidines as novel antifolates: synthesis and biological activities. Med Chem 1994;37:1169–76.
[116] Gangjee A, Devraj R, McGuire JJ, Kisliuk RL. Effect of bridge region variation on antifolate and antitumor activity of classical 5-substituted 2,4-diaminofuro[2,3-*d*] pyrimidines. J Med Chem 1995;38(19):3798–805.
[117] Gangjee A, Zeng Y, McGuire JJ, Kisliuk RL. Synthesis of classical, four-carbon bridged 5-substituted furo[2,3-*d*]pyrimidine and 6-substituted pyrrolo[2,3-*d*]pyrimidine analogues as antifolates. J Med Chem 2005;48:5329–36.

[118] Gewald K. Heterocyclen aus CH-Aciden Nitrilen. VII. 2-Aminothiophene aus R-oxo-mercaptanen und methylenaktiven. Nitrilen. Chem Ber 1966;98:3571–7.
[119] Gangjee A, Qiu Y, Kisliuk RL. Synthesis of classical and nonclassical 2-amino-4-oxo-6-benzyl-thieno[2,3-*d*]pyrimidines as potential thymidylate synthase inhibitors. J Heterocyclic Chem 2004;41:941–6.
[120] Gangjee A, Qiu Y, Li W, Kisliuk RL. Potent dual thymidylate synthase and dihydrofolate reductase inhibitors: classical and nonclassical 2-amino-4-oxo-5-arylthio-substituted-6- methylthieno[2,3-*d*]pyrimidine antifolates. J Med Chem 2008;51:5789–97.
[121] Taylor EC, Young WB, Chaudhari R, Patel M. Synthesis of a regioisomer of *N*-{4-[2-(2-amino-4(3*H*)-oxo-7*H*-pyrrolo[2,3-*d*]pyrimidin-5-yl)ethyl]benzoyl}-L-glutamic acid (LY231514), an active thymidylate synthase inhibitor and antitumor agent. Heterocycles 1993;36:1897–908.
[122] Itoh T, Mase T. A general palladium-catalyzed coupling of aryl bromides/triflates and thiols. Org Lett 2004;24:4587–90.
[123] Gangjee A, Li W, Kisliuk RL, Cody V, Pace J, Piraino J, Makin J. Design, synthesis, and X-ray crystal structure of classical and nonclassical 2-amino-4-oxo-5-substituted-6-ethylthieno[2,3-*d*]pyrimidines as dual thymidylate synthase and dihydrofolate reductase inhibitors and as potential antitumor agents. J Med Chem 2009;52:4892–902.
[124] Taylor EC, Liu B. A novel synthetic route to 7-substituted derivatives of the antitumor agent LY231514 (MTA). Tetrahedron Lett 1999;40:5291–4.
[125] Taylor EC, Liu B. A new and efficient synthesis of pyrrolo[2,3-*d*]pyrimidine anticancer agents: alimta (LY231514, MTA), homo-alimta, TNP-351, and some aryl 5-substituted pyrrolo[2,3-*d*]pyrimidines. J Org Chem 2003;68:9938–47.
[126] Rosowsky A, Vaidya CM, Bader H, Wright JE, Teicher BA. Analogues of *N*R-(4-amino-4-deoxypteroyl)-*Nä*-hemiphthaloyl-L-ornithine (PT523) modified in the side chain: synthesis and biological evaluation. J Med Chem 1997;40:286–99.
[127] Reynolds RC, Johnson CA, Piper JR, Sirotnak FM. Synthesis and antifolate evaluation of the aminopterin analogue with a bicyclo[2.2.2]octane ring in place of the benzene ring. Eur J Med Chem 2001;36(3):237–42.
[128] Vaidya CM, Wright JE, Rosowsky A. Synthesis and in vitro antitumor activity of new deaza analogues of the nonpolyglutamatable antifolate *N*α-(4-amino-4-deoxypteroyl)-*N*δ-hemiphthaloyl-L-ornithine (PT523). J Med Chem 2002;45:1690–6.
[129] Gangjee A, Vidwans AP, Vasudevan A, Queener SF, Kisliuk RL, Cody V, Li R, Galitsky N, Luft JR, Pangborn W. Structure-based design and synthesis of lipophilic 2,4-diamino-6-substituted quinazolines and their evaluation as inhibitors of dihydrofolate reductases and potential antitumor agents. J Med Chem 1998;41:3426–34.
[130] Gangjee A, Adair OO, Queener SF. Synthesis and biological evaluation of 2,4-diamino-6-(arylaminomethyl)pyrido[2,3-*d*]pyrimidines as inhibitors of *Pneumocystis carinii* and *Toxoplasma gondii* dihydrofolate reductase and as antiopportunistic infection and antitumor agents. J Med Chem 2003;46:5074–82.
[131] Corona P, Loriga M, Costi MP, Ferrari S, Paglietti G. Synthesis of N-(5,7-diamino-3-phenyl-quinoxalin-2-yl)-3,4,5-substituted anilines and N-[4[(5,7-diamino-3-phenylquinoxalin-2-yl)amino]benzoyl]-l-glutamic acid diethyl ester: evaluation of in vitro anti-cancer and anti-folate activities. Eur J Med Chem 2008;43(1):189–203.
[132] Sheng-Li C, Yu-Ping F, Yu-Yang J, Shi-Ying L, Guo-Yu D, Run-Tao L. Synthesis and in vitro antitumor activity of 4(3*H*)-quinazolinone derivatives with dithiocarbamate side chains. Bioorg Med Chem Lett 2005;15:1915–7.
[133] Alleca S, Corona P, Loriga M, Paglietti G, Loddo R, Mascia V, Busonera B, La Colla P. Quinoxaline chemistry. Part 16. 4-substituted anilino and 4-substituted phenoxymethyl pyrrolo[1,2-*a*]quinoxalines and *N*-[4-(pyrrolo[1,2-*a*]quinoxalin-4-yl)amino

and hydroxymethyl]benzoyl glutamates. Synthesis and evaluation of in vitro biological activity. Farmaco 2003;58(9):639–50.
[134] Gao T, Zhang C, Shi X, Guo R, Zhang K, Gu J, Li L, Li S, Zheng Q, Cui M, Cui M, Gao X, Liu Y, Wang L. Targeting dihydrofolate reductase: design, synthesis and biological evaluation of novel 6-substituted pyrrolo[2,3-d]pyrimidines as nonclassical antifolates and as potential antitumor agents. Eur J Med Chem 2019;178:329–40.
[135] Singla P, Luxami V, Paul K. Synthesis, in vitro antitumor activity, dihydrofolate reductase inhibition, DNA intercalation and structure-activity relationship studies of 1,3,5-triazine analogues. Bioorg Med Chem Lett 2016;26(2):518–23.
[136] Erickson JA, Jalaie M, Robertson DH, Lewis RA, Vieth M. Lessons in molecular recognition: the effects of ligand and protein flexibility on molecular docking accuracy. J Med Chem 2004;47(1):45–55.
[137] Zhou X, Lin K, Ma X, Chui WK, Zhou W. Design, synthesis, docking studies and biological evaluation of novel dihydro-1,3,5-triazines as human DHFR inhibitors. Eur J Med Chem 2017;125:1279–88.
[138] Hennequin LF, Thomas Boyle F, Michael Wardleworth J, Marsham PR, Kimbell R, Jackman AL. Quinazoline antifolates thymidylate synthase inhibitors: lipophilic analogues with modification to the C2-methyl substituent. J Med Chem 1996;39:695–704.
[139] Bavetsias V, Bisset GMF, Kimbell R, Boyle FT, Jackman AL. Synthesis of novel quinazoline-based antifolates with modified glutamate side chains as potential inhibitors of thymidylaye synthase and antitumor agents. Tetrahedron 1997;53(39):13383–96.
[140] Taylor EC, Dowling JE. Synthesis of a pyrimido[4,5-*b*]azepine analog of 5,10-dideaza-5,6,7,8-tetrahydrofolic acid (DDATHF). Bioorg Med Chem Lett 1997;7(4):453–6.
[141] Fales KR, Njoroge FG, Brooks HB, Thibodeaux S, Torrado A, Si C, Toth JL, Mc Cowan JR, Roth KD, Thrasher KJ, Frimpong K, Lee MR, Dally RD, Shepherd TA, Durham TB, Margolis BJ, Wu Z, Wang Y, Atwell S, Wang J, Hui YH, Meier TI, Konicek SA, Geeganage S. Discovery of N-(6-fluoro-1-oxo-1,2-dihydroisoquinolin-7-yl)-5-[(3R)-3-hydroxypyrrolidin-1-yl] thiophene-2-sulfonamide (LSN 3213128), a potent and selective nonclassical antifolate aminoimidazole-4-carboxamide ribonucleotide formyltransferase (AICARFT) inhibitor effective at tumor suppression in a cancer xenograft model. J Med Chem 2017;60(23):9599–616.
[142] Chong Y, Hwang I, Tavassoli A, Zhang Y, Wilson IA, Benkovic SJ, Boger DL. Synthesis and biological evaluation of alpha- and gamma-carboxamide derivatives of 10-CF3CO-DDACTHF. Bioorg Med Chem 2005;13:3587–92.
[143] Demko Z, Sharpless KB. An expedient route to the tetrazole analogues of α-amino acids. Org Lett 2002;4:2525–7.
[144] Itoh F, Yukishige K, Wajima M, Ootsu K, Akimoto H. Non-glutamate type pyrrolo[2,3-*d*]pyrimidine antifolates. I: synthesis and biological properties of pyrrolo[2,3-d]pyrimidine antifolates containingtetrazole congeners of glutamic acid. Chem Pharm Bull 1995;43:230–5.
[145] Thomae KG. British patent 807826; 1959 [*Chem. Abstr.* 1959, *53*, 12317e). Thomae, K. G. German Patent 1151806, 1963. (*Chem. Abstr.* 1964, *60*, 2974a].
[146] DeMartino JK, Hwang I, Xu L, Wilson IA, Boger DL. Discovery of a potent, nonpolyglutamatable inhibitor of glycinamide ribonucleotide transformylase. J Med Chem 2006;49(10):2998–3002.
[147] Curtin NJ, Barlow HC, Bowman KJ, Calvert AH, Davison R, Golding BT, Huang B, Loughlin PJ, Newell DR, Smith PG, Griffin RJ. Resistance-modifying agents. 11.(1) Pyrimido[5,4-d]pyrimidine modulators of antitumor drug activity. Synthesis and

structure-activity relationships for nucleoside transport inhibition and binding to alpha1-acid glycoprotein. J Med Chem 2004;47(20):4905–22.
[148] Liu Y, Zhang C, Zhang H, Li M, Yuan J, Zhang Y, Zhou J, Guo H, Zhao L, Du Y, Wang L, Ren L. Synthesis and antitumor activity of a novel series of 6-substituted pyrrolo[2,3-*d*]pyrimidines as potential nonclassical antifolates targeting both thymidylate and purine nucleotide biosynthesis. Eur J Med Chem 2015;93:142–55.

CHAPTER 6

Synthetic strategies for purine nucleoside analogs

1. Introduction

Purine nucleosides have significant importance and pivotal effects in medicinal chemistry. Following the enormous developments in the novel synthetic strategies of the purine nucleosides, we surveyed the most recent and up-to-date synthetic approaches of the purine nucleosides that reflect its comprehensive importance in the different fields.

Consequently, the novel insights of a broad array of novel nucleosides have been demonstrated.

Purine nucleoside analogs have unique and broad-spectrum curative effects. Various purine structural analogs that mimic the endogenous nucleosides are designed to target the main steps in the cellular metabolism. Purine nucleosides have integrated importance with the other classes of antimetabolites, pyrimidines [1], and antifolates [1,2]. Thus, well-recognized synthetic approaches have been set up for subsequent drug development and discovery.

Purine nucleosides are critical class of heterocyclic compounds that are widely distributed in all living organisms as constituents of nucleic acids and signal molecules [3]. They display a wide range of promising biological activities [4]. Crucial routes for the syntheses and development of new drugs are performed through targeting purine metabolic pathways [5]. In nucleic acid metabolism, modified nucleosides can be considered as therapeutic agents [6] and have many potential applications in biology and biotechnology [7]. Purine nucleosides are an important category of drugs in treating cancer [8–10]. PNA are clinically important anticancer drugs [11]. Purine nucleosides have an antileukemic activity, and they are used in the treatment of myelomonocytic [12], acute lymphoblastic [13,14], and acute myeloid leukemia [15]. Purine analogs are widely utilized not only as antileukemic agents, but also as cytotoxic agents to treat solid tumors. [11] The purine nucleoside analogs (PNA) form an important group of cytotoxic drugs active

New Strategies Targeting Cancer Metabolism
https://doi.org/10.1016/B978-0-12-821783-2.00002-9

in treating neoplastic diseases [16]. New purine nucleoside analogs (PNA) have been demonstrated to exhibit an antitumor activity in preclinical and clinical trials [17–23].

Novel PNAs comprising nelarabine, clofarabine, and Immucillin H have been introduced into clinical trials. These agents have various metabolic features, new mechanism of action, and are undergoing phase I-II clinical investigations for treating hematopoietic malignancies [24]. Many other novel purine nucleoside analogs are used in the treatment of hematological malignancies [25,26].

Clinical utility of purine nucleoside analog is indicated in treating many diseases [27,28]. Various derivatives of the purine nucleosides, which are used in medicine as immunosuppressors [29] has immunomodulatory and antiinflammatory properties [30,31]. Other purine nucleosides, such as inosine, which is a naturally occurring purine nucleoside with neurotrophic, antioxidant, and neuroprotective effects, are known to improve motor function in preclinical models of spinal cord injury (SCI) [32,33].

Purine nucleosides act as biomarkers of tissue ischemia and hypoxia as clinically applied in humans in vivo [34]. There is also clinical application of purine nucleoside analogs in the treatment of supraventricular tachycardia [35]. Additionally, PNA has antitrypanosomal activity [36] and other antimicrobial activities [37]. They also represent promising lead compounds in the development of clinically useful antiviral therapies [38]; many nucleoside analogs have been revealed to be potent inhibitors of human immunodeficiency virus type 1 (HIV-1) [39–41]. There are many purine nucleoside antibiotics utilized clinically to treat human DNA virus infections [42]. Many acyclic purine nucleoside analogs are used for therapy of herpes simplex virus (HSV) infection [43]. Various other analogs act as antiinfluenza virus agents [44], potent antihepatitis C (HCV) virus agents [45–81], antiproliferative agents [82], and pleiotropic agents for the potential treatment of Alzheimer's disease [83–85], and other derivatives act as selective cholinesterase inhibitors [86].

They have also fluorescence applications, which have drawn the interest of many chemists. For example, 7-deazapurine nucleosides [87] and 8-azapurine ribosides have many fluorescence properties [88].

The current research in the developments of the synthetic approaches has continued to expand in an intense pace. Up-to-date synthetic approaches of the purine nucleosides have been overviewed throughout this book.

Purine analogs are an important class of therapeutic agents that have many remarkable and comprehensive applications in biochemistry and biotechnology. They are used in the treatment of cancer, Alzheimer's disease, influenza virus, hematological malignancies, hepatitis C (HCV), and HIV. A wide variety of purine nucleosides with diverse biological activities have been synthesized. Besides the ribonucleosides, including free-, protected ribonucleosides, deoxy, and dideoxy derivatives, many nonribonucleosides were synthesized. The biological evaluation of the novel synthesized compounds was emphasized. Their synthetic procedures were also elaborated.

2. Examples of potent anticancer purine nucleoside analogs

Many purine nucleoside analogs were used in the treatment of cancer, such as carcinoma, adenocarcinoma, leukemia, and sarcoma, which may include hepatoma, liver cancer, breast cancer, esophageal carcinoma, pancreatic carcinoma, gastrointestinal cancer, bladder cancer, Kaposi's sarcoma, T-cell leukemia, squamous cell carcinoma (SCC) of the skin, ovarian cancer, pulmonary carcinoma, prostate cancer, skin cancer, gastric cancer, and Ewing's sarcoma. Additionally, they are used in the treatment of noncancerous proliferative disorders and benign tumors.

Accordingly, many invented compounds can be combined with other active ingredients, which are utilized in the treatment of proliferative diseases and disorders. For example, *N*-valproyl-9-(2-valproyloxy)ethoxymethylguanine (**1**) (Fig. 6.1) was synthesized by reacting $(MeCH_2CH_2)_2CHCOCl$ with acyclovir at 0°C under argon using DMAP and triethylamine in dichloromethane. The synthesized compounds were evaluated for the in vitro

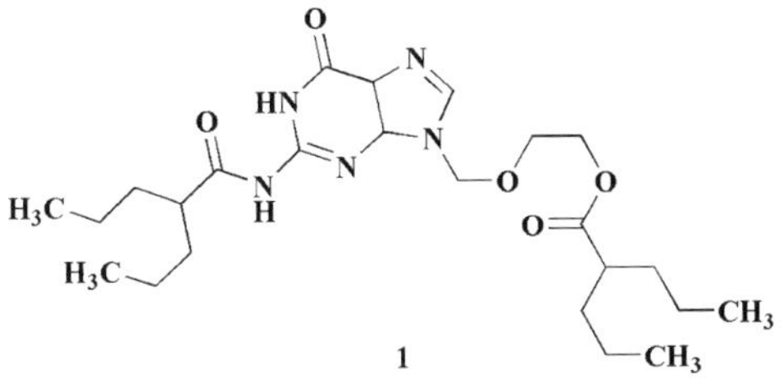

Fig. 6.1 *N*-Valproyl-9-(2-valproyloxy)ethoxymethylguanine (**1**).

antitumor viability against prostate carcinoma cells and human glioma, apoptosis in U251 cells, proteasome activity, and viral thymidine kinase activity [89].

Nucleoside phosphonates are provided for treatment, amelioration, or prevention of a variety of medical disorders associated with viral infections and/or cell proliferation. Along these lines, nucleoside phosphonate analog **2** (Fig. 6.2) was synthesized and estimated as antitumor agent [90].

The nucleotide **3** (Fig. 6.3) was synthesized and estimated in vitro as an antiviral and antitumor agent. It showed an active antiviral potency against Sapovirus, Norovirus, HSV-1, HSV-2, HIV-1, HIV-2, HCV, yellow fever, Dengue virus, and HBV [91].

The L-pyranosyl nucleoside derivative **4** (Fig. 6.4) tested in vitro indicated IC_{50} of 7.8 and 8.3 μg/mL against human tumor cell lines H578St (breast) and OM-1 (colon), as compared to <0.6 and 1.47, respectively, for 5-fluorouracil. It at 10 μg/mL indicated 94.1% survival of bone marrow cells versus 38.6% for 5-fluorouracil at 6 μg/mL [92].

Compound **5** (Fig. 6.5) and their pharmaceutically acceptable esters are used as antitumor agents [93].

Nucleosides **6** (Fig. 6.6) were synthesized from 9-β-D-arabinofuranosyl-6-mercaptopurine (II). Thus, treatment of the latter with $(EtCO)_2O$ in

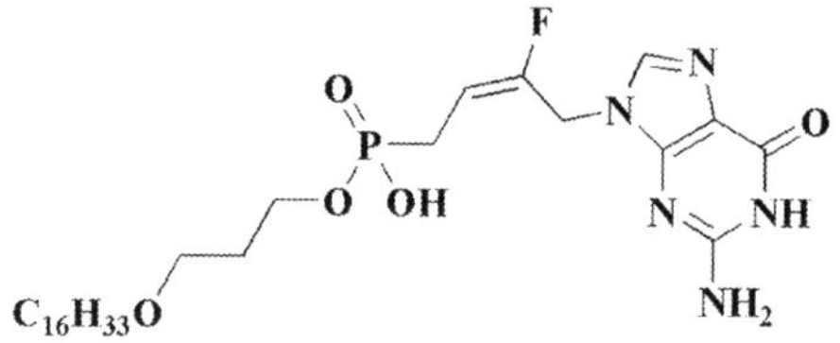

Fig. 6.2 Nucleoside phosphonate analog **2.**

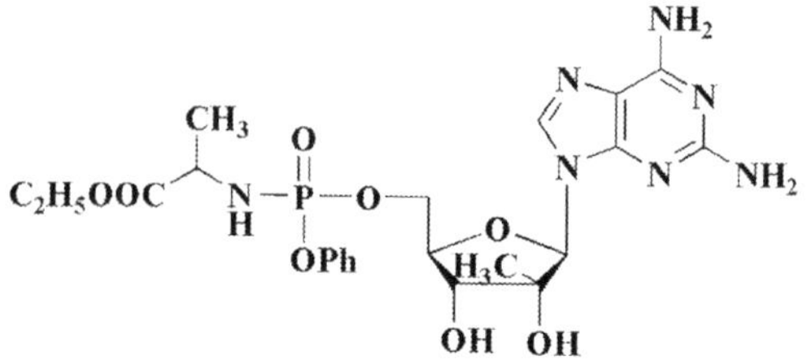

Fig. 6.3 Nucleotide **3.**

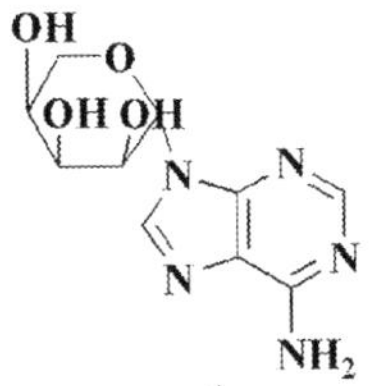

Fig. 6.4 L-Pyranosyl nucleoside derivative **4.**

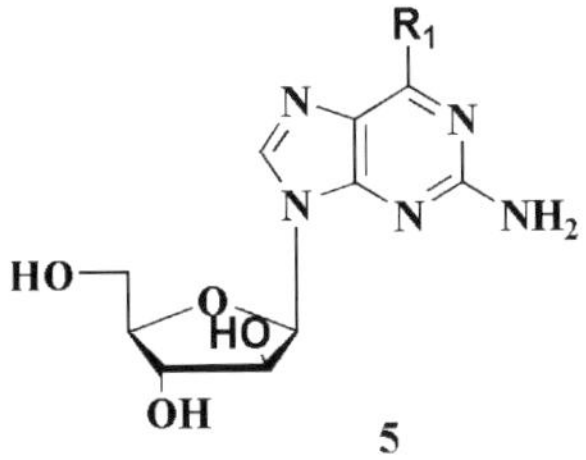

Fig. 6.5 Compound **5**; $R^1 = C_{1-5}$ alkoxy.

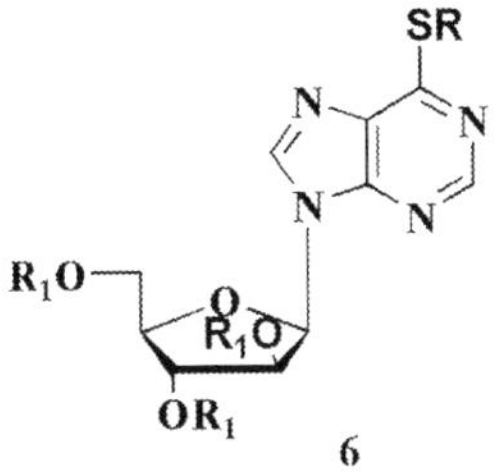

Fig. 6.6 Nucleosides **6**; R, $R^1 = H$, EtCO; H, HCO; NC, H; H_2NCOCH_2, H; HO_2CCH_2, H; MeO_2CCH_2, H; $PhCH_2$, H.

pyridine overnight at room temperature afforded 65% **6** (R = H, R^1 = EtCO). Anticarcinogenic (in vitro) data of **6** against leukemia P388 and mastocytoma P815-X2 are indicated [94].

Purine nucleosides are selective to A_3 adenosine receptors and are utilized for the treatment of inflammatory diseases and cancer. Also disclosed are pharmaceutical compounds comprising a compound of formula 7 (Fig. 6.7), an isomer, or its pharmacological acceptable salt as an active ingredient and a method for preventing or treating various diseases, including asthma, inflammation, heart diseases, cerebral ischemia, and cancer. Thus, (2*R*,3*S*,4*S*,5*R*)-2-[2-chloro-6-(3-iodobenzylamino)-purin-9-yl]-5-hydroxymethyltetrahydrothiophene-3,4-diol was synthesized

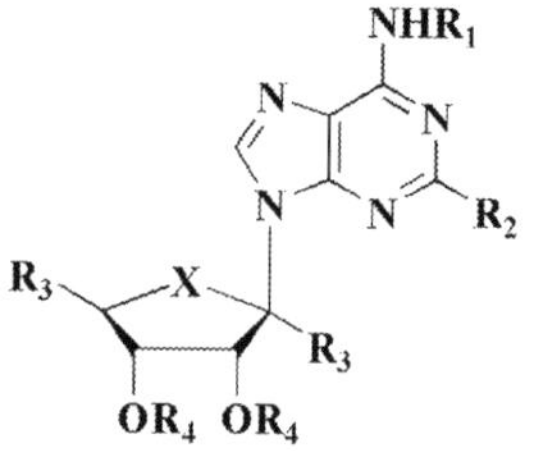

Fig. 6.7 Compound **7**; X is sulfur or oxygen; R^1 is hydrogen, alkyl, benzyl, halobenzyl, or phenylalkyl; R^2 is hydrogen, halogen, alkoxy, alkenyl, alkynyl, alkylthio, or thio; R^3 is hydrogen, hydroxyalkyl, alkoxycarbonyl; R^4 is hydrogen or alkyl.

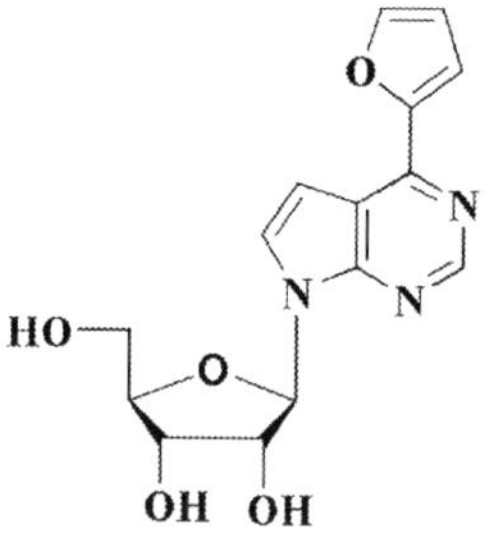

Fig. 6.8 7-Deazapurine nucleoside **8.**

and examined in mice as prodrug for the treatment of inflammation, asthma, heart diseases, cerebral ischemia, and cancer [95].

7-Deazapurine nucleoside **8** (Fig. 6.8) was prepared and tested in vitro as antitumor agent [96].

The nucleosides **9** (Fig. 6.9) were synthesized. Thus, cyclocondensation of (±)-(1α,4α)-4-[(2,5-diamino-6-chloro-4-pyrimidinyl)amino]-2-cyclopentenylcarbinol with $(EtO)_3CH$ in the presence of aqueous hydrochloric acid and ammonolysis of the resulting (±)-**9** (Y=N, X=Cl, Z=NH_2) in methanolic NH_3 at 75° in a bomb generated (±)-**9** (Y=N, X=Z=NH_2), which was deaminated by adenosine deaminase in 0.05 M phosphate buffer to afford (−)-I (Y=N, X=OH, Z=NH_2) (**9′**). Seventeen derivatives of compound **9** were synthesized and **9′** in vitro inhibited the infection of CEM cells with HIV-IIIB with an ED_{50} of 0.66 μg/ML [97].

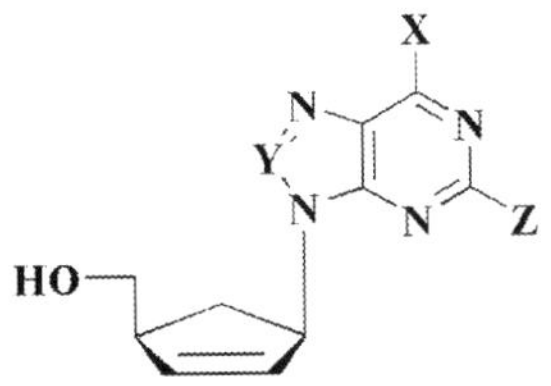

Fig. 6.9 Compound **9**; Y=CH, N; X=Cl, NH_2, SH, OH; Z=H, NH_2.

Thus, nucleoside analog **10** (Fig. 6.10) was synthesized and tested in vitro and in vivo as antitumor agent (EC_{50}=16–288 nM) and RNA polymerase I-III inhibitor [98].

Fluorinated purine nucleoside derivatives **11** (Fig. 6.11) were expected to be valuable as antitumor agents and antiviral agents [99].

7-Deazapurine nucleosides were used as DOT1 enzyme inhibitors and antitumor agents. In this way, nucleoside **12** (Fig. 6.12) was synthesized. The effect of DOT1L enzyme inhibitors on leukemia cell growth and viability is

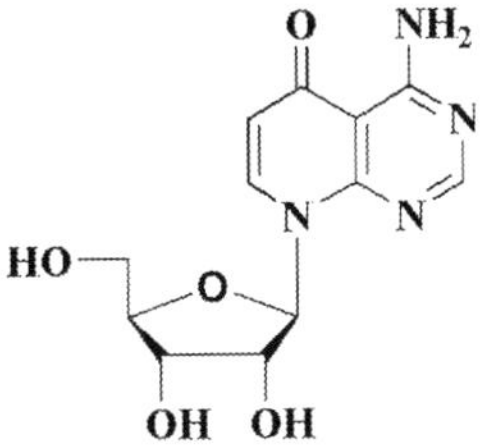

Fig. 6.10 Nucleoside analog **10.**

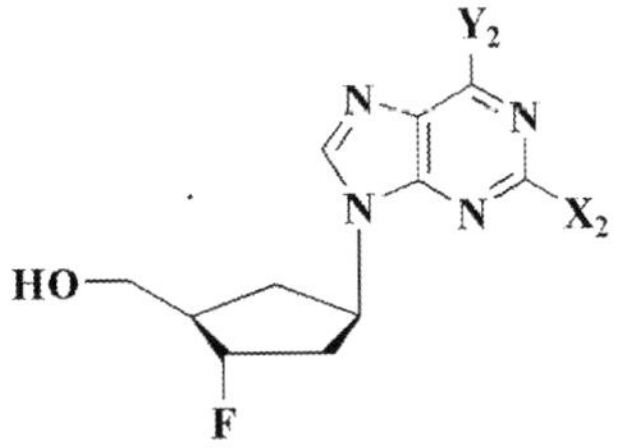

Fig. 6.11 Fluorinated purine nucleoside derivatives **11**; X^2, Y^2=H, halo, (un)substituted amino, (un)substituted OH.

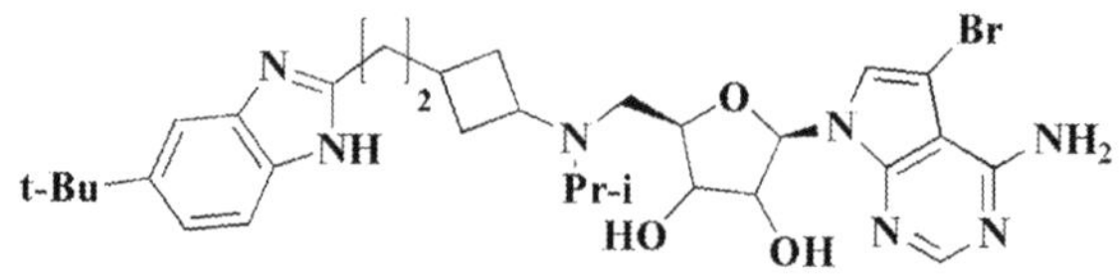

12

Fig. 6.12 7-Deazapurine nucleoside analog **12.**

estimated. Inhibition of DOT1L methyltransferase activity and gene expression in leukemia, wherein leukemia is acute lymphocytic leukemia, acute myeloid leukemia, or mixed lineage leukemia, is revealed [100].

Thus, nucleoside **13** (Fig. 6.13) was synthesized and estimated as an antitumor and antiviral agent; the viral infection is selected from the group comprising hepatitis C, hepatitis B, polio, human immunodeficiency virus, Coxsackie A and B, small pox, Echo, Rhino, Ebola, and West Nile virus. Compound **13** were investigated as HCV polymerase and protease inhibitors [101].

N^2-quinolyl- or isoquinolyl-substituted purine derivatives, and pharmaceutical compounds containing them, were prepared as antitumor agents. Thus, compound **14** were synthesized from 2-bromohypoxanthine. Compound **14** (R=H) (Fig. 6.14) indicated an antitumor potency at 40 mg/kg i.p. in mice inoculated with H_{22} tumor cells [102].

The 4-aza-steroid-purine nucleoside analog of formula **15** (Fig. 6.15) was invented useful in the treatment of cancer. Compound **15** was synthesized by the substitution of (2*S*,4a*S*,4b*R*,6a*R*,10a*R*,10b*S*,12a*S*)-2-(6-chloro-9*H*-purin-9-yl)tetradecahydro-10a,12a-dimethyl-2*H*-1-Benzopyrano[6,5-*f*]quinolin-8(4b*H*)-one and cyclic amine. The compound has significant inhibitory activities to cervical and cancer prostate cancer cell lines, and acts as antitumor agent [103].

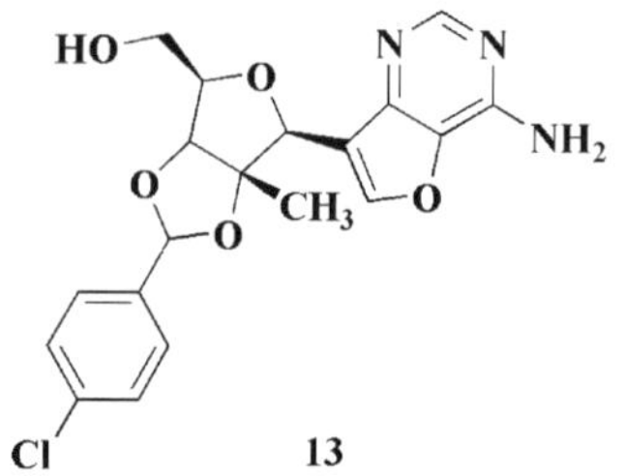

13

Fig. 6.13 Nucleoside **13.**

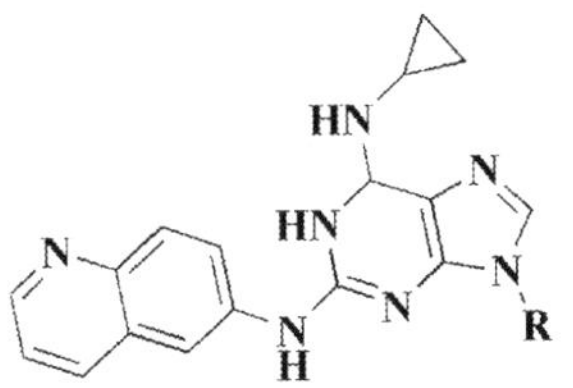

14

Fig. 6.14 Compound **14**; R = H, β-D-ribofuranosyl, β-D-2-deoxyribofuranosyl.

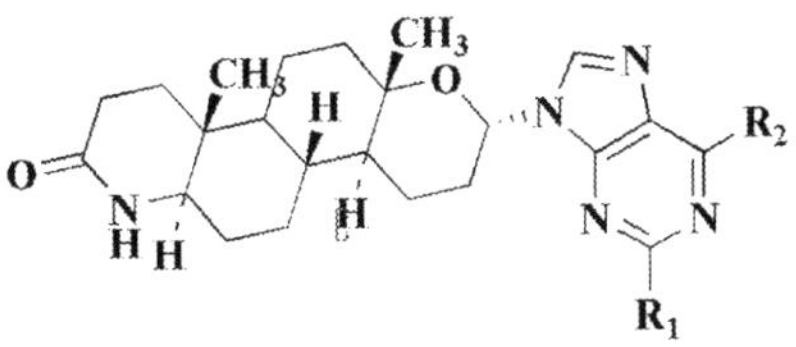

15

Fig. 6.15 4-Aza-steroid-purine nucleoside analog **15;** R^1 is H, Cl, or F; R^2 is pyrrolidyl, piperidinyl, morpholinyl, piperazinyl, or substituted piperazinyl.

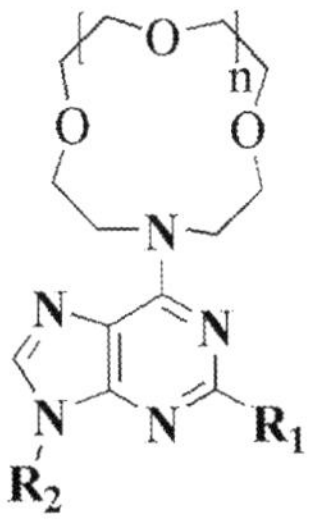

16

Fig. 6.16 6-Monoazacrown ether modified purine nucleoside compounds **16**; R^1 = halo, amino, diallylamino, etc.; R^2 = benzyl, substituted benzyl, etc.; n = 0, or any integer.

6-Monoazacrown ether-modified purine nucleoside compounds **16** (Fig. 6.16) with antitumor activity are obtained by reacting 9-R^2-substituted 6-chloropurine, 2-amino-6-chloropurine, 2,6-dichloropurine, 2-*N*,*N*-dimethyl allyl-6-chloropurine, 2-*N*,*N*-diallyl-6-chloropurine with diethanolamine in water under microwave radiation to access 6-*N*,*N*-di(2-hydroxyethyl) purine compound; then the further reaction of the 6-*N*,*N*-di(2-hydroxyethyl) purine with *p*-ethylene glycol di-*p*-toluenesulfonate, diethylene glycol *p*--toluenesulfonate, triethylene glycol di-*p*-toluenesulfonate and *p*-tetraethylene glycol di-*p*-toluenesulfonate in the presence of alkali in organic solvent for 6 h was accomplished [104]. Many other inventions of the purine nucleoside analogs were depicted [93,105–141].

3. Nucleoside analogs: Syntheses and biological evaluation

3.1 Ribonucleoside purine analogs

3.1.1 Free and protected purine ribonucleoside analogs

Impressive antitumor potency has been revealed with fludarabine phosphate against tumors that express *Escherichia coli* purine nucleoside phosphorylase (PNP) because of the liberation of 2-fluoro-7*H*-purin-6-amine in the tumor tissue. 6-Methylpurine (MeP) is another cytotoxic adenine derivative that does not display selectivity when administered systemically, and could be extremely valuable in a gene therapy approach to cancer treatment using *E. coli* PNP. The prototype MeP releasing prodrug 9-(2-deoxy-β-D-ribofuranosyl)-6-methylpurine (**17**) [MeP-dR] (Fig. 6.17) has elucidated good potency against tumors expressing *E. coli* PNP, but its antitumor potency is restricted because of toxicity resulting from the generation of MeP from gut bacteria [142].

Upon embarking on a medicinal chemistry program to identify a combination of nonhuman adenosine glycosidic bond-cleaving enzymes and nontoxic MeP prodrugs, the two best MeP-based substrates with M64V-*E. coli* PNP, a mutant that was tolerated the modification at the 5′-position of adenosine and its derivatives, were 9-(6-deoxy-α-L-talofuranosyl)-6-methylpurine (**25**) and [methyl(talo)-MeP-R] and 9-(α-L-lyxofuranosyl) 6-methylpurine (**29**) [lyxo-MeP-R]. The methyl (talo)-MeP-R and

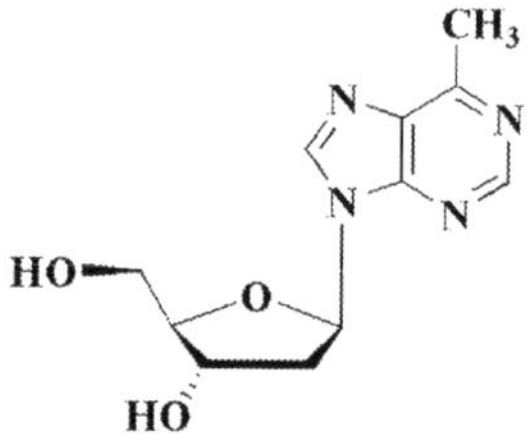

CH_3
N
N
N
N
HO
O
HO
17

Fig. 6.17 9-(2-Deoxy-β-D-ribofuranosyl)-6-methylpurine (**17**).

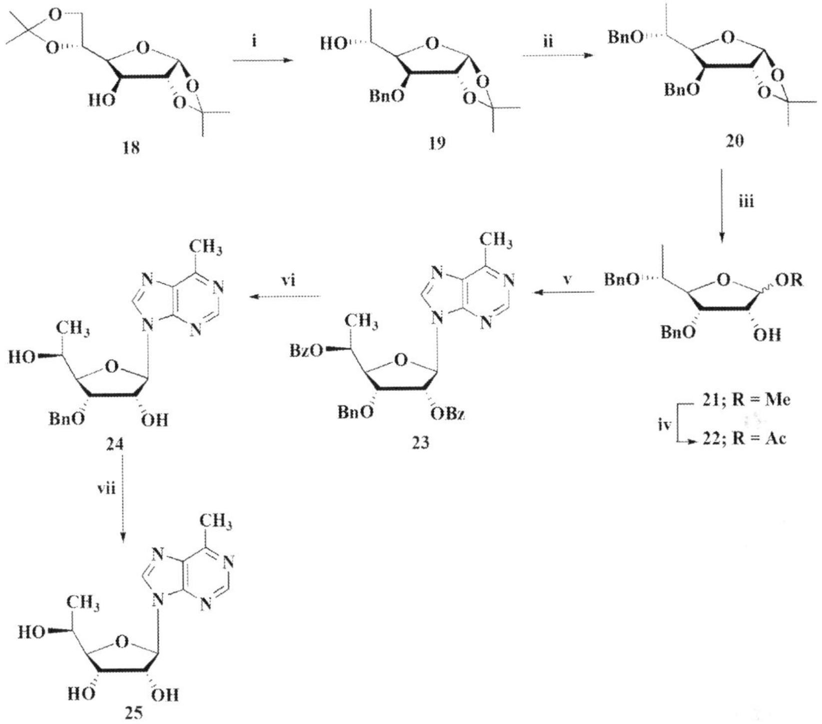

Scheme 6.1 Reagents and conditions: (i) [143]. (ii) TPP/DEAD/BzOH/THF/0°C/10 h; (iii) 20% HCl/CH_3OH-H_2O/6 h/RT/BzCl/dry pyr./30 min/RT; (iv) $Ac_2O/CH_3COOH/H_2SO_4$/4 h/RT; (v) 6-Mep/DCE/TMSCl/HMDS/3 h /80°C/$SnCl_4/CH_3CN$/−10°C, /10 min; (vi) NH_3/CH_3OH/48 h/RT; (vii) 10% Pd/C/C_2H_5OH/H_2/6 h/RT.

lyxo-MeP-R were synthesized (Scheme 6.1), and the estimation of their substrate potency with 4 enzymes not normally associated with cancer patients is reported [142].

The intraperitoneal pharmacokinetic (ip-PK) features of methyl(talo)-MeP-R were identified and its in vivo bystander potency has also been determined in mice bearing D54 tumors that express M64V PNP. The indicated good in vivo bystander potency of [methyl(talo)-MeP-R/M64V-*E coli* PNP combination shows that these agents could be utilized in treating cancer [142].

Synthesis of methyl(talo)-MeP-R, **25**. Vorbrüggen glycosylation of 6-alkyl/arylpurines and 1,2,3,5-tetra-Oacyl- ribofuranosides has been reported to selectively furnish 9-(β-D-ribofuranosyl)-6-alkyl/arylpurine nucleosides [144–148]. α-L-Talofuranoside derivative **22**, which is the pivotal intermediate required for the coupling reaction, could be accomplished by the inversion of the configuration at the C-5 position of 6-deoxy-D-allofuranose derivative **19**; then the application of the standard methods of glycosyl donor protection and activation was followed. 1,2:5,6-Di-*O*-isopropylidene-α-D-glucofuranose (**18**) was transformed to 6-deoxy-D-allofuranose analog **19** [149,150]. Further esterification of compound **19** under Mitsunobu reaction conditions afforded the corresponding 5-*O*-benzoate analog **20**. The latter compound 7 was then treated with hydrochloric acid, followed by benzoylation of the 2-hydroxyl group, which furnished the corresponding methyl glycoside analog **21**. Acetolysis of compound **21** afforded the corresponding 1-*O*-acetyl-α-L-talofuranose derivative **22**. Coupling of α-L-talofuranose analog **22** and silylated 6-methylpurine yielded the corresponding methylpurine **23**. The latter compound debenzoylated with methanolic ammonia afforded compound **24**. Pd/C-catalyzed hydrogenation of compound **24** yielded 9-(6-deoxy-β-D-talofuranosyl)-6-methylpurine (**25**) [142].

Synthesis of Lyxo-MeP-R, **29**. The glycosyl donor 1-*O*-acetyl-2,3,5--tri-*O*-benzoyl-L-lyxofuranose (**27**) was synthesized [151–153] starting from L-lyxose. The latter was treated with acetyl chloride to afford a mixture of the 1-*O*-melyxopyranoside and 1-*O*-melyxofuranoside derivatives in a 1:5 ratio. Through the crystallization of the mixture, the pure β-anomer of 1-*O*-melyxofuranoside was isolated. The methyl lyxofuranoside was benzoylated and then subsequent acetolysis of the corresponding methyl 2,3,4-tri-*O*-benzoyl-lyxofuranoside derivative furnished the corresponding 1-*O*-acetyl-2,3,5-tri-*O*-benzoyl-β-L-lyxofuranose (**27**) predominantly [153]. The silylated 6-methyl purine was coupled with compound **27** to yield the targeted 6-MeP-lyxoside analog **28**. The latter compound **28** was deprotected to afford the 9-(α-L-lyxofuranosyl)-6-methylpurine (**29**) (Scheme 6.2) [142].

By treatment of 6-(*N*,*N*-disubstituted aminomethyl)purine derivatives **30** and **31** with catalytic amounts of NaOMe, 6-(*N*,*N*-disubstituted aminomethyl) purine nucleosides **32** or **33** were obtained in good yields (Scheme 6.3) [154].

The corresponding sodium alcoholates were allowed to react with **34** to afford the targeted alkoxymethylpurine nucleosides.

Then the cleavage of the Tol-protecting groups furnished the free nucleosides **37**a and **38**a in reasonable yields in the case of NaOMe.

The 6-(*O*-substituted hydroxymethyl)- and 6-(*S*-substituted sulfanylmethyl)purine nucleosides **37**, **38** and **41**, **42** were generated in good yields via the treatment with NaOMe or aqueous NH_3 in the presence of

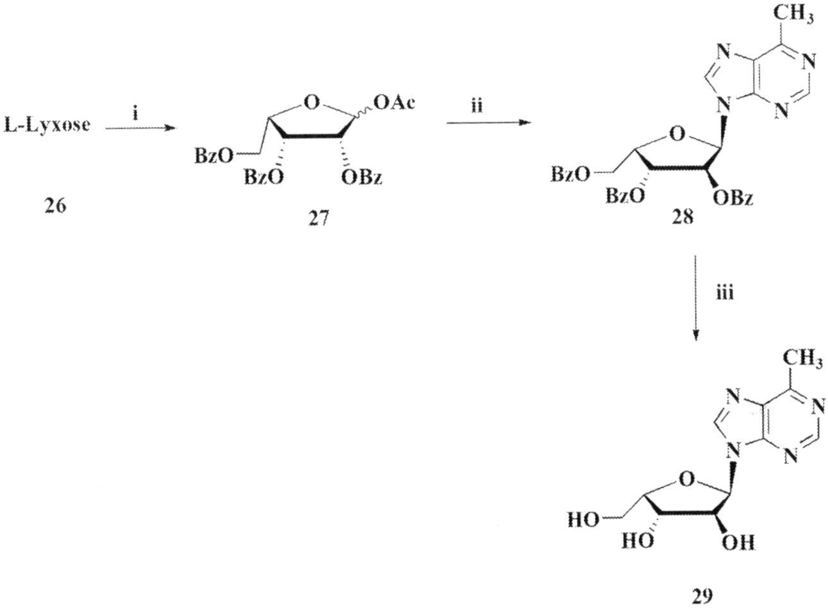

L-Lyxose → (i) → 27 → (ii) → 28 → (iii) → 29

26

Scheme 6.2 Reagents and conditions: (i) (i) H_2SO_4/CH_3OH/72 h/RT; (ii) BzCl/dry Pyr./30 min/0°C; (iii) $Ac_2O/CH_3COOH/H_2SO_4$/0°C/30 min/39.6%; (ii) 6-MeP/DCE/HMDS/TMSCl/3 h/80°C/$CH_3CN/SnCl_4$/-10°C/40 min; (iii) NaOMe/CH_3OH/2 h/RT.

methanol. The carboxylic acids **41**da and **42**da were provided when the thioglycolate methyl esters **39**d and **40**d were hydrolyzed using sodium methoxide, then aqueous NaOH (Scheme 6.4).

The compounds **32**, **33**, **37**, **38**, **41**, and **42** were subjected to biological activity screening. On the following cell cultures, mouse leukemia L1210 cells (ATCC CCL 219), human cervix carcinoma HeLaS3 cells (ATCC CCL 2.2), human promyelocytic leukemia HL60 cells (ATCC CCL 240), and human T lymphoblastoid CCRF-CEM cell line (ATCC CCL 119), the cytostatic activity in vitro was evaluated.

The cytostatic potencies of compounds were determined by evaluating the cell count in a hemocytometer. The cell viability was evaluated using XTT staining. The 2′-deoxyribonucleoside derivatives were not potent, while the ribonucleosides **32**, **37,** or **41** showed obvious antiproliferative effects at low mM concentrations [154].

Human HL-60 was the most susceptible cell line, while L1210 and CCRF-CEM cell lines showed little sensitivity to most of the compounds. None of the compounds were potent in HeLaS3 cell line. Compounds

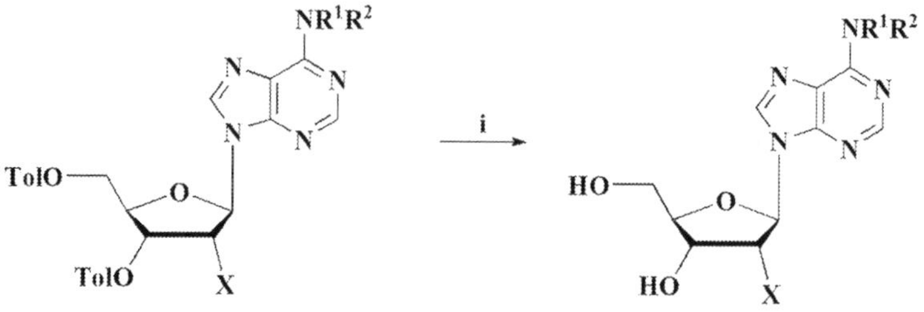

30; X = OTol
31; X = H

32; X = OH
33; X = H

R^1	R2	NR^1R^2
a, CH_3	Ac	m, piperidin-1-yl
b, Bn	Ac	n, morpholin-1-yl
c, cHex	Ac	o, thiazolidin-1-yl
d, CH_2COOEt	H	p, indolin-1-yl
e, CH_2COOH	H	q, indol-1-yl
f, CH_3	CH_3	r, pyrazol-1-yl
g, iPr	iPr	s, 4-pyrimidone-1-yl
h, CH_3	Bn	t, 2-pyrimidone-1-yl
i, CH_3	Ph	u, 2-pyrrolidinone-1-yl
j, cPr	Ac	v, Pro-OMe
k, CH_2CH_2OH	CH_2CH_2OH	w, Pro-OH
l, NHNHAc	H	X, N_3

Scheme 6.3 Reagents and conditions: (i) MeONa, MeOH, or NH_3, MeOH, 50°C.

32e–32h and **32j–32n** possessed cytostatic effects, while heteroarylmethyl and acetylated monoalkylaminomethyl derivatives were less potent. Additionally, 6-(methoxymethyl)-, 6-(methylsulfanyl)-, and 6-(benzylsulfanyl)-purine ribonucleosides **37a**, **41a**, and **41c** indicated obvious cytostatic potency. 6-(Sulfanylmethyl)purine ribonucleoside **41i**, the thia-analog of the highly

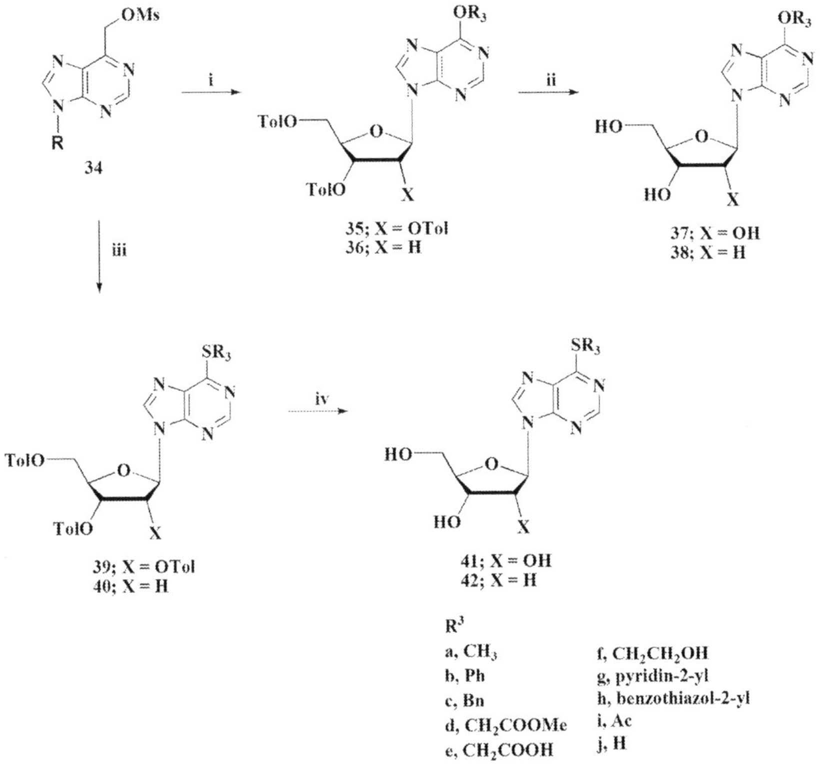

OMs
R
34
i
OR3
TolO
TolO
X
35; X = OTol
36; X = H
ii
OR3
HO
HO
X
37; X = OH
38; X = H
iii
SR3
TolO
TolO
X
39; X = OTol
40; X = H
iv
SR3
HO
HO
X
41; X = OH
42; X = H
R3
a, CH3
b, Ph
c, Bn
d, CH2COOMe
e, CH2COOH
f, CH2CH2OH
g, pyridin-2-yl
h, benzothiazol-2-yl
i, Ac
j, H

Scheme 6.4 Reagents and conditions: (i) R^3ONa. (ii) MeOH, or NH_3, MeOH. (iii) R^3SNa or R^3SH, iP_2NEt. (iv) MeONa, MeOH.

cytostatic 6-(hydroxymethyl)purine **37i**, showed a moderate cytostatic activity in the μM range. Interestingly, the azidomethyl derivative **32v** was the most potent antiproliferative compound from the series with IC_{50} values of 0.4–1.9 lM [154].

New ribonucleoside analogs **51–55** comprising a 4-substituted piperazine in the substituent at N6 were prepared (Scheme 6.5) and estimated for their cytotoxicity on HepG2, FOCUS, Huh7, MCF7 breast, Mahlavu liver, and HCT116 colon carcinoma cell lines. Based on a sulforhodamine B assay, the purine nucleoside derivatives were analyzed via an anticancer drug-screening method. Two nucleoside analogs with promising cytotoxic potencies **53** and **54** were then analyzed on the hepatoma cells. The piperazine derivative **53** exhibited the best antitumor potency, with IC_{50} values in range of 5.2 and 9.2 μM. Similar to previously described nucleoside derivatives, compound **53** also interferes with cellular ATP reserves, possibly

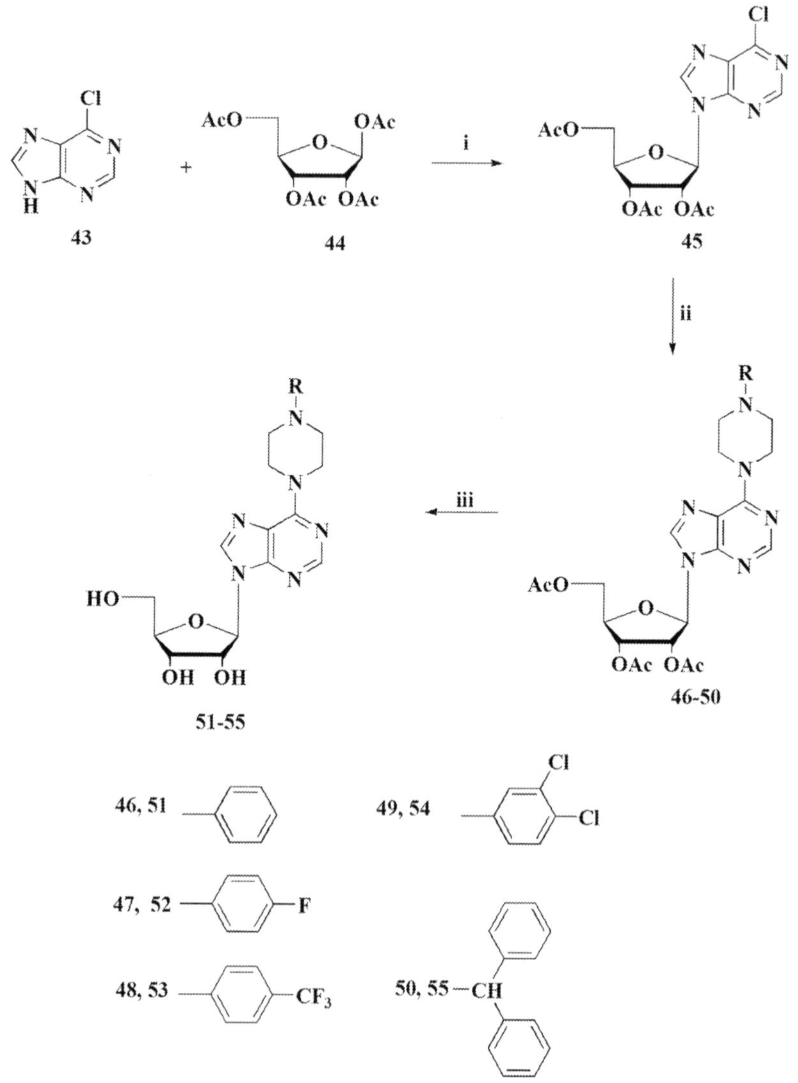

Scheme 6.5 Reagents and conditions: (i) silica gel 60/EtOAc/MW; (ii) piperazine/Et_3N/C_2H_5OH; (iii) NaOMe/CH_3OH.

through influencing cellular kinase activities. Additionally, the derivative **53** was indicated to induce senescence-associated cell death, as elucidated via the SAβ-gal assay. The senescence-dependent cytotoxic effect of compound **53** was also affirmed via phosphorylation of the Rb protein via p15INK4b overexpression in the existence of this analog [155].

3′-Deoxy-3β′-fluoro-3α′-C-hydroxymethyl purine **63d** nucleosides were prepared through the modified Corey-Link reaction with 3-oxo-D-glucofuranose analog to afford C-3 quaternary carbon with 3β′-fluoro-3α′-C-hydroxymethyl groups, in the sugar framework (Scheme 6.6). Antiproliferative potency of fluorinated nucleosides **63a–d** was estimated against L-132, HeLa, MDA-MB-231, and neural cell lines and observed to be moderately active [156].

2′,3′-Bis-*O*-tert-butyldimethylsilyl-5′-deoxy-5′-[*N*-(methylcarbamoyl) amino]-*N*6-(*N* phenylcarbamoyl)-adenosine (Scheme 6.7), a novel member

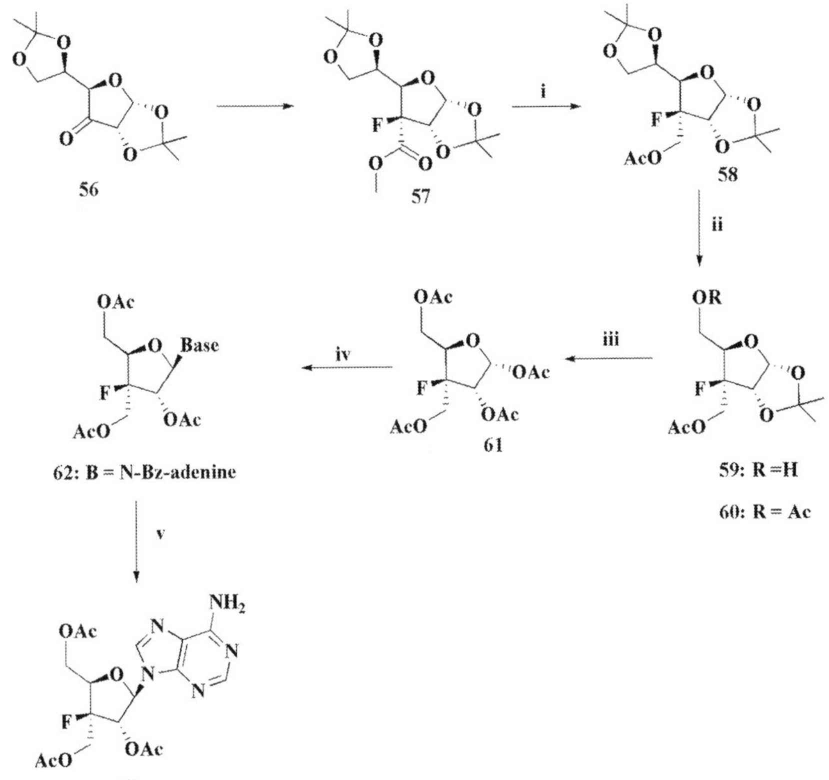

Scheme 6.6 Reagents and conditions: (i) (i) $NaBH_4/CH_3OH$, (ii) Ac_2O/pyridine/DCM/DMAP; (ii) (i) 65% $AcOH:H_2O$/50°C/3.5 h, (ii) $NaIO_4$/acetone/H_2O/RT; (iii) $NaBH_4/CH_3OH$/0°C; (iii) Ac_2O/pyridine/DCM/DMAP; (iv) (i) TFA: H_2O (3:2) 3 h, (ii) Ac_2O/pyridine/DMAP; (v) TMSOTf/BSA/CH_3CN/nucleobase/60°C/24 h (8a: 65%, 8b: 62%, 8c: 68%, 8d: 60%; (vi) NH_3/CH_3OH/55°C/1 h, (1a: 84%, 1b: 81%, 1c: 88%, 1d: 77%).

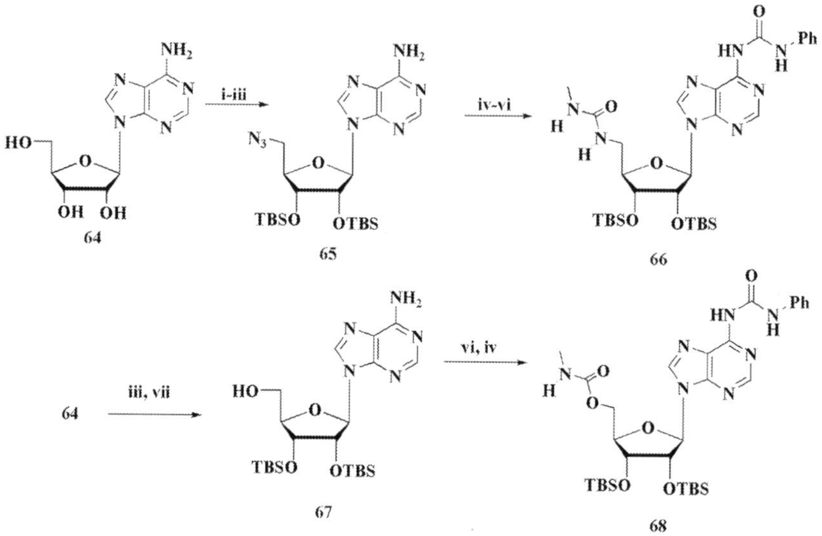

Scheme 6.7 Reagents and conditions: (i) $SOCl_2/CH_3CN$/RT/15 h; (ii) NaN_3/DMF/150°C/1 h; (iii) TBSCl/DMF/RT/16 h; (iv) PhN=C=O/CH_2Cl_2/RT/5d; (v) H_2/Pd–C/EtOAc; (vi) p-NO_2 $C_6H_4OCONHCH_3$/base/EtOAc/RT; (vii) TFA/H_2O (9:1)/RT/3 h.

of the *N* 6,5′-bis-ureidoadenosine class of anticancer nucleosides, is observed to display a broad-spectrum antiproliferative potency. A majority of the cell lines in the NCI-60 are inhibited with an average GI_{50} = 3.13 lM. Selective toxicity against human colon cancer cell lines (HCC-2998, COLO 205, HT29, KM12, HCT-116) was also displayed (LC_{50}'s = 6–10 lM) [157].

5′-*S*-(2 Acetamidoethyl)-6-*N*-[(4-substituted)benzyl]-5′-thioadenosine analogs, 5′-*S*-(2-aminoethyl)-6-*N*-(4-nitrobenzyl)-5′-thioadenosine (SAENTA), and 5′-*S*-[2-(6-aminohexanamido)]ethyl-6-*N*-(4-nitrobenzyl)-5′-thioadenosine (SAHENTA) were prepared through the SNAr displacement of fluoride from 6-fluoropurine intermediates with 4-(substituted)benzylamines. The fluorescein-5-yl isothiocyanate (FITC) was conjugated with the pendant amino groups of SAHENTA and SAENTA to afford fluorescent probes that bound at nanomolar concentrations specifically to human equilibrative nucleoside transporter 1 (hENT1) formed in recombinant form in model expression systems and in native form in cancer cell lines [158].

Cysteamine was trifluoroacetylated (Scheme 6.8) to afford **2**-(trifluoroacetamido)ethanethiol. 5′-Chloro-5′-deoxyadenosine [159] was treated with sodium thiolate derived from 2,2,2-trifluoro-*N*-(2-mercaptoethyl)acetamide to furnish 5′-thioether, which was acetylated and exposed to fluoro-diazotization to afford the fluoropurine **70**. SNAr displacement of fluoride from compound **70** with (4-nitrophenyl)methanamine and compound **71** generated **72a** and **72b**. Deprotection of the latter afforded compounds **73a** and **73b** and treated with FITC to furnish **74a** and **74b** in 56% and 95% yield. *Aminocaproic acid* was trifluoroacetylated to yield 6-(2,2,2-trifluoroacetamido)hexanoic acid, which was transformed into its

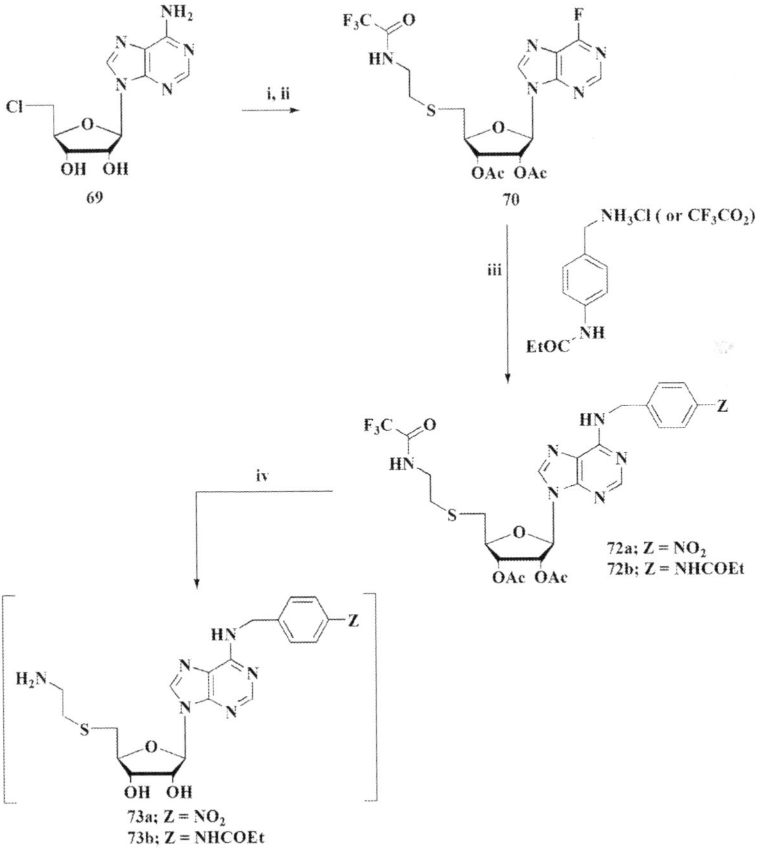

Scheme 6.8 Reagents and conditions: (i) 2,2,2-Trifluoro-*N*-(2-mercaptoethyl) acetamide, NaH/DMF/RT; (ii) Ac_2O/pyridine/0°C/$NaNO_2$/~55% HF-pyridine/−10 to −5° C; (iii) (4-nitrophenyl)methanamine, TEA/CH_3OH/RT; (iv) 30% NH_3/H_2O/CH_3OH/60°C;

(Continued)

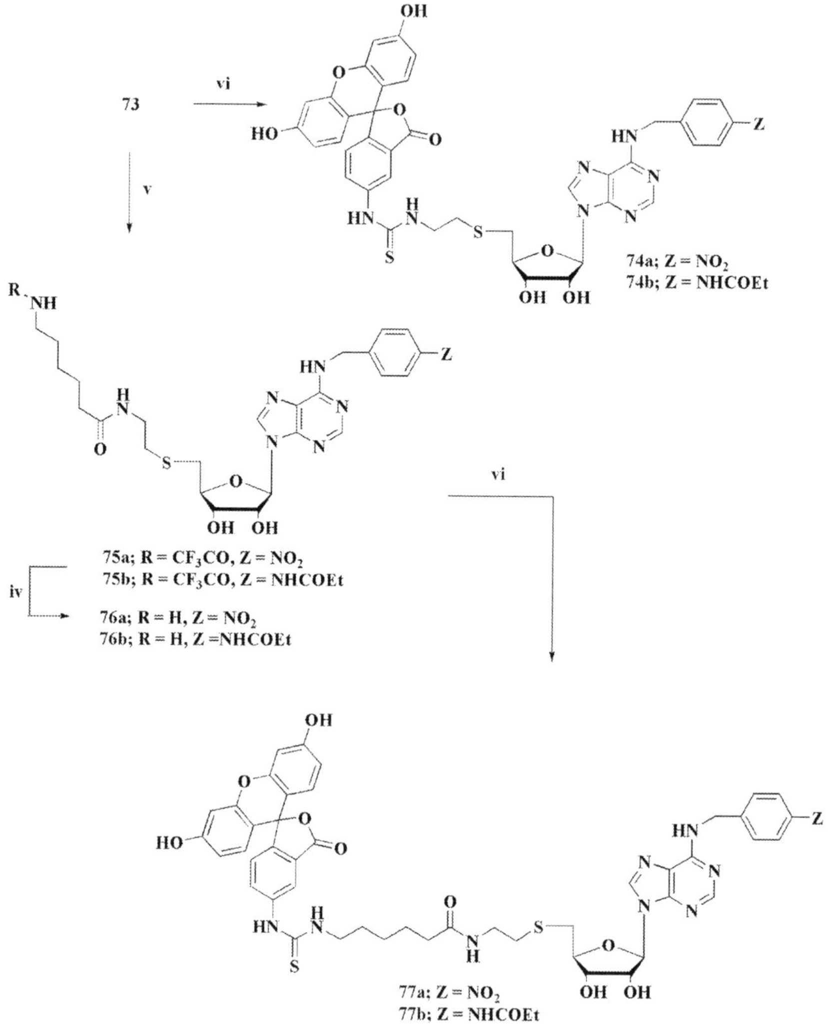

Scheme. 6.8, Cont'd (v) FITC/TEA/DMF/RT; (vi) 5-(2,2,2-trifluoroacetamido)pentyl 4-nitrobenzoate, TEA, DMF, RT.

nitrophenyl ester, 5-(2,2,2-trifluoroacetamido)pentyl 4-nitrobenzoate, with 4-nitrophenol and DCC. Compounds **73a** and **73b** were treated with **the** nitrophenyl ester to afford the 5′-thionucleoside derivatives **75a** and **75b**. The trifluoroacetyl group removal from **75a** and **75b** furnished compounds **76a** and **76b**, respectively. The treatment of the latter compounds with FITC/TEA/DMF generated compounds 7**7a** and 7**7b** in 86% and 89% yields [158].

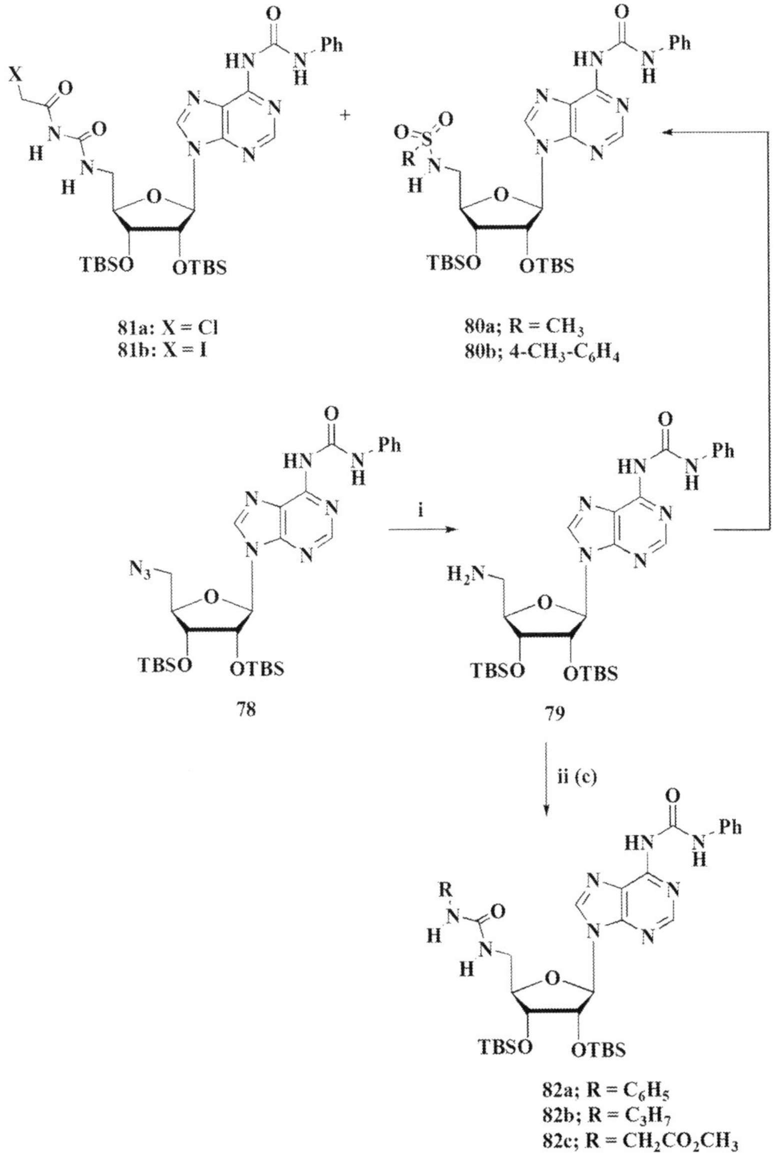

81a: X = Cl
81b: X = I

80a; R = CH_3
80b; 4-CH_3-C_6H_4

78

79

i

ii (c)

82a; R = C_6H_5
82b; R = C_3H_7
82c; R = $CH_2CO_2CH_3$

Scheme 6.9 Reagents and conditions: (i) H_2/Pd-C; (ii) PhC=N=O; (iii) *p*-NO_2-$C_6H_4O_2CNHR$; (iv) $ClCH_2CON=C=O$; (v) RSO_2Cl; (vi) NaI; (vii) $Cl_3CON=C=O$; (viii) SiOH/CH_3OH/CH_2Cl_2; (ix) CF_3COCl; (x) FCH_2COCl.

Additional analogs varying at the 5′-position were synthesized as depicted in Schemes 6.9 and 6.10. Compound **79** was synthesized through the hydrogenolysis of compound **78**, which could be accessed from 5′-chloro-5′-deoxyadenosine through a two-pot reaction sequence.

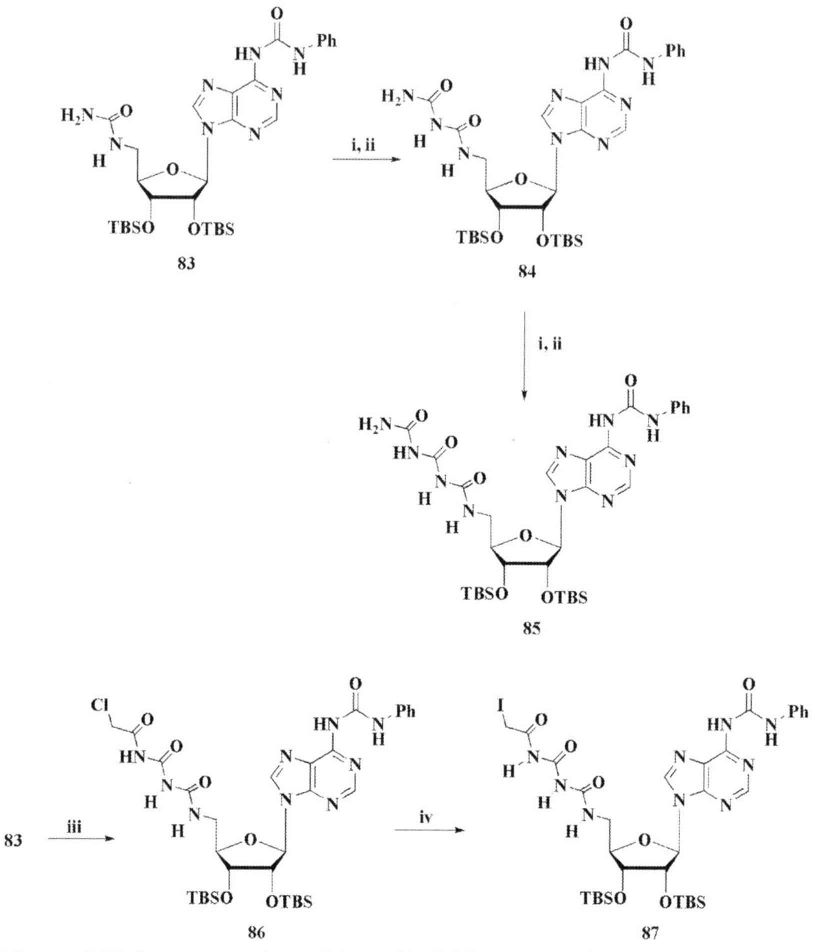

Scheme 6.10 Reagents and conditions: (i) $Cl_3CON{=}C{=}O$; (ii) SiOH, CH_3OH/CH_2Cl_2; (iii) $ClCH_2{-}CON{=}C{=}O$; (iv) NaI.

Compound **79** was treated with either the appropriate *N*-alkyl *p*-nitrophenyl carbamate or phenyl isocyanate to afford compounds **82a–c**. Acylation of compound **79** with chloroacetyl isocyanate furnished **81a**. Compound **81a** was transformed to **81b** using classical Finkelstein **87** conditions and afforded the targeted iodo product [160].

Compound **79** was treated using a modified Kočovský carbamylation [161] method to afford compound **83**. The latter compound **83** was treated with monofluoroacetyl chloride or trifluoroacetyl chloride to generate **89a** and **89b** (Fig. 6.18). Compound **83** could also be treated sequentially with the modified Kočovský carbamylation **79** method to generate compound **84**, then compound **85** (Scheme 6.10). Compound **86** was also derived from **83** after treating with 2-chloroacetyl isocyanate. Compound **86** was treated using Finkelstein [162,163] conditions to yield the iodo product **87** [160].

Compounds **82a–c**, **81a–b**, **80a–b**, **83**, and **84–87** were estimated for antiproliferative activity using human CD4+T-lymphocyte (CEM), murine leukemia L1210, and human cervix carcinoma (HeLa) cells. Compound **81** was found to be the most active analog in this assay, while compounds having sterically bulky 5′-amino substituents **82a–b** or **80b** were much less potent. The methanesulfonyl group in **80a** did not enhance upon the potency displayed by lead compound **88a** (Fig. 6.19) (although in comparison with **80b**, **80a** was markedly more potent).

Compounds **81a, 86,** and **87** were selected for multidose screening in the NCI-60. The average GI_{50} for **81a** was 1.08 ± 2.20lM. The average GI_{50} for compounds **86** and **87** was similar (4.58 ± 3.74lM and 3.35 ± 4.13lM). Overall selectivity patterns for **86** and **87** had a high correlation in the matrix COMPARE10 analysis for both single-dose and multidose experiments (single-dose growth percent correlation, $r = 0.88$; multidose GI_{50} correlation, $r = 0.85$). The high correlation between selectivity patterns for compounds **86** and **87** confirms the conclusion that these compounds show their antiproliferative effects through a shared mechanism. In contrast, the multidose matrix COMPARE analysis for **87** and **88a** afforded a significantly lower correlation (GI_{50} correlation, $r = 0.081$).

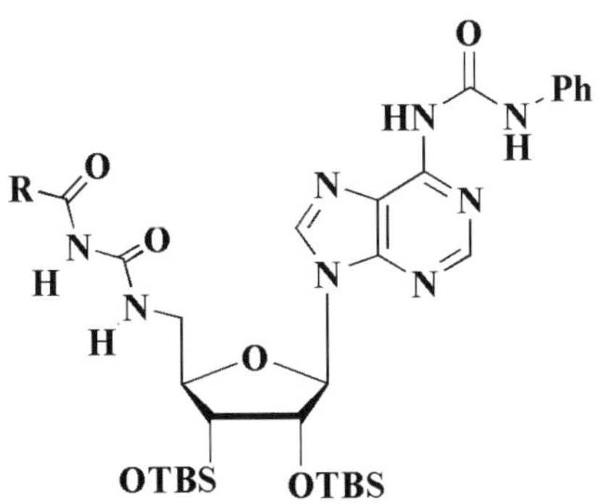

Fig. 6.18 Compound **89**.

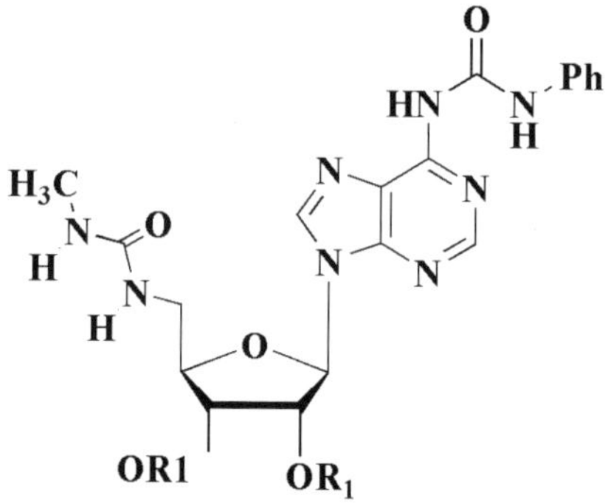

88 **a, t-$BuMe_2Si$**
b, Et_3Si
c, t-$BuPh_2Si$
d, $(iPr)_3Si$
e, H

Fig. 6.19 Compound **88**.

This latter study proposes that compounds **86** and **87** may target various biomolecular target(s) than **88a** (or interact differently with BMPR1b) [164,165], and these possibilities are currently under active investigation. Cell lines against which **86** and **87** demonstrated selective inhibition included melanomas MALME-3M and LOX IMVI, prostate cancer PC-3, ovarian cancer IGROV-1and non-small-lung cancer NCI-H522. Both **86** and **87** indicated a potent and selective potency against lung adenocarcinoma cell line NCI-H522, with compound **87** being 10-fold more potent (GI_{50}=96 nM and 9.7 nM for **86** and **87**, respectively) [160].

Efficient methods for the preparation of 5′-substituted 5′-amino-5′-deoxy-N^6-ureidoadenosine derivatives, e.g., I (R = iodo, chloro), are described. Compounds were screened for antiproliferative activity against a panel of murine and human cell lines (CEM, L1210, and HeLa) and/or against the NCI-60. The most potent derivative inhibited the lung adenocarcinoma cell line NCI-H522 at low nanomolar concentrations (GI_{50} = 9.7 nM). [160].

According to Vorbrüggen method and starting from 1,2,5,6-di-O-isopropylidene-α-D-allofuranose, different novel purine nucleoside analogs of allofuranose were synthesized using 1,2,3,5,6-pentaacetoxy-β-D-allofuranose as pivotal intermediate (Schemes 6.11 and 6.12). The allofuranosyl analogs and some acetyl derivatives were investigated for their cytotoxicity in vitro in three human cancer cell lines (MCF-7, HL-60, and Hela-229).

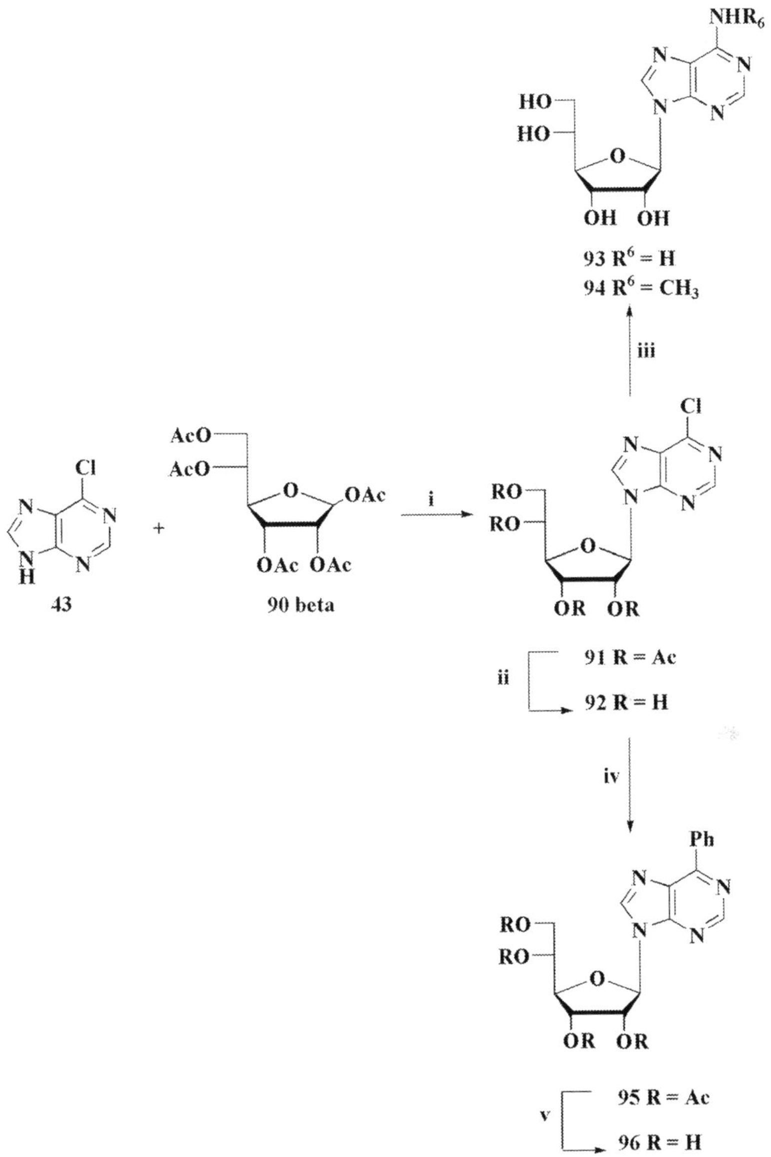

Scheme 6.11 Reagents and conditions: (i) Hexamethyldisilazane (HMDS)/TMSOTf/$(NH_4)_2SO_4$/CH_3CN/reflux (**91**); (ii) NH_3/CH_3OH/0°C; (iii) NH_4OH (25%)/CH_3CN/reflux (**93**); CH_3NH_2.HCl/TEA/CH_3OH (**94**); (iv) $PhB(OH)_2$/K_2CO_3/$Pd(PPh_3)_4$/toluene/100°C.

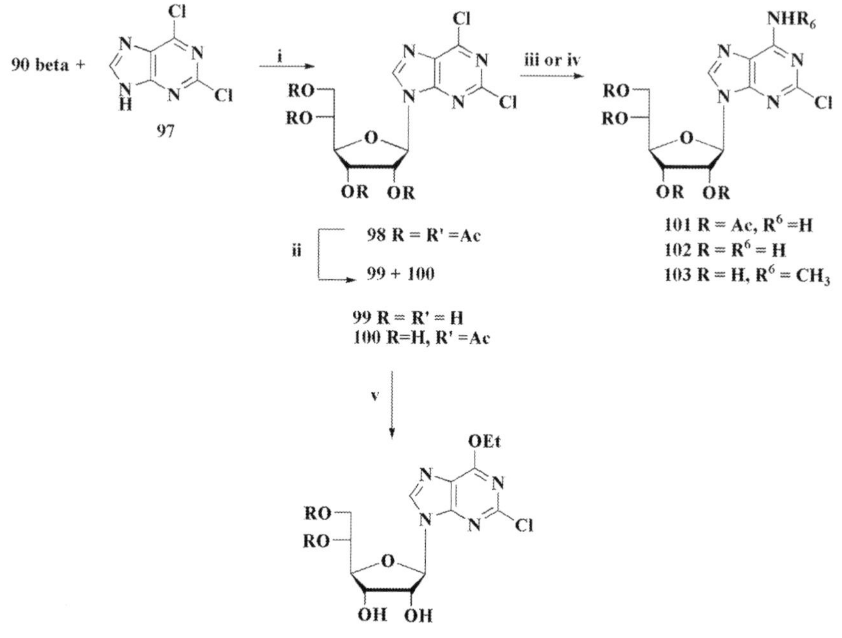

90 beta + 97 →(i) 98 →(iii or iv) 101–103

ii: 98 R = R' = Ac → 99 + 100

99 R = R' = H
100 R=H, R' =Ac

101 R = Ac, R^6 = H
102 R = R^6 = H
103 R = H, R^6 = CH_3

v → 104

Scheme 6.12 Reagents and conditions: (i) Hexamethyldisilazane (HMDS)/TMSOTf/$(NH_4)_2SO_4$/CH_3CN/reflux (91), 74% (**98**); (ii) dibutyltin oxide (DBTO)/toluene/100°C (**99**), 32% (**100**); (iii) NH_3 0.5 M/dioxane/RT (11); (iv) NH_4OH (25%)/CH_3CN/reflux (**93**), 60% (**102**);CH_3NH_2.HCl/TEA/CH_3OH (**94**), 62% (**103**); (v) K_2CO_3/CH_2Cl_2-C_2H_5OH/RT.

The 9-(2,3,5,6-tetra-*O*-acetyl-β-D-allofuranosyl)-2,6-dichloropurine **98** was the most active one on the three cell lines, being its potency against HL-60 cells similar to cisplatin [166].

Based on the Sonogashira coupling reaction, the C6-amino-C′-*N*-cyclopropyl carboxamido-C2-alkynylated purine nucleoside derivatives **115a–g** were synthesized (Schemes 6.13 and 6.14). Compound **113** is utilized as a pivotal intermediate for the preparation of the modified nucleoside analogs **115a–g**. The analogs **115a–g** are estimated for in vitro anticancer potency against Caco-2 cell and MDA-MB-231 lines. Screening data show that compounds **115b** and **115e** exhibited potency (IC_{50} = 7.5, 8.3 mg/mL) against Caco-2 and of 7.9 and 6.8 mg/mL respectively, against MDA-MB-231 than doxorubicin; thus, they have good potential for applying as anticancer agent [167].

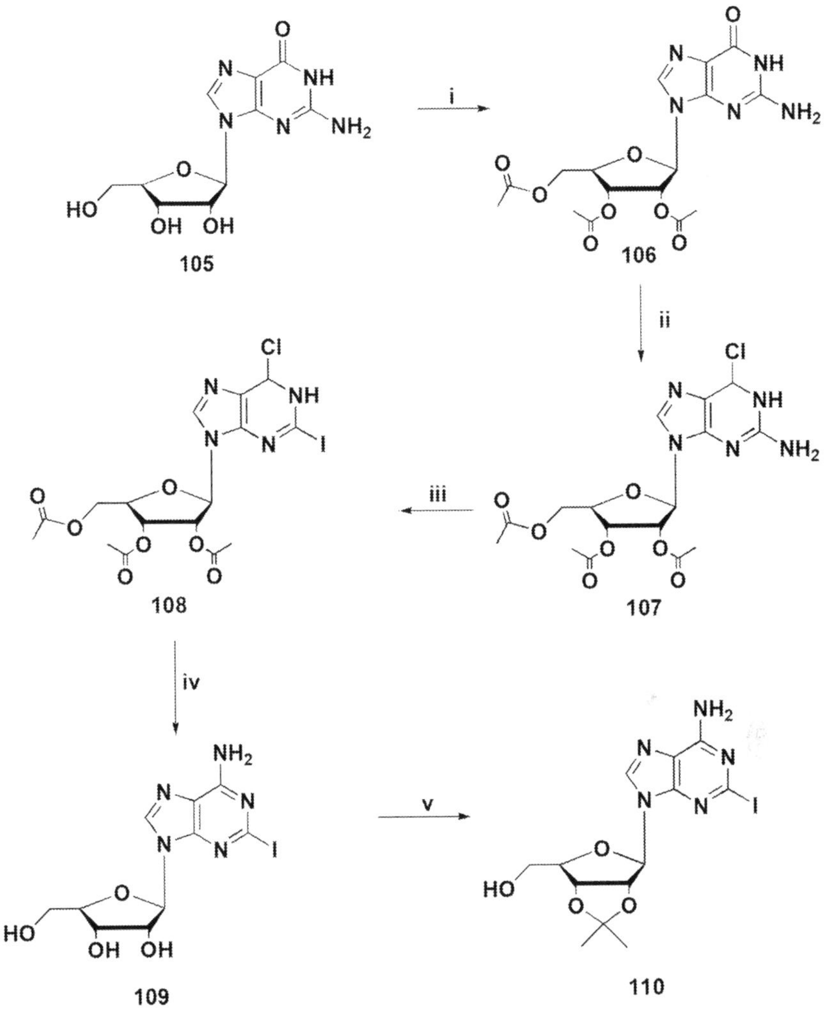

Scheme 6.13 Reagents and conditions: (i) Ac_2O/pyridine/DMF/75°C/4.15 h; (ii) $POCl_3$/tetramine chloride/*N,N*-dimethylaniline/CH_3CN/0°C to RT/16 h; (iii) 1,1-dimethylethyl nitrite/CH_2I_2/CH_3CN/80°C/4.50 h; (iv) Liquid NH_3/−78°C to RT/18 h; (v) PTSA/acetone dimethyl acetal/Acetone/RT/4 h/70%.

3′-Deoxy-3β′-fluoro-3α′-*C*-hydroxymethylpurine **105d** nucleosides were prepared as outlined in scheme. Antiproliferative potency of analogs **105a–d** was estimated against L-132, HeLa, MDA-MB-231, and neural cell lines and proved to have moderate potency [167].

Using the modified Corey-Link reaction, the desired compounds **105a–d** were synthesized starting from compound **107**, which was

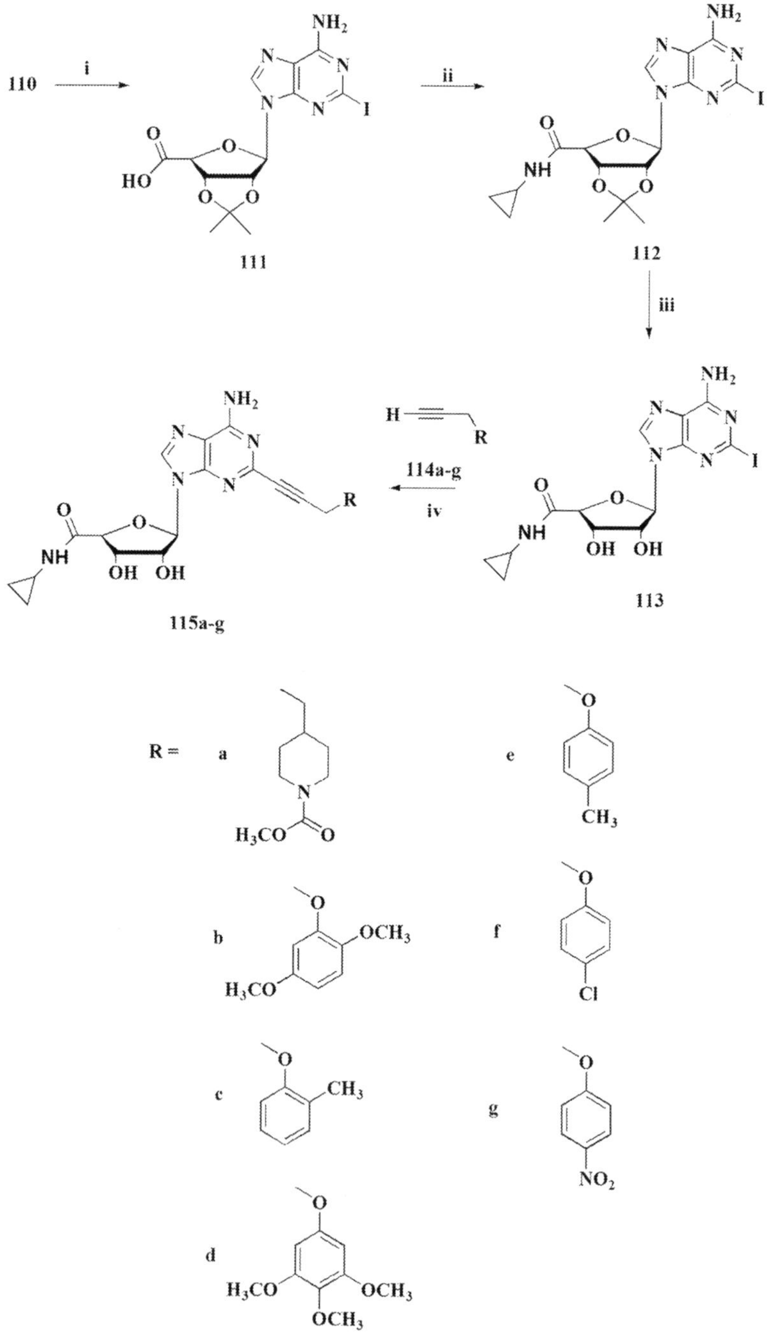

Scheme 6.14 Reagents and conditions: (i) KOH, $KMnO_4$, H_2O, rt., 72 h; (ii) HOBT, EDCI, DMAP, aminocyclopropane, DMF, Et_3N, rt., Overnight; (iii) 50% HCOOH, 80°C, 1.5 h; (iv) $Pd(PPh_3)_2Cl_2$, Copper(I) iodide, dry DMF, dry CH_3CN, dry TEA, rt., 24 h

synthesized from 1,2:5,6-di-*O*-isopropylidene-3-oxo-α-D-glucofuranose **2**. The compound **107** has requisite C3 quaternary carbon with 3-β-fluoro and 3-α-methylester functionalities, of which the ester group could be subsequently manipulated [168]. The C3-α-methylester in compound **107** was reduced followed by acetylation to yield the acetoxymethyl analog **108**. Further deprotection of 5,6-acetonide functionality in compound **108** furnished the diol, which on treatment with sodium periodate followed by sodium borohydride reduction yielded the acetoxymethyl-α-D-xylo-furanose **109** [167].

Acetylation of the C5-hydroxyl group in compound **109** was performed to access compound **110** that on reaction with TFA:H_2O (hydrolysis of 1,2-acetonide) followed by acetylation gave the tetra acetate **111** as an anomeric mixture. The latter was separately exposed to the Vorbruggen glycosylation reaction [169] with various nucleobases, such as *N*-Bz-cytosine, uracil, *N*-Bz-adenine, thymine, in the existence of TMSOTf and BSA in acetonitrile, that furnished nucleosides **112a–d**. The production of the bisomer is resulted from the neighboring group participation of C-2 acetyl group [170]. Subsequent deprotection of acetate groups in compound **112ad** with methanolic NH_3 yielded the corresponding hydroxymethyl nucleosides **105a–d**, respectively [167,171].

The synthesis of adenine analog of ç-(*Z*)-ethylidene-2,3-dimethoxybutenolide **121** involved hydration/dehydration of the C4dC5 in the precursor **118a**. Compound **118a** underwent a reverse Michael-type addition with water in the presence of ammonium hydroxide at high temperature, to give the hydrate **5**. In the presence of *S*-adenosyl-L-homocysteine hydrolase, compound **121** was also hydrated to generate compound **120** at 37°C. Butenolide **121** displayed an inhibitory property toward the enzyme. Such type II (enzyme-mediated addition of water across C4dC5) mechanism is the first example of "enzyme-substrate intermediate" inactivation of *S*-adenosyl-L-homocysteine hydrolase. In comparison with the type I mechanism-based inactivation, reduction of enzyme-bound NADP+ to NADPH was not indicated. Treating with hydrochloric acid, stereoselective dehydration of compound **120** occurred to yield the desired compound **121**. Hydration of compound **3a** was accomplished at ambient temperature in the presence of ammonium hydroxide to afford 6-chloropurine analog **119** (Scheme 6.15). Against the murine leukemia (P388) cell line, the adenine-containing ethylidene-2,3-dimethoxybutenolide **121** displayed an obvious selectivity in cytostatic potency [172].

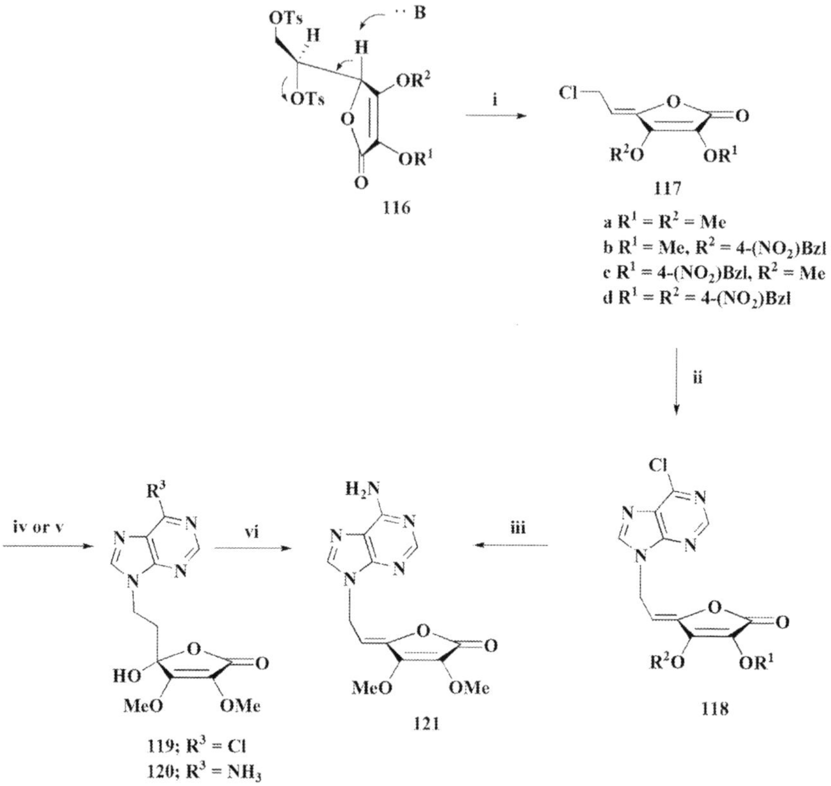

OTs H H ·· B OR² OTs O OR¹ O 116 i Cl O O R²O OR¹ 117
a R¹ = R² = Me
b R¹ = Me, R² = 4-(NO₂)Bzl
c R¹ = 4-(NO₂)Bzl, R² = Me
d R¹ = R² = 4-(NO₂)Bzl
ii
iv or v R³ N N N N HO O O MeO OMe vi H₂N N N N N O O MeO OMe 121 iii Cl N N N N O O R²O OR¹ 118
119; R³ = Cl
120; R³ = NH₃

Scheme 6.15 Reagents and conditions: (i) DBU/CH_3CN/*n*-$_{Bu4NCl}$/reflux/16 h (80% for **2a**, 79% for **2b**, 83% for **2c**, 81% for **2d**); (ii) 6-chloro-9*H*-purine/DMF/TEA/25°C/16 h, (80% for **3a**, 80% for **3b**, 70% for **3c**, 76% for **3d**); (iii) NH_3 (gas)/MeCN/pressure bottle/80°C/ 30.0 h/16%; (iv) For **4**: NH_4OH/MeCN/25°C/1.0 h; (v) For **5**: NH_4OH/MeCN/60°C/3.0 h; (vi) HCl (gas)/EtOAc or CH_3CN/reflux/2.0 h.

A novel silicon(IV) phthalocyanines (SiPcs), which are disubstituted axially with various nucleoside scaffolds, have been prepared (Scheme 6.16) and estimated for their singlet oxygen quantum yields ($\Phi\Delta$) and in vitro photodynamic potencies. The adenosine-substituted SiPc indicates a lower photosensitizing efficiency ($\Phi\Delta = 0.35$) than the uridine- and cytidine-substituted analogs ($\Phi\Delta = 0.42$–0.44), while the guanosine-substituted SiPc displays a weakest singlet oxygen generation efficiency ($\Phi\Delta$ value down to 0.03). On the other hand, replacing axial adenosines with chloro-modified adenosines and purines led to the increase of photogenerating singlet oxygen efficiencies of SiPcs. The formed SiPcs **122** and **127**, which comprise monochloro-modified adenosines and dichloro-modified purines,

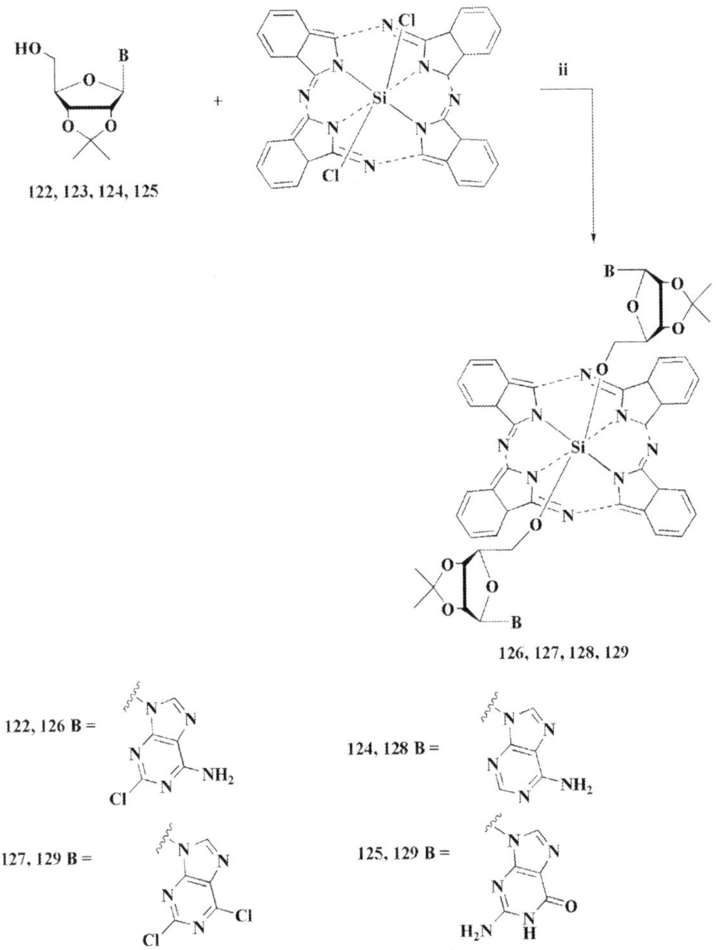

Scheme 6.16 Reagents and conditions: (i) PTSA, acetone, RT. (ii) NaH, toluene, reflux.

respectively, were found to be an efficient photosensitizers (ΦΔ of 0.42–0.44). Both compounds **122** and **127** present high photocytotoxicities against BGC823 and HepG2 cancer cells (IC_{50} values ranging from 9 nM to 33 nM). The photocytotoxicities of these two compounds are significantly higher than the well-known anticancer photosensitizer, chlorin e6 (IC_{50} = 752 nM against HepG2 cells) in the same condition. As indicated by confocal microscopy, for both cell lines, compound **122** can essentially bind to mitochondria, while compound **127** is just partially localized in mitochondria. Additionally, the two compounds induce cell death of HepG2 cells likely via apoptosis [173].

A new series of pyrazolopyrimidine thioglycoside analogs were prepared (Scheme 6.17). Nanoformulation of compound **135** into chitosan nanoparticles elucidated an anticancer potency and can be utilized as a drug delivery system; however, subsequent investigations are still needed.

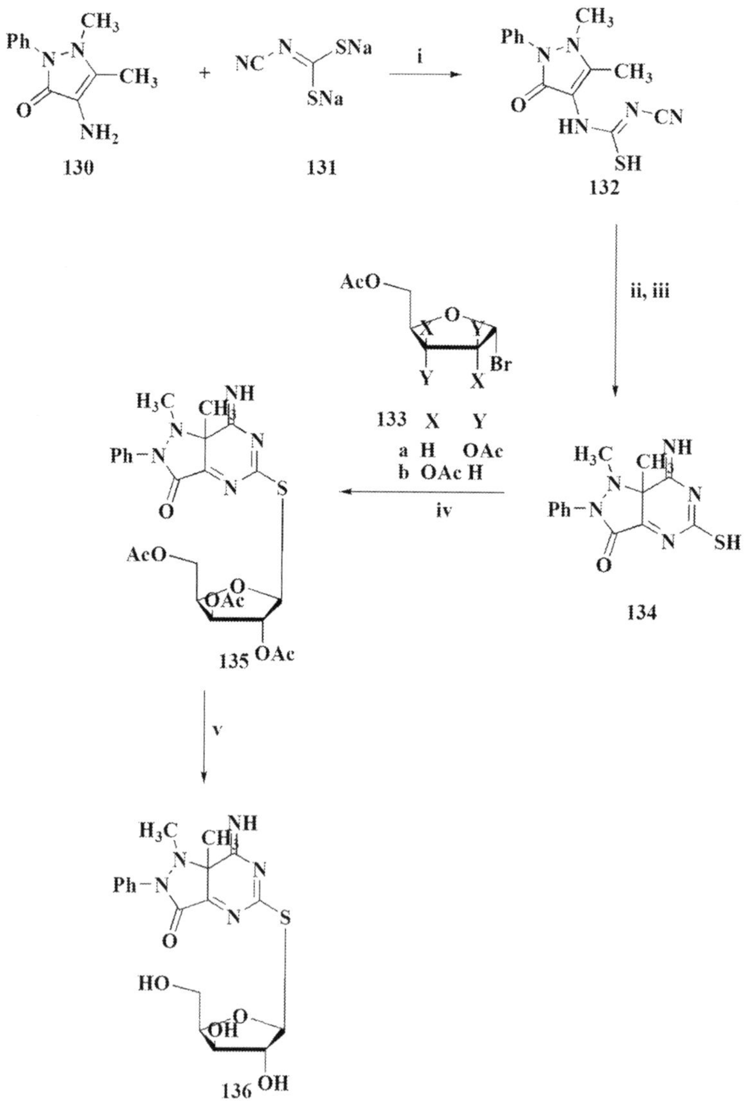

Scheme 6.17 Reagents and conditions: (i) AcOH, reflux, 5 min. (ii) EtOH, reflux. (iii) HCl. (iv) KOH, acetone, RT. (v) MeOH/NH_3/RT/10 min.

The nanopreparation was accessed via the encapsulation of compound **135** into chitosan nanoparticles. Treatment of Mcf-7 and Huh-7 indicated that compound **135** was the most cytotoxic compound on both cancer cell lines where IC_{50} was 12.203 (4.8812 μg/mL) and 24.59 (9.836 μg/mL) on Mcf-7 and Huh-7. But IC_{50} of the nanopreparation was 30.68 and 37.19 μg/mL on Mcf-7 and Huh-7, showing its aggressiveness on human breast cancer cells as affirmed by DNA fragmentation assay and theoretically via the CompuSyn tool [174].

The 7-deazapurine nucleoside antibiotic tubercidin was transformed into its 4-*N*-benzyl and 4-*N*-(4-nitrobenzyl) derivatives through alkylation at *N*3 followed by fluoro-diazotization and SNAr displacement of the 4-fluoro group using a benzylamine or by a Dimroth rearrangement to the 4-*N* isomer. The 4-*N*-(4-nitrobenzyl) analogs of toyocamycin and sangivamycin antibiotics were synthesized via the alkylation approach (Schemes 6.18–6.20). Cross-membrane transport of labeled uridine by human equilibrative nucleoside transporter 1 (hENT1), which is a prototypical nucleoside transporter protein that is expressed on the cell surface of almost all human tissues, was inhibited to a weaker extent by the sangivamycin and 4-nitrobenzylated tubercidin derivatives than was found with 6-*N*-(4-nitrobenzyl) adenosine. Type-specific inhibition of cancer cell proliferation was indicated at micromolar concentrations with the 4-*N*-(4-nitrobenzyl) derivatives of toyocamycin and sangivamycin, and also with 4-*N*-benzyltubercidin. The acetyladenosine was treated with aryl isocyanates to generate the 6-ureido analogs but none of them displayed an inhibitory potency against cancer cell proliferation or hENT1 [175].

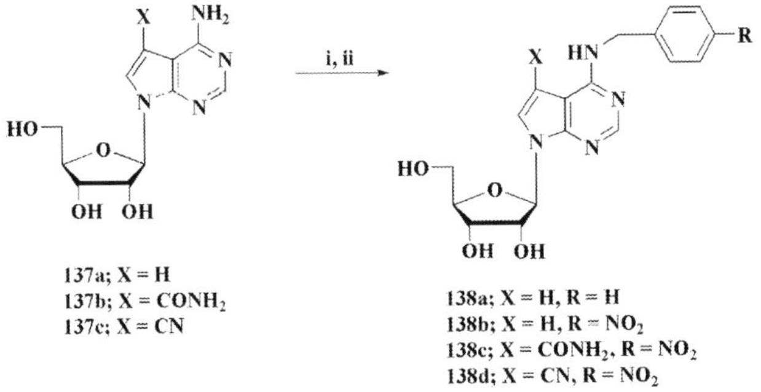

137a; X = H
137b; X = $CONH_2$
137c; X = CN

138a; X = H, R = H
138b; X = H, R = NO_2
138c; X = $CONH_2$, R = NO_2
138d; X = CN, R = NO_2

Scheme 6.18 Reagents and conditions: (i) $ArCH_2Br$/DMF/40–80°C/24–63 h; (ii) $(CH_3)_2NH$ (2 m in THF)/CH_3OH/reflux/20–32 h.

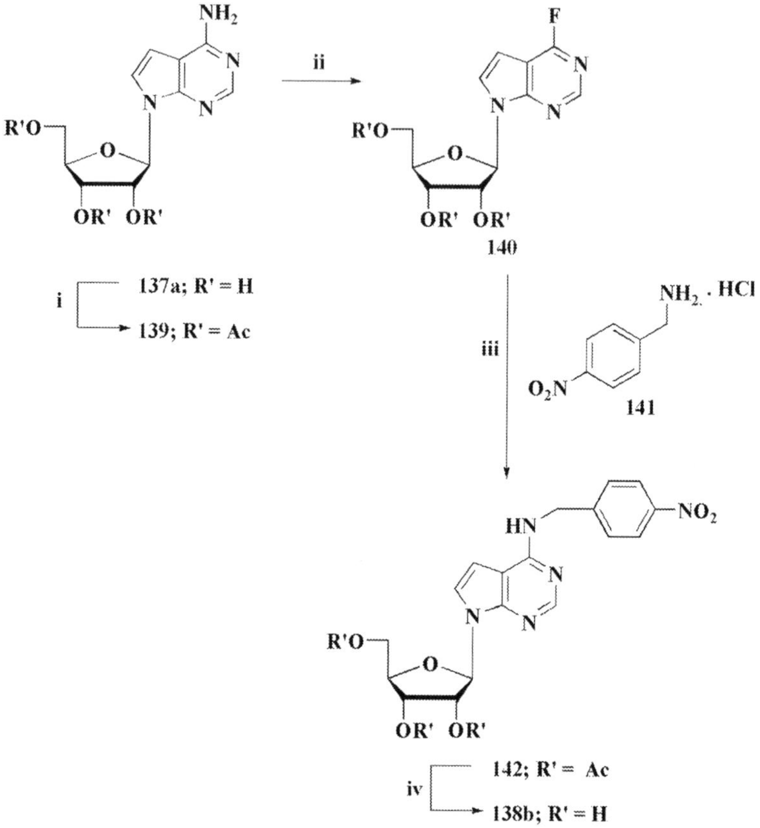

Scheme 6.19 Reagents and conditions: (i) Ac_2O/Py/0°C—RT/21 h; (ii) Py·HF (55% HF)/ $NaNO_2$/−10°C/0.25 h; (iii) TEA/CH_3OH/RT/5 h; (iv) NH_3/CH_3OH/RT/20 h.

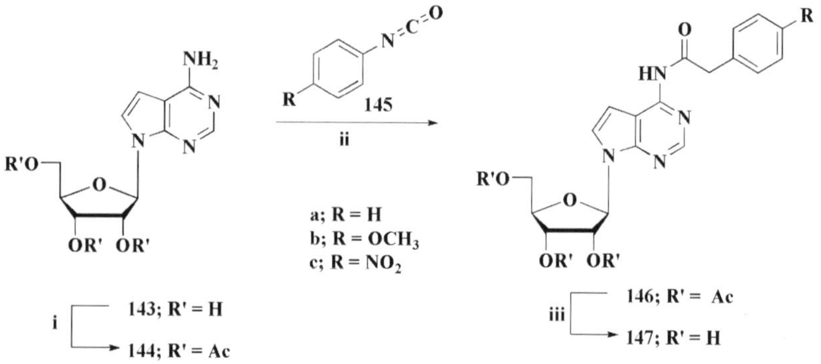

Scheme 6.20 Reagents and conditions: (i) Ac_2O/Py /0°C—RT/24 h; (ii) CH_2Cl_2/RT / 24–68 h; (iii) NH_3 /CH_3OH/RT/2 h.

Inhibition of the proliferation of human T-lymphocyte (CEM), murine leukemia (L1210) cells, prostate adenocarcinoma (PC-3), cervix carcinoma (HeLa), and kidney carcinoma (Caki-1) cells by the 7-deazaadenine and adenine nucleoside derivatives was estimated. No inhibitory potency was indicated with the 6-ureido derivatives **147a**, **147b**, and **147c** in any of the tumor cell lines. Significant inhibition of the proliferation of HeLa cells (IC_{50}: 7.4 mm) and PC-3 cells (IC_{50}: 0.92 mm) by 4-*N*-benzyltubercidin (**138a**) was shown but no marked potency was observed with its nitrobenzyl derivative and the analogous subsequent treatment of compound **144** with the isocyanates furnished compounds **147b** and **147c** [175].

Coupling 3′-C-ethynyl-D-ribofuranose sugar moiety with modified purine nucleobase motif afforded novel antitumor agent. The resulting nucleosides were probed for their ability to inhibit tumor cell proliferation. C7-substituted 7-deazapurine nucleosides elicited an antiproliferative potency. Their activity spectrum was tested in the NCI-60 tumor cell line panel showing potency against a large variety of solid tumor-derived cell lines. Analog **162**, equipped with a 7-deaza 7-chloro-6-amino-purin-9-yl base, was estimated in a metastatic breast tumor (MDA-MB-231-LM2) xenograft model. It inhibited tumor growth and reduced the formation of lung metastases as shown by BLI in vivo analysis [176].

The desired nucleoside analogs were synthesized using Vorbrüggen glycosylation using either the benzoate (**148**, Scheme 6.21) [177] protected sugar precursors (purine vs. 7-deazapurine analogs). Using the benzoate-protected sugar derivative **148** [177] enhances the coupling yields, [178] but the isolated glycosylation products were accompanied by some residues of the glycosyl donor degradation products. Decomposition was resulted upon treating the intermediates **149–152** with methanolic ammonia [178] (or ammonium hydroxide [179]) at high temperatures. Subsequently, an alternative method to introduce the 6-amino group was performed [180,181] Nucleophilic displacement of the 6-chloride with sodium azide furnished the corresponding azide analogs (**153–156**) efficiently. Staudinger reduction generated the corresponding 6-amino derivatives. Further deprotection furnished compounds (**161**, **162**, **163**, **164**) [176].

The synthesis of novel nucleoside platinum complexes (**174a–e**-(*R*) and **174a-e**-(*S*)) was reported, in which a cisplatin-like unit is attached to 7-deazapurine riboside through an alkyl chain bearing a diaminopropyl group, installed at the purine C6-position. The length of alkyl chains has

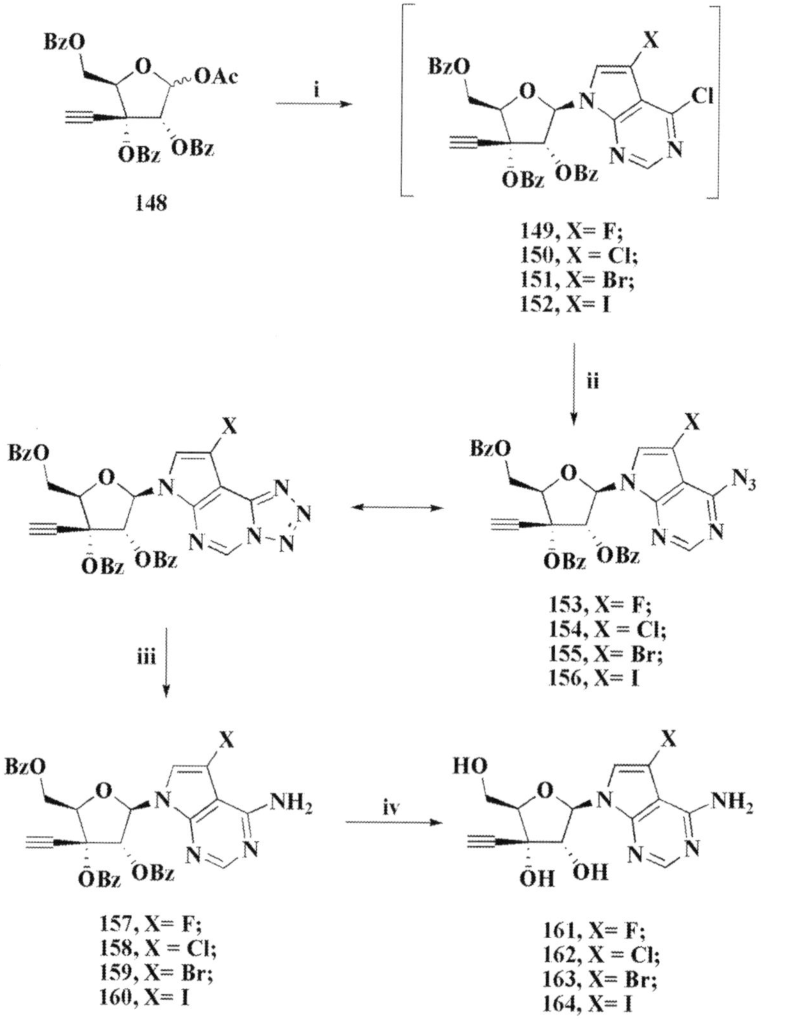

Scheme 6.21 Reagents and conditions: (i) 4-chloro-5-halo-7*H*-pyrrolo[2,3-*d*]pyrimidine (F[32, 33], Cl[34], Br[34], I[34])/CH_3CN/TMSOTf/80°C; (ii) NaN_3/DMF/65°C; (iii) (i) 1.0 M PMe_3 in THF/THF; (ii) aq. HOAc/CH_3CN/65°C; (iv) 7 N NH_3/CH_3OH.

been varied from 1 to 6 carbon atoms, and the resulting pending primary amines have been coupled both to (*R*)- and (*S*)-2,3-diaminopropanoic acids (DAPA), thus generating the chelating diamines **173a–e**-(*R*) and **173ae**-(*S*), then exposed to the platination reactions [182].

The platinum complexes **174a–e-**(*R*) and **174a–e-**(*S*) were synthesized as accomplished in Scheme 6.22. Compound **165** was allowed to react with the mono-Boc-protected diaminoalkanes **166a–e** to yield compounds **167a–e**. The Boc group was removed using trifluoroacetic acid, and then the TFA salts of nucleosides **168a–e** were treated with DIPEA to deprotonate the primary amine groups and then allowed to react with the

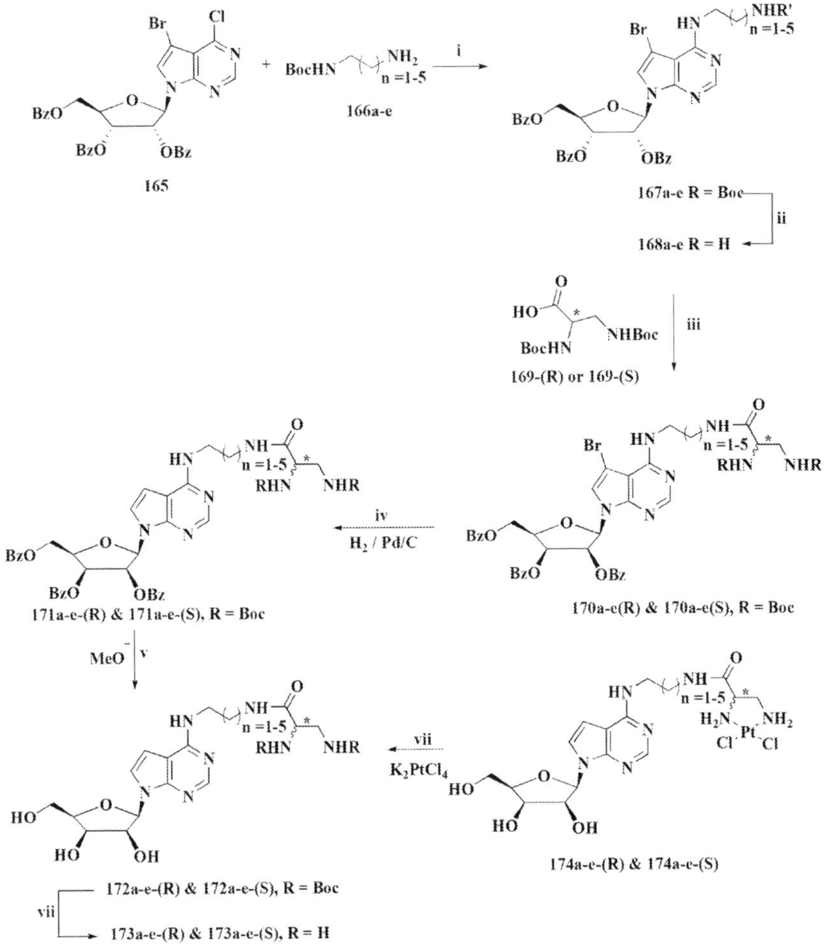

Scheme 6.22 Reagents and conditions: (i) **166a–e**/C_2H_5OH/5 h/reflux; (ii) TFA/CH_2Cl_2, 1:1 (v/v)/1 h/RT.; (iii) (*R*)-**169** or (*S*)-**169**'/ HOBt/EDC/DIPEA/16 h/RT; (iv) CH_3OH/AcONa/ H_2/Pd/C/2 h/ RT.; (v) CH_3ONa/CH_3OH/1 h/RT; (vi) (a) TFA/CH_2Cl_2, 1:1 (v/v)/1 h/RT; (b) Dowex-OH- resin/H_2O; (vii) K_2PtCl_4/ H_2O/CH_3OH 1:1, 16 h/RT in the dark.

enantiopure bis-Boc-protected propanoic acids **169-**(*R*) and **169-**(*S*) using the activating agents of the carboxylic function, thus furnishing the nucleosides **170a–e-**(*R*) and **170a–e-**(*S*) (80%–96%) [182].

Then the C7-brominated carbon was reduced to furnish compound **171**. Under basic conditions, the removal of the benzoyl groups from the ribose scaffold on nucleosides **171** generated the derivatives **172a–e-**(*R*) and **172ae-**(*S*). Further acidic treatment of compounds **172** resulted in the quantitative Boc removal, generating the compounds **173a–e-**(*R*) and **173a–e-**(*S*) in the form of TFA salts. Treatment of the latter salts **173** with a Dowex- OH-resin furnished the desalted amino functions that are cable to coordinate the platinum center. The free amino groups on the diaminopropionic scaffold in compounds **173** were deprotected and carried out through treatment with tertiary amines (Et_3N, DIPEA), even if the following platination reaction furnished very low yields of the platinum complexes **174** using these conditions [182].

The platinum complexes **174a–e-**(*R*) and **174a–e-**(*S*) were afforded through reacting K_2PtCl_4 with the diamines **173**, in H_2O/MeOH solutions. The platinated complexes **174** were furnished in 31%–55% yield.

Preliminary antiproliferative data on platinum complexes **174a**–e- (*R*) and **174a–e**-(*S*) have elucidated that all compounds, although less potently than cisplatin, could inhibit the proliferation of the A549 and Cal27 human cell lines in a dose-dependent manner. The A549 cells were less sensitive than the Cal27 to all the examined platinum complexes, which was explained through the genomic differences between the two cell lines, which was caused due to the some specific membrane transporters alterations, resulting in reduced drug uptake and/or enhanced drug efflux [182].

As part of an effort to explore the mechanism of potent, broad-spectrum anticancer and antiviral activities of a number of ring-expanded ("fat") nucleosides that was reported, a representative "fat" nucleoside 4,6-diamino-8-imino-8*H*-1-b-d-ribofuranosylimidazo[4,5-*e*][1,3]diazepine (**175**) was transformed to its 5′-triphosphate derivative (**176**) (Scheme 6.23), and biochemically screened for possible inhibition of nucleic acid polymerase activity, using synthetic DNA templates and the bacteriophage T7 RNA polymerase as a representative polymerase. The results propose that compound **176** is a moderate inhibitor of T7

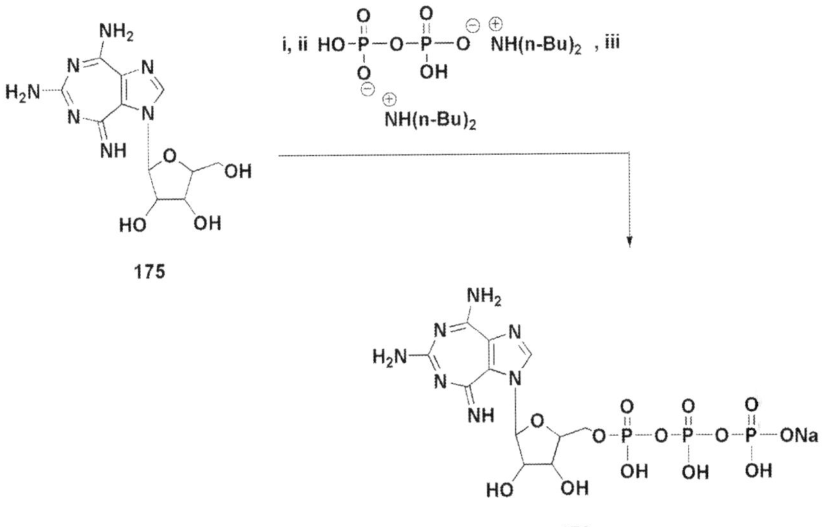

175

176

Scheme 6.23 Reagents and conditions: (i) $POCl_3$, $(OMe)_2PO$. (ii) $N(n\text{-}Bu)_3$, DMF. (iii) NaI, acetone.

RNA polymerase and that the 5′-triphosphate moiety of compound **176** appears to be essential for inhibition as compound **175** alone failed to inhibit the polymerase reaction [183].

3.1.2 Deoxyribonucleoside purine analogs

The synthesis of 2′-deoxy-2′-methylidine-β-D-arabinofuranosyladenosine compounds was outlined.

Guanosine has been converted to guanosine triacetate **178** by using acetic anhydride. The 2-amino-4-chloro-(9*H*)purine nucleoside **179** was yielded as a result of the reaction of guanosine triacetate **178** with thionyl chloride. After adding sodium ethoxide to compound **179**, the 4-ethoxy derivative **180** was obtained (Scheme 6.24).

The 3′,5′-hydroxyl groups in compound **180** were selectively protected using tetraisopropyldisiloxane-1,3-diyl (TIPDS) group to yield **181**. The latter was then transformed to the corresponding 2′-keto analog **182** via Swern oxidation by oxalyl chloride (Scheme 6.24) [184].

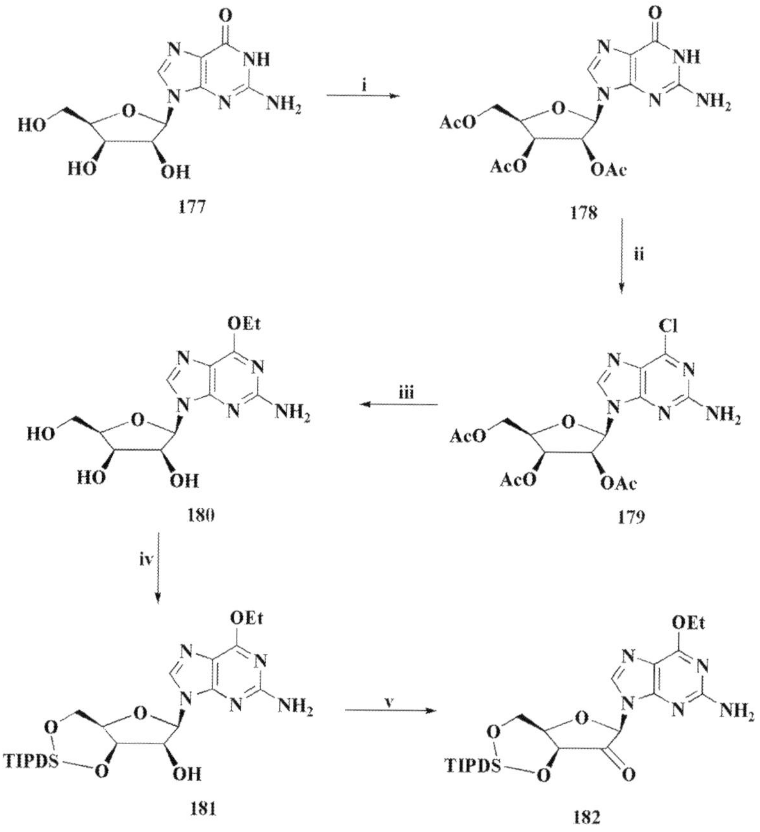

Scheme 6.24 Reagents and conditions: (i) Ac_2O. (ii) $SOCl_2$. (iii) NaOEt. (iv) DTIPDS. (v) Oxaylyl chloride.

The 2′-keto derivative **182** was allowed to react with triphenylphosphonium bromide and butyllithium to yield the corresponding 2′-deoxy-2′-hydroxy-2′-methyltriphenylphosphenium bromide **183**. The latter was treated with sodium hydride in tetrahydrofuran to afford 2′-methylidene compound **184** (78% yield).

The reaction of compound **184** with tetrabutylammonium fluoride (TBAF) in tetrahydrofuran furnished 2-amino-4-ethoxy-9-(2′-deoxy-2′-methylidene-β-D-ribofuranosyl)purine **185**.

Compound **185** reacted with NH_3 in MeOH and/or NH_2OH in MeOH to afford the corresponding 2,4-diamino and 2-amino-4-hydroxylamino analogs **186** (Scheme 6.25).

The 2-chloro compounds **170** were synthesized from the acetylated compound **187** using TNB and $SbCl_3$ in CH_2Cl_2. Similarly, the

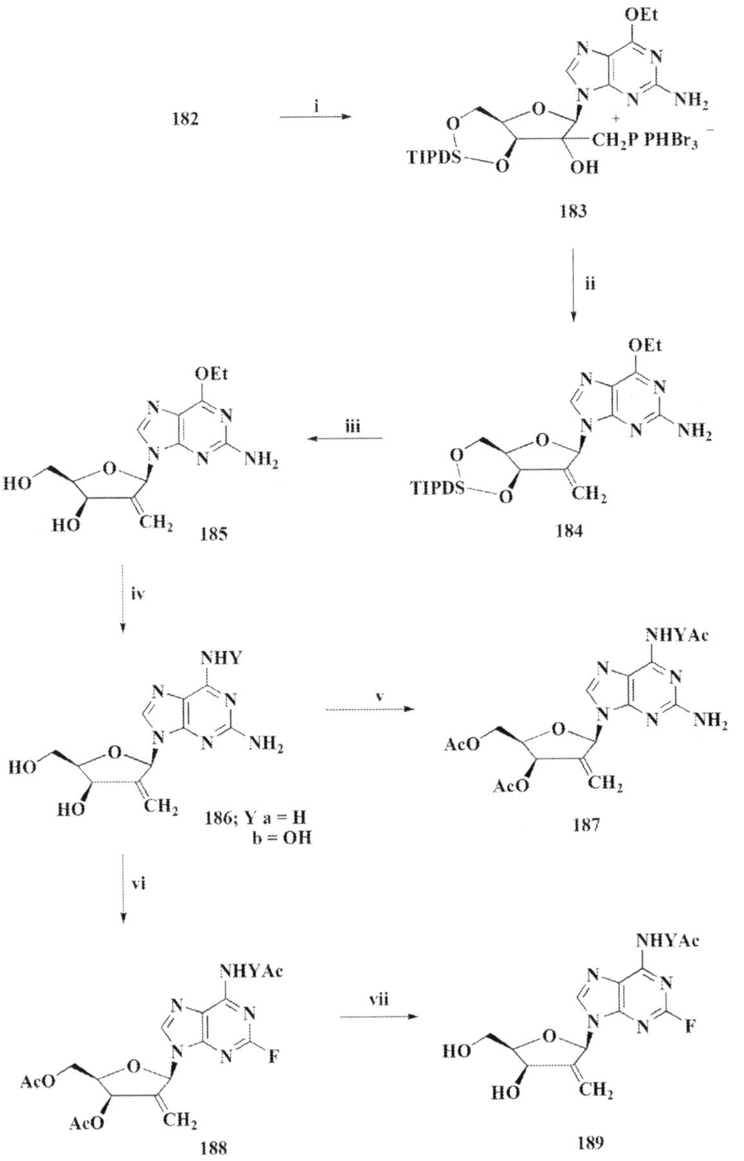

Scheme 6.25 Reagents and conditions: (i) Ph_3PCH_3Br. (ii) NaH. (iii) TBAF. (iv) NH_3/MeOH or NH_2OH/MeOH. (v) Ac_2O. (vi) $NaNO_2{\cdot}HBF_4$. (vii) NH_3/MeOH.

2-bromo analogs **188** were furnished using $SbBr_3$ and TNB in CH_2Br_2. Compounds **187** in HBF_4 were allowed to react with $NaNO_2$ solution to obtain the 2-flouro derivatives **188**. In addition, compounds **170** and **68** were generated by starting with compound **187** (Scheme 6.25) [184].

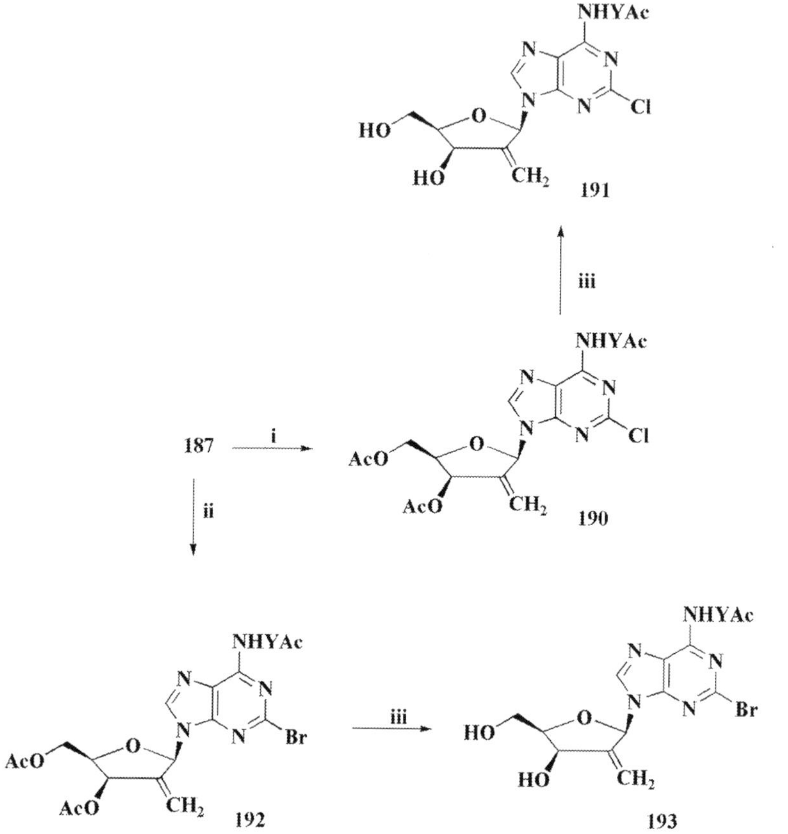

Scheme 6.26 Reagents and conditions: (i) $SbCl_3$, TBN. (ii) $SbBr_3$, TBN. (iii) NH_3, MeOH.

Deacetylation of compounds **190**, **192**, and **188** using methanolic ammonia afforded the 2-halogenated derivatives **191**, **173**, and compound **189** (Scheme 6.26).

Compounds **189a,b**, **191a,b**, and **193a,b** were exposed to in vitro antileukemic and antiviral potency upon a novel L1210 cell line that is twofold resistant to both deoxyadenosine and hydroxyurea, which was grown and recognized. The novel compounds indicated an active antileukemic potency [184].

The synthesis of the 6-methylpurine nucleosides with substitutions at 5′-position has been accomplished in Scheme 6.27. These compounds bear a 5′-heterocycle such as imidazole or a triazole with a two-carbon chain, and an amine, ether, or thio ether. To extend the SAR study of 6-methyl purine and 2-fluoroadenine nucleosides, their corresponding α-linker nucleosides

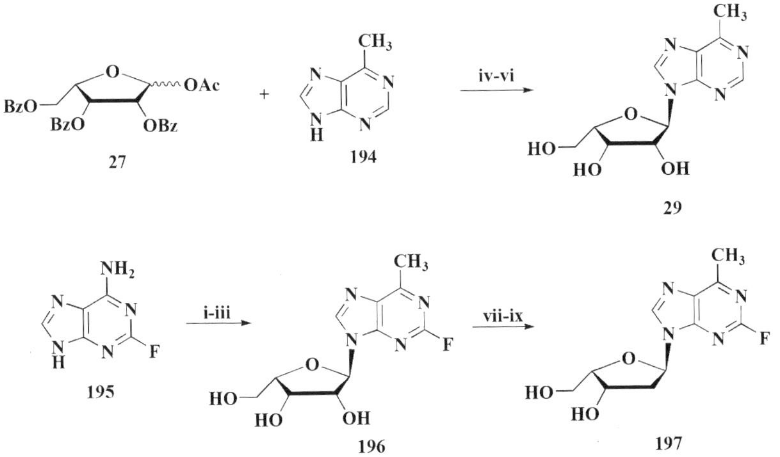

Scheme 6.27 Reagents and conditions: (i) HMDS/$(NH_4)_2$ SO_4/reflux/31 h; (ii) $SnCl_4$/AcCN/−10°C to RT/3–5 h; (iii) KOH/C_2H_5OH/RT/5 h; (iv) TMSCl/HMDS/ethylene dichloride/reflux/2 h; (v) NaOMe/CH_3OH/5°C/2 h; (vi) $TIPDSCl_2$/pyridine/RT/overnight; (vii) ImC(S)Im/DMAP/AcCN/80°C/overnight; (viii) ButOOBut/TTMSS/toluene/reflux/2 h; (ix) Et_4NFxH_2O/AcCN/RT/1 h.

with L-lyxose and L-xylose were also prepared. All of these compounds have been estimated for their substrate potency with *E. coli* PNP [144].

The synthesis of the 2′-deoxy expanded nucleoside derivatives where a heteroaromatic thiophene spacer ring has been localized in between the pyrimidine and imidazole ring systems of the natural purine moiety was accomplished in Schemes 6.28–6.30. The expanded 2′-deoxy-guanosine and -adenosine tricyclic derivatives were synthesized, and the preliminary biological results are also demonstrated.

The tricyclic targets were estimated for their potential cytostatic potency. Moderate inhibition was found for compounds **202** and **207** against the proliferation of three cancer cell lines. The compounds were somewhat more cytostatic against human embryonic lung fibroblasts (HEL). In all cases, compound **207** was twofold more inhibitory to cell proliferation than compound **202** [185].

Two synthetic methods to achieve the 2′-deoxynucleosides could be envisioned; one, a linear route stemming directly from a 2′-deoxy sugar, or two, treatment of the ribofuranose guanosine tricyclic nucleoside [186–189] with the Barton deoxygenation protocol 28,29 to furnish the 2′-deoxy target. The latter was initially chosen since a small amount of the ribose nucleosides were already in hand and could be used immediately. In preparation for the Barton deoxygenation, the protection of exocyclic amine group on the guanosine nucleoside with

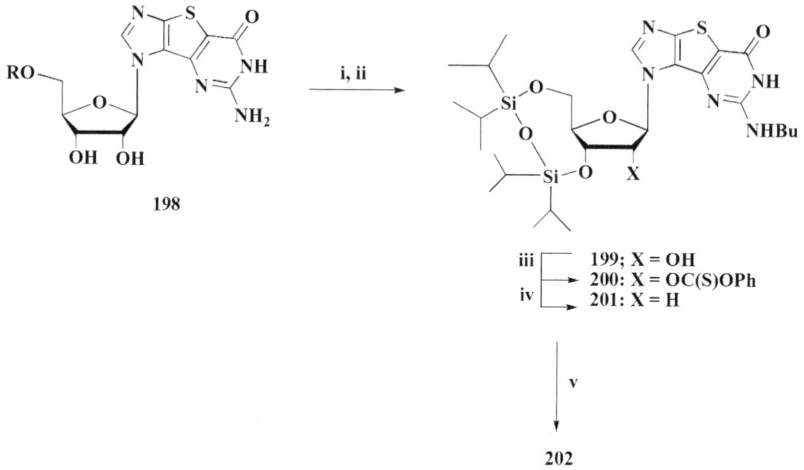

Scheme 6.28 Reagents and conditions: (i) TMSCl/pyridine then 2-methylpropanoyl *chloride*/24 h; (ii) 1,1,3,3-tetraisopropyl-1,3-dichlorodisiloxane/pyridine/RT/24 h; (iii) Phenyl chlorothionocarbonate/DMAP/RT/24 h; (iv) AIBN/Bu_3SnH/toluene/reflux/6 h; (v) (i) 1 M TBAF/THF/RT/4 h; (ii) NH_3/CH_3OH/RT.

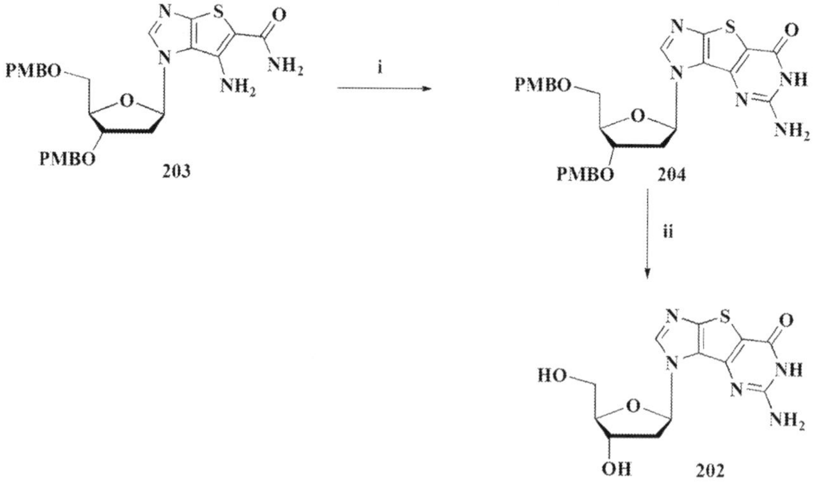

Scheme 6.29 Reagents and conditions: (i) (i) KSC(S)OEt/DMF/reflux/4 h; (ii) H_2O_2/CH_3OH/0°C/2 h; (iii) NH_3/CH_3OH/125°C/18 h; (ii) $BF_3.OEt_2$/EtSH/CH_2Cl_2.

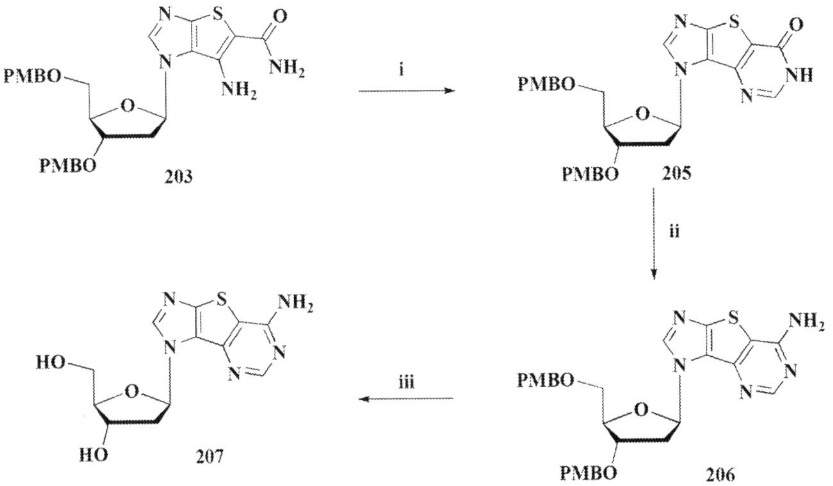

Scheme 6.30 Reagents and conditions: (i) $CH(OEt)_3$/4 Å molecular sieves; (ii) (i) DMAP/TPSCl/Et_3N/CH_3CN/3 h; (ii) NH_3/18 h/RT; (iii) $BF_3.OEt_2$/C_2H_5SH/CH_2Cl_2.

2-methylpropanoyl *chloride was* followed by selective bis-protection of the 3′- and 5′-hydroxyl groups with 1,3-dichlorotetraisopropyldisiloxane to yield **199** for the two steps. Using Barton deoxygenation methods, [190,191] the free 2′-hydroxyl group was then treated with *O*-phenyl carbonochloridothioate to afford the thiocarbonate **200**. Treatment with tributyltin hydride and AIBN then furnished compound **201** [185].

The silyl groups were deprotected and then the removal of isobutyryl group was followed to afford the desired compound **202** from **201** (Scheme 6.28). The synthesis of tricyclic intermediate **204** (Scheme 6.29) was subsequently accomplished. The first step of the ring closure method was modified and the performed potassium ethylxanthic acid salt was utilized instead of carbondisulphide and sodium hydroxide in methanol [185].

Deprotection of the PMB groups proved to be quite problematic despite the availability of various deprotection strategies [192–197]. Different conditions were employed, comprising the standard deprotection protocol using ammonium cerium (IV) nitrate (CAN). Unfortunately, CAN led to a marked glycosidic bond cleavage. In most cases, either no product was found or the product yields proved to be less than 10%. Ultimately, the original Lewis acid-mediated deblocking method was utilized, which afforded **202** (Scheme 6.29) [185].

Analogous to the synthetic route to the guanosine analog, the expanded adenosine nucleoside **207** could be approached from the same two routes, and since the synthesis of both **202** and **207** was being pursued concurrently, both routes were tried for comparison.

As was the case with the guanosine derivative, the route involving the Barton deoxygenation was accompanied by low yields and difficulties with different purifications; thus, the route utilizing the 2′-deoxy sugar was provided. Starting with the bicyclic intermediate **203**, ring closure to the tricyclic base was achieved with triethyl orthoformate and molecular sieves to afford the inosine intermediate **204** (Scheme 6.30) [185].

The desired compound **211** from the analog **208** synthesized through the bromination of 2′-deoxyadenosine [198,199] was outlined. Phosphorylation was performed. After chromatographic purification, a mixture containing mainly monophosphate **209** (89%) but also comprising a small amount of the analog **208** (9%) and a lesser quantity of diphosphate **210** was performed forward. Reaction of chloromethyl pivalate with the mixture furnished the targeted compound **211**. The inability of nucleosides to be initially activated to the monophosphate level via a nucleoside kinase or other activating enzyme is believed to be due to the lack of cytotoxicity and this is considered to be the main reason. In a quest of different nucleosides that might be useful anticancer agents, the usage of monophosphate prodrugs was examined to investigate if any improved cytotoxicity might be observed for the prodrugs of candidate nucleosides that possess little or no cytotoxicity. For this reason, 5′-bis (pivaloyloxymethyl) phosphate prodrugs of two weakly cytotoxic compounds, 8-bromo-2′-deoxyadenosine **208** and 8-aza-2′-deoxyadenosine **202**, have been synthesized (Schemes 6.31 and 6.32). The prodrug **211** were estimated for their cytotoxicity in CEM cells and were observed to have a marked enhanced cytotoxicity comparable to the corresponding nucleosides [198].

3.2 Nonribonucleoside analogs

3.2.1 Thiofuranoside analogs

4′-Thio-lxylofuranosyl analogs were synthesized and estimated as potential anticancer agents. A versatile sugar intermediate for direct coupling with the purine scaffold is also prepared. The desired compounds were tested in a variety of human cancer cell lines in vitro. The synthesis of the carbohydrate precursor 1-*O*-acetyl-2,3,5-tri-*O*-benzyl-4-thio-L-xylofuranose (**225**) and corresponding purine nucleosides is accomplished [200].

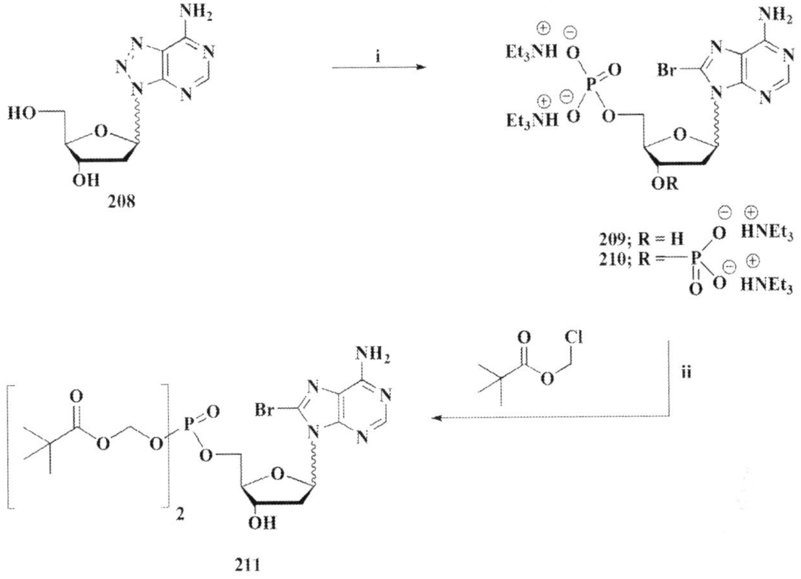

NH2
N N N N N
HO O
OH
208

i

Et3NH O
O P=O
Et3NH O O
NH2
Br N N N N
O
OR
209; R = H
210; R = P O HNEt3 O HNEt3
O

O Cl
O
ii

NH2
O P=O Br N N N N
O O O
O
2 OH
211

Scheme 6.31 Reagents and conditions: (i) $POCl_3$/TEA/PO(OMe)$_3$. (ii) $Me_2CCOOCH_2Cl$, TEA.

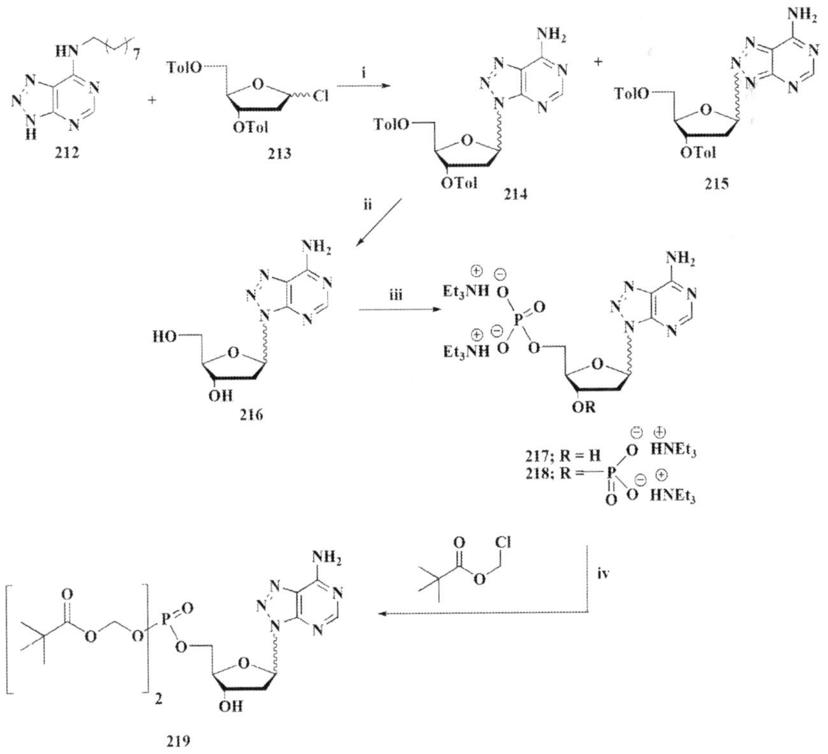

HN 7
N N N N N H
212
+
TolO O Cl
OTol
213
i
NH2
N N N N N
TolO O
OTol
214
+
NH2
N N N N N
TolO O
OTol
215

ii

NH2
N N N N N
HO O
OH
216

iii

Et3NH O
O P=O
Et3NH O O
NH2
N N N N N
O
OR
217; R = H
218; R = P O HNEt3 O HNEt3
O

O Cl
O
iv

NH2
O P=O N N N N N
O O O
O
2 OH
219

Scheme 6.32 Reagents and conditions: (i) DBU/MeCN. (ii) NH_3/EtOH. (iii) $POCl_3$/TEA/PO (OMe)$_2$. (iv) $Me_2CCOOCH_2Cl$, TEA.

According to a previously used method for the synthesis of other 4-thiofuranose precursors [201–204], the main intermediate **225** has been developed. Transformation of d-arabinose to methyl 2,3,5-tri-*O*-benzyl-D-arabinofuranoside **(222)** was demonstrated. Transformation to dibenzyl dithioacetal **223** utilizing stannic chloride and benzyl mercaptan was accomplished after chromatographic purification in 48% yield. Single inversion was involved as a result of the cyclization at C-4 [204]; thus, transforming the D-*arabino* to the L-*xylo* configuration was proceeded using iodine, imidazole, and triphenylphosphine. Finally, the benzylthio group at C-1 was replaced by an acetoxy group, involved the treatment of compound **224** with mercuric acetate in acetic acid. 25% is the yield of compound **225** from **220**, resulted from four column purifications, and generated a c. 1:1 mixture of α,β anomers [200].

Purine nucleoside derivatives were synthesized through the coupling of 2,6-dichloropurine and compound **225**. A Lewis acid-catalyzed reaction using stannic chloride in acetonitrile was proved to be an effective method to access this coupling, and α and β anomers of compound **226** were furnished in 30% and 25% yields after chromatographic purification and separation. Treating with ethanolic NH_3 to afford the blocked 2-chloroadenine nucleosides **227 α** and **227 β**, frequent removal of the *O*-benzyl groups was accomplished with boron trichloride to give the targeted nucleoside **228 α** (55%) and **228 β** (45%). Compounds **226 α** and **226 β** were treated with sodium azide in the presence of ethanol at reflux to furnish the corresponding diazido intermediates **229 α** and **229 β**, which were then reduced to yield the blocked diaminopurine nucleosides **230 α** and **230 β** in 80% and 82% yield. Deblocking of compounds **230 α** and **230 β** generated the diamino derivatives **231 α** (71%) and **231 β** (75%). The transformation of **231 β** to the corresponding guanine nucleoside **232** (45%) was accessed by treatment with adenosine deaminase. Slow deamination was occurred, even with a high concentration of enzyme. On the other hand, when 9-(4-thio-α-L-xylofuranosyl)-9*H*-purine-2,6-diamine (**231 α**) was treated with adenosine deaminase for several days, it was found that it was resistant to deamination (Schemes 6.33 and 6.34) [200].

2′-Deoxy-2′-fluoro-4′-thioarabinofuranosyl purine nucleosides were synthesized, and antitumor activities were evaluated.

2-Chloroadenine β-**247** and 2-fluoroadenine β-**241**, which were designed as potential antitumor agents, demonstrated little or no potency against both KB and CCRF-HSB-2 cells.

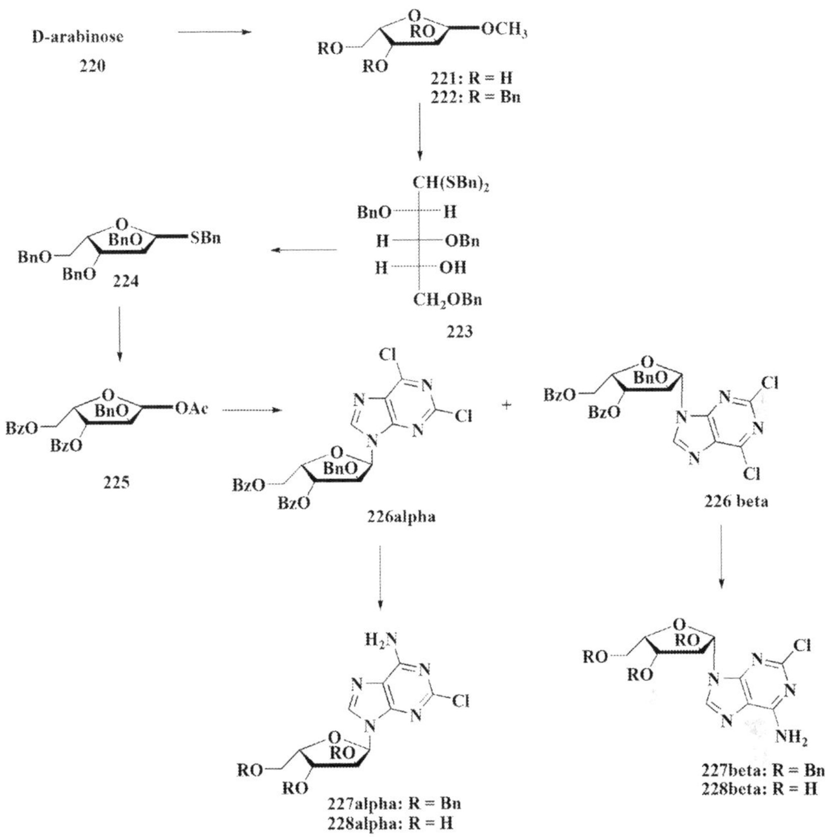

Scheme 6.33 Reagents and conditions: (i) stir 5 h in 0.5% hydrogen chloride/CH_3OH/RT/ Amberlite IRA-400 OH anion-exchange resin/purified by silica gel chromatography ($CHCl_3$/MeOH). (ii) Dry THF/NaH (60% dispersion in mineral oil/stir for 15 min under N_2/tetrabutylammonium iodide/benzyl bromide/stir 3 days RT/purified by silica gel chromatography (cyclohexane/EtOAc). (iii) Dichloromethane/benzyl mercaptan/ stannic chloride/stir RT overnight/5% aq. $NaHCO_3$/dichloromethane /silica gel chromatography (cyclohexane/EtOAc). (iv) dry 2:1 toluene/acetonitrile/ triphenylphosphine/iodine/imidazole/stir at 90°C for 24 h, purified by silica gel chromatography (cyclohexane/EtOAc, (v) mercuric acetate/acetic acid/stir at RT for 2 h/dichloromethane/water/saturated aqueous $NaHCO_3$/5% aq. KCN sol./ Na_2SO_4. (vi) 2,6-dichloropurine in acetonitrile RT/ stannic chloride/dichloromethane/over 1 min/stir 2 h/dichloromethane/$NaHCO_3$/($MgSO_4$). (vii) Ethanolic NH_3/at 50○C in a glass-lined stainless steel pressure vessel for 48 h/dichloromethane/1 M BCl_3/ CH_2Cl_2/−50°C/was stored at −20°C for 16 h/ice-cold CH_2Cl_2/aq. $NaHCO_3$/(pH 7–8)/ H_2O/CH_2Cl_2.

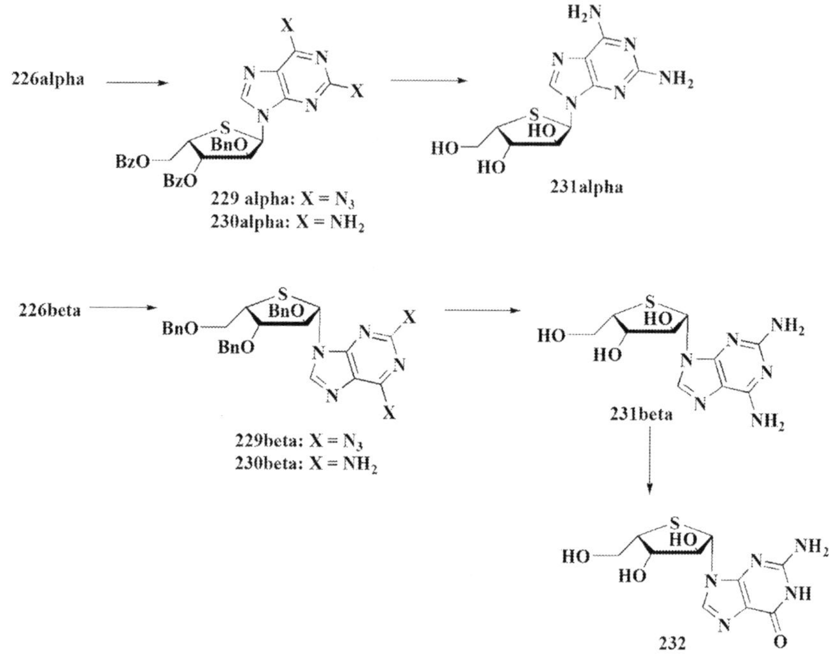

Scheme 6.34 Reagents and conditions: (i) Sodium azide, C_2H_5OH/reflux for 2 h/dichloromethane/H_2O/($MgSO_4$) and concentrated in vacuo. (ii) Dichloromethane/methanol/stannous chloride/stir 30 min/purification by silica gel chromatography ($CHCl_3$/CH_3OH). (iii) CH_2Cl_2/BCl_3 in CH_2Cl_2 at −50°C. at −20°C for 16 h. CH_2Cl_2/aq. $NaHCO_3$/(pH 7–8)/H_2O/CH_2Cl_2. (iv) H_2O/100 u of adenosine deaminase type VIII (40 μL)/stir for 72 h/boil for 3 min/charcoal.

The protecting group of **235a** and **236a** was deblocked to afford a mixture of α- and β-anomers, which was separated by ODS column chromatography to yield adenine analog **α-236a** and **β-235a**. The α- and β-anomers of diaminopurine analogs **235b** and **236b** were also accessed through the same method [205].

The mixture of the α- and β-anomers of **253a** and **236a** was treated with adenosine deaminase to favor the selective hydrolysis of the 6-amino group of **β-235a**. Thus, unreacted α-diamino derivatives **235** and **236** and the produced β-anomer of guanine analog **237** could be separated through simple ODS column chromatography (Scheme 6.35).

The 2-fluoroadenine analog was prepared from the diaminopurine derivative **234b** 16. Desilylation of the anomeric mixture of diamino analog **234b** was accomplished then and reprotection by an acetyl group was achieved to generate **238**. Further treatment with 1,1-dimethylethyl nitrite

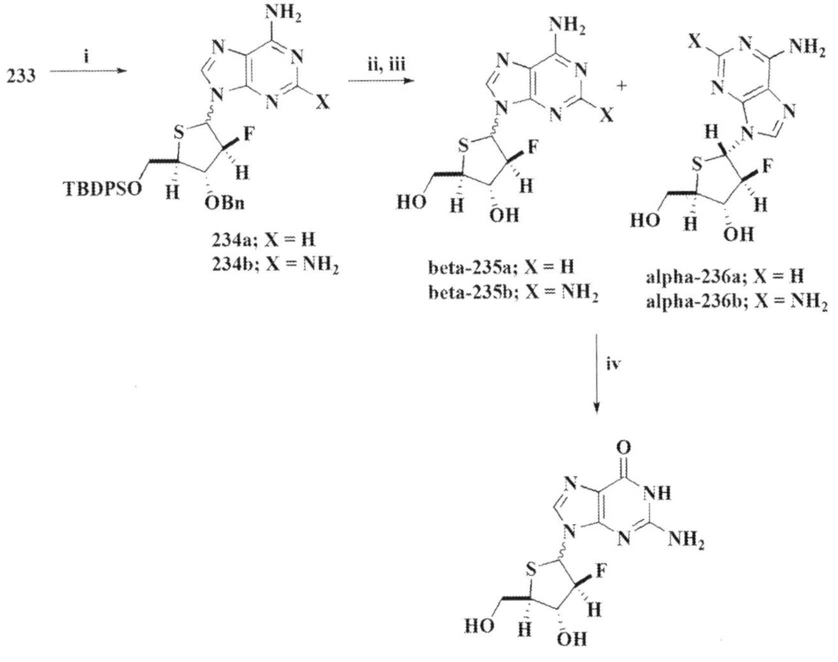

Scheme 6.35 Reagents and conditions: (i) 2,6-DAP or adenine/TMSOTf/CH_3CN; (ii) BCl_3/CH_2Cl_2/−78°C; (iii) $NH_4F.HF$/DMF/ODS column; (iv) adenosine deaminase/Tris-HCl buffer.

and 60% HF-pyridine afforded 2-fluoro derivative **239**, along with a 2,6-difluoro derivative in 58% and 20% yield. Compound **239** was deacetylated to give a mixture of α-**243** and β-**240**, which could be separated through chromatographic separation. The formed **α-243** was transiently protected by a TMS group to facilitate solubility in an organic solvent, and then debenzylation was performed using boron trichloride, then treatment with ammonium hydroxide in methanol was followed, to afford **α-244** (76% yield). On the other hand, the same reaction with compound **β-240** afforded the targeted **β-241** and a 5′-chlorinated derivative **β-242** in 56% and 36% yield (Scheme 6.36) [205].

A 2-chloroadenine analog was synthesized. The glycosylation reaction of intact 2,6-dichloropurine with compound **233** in the presence of trimethylsilyl trifluoromethanesulfonate (TMSOTf) selectively afforded a *N*9-adduct **245** in 69% yield (α:β = 1.5:1). The same reaction with Tin (IV) *chloride*, instead of TMSOTf, led to a decreased yield of compound

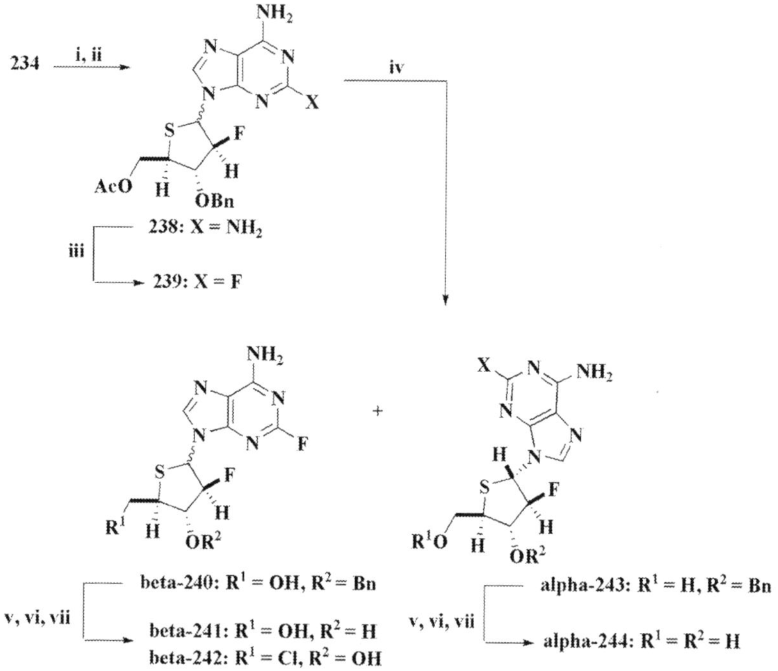

234 —i, ii→ 238: X = NH_2 —iii→ 239: X = F; —iv→ beta-240: R^1 = OH, R^2 = Bn + alpha-243: R^1 = H, R^2 = Bn

beta-240 —v, vi, vii→ beta-241: R^1 = OH, R^2 = H; beta-242: R^1 = Cl, R^2 = OH

alpha-243 —v, vi, vii→ alpha-244: R^1 = R^2 = H

Scheme 6.36 Reagents and conditions: (i) $NH_4F.HF/DMF$; (ii) $Ac_2O/TEA/CH_3CN$; (iii) 1,1-Dimethylethyl nitrite/60% HF-pyridine/0°C; (iv) NH_4OH/CH_3OH/ODS column; (v) TMSCl/CH_2Cl_2/BCl_3/−78°C.

245 (16%, $\alpha:\beta = 1.4:1$). After debenzylation, the α- and β-anomers of compound **246** were separated via chromatographic purification, were treated with ammonia in the presence of ethanol, then desilylation was followed to afford the α- and β-anomers of 2-chloroadenine analogs **249** (Scheme 6.37) [205].

3.2.2 Carbocyclic purine analogs

In a regio- and stereoselective manner, the 2′-β-substituted-6′-fluoro-cyclopentenyl purines **254** and **255** were prepared from D-ribose (Scheme 6.38). The C2-position of 6′-fluoro-cyclopentenyl nucleosides was functionalized and that was performed through regioselective protection of a hydroxyl group at the C3-position and stereoselective formation of C2-triflate followed through direct SN2 reaction with an azido or fluoro nucleophile. The carbocyclic nucleosides were estimated for their anticancer potencies in various tumor cell lines, but were proved to be neither toxic nor active.

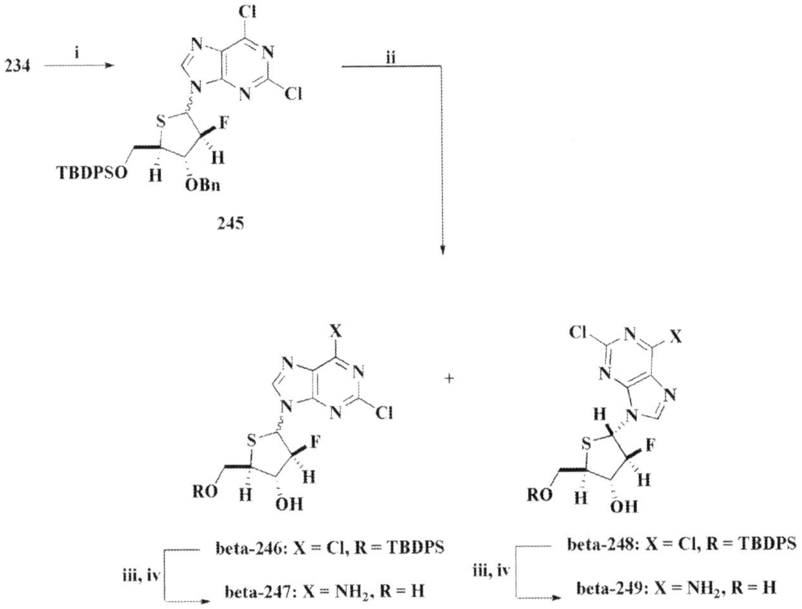

234 → i → 245 (TBDPSO, H, OBn, F, H; 2,6-dichloropurine) → ii → beta-246: X = Cl, R = TBDPS; beta-247: X = NH_2, R = H (iii, iv) + beta-248: X = Cl, R = TBDPS; beta-249: X = NH_2, R = H (iii, iv)

Scheme 6.37 Reagents and conditions: (i) 2,6-dichloro-9*H*-purine/TMSOTf/CH_3CN/0°C; (ii) BCl_3/CH_2Cl_2/−78°C; (iii) NH_3/C_2H_5OH/80°C; (iv) $NH_4F.HF$/DMF.

The 2′-β-substituted purine nucleoside derivatives **254a**, **255a**, and **255b** were furnished using the Mitsunobu reaction as the main step (Scheme 6.6). Under the standard Mitsunobu conditions, 6-chloropurine was condensed with glycosyl donors **250** and **251** to afford the N9-6-chloropurine analogs, which after purification, were treated with tert-butanolic ammonia solution to yield adenosine analogs **252** and **253,** respectively. Compounds **252** and **253** were treated with tribromoborane (10%) solution in methylene chloride followed by the pivaloyl group removal after treatment with the methanolic ammonia to generate nucleosides **254a** and **255a**. Further reduction of compound **255a** gave the 2′-β-aminofluoroneplanocin A derivative **255b** [206].

Based on the potent anticancer activity of 6′-fluorocyclopentenyl-cytosine **269** (Fig. 6.20) in phase II clinical trials for the treatment of gemcitabine-resistant pancreatic cancer, a systematic structure-activity relationship study of 6′-fluorocyclopentenylpurines **259, 261**, and **262** were performed to search new anticancer agents.

The phosphoramidate prodrug **268** of adenine derivative **260** was synthesized (Scheme 6.39) to identify if the anticancer potency relied upon the

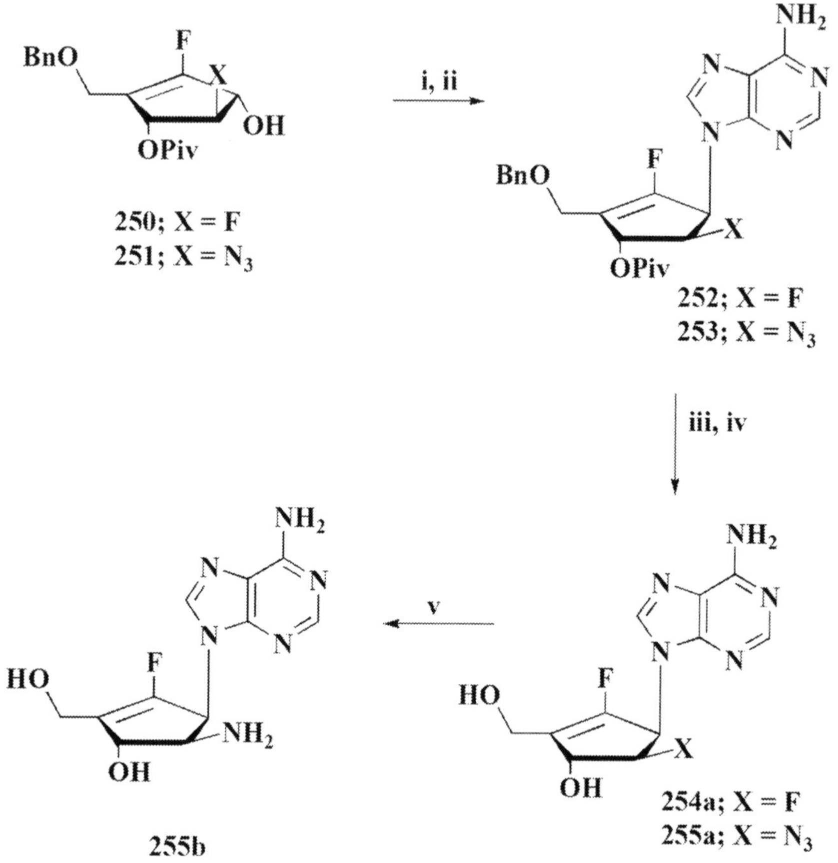

250; X = F
251; X = N_3

252; X = F
253; X = N_3

254a; X = F
255a; X = N_3

255b

Scheme 6.38 Reagents and conditions: (i) 6-Chloro-9H-purine/Diisopropyl azodicarboxylate (*DIAD*)/PPh_3/THF/RT/18 h; (ii) NH_3/tBuOH/120°C/sealed tube/12 h; (iii) BBr_3/CH_2Cl_2/−78°C/3 h; (iv) NH_3/CH_3OH/40°C/18 h; (v) PPh_3/NH_4OH/THF/H_2O/RT/18 h.

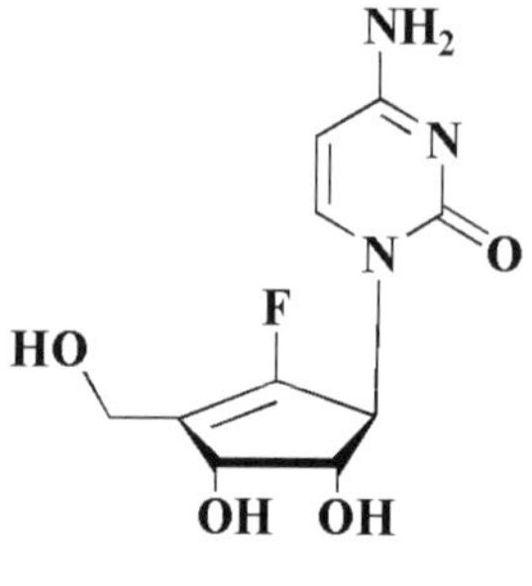

269

Fig. 6.20 6′-Fluorocyclopentenyl-cytosine **269**.

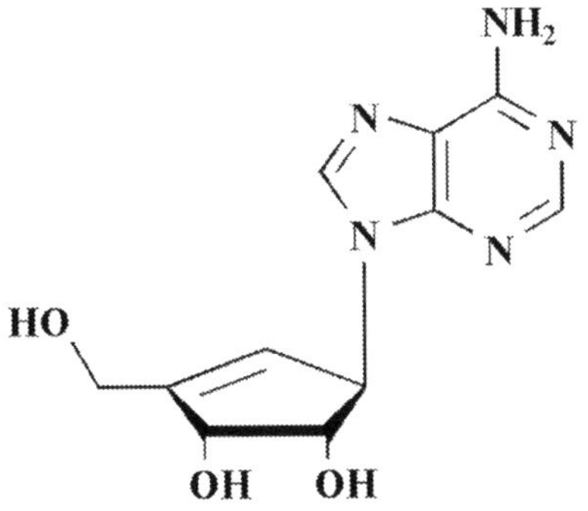

NH2 N N N N HO OH OH 270

Fig. 6.21 Neplanocin A (**270**).

inhibition of DNA and/or RNA polymerase in cancer cells and/or on the inhibition of *S*-adenosylhomocysteine (SAH) hydrolase. The adenine analog **260** and *N*6-methyladenine derivative **261** indicated an anticancer potency, indicating equipotent inhibitory potency as the positive control, neplanocin A (**270**) (Fig. 6.21), or Ara-C.

However, the phosphoramidate prodrug **268** revealed less anticancer potency than compound **260**, showing that it did not act as a RNA and/or DNA polymerase inhibitor. This result also elucidated that the anticancer potency of compound **260** relies on the inhibition of histone methyltransferase, resulting from strong inhibition of SAH hydrolase. The deamination of the *N*6-amino group, the introduction of the amino group at the C2 position, or the addition of the bulky alkyl group at the *N*6-amino group almost abolished the anticancer potency [207].

In order to determine if the anticancer potency depended on inhibiting DNA and/or RNA polymerase in cancer cells, the phosphoramidate prodrug **268** was prepared as accomplished in Scheme 6.40. The adenine analog **260** was protected as 2,3-acetonide **263**, which was treated with bis(tert-butoxycarbonyl)oxide (Boc_2O) to afford a mixture of mono-Boc- and di-Boc-adenine analogs **264** and **265**. In the presence of tert-butylmagnesium chloride, the mixture of **264** and **265** was treated with the phosphoramidate reagent (**266**) [207–209] and yielded the phosphoramidate analog **267**. Treating the latter with 50% HCOOH afforded the final phosphoramidate analog **268**.

Likewise, the anticancer potency of compound **261** is ascribed to the inhibition of SAH hydrolase ($IC_{50}=3.5\,\mu M$). Along these lines, it is

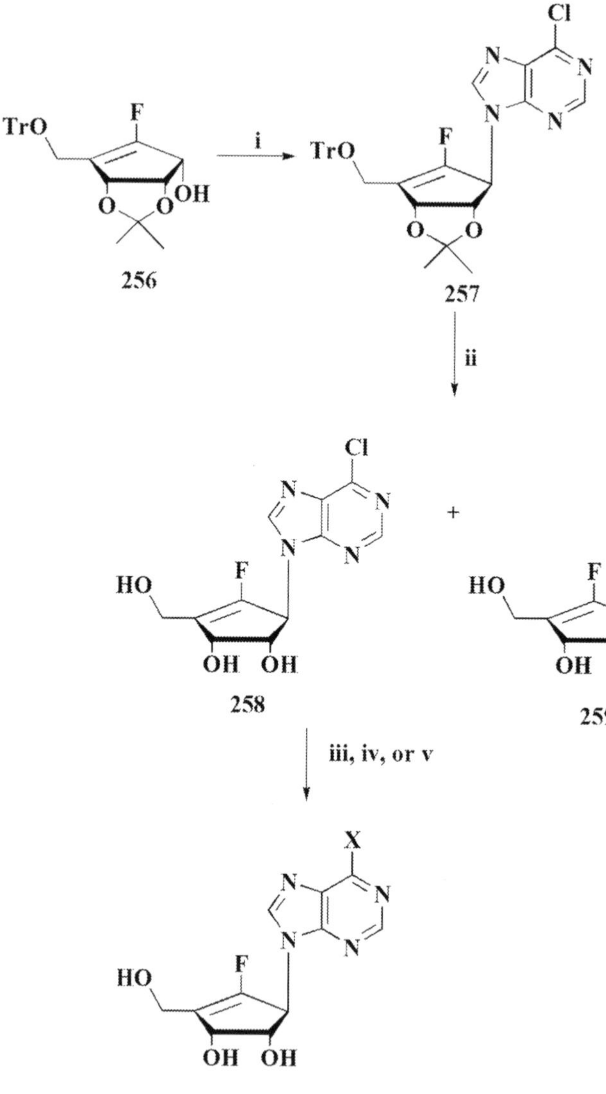

Cl N N N N TrO F OH O O 256 → i TrO F O O 257 → ii

HO F OH OH 258 + HO F OH OH O NH 259 → iii, iv, or v

X HO F OH OH

260, X= NH_2;
261, X = NHMe;
262, X= $NH_2CH(CH_2)_2$;

Scheme 6.39 Reagents and conditions: (i) 6-Chloro-9*H*-purine/PPh_3/DIAD/THF/0°C to RT, 15h; (ii) TFA/H_2O (1:1)/RT/15h; (iii) t-BuOH/NH_3/steel bomb/70–90°C/20h; (iv) CH_3NH_2/H_2O (40wt%), C_2H_5OH/sealed tube/RT/20h; (v) aminocyclopropane/TEA/C_2H_5OH/60–67°C/24h.

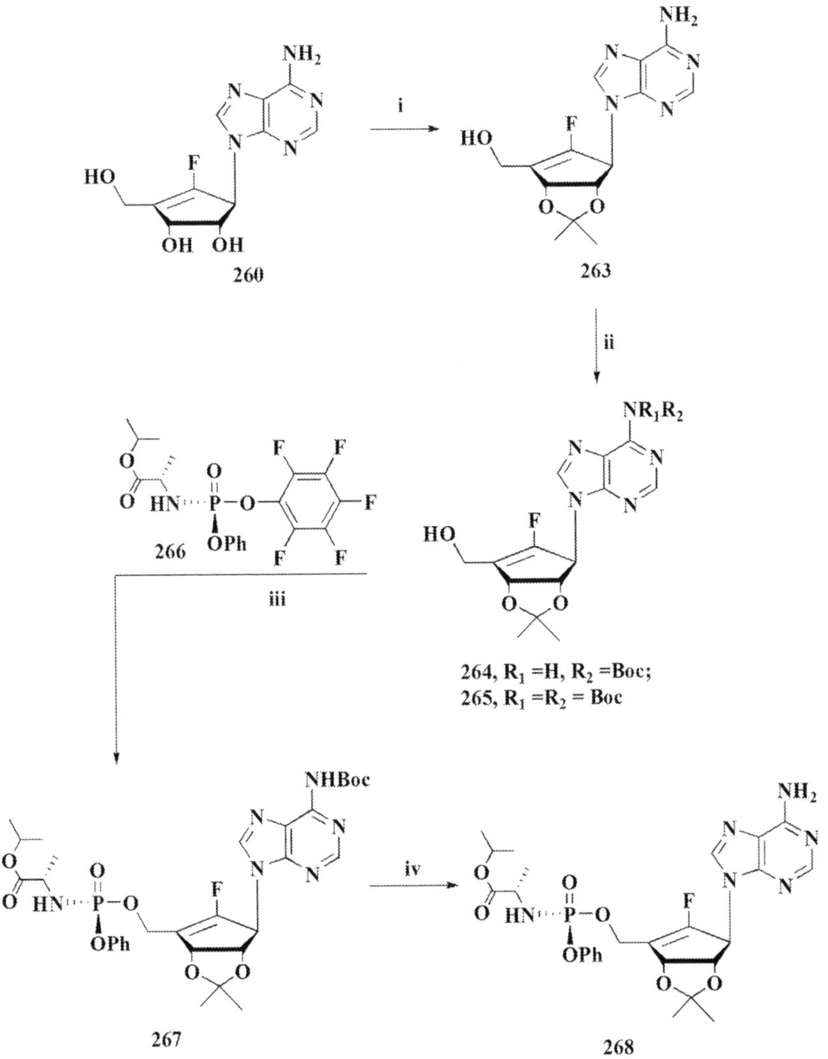

Scheme 6.40 Reagents and conditions: (i) Dimethyl ketone/concentrated H_2SO_4/6 h; (ii) (i) DMAP/HMDS/TMSOTf/75°C/2 h; (ii) Boc_2O/THF/RT/4 h; (iii) CH_3OH:TEA (5:1)/55°C/16 h; (iii) P,t-BuMgCl/molecular sieves/THF/0°C to RT/48 h; (iv) 50% HCOOH/RT/8 h.

proposed that the major pathway for the anticancer potency of compound **260** is the inhibition of histone methyltransferase, which is due to the strong inhibition (IC_{50} = 0.48 μM) of SAH hydrolase [207].

Using disk-diffusion assay [210,211], the derivatives **277** and **278** were evaluated in a series of tumor systems in vitro. The antitumor effect of the

2-fluoropurine Z-isomer **277a** exhibited some selectivity in comparison with normal fibroblast cells. There was little differentiation of the potency against solid tumors vs leukemia L1210. The E-isomer **278a** was cytotoxic across the board. Inactive potency of the chloro analogs **277b** and **278b** was observed.

2-Fluoro-, 2-chloro, and 2-oxopurine methylenecyclopropane nucleoside derivatives **277a**, **278a**, **277b**, and **278b** were prepared (Schemes 6.41 and 6.42). Fluoro analogs **277a** and **278a** are antitumor agents with limited selectivity [211].

Novel 1,2-di-substituted carbocyclic nucleosides with adenine, 6-chloropurine, and hypoxanthine bases were furnished by the construction of purine on the primary amino group of (±)-trans-2-aminocyclopentylmethanol.

Racemic compounds **281**, **282**, and **283** were prepared starting from (±)-trans-2-aminocyclopentyl-methanol (**279**), which was separated from

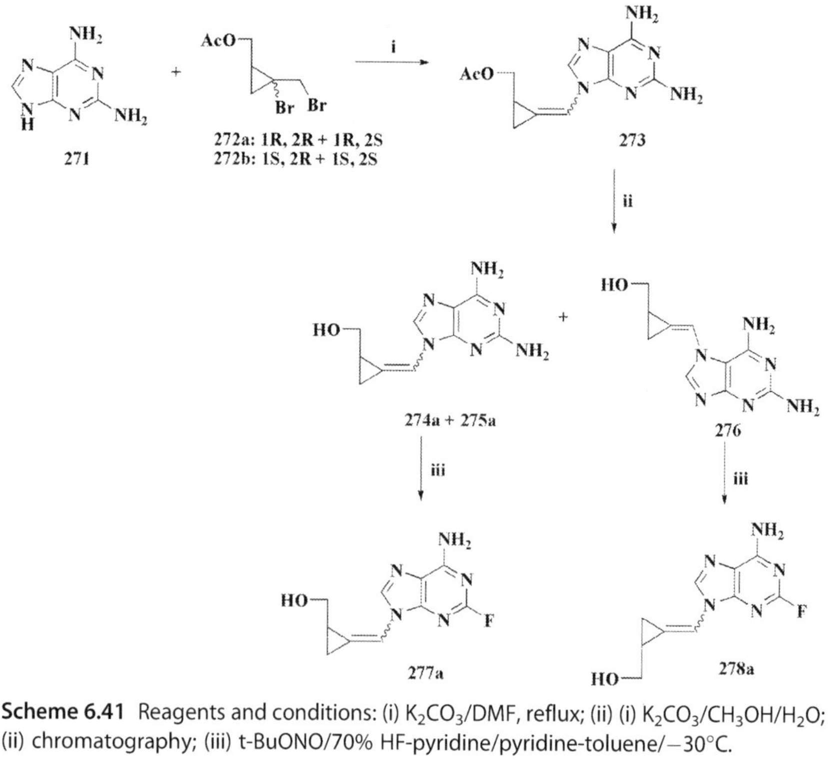

Scheme 6.41 Reagents and conditions: (i) K_2CO_3/DMF, reflux; (ii) (i) K_2CO_3/CH_3OH/H_2O; (ii) chromatography; (iii) t-BuONO/70% HF-pyridine/pyridine-toluene/−30°C.

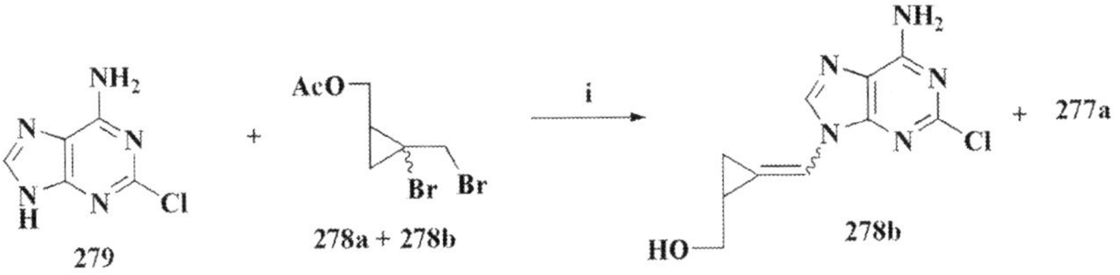

Scheme 6.42 Reagents and conditions: (i) (i) K_2CO_3/DMF, reflux; (ii) H_2O; (iii) chromatography.

a mixture of cis and trans isomers furnished through two steps from ethyl 2-oxocyclopentylcarboxylate. 5-Amino-4,6-dichloropyrimidine was condensed with the amine **279** to generate compound **280**. Ring closure with triethyl orthoformate in an acetic medium then yielded the 6-chloropurine **281**. Compound **281** was treated with methanolic ammonia to yield the adenine compound **283**, and with sodium hydroxide to access the hypoxanthine analog **282** (Scheme 6.43).

Assays comparing compounds **281–283** with Ara A revealed that the derivative **281** causes a 50% reduction in cell proliferation at concentrations of $39.2 \pm 8.0\,\mu M$ for Molt4/C8 human T-lymphocytes (Ar A: $11.9 \pm 7.3\,\mu M$), $56.1 \pm 3.0\,\mu M$ for murine leukemia cells L1210 (Ara A: $14.2 \pm 6.4\,\mu M$), and $19.0 \pm 4.2\,\mu M$ for CEM/0T cells (Ara A: $24.8 \pm 1.9\,\mu M$) [212].

Novel carbanucleosides were estimated with the model and examined for their antitumor activities on the proliferation of human T-lymphocyte cells (CEM/0 and Molt4/C8) and murine leukemia cells (L1210/0). The

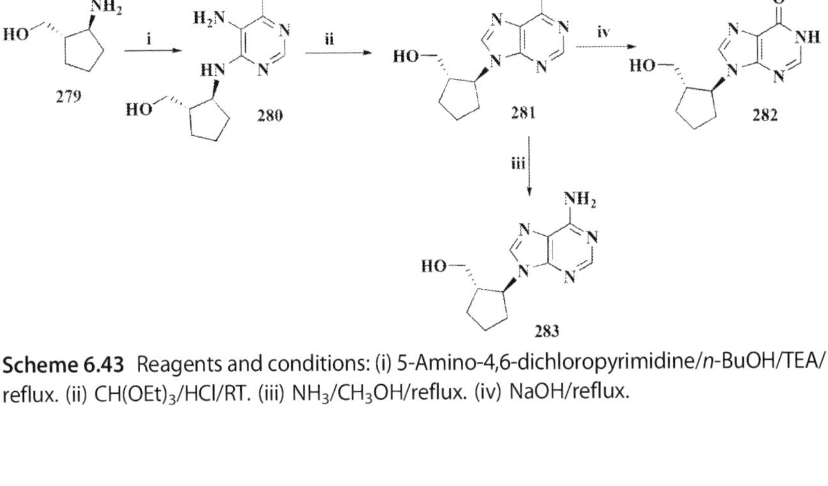

Scheme 6.43 Reagents and conditions: (i) 5-Amino-4,6-dichloropyrimidine/*n*-BuOH/TEA/reflux. (ii) $CH(OEt)_3$/HCl/RT. (iii) NH_3/CH_3OH/reflux. (iv) NaOH/reflux.

most interesting potency was identified for the compound **289a** (Scheme 6.44) (predicted probability = 80.2%; IC_{50} = 27.0, 27.2, and 29.4 lM), against the stated latter cellular lines as compared to the values of the Ara-A [213].

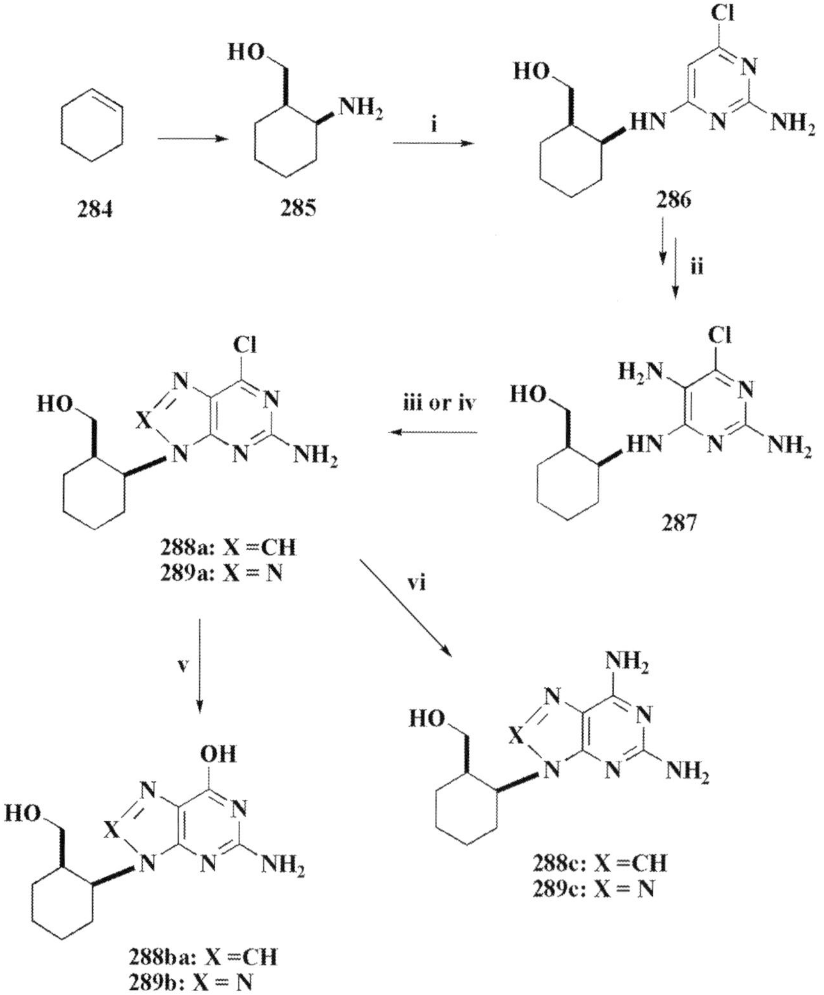

Scheme 6.44 Reagents and conditions: (i) 4,6-Dichloro-2-pyrimidinamine/TEA/n-BuOH/reflux/24 h; (ii) *p*-chloroaniline/$NaNO_2$/HCl/0°C; Zn/AcOH/C_2H_5OH/reflux/1 h; (iii) $CH(OEt)_3$/12 M HCl/reflux/12 h, (**288a**); (iv) $NaNO_2$/H_2O/AcOH, (**289a**); (v) 0.33 M NaOH/reflux/5 h/71% (288b), 92% (**289b**); (vi) NH_4OH/reflux (**288c** and **289c**).

The synthesis of the novel 6-substituted adenine nucleoside started from the intermediate **291** with the appropriate amines. The reaction of the 6-chloropurine intermediate **291** with sodium ethoxide in the presence of ethanol afforded the 6-ethoxypurine analog **295g** (Schemes 6.45 and 6.46) [214].

At a single high dose (10^{-5} M) in the full NCI 60 human tumor cell screen panel, the synthesized compounds were evaluated for their antitumor potency, and some of them showed a moderate activity.

The best results for antitumor potency were generated for compounds **292–293**, substituted at the 6 amino group with phenethyl **292e**, followed by the 6-dimethylaminopurine **293** derivative; a little lower potency was afforded for those substituted with 4-methylpiperazine **292h** and morpholine **292I** [214].

3.2.3 Glycoside purine analogs

Microwave-assisted synthesis of purine thioglycoside derivatives is outlined (Schemes 6.47 and 6.48). The main step of this syntheses is the synthesis of 7-mercaptopyrazolo[1,5-*a*]pyrimidine and sodium pyrazolo[1,5-*a*]pyrimidine-7-thiolate derivatives through the condensation of 2-(dimercaptomethylene)malononitrile or sodium 2,2-dicyanoethene-1,1-bis(thiolate) salts with

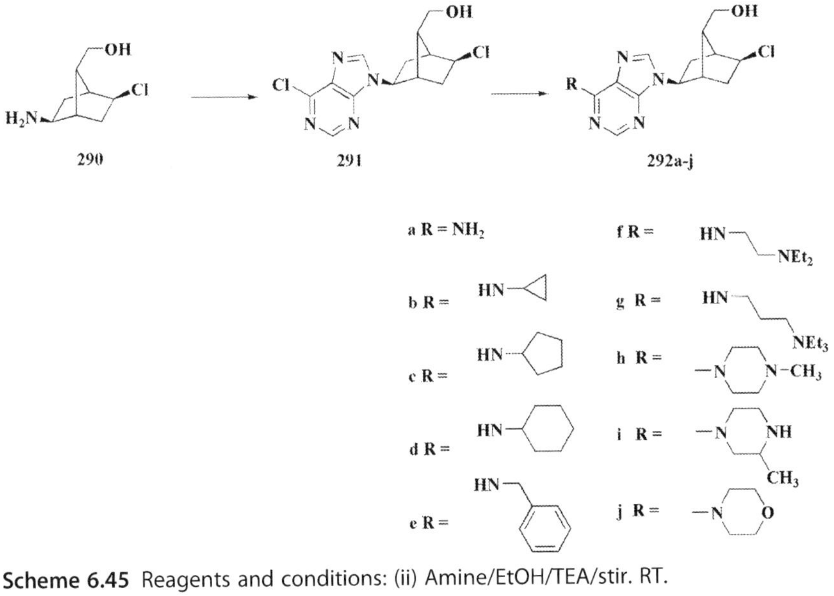

290 291 292a-j

a R = NH_2
b R = HN–cyclopropyl
c R = HN–cyclopentyl
d R = HN–cyclohexyl
e R = HN–benzyl
f R = HN–NEt_2
g R = HN–NEt_3
h R = –N(piperazine)N–CH_3
i R = –N(piperazine)NH, CH_3
j R = –N(morpholine)O

Scheme 6.45 Reagents and conditions: (ii) Amine/EtOH/TEA/stir. RT.

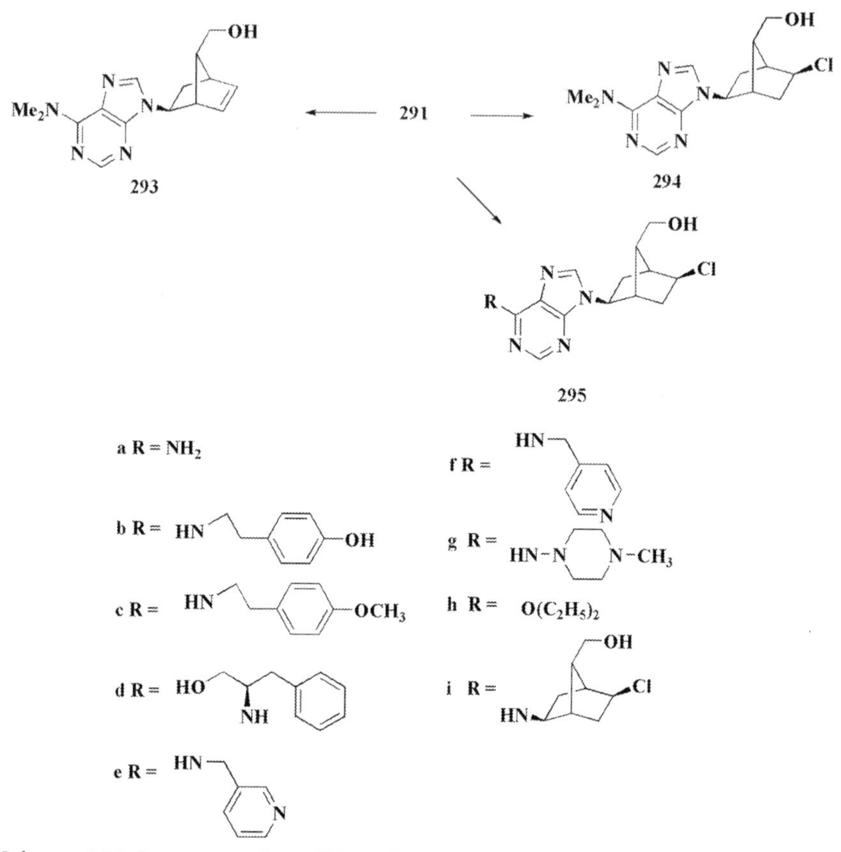

Scheme 6.46 Reagents and conditions: [215].

5-amino-1*H*-pyrazoles, using microwave irradiation. Subsequent coupling with halo sugars was carried out to afford the corresponding purine thioglycoside analogs. The targeted purines thioglycosides were estimated in vitro against colon (HCT116), lung (A549), prostate (PC3), and liver (HEPG2) cancer cell lines. Compounds (**300b, 300d, 300f**, and **9a–d**) displayed little activity toward the four cell lines. Compound **300a** displayed higher cytotoxicity on both colon (HCT116) and prostate (PC3), while it exhibited moderate potency on lung (A549) and inactivity on liver (HEPG2). In addition, compound **5c** displayed a moderate activity with LC_{50} values in range of 52.0–88.9 μM for almost all the cell lines [216].

The preparation of the *N*-glycosyl analogs comprising *N*-substituted glucuronamide scaffolds, as prospective mimetics of glycosyl phosphates

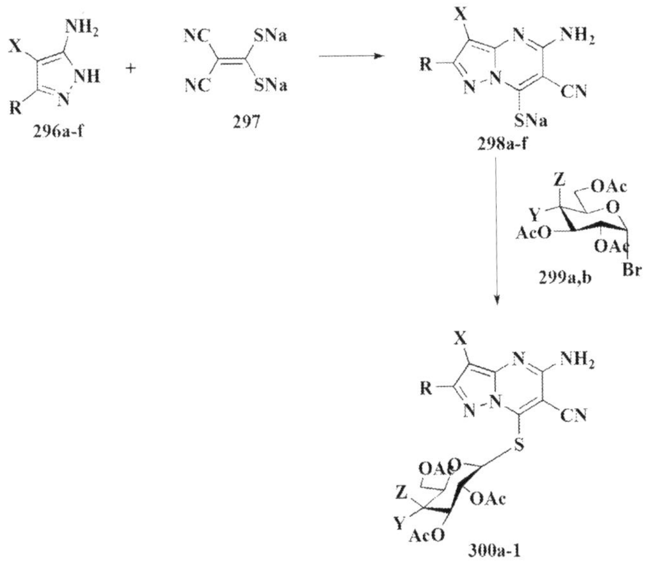

296, 298	X	R
a	CN	H
b	C_6H_5-N=N-	NH_2
c	4-MeO-C_6H_4-N=N-	NH_2
d	4-Me-C_6H_6-N=N-	NH_2
e	4-Br-C_6H_4-N=N-	NH_2
f	4-Br-C_6H_4-N=N-	NH_2

300	X	Y	Z	R
a	CN	OAc	H	H
b	C_6H_5-N=N-	OAc	H	NH_2
c	4-MeO-C_6H_4-N=N-	OAc	H	NH_2
d	4-Me-C_6H_6-N=N-	OAc	H	NH_2
e	4-Br-C_6H_4-N=N-	OAc	H	NH_2
f	4-Br-C_6H_4-N=N-	OAc	H	NH_2

300	X	Y	Z	R
g	CN	H	OAc	H
h	C_6H_5-N=N-	H	OAc	NH_2
i	4-MeO-C_6H_4-N=N-	H	OAc	NH_2
j	4-Me-C_6H_6-N=N-	H	OAc	NH_2
k	4-Br-C_6H_4-N=N-	H	OAc	NH_2
l	4-Br-C_6H_4-N=N-	H	OAc	NH_2

Scheme 6.47 Reagents and conditions: (i) MW/DMF /10 min. (ii) DMF/RT/8 h.

or of nucleotides, is accomplished in Scheme 6.49. These compounds contain *N*-anomerically-linked motifs that are surrogates of a phosphate group, as phosphoramidate or sulfonamide scaffolds. 1-Sulfonamido glucuronamides comprising *N*-benzyl, *N*-propargyl, or *N*-dodecyl carboxamide units were prepared via glycosylation of methanesulfonamide with tetra-*O*-acetyl glucuronamides. 1-Azido glucuronamides were furnished via microwave-assisted reactions of tetra-*O*-acetyl glucuronamides with trimethylsilyl azide and were subsequently transformed into *N*-glycosylphosphoramidates

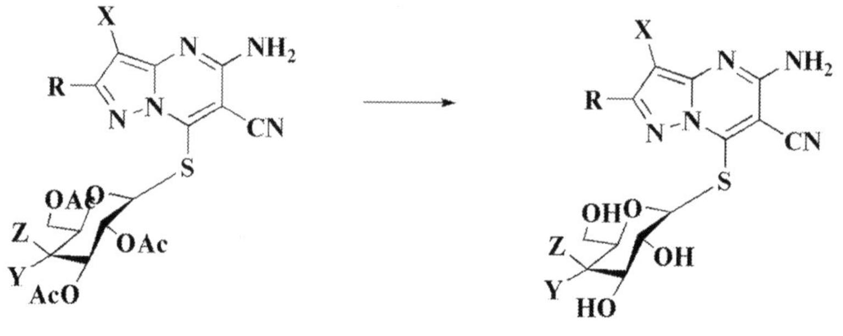

X
R
N
N
N
NH_2
CN
S
OAc O
Z
OAc
Y
AcO

X
R
N
N
N
NH_2
CN
S
OH O
Z
OH
Y
HO

300b, i

301a,b

300	X	Y	Z	R
b	C_6H_5-N=N-	OAc	H	NH_2
i	4-MeO-C_6H_4-N=N-	H	OAc	NH_2

301	X	Y	Z	R
a	C_6H_5-N=N-	OH	H	NH_2
b	4-MeO-C_6H_4-N=N-	H	OH	NH_2

Scheme 6.48 Reagents and conditions: (i) CH_3OH/NH_3/RT/10 min.

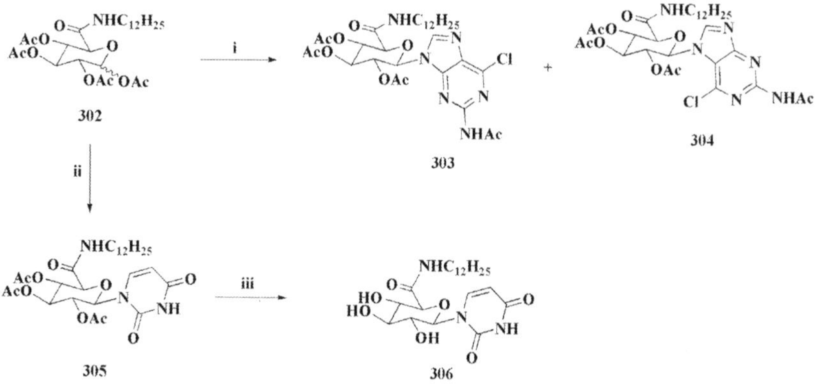

$NHC_{12}H_{25}$
AcO
AcO
O
OAc OAc
302

i

$NHC_{12}H_{25}$
AcO
AcO
O
OAc
N
N
Cl
N
N
NHAc
303

+

$NHC_{12}H_{25}$
AcO
AcO
O
OAc
N
N
N
Cl
N
NHAc
304

ii

$NHC_{12}H_{25}$
AcO
AcO
O
OAc
N
O
NH
O
305

iii

$NHC_{12}H_{25}$
HO
HO
O
OH
N
O
NH
O
306

Scheme 6.49 Reagents and conditions: (i) 2-NHAc-6-Cl-purine/TMSOTf/BSA/CH_3CN/65°C/MW, max. 150 W/40 min, 18% (303), 25% (304); (ii) Uracil/TMSOTf/BSA/CH_3CN/65°C/MW, max. 150 W, 40 min, 53%; (iii) NaOMe/CH_3OH/2 h, RT.

through the treatment with trimethyl phosphite. Potential glucuronamide-based nucleotide mimetics containing both a benzyltriazolylmethyl amide system at C-5, as nucleobase mimetics, and an anomeric sulfonamide/phosphoramidate group were prepared through the "click" cycloaddition of benzyl azide and N-propargyl glucuronamide derivatives. *N*-Dodecyl tetra-*O*-acetyl glucuronamides were transformed into purine and uracil nucleosides through *N*-glycosylation of the corresponding silylated nucleobases [217].

The purine nucleosides **303–304** displayed promising potencies against both cancer cell lines, strengthening the biological relevance of 2-acetamido-6-chloropurine nucleosides, as active anticancer agents [217–219]. The significant activity was identified for the N9-linked purine nucleoside **303** with GI_{50} values of 3.3 and 8.5 μM in MCF-7 and in K562 cells. The antiproliferative potency of compound **303** in MCF-7 cells like that of the clinically used 5-fluorouracil, while its GI_{50} value in K562 cells was about 12-fold more than that of imatinib. The *N*7-linked nucleoside **305** was two- to threefold less active than its *N*9 regioisomer toward the cancer cell lines. The *N*-dodecyl glucuronamide **302** did not reveal anticancer activities at concentrations less than 50 μL, revealing that the *N*-linked aglycon scaffolds are important for the potencies displayed via these *N*-glycosyl analogs. Additionally, the potent compounds exhibited in the majority of cases a slight selectivity to cancer cells than the nonmalignant fibroblasts. The purine *N*9-linked nucleoside **303** as well as the estimation of the impact of this compound on the potencies of caspases 3 and 7 indicated the induction of apoptosis in treated cells as the mechanism of cell death [217].

Novel purine nucleosides based on 6-chloropurine or 2-acetamido-6-chloropurine moieties attached to perbenzylated hexosyl (mannosyl, glucosyl, and galactosyl) scaffolds have been synthesized as accomplished in Scheme 6.50.

Their antitumor potential on lung, human melanoma, and colon adenocarcinoma and on ovarian carcinomas was investigated.

All compounds were examined in a sulforhodamine B (SRB) assay for their cytotoxicity and furnished micromolar GI_{50} values with order of magnitude comparable to structurally similar to cladribine. Additionally, the induction of apoptosis was demonstrated and cell cycle analysis was proceeded establishing a G2/M cell cycle arrest [218].

The nucleosides with the mannosyl moiety revealed the highest cytotoxic activity. Except compound **313**, a N7 CP nucleoside has GI_{50} values between 9 and 14 mM. The corresponding ACP analog **319** displayed GI_{50}

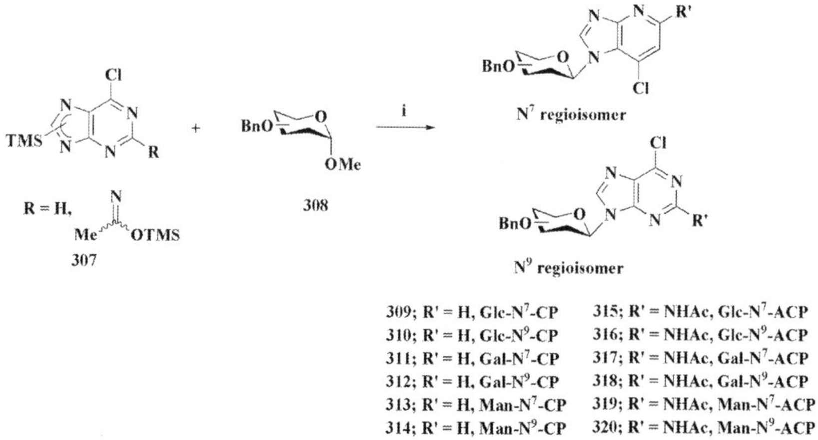

309; R' = H, Glc-N^7-CP
310; R' = H, Glc-N^9-CP
311; R' = H, Gal-N^7-CP
312; R' = H, Gal-N^9-CP
313; R' = H, Man-N^7-CP
314; R' = H, Man-N^9-CP
315; R' = NHAc, Glc-N^7-ACP
316; R' = NHAc, Glc-N^9-ACP
317; R' = NHAc, Gal-N^7-ACP
318; R' = NHAc, Gal-N^9-ACP
319; R' = NHAc, Man-N^7-ACP
320; R' = NHAc, Man-N^9-ACP

Scheme 6.50 Reagents and conditions: (i) TMSOTf/CH_3CN/65°C/2 h.

values of 1.4–2.2 mM and is the most active compound of this evaluation. This compound displayed a higher cytotoxicity than the antitumor drugs tamoxifen (GI_{50} from 7.6 to 9.7 mM [220]) or betulinic acid (GI_{50} from 11.0 to 14.9 mM [221,222]). There was no selectivity was determined when these substances were estimated on a nonmalignant cell line such as murine embryonic fibroblasts NiH 3T3 [218].

3.2.4 4-Azasteroidal purine analogs

The nucleophilic substitution of 6-chloro-purine with amines furnished novel C_6-amino-substituted 4-azasteroidal purine nucleoside analogs in high yields as described in Scheme 6.51 [223].

The synthesized novel compounds were examined for their anticancer potency in vitro against Hela, MCF-7, and PC-3 cell lines. Compounds **322**a,b,c showed an effective cytotoxicity with the IC_{50} values of 2.99 lM (PC-3), 2.84 lM (PC-3), and 2.69 lM (Hela) [223].

New *N*-glycoside derivatives with 4-azasteroid scaffold having sugar-like D ring were prepared through constructing the core dihydropyran ring embedded in 4-azasteroidal skeleton, which was synthesized from 4-aza-5a-androst-3,17-dione **323** as outlined in Scheme 6.52. Anticancer potency was observed for all of the analogs with purinyl moiety against human neuroblastoma (SK-N-SH), breast cancer (MCF-7), prostatic cancer (PC-3), and cervical cancer cell (HeLa). The study indicated the purinyl moiety linked to the pyran ring of **328a–d**, substituent at 6′-position of

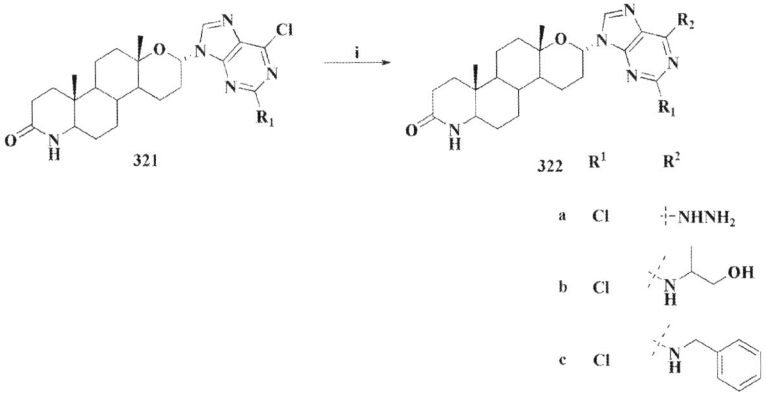

322	R[1]	R[2]
a	Cl	-NHNH$_2$
b	Cl	
c	Cl	

Scheme 6.51 Reagents and conditions: (i) Amines, Et_3N, MeOH, 5–20 min.

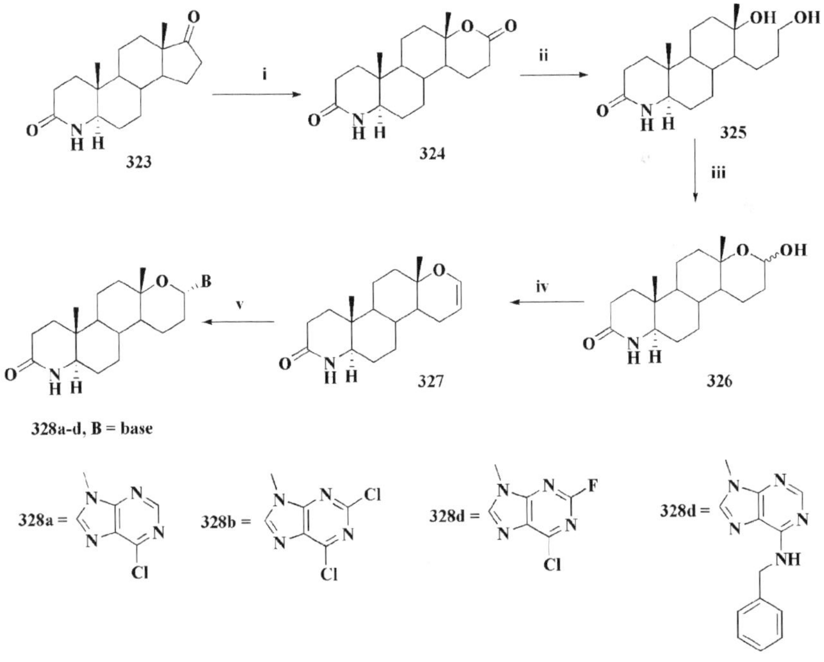

Scheme 6.52 Reagents and conditions: (i) Magnesium monoperoxyphthalate (MMPP)/CH_2Cl_2/H_2O/RT/96 h; (ii) $NaBH_4$/t-BuOH/MeOH/reflux/2 h; (iii) 2-iodoxybenzoic acid/DMSO/RT/2 h; (iv) mesyl/CH_2Cl_2/TEA/RT/20 min; (v) for **328a**, 6-chloro-9*H*-purine/TFA/EtOAc/reflux/24 h; for **328b**, 2,6-dichloro-9*H*-purine/TFA/EtOAc /reflux/24 h; for 6c, 6-chloro-2-fluoro-9*H*-purine/TFA/EtOAc/reflux/24 h; for **328d**, *N*-benzyl-9*H*-purin-6-amine/TFA/EtOAc/reflux/24 h; for 7,1,2,4-triazole/TFA/EtOAc/reflux/23 h.

purine base and introduction of a halogen atom at 2′-position of 6′-chloropurine, had an obvious impact on the estimated anticancer potency [224].

References

[1] Lansiaux A. Antimetabolites. Bull Cancer 2011;98(11):1263–74.

[2] Elgemeie GH, Mohamed RA. Microwave chemistry: synthesis of purine and pyrimidine nucleosides using microwave radiation. J Carbohydr Chem 2019;38(1):20–66.

[3] Luo L, He X, Shen Q, Li J, Shi X, Xie J, Li J, Chen G. Synthesis of (glycopyranosyltriazolyl)-purines and their inhibitory activities against protein tyrosine phosphatase 1B (PTP1B). Chem Biodivers 2011;8(11):2035–44.

[4] Cassera MB, Zhang Y, Hazleton KZ, Schramm VL. Purine and pyrimidine pathways as targets in plasmodium falciparum. Curr Top Med Chem 2011;11(16):2103–15.

[5] Marek R, Sklenář V. NMR studies of purines. Annu Rep NMR Spectrosc 2004;54:201–42.

[6] Xu Y, Narukulla R. Nucleotides and oligonucleotides: mononucleotides. Organophosphorus Chem 2014;43:117–45.

[7] Bergstrom DE. Unnatural nucleosides with unusual base pairing properties. Curr Protoc Nucleic Acid Chem 2009;37:1.4.1–1.4.32.

[8] Parker WB. Enzymology of purine and pyrimidine antimetabolites used in the treatment of cancer. Chem Rev 2009;109(7):2880–93.

[9] Parker WB, Secrist JA, Waud WR. Purine nucleoside antimetabolites in development for the treatment of cancer. Curr Opin Investig Drugs 2004;5(6):592–6.

[10] Huang L, Xu H, Yang Z, Zheng Y, Liu H. Synthesis and anticancer activity of novel C6-piperazine substituted purine steroid-nucleosides analogues. Steroids 2014;82:1–6.

[11] Galmarini CM, Popowycz F, Joseph B. Cytotoxic nucleoside analogues: different strategies to improve their clinical efficacy. Curr Med Chem 2008;15(11):1072–82.

[12] Kalinichenko. Synthesis of nucleosides 's analogues and their application as chemotherapeutic agents. Eurasian Chem Technol J 2013;15(3):189–94.

[13] Niitsu N. Possibility of the treatment for myelomonocytic leukemia by purine nucleoside analogs. Ketsueki Shuyoka 2004;49(4):383–90.

[14] Robak T. New purine nucleoside analogs for acute lymphoblastic leukemia. Clin Cancer Drugs 2014;1(1):2–10.

[15] Papayannidis C, Derenzini E, Iacobucci I, Curti A, Paolini S, Cilloni D, Baccarani M, Martinelli G. Successful combination treatment of clofarabine, cytarabine, and gemtuzumab-ozogamicin in adult refractory B-acute lymphoblastic leukemia. Am J Hematol 2009;84(12):849–50.

[16] Jabbour E, Short NJ, Ravandi F, Huang X, Xiao L, Garcia-Manero G, Plunkett W, Gandhi V, Sasaki K, Pemmaraju N, Daver NG, Borthakur G, Jain N, Konopleva M, Estrov Z, Kadia TM, Wierda WG, DiNardo CD, Brandt M, O'Brien SM, Cortes JE, Kantarjian H. A randomized phase 2 study of idarubicin and cytarabine with clofarabine or fludarabine in patients with newly diagnosed acute myeloid leukemia. Cancer 2017;123(22):4430–9.

[17] Robak T, Lech-Maranda E, Korycka A, Robak E. Purine nucleoside analogs as immunosuppressive and antineoplastic agents: mechanism of action and clinical activity. Curr Med Chem 2006;13(26):3165–89.

[18] Warrell RPJ, Berman E. Phase I and II study of fludarabine phosphate in leukemia: therapeutic efficacy with delayed central nervous system toxicity. J Clin Oncol 1986;4(1):74–9.
[19] Patel VF, Hardin JN, Mastro JM, Law KL, Zimmermann JL, Ehlhardt WJ, Woodland JM, Starling JJ. Novel acid labile COL1 trityl-linked difluoronucleoside immunoconjugates: synthesis, characterization, and biological activity. Bioconjug Chem 1996;7 (4):497–510.
[20] Ghias K, Ma C, Gandhi V, Platanias LC, Krett NL, Rosen ST. 8-Amino-adenosine induces loss of phosphorylation of p38 mitogen-activated protein kinase, extracellular signal-regulated kinase 1/2, and Akt kinase: role in induction of apoptosis in multiple myeloma. Mol Cancer Ther 2005;4(4):569–77.
[21] Korycka A, Blonski JZ, Robak T. Forodesine (BCX-1777, Immucillin H)—a new purine nucleoside analogue: mechanism of action and potential clinical application. Mini Rev Med Chem 2007;7(9):976–83.
[22] Robak T. New nucleoside analogs for patients with hematological malignancies. Expert Opin Investig Drugs 2011;20(3):343–59.
[23] Korycka A, Lech-Maranda E, Robak T. Novel purine nucleoside analogues for hematological malignancies. Front Anticancer Drug Discov 2010;1:219–40.
[24] Valton J, Guyot V, Marechal A, Filhol J, Juillerat A, Duclert A, Duchateau P, Poirot L. A multidrug-resistant engineered CAR T cell for allogeneic combination immunotherapy. Mol Ther 2015;23(9):1507–18.
[25] Lech-Maranda E, Korycka A, Robak T. Pharmacological and clinical studies on purine nucleoside analogs—new anticancer agents. Mini Rev Med Chem 2006;6 (5):575–81.
[26] Robak T, Korycka A, Kasznicki M, Wrzesien-Kus A, Smolewski P. Purine nucleoside analogues for the treatment of hematological malignancies: pharmacology and clinical applications. Curr Cancer Drug Targets 2005;5(6):421–44.
[27] Tobinai K. The role of purine analogs in the treatment of lymphoid malignancies and clinical trials in Japan. Ketsueki Shuyoka 2004;49(4):333–40.
[28] Berdis AJ. Chemotherapeutic intervention by inhibiting DNA polymerases. In: Kelley MR, editor. DNA repair in cancer therapy. Academic Press; 2012. p. 75–107.
[29] Allen DG, Barker MD, Cousins RPC. Preparation of purine nucleoside Analogs as Adenosine A2a Receptor agonists for the treatment of Inflammation. PCT Int. Appl.; 2007. WO 2007009757 A1 20070125.
[30] Pogosyan LG, Nersesova LS, Gazaryants MG, Mkrtchyan ZS, Meliksetyan GO, Akopyan ZI. Purine nucleoside phosphorylase inhibitors and clinical significance. Ukr Biokhim Zh 2008;80(5):95–104.
[31] Caciagli F, Ciccarelli R, Di Iorio P, Kleywegt S, Werstiuk ES, Rathbone MP, Vertes E. Purine nucleosides as anti-apoptotic agents for treating central nervous system diseases. PCT Int. Appl.; 2004. WO 2004022039 A2 20040318.
[32] Yamagiwa T, Shimosegawa T, Satoh A, Kimura K, Sakai Y, Masamune A. Inosine alleviates rat caerule in pancreatitis and pancreatitis-associated lung injury. J Gastroenterol 2004;39(1):41–9.
[33] Chung YG, Seth A, Doyle C, Tu DD, Estrada CR, Mauney JR, Adam RM, Franck D, Kim D, Benowitz LI, Tu DD, Estrada CR, Mauney JR, Sullivan MP, Adam RM. Inosine improves neurogenic detrusor overactivity following spinal cord injury. PLoS ONE 2015;10(11), e0141492.
[34] Fisher O, Benson RA, Imray CHE. The clinical application of purine nucleosides as biomarkers of tissue ischemia and hypoxia in humans in vivo. Biomark Med 2019;13 (11):953–65.

[35] Zimmerman MA, Kam I, Eltzschig H, Grenz A. Biological implications of extracellular adenosine in hepatic ischemia and reperfusion injury. Am J Transplant 2013;13 (10):2524–9.

[36] Sufrin JR, Rattendi D, Spiess AJ, Lane S, Marasco CJJ, Bacchi CJ. Antitrypanosomal activity of purine nucleosides can be enhanced by their conversion to O-acetylated derivatives. Antimicrob Agents Chemother 1996;40(11):2567–72.

[37] Tedder ME, Nie Z, Margosiak S, Chu S, Feher VA, Almassy R, Appelt K, Yager KM. Structure-based design, synthesis, and antimicrobial activity of purine derived SAH/ MTA nucleosidase inhibitors. Bioorg Med Chem Lett 2004;14(12):3165–8.

[38] Graci JD, Too K, Smidansky ED, Edathil JP, Barr EW, Harki DA, Galarraga JE, Bollinger JM, Peterson BR, Loakes D, Brown DM, Cameron CE. Lethal mutagenesis of picorna viruses with N-6-modified purine nucleoside analogues. Antimicrob Agents Chemother 2008;52(3):971–9.

[39] Gu Z, Wainberg MA, Nguyen-Ba N, L'Heureux L, De Muys J, Bowlin TL, Rando RF. Mechanism of action and in vitro activity of 1′,3′-dioxolanylpurine nucleoside analogs against sensitive and drug-resistant human immunodeficiency virus type 1 variants. Antimicrob Agents Chemother 1999;43(10):2376–82.

[40] Liu LJ, Kim E, Hong JH. Design and synthesis of novel threosyl-5′-deoxyphosphonic acid purine analogues as potent anti-HIV agents. Nucleosides Nucleotides Nucleic Acids 2012;31(5):411–22.

[41] Gonzalez-Moa MJ, Teijeira M, Teran C, Uriarte E, Pannecouque C, De Clercq E. Synthesis and anti-HIV activity of novel cyclopentenyl nucleoside analogues of 8-azapurine. Chem Pharm Bull 2006;54(10):1418–20.

[42] Wu P, Wan D, Xu G, Wang G, Ma H, Wang T, Gao Y, Qi J, Chen X, Zhu J, Li YQ, Deng Z, Chen W. An unusual protector-protege strategy for the biosynthesis of purine nucleoside antibiotics. Cell Chem Biol 2017;24(2):171–81.

[43] Xiang Y, Qian C, Xing G, Hao J, Xia M, Wang Y. Anti-herpes simplex virus efficacies of 2-aminobenzamide derivatives as novel HSP90 inhibitors. Bioorg Med Chem Lett 2012;22(14):4703–6.

[44] Meneghesso S, Vanderlinden E, Brancale A, Balzarini J, Naesens L, Mcguigan C. Synthesis and biological evaluation of purine 2′-fluoro-2′-deoxyriboside ProTides as anti-influenza virus agents. Chem Med Chem 2013;8(3):415–25.

[45] Chun B, Wang P, Hassan A, Du J, Tharnish PM, Murakami E, Stuyver L, Otto MJ, Schinazi RF, Watanabe KA. Synthesis and biological activity of 5′,9-anhydro-3-purine- isonucleosides as potential anti-hepatitis C virus agents. Nucleosides Nucleotides Nucleic Acids 2007;26(1):83–97.

[46] Yoo BN, Kim HO, Moon HR, Seol SK, Jang SK, Lee KM, Jeong LS. Synthesis of 2-C-hydroxymethylribofuranosylpuri
nes as potent anti-hepatitis C virus (HCV) agents. Bioorg Med Chem Lett 2006;16 (16):4190–4.

[47] Hwu JR, Lin S, Tsay S, De Clercq E, Leyssen P, Neyts J. Coumarin-purine ribofuranoside conjugates as new agents against hepatitis C virus. J Med Chem 2011;54 (7):2114–26.

[48] Shin YS, Jarhad DB, Jang MH, Kovacikova K, Kim G, Yoon JS, Kim HR, Hyun YE, Tipnis AS, Chang TS, van Hemert MJ, Jeong LS. Identification of 6′-β-Fluoro-homoaristeromycin as a potent inhibitor of chikungunya virus replication. Eur J Med Chem 2019;187, 111956.

[49] Klejch T, Keough DT, Chavchich M, Travis J, Skácel J, Pohl R, Janeba Z, Edstein MD, Avery VM, Guddat LW, Hocková D. Sulfide, sulfoxide and sulfone bridged acyclic nucleoside phosphonates as inhibitors of the *Plasmodium falciparum* and human 6-oxopurine phosphoribosyltransferases: synthesis and evaluation. Eur J Med Chem 2019;183, 111667.

[50] Ovadia R, Khalil A, Li H, De Schutter C, Mengshetti S, Zhou S, Bassit L, Coats SJ, Amblard F, Schinazi RF. Synthesis and anti-HCV activity of β-d-2′-deoxy-2′-α-chloro-2′-β-fluoro and β-d-2′-deoxy-2′-α-bromo-2′-β-fluoro nucleosides and their phosphoramidate prodrugs. Bioorg Med Chem 2019;27(4):664–76.
[51] Eletskaya BZ, Gruzdev DA, Krasnov VP, Levit GL, Kostromina MA, Paramonov AS, Kayushin AL, Muzyka IS, Muravyova TI, Esipov RS, Andronova VL, Galegov GA, Charushin VN, Miroshnikov AI, Konstantinova ID. Enzymatic synthesis of novel purine nucleosides bearing a chiral benzoxazine fragment. Chem Biol Drug Des 2019;93(4):605–16.
[52] Alexandre F, Rahali R, Rahali H, Guillon S, Convard T, Fillgrove K, Lai M, Meillon J, Xu M, Small J, Dousson CB, Raheem IT. Synthesis and antiviral evaluation of carbocyclic nucleoside analogs of nucleoside reverse transcriptase translocation inhibitor MK-8591 (4′-Ethynyl-2-fluoro-2′-deoxyadenosine). J Med Chem 2018;61(20):9218–28.
[53] Mitsuya H, Yamada K, Tomaya K, Ohno Y. 2'-Deoxy-7-deazapurine nucleoside derivative having antiviral activity. PCT Int. Appl.; 2018. WO 2018110591 A1 20180621.
[54] Andrei G, De Jonghe S, Groaz E, Herdewijn P, Luo M, Schols D, Snoeck R. Prodrugs of fluorinated acyclic nucleoside phosphonates. PCT Int. Appl.; 2018. WO 2018055071 A1 20180329.
[55] Wang G, Beigelman L, Deval J, Jekle CA. Preparation of substituted nucleosides, nucleotides and analogs thereof as antiviral agents. PCT Int. Appl.; 2018. WO 2018031818 A2 20180215.
[56] Chevrier F, Chamas Z, Lequeux T, Pfund E, Andrei G, Snoeck R, Roy V, Agrofoglio LA. Synthesis of 5,5-difluoro-5-phosphono-pent-2-en-1-yl nucleosides as potential antiviral agents. RSC Adv 2017;7(51):32282–7.
[57] Tichy M, Smolen S, Tloust'ova E, Pohl R, Ozdian T, Hejtmankova K, Liskova B, Gurska S, Dzubak P, Hajduch M, Hocek M. Synthesis and cytostatic and antiviral profiling of thieno-fused 7-deazapurine ribonucleosides. J Med Chem 2017;60(6):2411–24.
[58] Herdewijn P, De Jonghe S, Dumbre S, Liu C. Novel antiviral amino acid nucleoside phosphonate compounds, a process for their preparation, and their use for treating viral infections. PCT Int. Appl.; 2016. WO 2016174081 A1 20161103.
[59] Franchini S, Battisti UM, Sorbi C, Tait A, Cornia A, Jeong LS, Lee SK, Song J, Loddo R, Madeddu S, Sanna G, Brasili L. Synthesis, structural characterization and biological evaluation of 4'-C-methyl- and phenyl-dioxolane pyrimidine and purine nucleosides. Arch Pharm Res 2017;40(5):537–49.
[60] Hulpia F, Balzarini J, Schols D, Andrei G, Snoeck R, Van Calenbergh S. Exploring the purine core of 3'-C-ethynyladenosine (EAdo) in search of novel nucleoside therapeutics. Bioorg Med Chem Lett 2016;26(8):1970–2.
[61] Lin C, Sun C, Liu X, Zhou Y, Hussain M, Wan J, Li M, Li X, Jin R, Tu Z, Zhang J. Design, synthesis, and in vitro biological evaluation of novel 6-methyl-7-substituted-7-deaza purine nucleoside analogs as anti-influenza A agents. Antiviral Res 2016;129:13–20.
[62] McGuigan C, Serpi M, Slusarczyk M, Ferrari V, Pertusati F, Meneghesso S, Derudas M, Farleigh L, Zanetta P, Bugert J. Anti-flavivirus activity of different tritylated pyrimidine and purine nucleoside analogues. ChemistryOpen 2016;5(3):227–35.
[63] Zhou L, Zhang H, Tao S, Ehteshami M, Cho JH, McBrayer TR, Tharnish P, Whitaker T, Amblard F, Coats SJ, Schinazi RF. Synthesis and evaluation of 2,6-modified purine 2'-C-methyl ribonucleosides as inhibitors of HCV replication. ACS Med Chem Lett 2016;7(1):17–22.

[64] Naus P, Caletkova O, Perlikova P, Postova SL, Tloustova E, Hodek J, Weber J, Dzubak P, Hajduch M, Hocek M. Synthesis and biological profiling of 6- or 7-(het)aryl-7-deazapurine 4'-C-methylribonucleosides. Bioorg Med Chem 2015;23(23):7422–38.
[65] Schinazi F, Sivets GG, Detorio MA, McBrayer TR, Whitaker T, Coats SJ, Amblard F. Synthesis and antiviral evaluation of 2′,3′-dideoxy-2′,3′-difluoro-D-arabinofuranosyl 2,6-disubstituted purine nucleosides. Heterocycl Commun 2015;21(5):315–27.
[66] Dominguez-Perez B, Ferrer E, Figueredo M, Marechal J, Balzarini J, Alibes R, Busque F. Synthesis of novel nucleoside analogues built on a Bicyclo[4.1.0]heptane scaffold. J Org Chem 2015;80(19):9495–505.
[67] Solyev PN, Jasko MV, Kleymenova AA, Kukhanova MK, Kochetkov SN. Versatile synthesis of oxime-containing acyclic nucleoside phosphonates—synthetic solutions and antiviral activity. Org Biomol Chem 2015;13(44):10946–56.
[68] Khatri N, Lather V, Madan AK. Diverse models for anti-HIV activity of purine nucleoside analogs. Chem Cent J 2015;9:1–10.
[69] Niu H, Guo H, Wei T, Qu G, Wang D, Xie M, Chen J, Feng X, Hu L, Zhang J. Synthesis method of chiral noncyclic purine nucleoside analogue as antivirus drug. Faming Zhuanli Shenqing; 2015. CN 104557936 A 20150429.
[70] Perez-Toro I, Dominguez-Martin A, Choquesillo-Lazarte D, Vilchez-Rodriguez E, Gonzalez-Perez JM, Castineiras A, Niclos-Gutierrez J. Lights and shadows in the challenge of binding acyclovir, a synthetic purine-like nucleoside with antiviral activity, at an apical-distal coordination site in copper(II)-polyamine chelates. J Inorg Biochem 2015;148:84–92.
[71] Dejmek M, Sala M, Hrebabecky H, Dracinsky M, Procházková E, Chalupská D, Klíma M, Plačková P, Hájek M, Andrei G, Naesens L, Leyssen P, Neyts J, Balzarini J, Boura E, Nencka R. Norbornane-based nucleoside and nucleotide analogues locked in North conformation. Bioorg Med Chem 2015;23(1):184–91.
[72] Cho JH, Bondana L, Detorio MA, Montero C, Bassit LC, Amblard F, Coats SJ, Schinazi RF. Synthesis and antiviral evaluation of 2-amino-6-carbamoylpurine dioxolane nucleoside derivatives and their phosphoramidates prodrugs. Bioorg Med Chem 2014;22(23):6665–71.
[73] Novosjolova I, Barzdaine M, Mandrika I, Petrovska R, Klovins J, Bizdena E, Turks M. Synthesis of triazolylpurine nucleosides as antiviral agents. Latv; 2014. LV 14852 B 20140720.
[74] Or YS, Ma J, Wang G. Preparation of macrocyclic nucleoside amino acid phosphoramidate derivatives via ring closing olefin metathesis as antiviral agents. PCT Int. Appl.; 2013. WO 2013173759 A2 20131121.
[75] Quadrelli P, Mella M, Legnani L, Al-Saad D. From Cyclopentadiene to Isoxazoline-carbocyclic Nucleosides; synthesis of highly active inhibitors of influenza a virus H1N1. Eur J Org Chem 2013;2013(21):4655–65.
[76] Di Francesco ME, Avolio S, Pompei M, Pesci S, Monteagudo E, Pucci V, Giuliano C, Fiore F, Rowley M, Summa V. Synthesis and antiviral properties of novel 7-heterocyclic substituted 7-deaza-adenine nucleoside inhibitors of hepatitis C NS5B polymerase. Bioorg Med Chem 2012;20(15):4801–11.
[77] McGuigan C, Madela K, Bourdin C, Vernachio J, Chamberlain S. Preparation of amino acid-substituted purine nucleoside phosphoramidate and phosphorodiamidate derivatives as antiviral agents. PCT Int. Appl.; 2012. WO 2012092484 A2 20120705.
[78] Chang W, Naduthambi D, Nagarathnam D, Pamulapati GR, Ross BS, Sofia MJ, Zhang H. Preparation of purine nucleoside phosphoramidate as antiviral agent. PCT Int. Appl.; 2011. WO 2011123672 A1 20111006.
[79] Yang X, Niu H, Yuan T, Wang D, Guo H, Qu G, Xia R, Wang F, Li T, Ma G, et al. Method for preparation of chiral non-cycle purine nucleoside derivatives with antiviral activity. Faming Zhuanli Shenqing; 2011. CN 102161661 A 20110824.

[80] Amblard F, Fromentin E, Detorio M, Obikhod A, Rapp KL, McBrayer TR, Whitaker T, Coats SJ, Schinazi RF. Synthesis, antiviral activity, and stability of nucleoside analogs containing tricyclic bases. Eur J Med Chem 2009;44(10):3845–51.
[81] Gunic E, Hong Z, Girardet J. Preparation of 3′,5′-cyclic nucleoside analogs as antiviral agents for treatment of HCV. PCT Int. Appl.; 2007. WO 2007027248 A2 20070308.
[82] Fernndez-Cureses G, de Castro S, Jimeno M, Balzarini J, Camarasa M. Design, synthesis, and biological evaluation of unconventional aminopyrimidine, aminopurine, and amino-1,3,5-triazine methyloxynucleosides. Chem Med Chem 2015;10 (2):321–35.
[83] Kallitsakis MG, Yaez M, Soriano E, Marco-Contelles J, Hadjipavlou-Litina DJ, Litinas KE. Purine homo-*N*-nucleoside+coumarin hybrids as pleiotropic agents for the potential treatment of Alzheimer's disease. Future Med Chem 2015;7(2):103–10.
[84] Marcelo F, Silva FVM, Goulart M, Justino J, Sinay P, Bleriot Y, Rauter AP. Synthesis of novel purine nucleosides towards a selective inhibition of human butyrylcholinesterase. Bioorg Med Chem 2009;17(14):5106–16.
[85] Tumini E, Porcellini E, Chiappelli M, Conti CM, Beraudi A, Poli A, Caciagli F, Doyle R, Conti P, Licastro F. The G51S purine nucleoside phosphorylase polymorphism is associated with cognitive decline in Alzheimer's disease patients. Hum Psychopharmacol 2007;22(2):75–80.
[86] Schwarz S, Csuk R, Rauter AP. Microwave-assisted synthesis of novel purine nucleosides as selective cholinesterase inhibitors. Org Biomol Chem 2014;12(15):2446–56.
[87] Seela F, Schweinberger E, Xu K, Sirivolu VR, Rosemeyer H, Becker E. 1,*N*6-Etheno-2′-deoxytubercidin and pyrrolo-C: synthesis, base pairing, and fluorescence properties of 7-deazapurine nucleosides and oligonucleotides. Tetrahedron 2007;63 (17):3471–82.
[88] Stachelska-Wierzchowska A, Wierzchowski J, Wielgus-Kutrowska B, Mikleusevic G. Enzymatic synthesis of highly fluorescent 8-azapurine ribosides using a purine nucleoside phosphorylase reverse reaction: variable ribosylation sites. Molecules 2013;18(10):12587–98.
[89] Nudelman A, Rephaeli A. Novel derivatives of purinic and pyrimidinic antiviral agents and use thereof as potent anticancer agents. PCT Int. Appl.; 2008. WO 2008120205 A1 20081009.
[90] Dong S, Wuts PGM. Preparation of nucleoside phosphonates and analogs as antiviral and antitumor agents. PCT Int. Appl.; 2011. WO 2011031567 A1 20110317.
[91] Cho JH, Coats SJ, Schinazi RF, Zhang H, Zhou L. Purine nucleoside monophosphate prodrugs for treatment of cancer and viral infections. PCT Int. Appl.; 2010. WO 2010091386 A2 20100812.
[92] Weis AL, Goodhue CT. Preparation of L-pyranosyl nucleosides for treating various diseases in mammals. PCT Int. Appl.; 1996. WO 9612728 A1 19960502.
[93] Krenitsky TA, Averett DR, Wilson JD, Moorman AR, Koszalka GW, Chamberlain SD, Porter D, Wolberg G. Preparation of 2-amino-6-alkoxy-9-(β-D-arabinofuranosyl)-9H-purines and esters as antitumor agents. PCT Int. Appl.; 1992. WO 9201456 A1 19920206.
[94] Anon. Purine nucleosides. Jpn. Kokai Tokkyo Koho; 1980. JP 55164698 A 19801222.
[95] Jeong LS, Jacobson KA, Moon HR, Kim HO. Preparation of purine nucleosides as prodrugs for treating asthma, inflammation, cerebral ischemia, heart diseases, and cancer. PCT Int. Appl.; 2004. WO 2004038006 A2 20040506.
[96] Hocek M, Naus P. Preparation of cytostatic 7-deazapurine nucleosides as antitumor agents. PCT Int. Appl.; 2009. WO 2009089804 A1 20090723.
[97] Vince R, Hua M. Preparation of optically active carbocyclic purine 2′,3′-didehydro-2′,3′-dideoxynucleosides and their analogs as antiviral and antitumor agents. U.S; 1990. US 4931559 A 19900605.

[98] Wang G, Girardet J, Gunic E, Lau J, Pietrzkowski Z, Hong Z. Preparation of pyrido [2,3-d]pyrimidine and pyrimido[4,5-d]pyrimidine nucleosides as antitumor agents and RNA polymerase inhibitors. PCT Int. Appl.; 2002. WO 2002003997 A1 20020117.
[99] Torii T, Onishi T, Izawa K. Preparation of fluorinated purine nucleoside derivatives and their intermediates. Jpn. Kokai Tokkyo Koho; 2006. JP 2006052182 A 20060223.
[100] Chesworth R, Kuntz KW. Preparation of substituted 7-deazapurine nucleosides as DOT1 enzyme inhibitors and antitumor agents. PCT Int. Appl.; 2014. WO 2014152261 A1 20140925.
[101] Babu YS, Chand P. Preparation of tetrahydrofuro-[3,4-*d*]-dioxolane nucleosides for use in the treatment of viral infections and cancer. PCT Int. Appl.; 2008. WO 2008141079 A1 20081120.
[102] Wu Z. Preparation of quinolyl or isoquinolyl substituted purine derivatives as antitumor agents. PCT Int. Appl.; 2006. WO 2006133611 A1 20061221.
[103] Huang L, Liu H, Wang Y, Zheng Y. 4-Aza-steroid-purine nucleoside analog useful in treatment of cancer and its preparation. Faming Zhuanli Shenqing; 2014. CN 103483414 A 20140101.
[104] Guo H, Qu G, Wu J, Wang D, Yang X, Zhang X, Niu H, Xia R, Wang F, Cai Y, et al. 6-Monoazacrown ether modified purine nucleoside compounds with antitumor activity and the preparation method thereof. Faming Zhuanli Shenqing; 2009. CN 101544641 A 20090930.
[105] Huang L, Wu Z, Zhao J, Xu H. Steroidal purine nucleoside analogs containing 1,2,3-triazoles useful in treatment of cancer and its preparation. Faming Zhuanli Shenqing; 2017. CN 107236016 A 20171010.
[106] Cho JH, Coats SJ, Schinazi RF, Zhang H, Zhou L. Purine nucleoside monophosphate prodrugs for treatment of cancer and viral infections. U.S. Pat. Appl. Publ.; 2012. US 20120040924 A1 20120216.
[107] Bourderioux A, Hocek M, Naus P. Novel 7-deazapurine nucleosides for therapeutic uses. U.S. Pat. Appl. Publ.; 2015. US 20150218201 A1 20150806.
[108] Konstantinova ID, Eletskaya BZ, Dorofeeva EV, Berzina MY FIV, Lutonina OI, Miroshnikov AI. Method for obtaining 9-(beta-D-arabinofuranosyl)-6-(Nα-L-serylamido)-2-chloropurine. Russ; 2015. RU 2563257 C1 20150920.
[109] Zhang S, Han M, Gao Z, Cheng X, Zhao R. Antitumor agent prepared from deoxy purine nucleoside combined with other nucleoside or base, and preparation method and application thereof. Faming Zhuanli Shenqing; 2015. CN 104622887 A 20150520.
[110] Pozzoli CG, Canevari V, Brusasca M, Menna L, Curti M. Synthesis of antitumor and antiviral nucleosides via coupling reaction. U.S. Pat. Appl. Publ.; 2014. US 20140135490 A1 20140515.
[111] Jin G, Jiang W, Gao D, Wang Z. Synthesis of purine compd. conjugates and prepns. as immunomodulator and for prodn. of antibody, antiviral, antidiabetic, antitumor and other biopharmaceutical. Faming Zhuanli Shenqing; 2013. CN 103467590 A 20131225.
[112] Weinberg JB, Friedman DR, Cianciolo GJ, Rizzieri DA. Method of inducing apoptosis/cell death in leukemia cell using a purine nucleoside analogue. U.S. Pat. Appl. Publ.; 2011. US 20110319360 A1 20111229.
[113] Bourderioux A, Hocek M, Naus P. Novel 7-deazapurine nucleosides for therapeutic uses. PCT Int. Appl.; 2010. WO 2010121576 A2 20101028.
[114] Claiborne CF, Critchley S, Langston SP, Olhava EJ, Peluso S, Weatherhead GS, Vyskocil S, Visiers I, Mizutani H, Cullis C. Preparation of carbocyclic purine nucleoside analogs as antitumor agents and inhibitors of E1 activating enzymes. PCT Int. Appl.; 2008. WO 2008019124 A1 20080214.

[115] Langston SP, Olhava EJ, Vyskocil S. Preparation of purine nucleoside derivatives as antitumor agents and inhibitors of E1 activating enzymes. PCT Int. Appl.; 2007. WO 2007092213 A2 20070816.
[116] Trampota M, Murphy RB. Preparation of cubane nucleoside analogs as antiviral and antitumor agents. PCT Int. Appl.; 2007. WO 2007059330 A2 20070524.
[117] Ealick SE, Parker WB, Secrist III JA, Sorscher EJ. Mutant purine nucleoside phosphorylase proteins and cellular delivery thereof. U.S. Pat. Appl. Publ.; 2005. US 20050214901 A1 20050929.
[118] Feingold JM. Anti-CD33 cytotoxic conjugate combination with anthracycline or pyrimidine or purine nucleoside analog for the treatment of acute leukemia and myelodysplastic syndrome. PCT Int. Appl.; 2004. WO 2004043461 A1 20040527.
[119] Gupta PK, Munk SA. Process for the regioselective and stereoselective preparation of 9-β-anomeric nucleoside analogs as antitumor and antiviral prodrugs. U.S. Pat. Appl. Publ.; 2004. US 20040039190 A1 20040226.
[120] An H, Gunic E, Koh Y, Chen H, Barawkar D, Zhang W, Girardet J, Rong F, Hong Z. Solid phase combinatorial synthesis of substituted purine nucleoside libraries as potential antiviral and antitumor agents. PCT Int. Appl.; 2003. WO 2003051881 A1 20030626.
[121] Morris PE, Montgomery JA, Babu YS. Preparation of imminoribitol nucleosides as antitumor agents and purine nucleoside phosphorylase (PNP) inhibitors. U.S. Pat. Appl. Publ.; 2001. US 20010053784 A1 20011220.
[122] Weis AL, Pulenthiran K, Gero AM. Preparation of nucleoside analogs as parasiticides and antitumor agents. U.S; 2001. US 6242428 B1 20010605.
[123] Watanabe KA, Choi W. Synthesis of 2′-deoxy-L- nucleosides as antiviral and antitumor agents. PCT Int. Appl.; 2001. WO 2001034618 A2 20010517.
[124] Jiang J, Lee-Ruff E, Wan W. Preparation of optically active nucleosides from chiral cyclobutanones as antitumor and antiviral agents. Can. Pat. Appl.; 1995. CA 2113065 A1 19950708.
[125] Hamamoto T, Miyashita T. Manufacture of purine arabinosides with purine- nucleoside phosphorylase and uridine phosphorylase. Jpn. Kokai Tokkyo Koho; 1998. JP 10286097 A 19981027.
[126] Saneyoshi M, Eita T, Ishisaki K, Yokoyama Y, Yoshida N. Preparation of 3′-amino-3′-deoxypurine ribonucleosides as anticancer drugs. Jpn. Kokai Tokkyo Koho; 1998. JP 10007695 A 19980113.
[127] Haraguchi K, Tanaka H, Myasaka S. Method for producing 1′-carbon chain-substituted purine nucleoside derivatives. Jpn. Kokai Tokkyo Koho; 1996. JP 08301895 A 19961119.
[128] Izawa T, Kato K. Preparation of 2,3-bis(hydroxymethyl)azetidine derivatives as antiviral and anticancer agents and intermediates thereof. Jpn. Kokai Tokkyo Koho; 1995. JP 07242668 A 19950919.
[129] Sorscher EJ, Parker WB, Bennett LLJ. Purine nucleoside phosphorylase gene therapy for human malignancy. PCT Int. Appl.; 1995. WO 9507718 A2 19950323.
[130] Grindley GB, Grossman CS, Hertel LW, Kroin JS. 2′-deoxy-2′,2′-difluoro-(2,6,8--substituted) purine nucleosides having anti- viral and anti-cancer activity and intermediates. Eur. Pat. Appl.; 1993. EP 576227 A2 19931229.
[131] Sugimura H, Yamada K, Nagahata T, Narita K, Shiozawa A. Preparation of [tris(hydroxymethyl)cyclobutyl]purine and -pyrimidine nucleoside derivatives as antiviral and anticancer agents. Jpn. Kokai Tokkyo Koho; 1993. JP 05001042 A 19930108.
[132] Ichikawa Y, Sugawara Y, Sugimura H, Narita K, Shiozawa A. Preparation of 3-phosphonoalkyl-1-cyclobutyl purine and pyrimidine derivatives as antitumor and antiviral agents. Jpn. Kokai Tokkyo Koho; 1992. JP 04091094 A 19920324.

[133] Vince R, Hua M. Preparation of optically active isomers of dideoxycarbocyclic nucleosides as antiviral and antitumor agents. U.S; 1990. US 4950758 A 19900821.
[134] Yamamura S, Nishiyama S, Ogiya T, Kato K, Minami T, Takita T. Preparation of 3-hydroxy-4-(hydroxymethyl)-erythro-oxetany purine or pyrimidine nucleoside derivatives and their preparation. Jpn. Kokai Tokkyo Koho; 1990. JP 02209886 A 19900821.
[135] Morisawa Y, Nakayama T, Yasuda A, Uchida K, Sasaki T. 3'-Deoxy-3′-fluoropurine nucleoside derivatives as antitumor agents. Jpn. Kokai Tokkyo Koho; 1989. JP 01029316 A 19890131.
[136] Marquez VE, Driscoll JS. Preparation of cyclopentenylpyrimidine and –purine derivatives as antiviral and antitumor agents. U. S. Pat. Appl.; 1989. US 307115 A0 19890715.
[137] Amundsen AR, Stern EW, Hollis LS. Platinum triamine complexes with nucleotides or nucleosides as antitumor agents. Eur. Pat. Appl.; 1989. EP 303437 A2 19890215.
[138] Takeda Y, Mizutani T, Ueno A, Hirose K, Tanahashi E, Nishikawa S. Preparation of 5′-deoxy-5′-fluoromethylthio ribonucleosides and antitumor agents containing them. Jpn. Kokai Tokkyo Koho; 1988. JP 63215692 A 19880908.
[139] Iigo M, Miwa M, Nitta K. Antitumor composition. Eur. Pat. Appl.; 1986. EP 189755 A1 19860806.
[140] West CR, Hong C. Cytotoxic nucleoside-corticosteroid phosphodiesters. PCT Int. Appl.; 1981. WO 8100410 A1 19810219.
[141] Ueda T. 6-Dicyanomethylpurine nucleoside. Jpn. Kokai Tokkyo Koho; 1980. JP 55033434 A 19800308.
[142] Hassan AEA, Abou-Elkhair RAI, Parker WB, Allan PW, Secrist III JA. 6-methylpurine derived sugar modified nucleosides: synthesis and *in vivo* antitumor activity in D54 tumor expressing M64V-*Escherichia coli* purine nucleoside phosphorylase. Eur J Med Chem 2016;108:616–22.
[143] Hong JS, Waud WR, Levasseur DN, Townes TM, Wen H, McPherson SA, Moore BA, Bebok Z, Allan PW, Secrist JA, Parker WB, Sorscher EJ. Excellent in vivo bystander activity of fludarabine phosphate against human glioma xenografts that express the Escherichia coli purine nucleoside phosphorylase gene. Cancer Res 2004;64:6610–5.
[144] Silamkoti AV, Allan PW, Hassan AEA, Fowler AT, Sorscher EJ, Parker WB, Secrist JA. Synthesis and biological activity of 2-fluoro adenine and 6-methyl purine nucleoside analogs as prodrugs for suicide gene therapy of cancer. Nucleosides Nucleotides Nucleic Acids 2005;24:881–5.
[145] Parker WB, Allan PW, Ealick SE, Sorscher EJ, Hassan AEA, Silamkoti AV, Fowler AT, Waud WR, Secrist JA. Design and evaluation of 5 ’-modified nucleoside analogs as prodrugs for an *E. coli* purine nucleoside phosphorylase mutant. Nucleosides Nucleotides Nucleic Acids 2005;24:387–92.
[146] Ellermann M, Paulini R, Jakob-Roetne R, Lerner C, Borroni E, Roth D, Ehler A, Schweizer WB, Schlatter D, Rudolph MG, Diederich F. Molecular recognition at the active site of catechol-*O*-methyltransferase: adenine replacements in bisubstrate inhibitors. Chem A Eur J 2011;17:6369–81.
[147] Hocek M, Silhar P, Shih IH, Mabery E, Mackman R. Cytostatic and antiviral 6-arylpurine ribonucleosides. Part 7: synthesis and evaluation of 6-substituted purine L-ribonucleosides. Bioorg Med Chem Lett 2006;16:5290–3.
[148] Marasco CJ, Pera PJ, Spiess AJ, Bernacki R, Sufrin JR. Improved synthesis of β-D-6-methylpurine riboside and antitumor effects of the β-D- and α-D-anomers. Molecules 2005;10:1015–20.
[149] Whistler RL, Lake WC. 5-Thio-α-D-glucopyranose via conversion of a terminal oxirane ring to a terminal thiirane ring. Methods Carbohyd Chem 1972;6:286–91. Thiosugars [50].

[150] Yoshikawa M, Okaichi Y, Cha BC, Kitagawa I. Synthesis of (−)-aristeromycin from D-glucose. Tetrahedron 1990;46:7459–70.
[151] Guthrie RD, Smith SC. An improved preparation of 1,2,3,5-tetra-O-acetyl-β-D-ribofuranose. Chem Ind 1968;17:547–8.
[152] Migawa MT, Girardet JL, Walker JA, Koszalka GW, Chamberlain SD, Drach JC, Townsend LB. Design, synthesis, and antiviral activity of alpha-nucleosides: D- and L-isomers of lyxofuranosyl- and (5-deoxylyxofuranosyl)benzimidazoles. J Med Chem 1998;41:1242–51.
[153] Smrz M, Farkas J. Nucleic acid components and their analogs. CXVII. Synthesis of 1-α-L-lyxofuranosylthymine. Collect Czechoslov Chem Commun 1968;33:3803–8.
[154] Silhar P, Hocek M, Pohl R, Votruba I, Shih I, Mabery E, Mackman R. Synthesis, cytostatic and anti-HCV activity of 6-(N-substituted aminomethyl)-, 6-(O-substituted hydroxymethyl)- and 6-(S-substituted sulfanylmethyl)purine nucleosides. Bioorg Med Chem 2008;16(5):2329–66.
[155] Tuncbilek M, Bilget GE, Onder T, Cetin AR. Synthesis of novel 6-(4-substituted piperazine-1-yl)-9-(β-D-ribofuranosyl)purine derivatives, which lead to senescence-induced cell death in liver cancer cells. J Med Chem 2012;55(7):3058–65.
[156] Bhuma N, Burade SS, Bagade AV, Kumbhar NM, Kodam KM, Dhavale DD. Synthesis and anti-proliferative activity of 3′-deoxy-3′-fluoro-3'-Chydroxymethyl-pyrimidine and purine nucleosides. Tetrahedron 2017;73:6157–63.
[157] Shelton JR, Burt SR, Peterson MA. A broad spectrum anticancer nucleoside with selective toxicity against human colon cells in vitro. Bioorg Med Chem Lett 2011;21:1484–7.
[158] Robins MJ, Peng Y, Damaraju VL, Mowles D, Barron G, Tackaberry T, Young JD, Cass CE. Improved syntheses of 5′-S-(2-aminoethyl)-6-N-(4-nitrobenzyl)-5′-thioadenosine (SAENTA), analogues, and fluorescent probe conjugates: analysis of cell-surface human equilibrative nucleoside transporter 1 (hENT1) levels for prediction of the antitumor efficacy of gemcitabine. J Med Chem 2010;53:6040–53.
[159] Robins MJ, Hansske F, Wnuk SF, Kanai T. Nucleic acid related compounds. 66. Improved syntheses of 5'-chloro-5′-deoxy- and 5'-S-aryl(or alkyl)-5′-thionucleosides. Can J Chem 1991;69:1468–74.
[160] Shelton JR, Balzarini J, Peterson MA. Discovery of a nanomolar inhibitor of lung adenocarcinoma in vitro. Bioorg Med Chem Lett 2014;24:5107–10.
[161] Shelton JR, Peterson MA. Efficient synthesis of 5′-O(N)-carbamyl and -polycarbamyl nucleosides. Tetrahedron Lett 2013;54(50):6882–5.
[162] Finkelstein H. Darstellung organischer Jodide aus den entsprechenden Bromiden und Chloriden. Ber Dtsch Chem Ges 1910;43:1528–32.
[163] Smith MB, March J. Advanced organic chemistry: reactions, mechanisms, and structure. 6th ed. New York: Wiley-Interscience; 2007. p. 574–5.
[164] Garuti L, Roberti M, Bottegoni G. Irreversible protein kinase inhibitors. Curr MedChem 2011;18(20):2981–94.
[165] Liu QS, Sabnis Y, Zhao Z, Zhang TH, Buhrlage SJ, Jones LH, Gray NS. Developing irreversible inhibitors of the protein kinase cysteinome. Chem Biol 2013;20(2):146–59.
[166] Besada P, Costas T, Teijeira M, Teran C. Synthesis and cytostatic activity of purine nucleosides derivatives of allofuranose. Eur J Med Chem 2010;45(12):6114–9.
[167] Malthum S, Polkam N, Allaka TR, Chepuri K, Anireddy JS. Synthesis, characterization and biological evaluation of purine nucleoside analogues. Tetrahedron Lett 2017;58:4166–8.
[168] Fairlamb IJS. Palladium catalysis in synthesis: where next? Tetrahedron 2005;61(41):9661–2.
[169] Hocek M. Syntheses of purines bearing carbon substituents in positions 2, 6 or 8 by metal- or organometal-mediated C−C bond-forming reactions. Eur J Org Chem 2003;2003:245–54.

[170] Chinchilla R, Najera C. The Sonogashira reaction: a booming methodology in synthetic organic chemistry. Chem Rev 2007;107:874–922.

[171] Kolb CH, Sharpless BK. The growing impact of click chemistry on drug discovery. Drug Discov Today 2003;8(24):1128–37.

[172] Hakimelahi GH, Mei N, Moosavi-Movahedi AA, Davari H, Hakimelahi S, King K, Hwu JR, Wen Y. Synthesis and biological evaluation of purine-containing butenolides. J Med Chem 2001;44:1749–57.

[173] Zheng B, Shen X, Zhao D, Cai Y, Ke M, Huang J. Silicon(IV) phthalocyanines substituted axially with different nucleoside moieties. Effects of nucleoside type on the photosensitizing efficiencies and in vitro photodynamic activities. J Photochem Photobiol B Biol 2016;159:196–204.

[174] Abu-Zaied MA, Loutfy SA, Hassan AE, Elgemeie GH. Novel purine thioglycoside analogs: synthesis, nanoformulation and biological evaluation in in vitro human liver and breast cancer models. Drug Des Devel Ther 2019;13:2437–57.

[175] Rayala R, Theard P, Ortiz H, Yao S, Young JD, Balzarini J, Robins MJ, Wnuk SF. Synthesis of purine and 7-deazapurine nucleoside analogues of 6-*N*-(4-nitrobenzyl) adenosine; inhibition of nucleoside transport and proliferation of cancer cells. ChemMedChem 2014;9(9):2186–92.

[176] Hulpia F, Noppen S, Schols D, Andrei G, Snoeck R, Liekens S, Vervaeke P, Van Calenbergh S. Synthesis of a 3'-C-ethynyl-β-D-ribofuranose purine nucleoside library: discovery of C7-deazapurine analogs as potent antiproliferative nucleosides. Eur J Med Chem 2018;157:248–67.

[177] Hattori H, Tanaka M, Fukushima M, Sasaki T, Matsuda A. Nucleosides and nucleotides.158. 1-(3-*C*-ethynyl-*β*-d-ribo-pentofuranosyl)-cytosine, 1-(3-*C*-Ethynyl-*β*-d-ribopentofuranosyl)uracil, and their nucleobase analogues as new potential multifunctional antitumor nucleosides with a broad spectrum of activity. J Med Chem 1996;39:5005–11.

[178] Seela F, Ming X. 7-functionalized 7-deazapurine *β*-D and *β*-L-ribonucleosides related to tubercidin and 7-deazainosine: glycosylation of pyrrolo[2,3-*d*]pyrimidines with 1-*O*-acetyl-2,3,5-tri-*O*benzoyl-*β*-D or *β*-L-ribofuranose. Tetrahedron 2007; 63:9850–61.

[179] Peng X, Seela F. An efficient synthesis Of 7-functionalized 7-deazapurine β-D- or *β*-L-ribonucleosides: glycosylation of pyrrolo[2,3-*d*]pyrimidines with 1-O-Acetyl-2,-3,5-tri-*O*-benzoyl-D-Or L-ribofuranose. Nucleosides Nucleotides Nucleic Acids 2007;26:603–6.

[180] Yu W, Chory EJ, Wernimont AK, Tempel W, Scopton A, Federation A, Marineau JJ, Qi J, Barsyte-Lovejoy D, Yi J, Marcellus R, Iacob RE, Engen JR, Griffin C, Aman A, Wienholds E, Li F, Pineda J, Estiu G, Shatseva T, Hajian T, Al-awar R, Dick JE, Vedadi M, Brown PJ, Arrowsmith CH, Bradner JE, Schapira M. Catalytic site remodeling of the DOT1L methyltransferase by selective inhibitors. Nat Comm 2012;3:1288.

[181] Campeau LC, O'Shea PD. Chemoselective Staudinger strategy in the practical, fit for purpose, gram-scale synthesis of an HCV RNA polymerase inhibitor. Synlett 2011;2011:57–60.

[182] D'Errico S, Borbone N, Piccialli V, Di Gennaro E, Zotti A, Budillon A, et al. Synthesis and evaluation of the antitumor properties of a small collection of Pt(II) complexes with 7-deazaadenosine as scaffold. Eur J Org Chem 2017;2017(33):4935–47.

[183] Bretner M, Beckett D, Sood RK, Baldisseri DM, Hosmane RS. Substrate/inhibition studies of bacteriophage T7 RNA polymerase with the 5′-triphosphate derivative of a ring-expanded ('Fat') nucleoside possessing potent antiviral and anticancer activities. Bioorg Med Chem 1999;7:2931–6.

[184] Youssef KM, Lien EJ. Design and synthesis of potential ribonucleotide reductase enzyme (RNR) inhibitors as antileukemic and/or antiviral 2′-deoxymethylene nucleosides. Future J Pharm Sci 2015;1(2):42–9.
[185] Wauchope OR, Johnson C, Krishnamoorthy P, Andrei G, Snoeck R, Balzarini J, Seley-Radtke KL. Synthesis and biological evaluation of a series of thieno-expanded tricyclic purine 2′-deoxy nucleoside analogues. Bioorg Med Chem 2012;20 (9):3009–15.
[186] Seley KL. In: Chu CK, editor. Recent advances in nucleosides: chemistry and chemotherapy. Amsterdam: Elsevier Science; 2002. p. 299.
[187] Zhang Z, Wauchope OR, Seley-Radtke KL. Mechanistic studies in the synthesis of a series of thieno-expanded xanthosine and guanosine nucleosides. Tetrahedron 2008;64:10791.
[188] Seley KL, Zhang L, Hagos A, Quirk SJ. "Fleximers". Design and synthesis of a new class of novel shape-modified nucleosides(1). J Org Chem 2002;67(10):3365–73.
[189] Seley KL, Januszczyk P, Hagos A, Zhang L, Dransfield DT. Synthesis and antitumor activity of thieno-separated tricyclic purines. J Med Chem 2000;43:4877–83.
[190] Barton DHR, Ok JD, Jaszberenyi JC. On the mechanism of deoxygenation of secondary alcohols by tin hydride reduction of methyl xanthates and other thiocarbonyl derivatives. Tetrahedron Lett 1990;31(28):3991–4.
[191] Lopez RM, Hays DS, Fu GC. Bu_3SnH-catalyzed Barton – McCombie deoxygenation of alcohols. J Am Chem Soc 1997;119:6949–50.
[192] Horita K, Yoshioka T, Tanaka T, Oikawa Y, Yonemitsu O. On the selectivity of deprotection of benzyl, mpm (4-methoxybenzyl) and dmpm (3,4-dimethoxybenzyl) protecting groups for hydroxy functions. Tetrahedron 1986;42:3021–8.
[193] Oikawa Y, Yoshioka T, Yonemitsu O. Specific removal of o-methoxybenzyl protection by DDQ oxidation. Tetrahedron Lett 1982;23:885–8.
[194] Wright JA, Yu J, Spencer JB. Sequential removal of the benzyl-type protecting groups PMB and NAP by oxidative cleavage using CAN and DDQ. Tetrahedron Lett 2001;42:4033–6.
[195] Yu W, Su M, Gao X, Yang Z, Jin Z. A facile chemoselective deprotection of the **p** -methoxybenzyl group. Tetrahedron Lett 2000;41:4015.
[196] Yan L, Kahne D. p-methoxybenzyl ethers as acid-labile protecting groups in oligosaccharide synthesis. Synlett 1995;1995(SI):523–4.
[197] Oriyama T, Yatabe K, Kawada Y, Koga G. Direct conversion of **p** -Methoxybenzyl ethers into silyl-protected alcohols by the action of Trialkylsilyl Trifluoromethanesulfonate and Triethylamine. Synlett 1995;1995(1):45–6.
[198] Rose JD, Parker WB, Someya H, Shaddix SC, Montgomery JA, Secrist III JA. Enhancement of nucleoside cytotoxicity through nucleotide prodrugs. J Med Chem 2002;45:4505–12.
[199] Ettner N, Haak U, Niederweis M, Hillen W. Synthesis of 8-bromo and 8-azido-2′-deoxyadenosine-5′-*O*-(1-thiophosphate). Nucleosides Nucleotides 1993;12:757–71.
[200] Tiwari KN, Messini L, Montgomery JA, Secrist III JA. Synthesis and biological activity of 4′-Thio-L-xylofuranosyl purine nucleosides. Nucleosides Nucleotides Nucleic Acids 2005;24(10 – 12):1895–906.
[201] Walker RT. 4′-Thio-2′-deoxyribonucleosides, their chemistry and biological properties—a review. Spec Publ R Soc Chem 1997;198:203–37.
[202] Whistler RL, Nayak UG, Perkins AWJ. Anomeric methyl 4-thio-D-arabinofuranosides. J Org Chem 1970;35:519–21.
[203] Fu YL, Bobek M. An alternative synthesis of anomeric methyl 2-Deoxy-4-thio-D-erythro-pentofuranosides. J Org Chem 1976;41:3831–4.

[204] Yoshimura Y, Watanabe M, Satoh H, Ashida N, Ijichi K, Sakata S, Machida H, Matsuda A. A facile, alternative synthesis of 4′-thioarabinonucleosides and their biological activities. J Med Chem 1997;40:2177–83.
[205] Yoshimura Y, Kitano K, Yamada K, Sakata S, Miura S, Ashida N, Machida H. Synthesis and biological activities of 2'-deoxy-2′-fluoro-4′ thioarabinofuranosylpyrimidine
and -purine nucleosides. Bioorg Med Chem 2000;8:1545–58.
[206] Nayak A, Sahu PK, Song J, Leea SK, Jeong LS. Regio- and stereoselective synthesis of 2′-β-substituted-fluoroneplanocin a analogues as potential anticancer agents. Org Biomol Chem 2015;13(35):9236–48.
[207] Yoon JS, Jarhad DB, Kim G, Nayak A, Zhao LX, Yu J, Kim HR, Lee JY, Mulamoottil VA, Chandra G, Byun WS, Lee SK, Kim YC, Jeong LS. Design, synthesis and anticancer activity of fluorocyclopentenyl-purines and -pyrimidines. Eur J Med Chem 2018;155:406–17.
[208] Siddiqui AQ, Balatore C, McGuigan C, Pathirana TN, Balzarini J, De Clercq E. The presence of substituents on the aryl moiety of the aryl phosphoramidate derivative of d4T enhances anti-HIV efficacy in cell culture: a structure – activity relationship. J Med Chem 1999;42:393–9.
[209] Ross BS, Reddy PG, Zhang HR, Rachakonda S, Sofia MJ. Synthesis of diastereomerically pure nucleotide phosphoramidates. J Org Chem 2011;76:8311–9.
[210] Corbett TH, Valeriote FA, Polin L, Panchapor C, Pugh S, White K, Lowichik N, Knight J, Bissery MC, Wozniak A, LoRusso P, Biernat L, Polin D, Knight J, Biggar S, Looney D, Demchik L, Jones J, Jones L, Blair S, Palmer K, Essenmacher S, Lisow L, Mattes KC, Cavanaugh PF, Rake JB, Baker L. In: Valeriote FA, Corbett TH, Baker LH, editors. Cytotoxic anticancer drugs: models and conceptsfor drug discovery and development. Boston/Dordrecht/London: Kluwer Academic; 1992. p. 33.
[211] Qin X, Chen X, Wang K, Polin L, Kern ER, Drach JC, Gullen E, Cheng Y, Zemlick J. Synthesis, antiviral, and antitumor activity of 2-substituted purine methylenecyclopropane analogues of nucleosides. Bioorg Med Chem 2006;14:1247–54.
[212] Teran C, Santana L, Teijeira M, Uriarte E, De Clercq E. Design, synthesis, conformational analysis and biological activities of purine-based 1,2-di-substituted carbocyclic nucleosides. Chem Pharm Bull 2000;48(2):293–5.
[213] Gonza´lez-Dı´az H, Vin˜ D, Santana L, de Clercq E, Uriarte E. Stochastic entropy QSAR for the in silico discovery of anticancer compounds: prediction, synthesis, and in vitro assay of new purine carbanucleosides. Bioorg Med Chem 2006;14:1095–107.
[214] Tănase CI, Drăghici C, Cojocaru A, Galochkina AV, Orshanskaya JR, Zarubaev VV, Shova S, Enache C, Maganu M. New carbocyclic *N*6-substituted adenine and pyrimidine nucleoside analogues with a bicyclo[2.2.1]heptane fragment as sugar moiety; synthesis, antiviral, anticancer activity and X-ray crystallography. Bioorg Med Chem 2015;23:6346–54.
[215] Ta ˘nase CI, Dra ˘ghici C, Teodor Ca ˘proiu M, Shova S, Mathe C, Cocu FG, Enache C, Maganu M. New carbocyclic nucleoside analogues with a bicyclo[2.2.1]heptane fragment as sugar moiety; synthesis, X-ray crystallography andanticancer activity. Bioorg Med Chem 2014;22:513–22.
[216] Elgemeie G, Abu-Zaied M, Hebishy A, Abbasa N, Hameda M. A first microwave-assisted synthesis of a new class of purine and guanine thioglycoside analogs. Nucleosides Nucleotides Nucleic Acids 2016;35(9):459–78.
[217] Xavier NM, Porcheron A, Batista D, Jorda R, Řezníčková E, Kryštof V, Oliveira MC. Exploitation of new structurally diverse D-glucuronamide-containing N-glycosyl compounds: synthesis and anticancer potential. Org Biomol Chem 2017;15:4667.

[218] Schwarz S, Siewert B, Csuk R, Rauter AP. New antitumor 6-chloropurine nucleosides inducing apoptosis and G2/M cell cycle arrest. Eur J Med Chem 2015;90:592–602.
[219] Xavier NM, Goncalves-Pereira R, Jorda R, Řezničkova E, Kryštof V, Oliveira MC. Synthesis and antiproliferative evaluation of novel azido nucleosides and their phosphoramidate derivative. Pure Appl Chem 2017;89(9):1267–81.
[220] Csuk R, Albert S, Siewert B, Schwarz S. Synthesis and biological evaluation of novel (E) stilbene-based antitumor agents. Eur J Med Chem 2012;54:669–78.
[221] Csuk R, Barthel A, Schwarz S, Kommera H, Paschke R. Synthesis and biological evaluation of antitumor-active gamma-butyrolactone substituted botulin derivatives. Bioorg Med Chem 2010;18:2549–58.
[222] Csuk R, Nitsche C, Sczepek R, Schwarz S, Siewert B. Synthesis of antitumor active betulinic acid-derived hydroxypropargylamines by copper-catalyzed mannich reactions. Arch Pharm Chem Life Sci 2013;346:232–46.
[223] Huang L, Xu H, Yao Z, Wang Y, Liu H. Synthesis and biological evaluation of novel C_6-amino substituted 4-azasteroidal purine nucleoside analogues. Bioorg Med Chem Lett 2014;24:973–5.
[224] Huang L, Wang Y, Xu G, Zhang X, Zheng Y, He H, Fu W, Liu H. Novel 4-azasteroidal *N*-glycoside analogues bearing sugar-like D ring: synthesis and anticancer activities. Bioorg Med Chem Lett 2011;21:6203–5.

CHAPTER 7

Synthetic strategies for pyrimidine nucleoside analogs

1. Introduction

Pyrimidine nucleosides furnish versatile building blocks for pharmacological targets. They display fundamental and comprehensive roles in the growing area of medicinal chemistry. Various strategies for the synthesis of the many pyrimidines nucleosides have been exhibited throughout this chapter to access novel and potent pharmacological agents. Many of the synthesized pyrimidine nucleosides are potential chemotherapeutic agents, which provide beneficial impact on the current medicinal research. Pyrimidine analogs are fundamental category of therapeutic agents having many remarkable and comprehensive applications in biochemistry and biotechnology. They possess a promising and a wide spectrum of biological activities. They are used in the treatment of cancer, tumors, hematological malignancies, influenza virus, hepatitis C (HCV), HIV, and many bacterial diseases. A wide variety of pyrimidine nucleosides with diverse and tremendous biological activities have been reported.

Pyrimidine nucleosides are important chemical architectures, have pivotal roles in medicinal chemistry, and have wide pharmacological applications. They have great significance which exhibit a variety of clinical applications. They constitute a fundamental category of natural biologically active compounds [1]. Pyrimidine nucleosides have integrated metabolic importance with purines and antifolates.

Consequently, they are a significant class of heterocyclic compounds having diverse pharmacological activities, including anticancer [2–6], antitumor agents [6–11], antibacterial [12–16], antileishmanial [17–19], antinociceptive effects [20], anti-HCV [21–25], anti-HIV [26], antiinfluenza [27], antirubella [28], antivaricella zoster virus agents [29], anticoronavirus infections [30–34], and many other antiviral agents [35–105]. They are potential chemotherapeutic [106] and effective agents in treating many hematological disorders [107]. Additionally, they have various fluorescence applications, which have drawn the interest of many chemists [108–111].

New Strategies Targeting Cancer Metabolism
https://doi.org/10.1016/B978-0-12-821783-2.00010-8

Thus, well-recognized synthetic routes have been set up for subsequent drug development. It is worth noting that there are a huge series of pyrimidine nucleosides that have been synthesized [112–123].

Besides the ribonucleosides, including, free-, protected ribonucleosides, deoxy derivatives, phosphonated derivatives, and pyranose nucleosides, many modified nucleosides were also synthesized. The biological evaluation of the novel synthesized compounds was emphasized, and their synthetic procedures were also elaborated.

Crucial approaches for developing the syntheses of novel drugs are exhibited via targeting pyrimidine metabolic pathways. Following the recent developments in the synthetic strategies of the pyrimidine nucleosides, the most novel and up-to-date synthetic routes of the anticancer pyrimidine nucleoside analogs that reflect its significant importance were surveyed.

2. Examples of potent anticancer pyrimidine nucleoside analogs

Many inventions of the anticancer pyrimidine nucleoside analogs were depicted [124–145]. Examples of the potent nucleosides were outlined:

Pyridopyrimidine nucleoside derivatives **1** (Fig. 7.1) were synthesized as anticancer agents. $CH_2{=}CHCH_2NH_2$ and N^1-benzyl-2′,3′,5′-tri-*O*-benzoyl-6-chloroisouridine reacted to generate 95.0% 6-(allylamino)isouridine derivative (**2**) (Fig. 7.2), which was stirred with a suspension of palladium(II) chloride in aqueous dioxane, while bubbling oxygen into the mixture, to afford, after debenzoylation with sodium

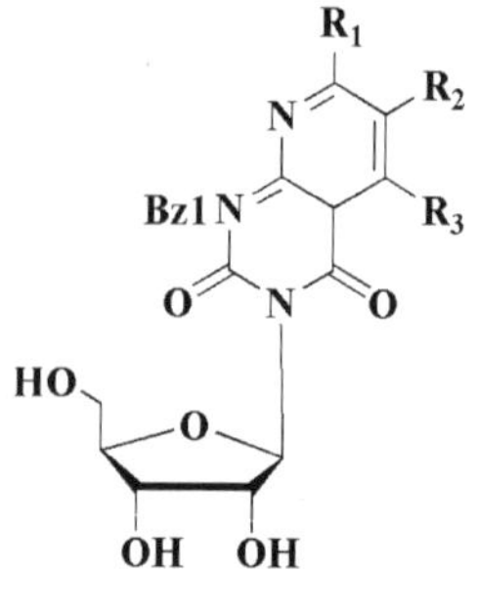

Fig. 7.1 Pyridopyrimidine nucleoside derivatives **1**; Bzl = $PhCH_2$; R^1-R^3 = H; or one of R^1-R^3 = lower alkyl and the rest of R^1-R^3 = H).

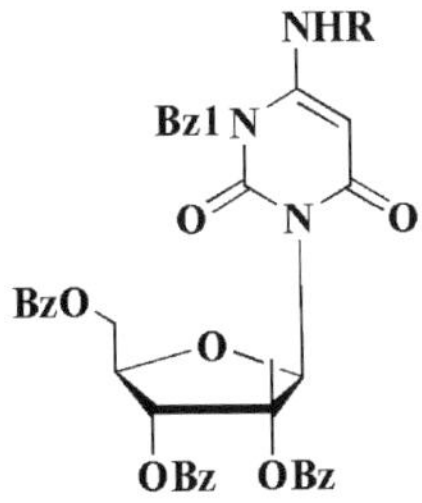

Fig. 7.2 6-(Allylamino)isouridine derivative **2**; R=$CH_2CH{=}CH_2$.

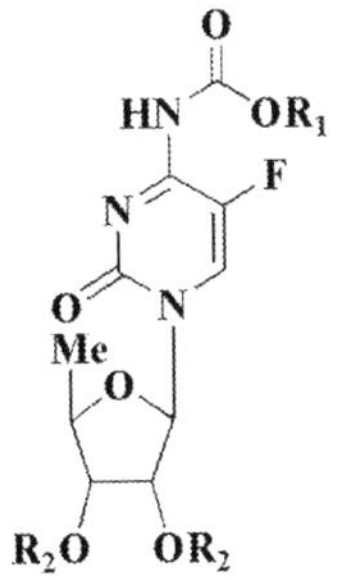

Fig. 7.3 Compound **3**; R^1=saturated or unsaturated, straight or branched hydrocarbon radical (wherein longest straight chain has 3–7C atoms), or $(CH_2)_nY$ (in which n=0–4 when Y=cyclohexyl, or n=2–4 when Y=C_{1-4} alkoxy or Ph); R^2=H or a radical easily hydrolyzable under physiological conditions.

methoxide/methanol, **1** (R^1-R^3=H). Compound **1** inhibited human HL-60 leukemia cells with IC_{50}'s of 23–65 μg/mL [146].

Compounds **3** (Fig. 7.3) and their solvates or hydrates are useful in the treatment of tumors. The compounds can be synthesized through reacting chloroformates R^1OCOCl with optionally protected N^4-unsubstituted 5′-deoxy-5-fluorocytidines. The compounds have improved pharmacokinetic profiles and less intestinal toxicity than the known compounds. For example, 5′-deoxy-5-fluorocytidine (5′-DFCR) was 2′,3′-di-*O*-acetylated with acetic anhydride in pyridine at 0°, and the product treated with *n*-Pr chloroformate in pyridine, to afford **3** (R^1=Pr, R^2=Ac). This was hydrolyzed by the addition of sodium hydroxide to a dichloromethane solution at ice temperature, furnishing **3** (R^1=Pr, R^2=H). The analogously synthesized **3** (R^1=Bu, R^2=H), a preferred compound, afford complete inhibition of growth of human colon cancer xenograft CXF280 in mice at a dose where intestinal toxicity was not observed, whereas the standard/metabolite 5-FU afforded only 58% inhibition at a toxic dose. Examples include preparations, formulations, acylamidase deacylation data, pharmacokinetics of selected **3** in monkeys, and additional antitumor and anticachexia data in mice [147].

Acylated derivatives of nonmethylated pyrimidine nucleosides (Fig. 7.4) are capable of attenuating damage to the hematopoietic system in animals receiving antineoplastic or antiviral chemotherapy. Oral administration of triacetyluridine ameliorated the hematological toxicity of 5-fluorouracil [148].

The fluorinated pyrimidine analogs were administered to treat solid tumors and hematological cancers. Thus, fluorinated pyrimidine nucleoside analog **5** (Fig. 7.5) was prepared through glycosylation reaction and estimated in vitro as antitumor agent and DNA methyltransferase inhibitor [149].

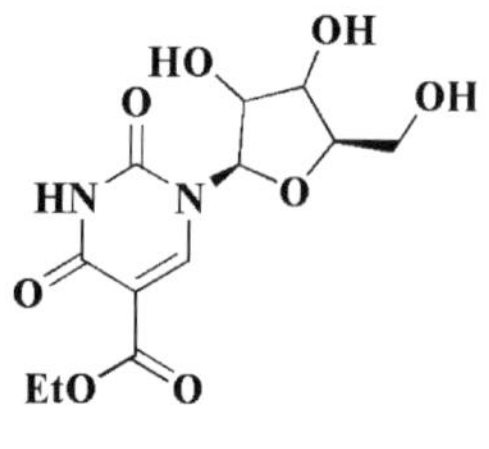

4

Fig. 7.4 Acylated derivative of nonmethylated pyrimidine nucleoside **4**.

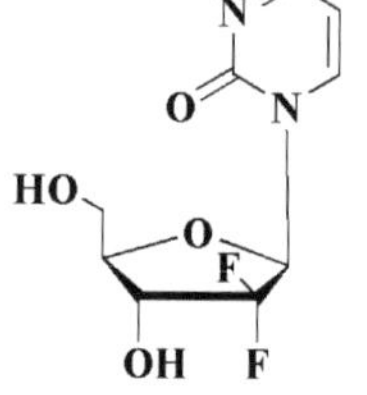

5

Fig. 7.5 Fluorinated pyrimidine nucleoside analog **5**.

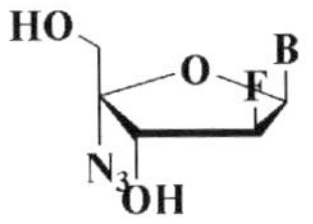

Fig. 7.6 The pyrimidine nucleoside derivatives **6** wherein B=a pyrimidine-containing base.

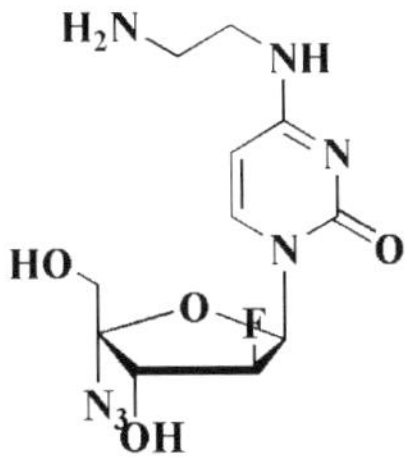

Fig. 7.7 Compound **7**.

The pyrimidine nucleoside derivatives **6** (Fig. 7.6) were synthesized as antiviral and antitumor agents. For example, compound **7** (Fig. 7.7) was prepared in a multistep synthesis. In biological test, **7** indicated an inhibitory potency with EC_{50} of 0.093 μM against HIV-1 virus [150].

Pyrimidine nucleoside analogs **8** (Fig. 7.8) were synthesized, as targeted mechanism-based modulators of cell cycle checkpoints. Cancers and/or malignancies can be treated by the administration of a cell cycle checkpoint modulator of the invention. Combinations of the cell cycle checkpoint modulator with a checkpoint kinase inhibitor to give synergistic apoptosis in cancer cells were disclosed. Thus, nucleoside **9** (Fig. 7.9) was synthesized and estimated as kinase inhibitor and anticancer agent, wherein the cancer is selected from the group comprising cancer of the breast, colon, bladder, liver, kidney, non-small-cell lung cancer, small-cell lung cancer, lung, neck esophagus and head, gall bladder, ovary, prostate, cervix thyroid, and skin, squamous cell carcinoma; leukemia, chronic lymphocytic leukemia, acute lymphocytic leukemia, acute lymphoblastic leukemia, B-cell lymphoma, T-cell lymphoma, non-Hodgkin's lymphoma, Hodgkin's lymphoma, mantle cell lymphoma, hairy cell lymphoma, and Burkitt's lymphoma, chronic and acute myelogenous leukemia, promyelocytic leukemia and myelodysplastic syndrome rhabdomyosarcoma, fibrosarcoma; neuroblastoma, astrocytoma, melanoma and glioma, osteosarcoma, teratocarcinoma, thyroid cancer, and Kaposi's sarcoma [151].

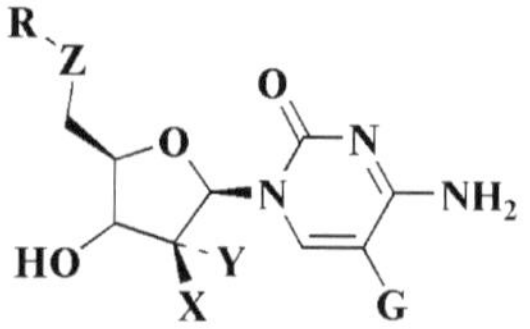

Fig. 7.8 Pyrimidine nucleoside analogs **8**; wherein G is H, halo, alkyl; X and Y are independently H, F, OR^2, alkyl; Z is O, NR^2, S, CR^2R^3, SO_2; R is alkenyl, alkyl, aryl, cycloalkyl, heteroaryl, alkylaryl, alkylheteroaryl, alkylheterocyclyl, heterocyclyl, alkyl-$C(O)NR^2R^3$ and alkyl-$C(O)R^2$, wherein each of said aryl, alkyl, heteroaryl and heterocyclyl can be (un)substituted with one or more groups which can be different or the same, each substituent being independently selected from the group comprising cyano, halo, hydroxy, alkynyl, alkenyl, alkyl, aryloxy, alkoxy, alkylthio, aryl, arylthio, heterocyclyl, heteroaryl, cycloalkenyl, cycloalkyl, NR^2R^3, OR, $C(O)OR^2$, $C(O)R^2$, and $C(O)NR^2R^3$; R^2 is H, heteroaryl, alkyl, aryl, wherein each of said aryl, alkyl, and heteroaryl can be unsubstituted or optionally independently substituted with one or more groups which can be different or the same and are independently selected from the group comprising cyano, halo, hydroxy, alkenyl, alkyl, alkynyl, aryloxy, alkoxy, arylthio, alkylthio, aryl, cycloalkyl, heterocyclyl, NR^2R^3, cycloalkenyl -OR^2, $C(O)NR^2R^3$, $C(O)R^2$; R^3 is aryl, alkyl, heteroaryl, cycloalkyl, heterocyclyl, cycloalkenyl, NR^2R^3, OR^2, $C(O)NR^2R^3$, and $C(O)R^2$.

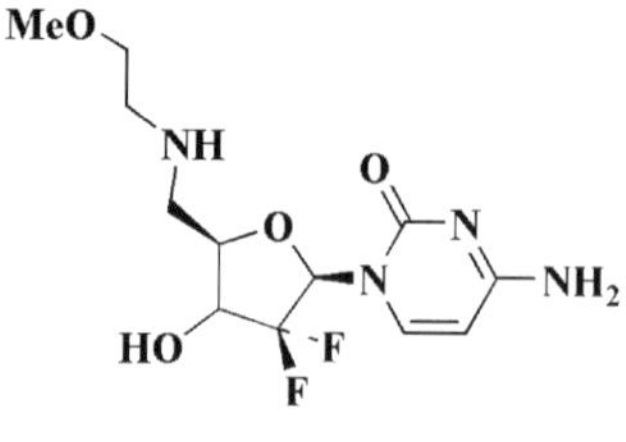

Fig. 7.9 Nucleoside analog **9**.

Compounds **10** were synthesized. Thus, the nucleoside **11** [R^1 = 1,2,4-triazol-1-yl, R = $SiMe_2CMe_3$] (Fig. 7.10) was treated with hydroxylamine and deblocked to afford compound **11** (Fig. 7.11), which had an IC_{50} against HSV-1 of 0.7 μg/mL and an IC_{50} against human leukemia cells of 0.086 μg/mL [152].

Gemcitabine compounds **12** (Fig. 7.12) were prepared and estimated as antitumor agents. Nucleoside **13** (Fig. 7.13) was prepared and estimated for oral activity against human HCT116 colorectal adenocarcinoma in female nude mice and indicated tumor growth inhibition (TGI) of 86.7% as compared to a TGI of 85.6% for gemcitabine [153].

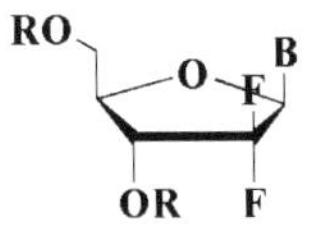

Fig. 7.10 Compound **10**; B = Pyrimidine, imidazopyrimidine, tetrazolopyrimidine, triazinopyrimidine, triazolopyrimidine.

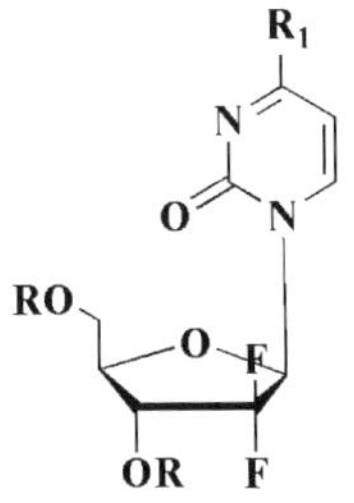

Fig. 7.11 Compound **11**; R = H, R^1 = NHOH.

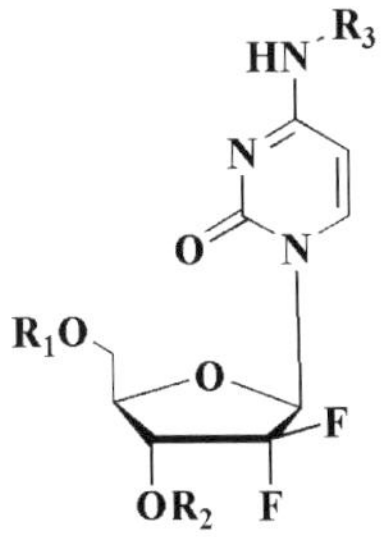

Fig. 7.12 Gemcitabine compounds **12**; wherein R^1 and R^2 are independently selected from H, —C(=O)—$(CH_2)_2$-aryl, and —C(=O)—$(CH_2)_n$—C(=O)—NH-aryl; $n = 2$–6; R^3 is H, —C(=O)—O—R^4; and R^4 is alkyl, alkynyl, alkenyl, arylalkyl, heteroalkyl, cycloheteroalkyl.

Novel pyrimidine compound represented by formula as shown in **14** (Fig. 7.14) and **15** (Fig. 7.15) was invented. Experimental results indicate that the compound can inhibit the growth of tumor cell, can be utilized for the manufacture of antitumor drug, and has safety and good effectiveness. And vascular stimulation test affirms that the compound does not have irritability or hemolysis, and can be prepared into the injection liquid for clinical use [154].

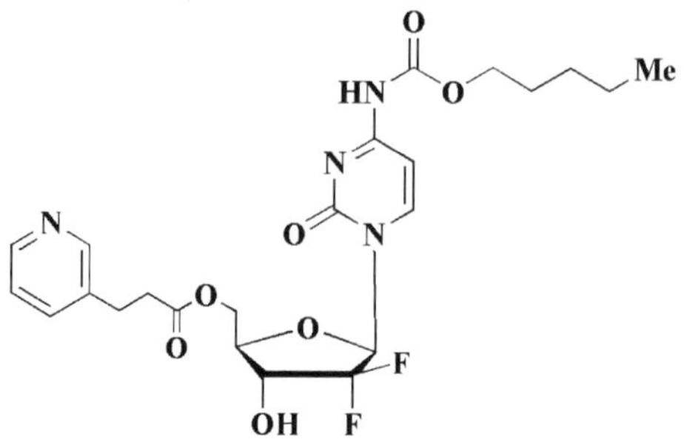

13

Fig. 7.13 Nucleoside analog **13**.

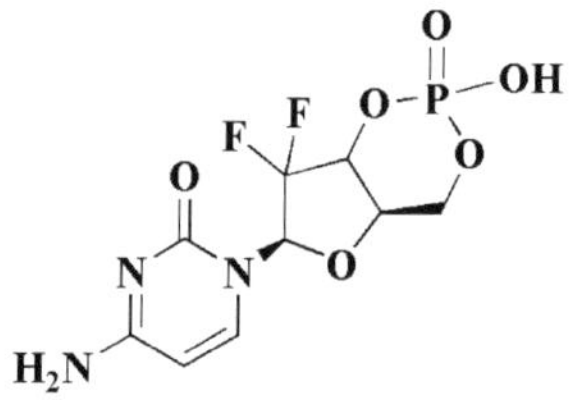

14

Fig. 7.14 Compound **14**.

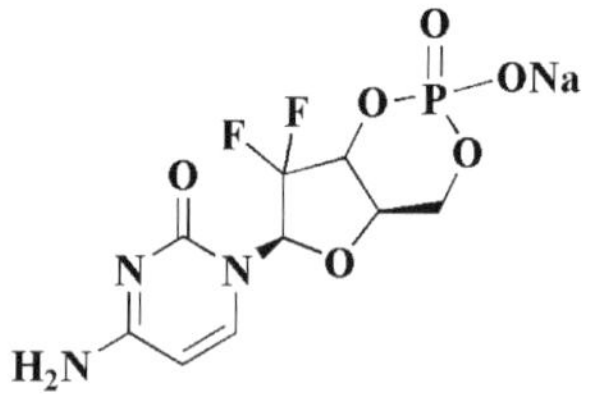

15

Fig. 7.15 Compound **15**.

The thiolated pyrimidine mononucleotides and pyrimidine mononucleosides (Fig. 7.16) were invented as therapeutic agents. The therapeutic applications of 4-thiouridine, its analogs, ribo- and deoxyribonucleotide derivatives and their various further derivatives, as well as these compounds for their utility in the therapeutic applications, were disclosed. These compounds are utilized as antiproliferative and/or antiviral agents. Preparation of 4-thiouridine monophosphate is reported [155].

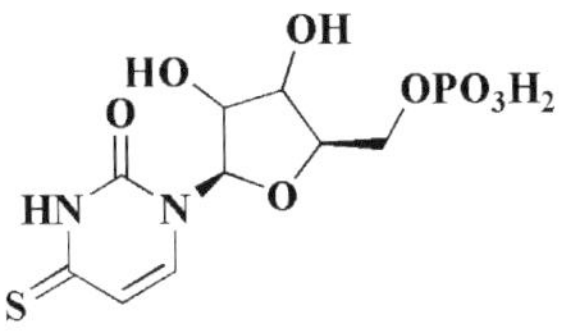

16

Fig. 7.16 Thiolated pyrimidine derivative **16**.

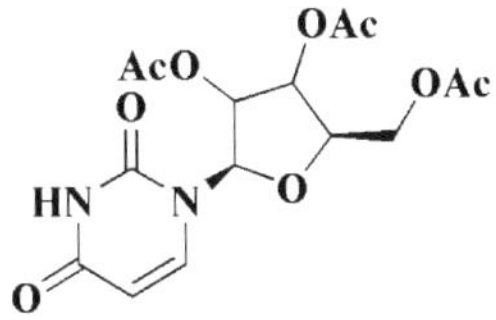

17

Fig. 7.17 Pyrimidine-based nucleoside **17**.

The administration of a pyrimidine-based nucleoside (Fig. 7.17), e.g., triacetyluridine, is provided for the treatment of mitochondrial disorders. Also provided are methods of eliminating or reducing symptoms associated with mitochondrial disorders. Mitochondrial disorders particularly appropriate for treatment include those attributable to a deficiency of one or more pyrimidines [156].

Methods for the treatment and prevention of toxicity due to antiviral and chemotherapeutic agents were invented. Thus the acylated derivatives of nonmethylated pyrimidine nucleosides were accomplished (Fig. 7.18). These compounds are capable of attenuating damage to the hematopoietic system in animals receiving antineoplastic or antiviral chemotherapy [157].

A synergistic antitumor pharmaceutical component contained interleukin-12 and a pyrimidine nucleoside derivative (Fig. 7.19) as well

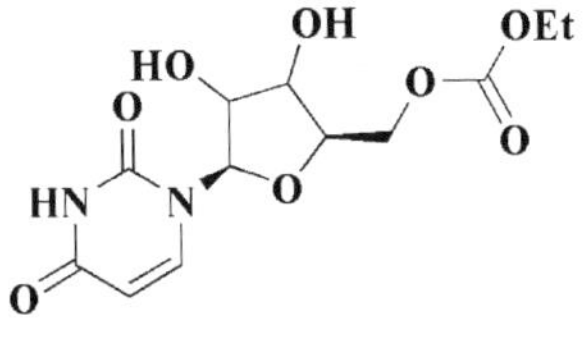

18

Fig. 7.18 Acylated derivative of nonmethylated pyrimidine nucleosides **18**.

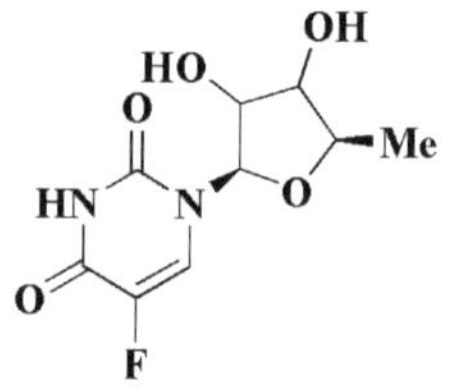

19

Fig. 7.19 Pyrimidine nucleoside derivative **19**.

as a solvate or hydrate that is transformed into fluorouracil or its derivative, and a pharmaceutically acceptable carrier. An injectable solution contained doxifluridine 1000, NaCl 41.4, NaH_2PO_4 16.2, and Na_2HPO_4 36.7, polysorbate 4 mg, and interleukin-12 50 μg. The solution was adjusted to pH 7.0 and water was added to make up the total volume of the solution to 20 mL [158].

A patient affected with a neoplastic disease is administered an effective antineoplastic amount of ionizing or nonionizing radiation, or of a DNA-reactive chemotherapeutic agent in conjunction with a sensitizing amount of compound (**20**) (Fig. 7.20) or a salt thereof. Thus, 2′-deoxy-2′-fluoromethylenecytidine (**20**, Y=H) radiosensitized HeLa cells to x-irradiation. [159].

2′-Deoxy-5-fluorouridine derivatives **21** (Fig. 7.21), which have an excellent antitumor potency with low toxicity and high therapeutic indexes and maintain a low blood concentration of 2′-deoxy-5-fluorouridine for a long period of time, are prepared. Thus, phosphorous oxychloride was added dropwise to a solution of 1-methylimidazole in acetonitrile and stirred for 10 min under ice-cooling, followed by adding a solution of 5′-

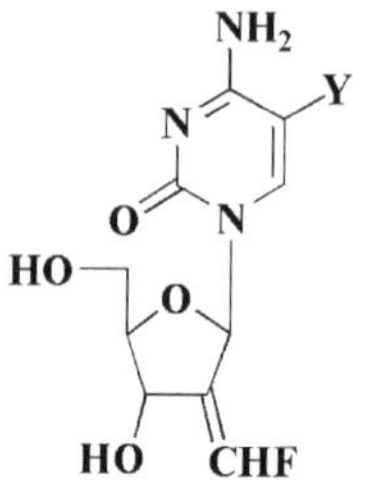

20

Fig. 7.20 2′-Deoxy-2′-fluoromethylenecytidine **20**; Y=H, C_{1-4} alkyl, C_{1-4} alkoxy.

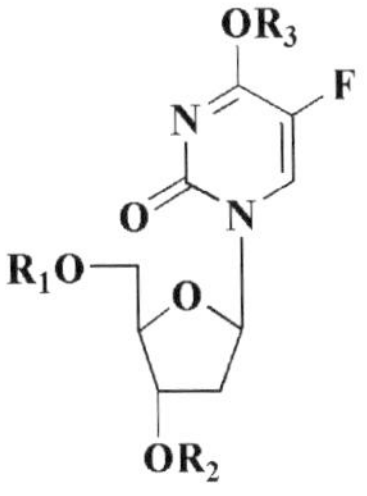

21

Fig. 7.21 2′-Deoxy-5-fluorouridine derivatives **21**; one of R_1 and R_2 represents hydrogen or a group hydrolyzable in vivo and the other represents CH_2Ph, which may be substituted by halogen or CF_3 on the benzene ring; $R^3=CH_2Ph$ or Ph optionally having a halogen as a substituent on the benzene ring, lower alkenyl, pH, lower alkyl optionally having OH, lower alkoxy, di(lower alkyl)amino, furanyl, thienyl, or pyridyl as a substituent.

O-acetyl-3′-*O*-(4-chlorobenzyl)-2′-deoxy-5-fluorouridine in acetonitrile. To the resulting mixture, ethanol was added, followed by adding triethylamine to afford a precursor **21** ($R^1=Ac$, $R^2=$4-chlorobenzyl, $R^3=Et$), which was treated with sodium ethoxide in ethanol to yield compound **21** ($R^1=H$, $R^2=$4-chlorobenzyl, $R^3=Et$). **21** ($R^1=H$, $R^2=$2,4-dichlorobenzyl, $R^3=Pr$) at 1.0 mL 0.5% solution/100 g per day for 7 consecutive days, which indicated an ED_{50} of 3 mg/kg/day and therapeutic index of 5.3 in rats *vs*. 0.8 mg/kg/day and 0.88, respectively, for 2-deoxy-5-fluorouridine (continuous i.v.) [160].

Compounds **22** (Fig. 7.22), useful as intermediates for antitumor 2′-substituted pyrimidine nucleosides, are prepared via oxidation of the OH group

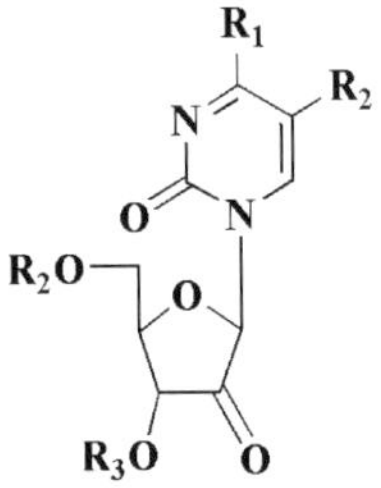

22

Fig. 7.22 Compound **22**; $R^1=$alkoxy, OH, amino, acylamino, (dialkylamino) methyleneimino; $R^2=$halo, H, alkyl, haloalkyl, halovinyl, alkenyl, alkynyl; $R^3=$ether- or silyl-type protecting group.

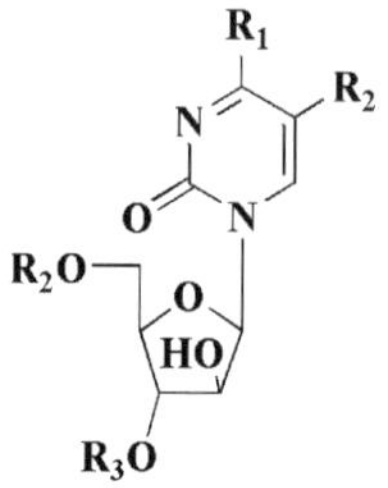

23

Fig. 7.23 1-β-D-arabinofuranosylpyrimidine derivatives **23**.

of the 1-β-D-arabinofuranosylpyrimidine derivatives **23** (Fig. 7.23). Thus, 3′,5′-bis-*O*-tetraisopropyldisiloxane-1,3-diyl-β-D-arabinofuranosyluracil was treated with HOAc and chromic anhydride in dichloromethane to afford the compound 1-(3,5-*O*-tetraisopropyldisiloxane-1,3-diyl-β-D-erythropentofuran-2-ulosyl)uracil. The transformation of compounds **22** to the antitumors 2′-substituted pyrimidine nucleosides is also reported [161].

Compounds **24** (Fig. 7.24), esters and prodrugs thereof, pharmaceutical components containing them, their uses in medicine, and their preparations are disclosed. **24** are useful as uridine phosphorylase inhibitors; e.g., the IC_{50} of **24** ($R^1=CH_2OCH_2CH_2OH$, $R^2=O$, $R^3=CH_2$, $R^4=OCHMe_2$, $R^5=R^6=H$) was <0.05 mM. Compounds **24** also minimize the toxicity of neoplastic pyrimidine nucleosides and potentiate the efficacy of antineoplastic drugs [162].

Compounds **25** (Fig. 7.25), which indicate antitumor or antiviral potency, are synthesized by treating nucleosides **26** (Fig. 7.26) (R^1 = same as **25**; R^5 = acyl; R^6 = protective group) with organosilicon compounds in the presence of Lewis acids. Compounds **26** were treated with allyltrimethylsilane and Tin(IV) chloride in dichloromethane at ≤−70° for 7 h to afford 74% **25** ($R^1=H$, $X^1=CH_2OSiPh_2CMe_3$, X^2 = allyl) and 5% I ($R^1=H$, X^1 = allyl, $X^2=CH_2OSiPh_2CMe_3$ [163].

The nucleosides **27** (Fig. 7.27) were synthesized as neoplasm inhibitors. Thus, compound **27** [$R^1=NC(CH_2)_{10}CO$, R^2-$R^4=H$, $R^5=CN$] was synthesized and indicated 97.3% of tumor growth inhibition of the fibrosarcoma cells [164].

5-Substituted pyrimidine derivatives of conformationally locked nucleoside analogs **28** (Fig. 7.28) were synthesized, wherein B is a 5-substituted

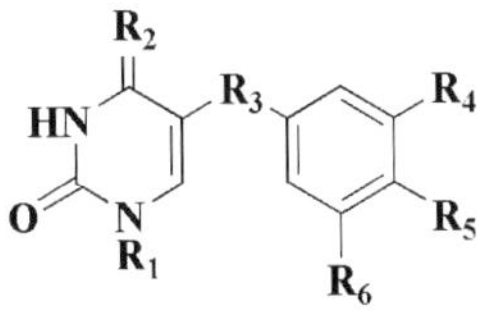

24

Fig. 7.24 Compound **24**; R^1 = H, C_{1-8} straight- or branched-chain alkyl, C_{2-6} alkenyl, (C_{1-3} alkyl-C_{3-6} cycloalkyl-C_{1-3} alkyl) optionally substituted by 1 or 2 substituents selected from OR^8, NR^8R^9 (R^8, R^9 = H, C_{1-6} straight or branched-chain alkyl, aralkyl); CH_2ZR^{10}, ZCH_2R^{10}, $CH_2ZR^{10a}Zr^{10}$ (R^{10a} = C_{1-6} branched- or straight-chain alkylene, R^{10} = C_{1-6} straight- or branched chain alkyl) each of R^{10a} and R^{10} being optionally substituted by 1 or 2 substituents independently selected from OR^8 and NR^8R^9 (same R^8, R^9), Z = O, S, CH_2O, CH_2S; R^2 = O, S; R^3 = O, S, SO, SO_2, NR^8, CO, C_{1-6} straight- or branched-chain alkanediyl; R^4 = H, C_{1-4} straight- or branched-chain alkyl, halogen, -OR^{11} (R^{11} = C_{1-4} branched- or straight-chain alkyl optionally substituted by aryl, halogen, C_{3-6} cycloalkyl, (C_{1-3} alkyl-C_{3-6} cycloalkyl), C_{2-6} alkenyl or C_{2-6} alkynyl), methylenedioxy, CX_3 (X = halogen), NO_2, CN; R^5 = H, halogen, OR^{11}; R^6 = H, Y-Ar-$R^7_{(m)}$ (Y = O, S, SO, SO_2, NR^8, C:O, alkanediyl), Ar = Ph, naphthyl, m = 1–3 and R^7 = R^8, CO_2R^8, COR^8, $CONR^8R^9$, $R^{8a}OR^8$ (R^{8a} = C_{1-6} straight- or branched-chain alkyl, aralkyl), CN, CX_3, OR^8, OCX_3, SR^8, SO_2R^8, F, Cl, Br, iodo, or a combination thereof, etc., provided that when R^1 = H, $CH_2OCH_2CH_2OH$ or $CH_2OCH(CH_2OH)_2$, R^2 = O, R^3 = CH_2 then R^4, R^5, and R^6 are other than OMe, OEt, OCH_2PH, or $OCHMe_2$.

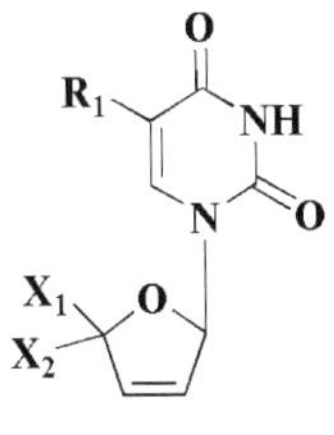

25

Fig. 7.25 Compound **25**; R^1 = H, halo, lower alkyl; (X^1, X^2) = (R^2, CH_2OR^3), (CH_2OR^3, R^2); R^2 = allyl, 2-alkylallyl, cycloalkanon-2-yl, R^4CH_2, cyano; R^3 = H, protective group; R^4 = acyl.

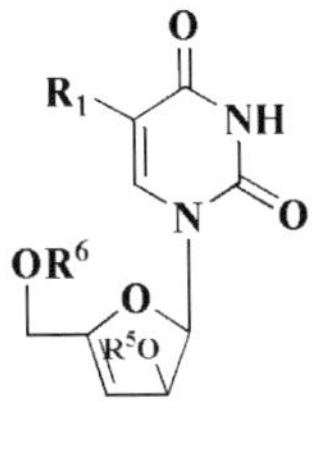

26

Fig. 7.26 Compound **26**; R^1 = H, R^5 = Ac, R^6 = $SiPh_2CMe_3$.

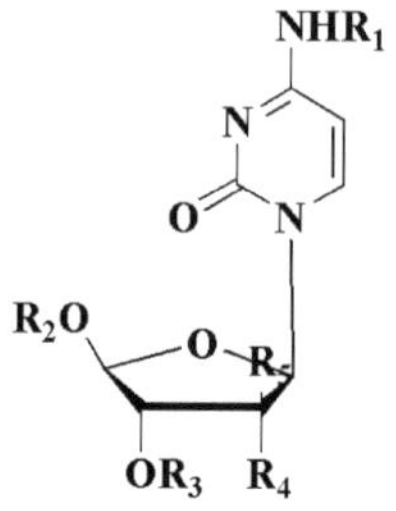

27

Fig. 7.27 Compound **27**; R^1-R^4=H; R^1-R^3=(un)substituted alkanoyl, alkenylcarbonyl, R^4=CN, R^5=H; R^4=H, R^5=CN.

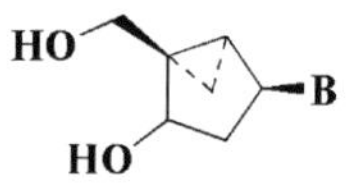

28

Fig. 7.28 5-Substituted pyrimidine derivatives of conformationally locked nucleoside analogs **28**; B is a 5-substituted derivatives of uracil, except when the 5-substituent is cytosine, or Me.

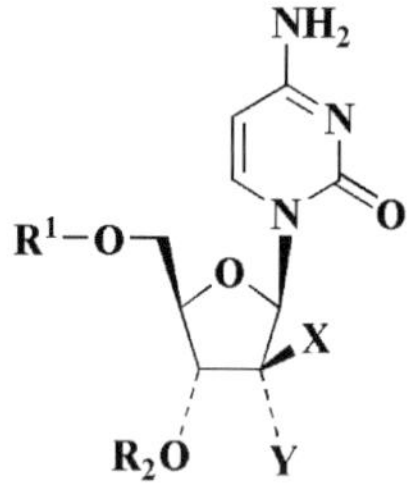

29

Fig. 7.29 Compound **29**; one of X and Y is a cyano, while the other is a hydrogen atom; one of R^1 and R^2 is a hydrogen atom, a carbonyl possessing an alkyl monosubstituted with amino, a group of the formula $(R^3)(R^4)(R^5)$Si— while the other is a silyl of the formula $(R^6)(R^7)(R^8)$Si-; R^1 and R^2 together represent a group of the formula —Si(R^9)(R^{10})— and may form a six-membered cyclic group; R^3-R^8=alkyl, cycloalkyl, aryl, etc.; R^9, R^{10}=(un)substituted alkyl.

derivatives of uracil, except when the 5-substituent is cytosine, or Me, and to the use of these derivatives as anticancer and antiviral agents [165].

Compounds **29** (Fig. 7.29) and their salts were synthesized. For example, reaction of 2′-cyano-2′-deoxy-1-β-D-arabinofuranosylcytosine with tert-butyldimethylsilyl chloride afforded compound **30** (Fig. 7.30). In tumor

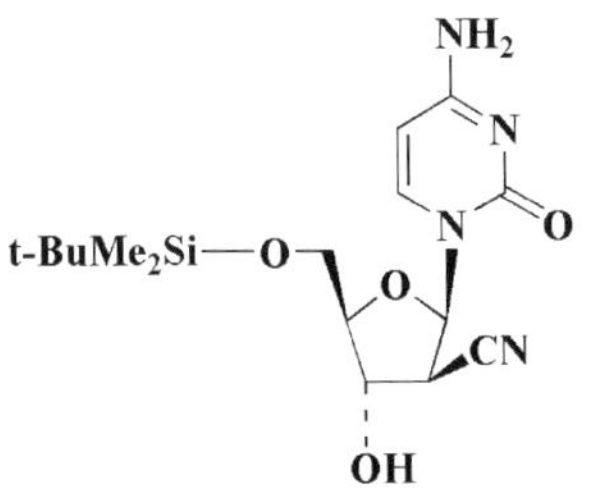

Fig. 7.30 Compound **30**.

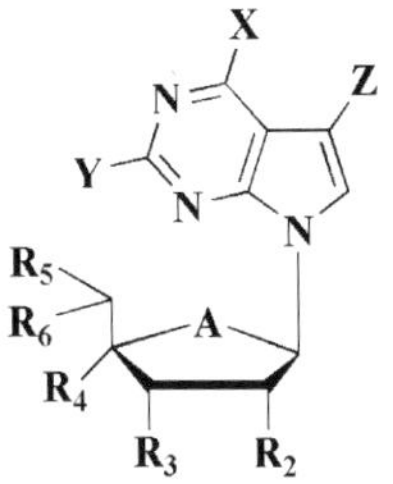

Fig. 7.31 Pyrrolo[2,3-*d*] pyrimidine nucleoside analogs **31**; wherein A is S, O, or CH_2; X is NH_2, OH, or H; Y is halogen, H, or NH_2; Z is selected from the group comprising halogen, H, OH, R, OR, SR, SH, CN, NHR, NR^2, NH_2, $C(O)NH_2$, COOR, COOH, CH_2NH_2, and C(=NH)NH_2 and C(=NOH)NH_2, where R is alkenyl, alkyl, aralkyl, or alkynyl; R^2 and R^3 are independently selected from the group comprising F, H, and OH; R^4 is selected from the group consisting of a hydrogen, an alkyl, an alkynyl, an alkenyl, and an aralkyl, wherein R^4 optionally has at least one of a hetero-atom and a functional group; R^5 is OH, $P(O)(OH)_2$, $OP(O)(OH)_2$, $P(O)(OR')_2$ or $OP(O)(OR')_2$, wherein R′ is a masking group; and R^6 is selected from the group comprising an alkyl, an alkynyl, an alkenyl, and an aralkyl, wherein R^6 has at least two carbon atoms, and optionally has at least one of a hetero-atom and a functional group, possessing substituents at the C5′ and C4′ positions of the ribofuranose moiety.

proliferation inhibition assays by oral administration, compound **30** displayed the inhibitory activity of 75% [166].

Pyrrolo[2,3-*d*] pyrimidine nucleoside analogs **31** (Fig. 7.31) are furnished. Contemplated components display, among other things, anticancer and immunomodulating effects at reduced cytotoxicity. Thus, compound **31** (A=O; R^2-R^4=OH; R^5=R^6=Me; Z=CN; X=NH_2; Y=H) was prepared and estimated for its immunomodulating effect at reduced cytotoxicity as antitumor. Inhibition of vascular endothelial growth factor

(VEGF) release in HTB 81 cells treated with **31** and inhibition of IL-8 release in HTB 81 cells at 0–50 μM are reported [167].

Compounds (**32**; B=Q, Q^1; R^1, R^2, R^3=H, HO-protecting group) (Fig. 7.32), which have sufficient effectiveness against viral infection and cancers, are prepared. Thus, 2′,3′,5′-tri-O-tert-butyldimethylsilyl-2-thio-5-iodouracil arabinoside was chlorinated by phosphorous oxychloride in the presence of triethylamine in acetonitrile followed by ammonolysis with aqueous ammonia in 1,4-dioxane and desilylation with tetra-n-butylammonium fluoride in tetrahydrofuran to afford compound (**33**) (Fig. 7.33). Compound **33** indicated an IC_{50} of 0.11, 1.40, 3.00, and 1.50 μg/mL against K562, B16F1, A549, and A375 cancer cells, respectively, and also inhibited the proliferation of herpes simplex 1 virus in host RPMI8226 cells with an EC_{50} of 1.22 μg/mL [168].

Compound **34** (Fig. 7.34), useful as an intermediate for an antitumor agent (**35**) (Fig. 7.35), is prepared via oxidation of a cytidine derivative

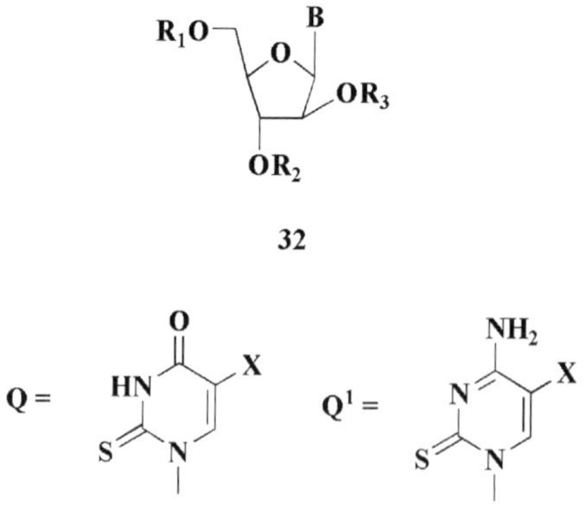

R_1O B O OR_3 OR_2

32

Q = O HN X S N

Q^1 = NH_2 N X S N

Fig. 7.32 Compounds (**32**; B=Q, Q^1; R^1, R^2, R^3=H, HO-protecting group).

NH_2 N X S N HO O OH OH

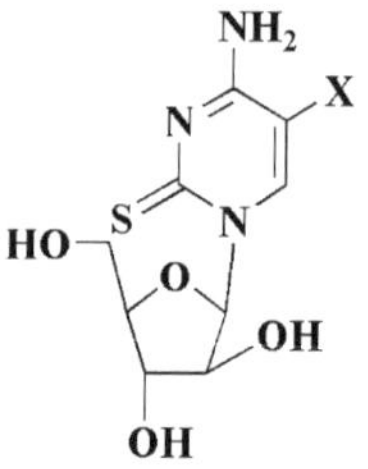

33

Fig. 7.33 Compound **33**.

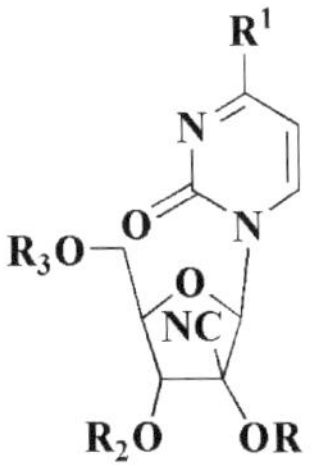

Fig. 7.34 Compound **34**; R=Q; R^1=(un)protected NH_2; R^2 and R^3 are separated or together a HO-protective group.

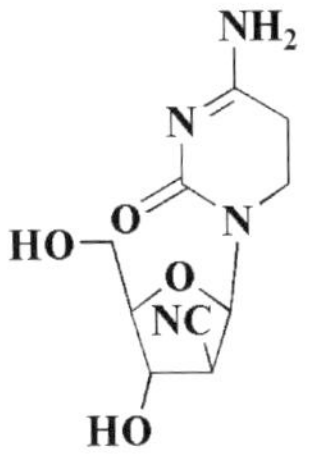

Fig. 7.35 Compound **35**.

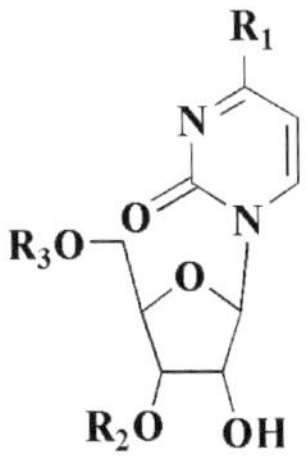

Fig. 7.36 Cytidine derivative **36**; R^1=Ac, R^2R^3=Q^1.

(**36**; R^1-R^3=same as above) (Fig. 7.36) with an oxidizing agent in the presence of an alkali metal halide and TEMPO (2,2,6,6-tetramethylpiperidinyloxy free radical) to a deoxycytidine (**37**; R^1-R^3=same as above) (Fig. 7.37), reaction of acetone cyanohydrin with the ketone, and esterification of the resulting cyanohydrin **34** (R=H; R^1-R^3=same as above) (Fig. 7.34) with 2-naphthyl chlorothioformate. Thus, cytidine derivative **36** (R^1=Ac, R^2R^3=Q^1) (Fig. 7.36) was dissolved

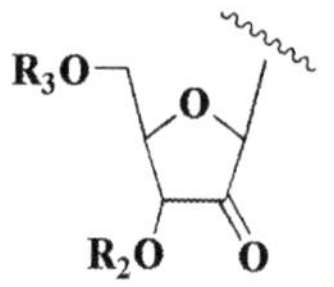

37

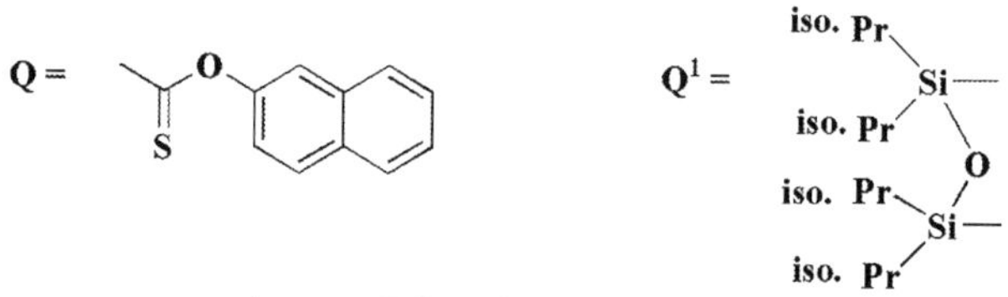

Q =

Q^1 =

Fig. 7.37 Compound **37**; $R^1 = Ac$, $R^2R^3 = Q^1$.

in CH_2Cl_2-H_2O and TEMPO and potassium bromide were added followed by cooling the mixture, adding 9% aqueous sodium hypochlorite solution (adjusting pH by adding sodium bicarbonate), and stirring the resulting mixture to furnish a ketone **37** ($R^1 = Ac$, $R^2R^3 = Q^1$) (Fig. 7.37). The latter compound was dissolved in dichloromethane, and acetone cyanohydrin and potassium dihydrogen phosphate buffer were added to yield a cyanohydrin **34** ($R = H$, $R^1 = Ac$, $R^2R^3 = Q^1$), which was esterified by 2-naphthyl chlorothioformate in the presence of 4-dimethylaminopyridine and triethylamine in toluene under cooling to afford compound **34** ($R = Q$, $R^1 = Ac$, $R^2R^3 = Q^1$) [169].

1′-C-Substituted pyrimidine nucleosides (**38**) (Fig. 7.38) and 1′-C-substituted pyrimidine 2,2′- nucleosides (**39**), which have biological

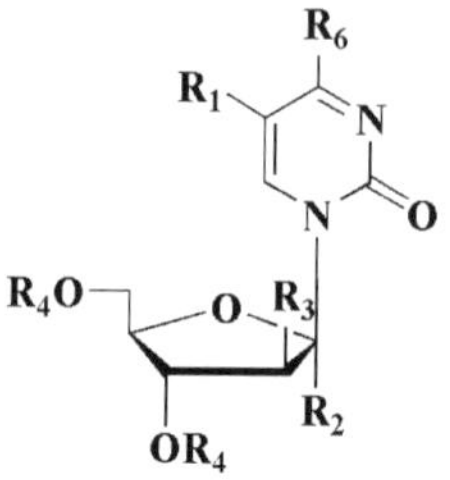

38

Fig. 7.38 1′-C-substituted pyrimidine nucleosides (**38**); $R^1 = H$, halo, lower alkyl; $R^2 =$ lower alkyl, allyl, 2-arylallyl, alkynyl, acylmethyl, cycloalkanon-2-yl, cyano, $CONH_2$; $R^3 =$ halo, H, OH, acyloxy; $R^4 = H$, HO-protecting group; $R^6 = OH$, NH_2.

activities such as antitumor activity, are prepared by reacting 1′,2′-didehydro-2′-deoxy-pyrimidine nucleosides (**40**) (Fig. 7.40) with an organic acid and a halogenating agent for acyloxylation and halogenation and reacting the resulting 1′-acyloxy pyrimidine nucleosides (**41**) (Fig. 7.41) with an organometallic compound to introduce a 1′-C substituent on the sugar moiety followed by optional removing HO-protecting group of the sugar HO groups or substituting with other protecting groups or amination at the 4-position of the base to yield **38** (Fig. 7.38). Compound **38** are transformed into **39** (Fig. 7.39) by treatment with a desilylating agent. Thus, triethylamine was added to a solution of pivalic acid in diethyl ether and stirred for 30 min, followed by successively adding **40** (R^1=H, Z=Me_3SiMe_2) (Fig. 7.40) and *N*-bromosuccinimide, and the resulting mixture was allowed to stirred at room temperature for 30 min to generate **41** (Z=Me_3SiMe_2, R^1=H, R^3=Br, R^5=O_2CCMe_3) (Fig. 7.41). Allyltrimethylsilane was added to a solution of V in dichloromethane and cooled to −40°, followed by adding $SiCl_4$ solution, and the mixture was warmed to

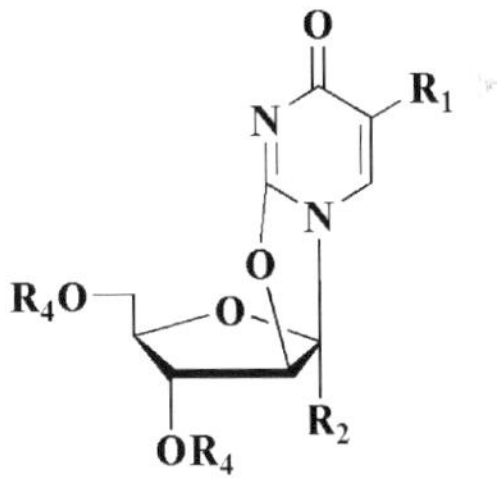

39

Fig. 7.39 Compound **39**; R^1, R^2, R^4=same as above.

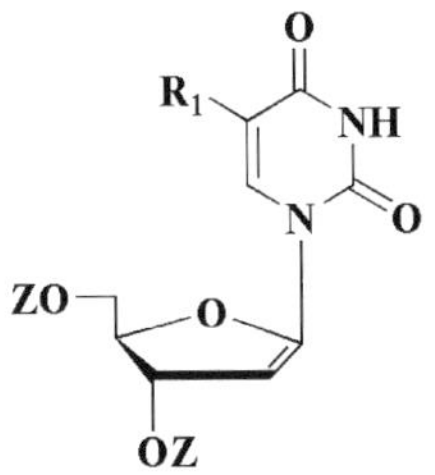

40

Fig. 7.40 Compound **40**; R^1=H, halo, lower alkyl; Z=silyl HO-protecting group.

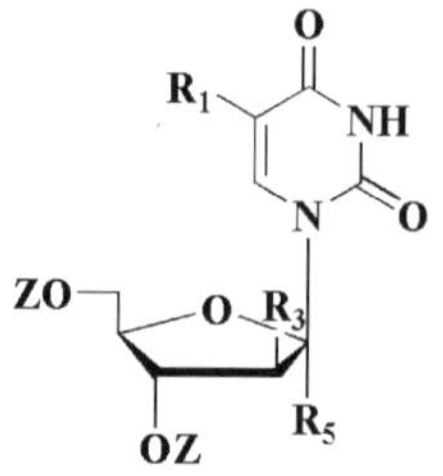

41

Fig. 7.41 Compound **41**; R[1], Z = same as above; R^3 = halo, H, OH, acyloxy; R^5 = acyloxy.

−20° over 2 h to afford **38** (R^1 = H, R^2 = allyl, R^3 = Br, R^4 = Me_3SiMe_2, R^6 = OH) (Fig. 7.38), which was stirred with Bu_4NF in tetrahydrofuran to afford **39** (R^1 = H, R^2 = allyl, R^4 = H) (Fig. 7.39) and then acetylated by Ac_2O in pyridine to yield **39** (R^1 = H, R^2 = allyl, R^4 = Ac). Similar coupling with 1-phenylallyltrimethylsilane, isopropenyloxytrimethylsilane, Me_3SiCN, 1-(trimethylsiloxy)cyclopentene, and 1-phenyl-1-(trimethylsiloxy)ethylene in the presence of $SnCl_4$ furnished **38** (R^2 = 1-phenylallyl, cyano, CH_2COMe, cyclopentanon-2-yl, and CH_2COPh; R^1 = H, R^3 = Br, R^4 = Me_3SiMe_2, R^6 = OH), respectively [170].

Pyrimidine nucleoside sulfamates **42** (Fig. 7.42) were prepared and utilized as antitumor agents and Atg7 enzyme inhibitors. Thus, nucleoside **43** (Fig. 7.43) was prepared and examined in vitro as Atg7 enzyme inhibitor [171].

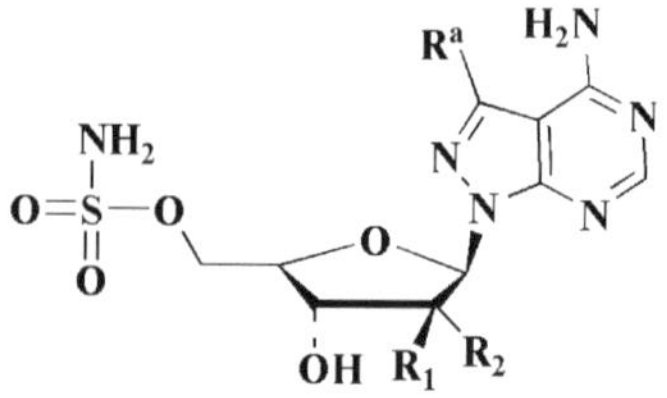

42

Fig. 7.42 Pyrimidine nucleoside sulfamates **42**; wherein R^1 is H and R^2 is H; or R^1 is H and R^2 is —OH; or R^1 is F and R^2 is H; R^a is H, iodo, CN, CO_2CH_3, $C(O)CH_3$, $C(S)NH_2$, $C(O)NH_2$, $S(O)CH_3$, fluoro-aliph., O-fluoro-aliph., S-fluoro-aliph.: or a pharmaceutically acceptable salt thereof.

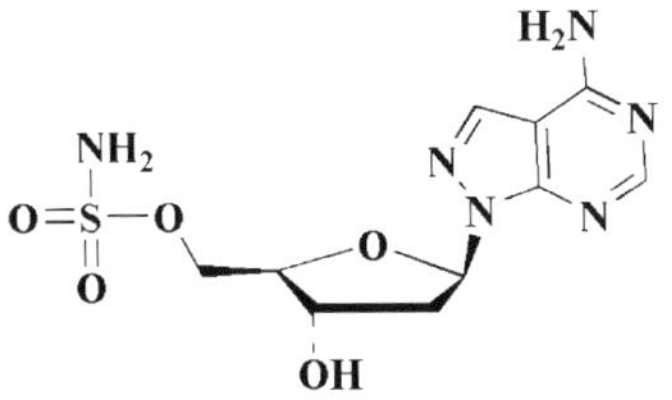

Fig. 7.43 Compound **43**.

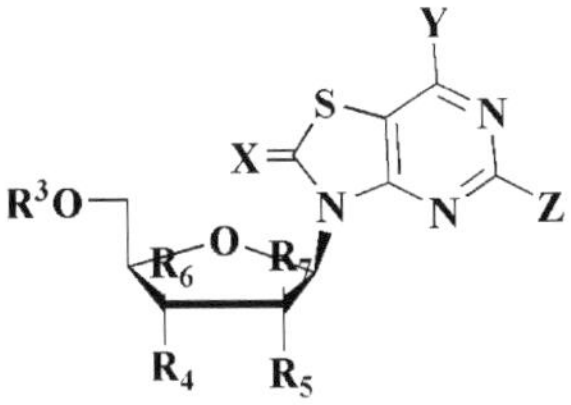

Fig. 7.44 Compound **44**; R^4-R^7 = H, OH, C_{1-18} acyloxy; R^3 = H, C_{1-18} acyl, $(HO)_2P(O)$; or R^5, R^7 = H, OH, R^6 = H and R^3R^4 = (HO)P(O)O; X = O, S; Y = OH, SH, NH_2, halo; Z = H, NH_2, OH, Cl, Br.

Compounds **44** (Fig. 7.44), specifically 5-amino-3-β-D-ribofuranosylthioazolo[4,5-*d*] pyrimidine-2,7(6*H*)-dione, or their pharmaceutically acceptable salts, which are useful for treating tumors, inhibiting tumor metastasis, stimulating the immune system, and enhancing natural killer immune cells, lymphocyte cells, and macrophage cells in a mammalian host, are prepared [172].

There are described phenylboronic acid compounds. Pyrimidine nucleosides **45** (Fig. 7.45) were also prepared as intermediates for oligonucleotide analogs. These compounds may be selectively activated to generate active anticancer agents in tumor cells, such as non-small-cell lung cancer, leukemia, CNS cancer, colon cancer, ovarian cancer, melanoma, prostate cancer, renal cancer, and breast cancer [173].

The compounds **46** (Fig. 7.46), useful as antitumor agents, are synthesized by reaction of 3′-acetylene-containing pyrimidine nucleosides **47** (Fig. 7.47) with a R^3-containing organocopper reagent [174].

Compounds **48** (Fig. 7.48), **49** (Fig. 7.49) and their pharmaceutically acceptable salts were synthesized. A solution of 1-[3,5-*O*-(1,1,3,3-tetraisopropyldisiloxanediyl)-β-D-erythropentofuran-2-urosyl]thymine in Et_2O-H_2O (2:1) was treated with sodium cyanide and sodium bicarbonate to afford

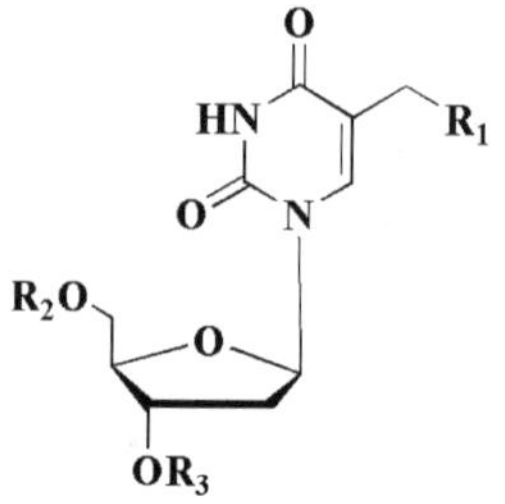

45

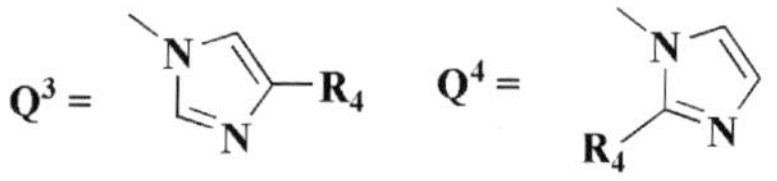

Fig. 7.45 Pyrimidine nucleoside derivatives **45**; $R^1 = Q^3, Q^4$; $R^2 = H$, a hydroxy-protecting group; $R^3 = H$, $P(N(CH(CH_3)_2)_2)(OCH_2CH_2CN)$; $R^4 =$ an electron-withdrawing group.

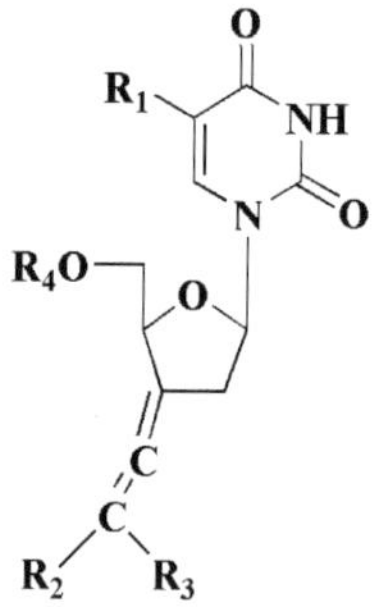

46

Fig. 7.46 Compound **46**; $R^1 = H$, lower alkyl; $R^2 =$ aryl, lower alkyl; $R^3 =$ aryl, lower alkyl; $R^4 = H$, HO-protective group.

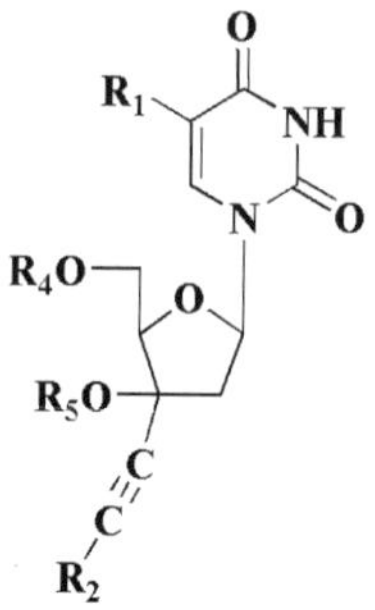

47

Fig. 7.47 3′-Acetylene-containing pyrimidine nucleosides **47**; $R^1 = H$, halo, lower alkyl; $R^2 =$ aryl, lower alkyl, silyl; $R^4 = H$, HO-protective group; $R^5 =$ acyl, sulfonyl.

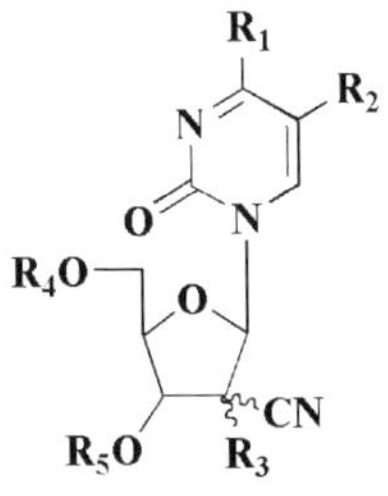

48

Fig. 7.48 Compound **48**.

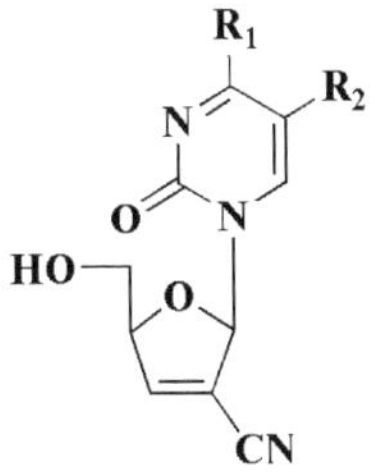

49

Fig. 7.49 Compound **49**; R^1 = OH, (substituted) amino; R^2 = H, alkyl; R^3 = H, OH; R^4, R^5 = H; or R^4R^5 = SiR^6R^7-O-SiR^8R^9; R^6-R^9 = alkyl.

1-[2′-cyano-3′,5′-*O*-(1,1,3,3-tetraisopropyldisiloxanediyl)-β-D-ribofuranosyl]thymine. 1-(2′-Cyano-2′-deoxy-β-D-arabinofuranosyl)cytosine-HCl had an IC_{50} of 0.21 μg/mL against L1210 cells in an in vitro study [175].

The toxicity of antineoplastic and antiviral agents, resulting from their damage to the mucosal tissue or the hematopoietic system, is prevented or treated with acylated derivatives of nonmethylated pyrimidine nucleosides (Fig. 7.50). These derivatives may themselves be antiviral,

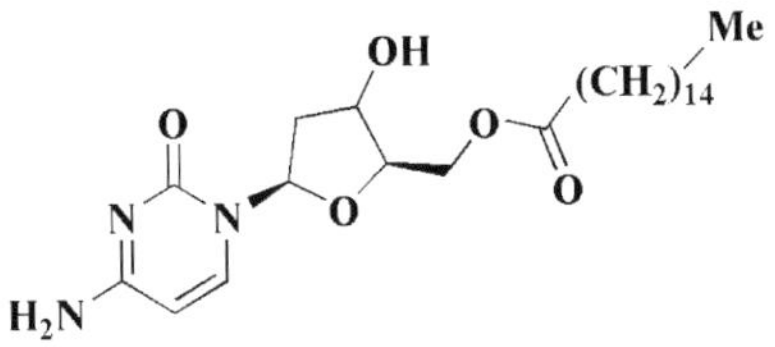

50

Fig. 7.50 Acylated derivative of nonmethylated pyrimidine nucleosides **50**.

antineoplastic, or antimalarial agents; they may be administered together with inhibitors of uridine phosphorylase, of cytidine deaminase, or of nucleotide transport. Thus, oral administration of triacetyluridine (500 mg/kg 8 times in 2 days) rescued mice from the hematololgical toxicity of 5-fluorouracil (150 mg/kg i.p.), as shown by leukocyte and platelet counts [176].

A method of using androgen receptor- and/or butyrylcholinesterase-targeted radiolabeled compounds, e.g., cycloSalingenyl pyrimidine nucleoside monophosphates, for the treating and diagnosing of cancer is reported. Compounds **51** (Fig. 7.51) were synthesized. The prepared compounds were estimated for BChE inhibitory activity and antitumor potency [177].

Compounds **52** (Fig. 7.52) were prepared as neoplasm inhibitors, which may be utilized by themselves or in conjunctive therapy with 5-fluorouracil. Thus, Me_3SiCH_2OH in dichloromethane and then triethylamine were added to a −79° mixture of $(COCl)_2/Me_2SO$ in dichloromethane to afford a solution of aldehyde; this was treated with a mixture prepared from

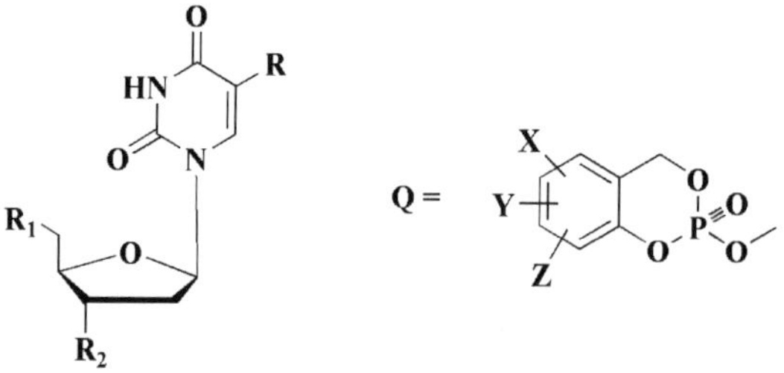

51

Fig. 7.51 Compound **51**; R = halo, radiohalo, aryl, alkyl, trialkylstannyl, etc.; R^1 = OH, Q; R^2 = halo, radiohalo, alkanoate, alkoxy, and rogen receptor binding ligand, Q, etc.; X = H, F, Cl, alkyl, alkoxy; Y, Z = H, alkyl, aryl, aryloxy.

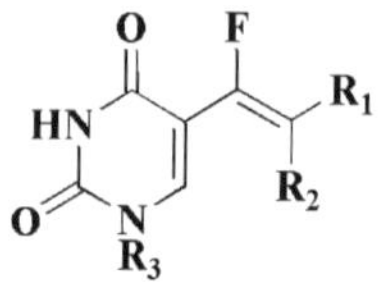

52

Fig. 7.52 Compound **52**; R^1, R^2 = H, alkyl, pH group unsubstituted or substituted with 1–3 alkyl or alkoxy groups; R^3 = H, ribosyl, 2′-deoxyribosyl, arabinosyl.

FCH_2SO_2Ph, $LiN(CHMe_2)_2$, and di-Et chlorophosphate to yield 77% (*E*,*Z*)-$Me_3SiCH{:}C(F)SO_2Ph$. This was refluxed for 24h in PhMe with AIBN and tributyltin hydride to afford 91% $Me_3SiCH{:}C(F)SnBu_3$.

The latter was heated with 5-iodo-2′-deoxyuridine and $Pd(Ph_3P)_4$ in dimethylformamide to yield the coupling product, which was protodesilylated with $(COOH)_2$ in methanol to afford compound **53** (Fig. 7.53). Compound **53** indicated $IC_{50}=9.4\,\mu M$ against KB cells, and a mixture of **53** and 5-fluorouracil revealed $IC_{50}=0.70\,\mu M$ [178].

6-(Halovinyl) pyrimidine nucleosides (**54**) (Fig. 7.54) are cyclized by a reducing agent to afford spirotype pyrimidine cyclonucleosides (**55**) (Fig. 7.55), which have antitumor and antiviral agents. Thus, *O* ^{2},2′-anhydrouridine derivative (**56**) (Fig. 7.56) was formylated by Me formate in the presence of $(Me_2CH)_2NLi$ in tetrahydrofuran to furnish 6-formyluridine derivative (**57**) (Fig. 7.57), which underwent Wittig reaction with [bromo(methoxycarbonyl)methylene]triphenylphosphorane in dimethylformamide to yield 6-(bromovinyl)uridine derivative **54**

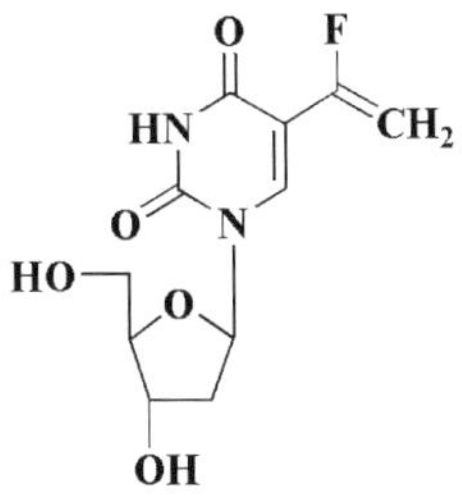

53

Fig. 7.53 Compound **53**.

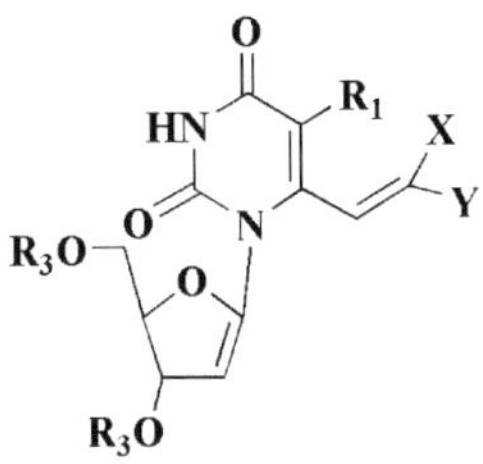

54

Fig. 7.54 6-(Halovinyl) pyrimidine nucleosides (**54**) R^1 = halo, H, lower alkyl; R^3 = HO-protective group, H; one of X and Y = halo and the other = halo, H, acyl, or alkoxycarbonyl.

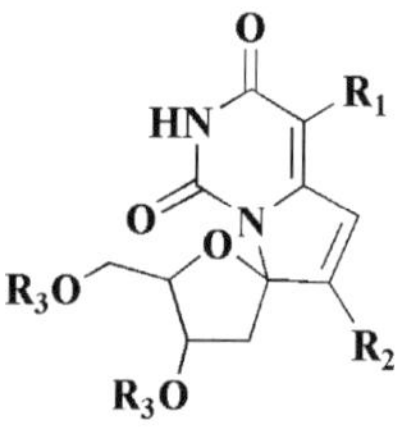

Fig. 7.55 Spiro-type pyrimidine cyclonucleosides (**55**) R^1, R^3 = same as above; R^2 = H, halo, acyl, or alkoxycarbonyl.

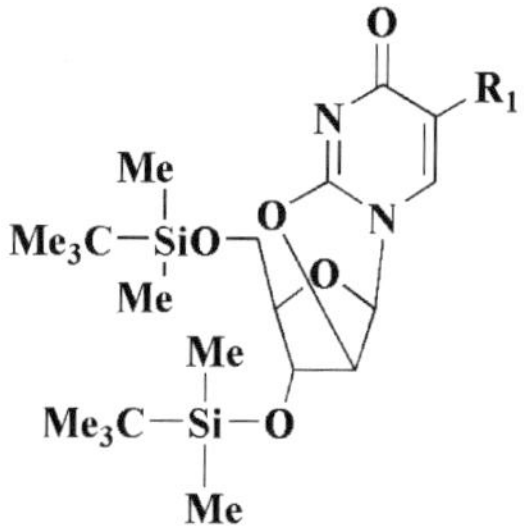

Fig. 7.56 *O*^{2},2′-Anhydrouridine derivative (**56**).

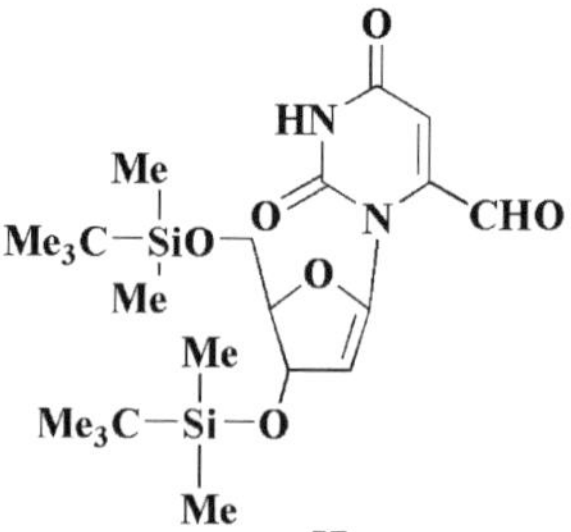

Fig. 7.57 6-Formyluridine derivative (**57**).

(R^1 = H, R^3 = Me_3CSiMe_2, X = Br, Y = CO_2Me). The latter compound was treated dropwise with a solution of tributyltin hydride and azobisisobutyronitrile in benzene to afford compound **55** (R^1 = H, R^2 = MeO_2C, R^3 = Me_3CSiMe_2). Also prepared was **55** (R^1 = H, R^2 = Br, R^3 = Me_3CSiMe_2) by cyclization of **54** (R^1 = H, R^3 = Me_3CSiMe_2, X = Y = Br [179].

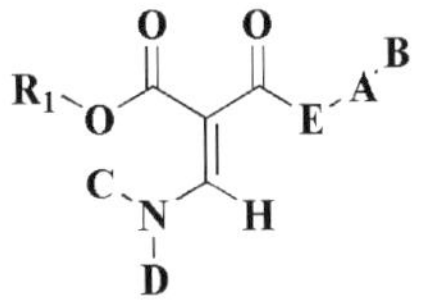

58

Fig. 7.58 Compound **58**; R^1 = H or (un)substituted alkyl; A = an spacer; B = group capable of binding to proteins; E = O or a single bond; (C)(D)N = a residue of an active drug having an amino group; N = the nitrogen atom of the amino group, which is a primary amine, a secondary amine, or a cyclic amine.

Compounds **58** (Fig. 7.58) or pharmaceutically acceptable salts were invented. The active drug having an amino group is an anticancer agent, an antimetabolic agent, a kinase inhibitor, a CDK inhibitor, an anthracycline anticancer agent, a FGFR inhibitor, a HDAC inhibitor, a PI3K inhibitor, a PARP inhibitor, a TLR7 agonist, a BTK inhibitor, or a topoisomerase inhibitor, preferably 5′-deoxy-5-fluorocytidine, gemcitabine, palbociclib, amrubicin, tucidinostat, imatinib, or crizotinib. After administration into an organism, the compounds **58** rapidly bind to blood albumin, persistently circulate systemically, and efficiently accumulate in tumor tissue due to the enhanced permeability and the retention (EPR) effect, and then selectively release an active drug in tumor cells. The compounds **58** display a high concentration in tumor tissue, and an excellent antitumor effect, and provide excellent anticancer agent while producing few adverse effects to normal cells.

When compound **59** (Fig. 7.59) (R = Na) at 10 mg/5 mL/kg i.v. was administered to male ICR mouse, the serum concentration of compound **59** (R = Na) was 174, 45, 89, and 35 μM after 0.25, 1, 8, 24, and 48 h, respectively [180].

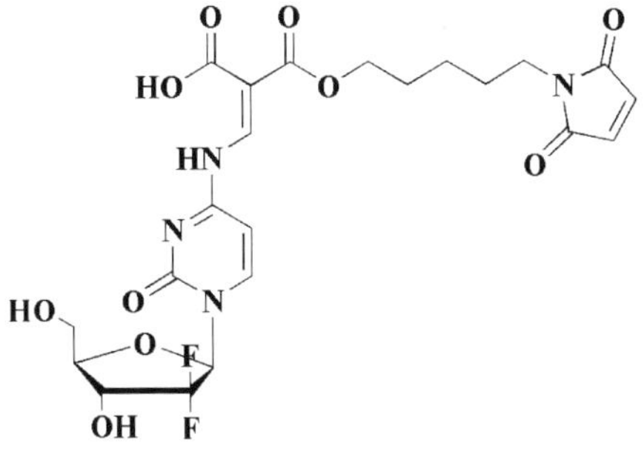

59

Fig. 7.59 Compound **59**.

5′-Deuterated nucleoside and nucleotides, and modifications thereof were prepared for use in medical therapies, including as antitumor, antiviral, and antineoplastic agents (Fig. 7.60). Thus, nucleotide **61** (Fig. 7.61) was prepared. In one embodiment, compounds, methods, and uses are provided for the treatment of hepatitis C, HSV, RSV, and other viral diseases in a host, including a human [181].

1-(2-Deoxy-2-fluoro-4-thio-β-D-arabinofuranosyl)cytosines (**62**; R = H, PO_3H_2) (Fig. 7.62), having an excellent antitumor potency, are prepared.

The latter compound in vivo at 3 mg/kg i.v. per day for 10 days decreased the tumor weight in mice transplanted with mouse sarcoma

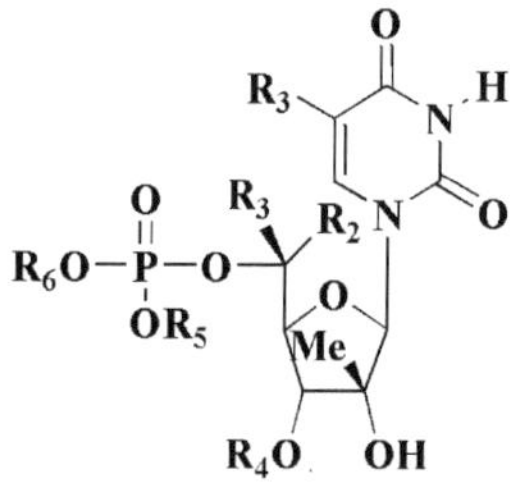

Fig. 7.60 Compound **60**; wherein R^1 and R^2 are independently deuterium, hydrogen, or C(*H*)m(D)n; and at least one of R^1 or R^2 is deuterium; R^3 is hydrogen, deuterium, halogen (F, Cl, Br, or I), C(H)*m*(D)*n*; or alkyne; wherein the R^3 alkyne and the C_4-Oxygen of the pyrimidine can form a heterocyclic ring; R^4 is hydrogen, deuterium, C(H)*m*(D)*n*, acyl or phosphate; R^5 is hydrogen, deuterium, alkyl, cycloalkyl, aryl, heteroaryl, heterocyclic, alkyl(heterocycle), alkyl(heteroaryl), alkylaryl; R^6 is hydrogen, deuterium, alkyl, cycloalkyl, aryl, heteroaryl, heterocyclic, alkyl(heterocycle), alkylheteroaryl, alkylaryl.

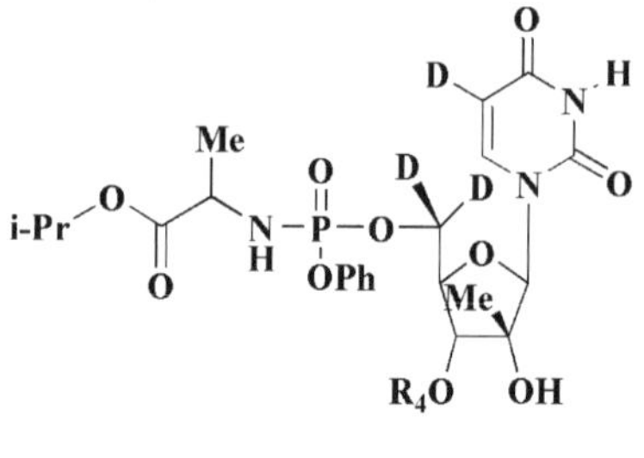

Fig. 7.61 Nucleotide **61**.

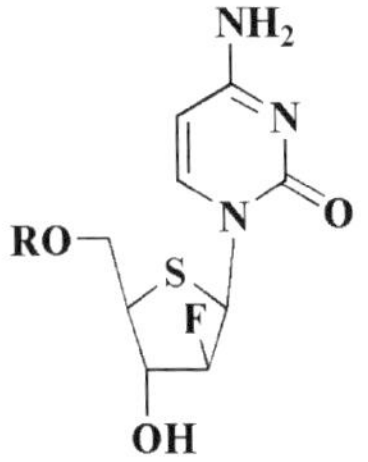

Fig. 7.62 1-(2-Deoxy-2-fluoro-4-thio-β-D-arabinofuranosyl)cytosines **62**; R = H, PO_3H_2.

S-180 by 83% compared to control animals. Pharmaceutical formulations, e.g., tablet containing **62**, were reported [182].

It is intended to provide a novel pyrimidine nucleoside compound in which anticancer efficacy and cytotoxicity are balanced. Disclosed is a compound represented by a general formula (**63**) (Fig. 7.63). A pharmaceutical component, especially an antitumor component containing the same, is also disclosed. For example, 1-[5′-*O*-tert-butyldimethylsilyl-2′-deoxy-4′-thio-1-β-D-ribofuranosyl]-5-fluorouracil showed an antitumor effect in KM20C human colon cancer cell-bearing mice with an inhibition rate (IR) of 73.2% [183].

A pyrimidine nucleoside compound represented by formula (**64**) (Fig. 7.64) is synthesized. A pharmaceutical component and an antitumor agent containing the pyrimidine nucleoside compound or its salt with a pharmaceutically acceptable carrier are also disclosed. For example, oral administration of 1-[2′-deoxy-3′-*O*-cyclohexanecarbonyl-4′-thio-1-β-

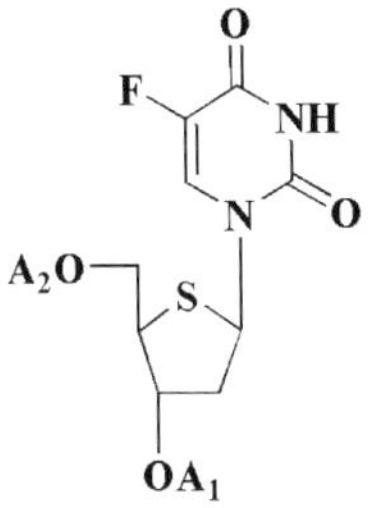

Fig. 7.63 Compound **63**, wherein at least one of A^1 and A^2 represents (un)substituted C_{1-12} alkyl- or C_{6-14} aryl-silyl group, or A^1 and A^2 may together form a Si-containing ring.

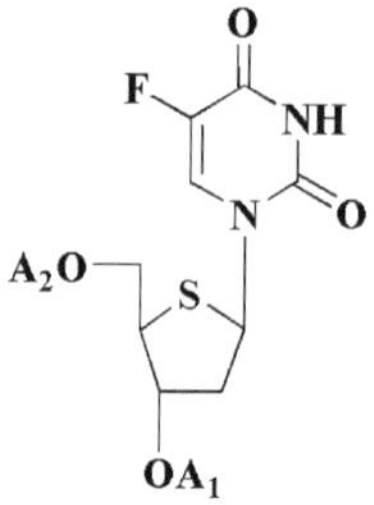

64

Fig. 7.64 Compound **64**; one of A^1 and A^2 is H or R^1CO and the other is R^2CO, one of A^1 and A^2 is H or $R^3(OCH_2)_m$ and the other is $R^4(OCH_2)_{m'}$, when one of A^1 and A^2 is H, $n=1$ (R^1, R^2 = (un)substituted C_{1-6} alkyl, (un)substituted C_{3-7} cyclic alkyl, (un)substituted C_{6-14} aryl; m, $n=0,1$; R^3, $R^4=C_{7-15}$ aralkyl), or its salt.

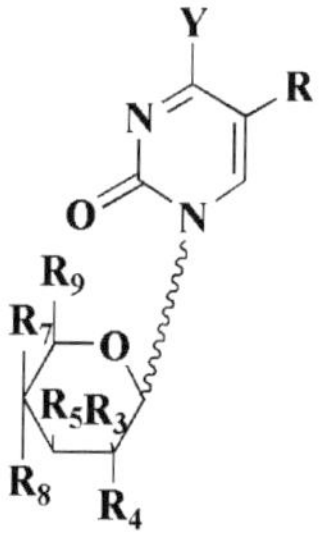

65

Fig. 7.65 Pyrimidine monosaccharides **65**; R = halo; Y = H, NH_2, OH or SH; R_3 or R_4 = H, with the other being H, OH, OAc or NHAc; R_5 = OH or OAc; either R_7 or R_8 = H, with the other being OH or OAc; R_9 = H, CH_2OH or CH_2OAc; with the proviso that when R_4 = OH, OAc or NHAc, then R_8 = H.

D-ribofuranosyl]-5-fluorouracil for 14 days inhibited tumor growth in KM20C tumor-bearing mice with an inhibition rate of 75.4% [184].

Pyrimidine monosaccharides **65** (Fig. 7.65) and enantiomers of such compounds were synthesized, e.g., 1-α- and 1-β-D-galactopyranosyl-5-fluorouracil were yielded from 5-fluorouracil and peracetylgalactose. The final compounds significantly slowed the rate of tumor growth in mice without causing toxic side effects (1300 mg/kg [185].

Compounds **66** (Fig. 7.66) are prepared as antitumor and antiviral agents, particularly potential anti-HIV agents, are synthesized. Thus, adenine and potassium carbonate were suspended in dimethylformamide and after stirring, 18-crown-6 ether and Me 2,3-anhydro-4,6-O-benzylidene-α-

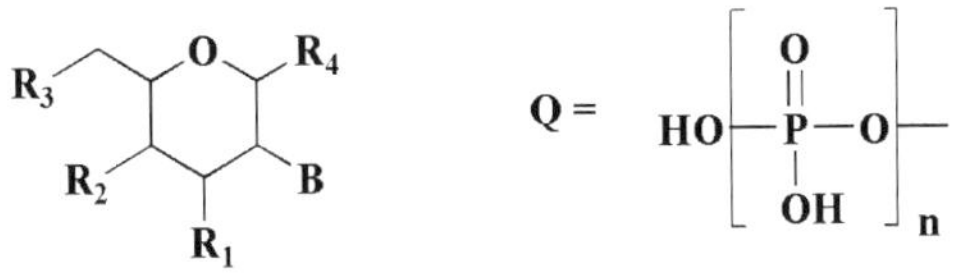

66

Fig. 7.66 Compound **66**; B = adenine, guanine, thymine, uracil, cytosine, xanthine, hypoxanthine, 5-methylcytosine, 4-ethoxy-5-methyl-2-oxopyrimidine, 5-methyl-2-oxopyrimidine, 4-isopropoxy-5-methyl-2-oxopyrimidine; R^1, R^2 = H, OH; or R^1R^2 = bond; R^3 = Q wherein n = 0,1,3; R^4 = H, lower alkoxy) or ethers, pharmacological acceptable esters, or salts thereof.

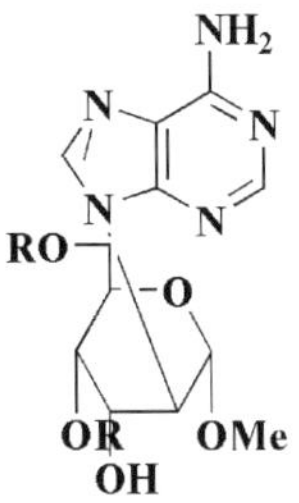

67

Fig. 7.67 Adenylaltropyranoside derivative **67**; RR = CHPh.

D-allopyranoside were added followed by stirring the resulting mixture to yield adenylaltropyranoside derivative (**67**; RR = CHPh) (Fig. 7.67) [186].

α- and β-L-Pyranosyl nucleosides **68** (Fig. 7.68), which are useful as anticancer, antifungal, antiviral, antibacterial, and/or antiparasitic agents in mammals, are prepared. Thus, 10 mmol silylated N^6-benzoyladenosine (**69**) (Fig. 7.69) and 1,2,3,4-tetra-*O*-acetyl-L-ribopyranose were treated with $CF_3SO_3SiMe_3$ in acetonitrile and the resulting clear solution was refluxed for 14 h to afford, after deprotection with methanolic ammonia, compound (**70**) (Fig. 7.70). Compound **70** in vitro indicated an IC_{50} of 8.3 and 7.8 μg/mL against human tumor cell lines OM-1 (colon) and H578St (breast), respectively, as compared to 1.47 and <0.6, respectively, for 5-fluorouracil. It at 10 μg/mL revealed 94.1% survival of bone marrow cells vs. 38.6% for 5-fluorouracil at 6 μg/mL [187].

The nucleosides **71** (Fig. 7.71), which selectively inhibit DNA synthesis, are synthesized from (*S*)-glycerin derivative **71** (X = OH, R = Ph, R^1 = Ac). Thus, tosylation of the latter with *p*-tosyl chloride in pyridine under ice-cooling followed by deacetylation with methanolic ammonia and

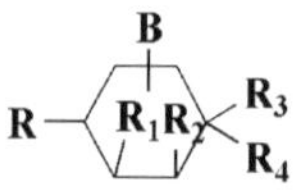

68

Fig. 7.68 α and β-L-Pyranosyl nucleosides **68**; B = (un)substituted naturally occurring nucleoside base such as A, G, C, U, T, or hypoxanthine, provided that when the base is a pyrimidine, the atom at position 4 of the base can be S and further provided that when the base is a purine, the atom at position 6 of the base may be S; R = OR^5; wherein R^5 = COR^6, H, $P(O)_nR^7R^8$; wherein R^6 = (un)substituted C_{1-5} alkyl or aromatic ring structure; R^7, R^8 = H, C_{1-5} alkyl; n = 2,3; R^1, R^2 = H, mono- or dihalo, B, group listed in R; R^3, R^4 = B, H, group listed in R; provided that only one of R^1-R^4 can be B and further provided that when R and R^1 are each R^2 = H, OH, and R^3 = B, then B cannot be thymine; and when R and R^1 are each OH, R^2 = H, and R^4 = B, then B cannot be thymidine.

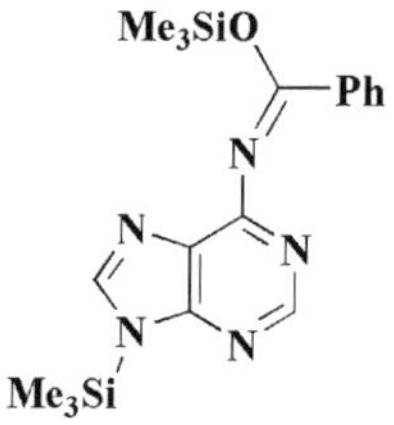

69

Fig. 7.69 Silylated N^6-benzoyladenosine (**69**).

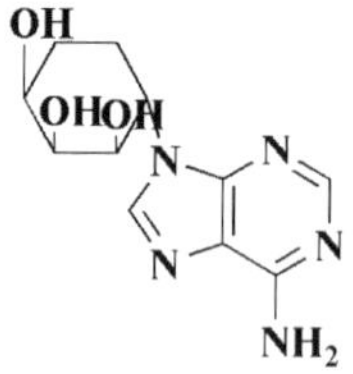

70

Fig. 7.70 Compound **70**.

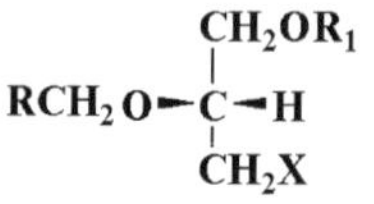

71

Fig. 7.71 Nucleosides **71**; R = heterocyclic base residue; R^1 = OH; X = halo, tosyloxy, N_3, NH_2.

hydrogenolysis over 5% Pd/C yielded (*R*)-(−)-3-tosyloxy-1,2-propane diol. The latter was cycloacetalized with refluxing trioxane containing 40% aqueous sulfuric acid yielded (*R*)-dioxolane derivative (**72**; Ts = *p*-$MeC_6H_4SO_2$) (Fig. 7.72), which was transformed to **71** (X = TsO, R = R^1 = AcO) with $Ac_2O/ZnCl_2$. The latter condensed with bistrimethylsilylthymine in the presence of tin tetrachloride in dichloromethane afforded **71** (X = TsO, R = Q, R^1 = Ac, R^2 = Me). The latter deacetylated with 25% aqueous methanolic ammonia followed by heating with sodium azide in dimethylformamide at 100° for 3 h yielded **71** (X = N_3, R = Q, R^1 = H, R^2 = Me). **71** (X = N_3, R = Q, R^1 = H, R^2 = F) was similarly prepared [188].

Compounds **73** (Fig. 7.73), useful as antiviral and anticancer agents, are prepared. Mesylation of alc. **74** (Fig. 7.74) with Et_2N and $MeSO_2Cl$ in dichloromethane afforded mesylate **74** (Z^1 = $MeSO_3$, R^3 = PhCO), which

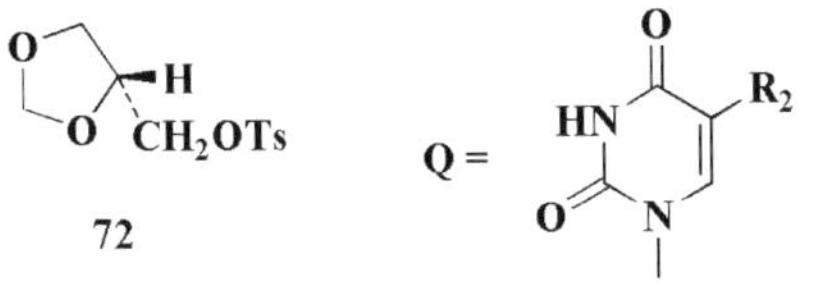

Fig. 7.72 (*R*)-Dioxolane derivative **72**; Ts = p-$MeC_6H_4SO_2$.

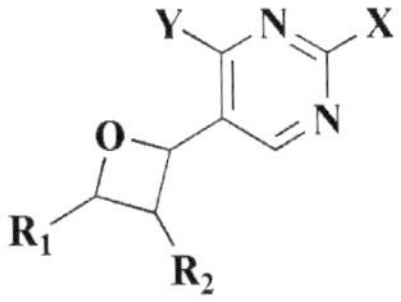

Fig. 7.73 Compound **73**; R^1 = alkyl, H, CH_2OR^3 (wherein R^3 = H, alkyl); R^2 = alkyl, H, OR^3, CH_2OR^3; X,Y = H, alkyl, OR^3, halo, NH_2.

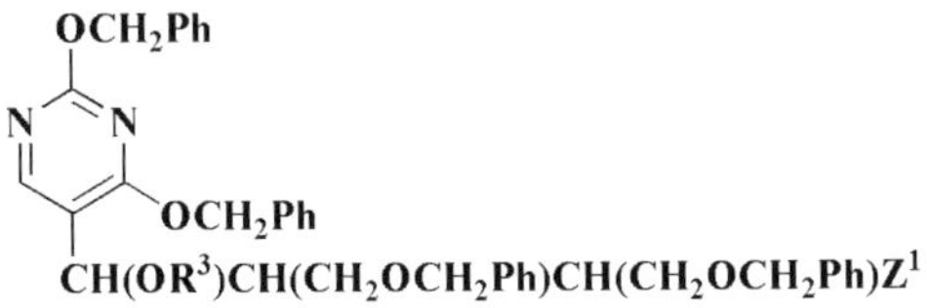

Fig. 7.74 Compound **74**; R^3 = PhCO, Z^1 = OH.

was hydrolyzed with sodium carbonate in methanol/tetrahydrofuran to afford **74** (R^3 = H, Z^1 = $MeSO_3$). A solution of the latter in tetrahydrofuran was added to a suspension of 60% sodium hydride in tetrahydrofuran with stirring at room temperature under nitrogen to yield β-D-erythro-I (R^2 = R^1 = $PhCH_2OCH_2$, Y = X = $PhCH_2O$ [189].

Compounds [(1*R*,2*R*,3*S*)-**75**] (Fig. 7.75), useful as antiviral agents against cytomegalovirus, herpes simplex virus (HSV), immunodeficiency virus, or hepatitis virus B, are prepared using cyclobutane intermediates (**76**) (Fig. 7.76), which in turn are prepared by cycloaddition of R^2CH:$CHCOQ^4$ to R^3CH:$C(SR^1)_2$. Thus, stereoselective cycloaddition of 3-[(*E*)-3-(methoxycarbonyl)propenoyl] oxazolidin-2-one to CH_2:$C(SMe)_2$ in the presence of HA molecular sieves and a Lewis acid complex furnished from $(Me_2CHO)_2TiCl_2$ and (2*S*,3*S*)-2,3-*O*-(1-phenylethylidene)-1,1,4,4-tetraphenyl-1,2,3,4-butanetetraol in 1,3,5-trimethylbenzene afforded (2*S*,3*S*)-(−)-**76** (R^1 = Me, R^2 = MeO_2C, R^3 = H, R = 2-oxooxazolidin-3-yl), which was transformed into (1*S*,2*S*,3*S*)-(+)-2,3-bis(tert-butyldiphenylsilyloxymethyl)-1-(methanesulfonyloxy)cyclobutane

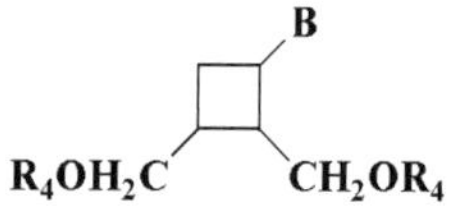

75

Fig. 7.75 Compound **75**; B = purine and pyrimidine base, i.e., Q-Q^3; R^4 = H, protecting group; Y^1 = alkyl, H; Y^2 = NH_2, H, halo; Y^3, Y^4 = NH_2, H.

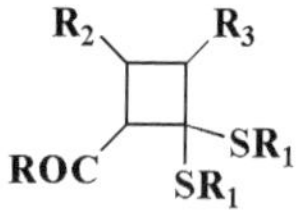

76

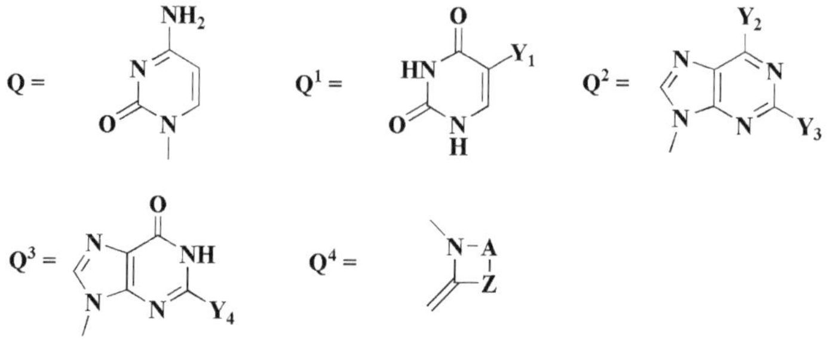

Fig. 7.76 Cyclobutane intermediates (**76**); R = Q^4; R^1 = C_{1-5} alkyl, aralkyl; or R^1R^1 = C_{2-3} alkylene; R^2 = H, C_{1-5} alkyl, alkoxy, or aralkyloxy; A = branched or straight C_{2-5} alkylene; Y = S, O; Z = S, O, or (un)substituted CH_2.

on several steps. The latter was condensed with 2-amino-6-(2-methoxyethoxy) purine in dimethylformamide containing sodium hydride followed by deprotection with 4N HCl/dioxane afforded (1*R*,2*R*,3*S*)-(+)-**75** (R^4 = H, B = guanin-9-yl), which was more efficient than acyclovir in reducing the mortality rate of mice infected with HSV-2 [190].

Cubane nucleoside analogs **77** (Fig. 7.77) were synthesized and are useful as anticancer and antiviral agents. The invention also provides pharmaceutical components containing one or more cubane nucleoside analogs and one or more pharmaceutically acceptable carriers. The invention also provides methods for the treatment of viral infections and cancer in mammals. Thus, cubane nucleoside analog **78** (Fig. 7.78) was synthesized and estimated in vitro as antitumor and antiviral agent [191].

Nucleoside phosphonates and analogs **79** (Fig. 7.79) and **80** (Fig. 7.80) were synthesized and estimated as antitumor and antiviral agents. Compounds and components are provided for treatment, prevention, or amelioration of a variety of medical disorders associated with viral infections and/or cell proliferation. Thus, nucleoside phosphonate analog **81** (Fig. 7.81) was synthesized and examined as antitumor and antiviral agent [192].

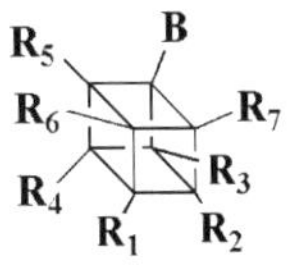

77

Fig. 7.77 Cubane nucleoside analogs **77**; R^1-R^7 are independently H, CN, halogen, azido, OH, COOH, halo-alkoxy, halo-alkyl, alkyl, alkynyl, alkenyl, alkoxy, alkyl-ester, alkylamino, cycloalkyl; B is optionally substituted pyrimidine or purine nucleobase.

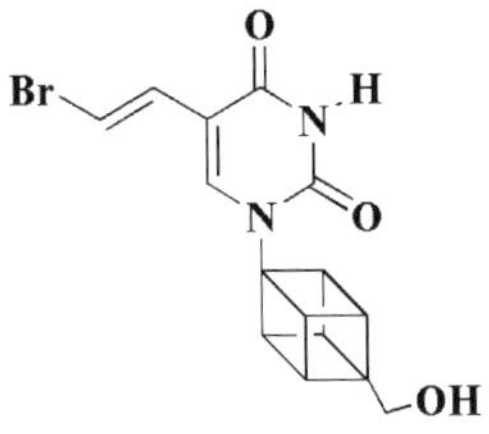

78

Fig. 7.78 Cubane nucleoside analog **78**.

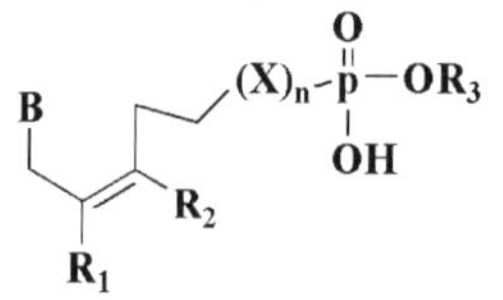

79

Fig. 7.79 Nucleoside phosphonate analog **79**; wherein B is a pyrimidine or purine nucleobase or an analog; R^1 and R^2 are independently H, alkyl, aryl, arylalkyl, R^4CO_2-, R^5O-, $R^6S(O)_m$, halogen, heteroalkyl, N_3, heteroaryl, heteroarylalkyl; R^1R^2 together with the carbon atom to which they are attached forms a cycloheteroalkyl, cycloalkyl; X is CH_2; *n* is 0–1; *m* is 0–2; R^3 is H, monovalent cation, lipophilic group; R^4-R^6 are independently alkyl, aryl, arylalkyl, heteroalkyl, heteroaryl, and heteroarylalkyl.

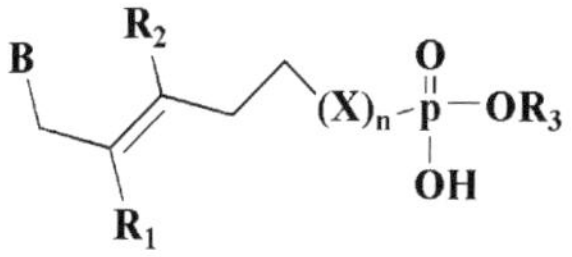

80

Fig. 7.80 Nucleoside phosphonate analog **80**; wherein B is a pyrimidine or purine nucleobase or an analog; R^1 and R^2 are independently H, alkyl, aryl, arylalkyl, R^4CO_2—, R^5O—, $R^6S(O)_m$, halogen, heteroalkyl, N_3, heteroaryl, heteroarylalkyl; R^1R^2 together with the carbon atom to which they are attached forms a cycloheteroalkyl, cycloalkyl; X is CH_2; *n* is 0–1; *m* is 0–2; R^3 is H, monovalent cation, lipophilic group; R^4-R^6 are independently alkyl, aryl, arylalkyl, heteroalkyl, heteroaryl, heteroarylalkyl.

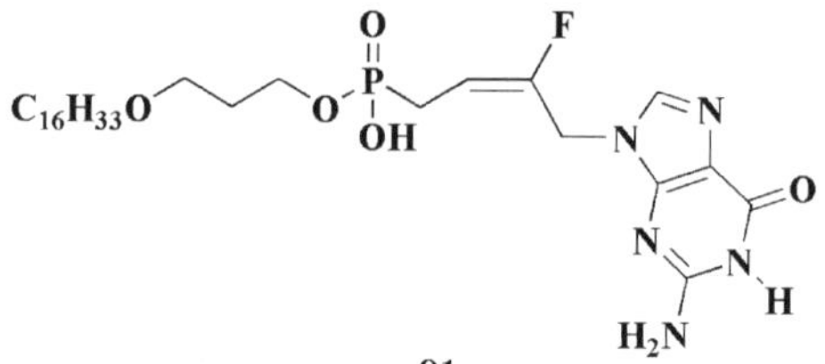

81

Fig. 7.81 Nucleoside phosphonate analog **81**.

3. Nucleoside analogs: Syntheses and biological evaluations

3.1 Ribonucleoside pyrimidine analogs

3.1.1 Free and protected pyrimidine ribonucleoside analogs

The synthesis of 5-[alkoxy-(4-nitro-phenyl)-methyl]-uridines was accomplished in Schemes 7.1 and 7.2. Compound **82** was refluxed with

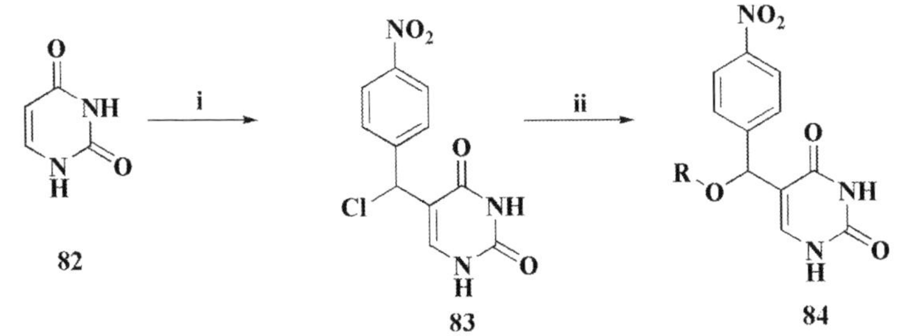

Scheme 7.1 Reagents and conditions:(i) 4-Nitrobenzaldehyde/concentrated HCl/reflux/4 h. (ii) Alcohol/reflux/4 h.

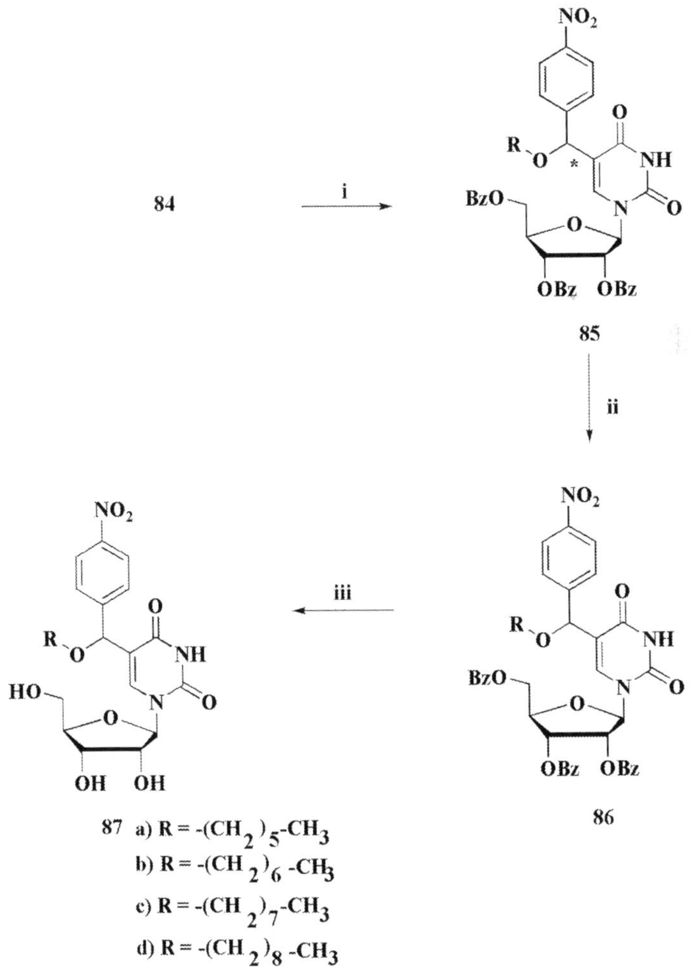

87 a) R = $-(CH_2)_5-CH_3$
b) R = $-(CH_2)_6-CH_3$
c) R = $-(CH_2)_7-CH_3$
d) R = $-(CH_2)_8-CH_3$

Scheme 7.2 Reagents and conditions: (i) Hexamethyldisilazane,/$(NH_4)_2SO_4$/trimethylsilyl trifluoro-methanesulfonate/1-*O*-acetyl-2,3,5-tri-*O*-benzoyl-β-D-ribofuranose/anhydrous 1,2-dichloroethane/RT/2 days. (ii) Separation of diastereoisomers. (iii) MeOH.

4-nitrobenzaldehyde in the presence of hydrochloric acid for 4 h to yield compound **83**, which was converted to 5-[alkoxy-(4-nitro-phenyl)-methyl]-uracils **84**, as shown in Scheme 7.1 [193].

Silylation of the starting compounds **84** is shown in Scheme 7.2. The reaction of the protected sugar with the silylated uracils in the presence of TMSOTf afforded the benzoylated ribonucleosides **85** that were in the form of mixture of two diastereoisomers **86a–d**. The latter diastereoisomers were isolated from each mixture using methanol in chloroform (0%–5%). Upon treatment of the ribonucleosides **86**a-d with methanolic ammonia, the nucleosides **87a–d** were yielded [193].

Under in vitro conditions, the synthesized nucleosides **87a–d** were examined for their cytotoxic activity against cancer cell lines, comprising drug-sensitive (K-562 and CEM), drug-resistant (K-562 TAX and CEM-DNR-B) cell lines, and A549 cells as representative of solid tumors.

Studies on the relationships of structure and the cytotoxic activities were performed. Through some cases of the transformation of bases to the nucleosides, there was a slight increase in the cytotoxic activity. The increase in the cytotoxic activity depended on the length of the alkyl chain [193].

To identify an orally available fluoropyrimidine having safety profiles and efficacy greatly enhanced over those of parenteral 5-fluorouracil (5-FU: **91**), a 5-FU prodrug was designed that would pass intact through the intestinal mucosa and be sequentially transformed to 5-FU by enzymes that are highly expressed in the human liver and then in tumors. Among a variety of the *N*4-substituted 5′-deoxy-5-fluorocytidine derivatives, a series of *N*4-alkoxycarbonyl derivatives were subjected to hydrolysis to generate the 5′-deoxy-5-fluoro-cytidine (5′-DFCR: **89**) specifically by carboxylesterase, which exists preferentially in the liver in monkeys and humans (Schemes 7.3 and 7.4). Particularly, derivatives having an *N*4-alkoxylcarbonyl moiety were the most susceptible to the human carboxylesterase. Those were then transformed to 5′-deoxy-5-fluorouridine (5′-DFUR: **90**) by cytidine deaminase highly expressed in the liver and solid tumors and finally to 5-FU by thymidine phosphorylase (dThdPase) preferentially located in tumors. When administered orally to monkeys, a derivative has the *N*4-alkoxylcarbonyl moiety with a C5 alkyl chain (capecitabine: **92**) (Fig. 7.82) and has the highest AUC and C_{max} for plasma 5′-DFUR. In tests with various human cancer xenograft models, capecitabine was more efficacious at wider dose ranges than either 5′-DFUR or 5-FUand was remarkably less toxic to the intestinal tract than the others in monkeys [194].

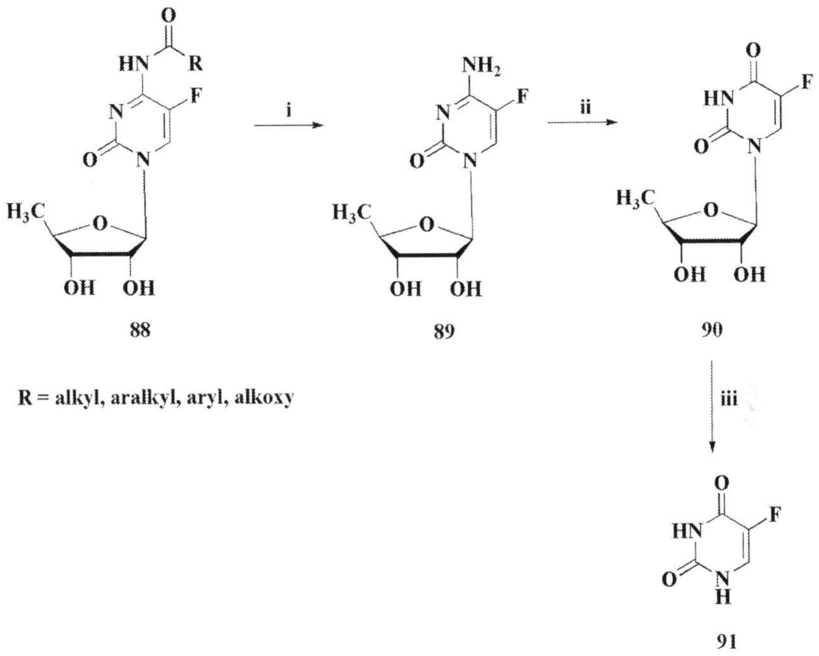

R = alkyl, aralkyl, aryl, alkoxy

Scheme 7.3 Reagents and conditions:(i) Carboxylesterase. (ii) Cytidine deaminase. (iii) Thymidine phosphorylase.

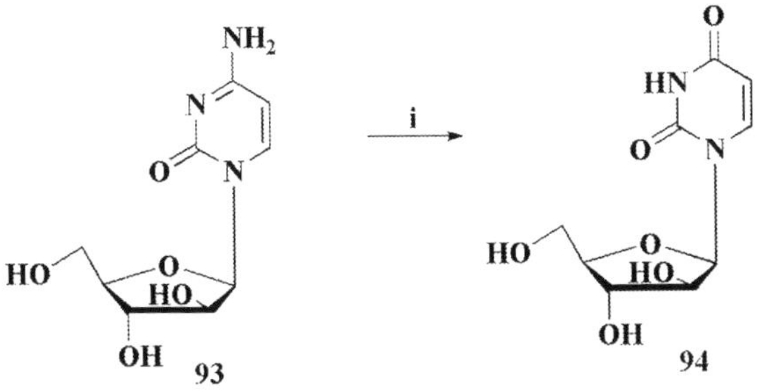

Scheme 7.4 Reagents and conditions:(i) Cytidine deaminase.

The synthesis of the *N*4-(substituted)-5′-DFCR (**98**) is accomplished in Scheme 7.5. *N*4-(Acyl or alkoxy-carbonyl)-5′-DFCR (**98**) derivatives were produced from 5-fluorocytosine (**95**) through three stages: first the glycosidation of compound **95** with 1,2,3-tri-*O*-acetyl-5-deoxyribose with stannic tetrachloride [195,196] or in situ generated trimethysilyl iodide [197], then the acylation of the *N*4-amino group of the resulting 5′-DFCR analog **96**

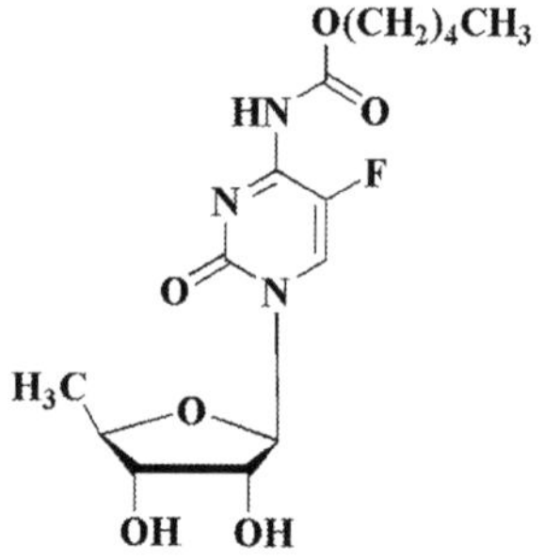

O(CH2)4CH3
HN O
N F
O N
H3C O
OH OH
92

Fig. 7.82 Capecitabine **92**.

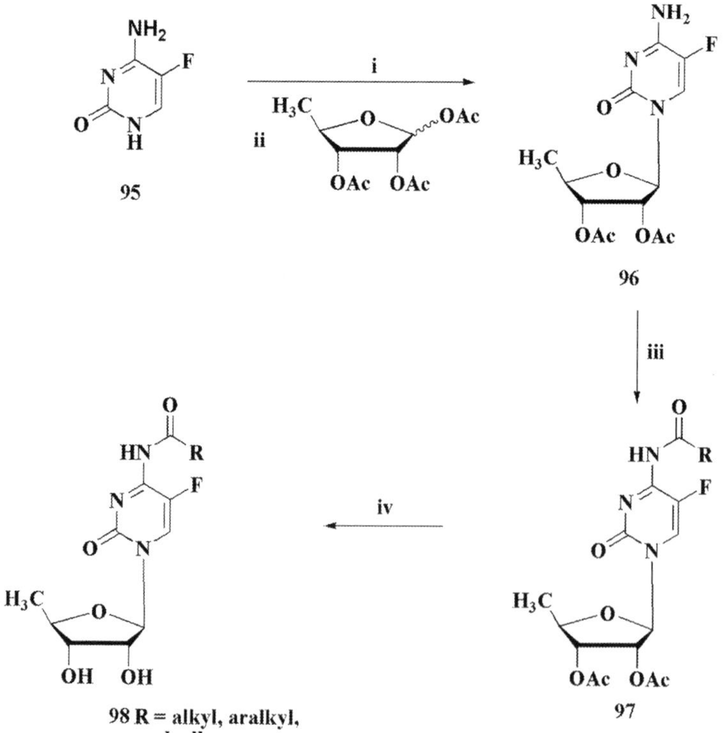

NH2
N F
O N H
95
i
ii
H3C O OAc
OAc OAc
NH2
N F
O N
H3C O
OAc OAc
96
iii
O
HN R
N F
O N
H3C O
OAc OAc
97
iv
O
HN R
N F
O N
H3C O
OH OH
98 R = alkyl, aralkyl, aryl, alkoxy

Scheme 7.5 Reagents and conditions: (i) HMDS/Toluene. (ii) $SnCl_4/CH_2Cl_2/15 \pm 20°C$ or $(CH_3)_3SiI/CH_3CN/0°C$. (iii) RCOCl/2 eq. pyridine/CH_2Cl_2. (iv) aqueous NaOH/MeOH.

with alkoxycarbonyl chlorides or acid chlorides, and in the final step the alkaline hydrolysis of the acetyl group of compound **97** at 0°C (Scheme 7.5) [194].

The apioarabinofuranosyl pyrimidine nucleosides were synthesized from the apiotriacetate **99**, which was prepared from D-ribose. Stereoselective-base condensation reaction furnished the β-apioarabinofurnosyl uredines [198]. Deprotection followed by the regioselective formation of 2,2′ anhydronucleosides and chemoselective nucleophilic ring opening was preformed to afford the apioarabinofurnosyl uracil nucleosides, which were converted to apioarabinofurnosyl cytosine nucleosides. Compounds **102a–c** and **105a–c** were estimated in colon, breast, and ovarian cancer cell lines. Among them, compounds **102a**, **102b**, and **105b** specifically prevent the growth of luminal A breast cancer cell line MCF7 than the triple-negative breast cancer cell line MDA-MB-231, showing that analogs of these compounds might be useful in treating the luminal A specific breast cancer. Using silylated uracil derivatives, the apiotriacetate **99** was investigated to glycosidic condensation reaction under Vorbrüggen condition to yield β-anomericapio-pyrimidine nucleosides **100a–c**. The latter apio-nucleosides **100a–c** were treated with a methanolic ammonia solution followed by the reaction with 1,1′ thiocarbonyl diimidazole [198,199] to afford cyclized 2,2′-anhydrouridine compounds **101a–c**. The chemoselective opening of the 2, 2′-anhydrobond of **101a–c** by nucleophilic attack at the C2 of uracil, followed by silyl group deprotection, furnished the apioarabino uridine derivatives **102a–c**. The uracil group of compound **102a–c** was transformed to the cytosine through conventional method [200], as depicted in Scheme 7.6 [198].

Protection of compounds **102a–c** by acetate using acetic anhydride in the presence of pyridine accessed peracetyl compounds **103a–c**. The latter compounds were then treated with phosphorous oxychloride and 1,2,4-triazole in the presence of triethylamine to afford triazole intermediates **104a–c**. The triazole intermediates **104a–c** were successively treated with 25% aqueous ammonia solution to generate the apio-cytarabine compounds **105a–c** (apio-ara-C) (Scheme 7.7) [198].

Cytotoxic activity of the compounds **102a–c** and **105a–c** was estimated in breast (MDA-MB-231, MCF7), ovarian cancer (OVCAR8), and colon (HCT116) cell lines utilizing the standard 3-(4,5-dimethylthiazol-2-yl)-2,5-diphenyltetrazolium bromide (MTT) cell viability assay [198,201]. Results showed that these compounds have a minimal cytotoxic activity in breast cancer cell lines than in the ovarian cancer and colon cell lines. Compounds **102a**, **102b**, and **105b** indicated a moderate cytotoxic effect

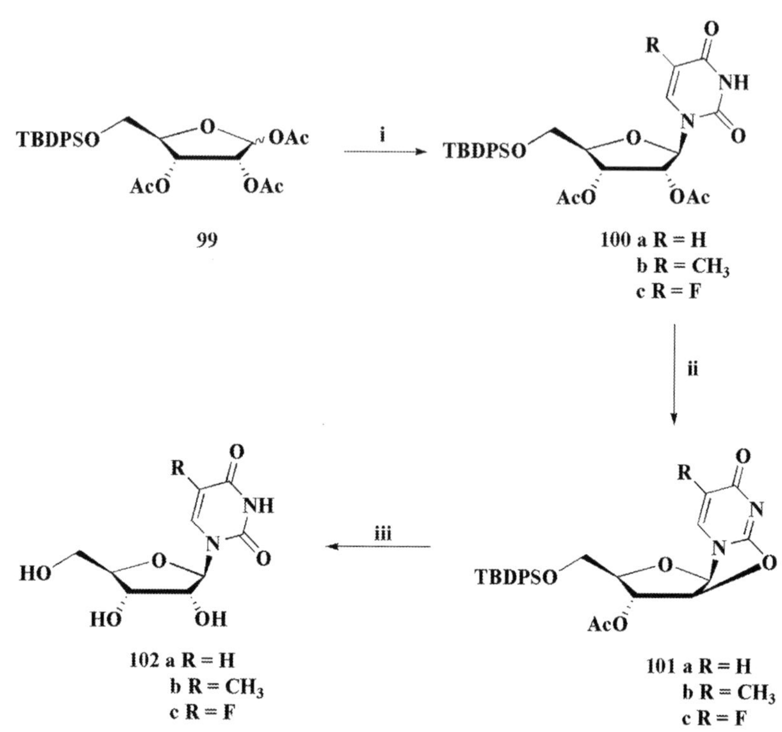

Scheme 7.6 Reagents and conditions: (i) Nucleobase/HMDS/$(NH_4)_2SO_4$/reflux/12 h/DCE/TMSOTf/0 °C/3 h; (ii) (a) CH_3OH/NH_3/RT/8 h; (b) 1,10thiocarbonyl diimidazole/TEA/toluene/reflux/24 h; (iii) CH_3OH/NaOH/RT/5 h.

in MCF7 cells as compared to other cell lines. However, their 50% growth inhibitor concentration is above 100 μM [198].

Nucleobase-modified analogs of the anticancer compounds 3′-C-ethynylcytidine (ECyd) and 3′-C-ethynyluridine (EUrd) were designed to overcome the strict substrate specificity of the activating uridine–cytidine kinase. ECyd, EUrd, and target nucleosides were accessed utilizing a short convergent synthetic route using diacetone-α-D-glucose [202].

Toward ECyd, EUrd, and nucleobase-modified analogs, a convergent strategy was proceeded, where a protected 3-C-ethynyl-D-ribo-pentofuranose as glycosyl donor reacted with a series of persilylated nucleobases. Of three synthetic routes reported to date leading to ECyd [202–204], two proceed through convergent strategies. A slightly different convergent strategy was reported in which peracetylated furanose **108** was identified as the key intermediate. Known furanose **106** [205] accessed from the diacetone-α-D-glucose, was utilized as starting material. Chemoselective

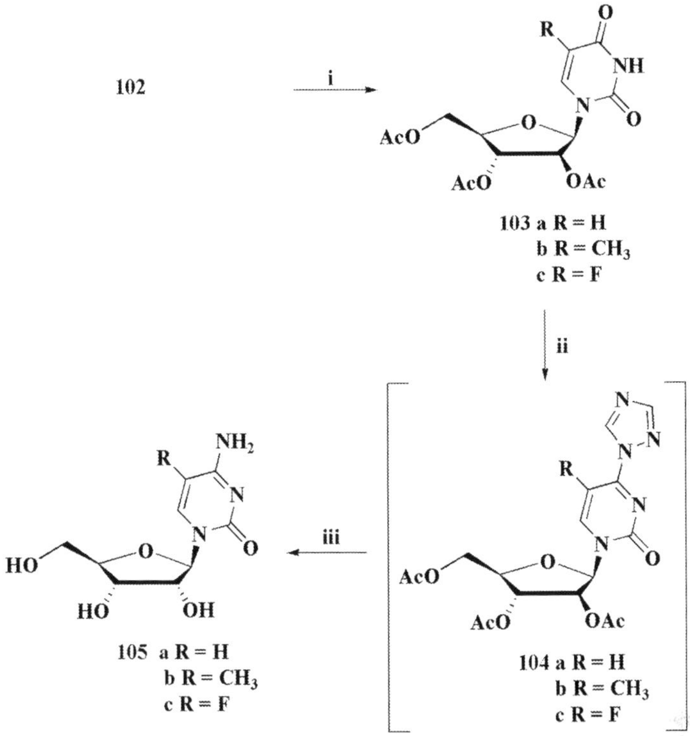

102 —i→ 103 a R = H, b R = CH_3, c R = F —ii→ [104 a R = H, b R = CH_3, c R = F] —iii→ 105 a R = H, b R = CH_3, c R = F

Scheme 7.7 Reagents and conditions:(i) Ac_2O/py/reflux/12 h; (ii) $POCl_3$/1,2,4 triazole/Et_3N/MeCN/20 h; (iii) 25% aqueous ammonia solution/1,4-dioxane/RT/15 h; then methanolic ammonia/RT/12 h.

cleavage of the 5,6-*O*-isopropylidene group of furanose **106**, followed by oxidative cleavage of the vicinal diol and reduction of the resulting aldehyde generated the diol **107**. The diol **107** was diacetylated, then the isopropylidene cleavage and peracetylation of the resulting anomeric diol were followed to yield glycosyl donor **108**. Glycosylation of furanose **108** with trimethylsilyl triflate (TMSOTf) and a series of persilylated nucleobases (obtained in situ from *N*,*O*-bis(trimethylsilyl)acetamide, BSA, and the corresponding nucleobase) [206] gave β-configured nucleosides **109** through an chimeric assistance of the *O*2-acetyl group and analogs. Reaction of glycosyl donor **108** with 4-thiouracil or persilylated 2-thiocytosine generated S-linked ribosides. A similar glycosylation has only been observed during the synthesis of S-β-D-glucuronides. Deprotection of nucleosides **109** afforded the nucleosides **110** (Scheme 7.8).

Nucleosides **110** were estimated against prostate cancer (PC-3) and human adenocarcinoma breast cancer (MCF-7) cell lines. ECyd exhibited

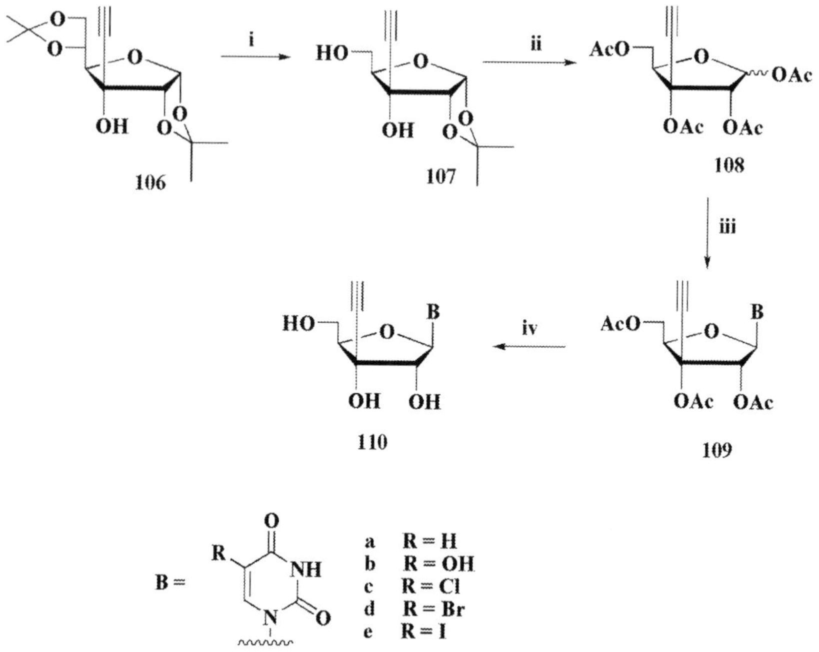

Scheme 7.8 Reagents and conditions: (i) 80% aqueous AcOH/RT/$NaIO_4$/MeOH/H_2O (3:1, v/v)/0°C to RT; $NaBH_4$/MeOH/0°C to RT/84%; (ii) Ac_2O/DMAP/pyridine/RT; 80% aqueous TFA/0°C to RT; Ac_2O/DMAP/pyridine/RT/74%; (iii) nucleobase/BSA/TMSOTf/CH_3CN; (iv) NH_3/MeOH.

very active anticancer potency with IC_{50} values of 0.15 nM and 2.2 nM against PC-3 and MCF-7 cells. 5-Iodouridine analog **110e**, the most active inhibitor of tumor cell proliferation among the nucleobase-modified nucleosides, exhibited IC_{50} values of 160 nM (PC-3) and 35 nM (MCF-7). Minimizing the size of the 5-substituent by the introduction of a chlorine or bromine atom leading to a systematic drop in potency. The corresponding 5-fluoro-EUrd derivative, bearing the smallest halogen, is known to be only **19–48** fold less potent than EUrd against L1210 and KB cells [202]. The substrate preference of UCK2 is decreased in the order: 5-fluorouridine >5-bromouridine >5-iodouridine [207]. The very weak anticancer potency of 6-azauridine compared to 5-iodouridine **110e**, and the lack of potency of 3-deazauridine derivative and 5-hydroxyuridine analog, is also interesting considering that the corresponding nucleosides without the C3′-ethynyl modification are remarkably better substrates for UCK2 than for 5-iodouridine [202].

The 5′-O-silyl- and trityl-protected nucleoside derivatives were synthesized as accomplished in Schemes 7.9–7.11.

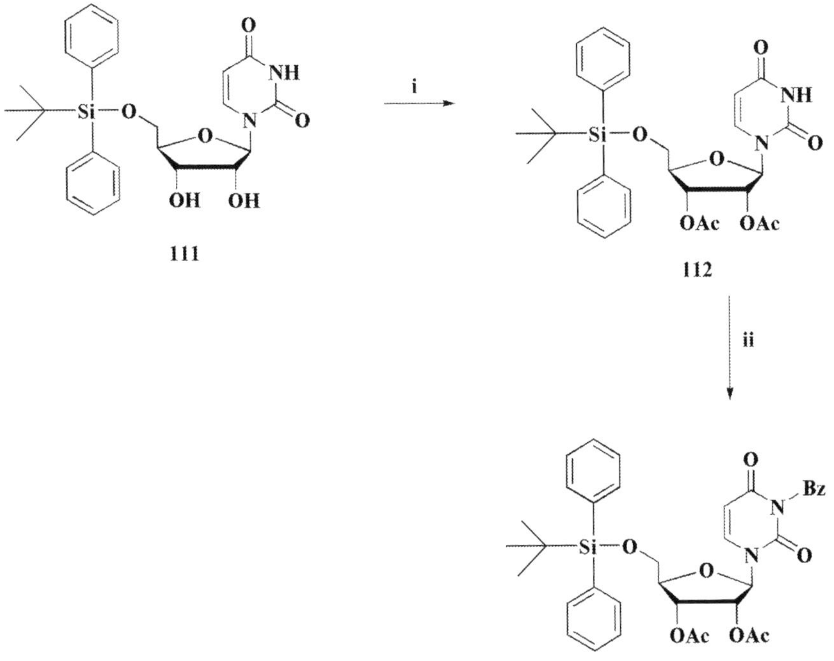

Scheme 7.9 Reagents and conditions: (i) Py/Ac_2O/RT/18 h. (ii) Hung's base/BzCl/py/RT/2 h.

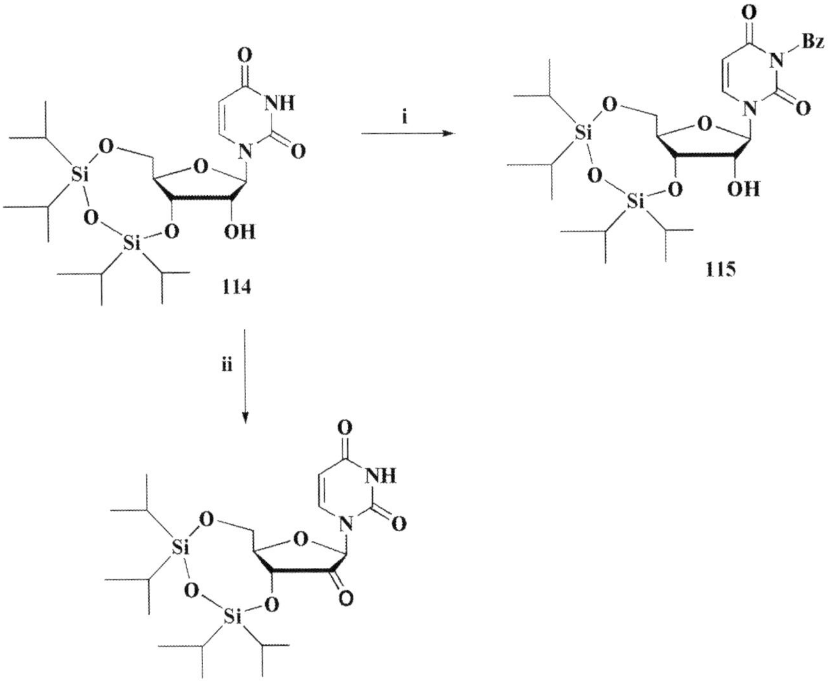

Scheme 7.10 Reagents and conditions: (i) Na_2CO_3/TBAB/BzCl/CH_2Cl_2/H_2O/RT/2 h. (ii) CrO_3/py/Ac_2O/CH_2Cl_2/RT/2 h.

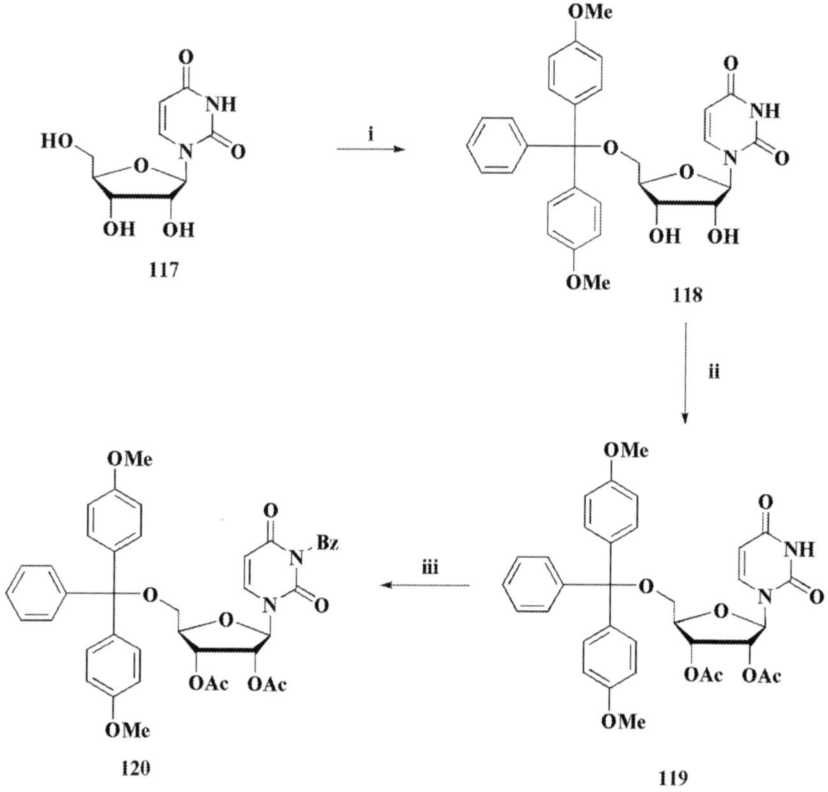

Scheme 7.11 Reagents and conditions: (i) DMTCl/DMAP/py/60°C/18 h. (ii) Ac_2O/py/RT/18 h. (iii) Hunig's base/BzCl/py/RT/2 h.

The TBDPS uridine analog **111** were protected as their acetyl derivatives to afford **112** followed by the protection of the base NH as the benzoyl derivative **113** (Scheme 7.9) [208].

The disila-**114** was readily transformed into the benzoyl derivative **115**, while the treatment of **114** with chromium(VI)oxide/pyridine/acetic anhydride (1:2:1) also yielded the keto-uridine derivative **116**. Uridine was treated with dimethoxytrityl chloride (DMTCl) to give the uridine derivative **118**. To afford additional compounds with modulated lipophilicity, this compound was transformed into the diacetyl derivative **119** and subsequently into the benzoyl-uridine derivative **120** (Scheme 7.10) [208].

The compounds were estimated at a fixed concentration of 100 lM against a small panel of tumor cell lines (K-562, HL-60, Caco-2, Jurkat, and HT-29). The entire set was also evaluated at varying concentrations against two human glioma lines (Hs683 and U373) to access GI_{50} values, with the best results (approximately 25 μM).

For the cell viability assay at 100 μM, compounds **111** and **118** reduced cell viability by 65%–95%, while for the GI_{50} determinations against two human glioma lines (Hs683 and U373), compounds **112**, **113**, **115**, **118**, and **119** furnished GI_{50} values in the range of 25–40 μM. This study showed that the nucleoside derivatives with the more lipophilic silyl/trityl groups had improved GI_{50} values [208].

The synthesis of 3′-deoxy-3′-fluoro-3′-*C*-hydroxymethyl-pyrimidine nucleosides **128a–d** from D-glucose was reported [209]. Compounds **128a–d** were synthesized starting from synthon **122**, which was prepared from 1,2:5,6-di-*O*-isopropylidene-3-oxo-α-D-glucofuranose **121** utilizing the modified Corey-Link reaction. The synthon **122** has requisite C3 quaternary carbon with 3-β-fluoro and 3-α-methyl-ester functionalities, of which the ester group is amicable for the further manipulation. Thus, reduction of C3-α-methylester in **122** followed by acetylation yielded 3β-fluoro-3α-acetoxymethyl derivative **123** [209]. Deprotection of 5,6-acetonide functionality in **123** afforded the diol that on treatment with $NaIO_4$ followed by reduction furnished 1,2-*O*-isopropylidene-3β-fluoro-3α-acetoxymethyl-α-D-xylo-furanose **124**. The C5-hydroxyl group in **124** was acetylated to get **125** that upon hydrolysis of 1,2-acetonide that was followed by acetylation generated tetra acetate **126** as ananomeric mixture. The latter was separately subjected to the Vorbruggen glycosylation reaction [210] with various nucleobases such as *N*-Bz-cytosine, uracil, thymine, *N*-Bz-adenine, in the presence of TMSOTf and BSA, that provided nucleosides **127a–d**. The exclusive formation of the β-isomer is due to the neighboring group participation of C-2 acetyl group. Further deprotection of acetate groups in **127a–d** with methanolic ammonia yielded the corresponding 3′-deoxy-3′-fluoro-3′-C-hydroxymethyl-nucleosides **128a–d** - (Scheme 7.12). The nucleosides **128a–d** revealed a moderate antiproliferative activity against the L-132, HeLa, MDA-MB-231, and neural cancer cell lines [209].

A series of theophylline containing 1,2, 3-triazoles with variant nucleoside derivatives have been synthesized [211].

The nucleosides **131**, **133**, and **134** were generated as depicted in Scheme 7.3. The uridine **129** was protected to access compound **130**. In the latter compound, the C-5 position of alcohol was transformed into nucleoside azide to afford compound **131** by a one-pot transformation involving two successive reactions. Upon alkylation, compound **132** was

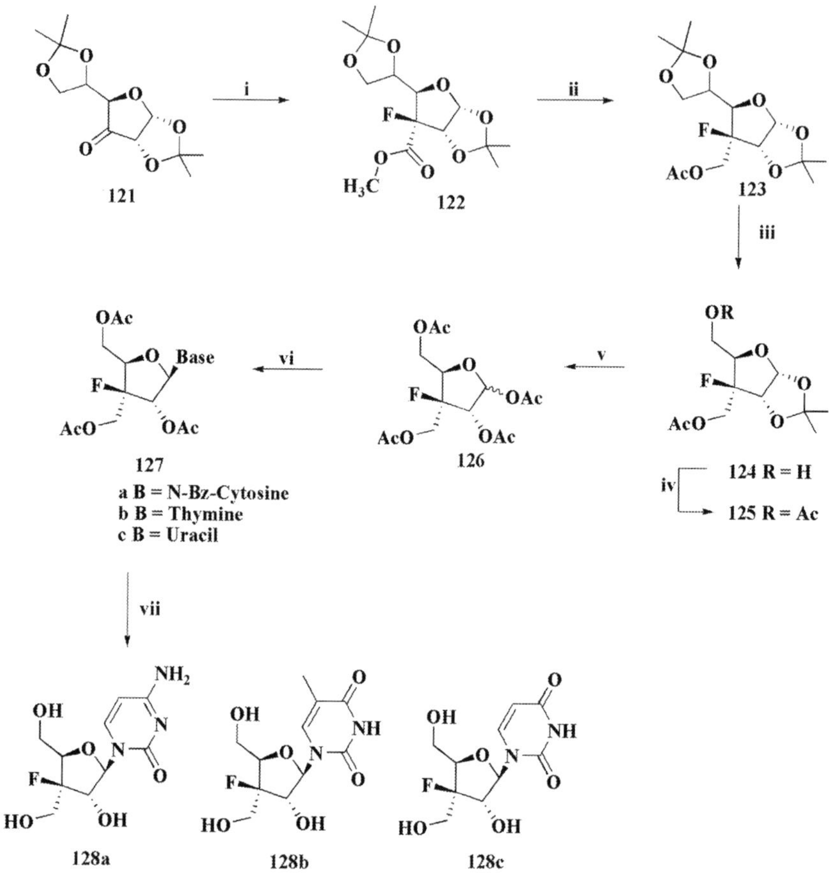

Scheme 7.12 Reagents and conditions: (i) (i) $NaBH_4/CH_3OH$. (ii) Ac_2O/pyridine/DMAP/DCM; (ii) (i) 65% $AcOH:H_2O$/50°C/3.5 h. (ii) $NaIO_4$/acetone/H_2O/RT (iii) $NaBH_4/CH_3OH$/0 °C; (iii) Ac_2O/pyridine/DMAP/DCM. (iv) (i) $TFA:H_2O$ (3:2) 3 h, (ii) Ac_2O/pyridine/DMAP. (v) BSA/TMSOTf/CH_3CN/nucleobase/60°C/24 h. (vi) NH_3/MeOH/55°C/1 h.

furnished. The acetal group was then removed from compound **132** through treating with TFA to generate compound **133**. Finally, the nucleoside compound **134** was synthesized from compound **133** treating with iodo methane in basic conditions (Scheme 7.13) [211].

The synthesis of theophylline containing 1,2,3-triazole nucleoside derivatives is outlined in Scheme 7.14. Under click chemistry conditions, the nucleoside compounds **131**, **133**, and **134** were condensed with theophylline acetylene compounds to afford the targeted solids [211].

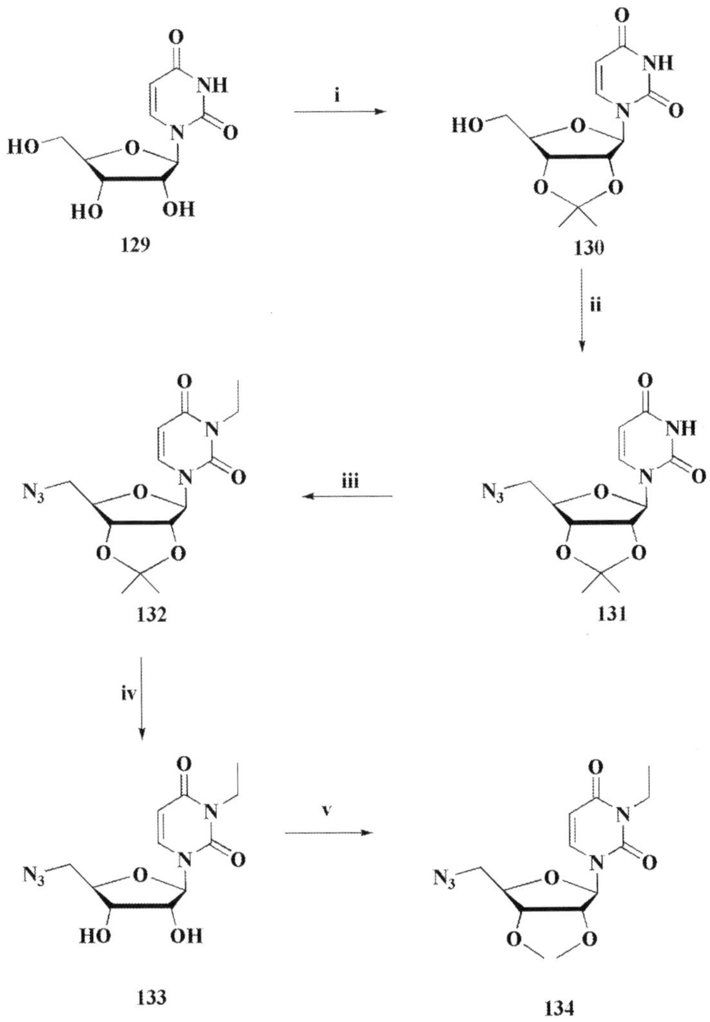

Scheme 7.13 Reagents and conditions: (i) Concentrated H_2SO_4/TEA/acetone/0°C to RT/4h; (ii) (i) $MeSO_2Cl$/Pyridine/CAN/0°C to RT/4h; (ii) NaN_3/DMF/90°C, 6h; (iii) Ethyl iodide/NaH/DMF/0°C to RT/3h; (iv) TFA/THF/H_2O/0°C to RT/10h; (v) CH_3I/NaH/DMF/0°C to RT/4h.

The theophylline-containing 1,2,3-triazole nucleoside hybrids were screened for an anticancer activity.

Computational docking and 2DQSAR were carried out utilizing MOE software to determine novel scaffolds. The data indicate that compound **136** displays an obvious cytotoxic effect on cancer cells such as lung (A549), colon (HT-29), breast (MCF-7), and melanoma (A375) with IC_{50} values

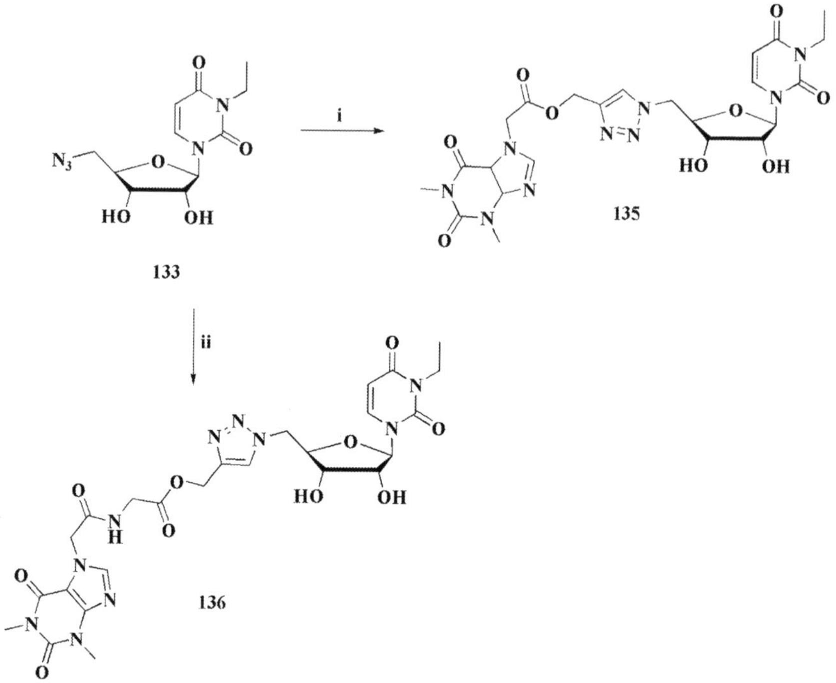

Scheme 7.14 Reagents and conditions: (i) Sodium ascorbate/$CuSO_4 \cdot 5H_2O$/EtOH:H_2O/RT/16h.

of 2.56, 2.19, 1.89, 4.89 mM and 3.57, 2.90, 2.10, and 5.81 mM, respectively. Compound **135** and **136** revealed good dock score and binding affinities with different therapeutic targets in cancer cell proliferation according to the docking studies. Additionally, these compounds have demonstrated an acceptable correlation with bioassay results in the regression plots assessed in 2D QSAR models [211].

A series of nucleosides having an aryl and hetaryl group in position 6 were synthesized (Schemes 7.15–7.18), and their biological potencies were evaluated through in vitro CDK2/Cyclin A2 and CDK1/Cyclin B1 kinase assay. The three xylocydine derivatives **145a**, **145b**, and **145c** displayed specific inhibitory potencies on CDK2/cyclin A2 with IC_{50} values of 4.6, 4.8, and 55 lM. Those compounds all induced G1/S-phase arrest in human epithelial carcinoma cell line (HeLa), and the results suggested that they may inhibit CDK2 activity in vitro. In addition, molecular modeling study, their docking into cyclin-dependent kinase 2 (CDK2) active site revealed high docking scores. These data reveal that those three compounds are good

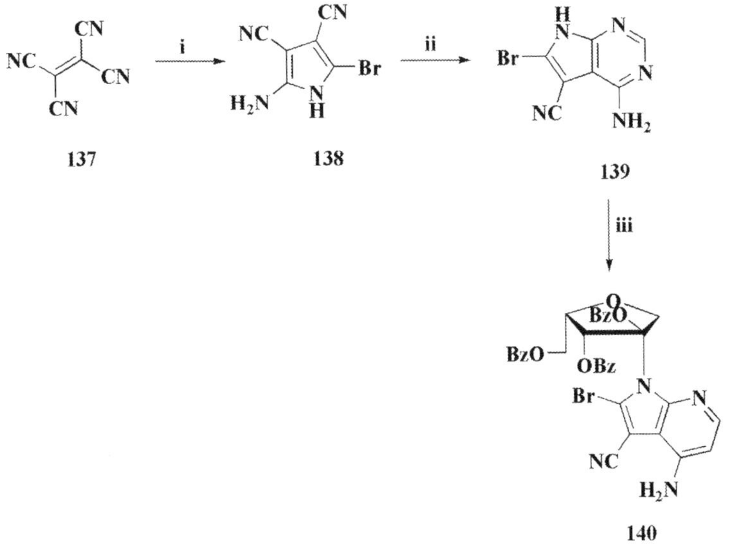

NC CN CN CN
137
i
NC CN Br H$_2$N N H
138
ii
H N N Br N NC NH$_2$
139
iii
O BzO BzO OBz N Br N NC H$_2$N
140

Scheme 7.15 Reagents and conditions: (i) HBr/HOAc/0 °C/2 h; (ii) HN=CHNH$_2$—CH$_3$COOH/CH$_3$CH$_2$OCH$_2$CH$_2$OH; (iii) BSA/CH$_3$CN/1-*O*-acetyl-2,3,5-tri-*O*-benzoyl-β-L-ribofuranose/TMSOTf/80°C.

OH HO O OH OH
141
i
O HO OCH$_3$ HO OH
142
ii
O BzO OAc BzO OBz
143
iii
140

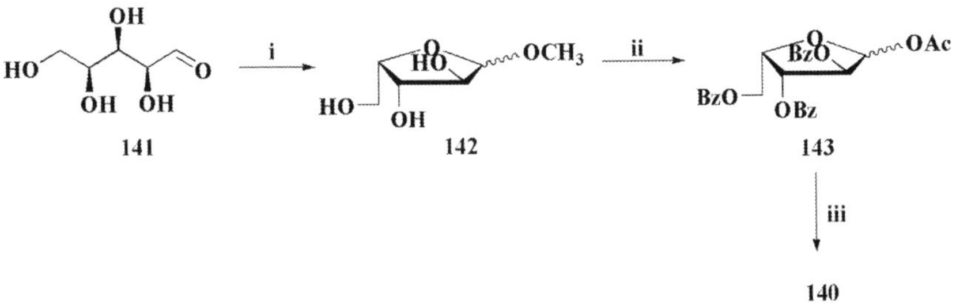

Scheme 7.16 Reagents and conditions: (i) CH$_3$OH/H$_2$SO$_4$; (ii) BzCl/pyridine/AcOH/Ac$_2$O; (iii) BSA/CH$_3$CN/1-*O*-acetyl-2,3,5-tri-*O*-benzoyl-β-L-ribofuranose/TMSOTf/80°C.

O BzO BzO OBz Br N N NC N H$_2$N
140
i
O BzO BzO OBz R N N NC N H$_2$N
ii
O HO HO OH R N N NC N H$_2$N

144 a R = thiophen-2-yl
b R = furan-2-yl
c R = 5-Bromothiophen-2-yl

145 a R = thiophen-2-yl
b R = furan-2-yl
c R = 5-Bromothiophen-2-yl

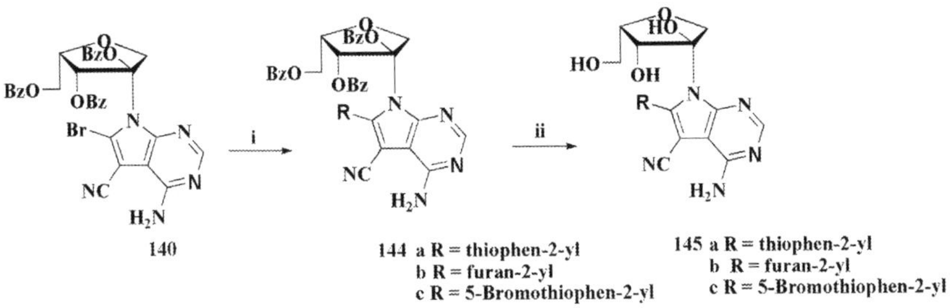

Scheme 7.17 Reagents and conditions: (i) R-B(OH)$_2$/Pd cat. (ii) CH$_3$ONa/CH$_3$OH.

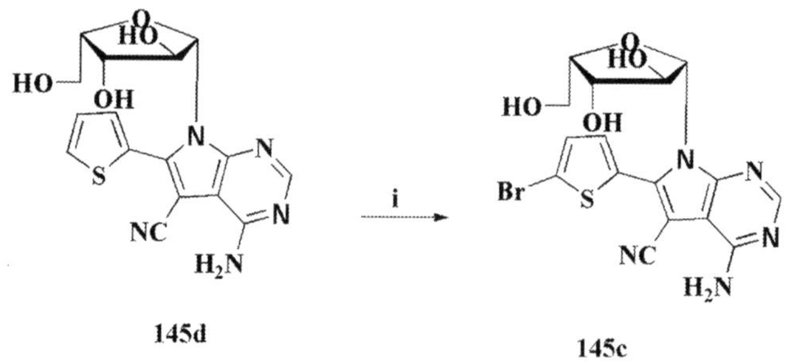

145d **145c**

Scheme 7.18 Reagents and conditions: (i) NBS/CH_3CN/0°C to RT/16 h.

inhibitors of CDK2 for studying this kinase signal transduction pathway in cell system [212].

3.1.2 Deoxyribonucleoside pyrimidine analogs

2′,2′-Fluoromethylidene nucleosides can be produced from the 3′,5′-di-*O*-protected nucleosides **146** using the procedure reported for the preparation of (*E*)-2′-deoxy-2′-dehydro-2′,2′-fluoromethylidenecytidine (**151**, X=NH_2, R=H). 3′,5′-*O*-(1,1,3,3-Tetraisopropyl-1,3-disiloxane-diyl)-nucleoside **147** was synthesized upon the treatment of **30** (X=N=$CHNMe_2$ or OEt) with 1,3-dichloro-1,1,3,3-tetraisopropyldisiloxane.

Further Swern oxidation of **147** affords compound **148**, which is treated with fluoromethylphenylsulfone and diethyl chlorophosphate in tetrahydrofurane at −70°C and the by lithium hexamethyldisilazane to give compound **149**. Then X′ can be deprotected using methanolic ammonia. Compound **151** was afforded upon treating compound **150** with tributyltin hydride (Scheme 7.19) [213].

Compound **153** was afforded via the oxidation of the 1,2:5,6-di-*O*-isopropylidene-α-D-glucofuranose **152** with pyridinium dichromate and acetic anhydride in dichloromethane; then the subsequent stereoselective reduction was performed followed by tosylation with *p*--toluenesulfonylchloride in pyridine [214].

When compound **153** reacted with potassium fluoride in acetamide displacement of the tosyloxy group at *C*-3 by a fluorine atom. Additionally, the further selective removal of the 5,6-*O*-isopropylidene group yielded the 3-deoxy-3-fluoro-1,2-*O*-isopropylidene-α-D-glucofuranose **154** (Scheme 7.20) [214].

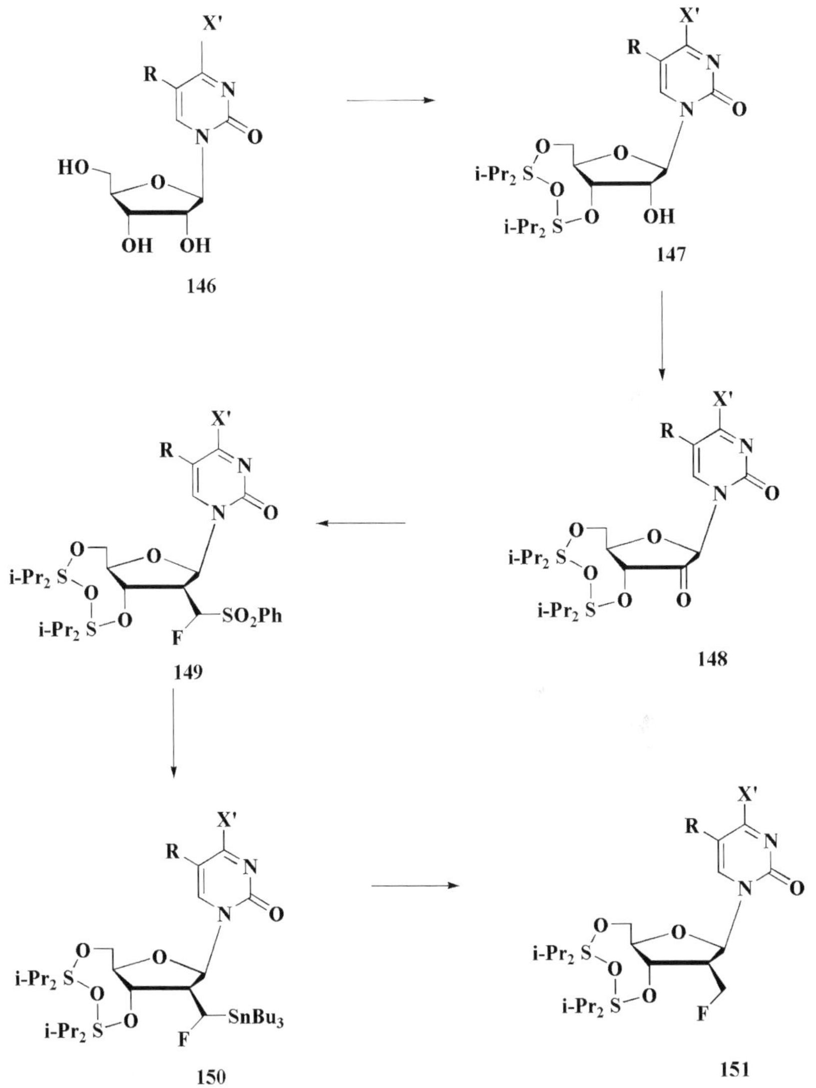

Scheme 7.19

Compound **155**a was afforded upon the periodate oxidation of compound **154** and the subsequent one-pot borohydride reduction of the resulting aldehyde. Sulfonylation of **155a** with *p*-toluenesulfonyl chloride in pyridine yielded compound **155b**. Tosylate **155b** was also treated with potassium thioacetate in *N*,*N*-dimethylformamide to afford the thioacetate **156**. Acetolysis of the latter in the presence of acetic anhydride, acetic acid,

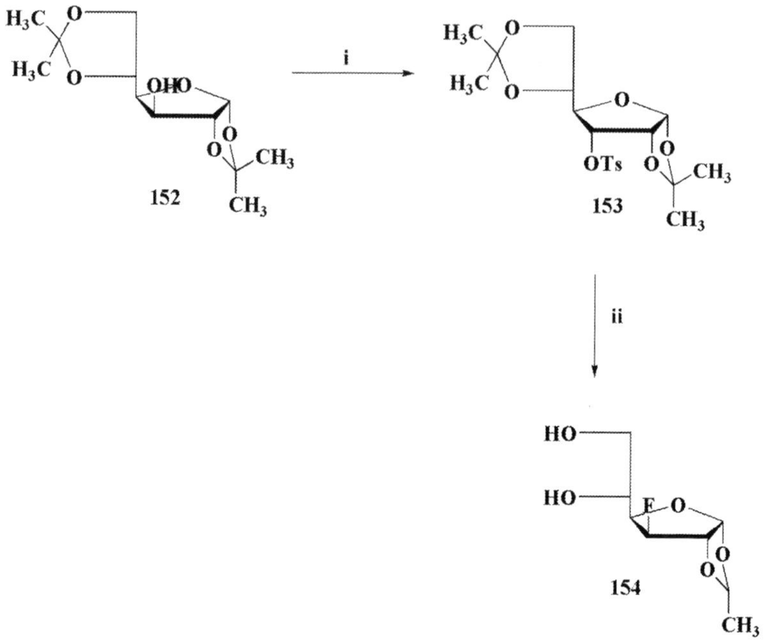

Scheme 7.20 Reagents and conditions: (i) PDC/CH_2Cl_2/Ac_2O/$NaBH_4$/CH_3OH/pyridine/TsCl. (ii) KF/AcOH (70%)/acetamide.

and sulfuric acid yielded a mixture of the anomeric diacetates **157** (α:β anomer = 1:2) (Scheme 7.21) [214].

The protected β-nucleosides1-(5-*S*-acetyl-3-deoxy-3-fluoro-5-thio-β-D-xylofuranosyl)nucleosides of thymine **158a**, uracil **158b**, and 5FU **158c** were afforded via the condensation of the anomeric mixture **157** with the silylated nucleic acid bases in the presence of trimethylsilyl trifluoromethanesulfonate (Scheme 7.15).

The cytotoxic effects of compounds **158** were evaluated. The most promising antitumor potency was indicated in the case of colon carcinoma treatment, where cytotoxic effect and growth inhibition were provided at low concentration in comparison with 5-fluorouracil (5FU) [214].

The reaction of 2,2′-anhydrouridine **159b** with a twofold excess of *p*--toluenesulfonylamide wasactivated with sodium hydride in dry DMF, so the *N* 2-tosylisocytidine **160** was produced which is then isolated by preparative chromatography (Scheme 7.22) [215].

4-Imino-*N* 2-tosylamino-1-(β-D-arabinofuranosyl)pyrimidine **161** was afforded when the 2,2′-anhydrocytidine hydrochloride **159a** reacted with

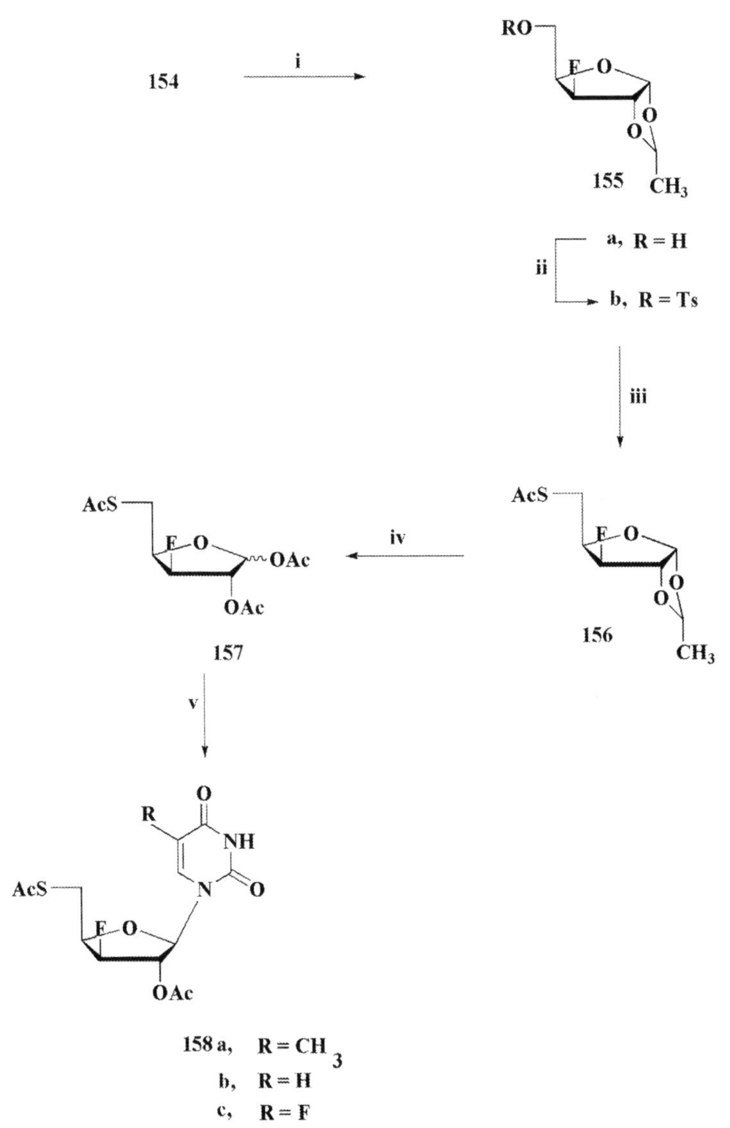

Scheme 7.21 Reagents and conditions: (i) $NaIO_4/NaBH_4/CH_3OH$. (ii) pyridine/TsCl. (iii) KSAc/DMF/100°C. (iv) $CH_3COOH/Ac_2O/H_2SO_4$. (v) Silylated base/$Me_3SiOSO_2CF_3$.

2 equivalents of the sodium salt of *p*-toluenesulfonamide in the presence of dry dimethylformamide at room temperature [215].

Similarly the sulfonamido derivative **161** was produced in 82% yield utilizing the sodium salt of 2-(aminosulfonyl)-*N*,*N*-dimethylnicotinamide as a

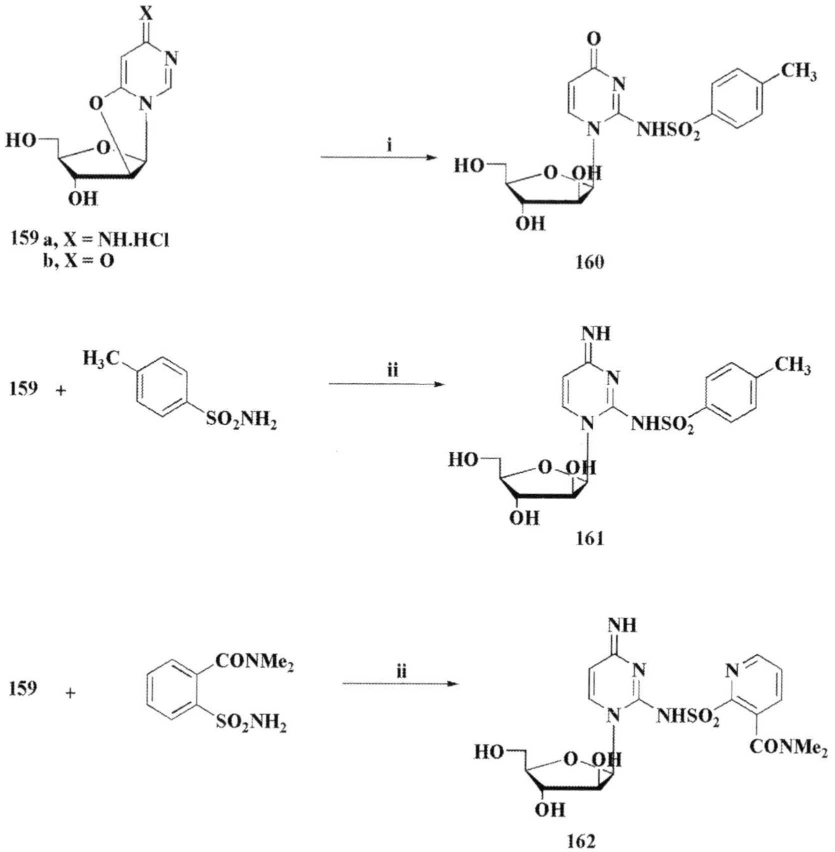

Scheme 7.22 Reagents and conditions: (i) *p*-toluenesulfonylamide/NaH/DMF/DBU/ CH_3CN. (ii) NaH/DMF.

nucleophile. The starting sulfonamide was synthesized through the reaction of 2-chlorosulfonyl-3-(*N*,*N*-dimethylaminocarbonyl) pyridine and methanolic ammonia (Scheme 7.22) [215].

Compound 5-methyl-*N* 2-tosyl-1-(2-deoxy-5-*O*-trityl-β-D-threo--pentofuranosyl)isocytosine **165** was yielded starting from compound **163**. The further reactions of the isocytosine **165** afforded compounds 5-methyl-*N* 2-tosyl-1-(2-deoxy-β-Dthreo-pentofuranosyl) isocytosine **166** and 5-methyl-*N* 2-tosylisocytosine **167**, respectively, as shown in Scheme 7.23 [215].

The effect of the *C*-2 sulfonamido pyrimidine compounds on human tumor cells growth is studied. Moderate (20%–50% inhibition) to weak

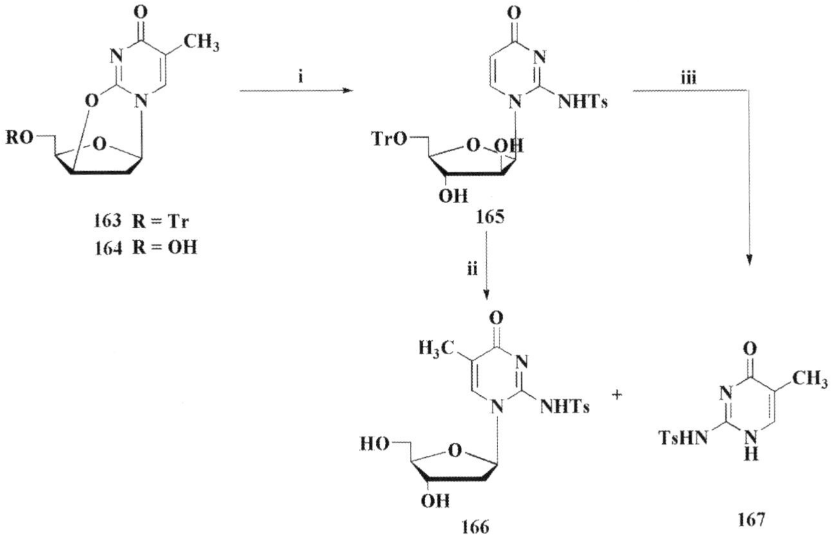

Scheme 7.23 Reagents and conditions: (i) *p*-toluenesulfonylamide/NaH/DMF/DBU/CH_3CN. (ii) TBAF/THF. (iii) $ZnBr_2/CH_2Cl_2$.

(0%–20% inhibition) activity of cytotoxicities was observed in all examined compounds [215].

Application of the cytidine derivatives **161** and **162** indicated similar effects on tumor cell growth in comparison with uridine analog **160**. The cell lines most sensitive to the tested compounds were human colon carcinoma (HT-29), human breast adenocarcinoma (MCF-7), and human hepatocarcinoma (Hep2). The most resistant cells were poorly differentiated cells from the lymph node metastases of colon carcinoma (SW 620) [215].

Azido-modified pyrimidine nucleosides were prepared as potential radiosensitizers; one class is 5-azidomethyl-2′-deoxyuridine (AmdU) and cytidine (AmdC), while the second class is cytidine (AvdC) and 5-(1-azidovinyl)-2′-deoxyuridine (AvdU). The addition of radiation produced electrons to C5-azido nucleosides resulted in the production of π-aminyl radicals followed by facile transformation to σ-iminyl radicals either through a bimolecular reaction involving intermediate α-azidoalkyl radicals in AmdU/AmdC or via tautomerization in AvdU/AvdC. AmdU reveals an efficient radiosensitization in EMT6 tumor cells [216].

Synthesis of 5-azidomethyl pyrimidine nucleosides and their 5′-triphosphates is outlined in Schemes 7.24 and 7.25 [215]. AmdU **171** was generated from thymidine **168** through the sequential silylation [216],

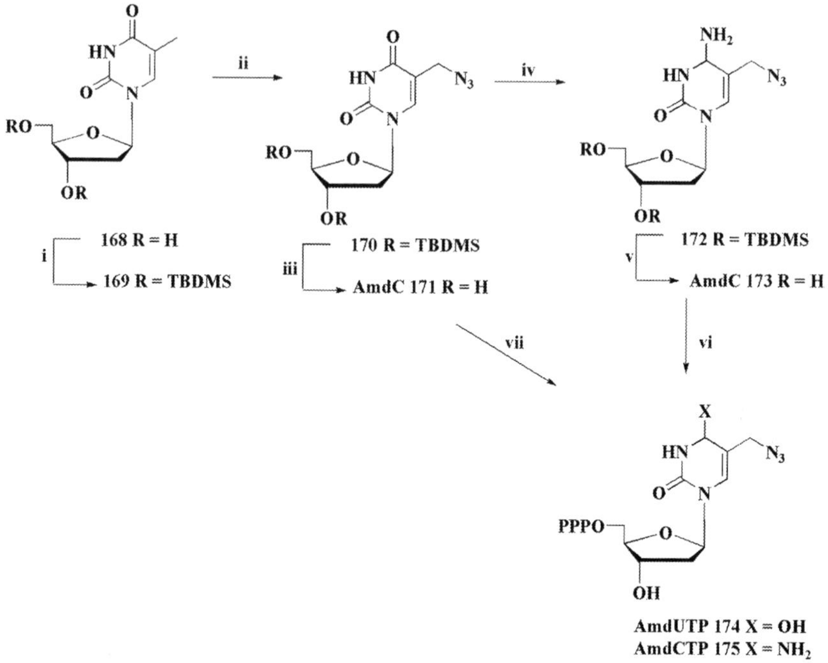

Scheme 7.24 Reagents and conditions: (i) TBDMSCl/imidazole/15 h. (ii) NBS/AIBN benzene/reflux/1 h/NaN_3/DMF/60°C/1 h. (iii) TBAF/THF/RT/4 h. (iv) TIPBSCl/TEA/DMAP/CH_2Cl_2/RT/1 h. aqueous NH_4OH/THF/RT/15 h. (v) TBAF/THF/RT/4 h. (vi) $PO(OMe)_3$/$POCl_3$/proton sponge 0°C/30 min./TBAPP/Bu_3N/DMF/0°C/2 min. (vii) $PO(OMe)_3$/proton sponge/$POCl_3$/0°C/30 min./TBAPP/Bu_3N/DMF/0°C/2 min.

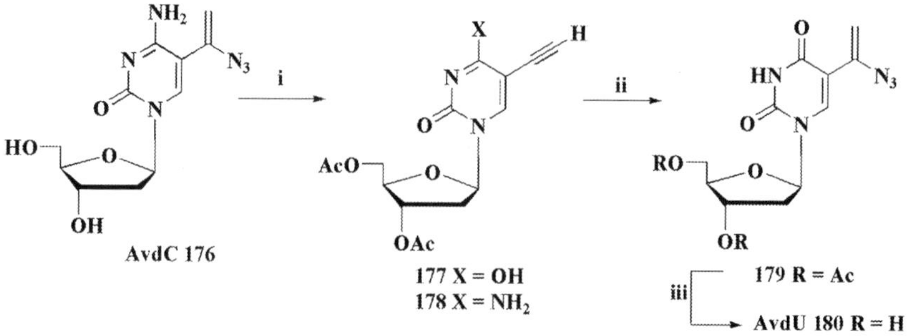

Scheme 7.25 Reagents and conditions: (i) $TMSN_3$/Ag_2CO_3 H_2O/DMF/80°C/1 h/NH_3/MeOH/0°C to RT/15 h. (ii) $TMSN_3$/Ag_2CO_3 H_2O/DMF/80°C/1 h. (iii) NH_3/MeOH 0°C to RT/15 h.

bromination with NBS, displacement of bromide with sodium azide, and desilylation [216–218]. Compound **170** was treated with 2,4,6-triisopropylbenzenesufonyl chloride (TIPBSCl) in the presence of TEA/DMAP followed by displacement of the resulting aryl sulfate with ammonium hydroxide and deprotection of compound **172** with TBAF furnished AmdC6 [216].

5-Azidovinyl pyrimidine nucleosides were synthesized as depicted in Scheme 7.25 [216].

AmdC and AmdU were triphosphorylated [219] to afford AmdCTP (23%) and AmdUTP (76%) [216]. Silver-catalyzed hydroazidation [220] of the protected 5-ethynyl-2′-deoxyuridine [221] **177** with $TMSN_3$ regioselectively furnished the α-azidovinyl **179** [222]. Compound **179** was deacetylated to yield AvdU **180** [223]. Analogous hydroazidation of the protected deoxycytidine [224] followed by deacetylation gave AvdC **176**. Transformation of the uracil base in **179** to a cytosine counterpart through treatment [225] of **179** with TIPBSCl and then ammonolysis was followed to afford **176** [226].

Reactions of *O*-t-butyldimethylsilyl-protected thymidine, 2′-deoxyuridine, and 3′-azidothymidine (AZT) with (benzotriazol-1-yloxy)tris(dimethylamino)phosphonium hexafluorophosphate (BOP) resulted in the activation of the C4 amide carbonyl through the formation of putative O4-(benzotriazol-1-yl) analogs. Further substitution with aryl and alkyl alcohols, thiols, and amines leads to facile functionalization at this position. Reactions with thiols and amines were conducted either as a two-step, one-pot transformation, or as a one-step conversion (Scheme 7.26). Reactions with alcohols were conducted as two-step, one-pot

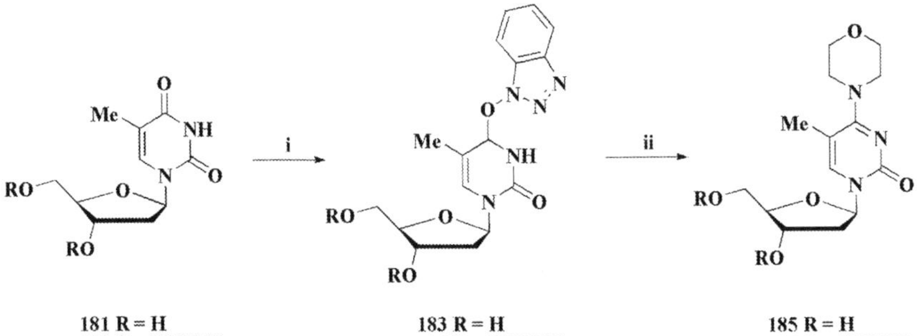

181 R = H
182 R = TBDMS

183 R = H
184 R = TBDMS

185 R = H
186 R = TBDMS

Scheme 7.26 Reagents and conditions: (i) THF, BOP, DBU, 3 h; (ii) morpholine, 12 h.

transformations. In the course of these investigations, the formation of 1-(4-pyrimidinyl)-1*H*-benzotriazole-3-oxide derivatives from the pyrimidine nucleosides was determined. However, these too underwent transformation to the targeted products. Products obtained from AZT were transformed to the 3′-amino derivatives via catalytic reduction. All products were assayed for their abilities to inhibit cancer cell proliferation and for antiviral activities [226].

The application of 2- and 4-chlorophenyl phosphoroditriazolides [227–230] for the phosphorylation of 5′-protected nucleosides has been provided in the phosphotriester synthesis of oligonucleotides [227].

4-Chlorophenyl *N*-alkyl phosphoramidates of 3′-azido-2′,3′-dideoxy-5-fluorouridine (**192**) were generated through the phosphorylation of 3′-azido-2′,3′-dideoxy-5-fluorouridine (**190**) with 4-chlorophenyl phosphoroditriazolide (**189**) as a phosphorylating agent followed by a reaction with the appropriate amine (Scheme 7.27). The phosphoramidates (**192**) were estimated for their cytotoxic activity in three human cancer cell lines: oral (KB), cervical (HeLa), and breast (MCF-7) using the sulforhodamine B (SRB) assay. The highest potency in all the investigated cancer cells was exhibited by phosphoramidate with the *N*-ethyl substituent, and its potency was several times than that of the parent nucleoside (AddFU). Also, phosphoramidate **192** ($R=CH_2C{\equiv}CH$) with the *N*-propargyl substituent displayed good potency in all the utilized cell lines. Moreover, phosphoramidate with the *N*-methyl substituent exhibited a moderate activity. However, phosphoramidates with a longer *N*-alkyl chain substituents were less active in all the cell lines used [227].

Phenstatin-stavudine derivatives were synthesized and biologically estimated as potential telomerase inhibitors (Scheme 7.28). The screening results revealed that some compounds had better anticancer potency in vivo and in vitro [231]. Compounds **196d** displayed potent activities against SMMC-7721, Hela, HepG2, and SGC-7901 cells. In vivo studies indicated that compound **196d** exhibited anticancer potency with inhibition tumor growth of HepG2 and S180 tumor-bearing mice. Compound **196d** indicated high inhibitory activity against telomerase. It indicated good antiproliferative activity against SGC-7901 cell with IC_{50} value of 0.77 mM via inducing cell cycle arrest at G2 phase. Annexin V/propidium iodide assay results showed that compound **196d** could induce apoptosis in SGC-7901 cells. The mitochondrial membrane potential assay revealed that the dissipation of MMP might participate in apoptosis induced by compound **196d** [231].

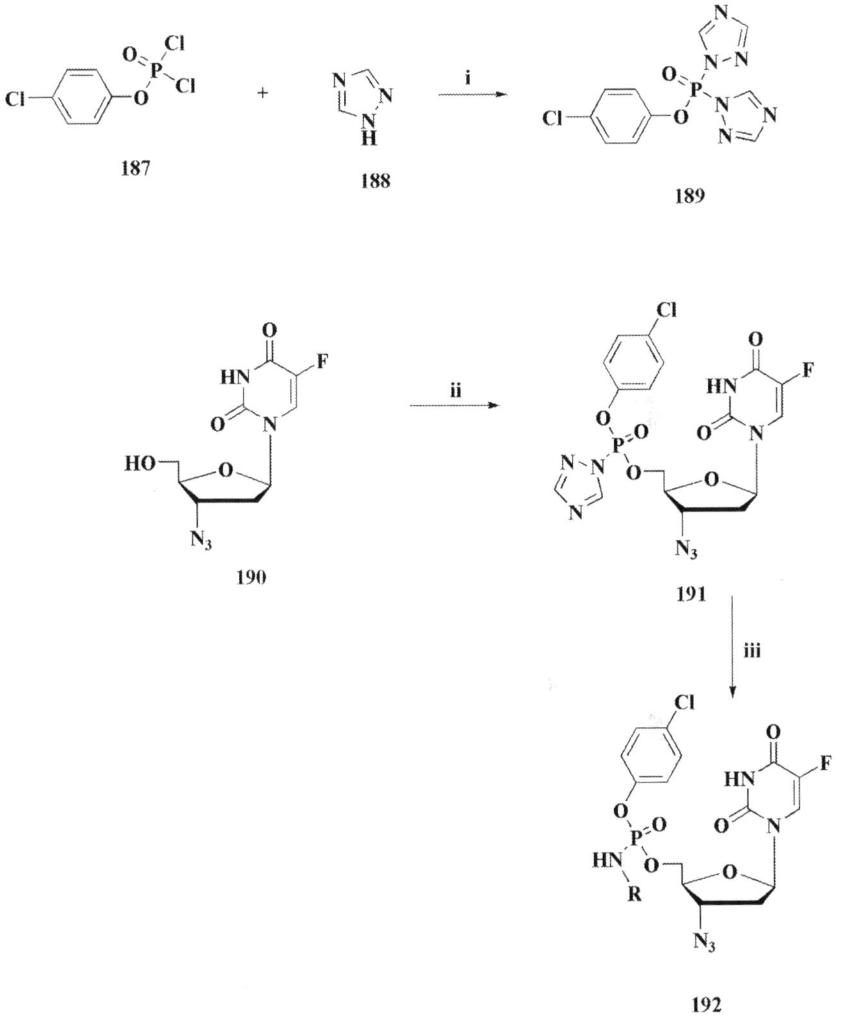

Scheme 7.27 Reagents and conditions: (i) NEt_3/CH_3CN/RT/30 min; (ii) **189**/pyridine/RT/1 h; (iii) R-NH_2/RT/1 h.

3.2 Nonribonucleoside pyrimidine analogs

3.2.1 Pyranose pyrimidine nucleosides

The production of the unsaturated dideoxyfluoroketopyranosyl nucleoside analogs attached to thymine, uracil, 5-fluorouracil, *N*4-benzoyl cytosine, and *N*6-benzoyl adenine, respectively, is outlined [232].

The condensation of compound **197** with silylated uracil, thymine, 5-fluorouracil, *N*4-benzoyl cytosine, and *N*6-benzoyl adenine in refluxing

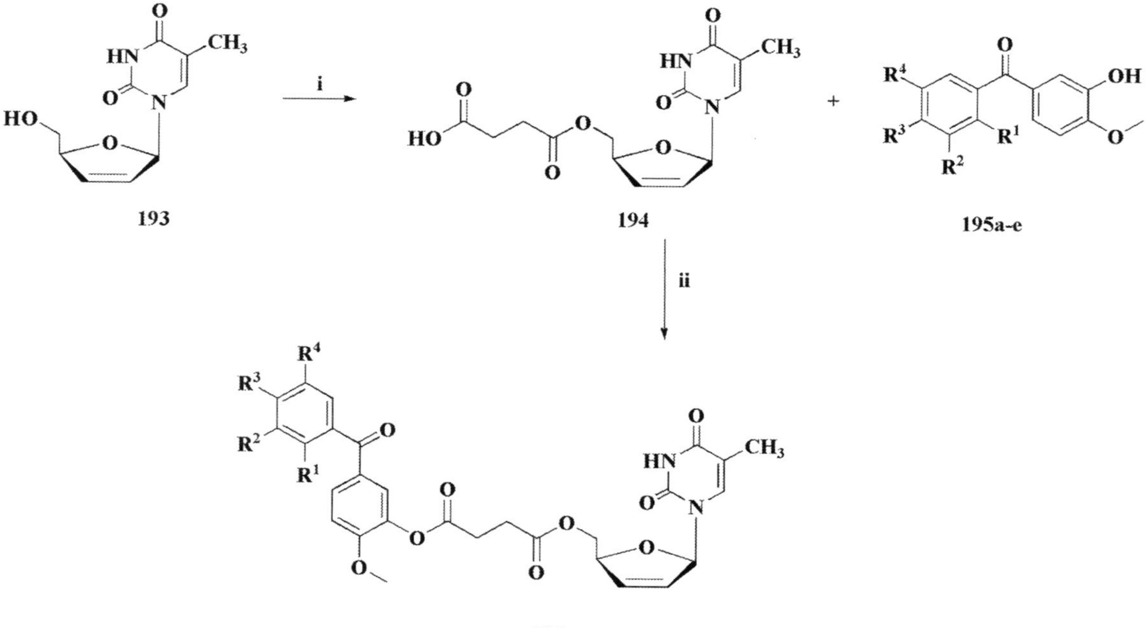

193 → (i) → 194 + 195a-e → (ii) → 196a-e

a R^1 = OMe, R^2 = H, R^3 = OMe, R^4 = OMe
b R^1 = H, R^2 = H, R^3 = F, R^4 = H
c R^1 = H, R^2 = H, R^3 = NO_2, R^4 = H
d R^1 = OMe, R2 = H, R^3 = H, R^4 = H
e R^1 = Cl, R^2 = H, R^3 = H, R^4 = H

Scheme 7.28 Reagents and conditions: (i) Succinic anhydride/TEA/CH_2Cl_2/RT/12h; (ii) DCC/DMAP/CH_2Cl_2/RT/12h.

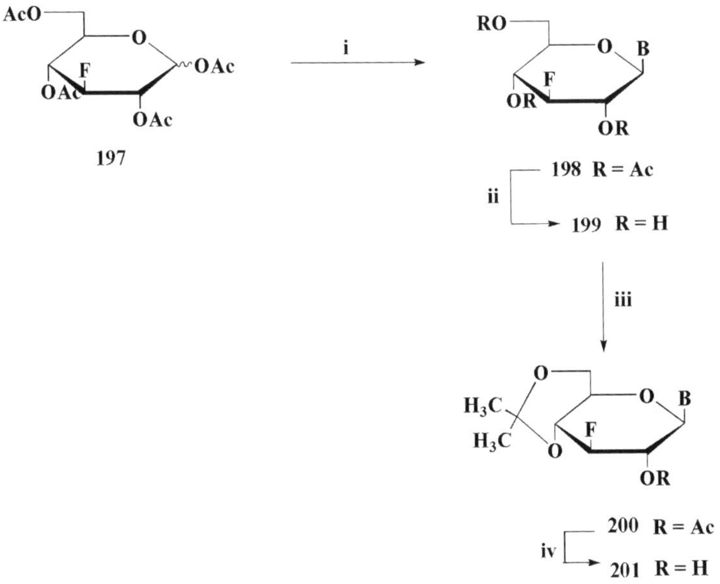

197 → (i) → 198 R = Ac; (ii) → 199 R = H; (iii) → 200 R = Ac; (iv) → 201 R = H

B = a: Uracil, b:5-Fluorouracil, c: Thymine, d:N -Benzoyl cytosine, e: N -Benzoyl adenine

Scheme 7.29 Reagents and conditions: (i) Silyated base/CH_3CN/tin chloride or trimethylsilyl trifluoromethane-sulfonate. (ii) NH_3/CH_3OH or C_2H_5OH/NaOH/pyridine/0°C/30 min/amberlite R 120 (H+) resin. (iii) *p*-Toluenesulfonic acid/2,2-dimethoxypropane/DMF. (iv) Acetic anhydride/pyridine.

acetonitrile yielded the nucleosides. Acetylation of compound **198** using 2,2-dimethoxypropane in *N,N*-dimethylformamide provides the 4′,6′-*O*-isopropylidene derivatives **200a**–**e** (Scheme 7.29) [232].

Acetylation of the free hydroxyl group and deisopropylidenation yielded the partially acetylated analogs **202a**–**e**. Replacement of the primary hydroxyl group of compounds **202a**–**e** by an iodine atom via iodine, triphenylphosphine, and imidazole afforded the targeted iodo derivatives **203a**–**e**, which were reduced to produce the deoxy nucleosides **204a**–**e**. Then the oxidation of the fluoro acetylated dideoxy precursors **204a**–**e** was carried out utilizing dimethyl sulfoxide/acetic anhydride system to yield compounds **205a**–**e** (Scheme 7.30) [232].

The compounds were tested for their cytostatic/cytotoxic activity. Compounds **144a**–**e** have an inhibitory activity against the proliferation of various tumor cell lines [232].

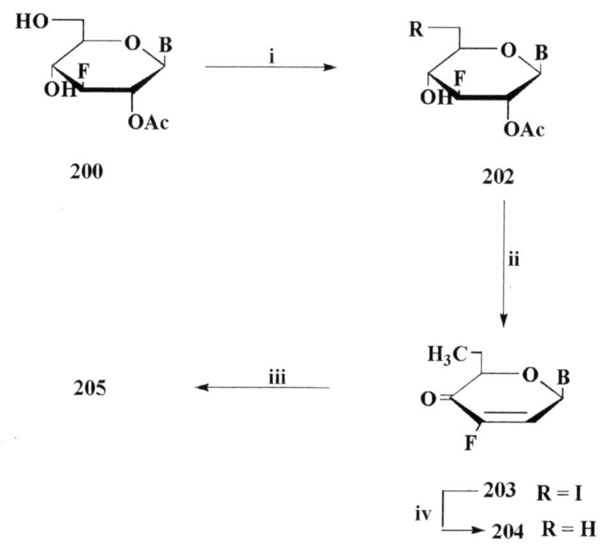

Scheme 7.30 Reagents and conditions: (i) 90% trifluoroacetic acid/CH_3OH/220°C/10 min. (ii) triphenylphosphine/iodine/imidazole/THF/80°C/1 h. (iii) Acetic anhydride/DMSO/100°C/10 min. (iv) H_2/10% Pd/C/triethylamine/ethyl acetate/ethanol/20°C/24 h.

3.2.2 Carbocyclic pyrimidine nucleosides

The biologically active fluoro-carbocyclic nucleosides were designed and synthesized as depicted in Scheme 7.31 [233].

Based on the anticancer potency of 6′-fluorocyclopentenyl-cytosine **212a** in phase IIa clinical trials for the treatment of gemcitabine-resistant pancreatic cancer, a systematic SAR study of 6′-fluorocyclopentenyl-pyrimidines **209a–e**, **212b–e**, which were prepared from D-ribose through a vinyl fluorination, was carried out to discover anticancer agents. The cytosine analog **212a** displayed the most anticancer potency among all of the tested compounds. The steric and electronic effects induced by halogen atoms, which were substituted at the C5 position, were detrimental to cellular phosphorylation of the final nucleosides, leading to weak anticancer activity. All of the synthesized pyrimidine nucleosides displayed much less active anticancer potency in vitro than the cytosine derivative **212a**, acting as DNA and/or RNA polymerase inhibitor, demonstrating that they could not be efficiently transformed to their triphosphates for anticancer potency. The addition of the bulky alkyl group at the *N*6-amino group, the deamination of the *N*6-amino group, or the introduction of

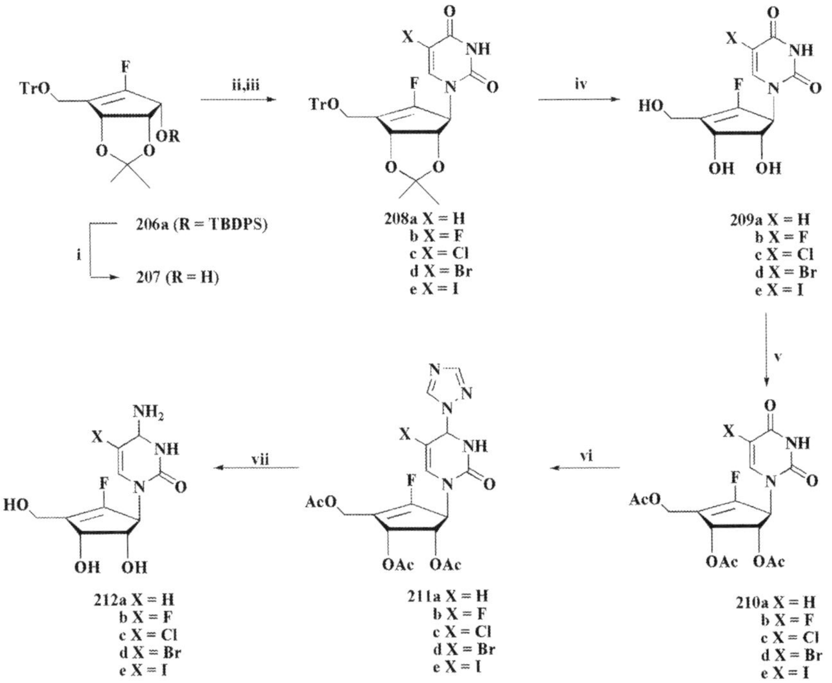

Scheme 7.31 Reagents and conditions: (i) TBAF/THF/0°C to RT/3 h. (ii) N^3-benzoyluracil or N^3-benzoyl-5-halouracil/PPh_3/DIAD/THF/0°C to RT/16 h. (iii) NH_3/$\mathbf{CH_3OH}$/RT/15 h. (iv) HCl/CH_3OH/RT/18 h. (v) Ac_2O/pyridine/RT/16 h. (vi) $POCl_3$/1,2,4-triazole/TEA/MeCN/RT/16 h. (vii) NH_4OH/H_2O/1,4-dioxane/RT/18 h/NH_3/MeOH/RT/3 h.

the amino group at the C2 position almost abolished the anticancer potency [233].

3.2.3 Modified pyrimidine nucleosides

The glycosyl donor, 4-selenoxide **214**, was yielded upon the oxidation of **213** with m-CPBA. The further Pummerer-type condensation of **214** with different pyrimidine bases as thymine, 5-halouracil, and uracil in the presence of trimethylsilyltrifluoromethane sulfonate and Et_3N afforded the β-anomers **215a–f**. In addition, the 4′-selenoribofuranosyl pyrimidines **216a–f** were furnished after the removal of the protecting groups of **215a–f** using 50% TFA (Scheme 7.32) [234].

The ribo analogs **216a–f** were subjected for the treatment with $TIPDSCl_2$ to furnish the 3′,5′-*O*-TIPDS-protected nucleosides **217a–f** - (Scheme 7.33). In order to prevent the production of O2,2′-anhydro

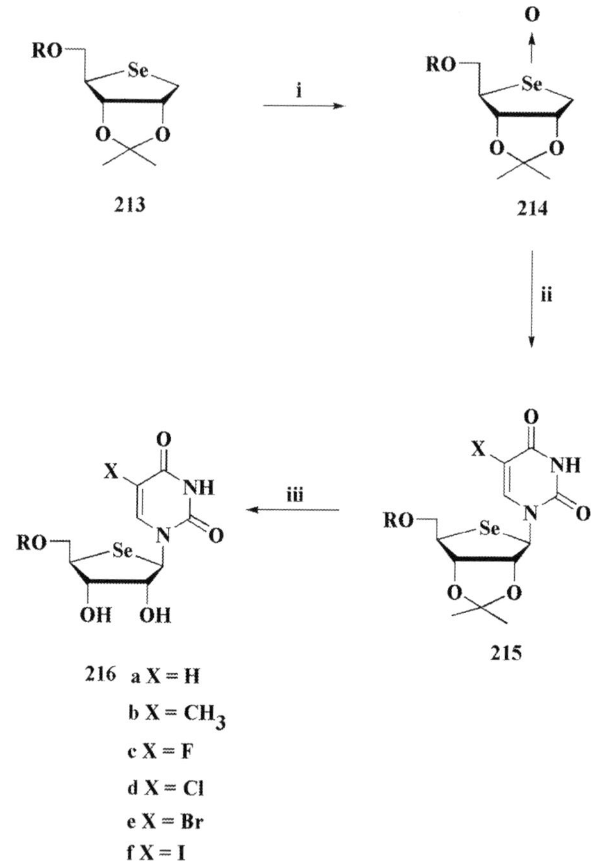

Scheme 7.32 Reagents and conditions: (i) mCPBA/CH_2Cl_2/−78°C/45 min. (ii) Base/Et_3N/TMSOT f/CH_2Cl_2/toluene/0°C to rt. (iii) THF/TFA/RT.

nucleosides during the Mitsunobu reaction with diphenylphosphorylazide (DPPA), the *N*-3 position of **217a–f** was selectively protected with electron-withdrawing benzoyl group in the presence of the 2′-hydroxyl group to afford **218a–f**. Treatment of **218a–f** with the DPPA under the Mitsunobu conditions yielded the targeted azidoarabinofuranosyl compounds **219a–f** [234].

Treatment of compound **219a–f** with 3HF.Et_3N afforded **220a–f**, which were then treated with methanolic ammonia to yield the pyrimidines **221a–f** (Scheme 7.34).

For the synthesis of the selenoarabinofuranosyl pyrimidines **229**, the same intermediates **217a–f** were transformed to **226**, which act as the

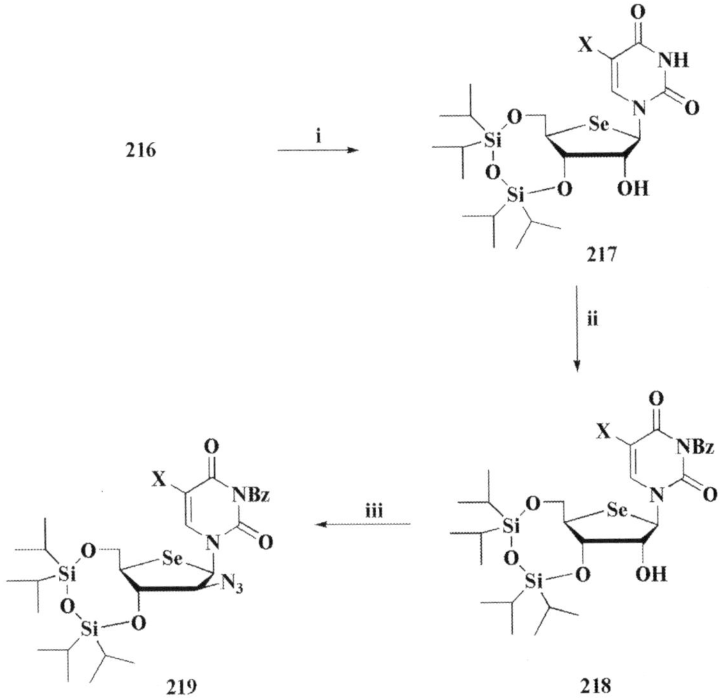

216 →(i) 217 →(ii) 218 →(iii) 219

Scheme 7.33 Reagents and conditions: (i) $TIPDSCl_2$, pyridine, RT. (ii) BzCl, Et_4N^+, Br^- in aqueous Na_2CO_3/CH_2Cl_2/RT for **a**, **c–f** and BzCl, Et_3N, CH_2Cl_2, RT.

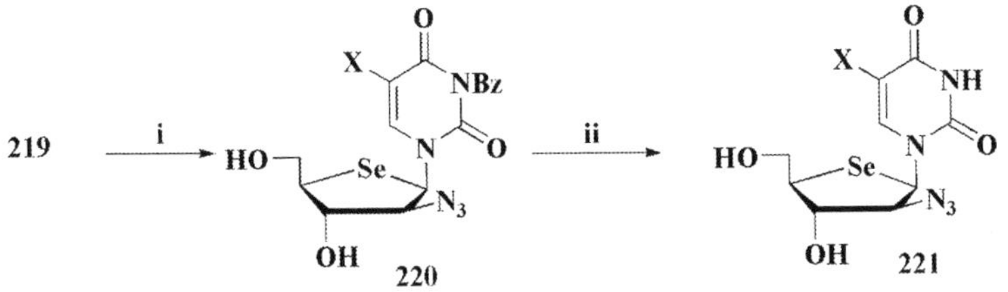

219 →(i) 220 →(ii) 221

Scheme 7.34 Reagents and conditions: (i) 3HF. Et_3N, Et_3N, THF, 0°C. (ii) NH_3/MeOH/RT.

substrates for DAST fluorination. Treatment of **217a–f** with MsCl/DBU afforded the *O*2,2′-anhydro nucleosides **222**. After the removal of the TIPDS groups of **222** with 3HF. Et_3N, the produced diols **223** were protected to afford **225**, respectively. Opening of the *O*2,2′-anhydro ring in **225** with sodium hydroxide yielded **226** with an arabino configuration at the 2′-position, which acts as the substrates for DAST fluorination (Scheme 7.35) [234].

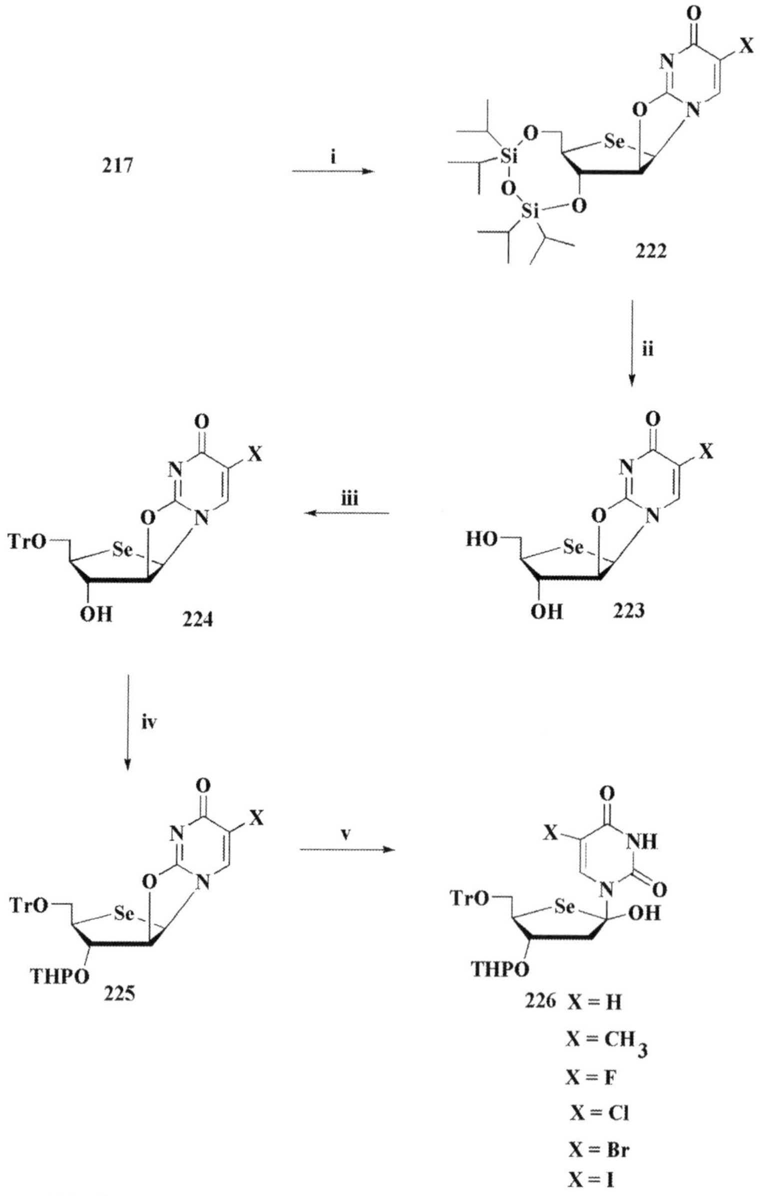

Scheme 7.35 Reagents and conditions: (i) MsCl/DMAP/Et_3N/MC/0°C/DBU/acetone/RT. (ii) 3HF.Et_3N/THF/0°C. (iii) TrCl/DMAP/pyridine/60°C. (iv) DHP/MC/RT. (v) NaOH/CH_3CN/RT.

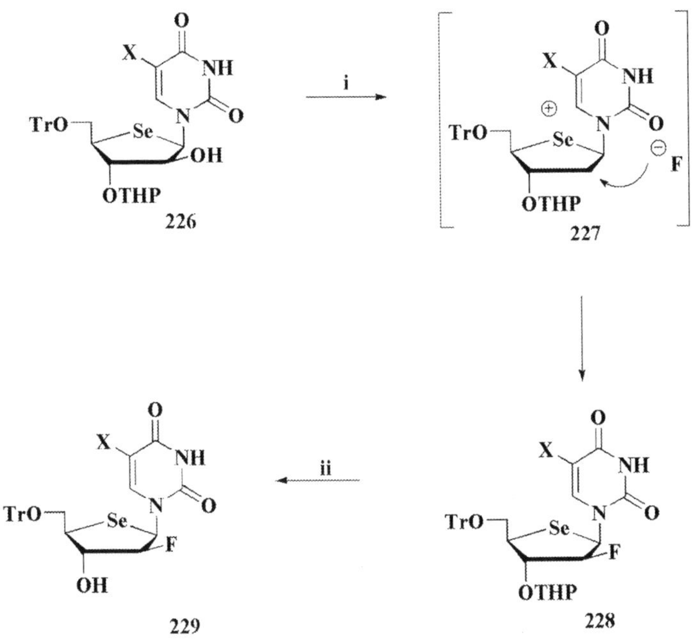

Scheme 7.36 Reagents and conditions: (i) DAST. (ii) CH_3COOH.

Treatment of **226** with DAST yielded the targeted 2′-fluoro analogs **228** with arabino configurations, which were synthesized via double SN_2 reactions as outlined in Scheme 7.36. The 2′-fluoro-4′-selenoarabinofuranosyl pyrimidines **229** were yielded upon the treatment of **228** with 80% acetic acid [234].

Compound **223a–e** were treated with sodium hydroxide to afford the 4′-selenarabinofuranosyl pyrimidines **230a–e**. The reaction of 5-iodo derivative **223f** under these set of conditions resulted in the production of the dehalogenated uracil compound **230f** instead of giving **231**. Thus, the formation of the 5-iodo derivative **231** was completed by the Mitsunobu reaction followed by the removals of all protecting groups, as outlined in Scheme 7.37 [234].

The uracil derivatives **221a**, **229a**, and **230a** were transformed to the corresponding cytosine derivatives **234a**, **234b**, and **234c** as illustrated in Scheme 7.38. The uracil analogs **221** were peracetylated to afford **232a**, **232b**, and **232c**. Treatment of **232a**, **232b**, and **232c** with 1,2,4-triazole,

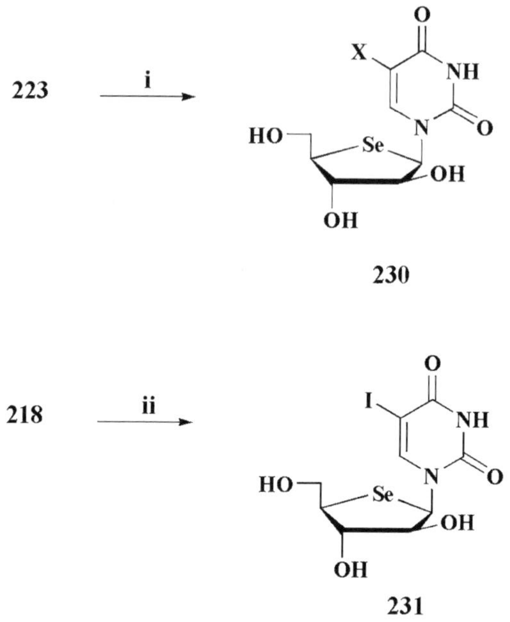

Scheme 7.37 Reagents and conditions: (i) 1 N/NaOH/CH_3CN. (ii) DIAD/PPh_3/BzOH/THF/0°C to RT/3HF·Et_3N/THF/°C/NH_3/MeOH/RT.

phosphorous oxychloride, and Et_3N yielded the corresponding triazole derivatives **233a**, **233b**, and **233c**. The latter compounds were then treated with ammonium hydroxide in methanolic ammonia and 1,4-dioxane to yield the cytosine derivatives **234a**, **234b**, and **234** [234].

The 2′-deoxy-4′-selenoarabinofuranosyl pyrimidines **221** and **234** were examined for cytotoxic effects in some human tumor cell lines. Among the evaluated compounds, the 2′-fluoro derivative **234b** was found to indicate the most anticancer potency and exhibited better anticancer potency than cytarabine [234].

A novel class of metallocene-nucleobase derivatives such as ferrocenyl-thymine-3,6-dihydro-2*H*-thiopyranes were synthesized as described in Schemes 7.39 and 7.40 [235].

The preparation of ferrocenyl-4-thiothymine-3,6-dihydro-2*H*-thiopyran **237** was started via the synthesis of the thioketone **236** (77% yield) through a thionation reaction of the ketone **235** and Lawesson's reagent. Further thermally induced hetero-Diels-Alder cycloaddition reaction of **236** with 2,3-dimethyl-1,3-butadiene afforded the complex **237** in 77% yield.

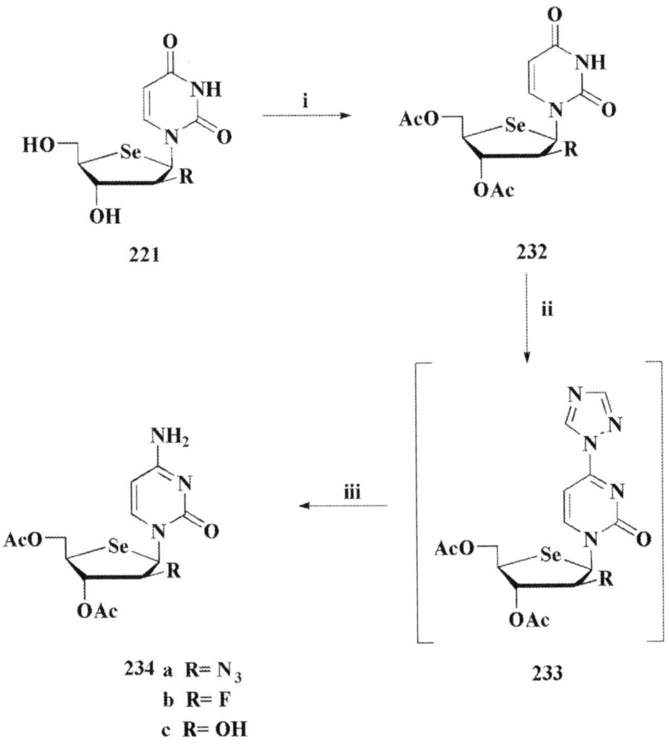

221 → (i) → 232 → (ii) → [233] → (iii) → 234 a R= N_3; b R= F; c R= OH

Scheme 7.38 Reagents and conditions: (i) Ac_2O/RT. (ii) 1,2,4-triazole, Et_3N, $POCl_3$/CH_3N. (iii) NH_4OH/1,4-dioxane/RT/NH_3/MeOH/RT.

To generate thioketone **239** (68% yield) as a sole isolated product, diphosphorus pentasulfide was used as thionation reagent. The cycloaddition reaction of thioketone **239** with 2,3-dimethyl-1,3-butadiene in the presence of lithium chloride resulted in the formation of ferrocenyl-thymine-3,6-dihydro-2*H*-thiopyran **240**, which is obtained in 65% yield [235].

The thymine derivative **240** exhibited a cytostatic activity at low micromolar levels; its thiothymine congener **237** indicated decreased or no cytostatic activity against cell lines tested. The apoptotic activity of complex **237** and **240** was evaluated using AO/EB staining. The complexes **237** and **240** have potential apoptosis-inducing properties on tumor cultures [235].

The synthesis of the isoxazolidinylfuropyrimidine **239** is accomplished. Thus, nitrone **241** and vinyl acetate **242** were allowed to react in anhydrous ether to afford adducts **243** and **244** in a relative ratio 1:4.2 in 90% yield (Scheme 7.41). According to Hilbert–Jones methodology, compound

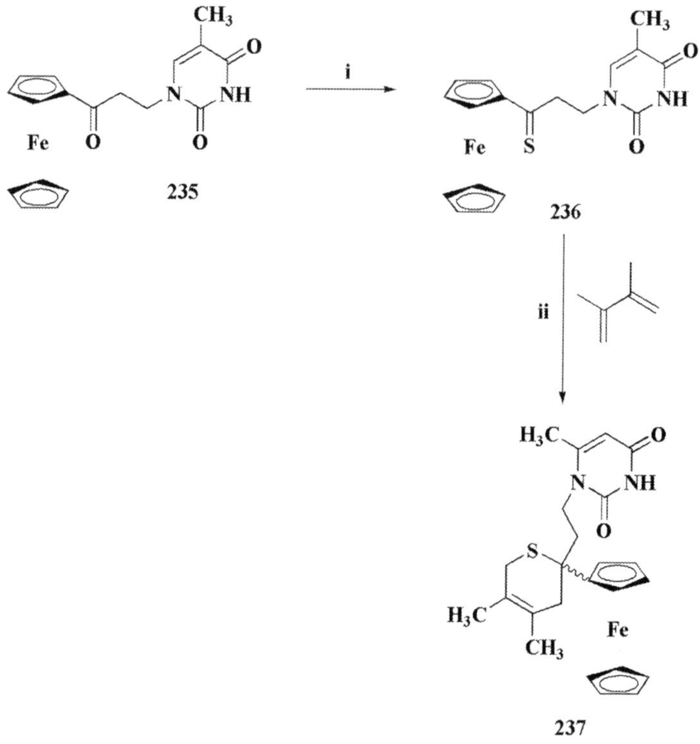

Scheme 7.39 Reagents and conditions: (i) P_2S_5/acetonitrile/reflux/40 min. (ii) Toluene/LiCl/reflux/20 h.

245 was afforded when the mixture of compounds **243** and **244** was reacted with silylated 5-iodouracil [236].

According to Sonogashira reaction, the coupling reaction of **245** with alkynes **246a–e** yielded alkynyl nucleosides **247a–e** in good to excellent yields (Scheme 7.42).

Reflux the latter nucleosides with CuI (1 equiv.) in a mixture of MeOH/TEA (3:2) for 4 h afforded compounds **248a–e** (80%–95% yield). The Pd-catalyzed coupling reaction of aryl acetylenes **246a–e** with nucleosides **245** afforded compounds **248a–e** in 85%–95% yield. The furopyrimidines **249a–e** containing the hydroxymethyl group at *C*-4′ were accessed by treating **248a–e** with TBAF [236].

The cytotoxicity of the examined compounds toward uninfected host cells, defined as the minimum cytotoxic concentration (MCC) that results in a microscopically detectable alteration of normal cell morphology, was

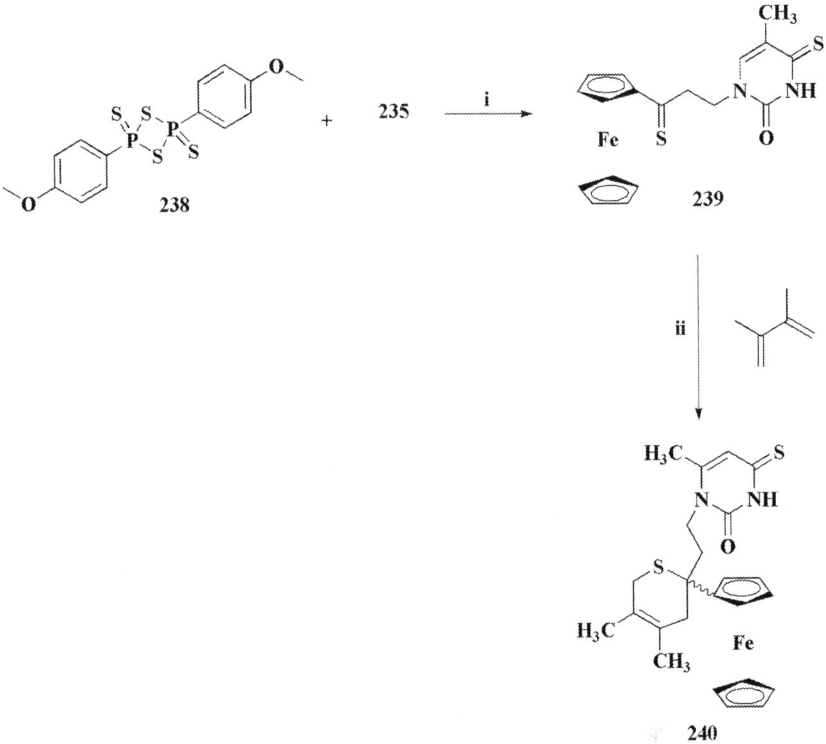

Scheme 7.40 Reagents and conditions: (i) Toluene/reflux/2 h. (ii) Toluene/90 °C/48 h.

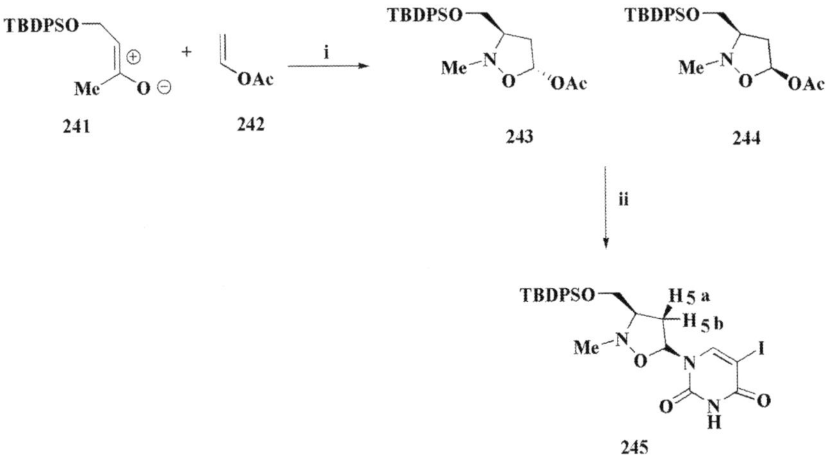

Scheme 7.41 Reagents and conditions: (i) Nitrone/vinyl acetate/ether/48/RT. (ii) MeCN/silylated 5-iodouracil/isoxazolidines **243** and **244**/TMSOTf/24 h/RT.

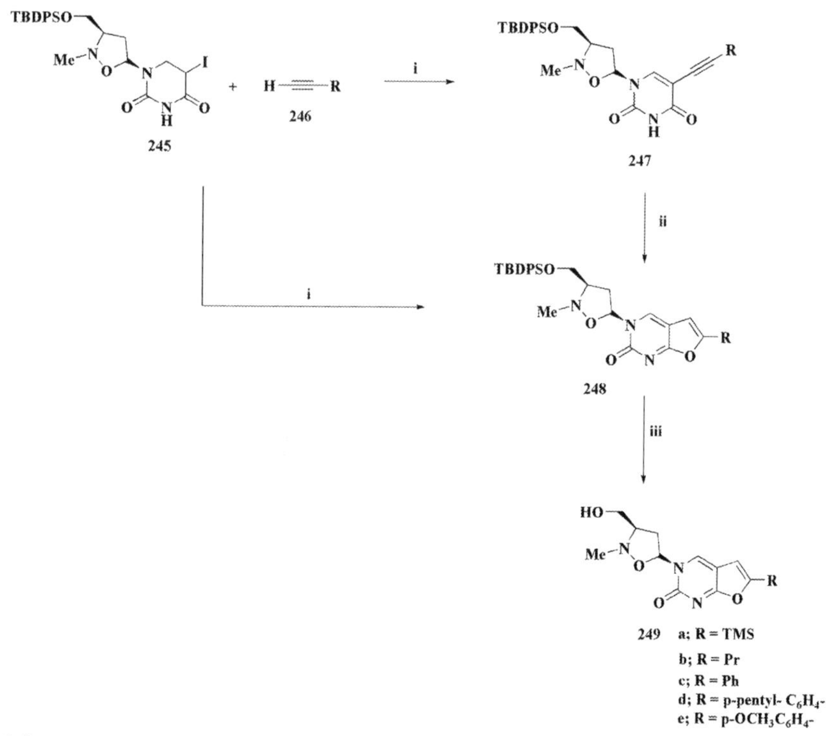

Scheme 7.42 Reagents and conditions: (i) DMF/TEA/Pd(PPh_3)$_4$/CuI/50°C/4h. (ii) CuI/MeOH/TEA (3:2). (iii) TBAF/RT/4–5h.

assayed. The compound **249d** exhibits the highest level of cell proliferation inhibitory potency with CC_{50} values of 0.7, 1.5, and 3 lM for BS-C-1, MRC-5, and Vero cells [236].

A mixture of diastereomers of the dioxolane derivatives **253** and **252** was obtained in yield (55%) upon the condensation of L-ascorbic acid **251** with -benzyloxy-2-propanone **250** with acetonitrile in the presence of *p*-toluene sulfonic acid. Oxidative degradation of the lactone ring of the diastereomers, and then subsequent oxidation of the resulting potassium salt **254** using sodium hypochlorite, catalyzed by benzyltriethylammonium chloride and ruthenium trichloride under controlled pH conditions, yielded the two dioxolane carboxylic acid isomers **255** and **256**. The latter compounds were then isolated. The transformation of the carboxyl group to the acetoxy group was done by oxidative decarboxylation of **255** and **256** using lead tetraacetate in acetonitrile in the presence of pyridine to afford the intermediates **257** and **258** [237].

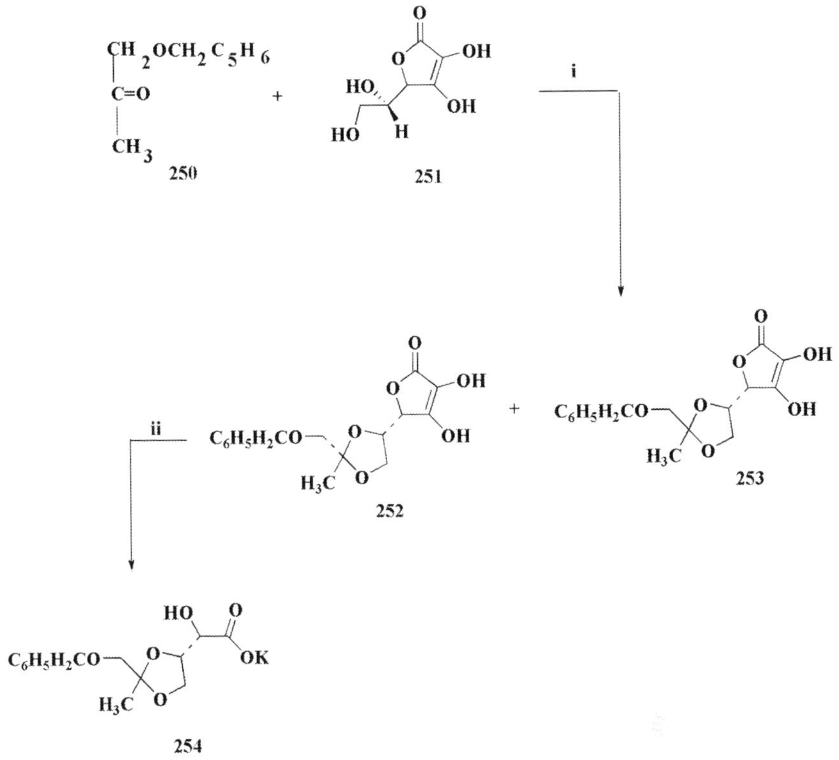

Scheme 7.43 Reagents and conditions: (i) TsOH/acetonitrile. (ii) 30% H_2O_2/K_2CO_3/H_2O/EtOH.

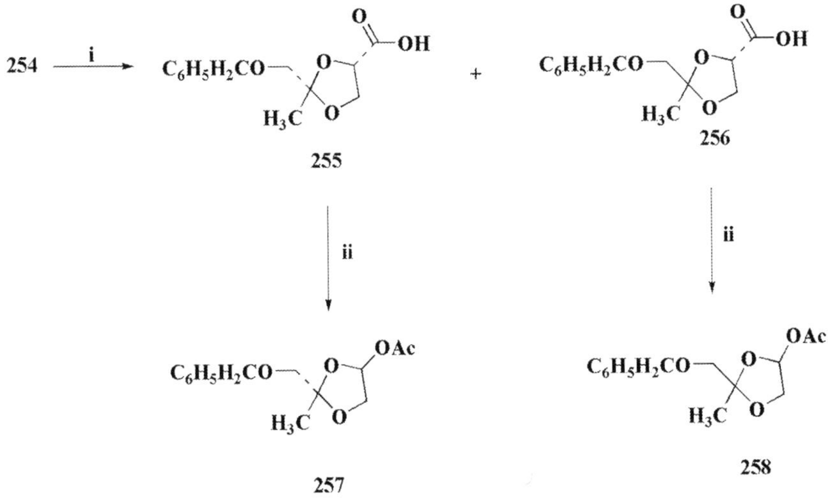

Scheme 7.44 Reagents and conditions: (i) NaOCl/$RuCl_3$/hydrate/benzyltriethylammonium chloride/H_2O/dichloroethane/acetonitrile/pH 8/HCl/dichloromethane flash chromatography. (ii) $Pb(OAc)_4$/pyridine/acetonitrile.

The reaction of the acetate **257** with silylated *N*-benzoyl-5-fluorocytosine or *N*-benzoylcytosine in the presence of the Lewis acid, trimethylsilyl trifluoromethanesulfonate in dichloroethane furnished a mixture of α- and β-nucleosides **259a, 260a**, **259b,** and **260b**, which were isolated using silica gel column chromatography. Coupling of the acetate **258** with *N*-benzoyl-5-cytosine or silylated *N*-benzoylcytosine, followed by separation and deprotection, yielded the corresponding compounds **263a**, **b** and the α-isomers **264a**, **b** (Schemes 7.43–7.45) [237].

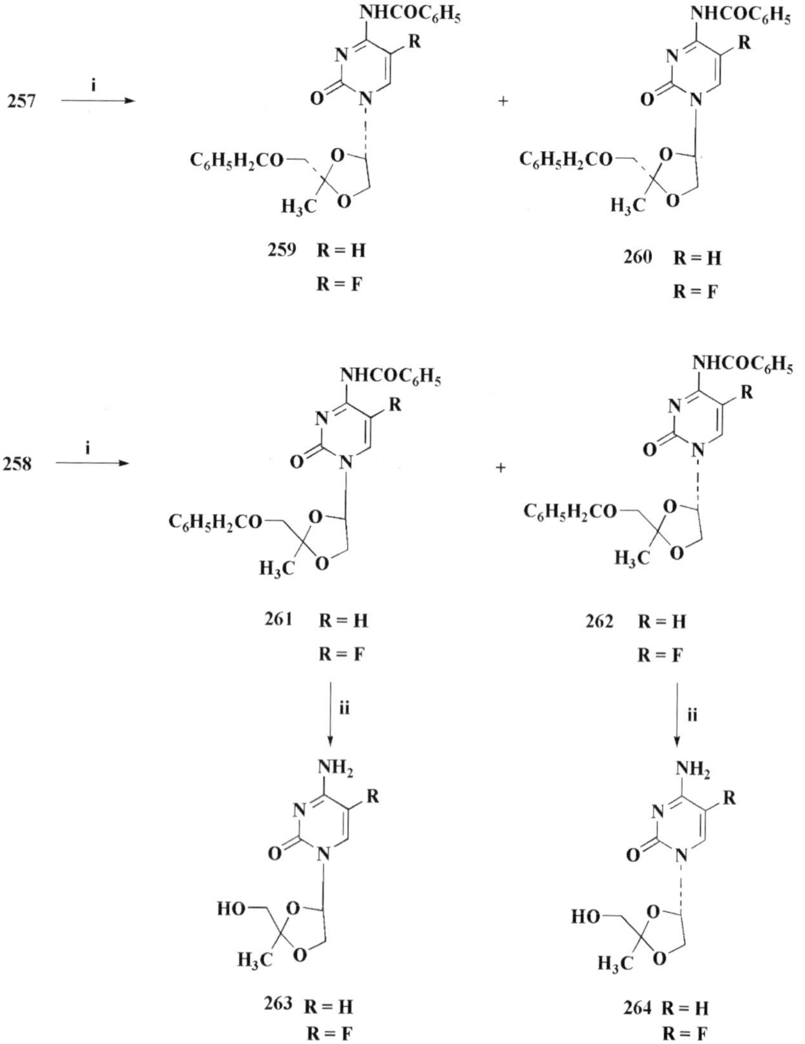

257 →(i) 259 R = H, R = F + 260 R = H, R = F

258 →(i) 261 R = H, R = F + 262 R = H, R = F

261 →(ii) 263 R = H, R = F

262 →(ii) 264 R = H, R = F

Scheme 7.45 Reagents and conditions: (i) 5-Fluorocytosine or cytosine/HMDS/reflux/6 h/TMSOTf/dichloroethane flash chromatography. (ii) NH_3/CH_3OH/PdO hydrate/cyclohexene/EtOH.

Compounds **263a**, **b**, and **264a**, **b** were tested in vitro for their cytotoxicities against the L1210 and P388 leukemias, the CCRF-CEM lymphoblastic leukemia, and the $B_{16}F_{10}$ melanoma cell lines. Compound **263a** showed IC_{50} values of 50, 10, 30, and 100 mM against L1210, P388, CCRF-CEM, and $B_{16}F_{10}$ cells. 5-Fluorocytosine analogs **263b** and **264b** indicated a weak or no potency up to 100 mM against these neoplastic cell lines [237].

References

[1] Lagoja IM. Pyrimidine as constituent of natural biologically active compounds. Chem Biodivers 2005;2(1):1–50.

[2] Kim S, Kim E, Oh C, Yoo KH, Hong JH. Synthesis and potent anti-leukemic activity of novel 5'-norcarbocyclic *C*-nucleoside. Bull Kor Chem Soc 2014;35(12):352–3508.

[3] Gartel AL. Novel anticancer compounds induce apoptosis in human tumor cells. Chin J Cancer 2008;27(7):675–6.

[4] Khalil A, Ishita K, Ali T, Tjarks W. *N*3-substituted thymidine bioconjugates for cancer therapy and imaging. Future Med Chem 2013;5(6):677–92.

[5] Bhat UG, Zipfel PA, Tyler DS, Gartel AL. Novel anticancer compounds induce apoptosis in melanoma cells. Cell Cycle 2008;7(12):1851–5.

[6] Alexander V, Song J, Yu J, Choi JH, Kim J, Lee SK, Choi WJ, Jeong LS. Synthesis and biological evaluation of 2′-substituted-4′-selenoribofuranosyl pyrimidines as antitumor agents. Arch Pharm Res 2015;38(6):966–72.

[7] Cappellacci L, Franchetti P, Petrelli R, Riccioni S, Vita P, Jayaram HN, Grifantini M. Purine and pyrimidine nucleoside analogs of 3′-*C*-methyladenosine as antitumor agents. Collect Czechoslov Chem Commun 2006;71(7):1088–98.

[8] Gunaga P, Moon HR, Choi WJ, Shin DH, Park JG, Jeong LS. Recent advances in 4′-thionucleosides as potential antiviral and antitumor agents. Curr Med Chem 2004; 11(19):2585–637.

[9] Meng W, Qing F. Fluorinated nucleosides as antiviral and antitumor agents. Curr Top Med Chem 2006;6(14):1499–528.

[10] Sriharsha SN, Ranganath Pai KS, Suhas, Shashikanth S, Chandra N, Prabhu KR. Synthesis, docking and anti-tumor activity of β-L-1,3-thiazolidine pyrimidine nucleoside analogues. Med Chem 2007;3(5):425–32.

[11] Matsuda A, Sasaki T. Antitumor activity of sugar-modified cytosine nucleosides. Cancer Sci 2004;95(2):105–11.

[12] Sriharsha SN, Satish S, Shashikanth S, Raveesha KA. Design, synthesis and antibacterial activity of novel 1,3-thiazolidine pyrimidine nucleoside analogues. Bioorg Med Chem 2006;14(22):7476–81.

[13] Johar M, Manning T, Kunimoto DY, Kumar R. Synthesis and in vitro antimycobacterial activity of 5-substituted pyrimidine nucleosides. Bioorg Med Chem 2005;13(24):6663–71.

[14] Rai D, Johar M, Manning T, Agrawal B, Kunimoto DY, Kumar R. Design and studies of novel 5-substituted alkynylpyrimidine nucleosides as potent inhibitors of mycobacteria. J Med Chem 2005;48(22):7012–7.

[15] Johar M, Manning T, Tse C, Desroches N, Agrawal B, Kunimoto DY, Kumar R. Growth inhibition of *Mycobacterium bovis*, *Mycobacterium tuberculosis* and *Mycobacterium avium* in vitro: effect of 1-β-D-2′-arabinofuranosyl and 1-(2′-deoxy-2′-fluoro-β-D-2′-ribofuranosyl) pyrimidine nucleoside analogs. J Med Chem 2007;50(15):3696–705.

[16] Srivastav NC, Manning T, Kunimoto DY, Kumar R. Studies on acyclic pyrimidines as inhibitors of mycobacteria. Bioorg Med Chem 2007;15(5):2045–53.

[17] Torrence PF, Fan X, Zhang X, Loiseau PM. Structurally diverse 5-substituted pyrimidine nucleosides as inhibitors of *Leishmania donovani* promastigotes in vitro. Bioorg Med Chem Lett 2006;16(19):5047–51.

[18] Zai W, Jiang P, Zhang X, Loiseau PM, Guo S, Fan X. Synthesis and antileishmanial activity of pyrimidine nucleoside-chalcone hybrids. Chin J Org Chem 2015; 35(6):1335–41.

[19] Hasan A, Satyanarayana M, Mishra A, Bhakuni DS, Pratap R, Dube A, Guru PY. Acyclic pyrazolo[3,4-*d*]pyrimidine nucleoside as potential leishmaniostatic agent. Nucleosides Nucleotides Nucleic Acids 2006;25(1):55–60.

[20] Shimizu T, Kimura T, Funahashi T, Watanabe K, Ho IK, Yamamoto I. Synthesis of N^3-substituted uridine and related pyrimidine nucleosides and their antinociceptive effects in mice. Chem Pharm Bull 2005;53(3):313–8.

[21] Amador A, Benzaria S, Cretton-Scott E, D'Amours M, Mao J, Mathieu S, Moussa A, Bridges EG, Standring DN, Sommadossi J, Storer R, Gosselin G. Pharmacokinetics of valopicitabine (NM283), an efficient prodrug of the potent anti-HCV agent 2′-*C*-methylcytidine. J Med Chem 2006;49(22):6614–20.

[22] Pei X, Choi WJ, Kim YM, Zhao LX, Jeong LS. Synthesis of 3'-*C*-hydroxymethyl-substituted pyrimidine and purine nucleosides as potential anti-hepatitis C virus (HCV) agents. Arch Pharm Res 2008;31(7):843–9.

[23] Ding Y, An H, Hong Z, Girardet J. Synthesis of 2'-beta-*C*-methyl toyocamycin and sangivamycin analogues as potential HCV inhibitors. Bioorg Med Chem Lett 2005;15(3):725–7.

[24] Pierra C, Benzaria S, Amador A, Moussa A, Mathieu S, Storer R, Gosselin G. NM 283, an efficient prodrug of the potent anti-HCV agent 2′-*C*-methylcytidine. Nucleosides Nucleotides Nucleic Acids 2005;24(5-7):767–70.

[25] Shi J, Du J, Ma T, Pankiewicz KW, Patterson SE, Hassan AEA, Tharnish PM, McBrayer TR, Lostia S, Stuyver LJ, Watanabe KA, Chu CK, Schinazi RF, Otto MJ. Synthesis and in vitro anti-HCV activity of β-D- and L-2′-deoxy-2′-fluororibonucleosides. Nucleosides Nucleotides Nucleic Acids 2005;24(5–7):875–9.

[26] Oh C, Kim J, Hong J. Simple synthesis and anti-HIV activity of novel 3′-vinyl branched apiosyl pyrimidine nucleosides. Nucleosides Nucleotides Nucleic Acids 2006; 25(8):871–8.

[27] Wang G, Wan J, Hu Y, Wu X, Prhavc M, Dyatkina N, Rajwanshi VK, Smith DB, Jekle A, Kinkade A, Symons JA, Jin Z, Deval J, Zhang Q, Tam Y, Chanda S, Blatt L, Beigelman L. Synthesis and anti-influenza activity of pyridine, pyridazine, and pyrimidine *C*--nucleosides as favipiravir (T-705) analogues. J Med Chem 2016;59(10):4611–24.

[28] Saladino R, Ciambecchini U, Maga G, Mastromarino P, Contic C, Bottad M. A new and efficient synthesis of substituted 6-[(2'-dialkylamino)ethyl] pyrimidine and 4-*N,N*-dialkyl-6-vinyl-cytosine derivatives and evaluation of their anti-rubella activity. Bioorg Med Chem 2002;10:2143–53.

[29] Chu CK. 5-(E)-Bromovinyl uracil analogues and related pyrimidine nucleosides as anti-varicella zoster virus agents and methods of use. US patent; 2013. US8492362 B1.

[30] Chu CK, Gadthula S, Chen X, Choo H, Olgen S, Barnard DL, Sidwell RW. Antiviral activity of nucleoside analogues against SARS-Coronavirus (SARS-CoV). Antiviral Chem Chemother 2006;17(5):285–9.

[31] Peters HL, Jochmans D, de Wilde AH, Posthuma CC, Snijder EJ, Neyts J, Seley-Radtke KL. Design, synthesis and evaluation of a series of acyclic fleximer nucleoside analogues with anti-coronavirus activity. Bioorg Med Chem Lett 2015;25(15):2923–6.

[32] Pyrc K, Jan Bosch B, Berkhout B, Jebbink MF, Dijkman R, Rottier P, van der Hoek L. Inhibition of human coronavirus NL63 infection at early stages of the replication cycle. Antimicrob Agents Chemother 2006;2000–8.

[33] Pruijssers AJ, Denison MR. Nucleoside analogues for the treatment of coronavirus infections. Curr Opin Virol 2019;35:57–62.
[34] Yates MK, Raje MR, Chatterjee P, Spiropoulou CF, Bavari S, Flint M, Soloveva V, Seley-Radtke KL. Flex-nucleoside analogues—novel therapeutics against filoviruses. Bioorg Med Chem Lett 2017;27(12):2800–2.
[35] Mantione D, Aizpuru OO, Memeo MG, Bovio B, Quadrelli P. 4-Heterosubstituted cyclopentenone antiviral compounds: synthesis, mechanism, and antiviral evaluation. Eur J Org Chem 2016;5:983–91.
[36] Chiacchio U, Rescifina A, Iannazzo D, Piperno A, Romeo R, Borrello L, Sciortino MT, Balestrieri E, Macchi B, Mastino A, Romeo G. Phosphonated carbocyclic 2′-oxa-3′-azanucleosides as new antiretroviral agents. J Med Chem 2007;50(15):3747–50.
[37] Wittine K, Benci K, Pavelic SK, Pavelic K, Mintas M. The novel unsaturated acyclic nucleoside analogues: cytostatic and antiviral activity evaluations. Nucleic Acids Symp Ser 2004;52:601–2.
[38] Rao JR, Schinazi RF, Chu CK. Enantioselective synthesis and antiviral activity of purine and pyrimidine cyclopentenyl *C*-nucleosides. Bioorg Med Chem 2007; 15(2):839–46.
[39] Zhu W, Gumina G, Schinazi RF, Chu CK. Synthesis and anti-HIV activity of L-β-3′-*C*-cyano-2′,3′-unsaturated nucleosides and L-3′-*C*-cyano-3′-deoxyribonucleosides. Tetrahedron 2003;59(34):6423–31.
[40] El-Essawy FA. Synthesis of novel acyclonucleosides analogs of pyridothienopyrimidine as antiviral agents. Nucleosides Nucleotides Nucleic Acids 2005;24(8):1265–76.
[41] Hockova D, Holy A, Masojidkova M, Andrei G, Snoeck R, De Clercq E, Balzarini J. 5-substituted-2,4-diamino-6-[2-(phosphonomethoxy) ethoxy]pyrimidines-acyclic nucleoside phosphonate analogues with antiviral activity. J Med Chem 2003;46 (23):5064–73.
[42] Kitano K, Machida H, Miura S. Syntheis of novel 4'-*C*-methyl-pyrimidine nucleosides and their biological activities. Bioorg Med Chem Lett 1999;9:827–30.
[43] Fan X, Zhang X, Zhou L, Keith KA, Kern ER, Torrence PF. Assembling a smallpox biodefense by interrogating 5-substituted pyrimidine nucleoside chemical space. Antiviral Res 2006;71(2–3):201–5.
[44] Merck C, INC. Fluorinated pyrrolo[2,3-*d*]pyrimidine N nucleosides for the treatment of RNA-dependent RNA viral infection. Patent; 2006. WO2006/65335A2.
[45] Fan X, Zhang X, Zhou L, Keith KA, Kern ER, Torrence PF. A pyrimidine-pyrazolone nucleoside chimera with potent in vitro anti-orthopoxvirus activity. Bioorg Med Chem Lett 2006;16(12):3224–8.
[46] McGuigan C, Balzarini J. Aryl furano pyrimidines: the most potent and selective anti-VZV agents reported to date. Antiviral Res 2006;71(2–3):149–53.
[47] Kumamoto H, Onuma S, Tanaka H, Dutschman GE, Cheng Y. 3′-Carbon-substituted pyrimidine nucleosides having a 2′,3′-dideoxy and 2′,3′-didehydro-2′,3′-dideoxy structure: synthesis and antiviral evaluation. Antiviral Chem Chemother 2006; 17(4):225–34.
[48] Procopio A, Alcaro S, De Nino A, Maiuolo L, Ortuso F, Sindona G. New conformationally locked bicyclic *N*,*O*-nucleoside analogues of antiviral drugs. Bioorg Med Chem Lett 2005;15(3):545–50.
[49] Amblard F, Aucagne V, Guenot P, Schinazi RF, Agrofoglio LA. Synthesis and antiviral activity of novel acyclic nucleosides in the 5-alkynyl- and 6-alkylfuro[2,3-*d*]pyrimidine series. Bioorg Med Chem 2005;13(4):1239–48.
[50] Janeba Z, Balzarini J, Andrei G, Snoeck R, De Clercq E, Robins MJ. Synthesis and biological evaluation of acyclic 3-[(2-hydroxyethoxy)methyl] analogues of antiviral furo- and pyrrolo[2,3-*d*]pyrimidine nucleosides. J Med Chem 2005;48(14):4690–6.

[51] Luoni G, McGuigan C, Andrei G, Snoeck R, De Clercq E, Balzarini J. Bicyclic nucleoside inhibitors of Varicella-Zoster virus: the effect of branching in the p-alkylphenyl side chain. Bioorg Med Chem Lett 2005;15(16):3791–6.
[52] Ying C, Holy A, Hocková D, Havlas Z, De Clercq E, Neyts J. Novel acyclic nucleoside phosphonate analogues with potent anti-hepatitis B virus activities. Antimicrob Agents Chemother 2005;49(3):1177–80.
[53] Gunaga P, Kim HO, Kim HJ, Chun MW, Jeong LS. Stereoselective synthesis of novel thioiso dideoxy nucleosides with exocyclic methylene as potential antiviral agents. Nucleosides Nucleotides Nucleic Acids 2005;24(5–7):1115–7.
[54] Ambrose A, Zemlicka J, Kern ER, Drach JC, Gullen E, Cheng Y. Phosphoralaninate pronucleotides of pyrimidine methylenecyclopropane analogues of nucleosides: synthesis and antiviral activity. Nucleosides Nucleotides Nucleic Acids 2005;24 (10–12):1763–74.
[55] Zhou S, Breitenbach JM, Borysko KZ, Drach JC, Kern ER, Gullen E, Cheng Y, Zemlicka J. Synthesis and antiviral activity of (Z)- and (E)-2,2-[bis(hydroxymethyl)cyclopropylidene]methylpurines and -pyrimidines: second-generation methylenecyclopropane analogues of nucleosides. J Med Chem 2004;47(3):566–75.
[56] Maslen HL, Hughes D, Hursthouse M, De Clercq E, Balzarini J, Simons C. -6-Azapyrimidine-2′-deoxy-4′-thionucleosides: antiviral agents against TK+ and TK- HSV and VZV strains. J Med Chem 2004;47(22):5482–91.
[57] Angell A, McGuigan C, Garcia SL, Snoeck R, Andrei G, De Clercq E, Balzarini J. Bicyclic anti-VZV nucleosides: thieno analogues bearing an alkylphenyl side chain have reduced antiviral activity. Bioorg Med Chem Lett 2004;14(10):2397–9.
[58] McGuigan C, Carangio A, Snoeck R, Andrei G, De Clercq E, Balzarini J. Synthesis and antiviral evaluation of some 3′-fluoro bicyclic nucleoside analogues. Nucleosides Nucleotides Nucleic Acids 2004;23(1-2):1–5.
[59] Grote M, Noll S, Noll B, Johannsen B, Kraus W. Syntheses of novel modified acyclic purine and pyrimidine nucleosides as potential substrates of herpes simplex virus type-1 thymidine kinase for monitoring gene expression. Can J Chem 2004;82(4):513–23.
[60] De Clercq E. Nucleoside analogues exerting antiviral activity through a non-nucleoside mechanism. Nucleosides Nucleotides Nucleic Acids 2004;23(1-2):457–70.
[61] Balzarini J, Pannecouque C, Naesens L, Andrei G, Snoeck R, De Clercq E, Hocková D, Holý A. 6-[2-(Phosphonomethoxy)alkoxy]-2,4-diaminopyrimidines: a new class of acyclic pyrimidine nucleoside phosphonates with antiviral activity. Nucleosides Nucleotides Nucleic Acids 2004;23(8–9):1321–7.
[62] Gawin R, De Clercq E, Naesens L, Koszytkowska-Stawinska M. Synthesis and antiviral evaluation of acyclic azanucleosides developed from sulfanilamide as a lead structure. Bioorg Med Chem 2008;16(18):8379–89.
[63] Gadthula S, Chu CK, Schinazi RF. Synthesis and anti-HIV activity of β-D-3′-azido-2′, 3′-unsaturated nucleosides and β-D-3′-azido-3′- deoxyribofuranosylnucleosides. Nucleosides Nucleotides Nucleic Acids 2005;24(10–12):1707–27.
[64] Abdel-Rahman AAH, El-Etrawy AS, Abdel-Megied AES, Zeid IF, El Ashry EH. Synthesis and antiviral evaluation of novel 2,3-dihydroxypropyl nucleosides from 2- and 4-thiouracils. Nucleosides Nucleotides Nucleic Acids 2008;27(12):1257–71.
[65] Chacko A, Qu W, Kung HF. Synthesis and in vitro evaluation of 5-[^{18}F]fluoroalkyl pyrimidine nucleosides for molecular imaging of herpes simplex virus type 1 thymidine kinase reporter gene expression. J Med Chem 2008;51(18):5690–701.
[66] Cheek MA, Dobrikov MI, Wennefors CK, Xu Z, Hashmi SN, Shen X, Shaw BR. Synthesis and properties of (alpha-P-borano)-nucleoside 5'-triphosphate analogues as potential antiviral agents. Nucleic Acids Symp Ser (Oxf) 2008;(52):81–2.
[67] Fogt J, Januszczyk P, Framski G, Onishi T, Izawa K, De Clercq E, Neyts J, Boryski J. Synthesis and antiviral activity of novel derivatives of 2'-beta-C-methylcytidine. Nucleic Acids Symp Ser (2004) 2008;52:605–6.

[68] Lee H, Kim WH, Park A, Kang J, Chun P, Bae JH, Jeong LS, Moon HR. Synthesis of pyrimidine analog of fluoroneplanocin A as potential anti-HCV agent. Nucleic Acids Symp Ser (Oxf) 2008;(52):607–8.
[69] Koszytkowska-Stawinska M, Kaleta K, Sas W, De Clercq E. Synthesis and antiviral properties of aza-analogues of acyclovir. Nucleosides Nucleotides Nucleic Acids 2007;26(1):51–64.
[70] Robins MJ, Nowak I, Rajwanshi VK, Miranda K, Cannon JF, Peterson MA, Andrei G, Snoeck R, De Clercq E, Balzarini J. Synthesis and antiviral evaluation of 6-(alkyl-heteroaryl)furo[2,3-*d*] pyrimidin-2(3*H*)-one nucleosides and analogues with ethynyl, ethenyl, and ethyl spacers at C6 of the furopyrimidine core. J Med Chem 2007;50 (16):3897–905.
[71] Aly YL, Pedersen EB, La Colla P, Loddo R. Novel synthesis and anti-HIV-1 activity of 2-arylthio-6-benzyl-2,3-dihydro-1*H*-pyrimidin-4-ones (Aryl S-DABOs). Synthesis 2007;13:1955–60.
[72] Zhang X, Amer A, Fan X, Balzarini J, Neyts J, De Clercq E, Prichard M, Kern E, Torrence PF. Synthesis and antiviral activities of new acyclic and "double-headed" nucleoside analogues. Bioorg Chem 2007;35(3):221–32.
[73] Migliore MD, Zonta N, McGuigan C, Henson G, Andrei G, Snoeck R, Balzarini J. Synthesis and antiviral activity of the carbocyclic analogue of the highly potent and selective anti-VZV bicyclo furano pyrimidines. J Med Chem 2007;50 (26):6485–92.
[74] Kasula M, Samunuri R, Chakravarty H, Bal C, Baba M, Jha AK, Sharon A. Regioselective synthesis of pyrazolo[3,4-*d*]pyrimidine based carbocyclic nucleosides as possible antiviral agent. Nucleosides Nucleotides Nucleic Acids 2016;35(1):43–52.
[75] Xu X, Wang J, Yao Q. Synthesis and quantitative structure-activity relationship (QSAR) analysis of some novel oxadiazolo[3,4-*d*]pyrimidine nucleosides derivatives as antiviral agents. Bioorg Med Chem Lett 2015;25(2):241–4.
[76] Shakya N, Vedi S, Liang C, Agrawal B, Kumar R. 4′-Modified pyrimidine nucleosides as potential anti-hepatitis C virus (HCV) agents. Lett Drug Des Discovery 2014; 11(7):917–21.
[77] Shakya N, Vedi S, Liang C, Yang F, Agrawal B, Kumar R. 4′-Substituted pyrimidine nucleosides lacking 5′-hydroxyl function as potential anti-HCV agents. Bioorg Med Chem Lett 2014;24(5):1407–9.
[78] Bialek-Pietras M, Olejniczak AB, Paradowska E, Studziska M, Suski P, Jabloska A, Lesnikowski ZJ. Synthesis and in vitro antiviral activity of lipophilic pyrimidine nucleoside/carborane conjugates. J Organomet Chem 2015;798:99–105.
[79] Ivanov MA, Aleksandrova LA. Bicyclic furano-, pyrrolo-, and thiopheno[2,3-*d*] derivatives of pyrimidine nucleosides: synthesis and antiviral properties. Russ J Bioorg Chem 2013;39(1):22–39.
[80] Coen S, Vuyyuru V, Balzarini D, Snoeck C, Chu A. Activity and mechanism of action of HDVD, a novel pyrimidine nucleoside derivative with high levels of selectivity and potency against gammaherpesviruses. J Virol 2013;87(7):3839–51.
[81] Ivanov A. Bicyclic furano[2,3-*d*] derivatives of pyrimidine nucleosides-synthesis and antiviral properties. Bioorg Khim 2013;39(1):26–45.
[82] Kim A, Hong JH. Enantiomeric synthesis of novel apiosyl nucleosides as potential antiviral agents. Bull Kor Chem Soc 2004;25(2):221–5.
[83] Kumar R. 5-(1-Substituted) alkyl pyrimidine nucleosides as antiviral (herpes) agents. Curr Med Chem 2004;11(20):2749–66.
[84] Kasula M, Balaraju T, Toyama M, Thiyagarajan A, Bal C, Baba M, Sharon A. A conformational mimetic approach for the synthesis of carbocyclic nucleosides as anti-HCV leads. ChemMedChem 2013;8(10):1673–80.

[85] Sun J, Duan R, Li H, Wu J. Synthesis and anti-HIV activity of triazolo-fused 2′,3′-cyclic nucleoside analogs prepared by an intramolecular huisgen 1,3-dipolar cycloaddition. Helv Chim Acta 2013;96(1):59–68.

[86] Hollecker L, Choo H, Chong Y, Chu CK, Lostia S, McBrayer TR, Stuyver LJ, Mason JC, Du J, Rachakonda S, Shi J, Schinazi RF, Watanabe KA. Synthesis of β-enantiomers of N^4-hydroxy-3′-deoxypyrimidine nucleosides and their evaluation against bovine viral diarrhoea virus and hepatitis C virus in cell culture. Antiviral Chem Chemother 2004;15(1):43–55.

[87] Choo H, Chong Y, Choi Y, Mathew J, Schinazi RF, Chu CK. Synthesis, anti-HIV activity, and molecular mechanism of drug resistance of L-2′,3′-didehydro-2′,3′-dideoxy-2′-fluoro-4′-thionucleosides. J Med Chem 2003;46(3):389–98.

[88] Yale University; The University of Georgia Research Foundation, Inc. 5-(E)-Bromovinyl uracil analogues and related pyrimidine nucleosides as anti-viral agents and methods of use; 2003. Patent US6653318B1.

[89] Chong Y, Gumina G, Mathew JS, Schinazi RF, Chu CK. l-2',3'-Didehydro-2',3'-dideoxy-3'-fluoronucleosides: synthesis, anti-HIV activity, chemical and enzymatic stability, and mechanism of resistance. J Med Chem 2003;46(15):3245–56.

[90] Jeannot F, Gosselin G, Mathe C. Synthesis and antiviral evaluation of 2'-deoxy-2'-*C*-trifiuoromethyl beta-D-ribonucleoside analogues bearing the five naturally occurring nucleic acid bases. Org Biomol Chem 2003;1(12):2096–102.

[91] Bidet O, McGuigan C, Andrei G, Snoeck R, De Clercq E, Balzarini J. Synthesis of unusual bicyclic nucleosides bearing an unsaturated side-chain, as potential inhibitors of varicella-zoster virus (VZV). Nucleosides Nucleotides Nucleic Acids 2003;22 (5–8):817–9.

[92] Luoni GM, McGuigan C, Andrei G, Snoeck R, De Clercq E, Balzarini J. Bicyclic nucleoside inhibitors of varicella-zoster virus: 5′-chloro and 3′-chloro derivatives. Nucleosides Nucleotides Nucleic Acids 2003;22(5–8):931–3.

[93] Carangio A, McGuigan C, Andrei G, Snoeck R, De Clercq E, Balzarini J. Bicyclic nucleoside inhibitors of varicella-zoster virus: synthesis and biological evaluation of 2′,3′-dideoxy-3′-fluoro and 2′-deoxy-xylo derivatives. Nucleosides Nucleotides Nucleic Acids 2003;22(5–8):935–7.

[94] McGuigan C, Brancale A, Andrei G, Snoeck R, De Clercq E, Balzarini J. Novel bicyclic furanopyrimidines with dual anti-VZV and -HCMV activity. Bioorg Med Chem Lett 2003;13(24):4511–3.

[95] Mcguigan C, Pathirana RN, Migliore M, Adak R, Luoni G, Jones AT, Diez-Torrubia A, Camarasa M, Velazquez S, Henson G, Verbeken E, Sienaert R, Naesens L, Snoeck R, Andrei G, Balzarini J. Preclinical development of bicyclic nucleoside analogues as potent and selective inhibitors of varicella zoster virus. J Antimicrob Chemother 2007;60(6):1316–30.

[96] Bouisset T, Counor C, Rabeson C, Pierra C, Storer R, Loi AG, Cadeddu A, Mura M, Musiu C, Liuzzi M, Loddo R, Bergelson S, Bichko V, Bridges E, Cretton-Scott E, Mao J, Sommadossi J, Seifer M, Standring D, Tausek M, Gosselin G, La Colla P. 2′-*C*-Methyl branched pyrimidine ribonucleoside analogues: potent inhibitors of RNA virus replication. Antiviral Chem Chemother 2007;18(4):225–42.

[97] Rashad AE, Ali MA. Synthesis and antiviral screening of some thieno[2,3-*d*]pyrimidine nucleosides. Nucleosides Nucleotides Nucleic Acids 2006;25(1):17–28.

[98] Fan X, Zhang X, Zhou L, Keith KA, Kern ER, Torrence PF. 5-(Dimethoxymethyl)-2′-deoxyuridine: a novel gem diether nucleoside with anti-orthopoxvirus activity. J Med Chem 2006;49(11):3377–82.

[99] Kim A, Hong J. Synthesis and antiviral evaluation of novel 5′-norcarboacyclic phosphonic acid nucleosides. Nucleosides Nucleotides Nucleic Acids 2006;25(8):941–50.

[100] Zhou W, Gumina G, Chong Y, Wang J, Schinazi RF, Chu CK. Synthesis, structure-activity relationships, and drug resistance of β-D-3′-fluoro-2′,3′-unsaturated nucleosides as anti-HIV agents. J Med Chem 2004;47(13):3399–408.
[101] Kumar R. 5-Bromo (or chloro)-6-azido-5,6-dihydro-2'-deoxyuridine and -thymidine derivatives with potent antiviral activity. Bioorg Med Chem Lett 2002;12:275–8.
[102] McGuigan C, Pathirana RN, Snoeck R, Andrei G, De Clercq E, Balzarini J. Discovery of a new family of inhibitors of human cytomegalovirus (HCMV) based upon lipophilic alkyl furano pyrimidine dideoxy nucleosides: action via a novel non-nucleosidic mechanism. J Med Chem 2004;47(7):1847–51.
[103] Duraffour S, Snoeck R, Krecmerova M, Van Den Oord J, De Vos R, Holy A, Crance J, Garin D, De Clercq E, Andrei G. Activities of several classes of acyclic nucleoside phosphonates against camelpox virus replication in different cell culture models. Antimicrob Agents Chemother 2007;51(12):4410–9.
[104] Kumar R, Semaine W, Johar M, Tyrrell DLJ, Agrawal B. Effect of various pyrimidines possessing the 1-[(2-hydroxy-1-(hydroxymethyl)ethoxy)methyl] moiety, able to mimic natural 2′-deoxyribose, on wild-type and mutant hepatitis B virus replication. J Med Chem 2006;49(12):3693–700.
[105] Prichard MN, Hartline CB, Harden EA, Daily SL, Beadle JR, Valiaeva N, Kern ER, Hostetler KY. Inhibition of herpesvirus replication by hexadecyloxypropyl esters of purine- and pyrimidine-based phosphonomethoxyethyl nucleoside phosphonates. Antimicrob Agents Chemother 2008;52(12):4326–30.
[106] Khatoon S, Yadav AK. Synthesis and antimicrobial screening of some new 4-imino-3,5,7-trisubstituted pyrido[2,3-*d*]pyrimidines and their ribofuranosides as potential chemotherapeutic agents. Phosphorus Sulfur Silicon Relat Elem 2004;179(2): 345–52.
[107] Szafraniec SI, Stachnik KJ, Skierski JS. New nucleoside analogs in the treatment of hematological disorders. Acta Pol Pharm 2004;61(3):223–32.
[108] Greco NJ, Tor Y. Synthesis and site-specific incorporation of a simple fluorescent pyrimidine. Nat Protoc 2007;2(2):305–16.
[109] Mata G, Luedtke NW. Synthesis and solvatochromic fluorescence of biaryl pyrimidine nucleosides. Org Lett 2013;15(10):2462–5.
[110] Vasilyeva SV, Kuznetsova AS, Khalyavina JG, Glazunova VA, Shtil AA, Gornostaev LM, Silnikov VN. Novel fluorescent pyrimidine nucleosides containing 2,1,3-benzoxadiazole and naphtho-[1,2,3-CD] indole-6 (2*H*)-One fragments. Nucleosides Nucleotides Nucleic Acids 2014;33(9):615–25.
[111] Mizuta M, Miyata K, Seio K, Santa T, Sekine M. Synthesis of fluorescent cyclic cytosine nucleosides and their fluorescent properties upon incorporation into oligonucleotides. Nucleic Acids Symp Ser 2006;(50):19–20.
[112] Vargas G, Escalona I, Salas M, Gordillo B, Sierra A. Synthesis and RT inhibitory activity evaluation of new pyrimidine-based Seco-nucleosides. Nucleosides Nucleotides Nucleic Acids 2006;25(3):243–57.
[113] Sunkara NK, Mosley SL, Seley-Radtke KL. A carbocyclic 7-deazapurine-pyrimidine hybrid nucleoside. Collect Czechoslov Chem Commun 2006;71(8):1161–8.
[114] Luisier S, Silhar P, Leumann CJ. Highly beta-selective, *N*-iodosuccinimide-mediated nucleosidation to bicyclo- and tricyclo-nucleosides. Nucleic Acids Symp Ser 2004;2008(52):581–2.
[115] Zare A, Hasaninejad A, Safinejad R, Moosavi-Zare AR, Khalafi-Nezhad A, Beyzavi MH, Miralai-Moredi M, Dehghani E, Kazerooni-Mojarrad P. Ionic liquid-accelerated Michael addition of pyrimidine and purine nucleobases to α,β-unsaturated esters: a rapid approach to carboacyclic nucleosides synthesis. ARKIVOC 2008;16:61–74.

[116] Elgemeie G, Abu-zaied M, Loutfy SA. 4-Aminoantipyrine in carbohydrate research: design, synthesis and anticancer activity of thioglycosides of a novel class of 4-aminoantipyrines and their corresponding pyrazolopyrimidine and pyrazolopyridine thioglycosides. Tetrahedron 2017;73(40):5853–61.
[117] Elgemeie G, Alkhursani SA, Mohamed RA. New synthetic strategies for acyclic and cyclic pyrimidinethione nucleosides and their analogues. Nucleosides Nucleotides Nucleic Acids 2019;38(1):1–76.
[118] Elgemeie G, Elnaggar D. Novel dihydropyridine thioglycosides and their corresponding dehydrogenated forms as potent anti-hepatocellular carcinoma agents. Nucleosides Nucleotides Nucleic Acids 2018;37(4):1–18.
[119] Elgemeie G, Fathy NM, Farag A, Alkhursani SA. Design, synthesis, molecular docking and anti-hepatocellular carcinoma evaluation of novel acyclic pyridine thioglycosides. Nucleosides Nucleotides Nucleic Acids 2018;37(3):1–13.
[120] Elgemeie G, Fathy NM, Farag A, Alkhursani SA. Antimetabolites: design, synthesis, and cytotoxic evaluation of novel dihydropyridine thioglycosides and pyridine thioglycosides. Nucleosides Nucleotides Nucleic Acids 2017;36(5):1–23.
[121] Elgemeie G, Salah AM, Abbas NS, Hussein HA, Mohamed RA. Nucleic acid components and their analogs: design and synthesis of novel cytosine thioglycoside analogs. Nucleosides Nucleotides Nucleic Acids 2017;36(2):1–12.
[122] Elgemeie G, Abu-zaied M, Azzam R. Antimetabolites: a first synthesis of a new class of cytosine thioglycoside analogs. Nucleosides Nucleotides Nucleic Acids 2016;35(4): 211–22.
[123] Elgemeie G, Abu-zaied M, Alsaid S, Hebishy A, Essa H. Novel nucleoside analogues: first synthesis of pyridine-4-thioglycosides and their cytotoxic evaluation. Nucleosides Nucleotides Nucleic Acids 2015;34(10):1–15.
[124] Baranowska-Kortylewicz J, Kortylewicz ZP. Targeted radiolabeled compounds and their use for the treatment and diagnosis of cancer. U.S. Pat. Appl. Publ; 2012. US 20120269725 A1 20121025.
[125] Pallua C. Cancer drug comprising pyrimidine nucleoside deriv. Ger. Offen.; 2016. DE 102014113936 A1 20160331.
[126] Daifuku R, Gall AS, Sergueev DS. Preparation of fluorinated pyrimidine nucleoside analogs as antitumor agents and DNA methyltransferase inhibitors. PCT Int. Appl.; 2014. WO 2014143051 A1 20140918.
[127] Pozzoli CG, Canevari V, Brusasca M, Menna L, Curti M. Synthesis of antitumor and antiviral nucleosides via coupling reaction. U.S. Pat. Appl. Publ.; 2014. US 20140135490 A1 20140515.
[128] Westwood R, Selkirk A. Capsules of antitumor pyrimidine nucleoside derivative CYC682. PCT Int. Appl.; 2007. WO 2007072061 A2 20070628.
[129] Feingold JM. Anti-CD33 cytotoxic conjugate combination with anthracycline or nucleoside or purine nucleoside analog for the treatment of acute leukemia and myelodysplastic syndrome. PCT Int. Appl.; 2004. WO 2004043461 A1 20040527.
[130] Takita T, Ohtsuka K, Numagami E, Harashima S. Crystalline forms of pyrimidine nucleoside derivative. U.S. Pat. Appl. Publ.; 2004. US 20040053883 A1 20040318.
[131] Binderup L. Combination of vitamin D analog and pyrimidine nucleoside analog. PCT Int. Appl.; 2002. WO 2002092062 A2 20021121.
[132] Takita T, Ohtsuka K, Numagami E, Harashima S. Preparation of crystal of pyrimidine nucleoside derivative. PCT Int. Appl.; 2002. WO 2002064609 A1 20020822.
[133] Weis AL, Pulenthiran K, Gero AM. Preparation of nucleoside analogs as parasiticides and antitumor agents. U.S.; 2001. US 6242428 B1 20010605.
[134] Von Borstel R, Bamat MK. Treatment of chemotherapeutic agent and antiviral agent toxicity with acylated pyrimidine nucleosides. U.S.; 1999. US 5968914 A 19991019.

[135] Vonborstel RW, Bamat MK. Methods of reducing toxicity of chemotherapeutic and antiviral agents with acylated non-methylated pyrimidine nucleosides. PCT Int. Appl.; 1996. WO 9640165 A1 19961219.
[136] Kassis AI, Adelstein SJ. Diagnosis and therapy of tumors with 5-radioiodo-2'-deoxyuridine or other radiohalogenated pyrimidine nucleoside. U.S.; 1994. US 5308605 A 19940503.
[137] Kawaguchi T, Nakajima S, Fukushima S. 5'-Deoxy-5-fluorouridine preparations with low toxicity. Jpn. Kokai Tokkyo Koho; 1993. JP 05213761 A 19930824.
[138] Sugimura H, Yamada K, Nagahata T, Narita K, Shiozawa A. Preparation of [tris(hydroxymethyl)cyclobutyl]purine and - pyrimidine nucleoside derivatives as antiviral and anticancer agents. Jpn. Kokai Tokkyo Koho; 1993. JP 05001042 A 19930108.
[139] Ichikawa Y, Sugawara Y, Sugimura H, Narita K, Shiozawa A. Preparation of 3-phosphonoalkyl-1-cyclobutyl purine and pyrimidine derivatives as antitumor and antiviral agents. Jpn. Kokai Tokkyo Koho; 1992. JP 04091094 A 19920324.
[140] Kassis AL, Adelstein SJ. Treatment of tumors with 5-radioiodo-2'-deoxyuridine. U.S.; 1991. US 5077034 A 19911231.
[141] Yamamura S, Nishiyama S, Ogiya T, Kato K, Minami T, Takita T. Preparation of 3-hydroxy-4-(hydroxymethyl)-erythro-oxetany purine or pyrimidine nucleoside derivatives and their preparation. Jpn. Kokai Tokkyo Koho; 1990. JP 02209886 A 19900821.
[142] Marquez VE, Driscoll JS. Preparation of cyclopentenylpyrimidine and -purine derivatives as antiviral and antitumor agents. U. S. Pat. Appl.; 1989. US 307115 A0 19890715.
[143] Amundsen AR, Stern EW, Hollis LS. Platinum triamine complexes with nucleotides or nucleosides as antitumor agents. Eur. Pat. Appl.; 1989. EP 303437 A2 19890215.
[144] Morton O. Therapeutic agents comprising nucleoside analogs and nucleoside analog cell uptake inhibitors for cancer treatment. Brit. UK Pat. Appl.; 1987. GB 2189391 A 19871028.
[145] West CR, Hong C. Cytotoxic nucleoside-corticosteroid phosphodiesters. PCT Int. Appl.; 1981. WO 8100410 A1 19810219.
[146] Hozumi M, Pponma Y, Ito T, Rafuitsuku JMO, Ishikawa I, Mizuno Y, Kawahara T, Ogura H. Preparation and testing of β-D-ribofuranosylpyridopyrimidinedione derivatives as anticancer agents. Jpn. Kokai Tokkyo Koho; 1989. JP 01143895 A 19890606.
[147] Arasaki MNR, Ishitsuka H, Kuruma I, Miwa M, Murasaki C, Shimma N, Umeda IIH. *N*-Oxycarbonyl-substituted 5'-deoxy-5-fluorocytidines as antitumor agents. Eur. Pat. Appl.; 1994. EP 602454 A1 19940622.
[148] Von Borstel RW, Bamat MK. Acylated pyrimidine nucleosides for treatment of toxicity from chemotherapeutic and antiviral agents. PCT Int. Appl.; 1994. WO 9426761 A1 19941124.
[149] Daifuku R, Gall A, Sergueev DS. Preparation of fluorinated pyrimidine nucleoside analogs as antitumor agents. U.S. Pat. Appl. Publ.; 2014. US 20140024612 A1 20140123.
[150] Chang J, An H, Yu X, Guo X. Preparation of pyrimidine nucleoside derivatives as antitumor and antiviral agents. Faming Zhuanli Shenqing; 2012. CN 102351931 A 20120215.
[151] Guzi TJ, Parry DA, Labroli MA, Dwyer MP, Paruch K, Rosner KE, Shen R, Popovici-Muller J. Preparation of pyrimidine nucleoside analogs as modulators of cell cycle checkpoints and their use in combination with checkpoint kinase inhibitors. PCT Int. Appl.; 2009. WO 2009061781 A1 20090514.

[152] Hertel LW, Kroin JS. 2'-deoxy-2',2'-difluoro-(4-substituted Pyrimidine) nucleosides having antiviral and anti-cancer activity and intermediates. Eur. Pat. Appl.; 1993. EP 576230 A1 19931229.

[153] Wu LI. Preparation of gemcitabine as antitumor prodrugs. U.S. Pat. Appl. Publ; 2014. US 20140134160 A1 20140515.

[154] Chen X. New pyrimidine compound as antitumor agent. Faming Zhuanli Shenqing; 2015. CN 104945437 A 20150930.

[155] Aradi J, Fesues L, Beck Z. Pyrimidine mononucleotide and mononucleoside compounds for use in therapy, particularly as antitumor and antiviral agents. PCT Int. Appl.; 2007. WO 2007083173 A2 20070726.

[156] Naviaux RK. Methods using pyrimidine -based nucleosides for treatment of mitochondrial disorders. PCT Int. Appl.; 2000. WO 2000050043 A1 20000831.

[157] Von Borstel RW, Bamat MK. Compositions of chemotherapeutic agent or antiviral agent with acylated pyrimidine nucleosides. U.S.; 1998. US 5736531 A 19980407.

[158] Mori K, Tanaka Y. Compositions of interleukin and pyrimidine nucleosides. PCT Int. Appl.; 1996. WO 9637214 A2 19961128.

[159] Snyder RD. Treatment of neoplastic diseases by conjunctive therapy with 2'-fluoromethylene derivatives of pyridine deoxyribonucleosides and radiation or chemotherapy. PCT Int. Appl.; 1996. WO 9601638 A1 19960125.

[160] Nomura M, Kajitani M, Yamashita J, Fukushima M, Shimamoto Y. Preparation of 2'-deoxy-5-fluorouridine derivatives as antitumor drugs. PCT Int. Appl.; 1995. WO 9518138 A1 19950706.

[161] Sakata S, Mori T. Preparation of 1-(β-D-erythropentofuran-2-ulosyl)pyrimidines as intermediates for antitumors. Jpn. Kokai Tokkyo Koho; 1994. JP 06192285 A 19940712.

[162] Kelley JL, Baccanari DP. Uracil derivatives as enzyme inhibitors. PCT Int. Appl.; 1994. WO 9401414 A1 19940120.

[163] Haraguchi K, Tanaka H, Myasaka S. 4'-Carbon-substituted pyrimidine nucleosides as pharmaceuticals and their preparation. Jpn. Kokai Tokkyo Koho; 1993. JP 05230058 A 19930907.

[164] Kaneko M, Hotoda H, Shibata T, Kobayashi T, Mitsuhashi Y, Matsuda A, Sasaki T. The synthesis of pyrimidine nucleosides as neoplasm inhibitors. Eur. Pat. Appl.; 1993. EP 536936 A1 19930414.

[165] Marquez VE, Russ PL. Preparation of 5-substituted pyrimidine derivatives of conformationally locked nucleoside analogs as antiviral and anticancer agents. PCT Int. Appl.; 2002. WO 2002008204 A2 20020131.

[166] Nomura M, Ono Y. Preparation of pyrimidine nucleoside compounds as antitumor agents. PCT Int. Appl.; 2006. WO 2006080509 A1 20060803.

[167] Tam R, Wang G, Lau J, Hong Z. Preparation and immunomodulating effects at reduced cytotoxicity of pyrrolo[2,3-*d*]pyrimidine nucleoside analogs as antitumors. U.S. Pat. Appl. Publ.; 2002. US 20020035077 A1 20020321.

[168] Miyoahi M, Kato H, Yokoyama Y, Nagata T, Yoshida M. Preparation of 2-thio-5-halogenopyrimidine arabinoside as anticancer and antiviral agents. Jpn. Kokai Tokkyo Koho; 1998. JP 10237092 A 19980908.

[169] Maruyama H, Ootsuka K, Morya K. Method for preparation of 2'-cyano-pyrimidine nucleoside derivative. Jpn. Kokai Tokkyo Koho; 1995. JP 07053586 A 19950228.

[170] Haraguchi K, Tanaka H, Myasaka S, Yoshimura J, Kano F. Preparation of 1'-C-substituted pyrimidine nucleosides and 2,2'-anhydronucleosides as antitumor agents. Jpn. Kokai Tokkyo Koho; 1995. JP 07109289 A 19950425.

[171] Adhikari S, Calderwood EF, England DB, Gould AE, Harrison SJ, Huang S, Ma L. Preparation of pyrimidine nucleoside sulfamates as antitumor agents and Atg7 enzyme inhibitors. PCT Int. Appl.; 2018. WO 2018089786 A1 20180517.

[172] Robins RK, Cottam HB. Preparation and testing of antiviral, antitumor, antimetastatic, and immune system enhancing 5-aminothiazolo[4,5-*d*]pyrimidine-2,7-(6*H*)-dione nucleoside and nucleotide analogs. PCT Int. Appl.; 1989. WO 8905649 A1 19890629.
[173] Peng X, Kuang Y, Cao S, Chen W, Wang Y. Preparation of phenylboronic acid derivatives as DNA-crosslinking anticancer agents. U.S. Pat. Appl. Publ.; 2013. US 20130045949 A1 20130221.
[174] Hayakawa H, Tanaka H, Myasaka S. Preparation of 3'-allene containing pyrimidine nucleosides as antitumor agents. Jpn. Kokai Tokkyo Koho; 1996. JP 08041054 A 19960213.
[175] Matsuda A, Sasaki T, Ueda T. Preparation of pyrimidine nucleoside derivatives as neoplasm inhibitors and pharmaceutical compositions containing them. PCT Int. Appl.; 1991. WO 9119713 A1 19911226.
[176] Von Borstel RW, Bamat MK. Treatment of chemotherapeutic agent and antiviral agent toxicity with acylated pyrimidine nucleosides. PCT Int. Appl.; 1993. WO 9301202 A1 19930121.
[177] Baranowska-Kortylewicz J, Kortylewicz ZP. Preparation of targeted radiolabeled pyrimidine nucleosides and their use for the treatment and diagnosis of cancer. PCT Int. Appl.; 2011. WO 2011079245 A1 20110630.
[178] McCarthy JR, Matthews DP, Sabol JS. Preparation of 5-(1-fluorovinyl)-1H-pyrimidine-2,4-dione derivatives as antineoplastic agents. PCT Int. Appl.; 1995. WO 9507917 A1 19950323.
[179] Kitsutaka A, Tanaka H, Myasaka S. Method for preparation of spiro-type pyrimidine cyclonucleosides. Jpn. Kokai Tokkyo Koho; 1994. JP 06256379 A 19940913.
[180] Shoji Y, Otake K, Morishita K, Kitao T. Preparation of anticancer agents modified by maleimide derivatives as protein-binding drugs for treatment of cancers. PCT Int. Appl.; 2019. WO 2019103050 A1 20190531.
[181] Deshpande M, Wiles JA, Hashimoto A, Phadke A. Preparation of stabilized nucleotides for medical treatment of hepatitis C, RSV, HSV and as antitumor agents. U.S. Pat. Appl. Publ.; 2016. US 20160016986 A1 20160121.
[182] Yoshimura Y, Kitano K, Miura S, Machida H, Watanabe M. Preparation of 1-(2-deoxy-2-fluoro-4-thio-β-D-arabinofuranosyl)cytosines as antitumor agents. PCT Int. Appl.; 1997. WO 9738001 A1 19971016.
[183] Tanaka Y, Nomura M, Kazuno H, Oguchi K. Silyl derivatives of pyrimidine nucleoside and antitumor agent comprising the same. Jpn. Kokai Tokkyo Koho; 2014. JP 2014189499 A 20141006.
[184] Tanaka Y, Nomura M, Kazuno H, Oguchi K. Novel pyrimidine nucleoside compound or its salt, and antitumor agent containing the same. Jpn. Kokai Tokkyo Koho; 2015. JP 2015024981 A 20150205.
[185] Courtney SM. Preparation of pyrimidine nucleosides as anti-tumor agents. U.S.; 1999. US 5945406 A 19990831.
[186] Waga T, Meguro H, Oorui H. Preparation of pyranose nucleoside derivatives as antiviral and antitumor agents. Jpn. Kokai Tokkyo Koho; 1994. JP 06263793 A 19940920.
[187] Weis AL, Goodhue CT. Preparation of L-pyranosyl nucleosides for treating various diseases in mammals. PCT Int. Appl.; 1996. WO 9612728 A1 19960502.
[188] Achinami K, Terao Y, Murata M, Nishio T, Akamatsu M, Kamimura M. Preparation of optically active acyclo nucleosides as virucides and antitumor agents. Jpn. Kokai Tokkyo Koho; 1990. JP 02009870 A 19900112.
[189] Niitsuma S, Yamanaka H, Sakamoto H. Preparation of 5-oxetanosylpyrimidine derivatives as anticancer and antiviral agents. Jpn. Kokai Tokkyo Koho; 1991. JP 03279381 A 19911210.

[190] Ichikawa Y, Narita A, Matsuo K, Aoyama K, Matsumura F, Nishiyama Y, Matsubara K, Nagahata T, Hoshino H, et al. Preparation of cyclobutane nucleoside analogs as virucides and anticancer agents. Eur. Pat. Appl.; 1990. EP 358154 A2 19900314.
[191] Trampota M, Murphy RB. Preparation of cubane nucleoside analogs as antiviral and antitumor agents. PCT Int. Appl.; 2007. WO 2007059330 A2 20070524.
[192] Dong S, PGM W. Preparation of nucleoside phosphonates and analogs as antiviral and antitumor agents. PCT Int. Appl.; 2011. WO 2011031567 A1 20110317.
[193] Brulíková L, Dzubák P, Hajdúch M, Lachnitová L, Kollareddy M, Kolár M, Bogdanová K, Hlavác J. Synthesis of 5-[alkoxy-(4-nitro-phenyl)-methyl]-uridines and study of their cytotoxic activity. Eur J Med Chem 2010;45:3588–94.
[194] Shimma N, Umeda I, Arasaki M, Murasaki C, Masubuchi K, Kohchi Y, Miwa M, Ura M, Sawada N, Tahara H, Kuruma I, Horii I, Ishitsuka H. The design and synthesis of a new tumor-selective fluoropyrimidine carbamate, capecitabine. Bioorg Med Chem 2000;8(7):1697–706.
[195] Niedballa U, Vorbru»ggen H. Synthesis of nucleosides. 9. General synthesis of *N*-‐glycosides. I. Synthesis of pyrimidine nucleosides. J Org Chem 1974;39:3654–60.
[196] Saneyoshi M, Inomata M, Fukuoka F. Synthetic nucleosides and nucleotides. XI.: Facile synthesis and antitumor activities of various 5-fluoropyrimidine nucleosides. Chem Pharm Bull 1978;26(10):2990–7.
[197] Matsuda A, Kurasawa Y, Watanabe KA. A simplified method for the synthesis of pyrimidine nucleosides. Synthesis 1981;1981(9):748.
[198] Sivakrishna B, Islam S, Santra MK, Pal S. Synthesis and cytotoxic evaluation of apioarabinofuranosyl pyrimidines. Drug Dev Res 2019;1–9.
[199] Jeong LS, Tosh DK, Choi WJ, Lee SK, Kang YJ, Choi S, Lee JH, Lee H, Lee HW, Kim HO. Discovery of a new template for anticancer agents: 2′-deoxy-2′-fluoro-4′-selenoarabinofuranosyl-cytosine (2′-F-4′-seleno-ara-C). J Med Chem 2009;52 (17):5303–6.
[200] Panda A, Islam S, Santra MK, Pal S. Lead tetraacetate mediated one pot oxidative cleavage and acetylation reaction: an approach to Apio and homologated Apio pyrimidine nucleosides and their anticancer activity. RSC Adv 2015;5:82450–9.
[201] Cole SPC. Rapid chemosensitivity testing of human lung tumor cells using the MTT assay. Cancer Chemother Pharmacol 1986;17(3):259–63.
[202] Hrdlicka PJ, Jepsen JS, Nielsen C, Wengel J. Synthesis and biological evaluation of nucleobase-modified analogs of the anticancer compounds 3′-C-ethynyluridine (EUrd) and 3′-C-ethynylcytidine (ECyd). Bioorg Med Chem 2005;13:1249–60.
[203] Nomura M, Sato T, Washinosu M, Tanaka M, Asao T, Shuto S, Matsuda A. Nucleosides and nucleotides. Part 212: practical large-scale synthesis of 1-(3-C-ethynyl-β-D-ribo-pentofuranosyl)cytosine (ECyd), a potent antitumor nucleoside. Isobutyryloxy group as an efficient anomeric leaving group in the Vorbrüggen glycosylation reaction. Tetrahedron 2002;58:1279–88.
[204] Ludwig PS, Schwendener RA, Schott HA. New laboratory scale synthesis for the anticancer drug 3′-*C*-ethynylcytidine. Synthesis 2002;16:2387–92.
[205] Kakinuma K, Iihama Y, Takagi I, Ozawa K, Yamauchi N, Imamura N, Esumi Y, Uramoto M. Diacetone glucose architecture as a chirality template. II.[1] Versatile synthon for the chiral deuterium labeling and synthesis of all diastereoisomers of chirally monodeuterated glycerol. Tetrahedron 1992;48:3763–74.
[206] Vorbrüggen H, Krolikiewicz K, Bennua B. Nucleoside syntheses, XXII[1] Nucleoside synthesis with trimethylsilyl triflate and perchlorate as catalysts. Chem Ber 1981;114:1234–55.
[207] Van Rompay R, Norda A, Linden K, Johansson M, Karlsson A. Phosphorylation of uridine and cytidine nucleoside analogs by two human uridine-cytidine kinases. Mol Pharmacol 2001;59(5):1181–6.

[208] Panayides JL, Mathieu V, Banuls LM, Apostolellis H, Dahan-Farkas N, Davids H, Harmse L, Rey ME, Green IR, Pelly SC, Kiss R, Kornienko A, van Otterlo WA. Synthesis and in vitro growth inhibitory activity of novel silyl- and trityl-modified nucleosides. Bioorg Med Chem 2016;24(12):2716–24.
[209] Bhuma N, Burade SS, Bagade AV, Kumbhar NM, Kodam KM, Dhavale DD. Synthesis and anti-proliferative activity of 3′-deoxy-3′-fluoro-3′-C-hydroxymethyl-pyrimidine and purine nucleosides. Tetrahedron 2017;73:6157–63.
[210] Niedballa U, Vorbruggen H. A general synthesis of pyrimidine nucleosides. Angew Chem Int Ed Engl 1970;9(6):461–2.
[211] Ruddarraju RR, Murugulla AC, Kotla R, Chandra Babu Tirumalasetty M, Wudayagiri R, Donthabakthuni S, Maroju R, Baburao K, Parasa LS. Design, synthesis, anticancer, antimicrobial activities and molecular docking studies of theophylline containing acetylenes and theophylline containing 1,2,3-triazoles with variant nucleoside derivatives. Eur J Med Chem 2016;123:379–96.
[212] Xiao C, Sun C, Han W, Pan F, Dan Z, Li Y, Song ZG, Jin YH. Synthesis of 6-(het) ary Xylocydine analogues and evaluating their inhibitory activities of CDK1 and CDK2 in vitro. Bioorg Med Chem 2011;19(23):7100–10.
[213] Shi J, Stuyver LJ, Watanabe KA. Modified fluorinated nucleoside analogues; 2003. EP 1480982 A2, WO2003068162A2.
[214] Tsoukala E, Agelis G, Dolinsˇek J, Botić T, Cencic A, Komiotisa D. An efficient synthesis of 3-fluoro-5-thio-xylofuranosyl nucleosides of thymine, uracil, and 5-fluorouracil as potential antitumor or/and antiviral agents. Bioorg Med Chem 2007;15:3241–7.
[215] Krizmanic I, Visnjevac A, Luic M, Glavas-Obrovac L, Zinic M, Zinic B. Synthesis, structure, and biological evaluation of *C*-2 sulfonamido pyrimidine nucleosides. Tetrahedron 2003;59(23):4047–57.
[216] Wen Z, Peng J, Tuttle PR, Ren Y, Garcia C, Debnath D, Rishi S, Hanson C, Ward S, Kumar A, Liu Y, Zhao W, Glazer PM, Liu Y, Sevilla MD, Adhikary A, Wnuk SF. Electron-mediated aminyl and iminyl radicals from C5 azido-modified pyrimidine nucleosides augment radiation damage to cancer cells. Org Lett 2018;20(23):7400–4.
[217] Neef AB, Luedtke NW. An azide-modified nucleoside for metabolic labeling of DNA. ChemBioChem 2014;15:789–93.
[218] Krim J, Taourirte M, Grünewald C, Krstic I, Engels J. Microwave-assisted click chemistry for nucleoside functionalization: useful derivatives for analytical and biological applications. Synthesis 2013;45(3):396–405.
[219] Kovács T, Ötvös L. Simple synthesis of 5-vinyl- and5-ethynyl-2′-deoxyuridine-5′-triphosphates. Tetrahedron Lett 1988;29(36):4525–8.
[220] Liu Z, Liao P, Bi X. General silver-catalyzed hydroazidation of terminal alkynes by combining TMS-N_3 and H_2O: synthesis of vinyl azides. Org Lett 2014;16(14): 3668–71.
[221] Liang Y, Suzol SH, Wen Z, Artiles AG, Mathivathanan L, Raptis RG, Wnuk SF. Uracil nucleosides with reactive group at C5 position: 5-(1-halo-2-sulfonylvinyl)uridine analogues. Org Lett 2016;18(6):1418–21.
[222] Hu B, DiMagno SG. Reactivities of vinyl azides and their recent applications in nitrogen heterocycle synthesis. Org Biomol Chem 2015;13:3844–55.
[223] Balzarini J, Andrei G, Kumar R, Knaus EE, Wiebe LI, De Clercq E. The cytostatic activity of 5-(1-azidovinyl)-2'-deoxyuridine (AzVDU) against herpes simplex virus thymidine kinase gene-transfected FM3A cells is due to inhibition of thymidylate synthase and enhanced by UV light (lambda = 254 nm) exposure. FEBS Lett 1995;373(1):41–4.
[224] Suzol SH, Howlader AH, Wen Z, Ren Y, Laverde EE, Garcia C, Liu Y, Wnuk SF. Pyrimidine nucleosides with a reactive (β-Chlorovinyl)sulfone or (β-Keto)sulfone

group at the C5 position, their reactions with nucleophiles and electrophiles, and their polymerase-catalyzed incorporation into DNA. ACS Omega 2018;3(4):4276–88.
[225] Fauster K, Hartl M, Santner T, Aigner M, Kreutz C, Bister K, Ennifar E, Micura R. 2′-azido RNA, a versatile tool for chemical biology: synthesis, X-ray structure, siRNA applications, click labeling. ACS Chem Biol 2012;7(3):581–9.
[226] Akula HK, Kokatla H, Andrei G, Snoeck R, Schols D, Balzarini J, Yang L, Mahesh Lakshman MK. Facile functionalization at the C4 position of pyrimidine nucleosides via amide group activation with (benzotriazol-1-yloxy)tris(dimethylamino)phosphonium hexafluorophosphate (BOP) and biological evaluations of the products. Org Biomol Chem 2017;15(5):1268.
[227] Lewandowska M, Ruszkowski P, Baraniak D, Czarnecka A, Kleczewska N, Celewicz L. Synthesis of 3′-azido-2′,3′-dideoxy-5-fluorouridine phosphoramidates and evaluation of their anticancer activity. Eur J Med Chem 2013;67:188–95.
[228] Stawinski J, Hozumi T, Narang SA, Bahl CP, Wu R. Arylsulfonyltetrazoles, new coupling reagents and further improvements in the triester method for the synthesis of deoxyribooligonucleotides. Nucleic Acids Res 1977;4(2):353–71.
[229] Agarwal KL, Riftina F. Chemical synthesis of a self-complementary octanucleotide, dG-G-T-T-A-A-C-C by a modified triester method. Nucleic Acids Res 1978;5:2809–23.
[230] Chattopadhyaya JB, Reese CB. Some observations relating to phosphorylation methods in oligonucleotide synthesis. Tetrahedron Lett 1979;20:5059–62.
[231] Shi JB, Chen LZ, Wang Y, Xiou C, Tang WJ, Zhou HP, Liu XH, Yao QZ. Benzophenone-nucleoside derivatives as telomerase inhibitors: design, synthesis and anticancer evaluation in vitro and in vivo. Eur J Med Chem 2016;124:729–39.
[232] Manta S, Tzioumaki N, Tsoukala E, Panagiotopoulou A, Pelecanou M, Balzarini J, Komiotis D. Unsaturated dideoxyfluoro- ketopyranosyl nucleosides as new cytostatic agents: a convenient synthesis of 2,6-dideoxy-3-fluoro-4-keto-β-D-glucopyranosyl analogues of uracil, 5-fluorouracil, thymine, *N*4-benzoyl cytosine and *N*6-benzoyl adenine. Eur J Med Chem 2009;44:4764–71.
[233] Yoon JS, Jarhad DB, Kim G, Nayak A, Zhao LX, Yu J, Kim HR, Lee JY, Mulamoottil VA, Chandra G, Byun WS, Lee SK, Kim YC, Jeong LS. Design, synthesis and anticancer activity of fluorocyclopentenyl-purines and -pyrimidines. Eur J Med Chem 2018;155:406–17.
[234] Kim J, Yu J, Alexander V, Choi JH, Song J, Lee HW, Kim HO, Choi J, Lee SK, Jeonga LS. Structure activity relationships of 2′-modified-4′-selenoarabinofuranosyl-pyrimidines as anticancer agents. Eur J Med Chem 2011;83:208–25.
[235] Skiba J, Karpowicz R, Szabo I, Therrien B, Kowalski K. Synthesis and anticancer activity studies of ferrocenyl-thymine-3,6- dihydro-2*H*-thiopyranes—a new class of metallocene-nucleobase derivatives. J Organomet Chem 2015;794:216–22.
[236] Romeo R, Giofrè SV, Garozzo A, Bisignano B, Corsaro A, Chiacchioc MA. Synthesis and biological evaluation of furopyrimidine *N*, *O*-nucleosides. Bioorg Med Chem 2013;21:5688–93.
[237] Liu M, Luo M, Mozdziesz DE, Lin T, Dutschman GE, Gullen EA, Cheng Y, Sartorelli AC. Synthesis and biological evaluation of L- and D-configurations of 2′,3′-dideoxy-4′-*C*-methyl-3′-oxacytidine analogues. Bioorg Med Chem Lett 2001;11:2301–4.

CHAPTER 8
Anticancer alkylating agents

1. Introduction

Alkylating agents are from the major classes of antineoplastic drugs that directly target and bind DNA [1]. They are from the oldest categories of anticancer drugs. The potency of these compounds differs based on bifunctional or monofunctional features that make either simple adducts, such as methylating agents and triazine derivatives, or intra- and interstrand cross-links in nucleic acids [2]. Alkylating agents are cytotoxic class of anticancer drugs [3,4]. Specific DNA-damaging drugs can kill tumor cells selectively due to the inability of tumor cells to repair some kinds of DNA damage [5] At various sites, the genomic DNA was alkylated [6–8]. Alkylating agents act during all phases of the cell cycle, directly on DNA, cross-linking the N-7-guanine residues, resulted in DNA strand breaks, resulted in abnormal base pairing, inhibition of cell division, and then lead to cell death [9]. Pyrrolobenzodiazepine (PBD) dimer is a DNA minor groove alkylator [10]. The cytotoxicity of many bifunctional alkylating agents was exerted through generating cross-links between the two complementary strands of DNA, called interstrand cross-links (ICLs) [11], which are considered among the most efficient antitumor agents in clinical use [12]. They have an antiproliferative activity in hepatoma cells. [13]. The most critical target of alkylating antineoplastic drugs belonging to the group of chloroethylating and methylating agents is DNA [14]. Minor groove binders (MGBs) are one of the most widely studied class of alkylating agents, which are characterized by a high level of sequence specificity [15]. Novel approaches to drug discovery have yielded candidate agents that are focused on "soft alkylation"—alkylators with higher target selectivity [16]. The alkylators are composed of the nitrogen mustards (melphalan, mechlorethamine, ifosfamide, cyclophosphamide, and chlorambucil), nitrosoureas (carmustine or BCNU, lomustine or CCNU, semustine or methyl-CCNU, and fotemustine), alkylalkane sulfonates (treosulfan and busulfan), aziridines (thiotepa), methylating agents (procarbazine, dacarbazine, and temozolomide), and the novel minor-groove binding agents (ectainascindin-473, tallimustine). These drugs alkylate DNA via the production of reactive intermediates that

New Strategies Targeting Cancer Metabolism
https://doi.org/10.1016/B978-0-12-821783-2.00003-0

attack nucleophilic sites. The group of platinum compounds consists of carboplatin, cisplatin [17–20], and satraplatin (JM216) and oxaliplatin. After binding of these agents to DNA, interstrand and intrastrand cross-links are furnished, which cause their cytotoxicity [21]. Alkylating agents comprising the chloroethylating nitrosoureas (CNU) lomustine, fotemustine, nimustine, and carmustine are considered from the first- and second-line therapeutics for malignant glioblastomas [22,23]. DNA-targeting anticancer medicines have been utilized in the clinic for more than 60 years [24,25]. Under physiological conditions, alkylating agents are able to covalently attach an alkyl group to a biomolecule. DNA alkylating agents interact with proliferating and resting cells in any phase of the cell cycle; they are more cytotoxic during the late G1 as well as the S phases as not adequate time to repair the damage before DNA synthesis occurs. Upon the attack, every electrophilic or nucleophilic species to DNA covalent bonds was furnished, and under physiological conditions, it need some nucleophiles (e.g., hydroxylamine, bisulfite, and hydrazine) that attack DNA bases. All oxygen and nitrogen atoms of these bases are nucleophiles, with the exemption of the nitrogen atoms implicated in the nucleoside bond (pyrimidines or N9 or N1 in purines); therapeutically advantageous drugs always act as carbon electrophiles [26]. Independent and related interactions govern the attraction between electrophiles and nucleophiles: electrostatic attraction between negative and positive charges and orbital overlap between the LUMO of the electrophile (orbital control) and the HOMO of the nucleophile. Highly electronegative oxygen atoms react under electrostatic control and are referred to as "hard nucleophiles," meaning they react with hard electrophiles. As the oxygen atoms of DNA bases are less, softer nucleophiles than nitrogen atoms, many therapeutically alkylating agents are soft electrophiles; mainly, DNA-alkylating compounds react at nitrogen sites in the subsequent rearrangement: N1 of adenine, N7 of guanine, N3 of thymine, and N3 of cytosine. Furthermore, diazonium salts, furnished from nitrosoureas and other antitumor agents, are exemplified as therapeutically hard electrophiles.

DNA alkylation is affected by steric factors, with nucleophilic sites in the double helix being less vulnerable to alkylation than those in the minor and major grooves [27]. Alkylation of DNA's bases has a significant impact on its dynamics and structure, resulting in a variety of outcomes. RNA transcription and DNA replication from the affected DNA molecule were prevented via alkylation. It also resulted in the fragmentation of DNA via hydrolytic reactions and through the action of repair enzymes during the removal of the alkylated bases. The mispairing of the nucleotides is induced via alkylation through alternating the normal hydrogen bonding between bases. Compounds susceptible of bis-alkylation can generate bridges between two complementary DNA strands "interstrand cross-linkage" or within a single DNA strand "intrastrand cross-linkage, limpet attachment," inhibiting their separation during DNA transcription or replication (Figs. 8.1–8.3). They can also resulted in cross-linking between

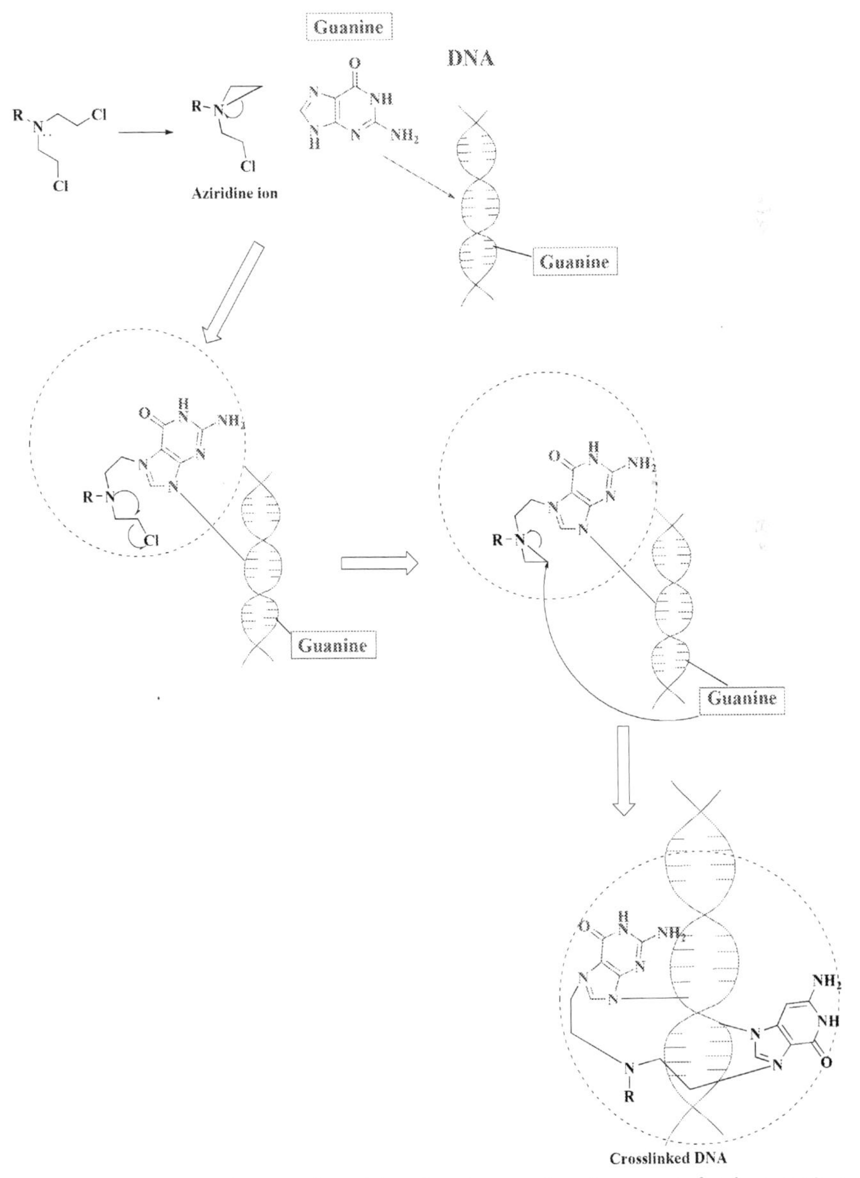

Fig. 8.1 Mechanism of action of mechlorethamines as alkylating agents for drugs acting on DNA.

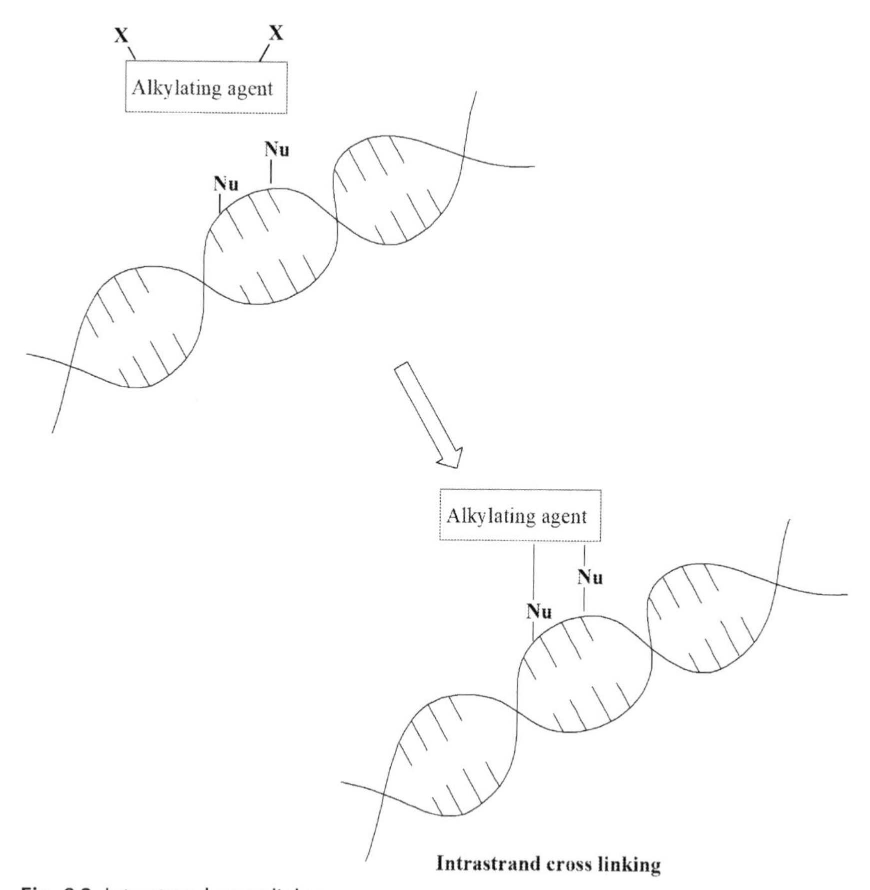

Fig. 8.2 Interstrand cross-linkage.

the associated proteins and DNA. The cytotoxicity of bifunctional alkylating chemicals is higher than that of monofunctional alkylating drugs. Additionally, cytotoxicity and the degree of interstrand cross-linking have a direct relationship [30,31].

2. Nitrogen mustards

Sulfur mustard (Fig. 8.4) is an irritant vesicant gas that was utilized in chemical warfare during World War I.

This resulted in systemic impacts which propose that this compound has an effect in treating cancer, while further investigations showed its toxicity on systemic use [32].

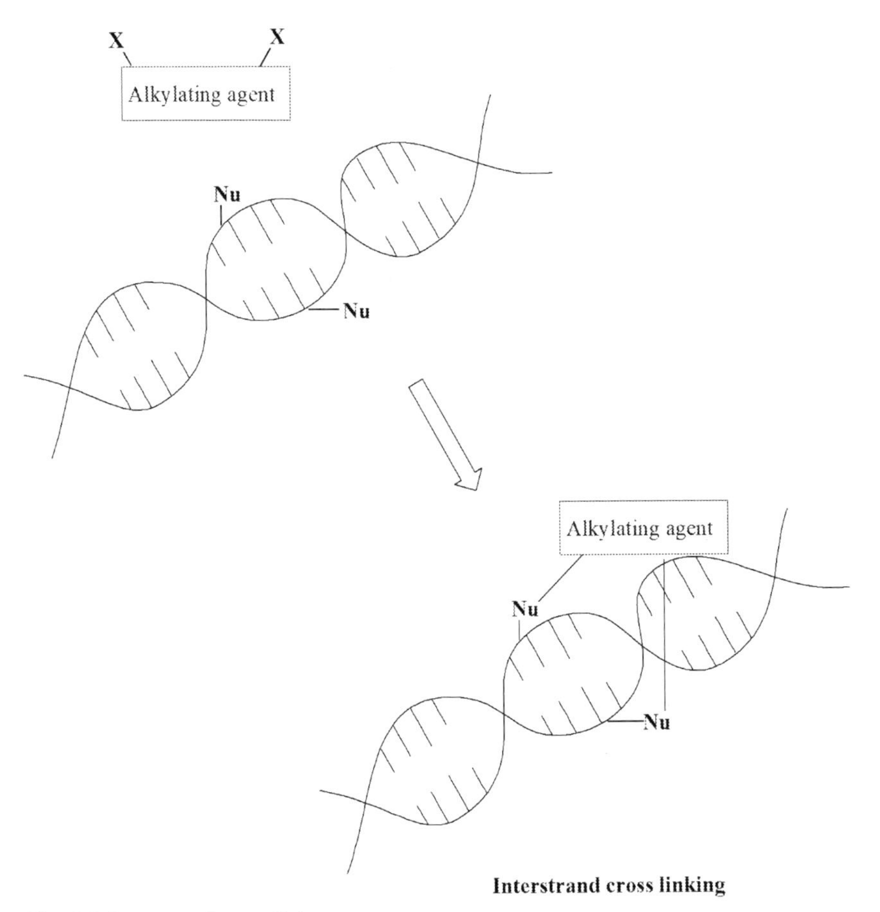

Fig. 8.3 Intrastrand cross-linkage.

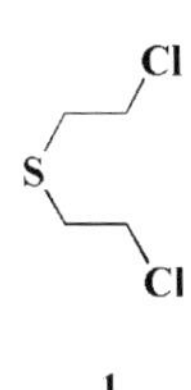

Fig. 8.4 Sulfur mustard.

Sulfur mustard vapor was emitted as a result of the sinking of a secret truck carrying this compound in Italy in 1943, which caused many deaths at that time. After investigating this disaster, mustard gas has been observed to cause lymphoid and myeloid suppression. This has suggested that this gas has an effect and is capable of killing divided cancerous cells. Upon an exploratory study of the impact of mustard gas on animals, mustine (chlormethine), a nitrogen analog of mustard gas, has been tested on animals and at another stage on humans, the results showed that it was effective as a treatment for lymphoma, including acute lymphoblastic leukemia and Hodgkin's lymphoma. It has been shown that the therapeutic effect of the nitrogen mustards may be limited due to marrow toxicity. However, mechloretamine (Mustargen) was then approved later on and is utilized as an antitumor agent for Hodgkin's lymphoma [33].

The mechloretamine gel (Valchlor) (Fig. 8.5) was approved for treating the stage IA/IB mycosis fungoides-type cutaneous T-cell lymphoma in 2013.

Chlorambucil (Leukeran) (Fig. 8.6) is considered active compound approved as an orally administered alkylating agent utilized for treating lymphomas and chronic lymphocytic leukemia.

Chlorambucil (*N*,*N*-bis(2-chloroethyl)-*p*-aminophenylbutyric acid; CLB) [34–36] is a bifunctional alkylating agent utilized as an antineoplastic drug and as an immunosuppressant [34,37]. Antineoplastic conjugate,

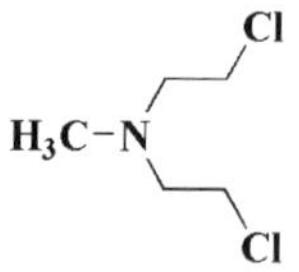

Cl
H_3C-N
Cl

2

Fig. 8.5 Mechloretamine.

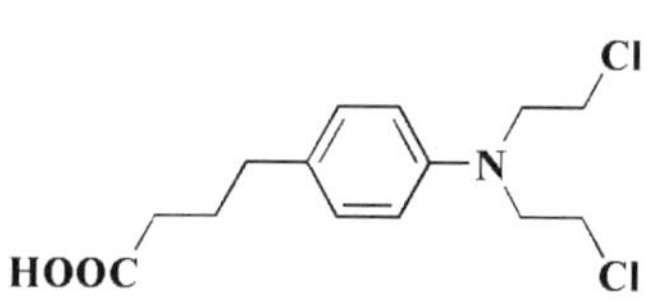

Cl
HOOC
N
Cl

3

Fig. 8.6 Chlorambucil.

chlorambucil-methionine, was estimated on the breast cancer MCF-7 cell line. This antineoplastic conjugate has indicated higher antineoplastic features and had not revealed abnormal toxicity. The conjugate indicated very good anti-cancer effects, which are the same as chlorambucil, but it had less toxicity as compared with chlorambucil. Based on the results, methionine-chlorambucil conjugate may be a better than chlorambucil alone for treating breast cancer [38]. The combination of chlorambucil plus rituximab seems more effective than alkylating agents or rituximab alone in gastric MALT lymphoma [39]. The cellular actions of CLB have remained poorly inspected [40].

Bendamustine (Ribomustin, Levact, Treandra) (Fig. 8.7) is a concerned nitrogen mustard in which the benzimidazole is present instead of the benzene ring. It is utilized in treating chronic lymphocytic leukemia, B-cell non-Hodgkin's lymphoma [41], as well as myeloma [41,42].

The butyric acid scaffold of bendamustine and chlorambucil modulates their metabolism as well as their reactivity as alkylating agents and their aqueous solubility [43].

Melphalan (Alkeran) (Fig. 8.8) is utilized as an antineoplastic drug [44], which comprises L-phenylalanine unit, and is utilized in the treatment of types of bone marrow tumors, breast or ovarian cancers, and multiple myeloma. Melphalan is considered to be the most agent widely utilized in

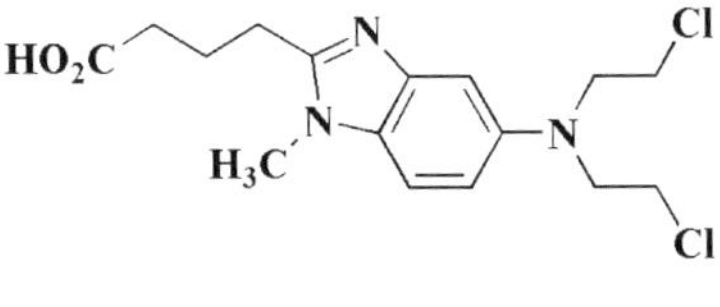

4

Fig. 8.7 Bendamustine.

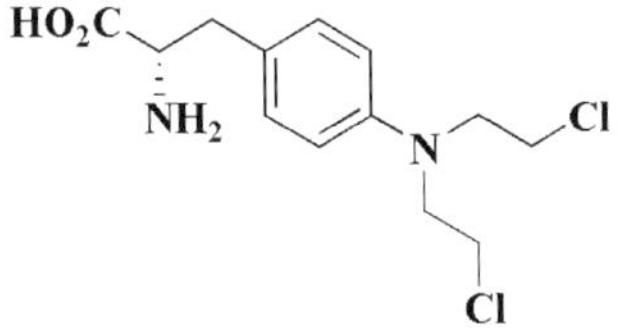

5

Fig. 8.8 Melphalan (L-phenylalanine mustard).

hematopoietic stem cell transplant [45] due to its antitumor impacts as a DNA alkylating agent and its myeloablative features.

Antineoplastic conjugates, such as the melphalan-flufenamide (melflufen), have efficacy as a new dipeptide prodrug of melphalan in multiple myeloma [46].

Other compounds designed on similar approaches are estramustine and uramustine (uracil mustard) (Fig. 8.9).

For the purpose of stabilizing melphalan for hydrolysis, compound design was developed called melphalan-flufenamide, which indicated portentous preclinical potency in various solid tumors [47], which moved the trend to the clinical examination of adult patients with advanced ovarian cancer, advanced solid tumors, or non-small-cell lung cancer [48]. Significantly, flufenamide-melphalan displays higher cytotoxicity than melphalan does in case of in vitro and in vivo. In vitro estimations indicate that following a rapid incorporation into the cytoplasm of tumor cell lines, intracellular hydrolysis led to the release of melphalan. Flufenamide-melphalan can thus be considered a melphalan prodrug, the aminopeptidase, e.g., aminopeptidase N (APN), is responsible for the enzymatic activation of melphalan [49]. In some malignancies, this enzyme is overexpressed and this is offered tumor selectivity [50].

Estramustine (Emcyt, Estracyt), comprising a β-estradiol unit joined to a nitrogen mustard portion through a carbamate bridge, was designed as a prodrug due to the electron-withdrawing effect of the carbonyl group, which makes the electron density of its nitrogen atom insufficient to trigger aziridinium formation.

Cyclophosphamide (Endoxana, Cytoxan) (Fig. 8.10), which is inactive as the electron-withdrawing effect of the P=O bond, stops the generation of aziridinium cations and was prospected to be activated via the

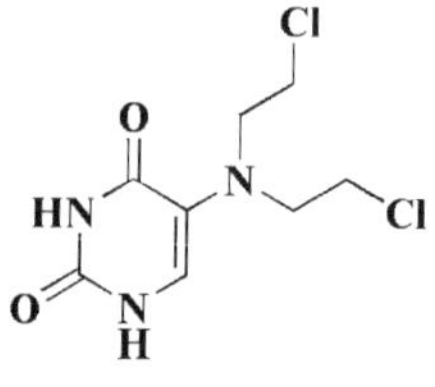

6

Fig. 8.9 Uramustine (uracil mustard).

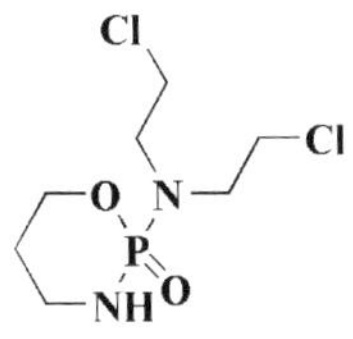

7

Fig. 8.10 Cyclophosphamide.

phosphoramide enzymatic hydrolysis on the basis of that some tumors comprise high levels of phosphoramidases.

This alkylating anticancer drug is utilized to treat leukemia, Hodgkin's disease, and other lymphomas, breast, lung, as well as ovarian malignancies, and is typically used in combination with other drugs. In multiple myeloma, it is particularly effective when used with dexamethasone and bortezomib [51]. Other applications include the treatment of Wegener's granulomatosis, lupus erythematosus, and severe rheumatoid arthritis; in smaller dosages, the drug also acts as an immunosuppressant.

Cyclophosphamide (CYP or CPA) is a synthetic alkylating agent utilized as an antineoplastic drug; it is chemically related to the nitrogen mustards as an anticancer and immunosuppressive drug [52,53] and is mostly utilized in rheumatoid arthritis, bone marrow transplantation, multiple sclerosis, lupus erythematosus, and neuroblastoma [54]. Oral cyclophosphamide is well tolerated and effective [55]. Cyclophosphamide (CPA) is an anticancer drug and alkylating agent (mutagen) [56]. It is also used in many pharmaceutical compounds in combination with the active ingredient as an immunostimulating lipid-A derivative in anticancer therapy [57]. In addition, CPA and ifosfamide (IFO) have a complementary mechanism of action; they are usually combined with other anticancer drugs in the chemotherapy of hematological malignancies and solid tumors to afford a synergistic or an additive anticancer effect [58].

Ifosfamide (Mitoxana) (Fig. 8.11) is utilized to treat lymphomas and sarcoma. The metabolism of ifosfamide differs from that of cyclophosphamide but they are similar in their structure.

Many other nitrogen mustards possess a remarkable cytotoxicity in inhibiting colon carcinoma (HCT-116), breast carcinoma (MX-1), human lymphoblastic leukemia (CCRF-CEM), and human non-small-cell lung cancer (H1299) cell growth in vitro [59–61].

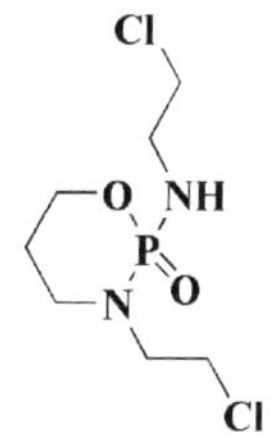

8

Fig. 8.11 Ifosfamide.

3. Aziridines

Aziridinium cation is the active species that is involved in DNA alkylation by nitrogen mustards, so various aziridines have also been evaluated as antitumor agents such as thiotepa (Thioplex), which is a sulfur analog of triethylenephosphoramide (TEPA) and triethylenemelamine (Tetramine) [62] (Figs. 8.12–8.14).

Thiotepa shows less stability in the acidic media that is prevalent in stomach. But it is intracavitary administrated when utilized in cases of bladder carcinoma. Thiotepa produces myelosuppression as a main toxicity effect. As a conditioning treatment prior to hematopoietic stem cell transplantation, it was designated as "orphan drug" by the U.S. Food and Drug Administration (FDA) and the European Medicines Agency (EMA) in 2007.

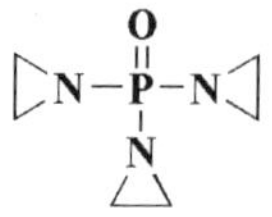

9

Fig. 8.12 TEPA (triethylenephosphoramide).

10

Fig. 8.13 Thiotepa.

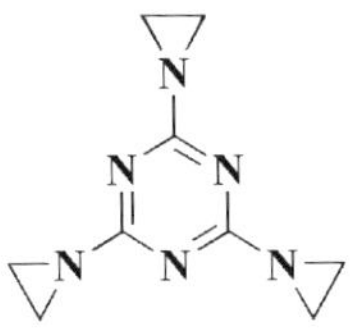

11

Fig. 8.14 TEM triethylenemlamine.

In hematological diseases it is also used as a conditioning treatment prior to autologous or allogeneic hematopoietic progenitor cell transplantation (HPCT).

Other antitumor compounds having two or three aziridine rings linked to a benzoquinone system can act as cross-linking agents and DNA bis-alkylators. They were synthesized in that design in order to be able to cross the blood–brain barrier due to their ionization and high lipophilicity.

Diaziquone (AZQ) and carboquone (carbazilquinone) (Figs. 8.15 and 8.16) have been tested in treating several cancers, but they are limited in their use because of their low effectiveness and toxicity.

Diaziquone (AZQ) [63] was received orphan drug status from the FDA in the early 1980s, but it revealed no obvious priority over pre-existing drugs.

In 1960, triaziquone (Oncovedex) (Fig. 8.17) was utilized clinically for treating cancer; however, as a result of its toxicity to the blood vessel walls and bone marrow, more significant and efficient agents have replaced it [64].

Apaziquone (EO9) (Fig. 8.18) is a bioreductive drug having only one azidirine scaffold, which indicates no potency in phase II clinical trials when administered intravenously due to poor penetration via the vascular tissue and rapid pharmacokinetic elimination.

But recent investigations indicated that after intravesical administration in patients with superficial transitional cell carcinoma of the bladder,

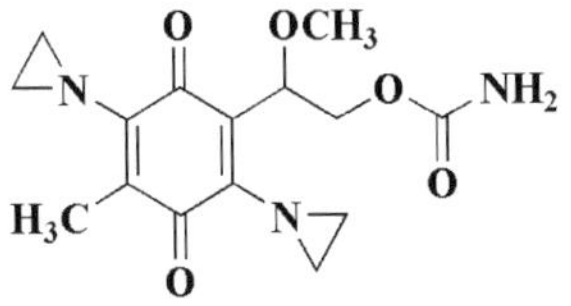

12

Fig. 8.15 Carboquone.

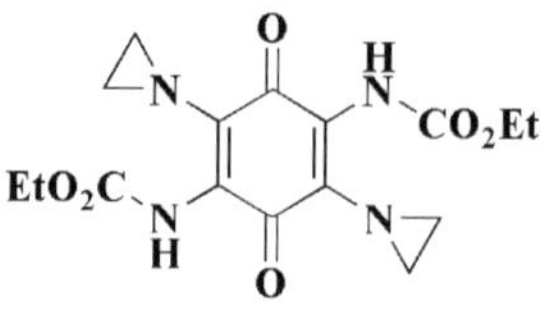

13

Fig. 8.16 Diaziquone (AZQ).

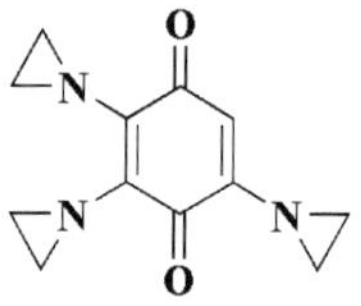

14

Fig. 8.17 Triaziquone.

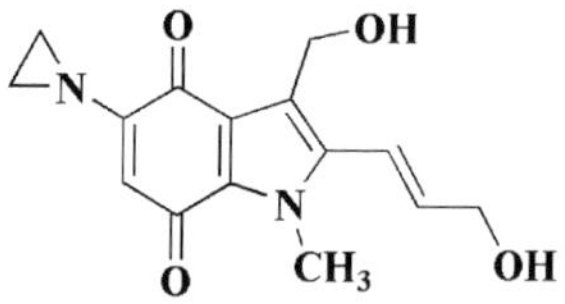

15

Fig. 8.18 Apaziquone (EO-9).

apaziquone showed potency, and this is affirmed by the FDA [65]. The related aziridinylquinone BZQ (Fig. 8.19) has also been subjected to clinical trials in humans [66].

DT-diaphorase (DTD, NQO1), which is two-electron reductase, is considered as target for antitumor compounds, and is in the nucleus of the cell, and its levels rise in many tumors, such as non-small-cell lung cancer [67]. It can activate the aziridinylbenzoquinones [68]; there has been much effort to furnish new agents that target this enzyme efficiently. DNA damage increases due to the presence of the electron donor groups that are present on the benzoquinone ring, as shown by SAR studies, while the electron-withdrawing or the presence of the sterically bulky groups at the C-6 position result in less potency or inactive compounds [69].

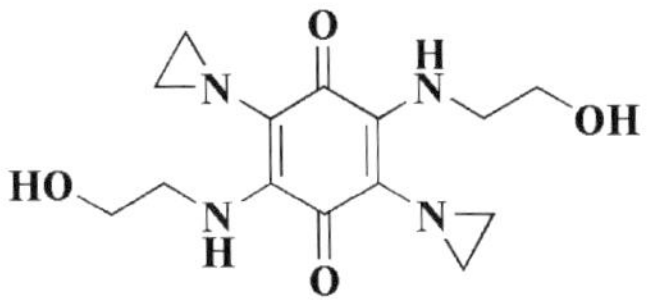

16

Fig. 8.19 BZQ.

Bioreductive alkylating agents, quinone-containing alkylating agents, need reduction of the quinone scaffold for activating their alkylating substituents [70]. Mitomycin C is the only bioreductive alkylating agent approved for use in the clinic; other quinone-containing alkylating agents such as diaziquone, porfiromycin, triaziquone, carbazilquinone, and EO9 have also been utilized in the clinic for treating cancer [71].

Like Me-DZQ (Fig. 8.20), the usefulness of this compound is limited in some cases due to its poor solubility; this has been partially overcome through introducing hydrophilic hydroxyl groups in the side chains.

RH1 (Fig. 8.21) [72] is a prime substrate for DTD, which has high DNA cross-linking potency and high antitumor activity in vitro and in vivo [73] with less toxicity in normal tissues. Phase I clinical studies were applied in

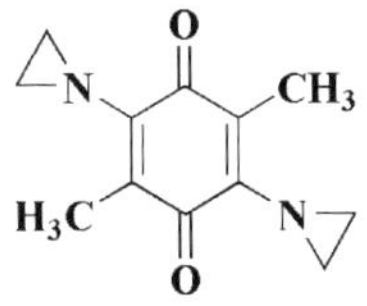

17

Fig. 8.20 Me-DZQ.

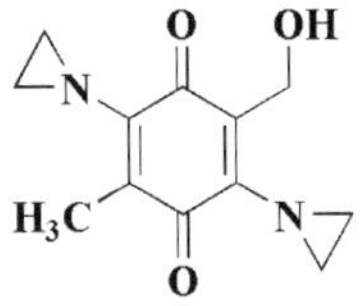

18

Fig. 8.21 RH1.

patients with solid tumors [74,75] although due to its limited aqueous solubility and stability, it will need a proper formulation for more proceeded clinical trials.

4. Epoxides

Diepoxybutane is considered as simple epoxide; it can cross-link DNA. It is nonenzymatically furnished from the class of methanesulfonate particularly from treosulfan (Ovastat). The latter can alkylate DNA at guanine bases as nitrogen mustards [76]. It is utilized in treating ovarian cancer. Also, it clearly has clinical activities against hematological malignancies and some solid tumors, and it is also utilized before stem cell transplantation [77] in bone marrow ablation and in the treatment of breast cancer and malignant melanoma. The study of the use of treosulfan is being carried out in combination with some other drugs for patients suffering from major thalassemia, which is considered as a nonmalignant genetic disorder [78]. The bromhydrine prodrug, the 1,6-dibromo analog of mannitol, "Mitobronitol:Myelobromol" is similar to treosulfan; double intramolecular nucleophilic displacement in the latter drug produces the diepoxide "DNA cross-linking reagent." In accelerated chronic granulocytic leukemia, it is utilized in the treatment of myelosuppression prior to allogenic bone marrow transplantation, revealing lower toxicity than busulfan [79]. VAL-083 (dianhydrogalactitol, dianhydrodulcitol) is a diastereomer of the bis-epoxide used in the treatment of lung cancer and chronic myelogenous leukemia (CML). Clinical tests have been performed on this drug for the brain cancer, particularly glioblastoma multiforme (GBM). Mixed epoxide-aziridine is considered as one of the known antitumor compounds. For example, azinomycin B (carzinophilin) is a complex natural achiroi and has an effective antitumor activity. It has been tested in a clinical study in humans with multiple types of cancer and has promising results [80,81].

5. Methanesulfonates

Methanesulfonate has the ability to remove a negative charge between three oxygen atoms, which is a characteristic that what makes it capable of being an efficient and good leaving group. So various compounds comprising 2 methanesulfonate groups segregated through a polymethylene chain were evaluated as antitumor agents, finding that the optimal potency is related to the compound with four carbon atoms busulfan. Busulfan (Myleran) is considered an important CML chemotherapeutic medication, and it has been replaced by imatinib, but it is also still used sometimes as it is affordable. In high-dose combination with cyclophosphamide, it is used in the treatment of patients who suffer from chronic granulocytic leukemia and need an allogeneic bone marrow transplant, but it causes partial toxicity to the pulmonary tissue, and this is restricted to its dosage. Hepsulfam is subjected to early clinical studies [82].

The anticancer drug busulfan (Fig. 8.22) is a cell cycle-nonspecific alkylating agent, which is used in treating chronic myelogenous leukemia (CML) [83] (Figs. 8.23 and 8.24).

6. Nitrosoureas

The antitumor potency of 1-methyl-1-nitrosourea was discovered; introducing a 2-chloroethyl chain on nitrogen atom with the nitroso group (CNUs) resulted in an increase in potency. The chloroethyl derivatives are effective in treating brain tumors, because they are lipophilic so they are able to cross the blood-brain barrier. This features resulted in the synthesis of various nitrosoureas, such as lomustine (CCNU) and its methyl analog semustine, nimustine (ACNU), carmustine (BCNU), and the water-soluble fotemustine (Fig. 8.25) and tauromustine (Fig. 8.26), but they are not widely used due to their toxicity.

From the *Streptomyces achromogenes* strain, a hydrophilic natural nitrosourea was isolated called streptozotocin (Zanosar) (Fig. 8.27). This compound was considered as a lead as the primary SAR studies purposed that hydrophilic nitrosoureas were less toxic and more active. ACNU, BCNU, streptozotocin, and lomustine were considered as the most clinically relevant nitrosoureas.

Lomustine (CeeNU, CCNU) is utilized in treating brain tumors; pancreatic, breast, lung, and ovarian cancer; melanoma; Hodgkin's lymphoma; and multiple myeloma (Figs. 8.28–8.30).

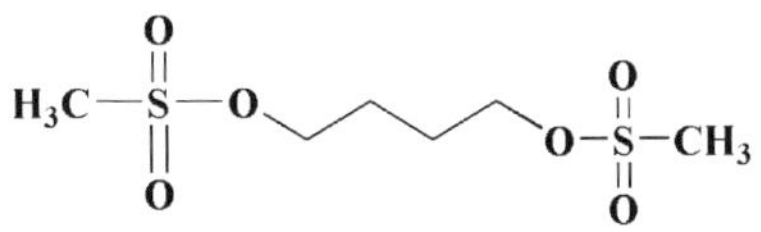

19

Fig. 8.22 Busulfan.

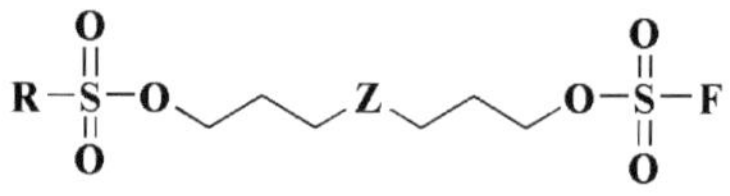

O
R—S—O Z O—S—F
O O
O

20

Fig. 8.23 R Z CH_3 NH Improsulfan NH_2 CH_2 Hepsulfam.

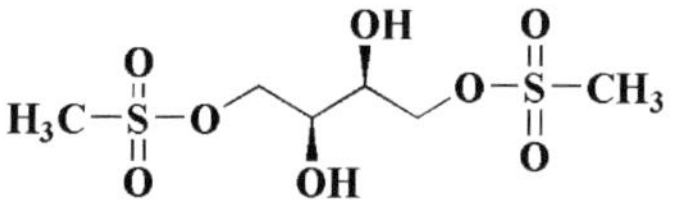

OH O
O
H_3C—S—O O—S—CH_3
O OH O

21

Fig. 8.24 Treosulfan.

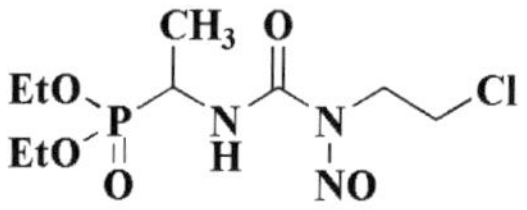

CH_3 O
EtO P N N Cl
EtO H
O NO

23

Fig. 8.25 Fotemustine.

O O
S Cl
$(H_3C)_2N$ N N
O H NO

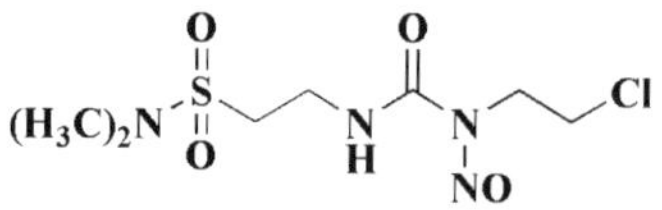

22

Fig. 8.26 Tauromustine.

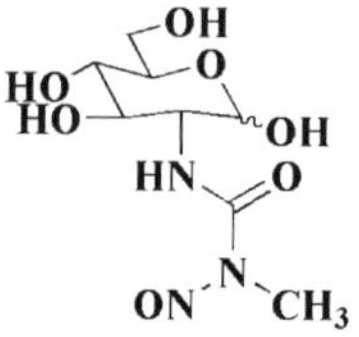

OH
O
HO
HO OH
HN O
N
ON CH_3

24

Fig. 8.27 Streptozotocin.

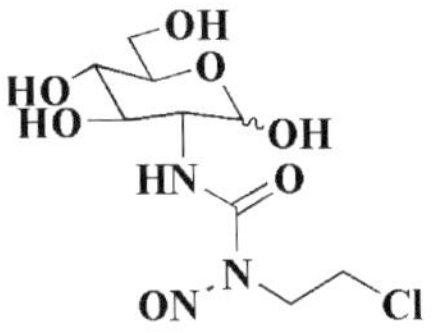

25

Fig. 8.28 Chlorozotocin.

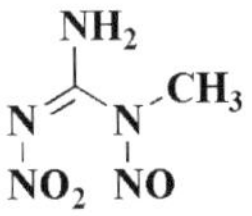

26

Fig. 8.29 1-Methyl-3-nitro-1-nitrosoguanidine.

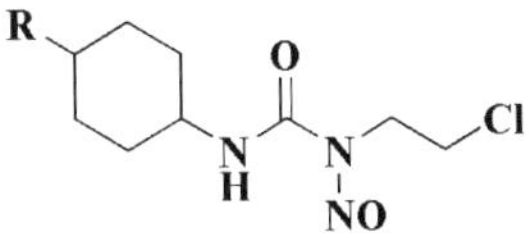

27

Fig. 8.30 Lomustine (CCNU, CeeNU) and semustine.

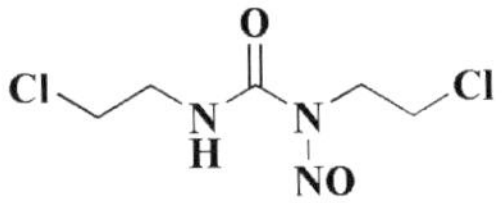

28

Fig. 8.31 Carmustine (BCNU).

Carmustine (BiCNU) (Fig. 8.31) is utilized in various forms of brain cancer (including glioblastoma, glioma, medulloblastoma, multiforme, and astrocytoma), lymphoma (Hodgkin's and non-Hodgkin's lymphoma), as well as multiple myeloma. With a reduced systemic toxicity, novel formulation of carmustine has been generated for local brain tumors treating.

Implantation of the formulated slow-release "wafer" dosage form (Gliadel Wafer; Carmustine in combination with Polifeprosan) into the resection cavity, which is left after the surgical removal of tumor [84]. It was approved by the FDA for use as an adjunct to surgery to prolong survival in patients with recurrent GBM for whom surgical resection is indicated.

Nimustine (Fig. 8.32) is utilized in combination with teniposide as a second- or third-line chemotherapy for recurrent glioblastoma [85]. After the discovery that streptozotocin was selectively toxic to β-cell of the pancreatic islets, it was assumed that this drug might be utilized in treating pancreatic cancer. The FDA approval of the drug is obtained for treating pancreatic islet cell cancer, and it is marketed as Zanosar; its utility is restricted to patients whose cancer cannot be treated through surgery. Concerning chlorozotocin, a phase II study indicated that it is potent against metastatic melanoma to the same degree as other chloroethylnitrosoureas used clinically [86], but without doing bone marrow toxicity.

1,3-Bis(2-chloroethyl)-1-nitrosourea (BCNU) is utilized in treating brain and other forms of cancer. Through forming DNA interstrand cross-links, BCNU can kill cells [87] (Fig. 8.33).

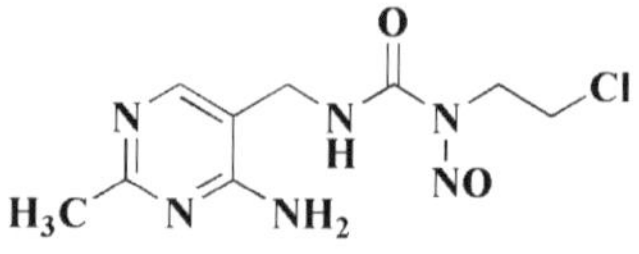

29

Fig. 8.32 Nimustine (ACNU).

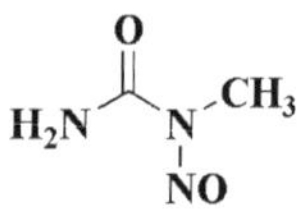

30

Fig. 8.33 *N*-Methyl-*N*-nitrosourea (MNU).

7. Triazenes

Dacarbazine (DTIC-Dome) (Fig. 8.34) is used in combination therapy for treating Hodgkin's lymphoma, metastatic malignant melanoma, sarcoma [88], and islet cell carcinoma of the pancreas. Initially, it was designed as an antimetabolite and as an analog of 5-aminoimidazole-4-carboxamide, which is an intermediate in purine biosynthesis. Its cytotoxic potency is related to the generation during its metabolism of methyldiazonium, which methylates DNA [89].

Dacarbazine has many impediments due to its toxicity and excessive hydrophilicity leading to slow absorption and is also orally incomplete; therefore, it is necessary to be administrated intravenously.

High photosensitivity is its disadvantage, i.e., it can be decomposed to 2-azahypoxanthine through an intermediate diazonium species. Thus, intravenous infusion bags of dacarbazine must be hidden from light. These reasons have motivated the design and preparation of the dacarbazine analogs, such as temozolomide (Temodal) (Fig. 8.35). At a later stage, a related compound, which was first termed azolastone and then mitozolomide, indicated a significant preclinical antitumor potency, but early phase I studies revealed that it induced irreversible thrombocytopenia. Its analog, temozolomide,

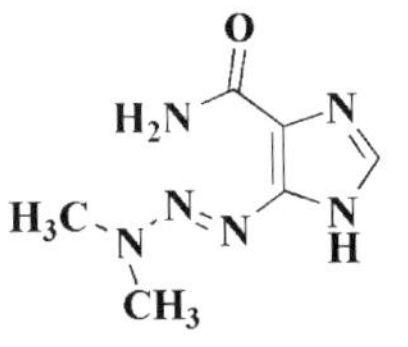

31

Fig. 8.34 Dacarbazine (DTIC).

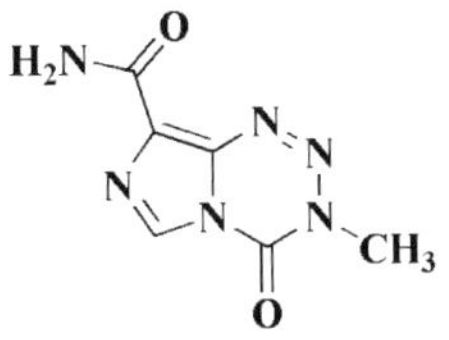

32

Fig. 8.35 Temozolomide.

was finally selected for clinical trials [90]. Temozolomide is considered one of the drugs that treat brain tumor called anaplastic astrocytoma, and is administered orally [91,92].

It is an effective anticancer drug utilized for treating different solid tumors, comprising melanoma and glioma [93–100]. It improves survival in patients with malignant gliomas such as glioblastoma [101]. Like dacarbazine, it is a prodrug that is transformed into the same intermediate (MTIC); however, in case of temozolomide, the bioactivation process comprises a non-enzymatic hydrolysis reaction, and then a spontaneous decarboxylation was followed. The disappearance of the hepatic activation is an advantage as the metabolic individual variation in patient microsomal potency is not to be taken into consideration. The bone marrow toxicity is considered the problem linked to the temozolomide administration.

8. Methylhydrazines

For advanced Hodgkin's disease, the use of procarbazine (Matulane) (Fig. 8.36) in combination therapy was affirmed. The unique agent procarbazine performs its functions with multiple mechanisms of action, and it does not interfere with other alkylating agents.

It allows the inhibition of the incorporation of small DNA precursors, as well as protein synthesis and RNA. Procarbazine can also directly damage DNA via a methylation reaction.

9. 1,3,5-Triazines: Hexamethylmelamine and trimetelamol

Altretamine (Hexalen, hexamethylmelamine) (Fig. 8.37), which is structurally similar to the aziridine derivative triethylenemelamine (TEM), was

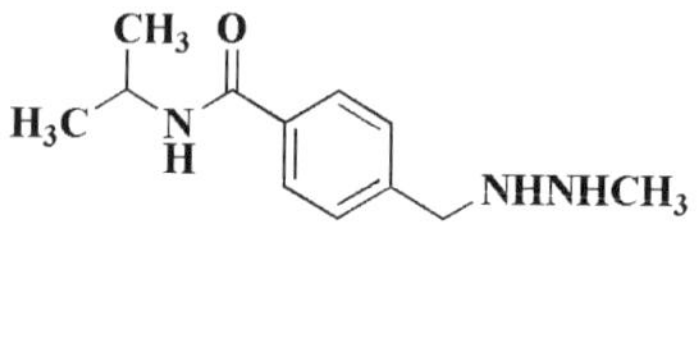

33

Fig. 8.36 Procarbazine.

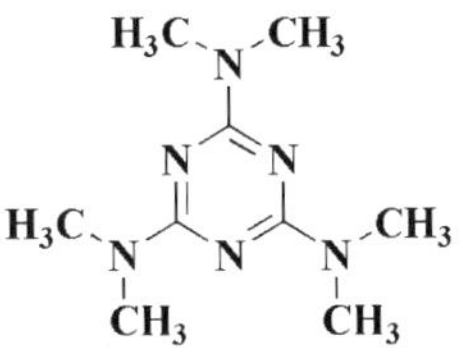

34

Fig. 8.37 Altretamine.

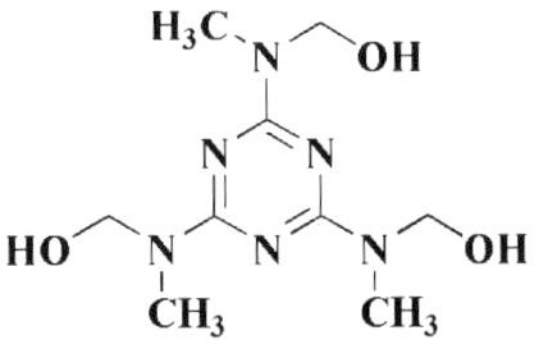

35

Fig. 8.38 Trimelamol.

studied as an antitumor agent although it was originally synthesized as a resin precursor. Its main therapeutic role is to treat recurrent ovarian cancer, following the first-line treatment with cisplatin, and it is also active in many other types of tumors [102].

The water-soluble compound trimelamol, which is considered a tris(hydroxymethyl) analog of altretamine, does not require metabolic activation, although it is less stable because of the existence of its carbinolamine species [103]. Treating refractory ovarian cancers using trimelamol (Fig. 8.38) was studied [104], but it had to be withdrawn from subsequent clinical investigations due to the formulation difficulties regarding its low stability.

10. Platinum complexes

Cisplatin (cisplatinum, cisdiamminedichloroplatinum II, CDDP) (Platinol) has become a substantial component in chemotherapy regimens for treating bladder and lung cancer, lymphomas, melanoma, and myelomas. Its utility is limited via severe dose-limiting side effects and acquired or intrinsic drug resistance. These side effects include neurotoxicity, nephrotoxicity, myelosuppression, and ototoxicity [105].

Cisplatin is a square-planar complex, possessing two inert ammonia molecules and two labile chlorines coordinated to the central Pt(II) atom, and it is *cis*-configurated. As a result of the very high toxicity of cisplatin and the presence of acquired or intrinsic drug resistance problems, developed derivatives have been synthesized to enhance the therapeutic index and its selectivity [106]. Many platinum-based drugs failed in clinical trials [107].

Oxaliplatin and carboplatin have received international marketing approval; lobaplatin, nedaplatin, and heptaplatin are approved in some nations, and currently, there are a few drugs in different phases of clinical trials (picoplatin and satraplatin). The research in this field has shifted toward drug delivery, as in case of aliposomal analog of cisplatin, lipoplatin [108].

Cisplatin analogs comprise tetragonal Pt(II) complexes, such as nedaplatin, oxaliplatin, carboplatin, SKI-2053R, and ZD-0473. Octahedral Pt(IV) complexes are also known, including iproplatin, satraplatin (JM 216), and tetraplatin [109]. The development of these cisplatin derivatives has indicated concerted demands that are required for being an anticancer drug, such as the electroneutrality to allow for the drug to pass through cell membranes, although the potent form is charged after ligand exchange, the existence of at least two good leaving groups (*cis* to one another), although trans complexes also indicate potency in some cases, and the existence of "inert" carrier ligands, usually nontertiary amine groups that increase adduct stabilization via H-bonding with bases.

In 1989 in combination with other chemotherapeutic agents, **carboplatin** (Paraplatin) (Fig. 8.39) was approved for the initial treatment of advanced ovarian cancer. It forms cross-links with guanine in DNA. Its mechanism of action is mainly identical to that of cisplatin. It causes substantially reduced nephrotoxicity at effective doses, as the dicarboxylate ligands facilitate its excretion.

Oxaliplatin (Eloxatin) (Fig. 8.40) was approved in Europe in 1999 and then gained the FDA approval later on [110]. Against several tumor cell

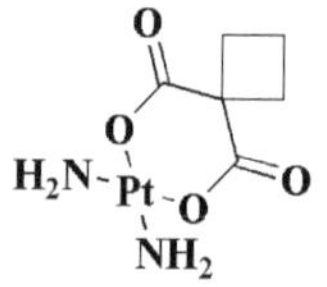

Fig. 8.39 Carboplatin.

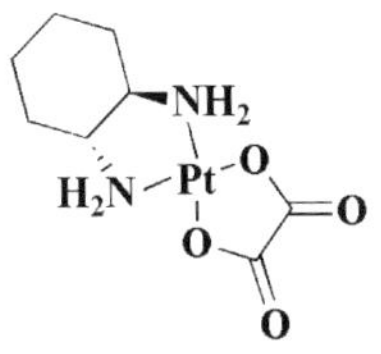

37

Fig. 8.40 Oxaliplatin.

lines, comprising some that are resistant to carboplatin and cisplatin, it revealed in vitro and in vivo efficacy.

The existence of the bulky diaminocyclohexane ring is thought to lead to the production of platinum-DNA adducts more efficient at blocking DNA replication than in case of cisplatin.

Oxaliplatin has a spectrum of potency variant from that of either carboplatin or cisplatin and lacks cross-resistance with them, proposing that it has various molecular targets [111]. It was the first platin-based drug to be potent against metastatic colorectal cancer in combination with folinic acid and fluorouracil [112].

Other platin-based drugs have been approved: lobaplatin, nedaplatin (Aqupla), and heptaplatin.

Nedaplatin (Fig. 8.41) is less toxic than cisplatin; however, it is only moderate in overcoming cisplatin resistance in a successful manner.

A mixture of the (*RRS*)- and (*SSS*)-diastereoisomers of 1,2-diammino-methyl-cyclobutaneplatinum(II) lactate, Lobaplatin (Fig. 8.42), was licensed for treating metastatic breast cancer, CML, and small cell lung cancer [113].

Heptaplatin (SKI-2053R) (Fig. 8.43), which has also less toxicity than the parent molecule [114], was affirmed for treating gastric cancers (Figs. 8.44 and 8.45).

ZD-0473 (picoplatin) (Fig. 8.46) was developed to overcome resistance caused by interactions with thiol-containing molecules. One of the amines

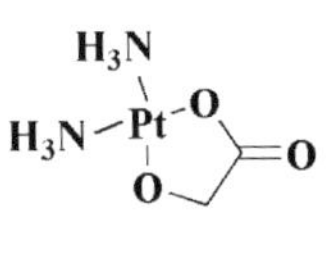

38

Fig. 8.41 Nedaplatin.

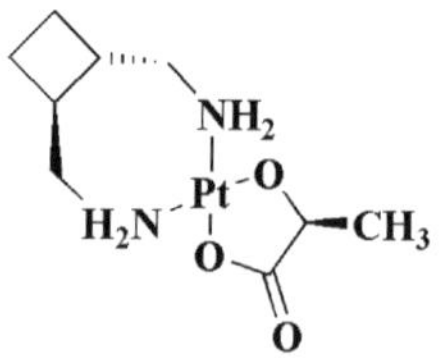

39

Fig. 8.42 Loboplatin.

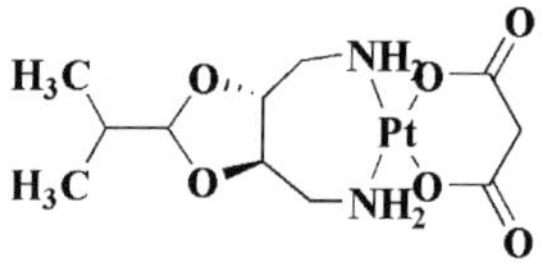

40

Fig. 8.43 Heptaplatin.

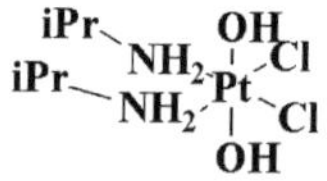

41

Fig. 8.44 Iproplatin.

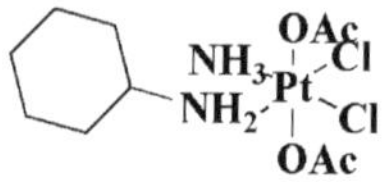

42

Fig. 8.45 Satraplatin (JM 216).

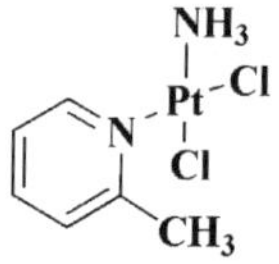

43

Fig. 8.46 Picoplatin.

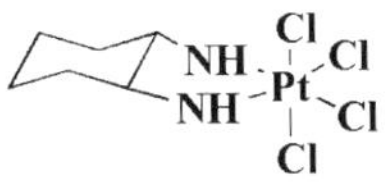

44

Fig. 8.47 Tetraplatin.

connected to Pt was replaced with a bulky methyl-substituted pyridine, allowing for a prolonged half-life [115,116] (Fig. 8.47).

Investigations to research novel orally administered analogs led to the octahedral Pt(IV) complex satraplatin (JM 216) (Fig. 8.45). The latter two drugs are still under clinical trials. Pt(IV) complexes appear to act by a different mechanism; it was suggested that reduction to the corresponding Pt(II) derivatives is necessary for potency. The active species for satraplatin is suggested to be compound JM 118 [117]. Unlike Pt(II) complexes, Pt(IV) compounds can reach the tumor site unchanged and then be converted to the biologically active Pt(II) species as they are highly stable in plasma. Some transplatinum complexes were shown to be effective against tumors resistant to cisplatin, owing to differences in the DNA binding of both types of complexes. Because of its stereochemistry, the trans isomer of cisplatin, known as TDDP (trans-diaminedichloroplatinum(II)), is unable to produce 1,2-intrastrand adducts; however, it does produce interstrand cross-links between complementary cytosine and guanine, as well as 1,3-intrastrand adducts, resulting in a different type of double helix conformational distortion [118]. Other types of trans-platinum antitumor compounds comprising 2 or 3 reactive platinum centers are trinuclear and dinuclear platinum(II) complexes, synthesized to afford long-range intrastrand and interstrand DNA cross-links, such as the triplatinum complexes BBR3610, diplatinum and triplatin (BBR3464) [119]. It possesses a wide range of antitumor potency and is now being tested in clinical trials [120].

11. Examples of potent anticancer alkylating agents

Compound **45** (Fig. 8.48) is utilized in treating brain cancer selected from a O6-methylguanine-DNA methyltransferase (MGMT)-pos. astrocytic brain tumor, primary CNS lymphoma, and metastatic brain cancer [121].

Melphalan derivative was synthesized through introducing DNA alkylation agent into *N*-13 site of evodiamine. Compound **46** (Fig. 8.49) was

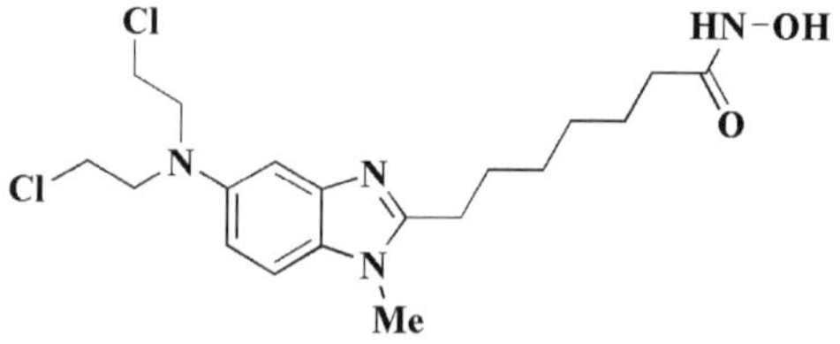

Fig. 8.48 Compound **45**.

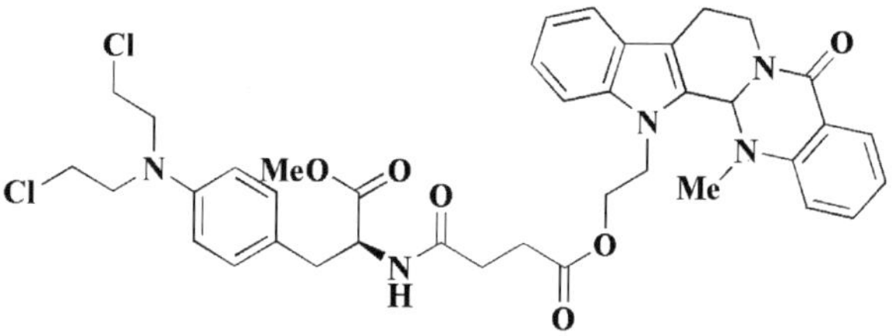

Fig. 8.49 Compound **46**.

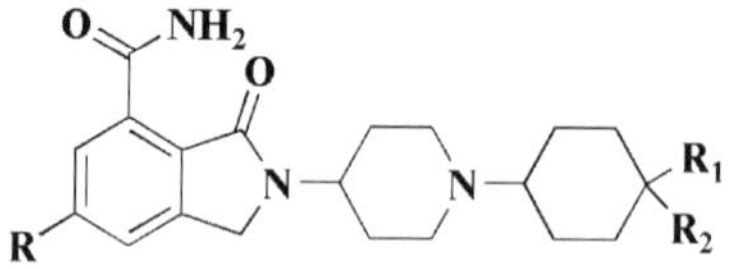

Fig. 8.50 Compound **47**.

synthesized that revealed antitumor potencies against cancer cell lines Bel-7402, HT-29, and THP-1 [122].

A therapeutic combination containing compound **47** (Fig. 8.50) and one or more antineoplastic agents selected consisting of an alkylating or alkylating-like agent, a topoisomerase II inhibitor, a topoisomerase I inhibitor, an antimetabolite agent, an antimitotic agent, and radiation was prepared [123].

Pharmaceutical components contain a pharmaceutically acceptable carrier and a prodrug, and are utilized in improving the chemotherapeutic treatment of tumor cells in a mammal, e.g., a human, with an antineoplastic

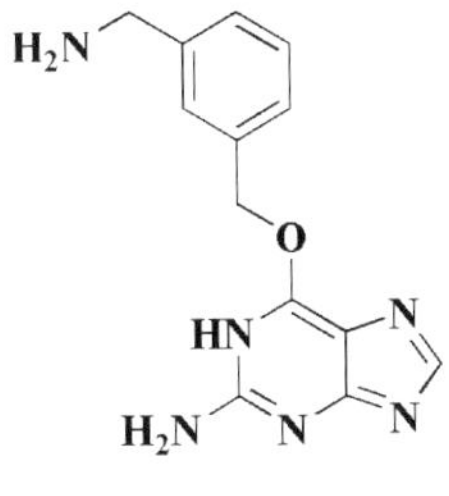

48

Fig. 8.51 Compound **48**.

alkylating agent that causes cytotoxic lesions at the *O*6-position of guanine (Fig. 8.51). So *O*6-benzyl-*N*2-[[[[4″-[(β-dglucopyranuronosyl)oxy]-3″ nitrophenyl]methyl]oxy]carbonyl]guanine monosodium salt was synthesized through coupling of 4-*O*-(2′,3′,4′-tri-*O*-acetyl-6′-methyl-β-D-glucopyranuronosyl)-3-nitrobenzyl alcohol with *O*6-benzyl-9-[(pivaloyl-oxy) methyl]guanine [124].

Compounds related to precolibactin, pharmaceutical components based upon these compounds, and methods of synthesis are utilized to access intermediates and compounds, which are alkylating agents [125] (Fig. 8.52).

Hypoxia-activated drug compounds possessing a structure of formula **50** (Fig. 8.53) are useful in treating cancer and other hyperproliferative diseases. Compound **50** is a bioreductive group as nitroimidazolyl, nitrothienyl, and nitrofuranyl, and pharmaceutically acceptable salts are reported. Compound **51** (Fig. 8.54) was synthesized through alkylating 2-nitrothiophene with 4-(bis(2-chloroethyl)amino)-3-methylphenyl chloromethylsulfonate. The

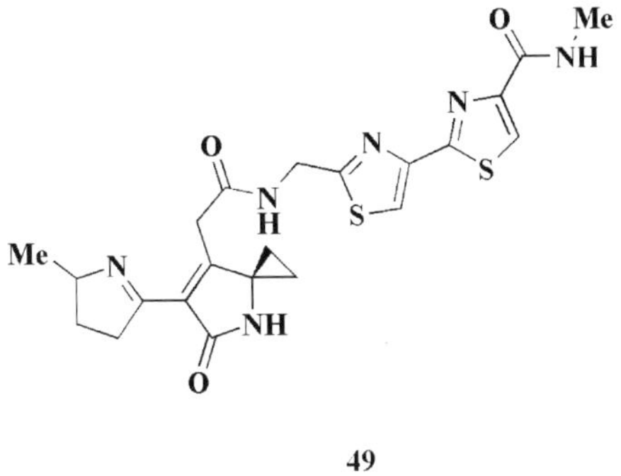

49

Fig. 8.52 Compound **49**.

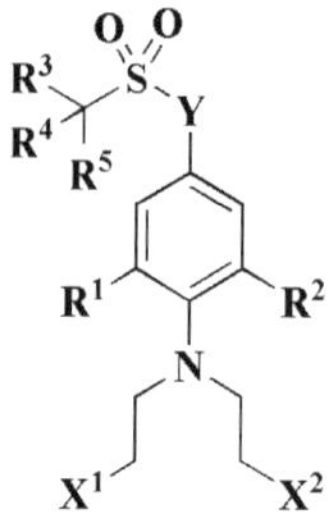

50

Fig. 8.53 Compound **50**; wherein X^1 and X^2 are independently I, Cl, Br and sulfonate; Y is NH and O and derivatives; R^1 and R^2 are independently halo, H, and (un)substituted C1–6 alkyl; R^3 and R^4 are independently H and C1–6 alkyl; R^5.

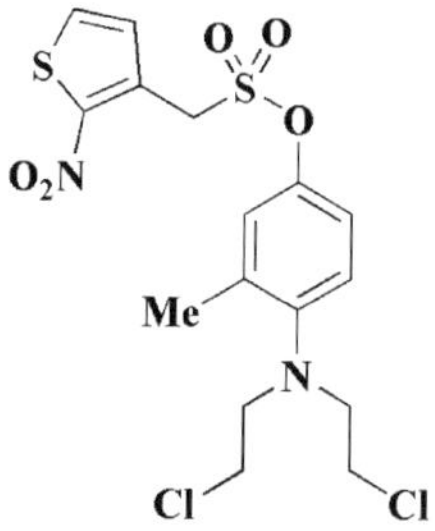

51

Fig. 8.54 Compound **51**.

compounds were estimated for their anticancer potency under normoxia and hypoxia conditions. It was determined that compound **50** reveals that these compounds are more cytotoxic under hypoxia conditions [126].

Substituted dihydropyridine analogs incorporating nitrogen mustard pharmacophore hybrids without spacer and with Et spacer were prepared. They were subjected to in silico ADME prediction data to check their drug-like features and estimated for their cytotoxicity against A 549 (lung), U 87 (glioblastoma), IMR-32 (neuroblastoma), and COLO 205 (colon) human cancer cell lines in vitro utilizing 3-(4,5-dimethylthiazole-2-yl)-2,5-diphenyl tetrazolium bromide (MTT) assay against docetaxel and chlorambucil. The compounds estimated a moderate-to-remarkable cytotoxic potency. The highest potency in all the investigated cancer cells was exhibited by compound **52** (Fig. 8.55). This may be a result of the less steric hindrance offered by compound **52** [127].

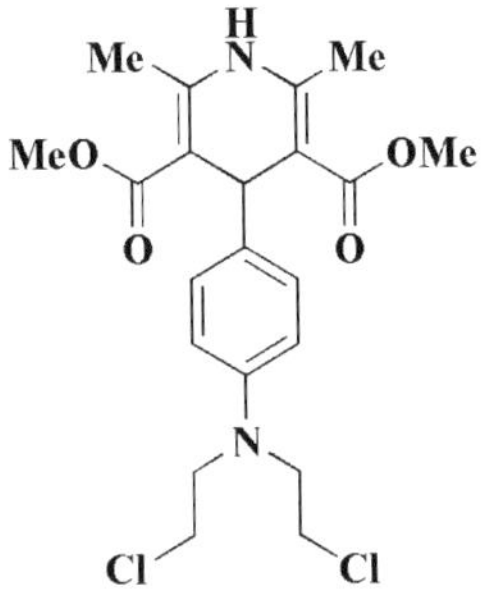

52

Fig. 8.55 Compound **52**.

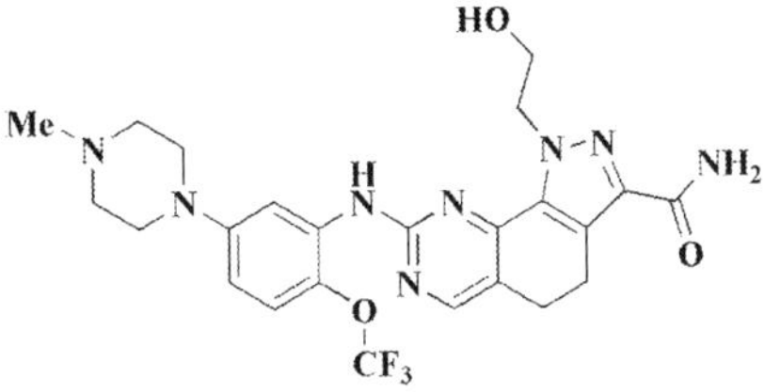

53

Fig. 8.56 Compound **53**.

A combination comprising compound **53** (Fig. 8.56) and one or more antineoplastic agents selected from the group comprising an antimetabolite agent, an alkylating or alkylating-like agent, an intercalating agent, a topoisomerase I or II inhibitor, a proteasome inhibitor, a kinase inhibitor, an antimitotic agent, and an antibody preventing a growth factor or its receptor, wherein active ingredients of the combination exist in each case in free form or in the form of a pharmaceutically acceptable salt or any hydrate or solvate thereof, is useful in treating tumors [128].

The synthesis of new acetoxymethyltriazene **54** (Fig. 8.57) designed to be a prodrug of multiple inhibitors of the epidermal growth factor receptor (EGFR) and a methyldiazonium species is depicted. Investigations with each of the expected metabolites revealed remarkable EGFR tyrosine kinase inhibitory potencies, and the released methyldiazonium was trapped with p-nitrobenzylpyridine. The ability of expected metabolites to damage genomic DNA in whole cells was accomplished [129].

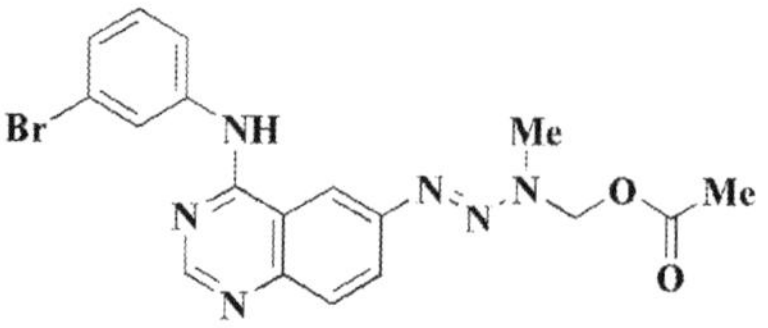

54

Fig. 8.57 Acetoxymethyltriazene derivative **54**.

A compound that comprises a proactive alkylating moiety-bearing bioreductive fragment that contains two double bonds, an electron-withdrawing group, and a linker joining the proactive alkylating moiety and the bioreductive scaffold, such that reduction of the bioreductive scaffold breaks the bond between the electron-withdrawing group and the linker and transforms the proactive alkylating agent into an active alkylating compound, is reported. Thus, 3-bis(2′-chloroethyl)amino-4-methoxybenzoic acid and 2-chloromethyl-3,5,6-trimethylbenzoquinone were refluxed with potassium carbonate and NaI in acetone to afford compound **55** (Fig. 8.58) [130].

Bis(hydroxymethyl) and corresponding bis(carbamate) of 8*H*-3aazacyclopenta[*a*]indene-1-yl and 5,10-dihydropyrrolo-[1,2-*b*]isoquinoline derivatives **56** and **57** were synthesized as DNA dialkylating agents to treat cancer (Figs. 8.59 and 8.60).

Compound **58** was prepared in a multistep synthesis that culminated in the reduction of the corresponding di-Me ester compound. Compound **58** had IC_{50} values of 0.076, 0.357, and 0.332, 0.357, 0.076 μM against HCT-116, MX-1, and human lymphoblastic leukemia cells. Against vinblastine-resistant and taxol-resistant leukemia cells, compound **58** had IC_{50} values of 0.0297 and 0.0663 μM, revealing no cross-resistance [131] (Fig. 8.61).

A pharmaceutical combination that contains $HO_2CC(R^1)(R^2)XC(Y)(Z)CO_2H$ (R^1, R^2=H, F, C1–6 alkyl; X=CR^1R^2, NR^3, O; R^3=H,

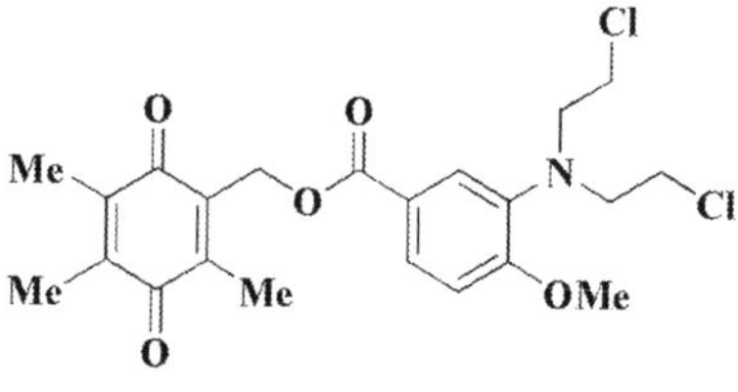

55

Fig. 8.58 Compound **55**.

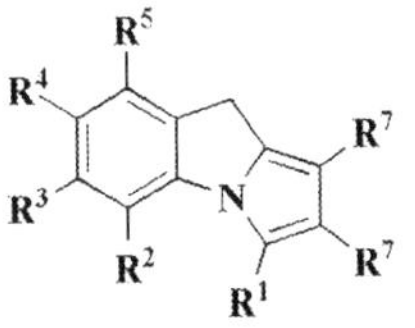

56

Fig. 8.59 Compound **56**; R^1 is C1-C5 alkyl, aryl, H, etc.; R^5, R^4, R^3, and R^2 are independently C1–C6 alkyl, H, halo, etc.; R^7 is $CH_2OCONHR^6$ or CH_2OH; R^6 is C1–C5 alkyl group or benzyl or (un)substituted Ph).

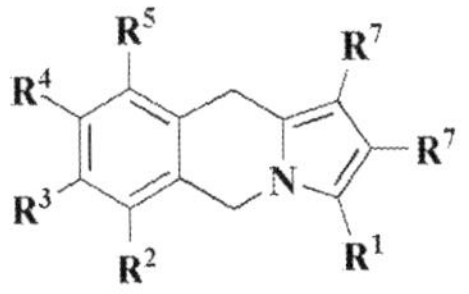

57

Fig. 8.60 Compound **57**; R^1 is C1-C5 alkyl, aryl, H, etc.; R^5, R^4, R^3, and R^2 are independently C1-C6 alkyl, H, halo, etc.; R^7 is $CH_2OCONHR^6$ or CH_2OH; R^6 is C1-C5 alkyl group or benzyl or (un)substituted Ph).

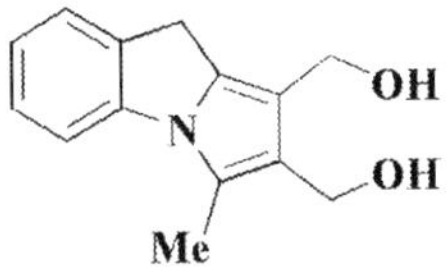

58

Fig. 8.61 Compound **58**.

C1–6 alkyl; Y, Z = H, NR^3, OH, SH, or Y and Z together are =O, =S), or compound **59** (Fig. 8.62), and an anticancer alkylating agent is reported. The combination is useful in treating cancer [132].

A method of treating cancer using a therapeutically effective amount of an alkylating agent as a substituted anthracycline compound and temozolomide is invented (Fig. 8.63) [133].

Pyrrole-imidazole polyamides may be screened for such biological potencies as accurate recognition of DNA sequences for anticancer potency for targeting DNA sequences specific to cancer cells. The *N*-[[4-[[(1-methy-l-4-acetamido-2-imidazolyl)carbonyl]amino]-1-methyl-2-pyrrolyl]

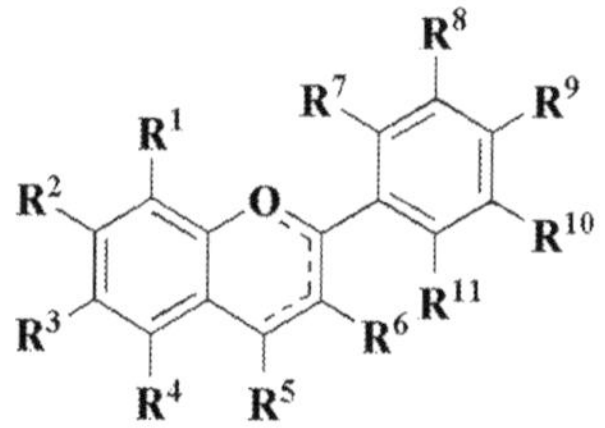

59

Fig. 8.62 Compound **59**; [R1-R11 = H, OH, (un)branched C1–6 alkyl, etc.; Q1 = Q2, Q3, Q4].

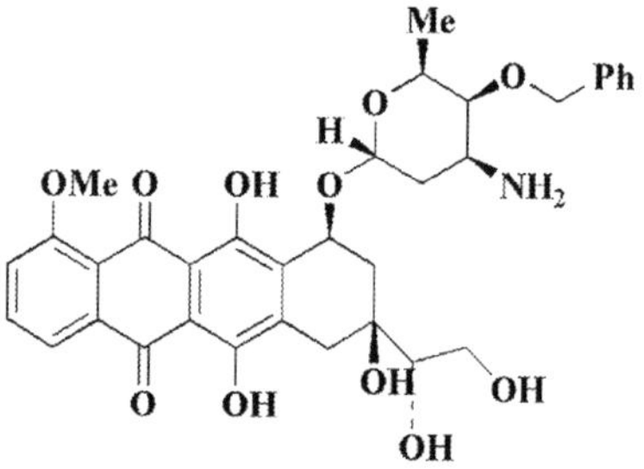

60

Fig. 8.63 Compound **60**.

carbonyl]-β-alanine (**61**; R = OH) (Fig. 8.64) was prepared utilizing Fmoc-β-alanine bound on a Wang resin, 4-[(9-fluorenylmethoxycarbonyl)amino]-1-methyl-2-pyrrole-2-carboxylic acid, and 4-[(9-fluorenylmethoxycarbonyl)amino]-1-methyl-2-imidazole-2-carboxylic acid, transferred into an imidazole ester **61** (R = imidazol-1-yl), and condensed with the segment A of Du-86 to afford **61** (R = Q). **61** (R = Q) in vitro renowned and alkylated 5′-TAAA-3′ of 5’-CTATAAAGA-3′ [134].

The synthesis of alkyl 2,3-diamino (e.g., **62**, R = R^1) and alkyl 2-amino-3-thioglucopyranosides (e.g., **62**, R = SBu-t) (Fig. 8.65) from alkyl 2-nitrosugar derivatives and amines or thiols is reported. The reaction occurs via 2-nitroalkene and with high stereoselectivity. The bonding of alkyl 2,3 diaminoglucopyranosides to chlorambucil through a spacer arm and an ester function is reported [135].

Many inventions revealed antitumor sustained-release injection containing alkylating agents as a component from the pharmaceutical components [136–148]. It has also many therapeutic applications [149–214].

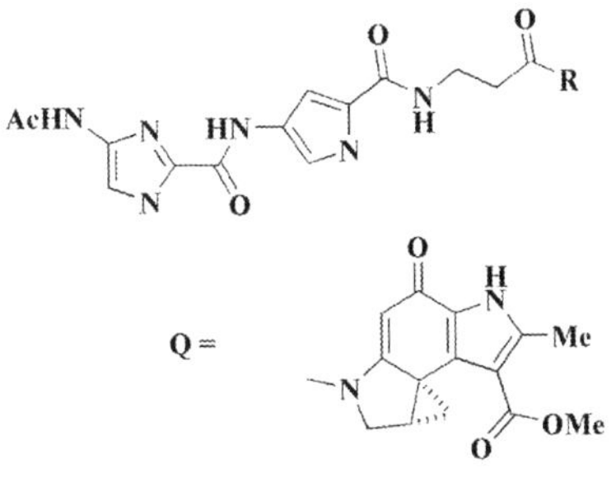

Fig. 8.64 Compound **61**.

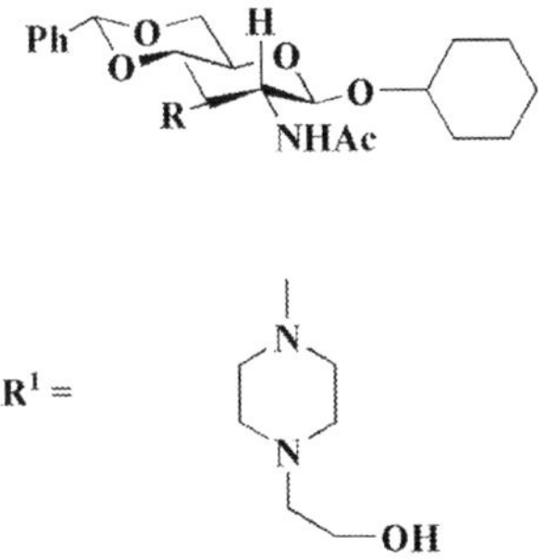

Fig. 8.65 Compound **62**.

12. Synthetic strategies and biological evaluations

12.1 Synthetic strategies of chlorambucil analogs

Amidine derivatives of chlorambucil **65**, where the chlorambucil moiety and 5-[4-(N-alkylamidino)phenyl]-2-furancarboxamide are attached by a $NH(CH_2)_2NH$ chain, were prepared (Scheme 8.1), and their cytotoxicity has been estimated against the growth of human breast cancer MCF-7 cells. The cytotoxicity of compound **65** was estimated using a MTT assay and inhibition of [3H]thymidine incorporation into DNA established that these conjugates were more potent than chlorambucil. These compounds can bind in the minor groove of DNA and indicate a moderate specificity for AT base pairs, and this was indicated using the ethidium displacement assay.

Compounds **65** were active topoisomerase II inhibitors ($IC_{50} = 10$ to $40\,\mu M$). The cytotoxicity of the compounds **65** correlates with their

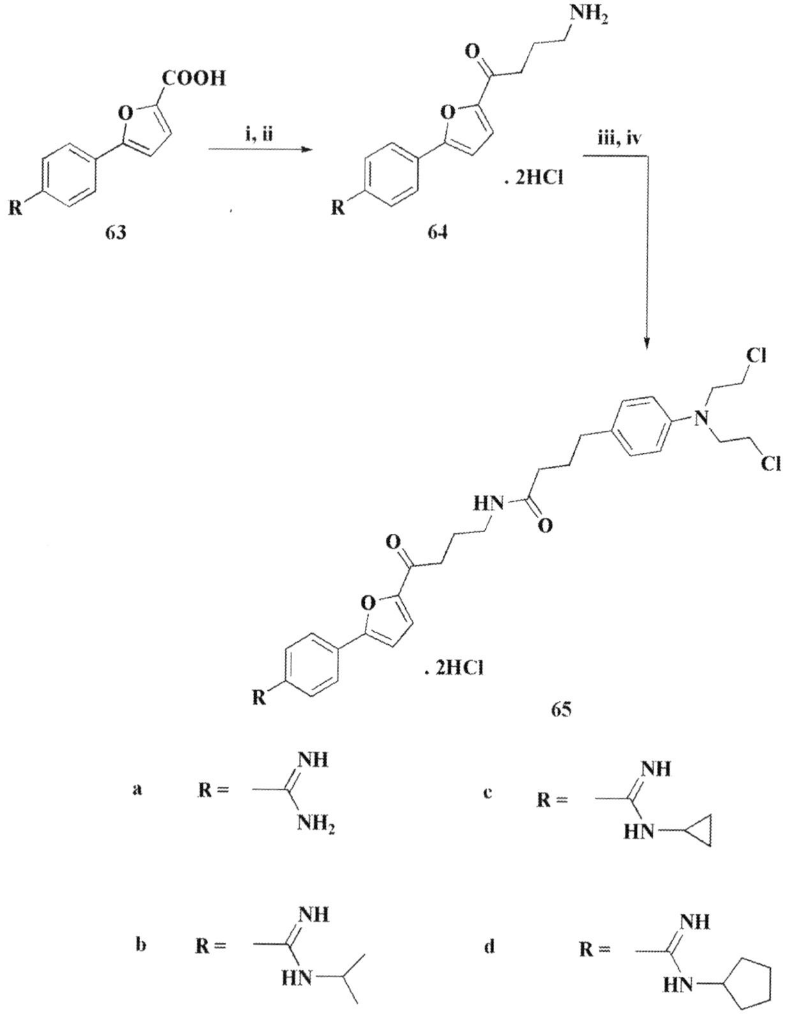

Scheme 8.1 Reagents and conditions: (i) $H_2NCH_2CH_2NH_2$/DCI. (ii) HCl. (iii) Chlorambucil/$N(C_2H_5)_3$/DCC. (iv) HCl.

DNA-binding affinities and their relative activity as topoisomerase II inhibitors. It was suggested from those data that the cytotoxic potency of compounds **65** may be because of the combined impacts of alkylation, DNA-minor groove binding, and that the substituted furan carboxamides **64** ligands are linkers that favor DNA targeting via the chlorambucil analogs [215].

Cyclic amidine derivatives of chlorambucil were prepared (Scheme 8.2) and investigated for cytotoxicity in breast cancer cell cultures and for the inhibition of topoisomerases I and II. Estimation of the cytotoxicity of these compounds utilizing a 3-(4,5-dimethylthiazol-2-yl)-

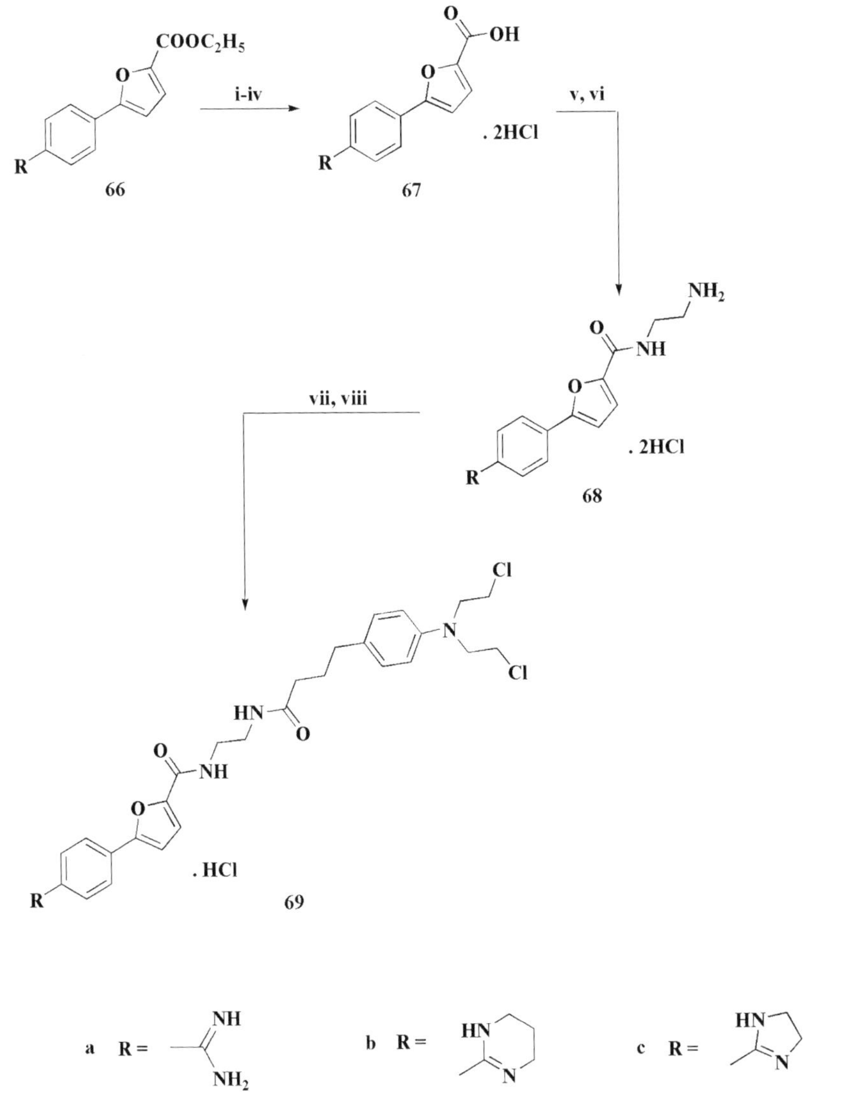

Scheme 8.2 Reagents and conditions: (i) C_2H_5OH/HCl. (ii) $H_2N(CH_2)n\text{-}NH_2$ or NH_3. (iii) LiOH $n=2,3$. (iv) HCl. (v) $H_2NCH_2CH_2NH_2$/DCl. (vi) HCl. (vii) Chlorambucil/$N(C_2H_5)_3$/DCC. (viii) HCl.

2,5-diphenyltetrazolium bromide assay and inhibition of [3*H*]-thymidine incorporation into DNA in both MCF-7 and MDA-MB-231 breast cancer cells demonstrated that these compounds were more potent than chlorambucil. Cyclic amidine analogs of chlorambucil are an active catalytic inhibitor of topoisomerase II but not topoisomerase I. The cytotoxicity and the highest degree of DNA binding in MCF-7 and MDA-MB-231 breast cancer cells were found for the compound, which has a 4,5-dihydro-1*H*-imidazol moiety [216].

Various tyrosine-chlorambucil conjugates have been synthesized. Thus, para-, meta-, and ortho-tyrosine-chlorambucil derivatives were prepared to afford novel anticancer drugs.

These novel derivatives were furnished by the subsequent effective synthetic procedure. All the chlorambucil-tyrosine hybrids were more efficient than chlorambucil. In vitro biological estimation on estrogen receptor-negative and estrogen receptor-positive breast cancer cell lines showed an improved cytotoxic potency for compounds with the phenol function present at the position meta. The preparation of all tyrosine–chlorambucil hybrid regioisomers was accomplished in Scheme 8.3 [217].

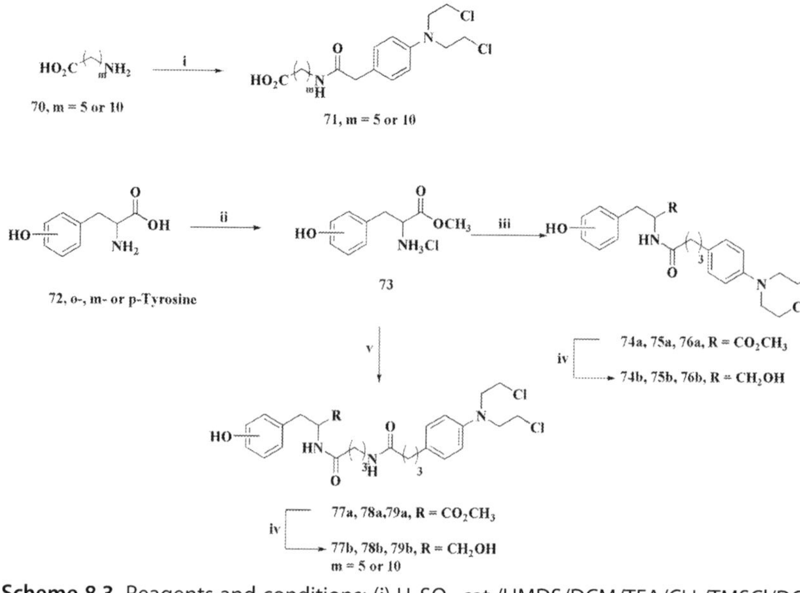

70, m = 5 or 10

71, m = 5 or 10

72, o-, m- or p-Tyrosine

73

74a, 75a, 76a, R = CO_2CH_3

74b, 75b, 76b, R = CH_2OH

77a, 78a, 79a, R = CO_2CH_3

77b, 78b, 79b, R = CH_2OH

m = 5 or 10

Scheme 8.3 Reagents and conditions: (i) H_2SO_4 cat./HMDS/DCM/TEA/CLL/TMSCl/DCC/HOBt/TEA/DMF. (ii) $CH_3OH/SOCl_2$. (iii) HOBt/CLL/DCC/TEA/DMF. (iv) $Et_2O/LiBH_4$/0°C to 22°C. (v) **71**, *m* = 5 or 10/DCC/HOBt/TEA/DMF.

D- and L-Tyrosine-chlorambucil derivatives were prepared as chemotherapeutic agents for treating breast cancer (Schemes 8.4-8.6). The compounds were generated via the efficient modifications of L- and D-tyrosine. The prepared compounds were estimated for their anticancer effectiveness in several hormone-independent and hormone-dependent (ER+ and ER−) breast cancer cell lines. The new derivatives indicated a remarkable in vitro anticancer potency as compared to chlorambucil. The impact of the stereochemistry of the tyrosine scaffold and the length of the spacer chain was revealed through structure–activity relationship

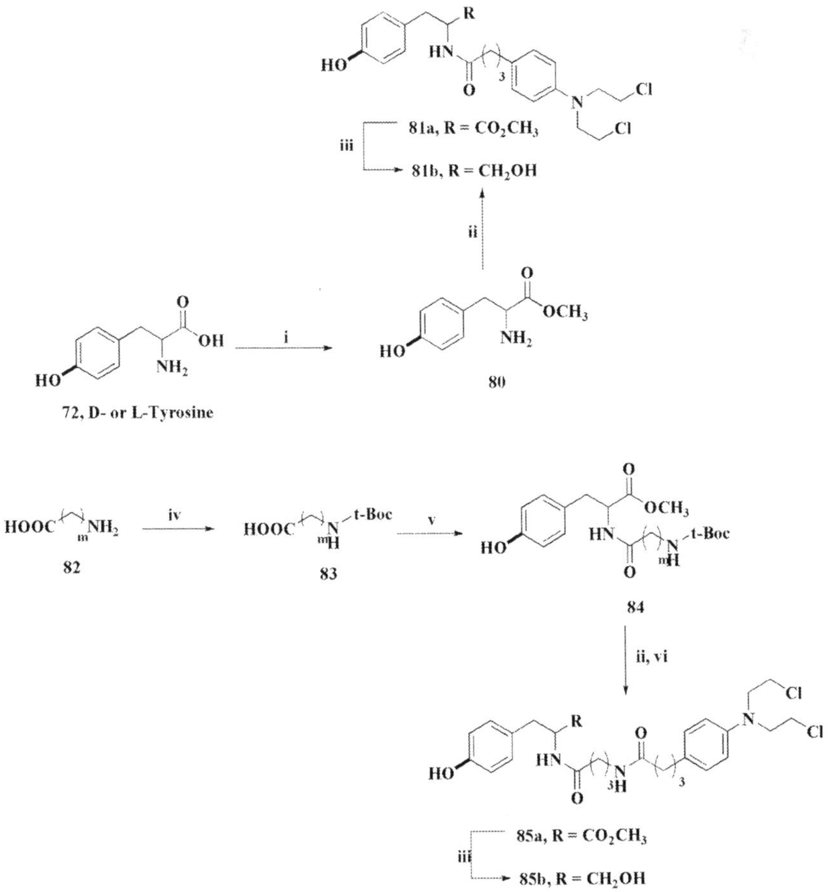

Scheme 8.4 Reagents and conditions: (i) $SOCl_2$/CH_3OH/reflux/3.5 h. (ii) chlorambucil/DCC/HOBt/TEA/DMF/22°C. (iii) Dry Et_2O/$LiBH_4$ excess/22°C. (iv) Boc-ON/H_2O/1,4-dioxane/TEA/22°C. (v) 6/DCC/HOBt/TEA/DMF/22°C. (vi) DCM/excess TFA/22°C/10 min.

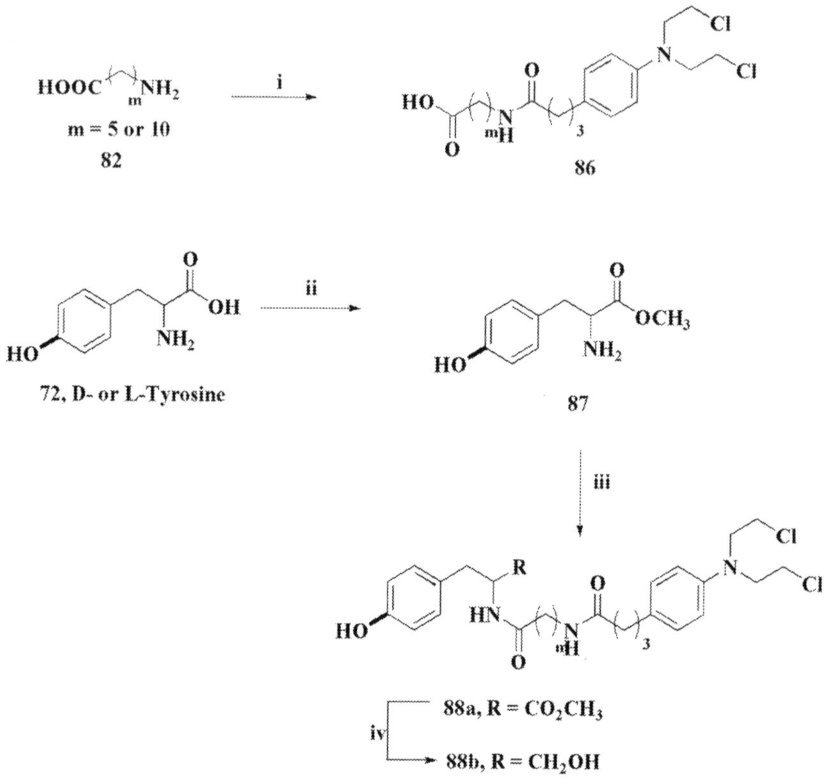

Scheme 8.5 Reagents and conditions: (i) H_2SO_4 cat./HMDS/DCM/TEA/CHL/TMSCl/DCC/HOBt/TEA/DMF. (ii) $SOCl_2/CH_3OH$/reflux/3.5 h. (iii) Chlorambucil-$NH(CH_2)mCO_2H/m=5$ or 10/DCC/HOBt/TEA/DMF/22°C. (iv) $Et_2O/LiBH_4$ excess/22°C.

(SAR). The L- and D-tyrosinol-chlorambucil analogs with ten carbon atoms spacer are selective toward MCF-7 (ER+) breast cancer cell line [218].

Novel carboxylic derivatives of netropsin and distamycin with chlorambucil scaffolds **94–97** have been furnished (Scheme 8.7). At the minor groove of DNA, these compounds can bind as indicated from the ethidium displacement assay. The increased affinity toward GC sequences of the carbocyclic lexitropsins with chlorambucil fragment **94–97** and the reduced affinity to AT pairs as compared with distamycin and netropsin were found and rationalized through molecular modeling techniques. In the standard cell line of the mammalian tumor MCF-7, compounds **94–97** indicated cytotoxic and antiproliferative impacts [219].

Perfluorinated CLB analogs **102–104** and compounds **98–101** were prepared (Scheme 8.8). An ester link with either a fluorinated alcohol

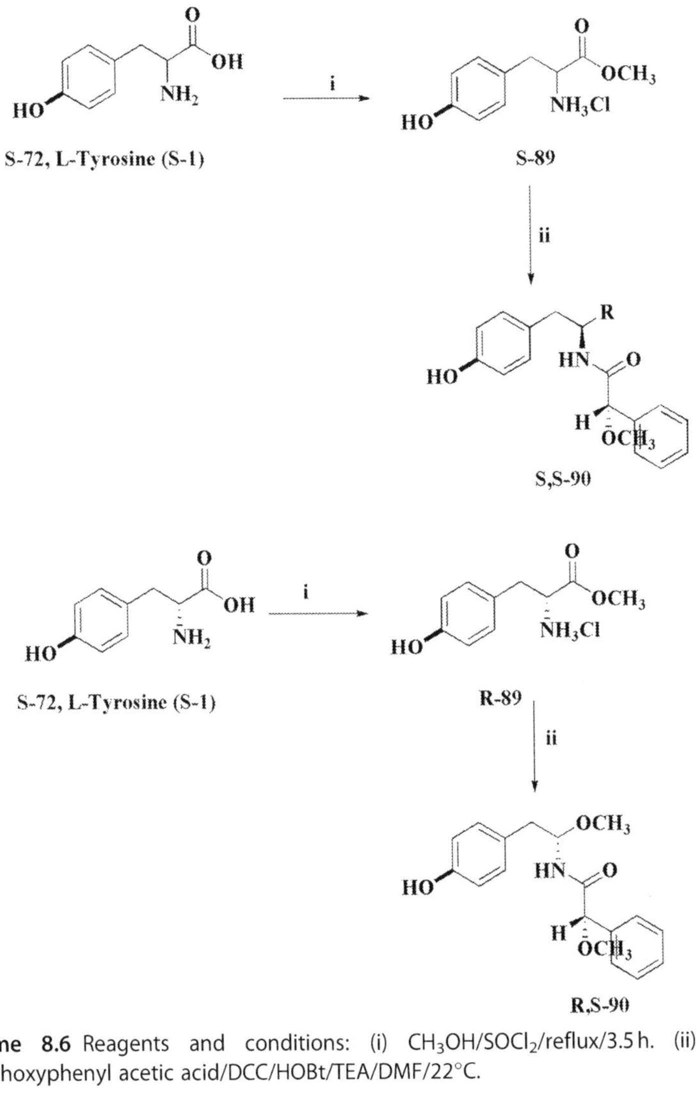

S-72, L-Tyrosine (S-1) → (i) → S-89 → (ii) → S,S-90

S-72, L-Tyrosine (S-1) → (i) → R-89 → (ii) → R,S-90

Scheme 8.6 Reagents and conditions: (i) $CH_3OH/SOCl_2$/reflux/3.5 h. (ii) *S*-(+)-a-methoxyphenyl acetic acid/DCC/HOBt/TEA/DMF/22°C.

or its hydrocarbon equivalent was coupled with CLB utilizing 4-dimethylaminopyridine (DMAP) and N,N′-dicyclohexylcarbodiimide (DCC) as accomplished in scheme. CLB derivatization with hydrophobic chains makes compounds **102–104** and **101** less soluble than CLB in water, with the solubility of compounds **104** and **101**. The solubility is affected greatly with temperature. The perfluorinated derivatives are the most lipophilic. The anticancer potency of CLB and its analogs was initially estimated

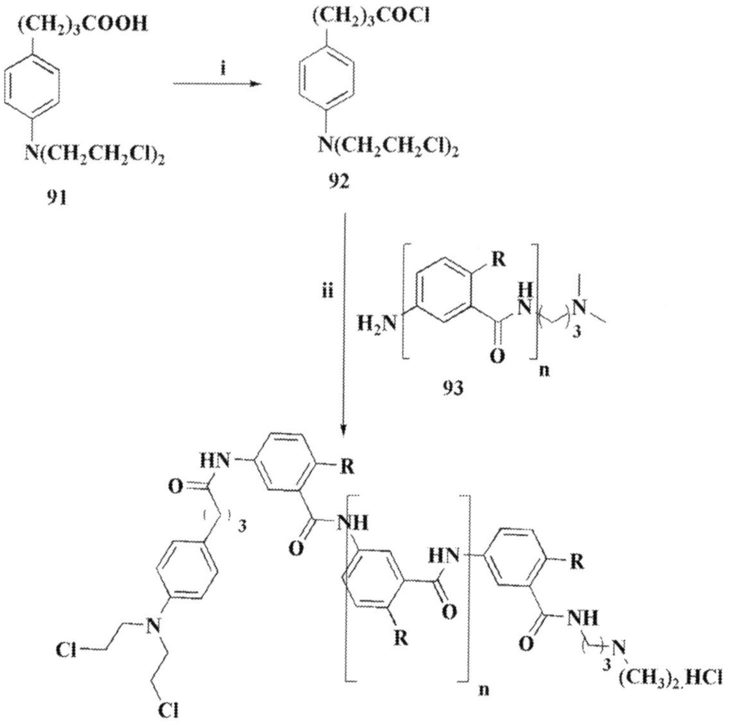

94; n = 1, R = H
95; n = 2, R = H
96; n = 1, R = OCH_3
97; n = 2, R = OCH_3

Scheme 8.7 Reagents and conditions: (i) THF/ethanedioyl dichloride/RT/2 h. (ii) DMAP/pyridine/20 h.

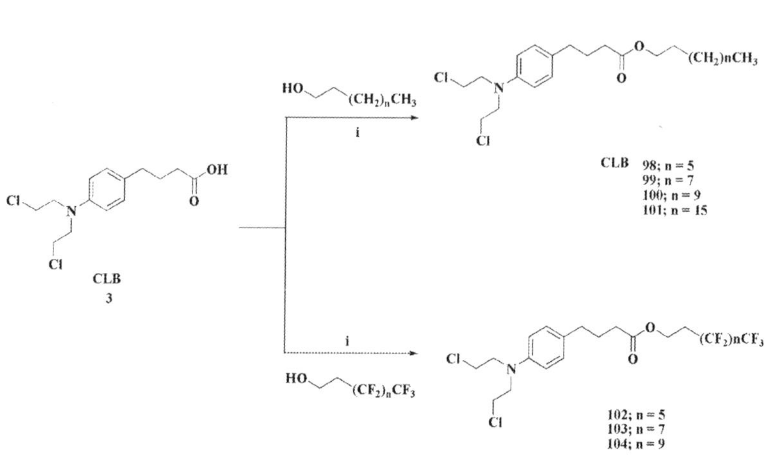

Scheme 8.8 Reagents and conditions: (i) DMAP/DCC.

in two human ovarian carcinoma cell lines (A2780 and its cisplatin-resistant variant A2780cisR). IC_{50} values were furnished on cells treated with the compounds and also under mild hyperthermia. To target tumor tissue, the anticancer drug modified with a perfluorinated chain can overstrain thermoresponsive behavior. The concept utilizes hyperthermia to trigger drug cytotoxicity inside the heated tumor cells. The fluorinated CLB analog **104** has revealed a thermoactive behavior in the tested cell lines, and, under mild hyperthermia, it is as cytotoxic as the parent drug, but does not indicate a remarkable toxicity [220].

Chlorambucil (CHL) analogs have been synthesized (Schemes 8.9 and 8.10), and their effects were estimated on leukemias L1210 and P388 in vivo and in normal human lymphocytes in vitro. Some compounds produced a submultiple toxicity, while the measured antileukemic activity was remarkably increased. The lactamization of the B-steroidal ring varnished the molecules more active; however, the corresponding 7-oxidized analogs established better in both leukemias examined. Regardless of the

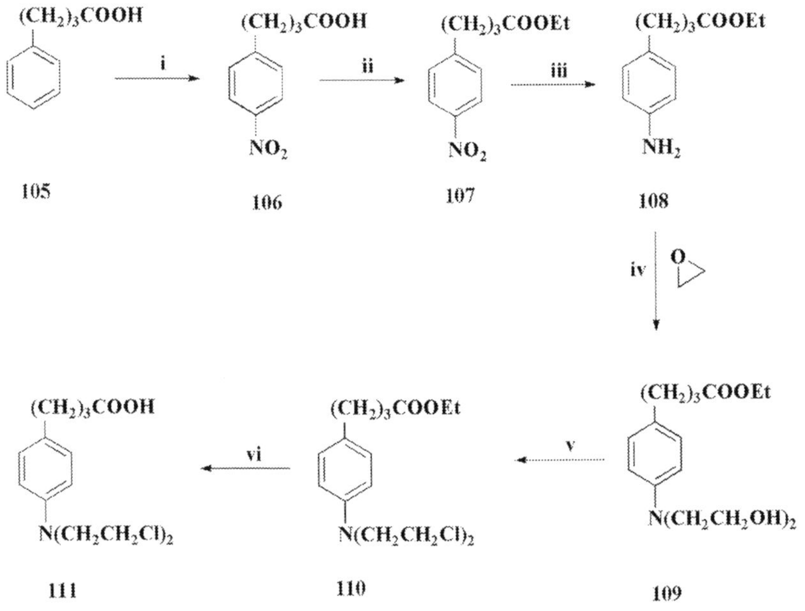

Scheme 8.9 Reagents and conditions: (i) CH_2Cl_2/ab.HNO_3. (ii) concentrated H_2SO_4/Ab. EtOH. (iii) H_2/Pd/C, CH_3OH. (iv) CH_3COOH. (v) $POCl_3$. (vi) concentrated HCl.

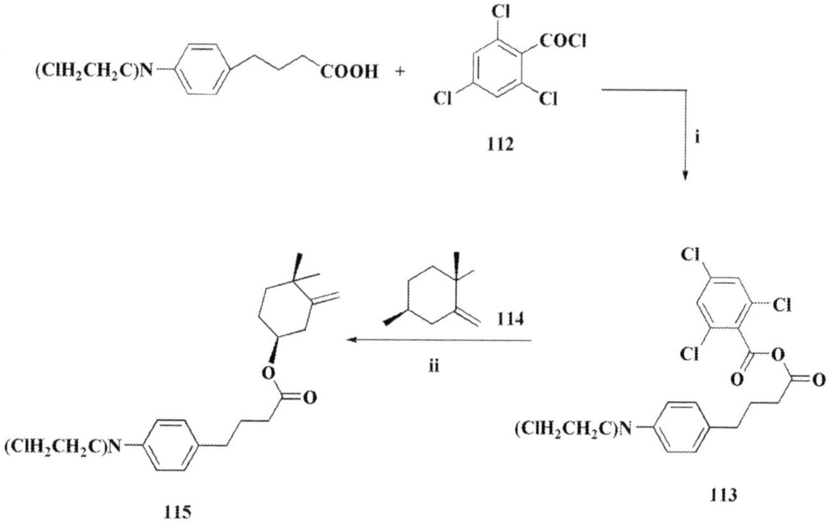

(ClH₂CH₂C)N–C₆H₄–(CH₂)₃–COOH + 112 →(i) 113 →(ii, 114) 115

Scheme 8.10 Reagents and conditions: (i) Benzene/NEt_3. (ii) benzene/4-DMAP.

configuration of the B-ring, the lactamization of the D-steroidal ring yielded active compounds [221].

The prodrug ether phospholipid conjugates were synthesized (Schemes 8.11–8.14). The lipids are synthesized from chlorambucil and possess C18 and C16 ether chains with phosphatidylglycerol or phosphatidylcholine head groups. All the prodrugs possess the ability to afford unilamellar liposomes and are hydrolyzed by phospholipase A2, leading to the release of chlorambucil. Liposomal formulations of prodrug lipids exhibited cytotoxicity toward MT-3, ES-2, and HT-29 cancer cell lines in the existence of phospholipase A2 (IC_{50} = 8–36 μM) [222].

A chlorambucil analog has been radiolabeled with [99mTc$(CO)_3$ $(H_2O)_3$] + core (Scheme 8.15), and its effectiveness as a tumor-targeting

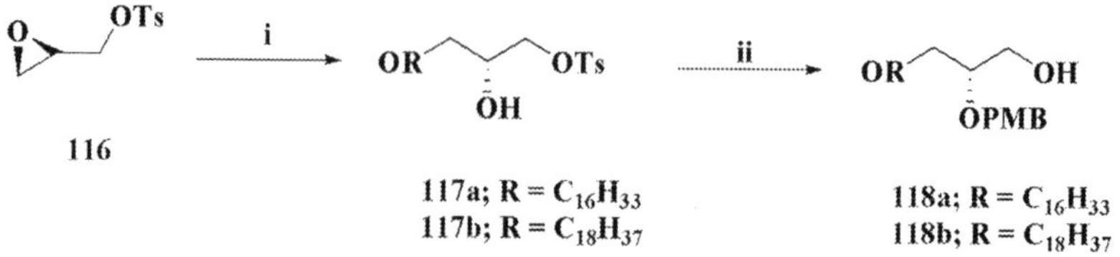

116 →(i) 117a; R = $C_{16}H_{33}$, 117b; R = $C_{18}H_{37}$ →(ii) 118a; R = $C_{16}H_{33}$, 118b; R = $C_{18}H_{37}$

Scheme 8.11 Reagents and conditions: (i) $C_{18}H_{37}OH$, $BF_3{\cdot}OEt_2$ or $C_{16}H_{33}OH/CH_2Cl_2$. (ii) PMBTCA/La(OTf)$_3$/toluene/CsOAc/DMF/DMSO/NaOMe/CH_3OH.

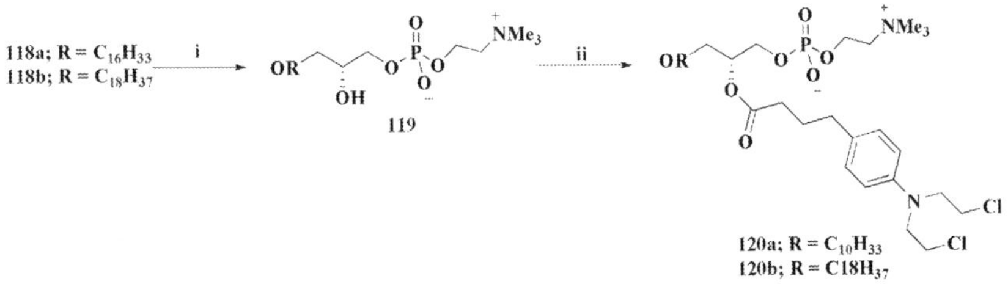

Scheme 8.12 Reagents and conditions: (i) $POCl_3$/TEA/CH_2Cl_2/pyridine/choline tosylate/DDQ/H_2O/CH_2Cl_2/H_2O. (ii) **3**/DMAP/DCC/1,2-dichloroethane or $CHCl_3$.

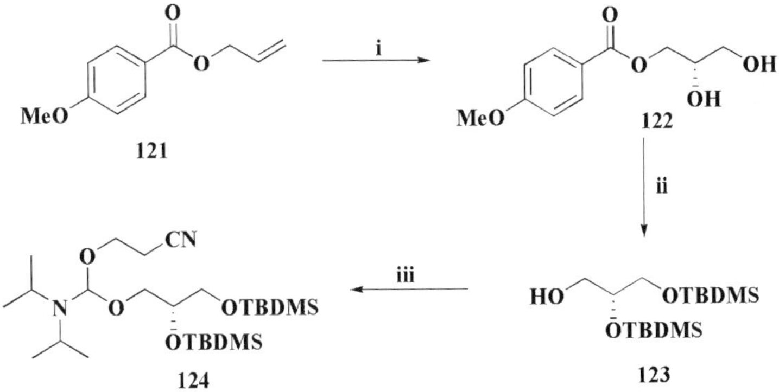

Scheme 8.13 Reagents and conditions: (i) $K_2OsO_4{\cdot}2H_2O$/$K_3Fe(CN)_6$/$(DHQD)_2$PHAL/K_2CO_3/*t*BuOH/H_2O. (ii) DIPEA/TBDMSOTf/CH_2Cl_2/DIBAL-H/CH_2Cl_2. (iii) CH_2Cl_2/DIPEA/2-cyanoethyl *N*,*N*-diisopropylchlorophosphoramidite.

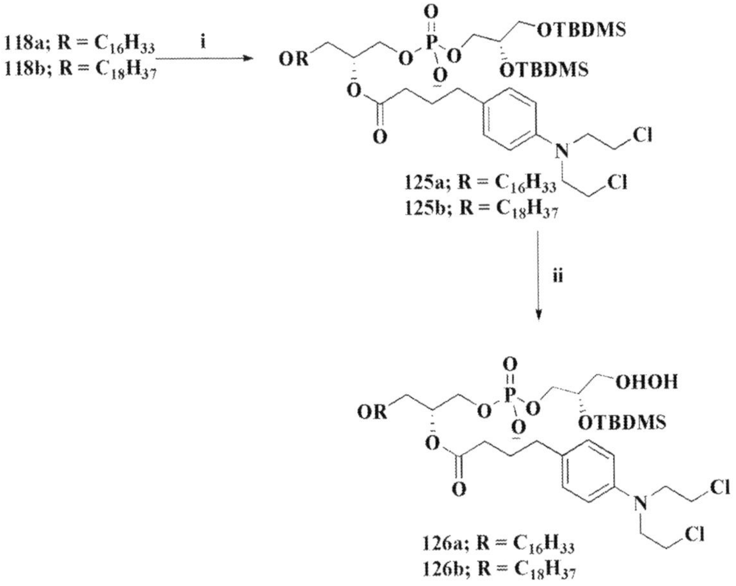

Scheme 8.14 Reagents and conditions: (i) **11**/tetraazacyclopentadiene/CH_2Cl_2/MeCN/*t*BuOOH./CH_2Cl_2/DDQ/H_2O/**3**/DMAP/EDCI/CH_2Cl_2/DBU/CH_2Cl_2. (ii) MeCN/HF/H_2O.

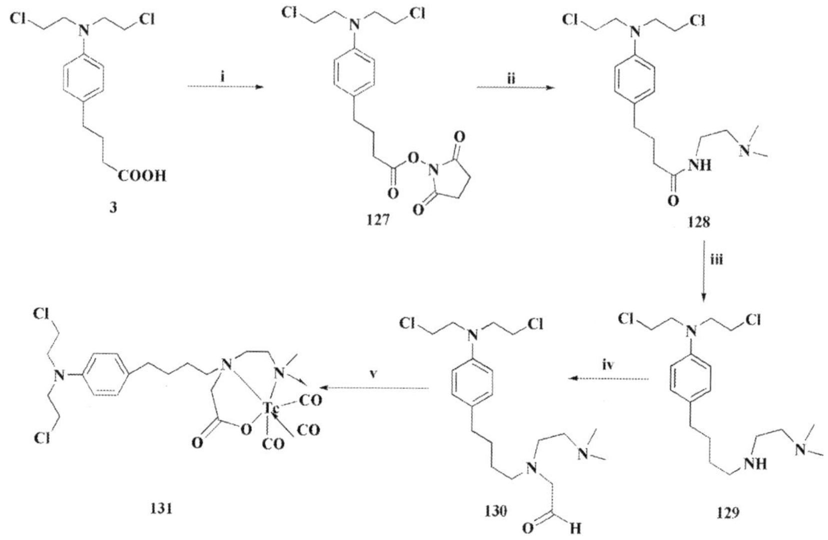

Scheme 8.15 Reagents and conditions: (i) TEA/NHS/ethyl acetate. (ii) TEA/N^1,N^1-dimethylethane-1,2-diamine/ethylacetate. (iii) LiIH$_4$/THF. (iv) $CH_3OH/BrCH_2COOH$. (v) $[99mTc(CO)_3(H_2O)_3]^+$.

agent has been investigated. The in vitro data in MCF-7 breast cancer cells indicated the inhibition of the radiolabeled complex in the presence of cold chlorambucil analog by 30% [223].

Chlorambucil amide derivatives involving chiral glycosyl glycerols derived from D-glucosamine were prepared via coupling chlorambucil scaffold to the amino group of □-amino-(□-1)-hydroxyalkyl 2-aylamino-4,6-*O*-benzylidene-2-deoxy-β-D-glucopyranosides, and further hydrolysis of the benzylidene group. From 2-acetamido-2-deoxy-D-glycose, the starting component was generated. The bonding of chlorambucil and 2,3,4,6-tetra-*O*-pivaloyl-β-D-glactopyranosylamine via the generation of an amide function is reported (Schemes 8.16–8.18) [224].

The protected intermediate sugars comprising the starting material to which chlorambucil would be bonded as amide or ester were synthesized from 2-acetamido-2-deoxy-D-glucose for the allo and gluco analogs, and from methyl α-D-glucopyranoside for the altro derivatives [225]. The preparation of compound **152** and compound **153** was carried out starting from compound **150** [226–229]. Compound **152** and its derivatives **154** and **155** [227] are the key intermediates for the preparation of the compounds in which chlorambucil bonds directly to C-3 of the sugar moiety through an ester function (Scheme 8.19). Removal of the benzylidene group in

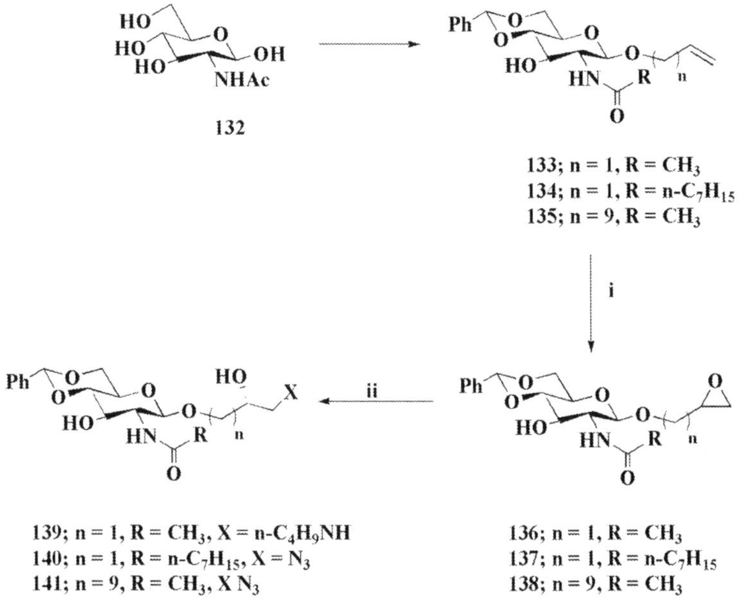

132
133; n = 1, R = CH3
134; n = 1, R = n-C7H15
135; n = 9, R = CH3
i
ii
136; n = 1, R = CH3
137; n = 1, R = n-C7H15
138; n = 9, R = CH3
139; n = 1, R = CH3, X = n-C4H9NH
140; n = 1, R = n-C7H15, X = N3
141; n = 9, R = CH3, X N3

Scheme 8.16 Reagents and conditions: (i) $CHCl_3$/M-CPBA. (ii) N$=$$NaN_3$ or $BuNH_2$/$LiClO_4$/CH_3CN.

139
140, 141
i
ii, iii
iv
142-144
145-147
142 & 145; n =1, R = CH3, R' = n-C4H9
143& 146; n = 1, R = n-C7H15, R' = H
144& 147; n = 9, R = CH3, R' = H

Scheme 8.17 Reagents and conditions: (i) THF/chlorambucil pentachlorophenol ester. (ii) H_2/Pd(C)/CH_3OH. (iii) Chlorambucil/DCC/DMAP/dioxane. (iv) HCl/CH_2Cl_2/CH_3OH.

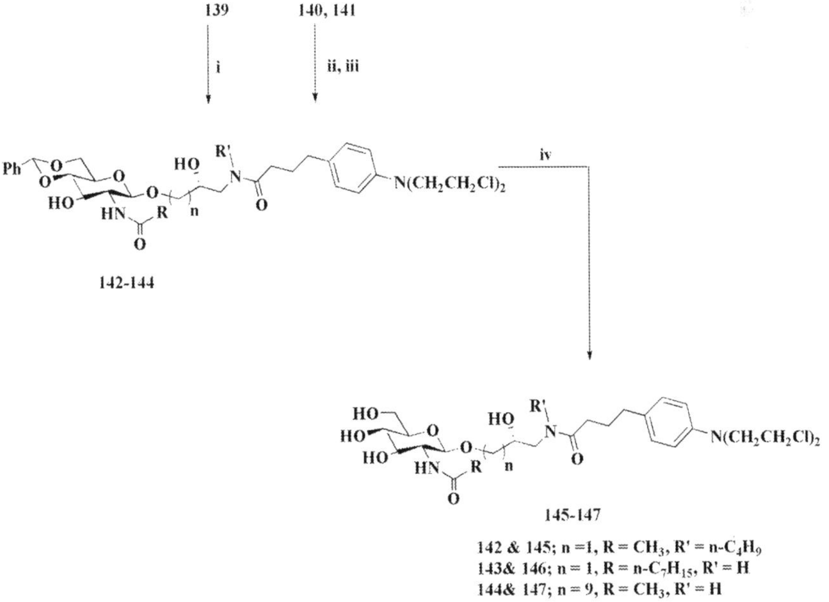

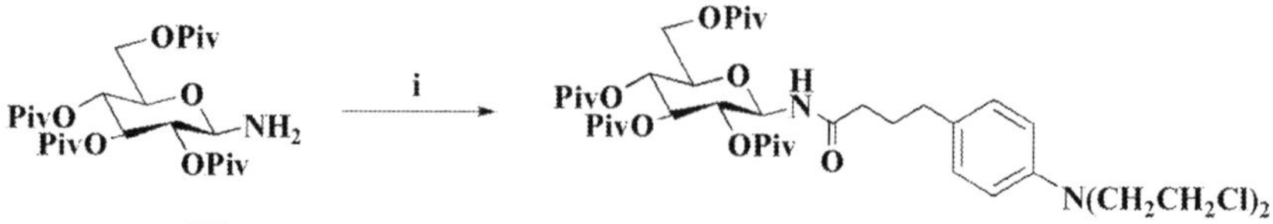

Piv = $(CH_3)_3CCO$

Scheme 8.18 Reagents and conditions: (i) Chlorambucil/DDC/DMAP/CH_2Cl_2.

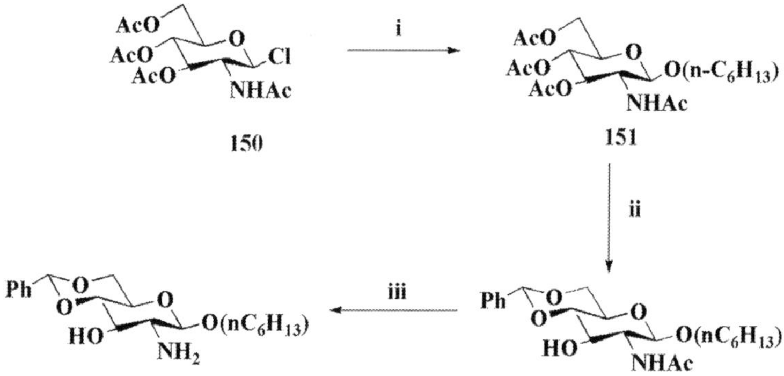

Scheme 8.19 Reagents and conditions: (i) $Hg(CN)_2$/n-$C_6H_{13}OH$. (ii) NaOMe/CH_3OH/$ZnCl_2$/PhCHO. (iii) EtOH/KOH.

compounds **156–158** affords diols **159–161**. The coupling of chlorambucil at C-2 of the sugar through an amide bond requires the synthesis of compound **153** (Scheme 8.19) and its *gluco* analogs **162–164** [230], as well as the *allo* analoe **165** [227] (Scheme 8.20). The reaction of compounds **153**, **162–165** with chlorambucil, activated by the pentachlorophenyl trichloroacetate procedure [231], in tetrahydrofuran affords the corresponding amides **166–170** (Scheme 8.21). Hydrolysis of the benzylidene group yields compounds **171–175** [225].

12.2 Synthetic strategies of melphalan analogs

New spin-labeled melphalan and chlorambucil compounds were synthesized through coupling 2,2,6,6-tetramethyl-1-piperidinyloxy (TEMPO) radicals with the alkylating agents (Schemes 8.22 and 8.23). The compounds were furnished via EDC ·HCl coupling reaction, in which the amino or

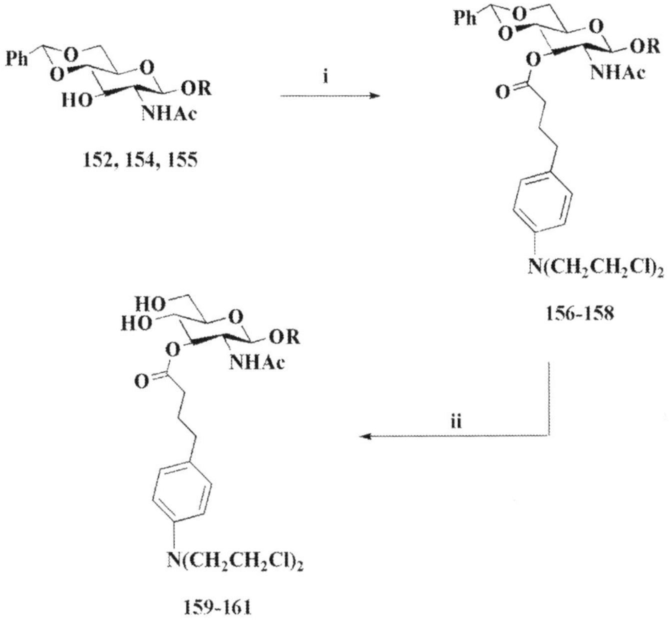

$R = n\text{-}C_6H_{13}, c\text{-}C_6H_{11}, n\text{-}C_8H_{17}$

Scheme 8.20 Reagents and conditions: (i) Chlorambucil/DCC/DMAP/THF. (ii) CH_2Cl_2/HCl/CH_3OH.

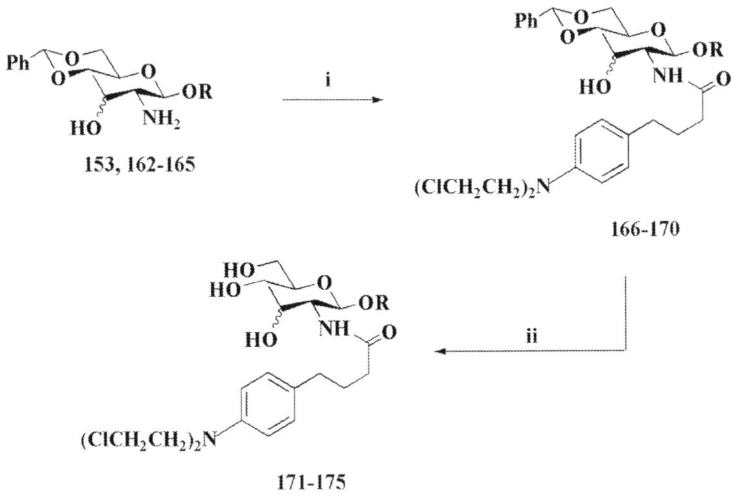

$R = n\text{-}C_6H_{13}$ gluco, o-C_6H_{11} gluco, n-C_8H_{17} glyco,
n-$C_{12}H_{25}$ gluco, n-$C_{12}H_{25}$ allo

Scheme 8.21 Reagents and conditions: (i) THF/Chlorambucil pentachlorophenol ester. (ii) CH_2Cl_2/HCL/CH_3OH.

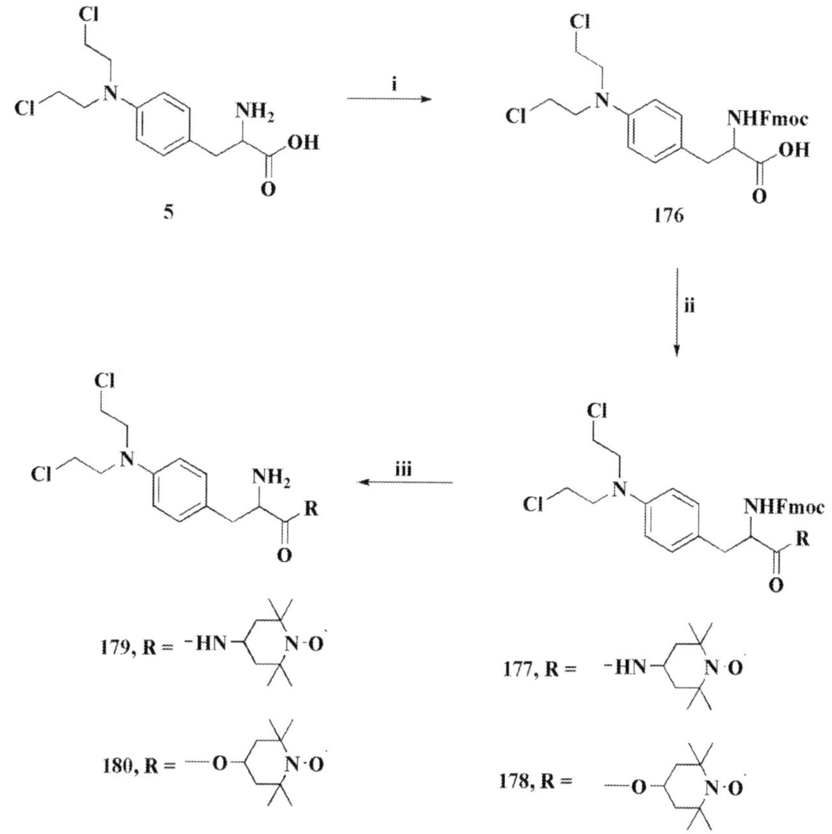

Scheme 8.22 Reagents and conditions: (i) Fmoc-Cl/$NaHCO_3$ solution/dioxane/0°C/8 h. (ii) EDC ·HCl/4-NH_2-TEMPO/HOBT/CH_2Cl_2/8 h/DMAP/EDC·HCl/4-OH-TEMPO/CH_2Cl_2/8 h. (iii) DMF/piperidine/RT/1 h.

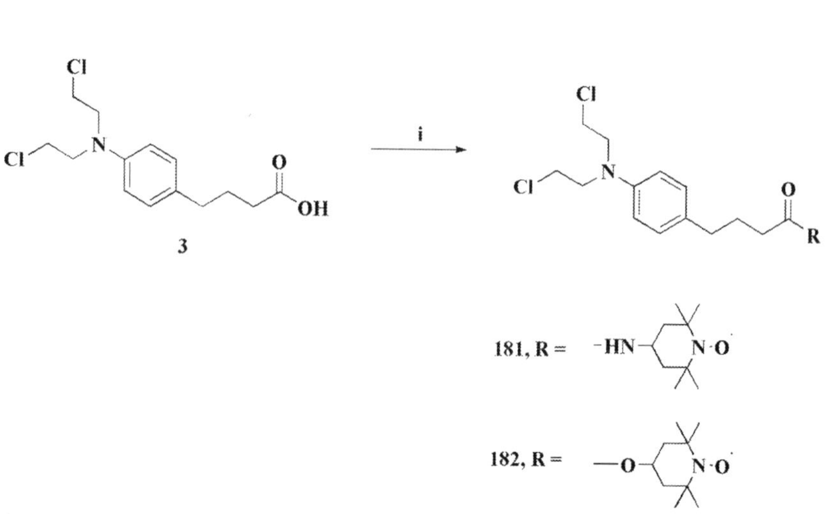

Scheme 8.23 Reagents and conditions: (i) HOBT/EDC·HCl/CH_2Cl_2/4-NH_2-TEMPO/8 h/EDC·HCl/4-OH-TEMPO/DMAP/CH_2Cl_2/8 h.

hydroxyl group of 4-hydroxyl-2,2,6,6-tetramethyl-1-piperidinyloxy (4-OH-TEMPO)/4-amino-2,2,6,6-tetramethyl-1-piperidinyloxy (4-NH_2-TEMPO) was conjugated with the carboxylic group of melphalan (**5**)/chlorambucil (**3**). Protection of the amino group of the melphalan using 9-fluorenylmethyl chloroformate (Fmoc-Cl) and then the *N*-(9-fluorenyl-methoxycarbonyl) (*N*-Fmoc) analogs were produced before the coupling reaction. Then, acylation or esterification reaction was followed. The chlorambucil and *N*-protected melphalan reacted with 4-NH_2-TEMPO in 4-OH-TEMPO in EDC·HCl·4-dimethylaminopyridine and EDC·HCl·1-hydroxybenzotriazole. Then, deprotection of the *N*-Fmoc analogs of the melphalan was carried out. After estimating their biological features in vitro, these compounds revealed much higher cytotoxic potency against human leukemia cell line K562 in vitro than their parent compounds [232].

Proline analogs of melphalan were synthesized as accomplished in Scheme 8.24. The D- and L-proline prodrugs of melphalan, designated as prophalan-D and prophalan-L, revealed a seven time higher rate of activation of prophalan-L comparable to that of the prophalan-D in SK-MEL-5 cell homogenates as affirmed by hydrolysis studies. Prophalan-L displayed cytotoxicity (GI_{50} 74.8 ıM) as compared to melphalan (GI_{50} 57.0 ıM) in SK-MEL-5 cells, while prophalan-D was ineffective, proposing that prolidase-specific activation to the parent drug may be fundamental for cytotoxic action. So melphalan prodrugs, as the prophalan-L, that are cleavable by prolidase show the potential for improved selectivity through facilitating cytotoxic potency only in cells overexpressing prolidase [233].

The amidine derivatives of melphalan were synthesized by following the steps illustrated in Scheme 8.25. These compounds are more potent than melphalan, and the evidence for this is the [3*H*] thymidine inhibition incorporation into DNA in MCF-7 and MDAMB-231 human breast cancer cells. These compounds were capable of binding in the minor groove-binding mode in AT sequences of DNA, and this was confirmed using the ethidium displacement assay. The cytotoxic features of the amide derivatives of melphalan are not correlated with the binding properties of the DNA, but are related to the cultured human breast cancer cells detected with the properties of topoisomerase II inhibitors [234].

Linear methoxy poly(ethylene glycol) was conjugated to melphalan drug (M-PEG) (Scheme 8.26). An ester linkage between drug and polymer was utilized in the coupling to afford a polymeric prodrug. By modifying melphalan to overcome the known aqueous solubility problem, a polymer-drug bioconjugate was accessed for parental and oral administrations. The

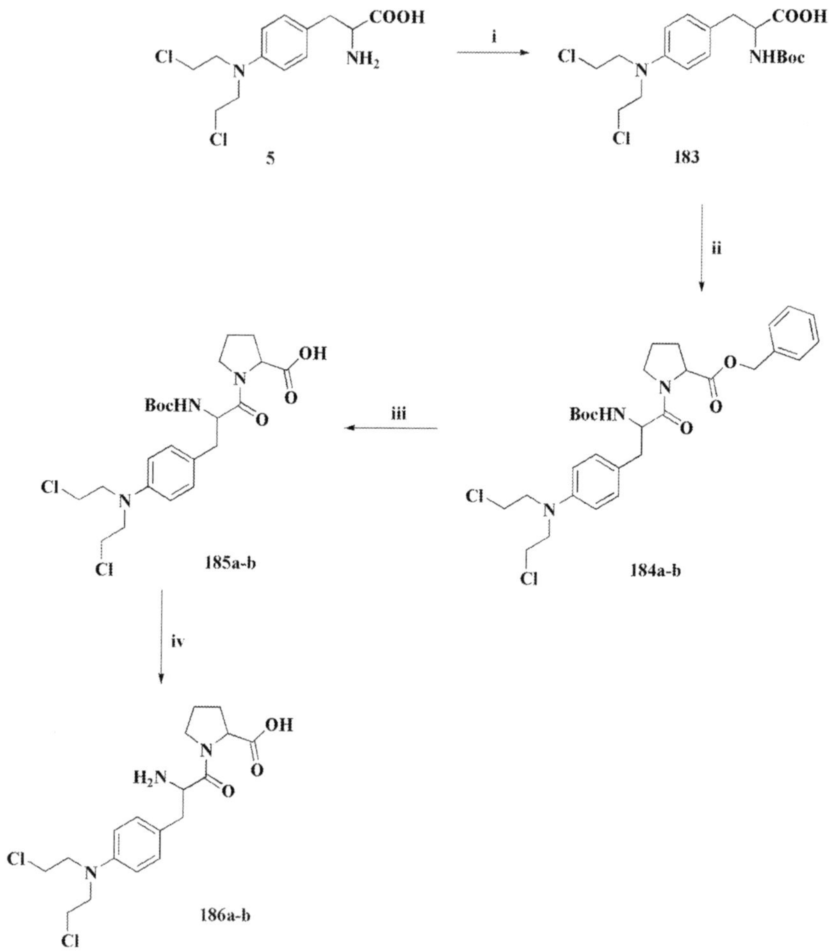

Scheme 8.24 Reagents and conditions: (i) NaOH/bis(tert-butoxycarbonyl)oxide/dioxane/H_2O. (ii) TEA/DCC/D-proline benzyl ester or L-proline benzyl ester. (iii) EtOH/10% Pd/C. (iv) dioxane/HCl(g).

aqueous solubility, conjugates stability, and hemolytic potency depend on the molecular weight of M-PEG. In vitro anticancer potency of melphalan and its conjugates was carried out with breast cancer MCF-7 cell lines. It indicates that LD_{50} concentration was higher (2 and 1.14 μm) for M-PEG 5000 and M-PEG 2000, comparable to that of pure melphalan (0.74 μm) [235].

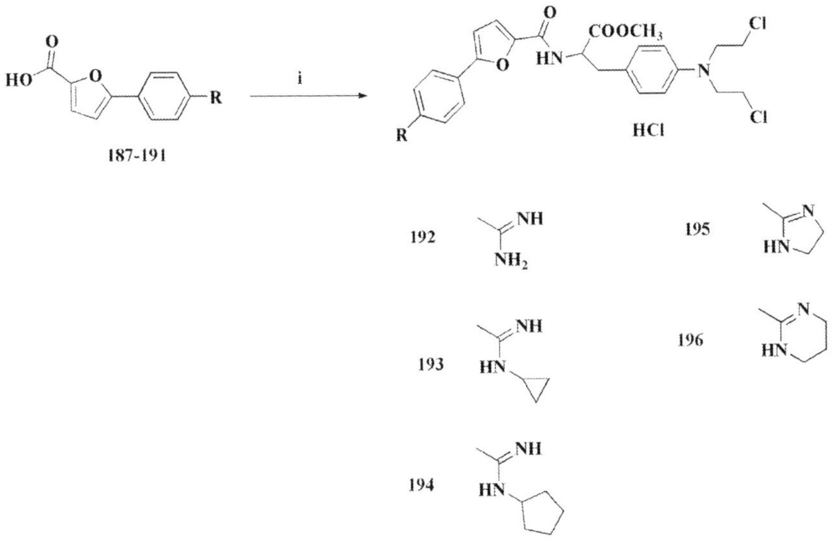

Scheme 8.25 Reagents and conditions: (i) melphalan methyl ester/HCl/DCl.

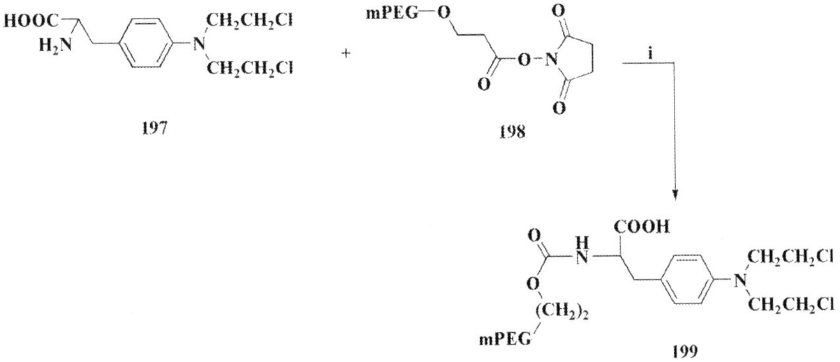

Scheme 8.26 Reagents and conditions: (i) CH_3OH /24hr.

The quaternary ammonium-melphalan conjugate was synthesized [236] [237]. Protection of the amine function of the melphalan was carried using di-tert-butyl dicarbonate in the presence of triethylamine. The further reaction with *N,N'*-dicyclohexylcarbodiimide, 3-(dimethylamino)-propylamine, and 1*H*-benzo[*d*][1,2,3]triazol-1-ol in methylene chloride yielded the amide **200**. Treating with methyl iodide in the presence of ethanol furnished the quaternary

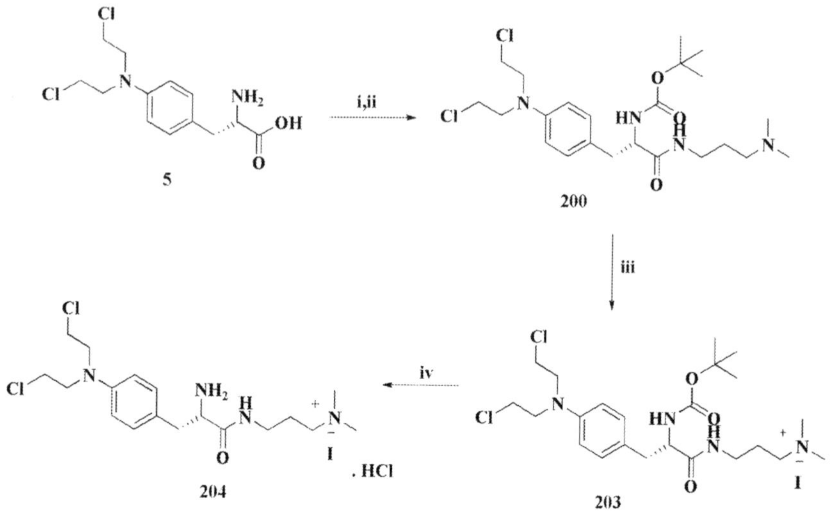

Scheme 8.27 Reagents and conditions: (i) TEA/di-tert-butyl dicarbonate/CH_3OH/RT. (ii) HOBt/DCC/CH_2Cl_2/RT. (iii) MeI/C_2H_5OH/3-(dimethylamino)propylamine/RT. (iii) MeI/ C_2H_5OH/RT. (iv) C_2H_5OH/HCl anhydrous dol./RT.

ammonium **203**. Removal of the t-Boc-protecting group yielded the QA analog **204** from compound **5** (Scheme 8.27) [236].

Malignant cartilaginous tumors are generally unresponsive to chemotherapy [238]. Based on its high binding capacity to proteoglycans, quaternary ammonium conjugated with melphalan was prepared, and its potential antitumor potency in chondrosarcoma was examined in vitro and in vivo. The differential in vitro antitumor potency of Mel-AQ, its ability to inhibit tumor cell growth in vivo, and its good tolerance in animals make Mel-AQ a candidate to develop an effective treatment for chondrosarcoma [236].

99mTc-Labeled tumor-imaging agents with SPECT, MFL that were directly labeled by 99mTc utilizing diethylenetriaminepentacetate acid as bifunctional chelating agent were developed. The synthesis of the new ligands was performed via the conjugation of DTPA to melphalan to afford bis-substituted DTPA-2MFL and monosubstituted DTPA-MFL. Radiolabeling was carried out to afford 99mTc-DTPA-2MFL and 99mTc-DTPA-MFL, which were stable at room temperature and hydrophilic (Scheme 8.28). The high initial tumor uptake with retention, good tumor/muscle ratios, and satisfactory scintigraphic images proposed the potential of 99mTc-DTPA-MFL and 99mTc-DTPA-2MFL for tumor

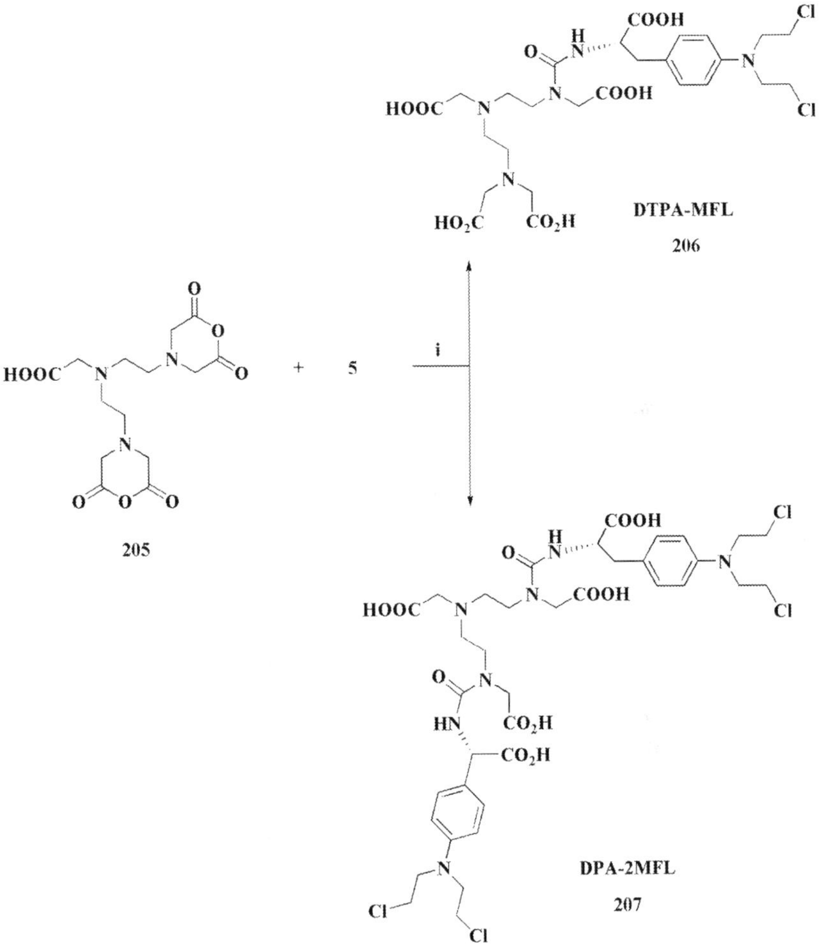

Scheme 8.28 Reagents and conditions: (i) H_2O/DIPEA.

imaging. Subsequent modification on the linker and/or 99mTc-chelate to enhance the tumor-targeting efficiency and in vivo kinetic profiles is in progress [237].

In order to increase melphalan and bendamustine antitumor potency and tumor selectivity, both compounds were integrated in SAR studies, including novel drug carrier systems. The synthesis of novel bivalent bendamustine and melphalan analogs were accomplished as outlined in Schemes 8.29 and 8.30. Two molecules each esterified with *N*-(2-hydroxyethyl)maleimide were joined through diamines with different chain lengths [238].

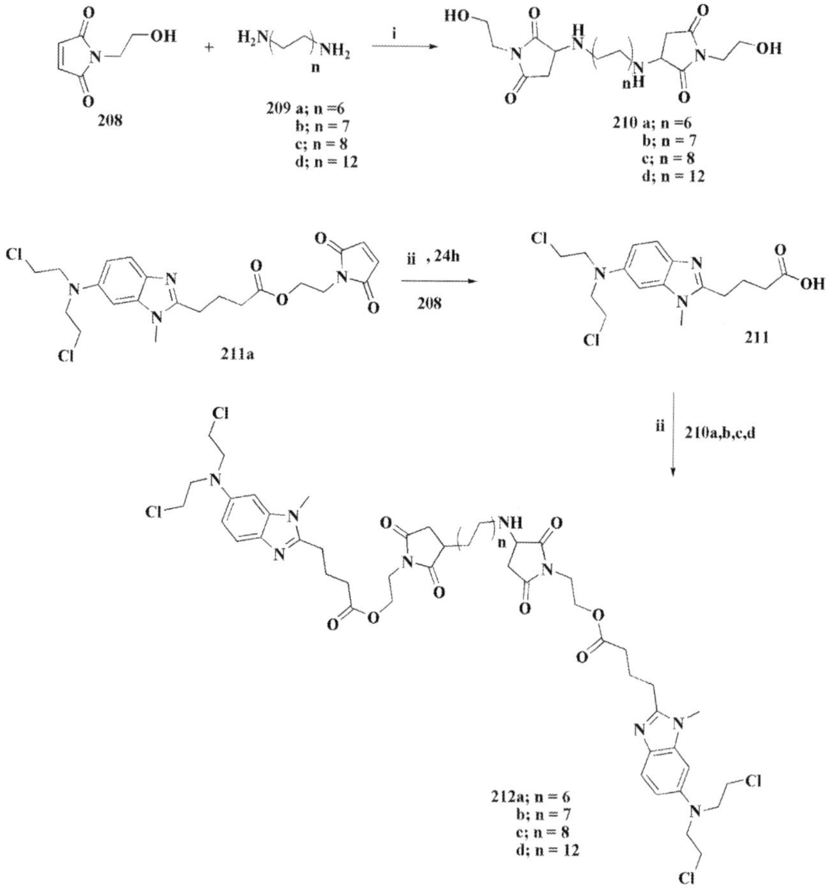

Scheme 8.29 Reagents and conditions: (i) $CHCl_3$/reflux. (ii) DCC/CH_2Cl_2/DMAP/RT.

It was supposed that these conjugates (**212a–d**, **217a–d**, and **218a–d**) cause cytotoxic effects preferred as a bivalent drug. The cytotoxicity of the novel compounds increased compared to melphalan and bendamustine as shown from the data concentration-dependent in vitro assays using the human MDA-MB-231 and MCF-7 breast cancer cell lines [238].

The amidine derivatives of alkylating antineoplastic compounds were synthesized (Scheme 8.31). This class is considered a novel category of cytotoxic minor groove binders and topoisomerase II inhibitors. The construction of a prodrug is done with lower hydrophobicity and cytotoxicity but is activated in cancer cells to overcome the toxicity of alkylating agents to normal tissue. Overexpression of prolidase in some neoplastic cells proposes that

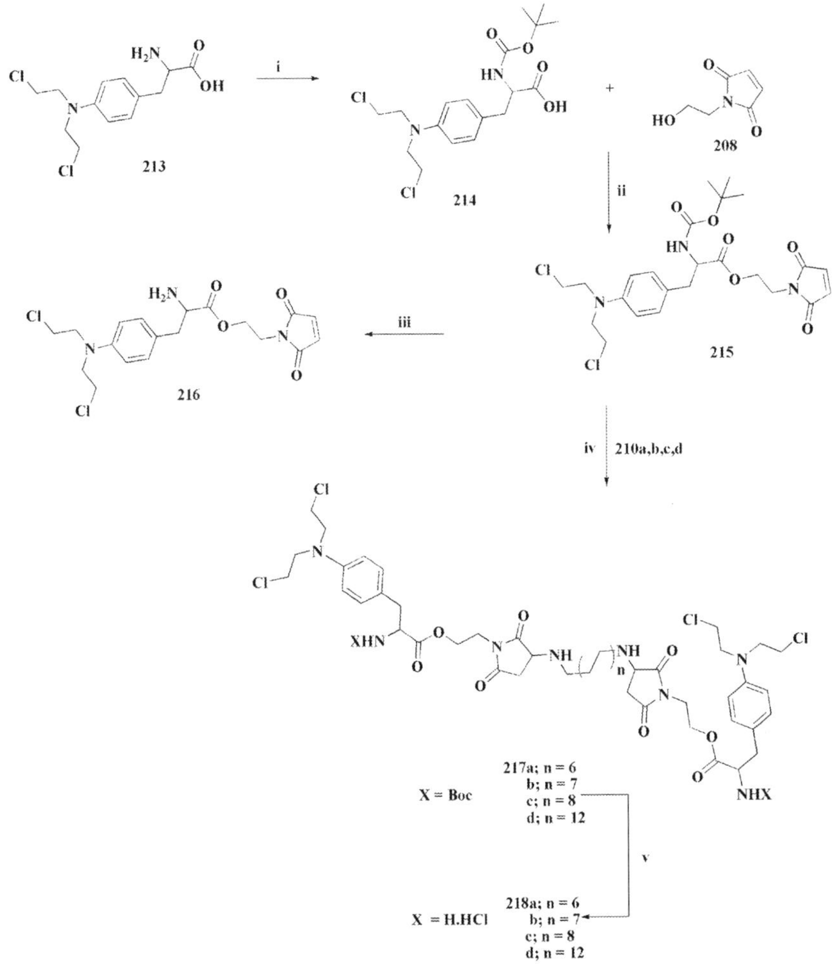

Scheme 8.30 Reagents and conditions: (i) TEA/Boc_2O/CH_3OH/RT/1 h; (ii) DMAP/CH_2Cl_2/DCC/RT/24h. (iii) THF/aq. HCl/RT/24h. (iv) $CHCl_3$/reflux. (v) THF/aq. HCl/RT.

the proline derivative of alkylating agents may act as a prolidase convertible prodrug. Various aspects of pharmacological actions of proline derivatives of melphalan as well as chlorambucil in breast cancer cells were compared. It was suggested that the prolidase could act as a target enzyme for the exquisite action of anticancer agent [239].

1,3,5-Tris(3-aminopropyl)benzene (**230**) and its derivative **235** were elected to design cytostatic drug-dendrimer conjugates to realize tumor cell accumulation through endocytosis as established for platinum complexes

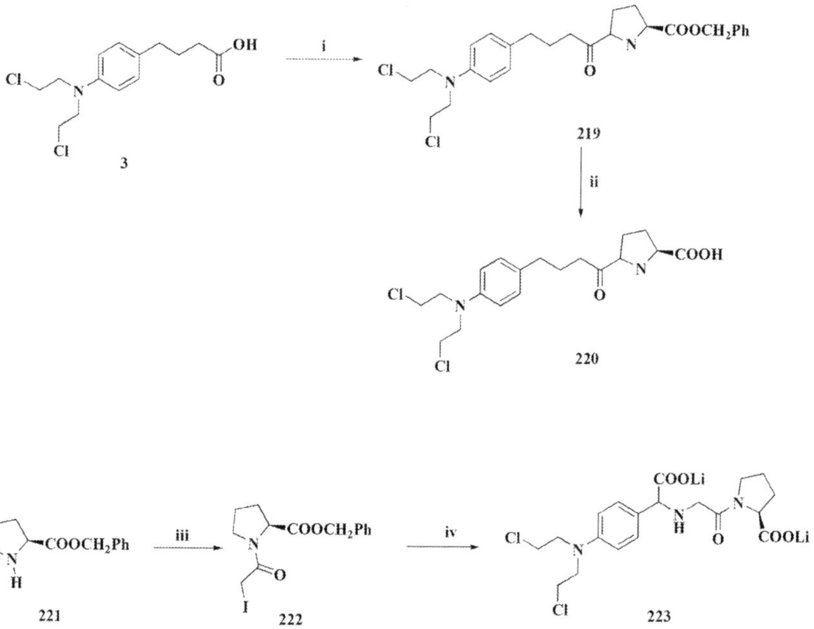

Scheme 8.31 Reagents and conditions: (i) L-proline benzyl ester/DCC. (ii) H_2/10% Pd/C. (iii) NaI/$ClCH_2COCl$. (iv) LiOH/melphalan.

(Schemes 8.32–8.37). The dendrimers act as carriers, and an *N*-(2-hydroxyethyl)maleimide spacer between carrier and drug should cause a selective release of the cytostatics in the tumor cells. Using the human MCF-7 and MDA-MB-231 breast cancer cell lines, cytotoxicity was demonstrated in vitro. The cytotoxicity caused by melphalan is due to the presence of its free amino group, and the potency is decreased as a result of the Boc protection. The cytotoxicity is independent of a derivation of the carboxylic group (spacer binding and dendrimers). Esterification of the *N*-(2-hydroxyethyl)maleimide spacer and bendamustine highly increased the hydrolytic stability of the N-lost scaffold, so in vitro antiproliferative impacts were shown [240].

The C-Mel, a cephalosporin carbamate analog of melphalan, was synthesized (Scheme 8.38). C-Mel was designed to release melphalan upon tumor-specific activation by targeted *â*-lactamase (bL). In vitro cytotoxicity assays with 3677 human melanoma cells established that C-Mel was 40-time less toxic than melphalan and was activated in an immunologically specific manner by L49-sFv-bL, a recombinant fusion protein that binds to the melanotransferrin antigen on some carcinomas and melanomas. L49-sFv-bL in

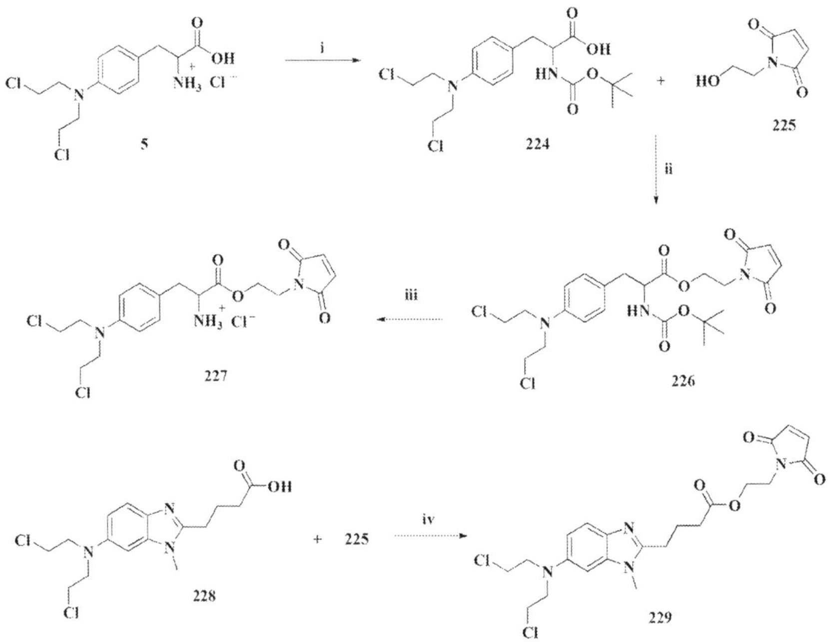

Scheme 8.32 Reagents and conditions: (i) TEA/Boc_2O/CH_3OH/RT/1 h. (ii) DMAP/CH_2Cl_2/DCC/RT/24 h. (iii) THF/aqueous HCl/RT/24 h. (iv) DMAP/CH_2Cl_2/DCC/RT/20 h.

combination with C-Mel led to regressions and cures of established subcutaneous 3677 tumors in nude mice. The effects were remarkably greater than those of melphalan, which did not lead to any long-term regressions in this tumor model [241].

The therapeutic impacts were comparable to those furnished in mice treated with the L49-sFv-bL/7-(4-carboxybutanamido)-cephalosporin mustard (CCM) combination. C-Mel may be more attractive than CCM for clinical development since the released drug is clinically affirmed [241].

A PBR ligand-melphalan conjugate (PBRMEL) was prepared (Schemes 8.39 and 8.40) and estimated for affinity for peripheral benzodiazepine receptors (PBRs) and cytotoxicity. PBR-MEL (**255**) was generated via the coupling reaction of compound **254** and compound **250**. On the basis of receptor-binding displacement assays in glioma cells and in rat brain, **255** had an appreciable binding affinity and displaced a prototypical PBR ligand, Ro 5-4864, with IC_{50} values between 289 and 390 nM. Compound **255** exhibited differential cytotoxicity to a variety of rat and human brain tumor cell lines. In some of the cell lines examined, including rat and human

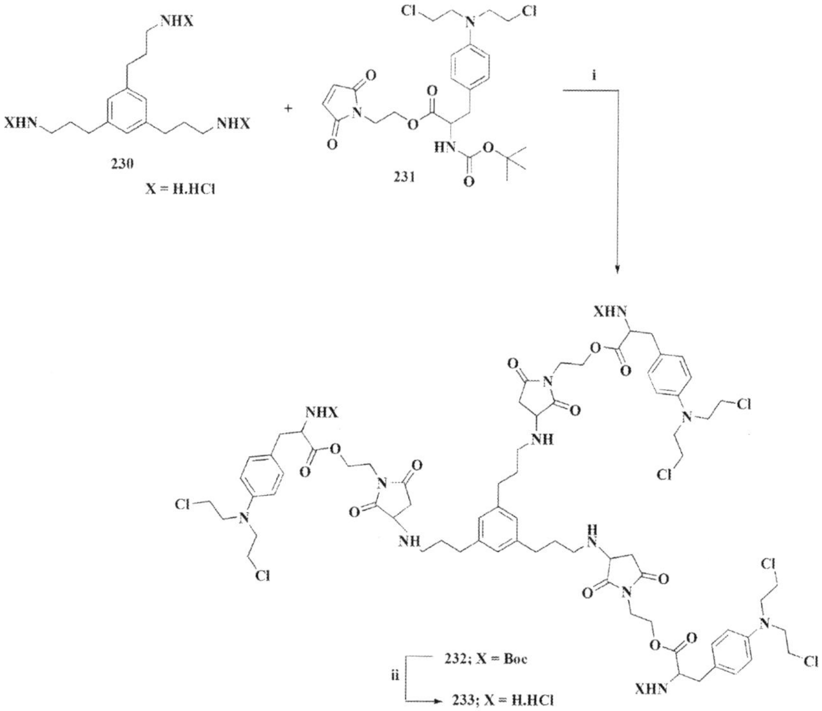

Scheme 8.33 Reagents and conditions: (i) DIPEA/$CHCl_3$/reflux/72 h. (ii) THF/aqueous HCl/RT.

melphalan-resistant cell lines, compound **255** established an appreciable cytotoxicity with IC_{50} values in the micromolar range, which were lower than that of melphalan alone. The improved potency of compound **255** may show increased membrane permeability, increased intracellular retention, or modulation of melphalan's mechanisms of resistance [242].

Water-soluble polyphosphoesters were utilized as polymer carriers of hydrochloride of *p*-bis(2-chloroethyl)amino-L-phenylalanine (melphalan hydrochloride), which is a multifunctional alkylating agent [243].

Under Atherton-To dd reaction conditions, the melphalan was chemically immobilized by covalent bonding to poly(oxyethylene *H*-phosphonate). As control, the melphalan-polymer complexes with ionic and hydrogen bonds were designed, basing on two other biodegradable polyphosphoesters: poly(methyloxyethylene phosphate and poly(hydroxyoxyethylene phosphate (Scheme 8.41).

The cytotoxic effect of the melphalan formulations was estimated on several tumor cell lines. The new polymer formulations indicated a

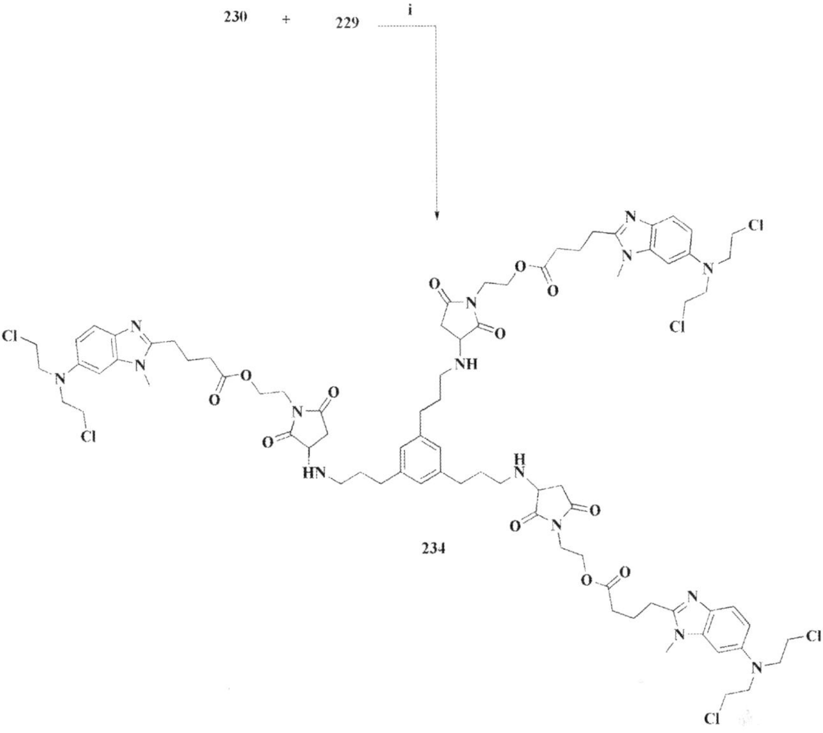

Scheme 8.34 Reagents and conditions: (i) $CHCl_3$/DIPEA/reflux/72 h.

concentration-dependent antitumoral activity, comparable to the impact of the unmodified melphalan. The melphalan-polymer conjugate was also examined in vivo in the human hepatocellular carcinoma HuH7 xenograft mouse model. Without rising side effects, the therapeutic efficiency of pure melphalan was improved [243].

The clinical application of melphalan for treating hematologic malignancies has been limited because of the lack of target specificity, its poor water solubility, and the rapid elimination. In trying to solve these problems, the syntheses of the *O,N*-carboxymethyl chitosan-peptide-melphalan conjugates were carried out. All polymeric prodrugs indicated a satisfactory water solubility. The peptide spacer and the molecular weight of *O,N*-carboxymethyl chitosan (*O,N*-CMCS) have a crucial role in controlling the drug diameter, content, and drug release characteristics of *O,N*-carboxymethylchitosan-peptide-melphalan conjugates. The investigations of in vitro drug release and cell cytotoxicity by MTT assay showing

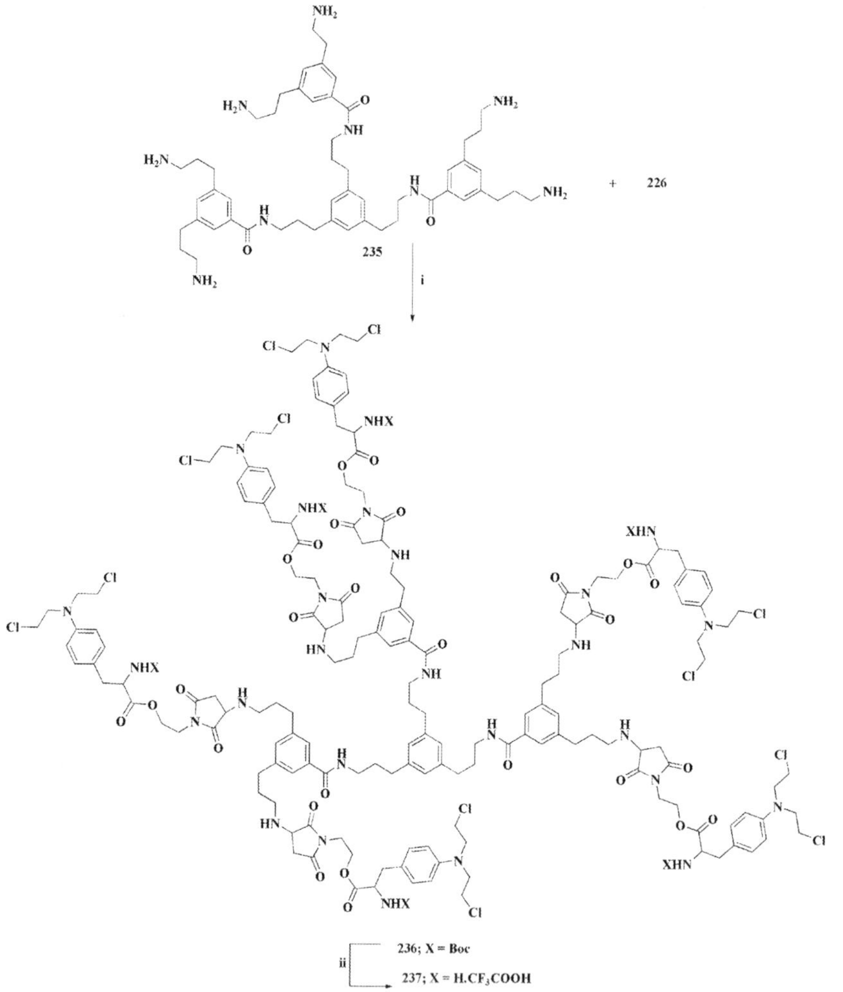

Scheme 8.35 Reagents and conditions: (i) DMF/CH_2Cl_2/−20°C/TBTU/30 min/CH_3OH/DIPEA/RT/20 h. (ii) RT/TFA/4 h.

that upon employing the polymeric conjugation strategy and utilizing the peptides glycylglycine(Gly-Gly) as a spacer, the conjugates will possess good cathepsin X-sensitivity and lower toxicity, and the drug-release behavior will be enhanced significantly (Schemes 8.42 and 8.43) [244].

The syntheses of melphalan-*O*-carboxymethyl chitosan (Mel-OCM-chitosan) conjugates were furnished. All conjugates indicated a satisfactory water solubility (160–217 times of Mel solubility). In vitro drug release behaviors through chemical and enzymatic hydrolysis were estimated.

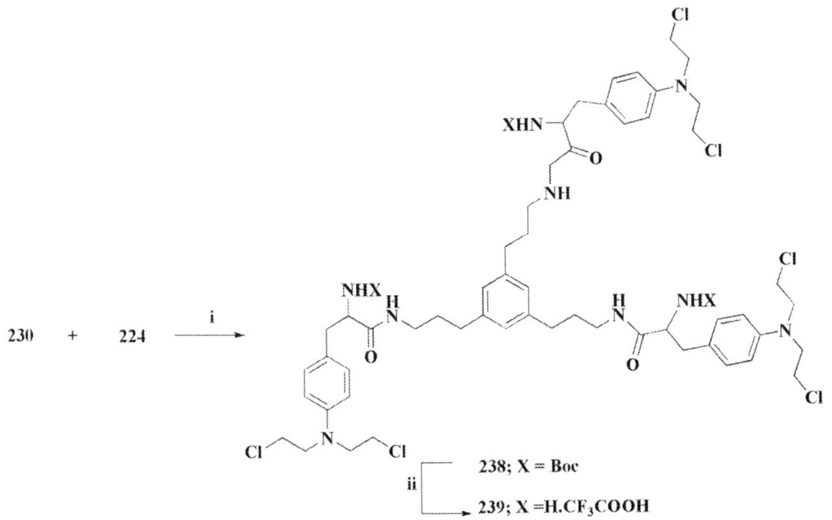

230 + 224 →(i)

ii: 238; X = Boc → 239; X = H.CF_3COOH

Scheme 8.36 Reagents and conditions: (i) DIPEA/chloroform/reflux/70 h. (ii) THF/aqueous HCl/RT.

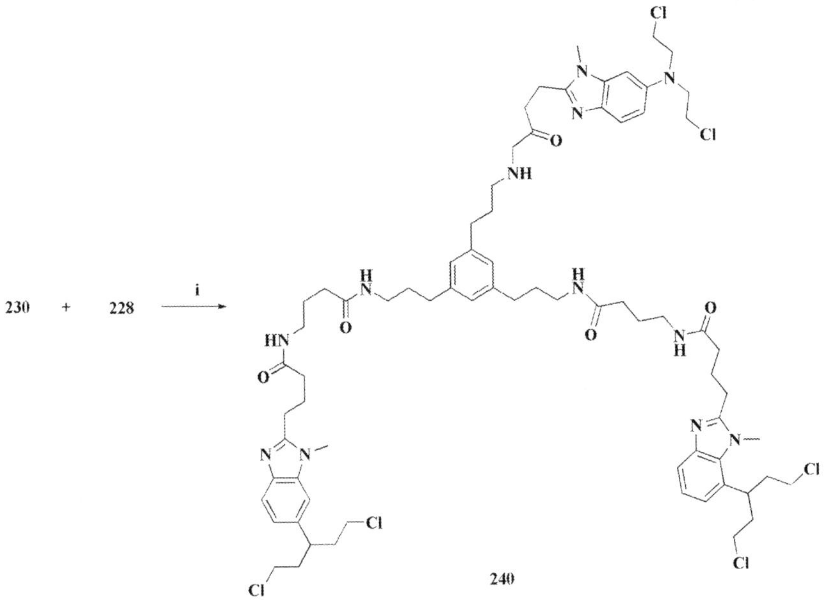

230 + 228 →(i) 240

Scheme 8.37 Reagents and conditions: (i) DMF/CH_2Cl_2/−20°C/TBTU/30 min/CH_3OH/DIPEA/RT/20 h.

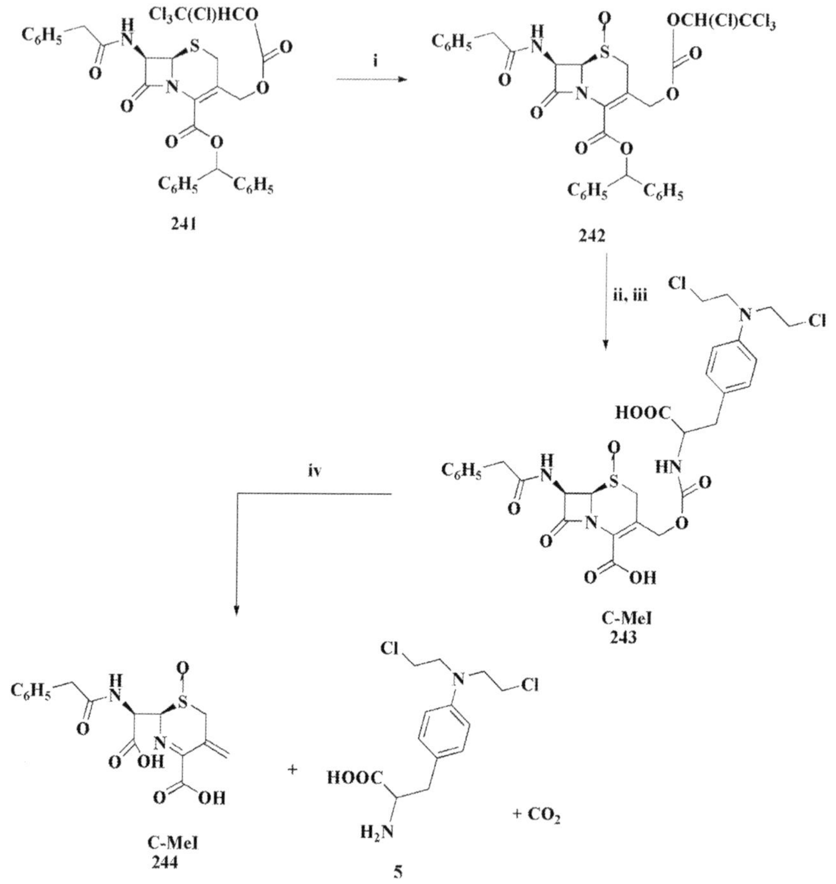

Scheme 8.38 Reagents and conditions: (i) *m*-Chloroperoxybenzoic acid/CH_2Cl_2. (ii) Melphalan/aq. $NaHCO_3$/THF. (iii) CH_2Cl_2/TFA-anisole. (iv) â-lactamase.

The prodrugs released Mel rapidly within papain and lysosomal enzymes of about 40%–75%, while those released in plasma and buffer of about 4%–5%, which proposed that the conjugates possess good plasma stability and the hydrolysis in both lysosomes and papain takes place mostly through enzymolysis. The spacers have an essential impact on the water solubility, drug content, drug release features, and cytotoxicity of Mel-OCM-chitosan conjugates. Cytotoxicity data using the MTT assay established that these conjugates had 52%–70% cytotoxicity against RPMI8226 cells in vitro compared to that of the free Mel, revealing the conjugates did not miss an anticancer potency of Mel (Scheme 8.44) [245].

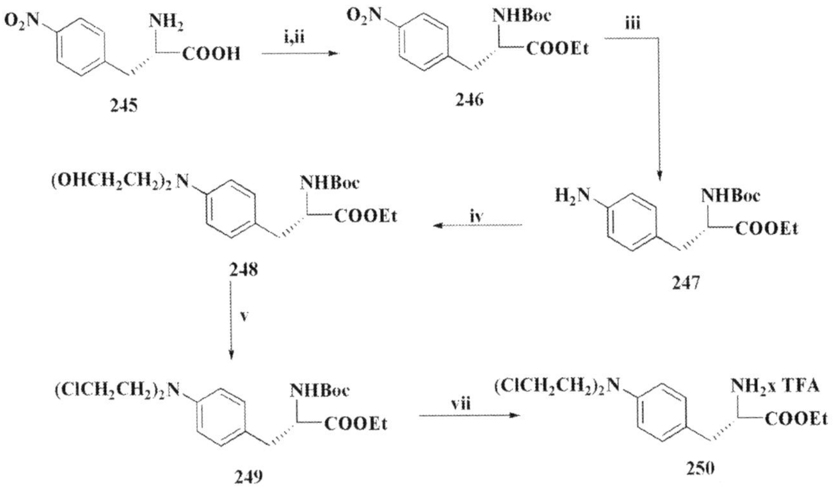

Scheme 8.39 Reagents and conditions: (i) HCl (g)/C_2H_5OH. (ii) Bis(tert-butoxycarbonyl) oxide/TEA. (iii) H_2/10% Pd/C/CH_3OH. (iv) 1,2-Epoxyethane/acetic acid/H_2O. (v) Ph_3P/CCl_4/CH_2Cl_2. (vii) TFA/methoxybenzene.

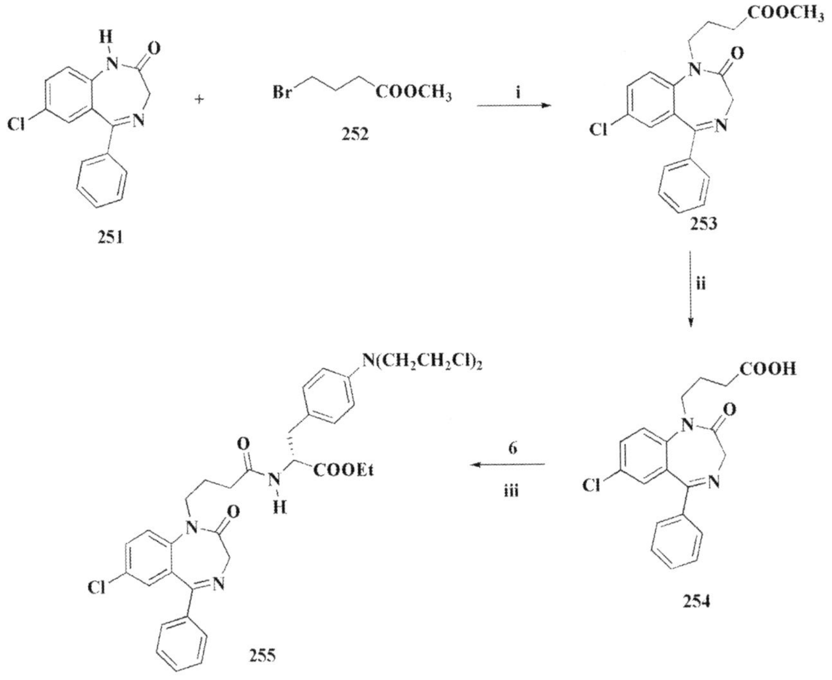

Scheme 8.40 Reagents and conditions: (i) DMF/NaH. (ii) CH_3OH/KOH. (iii) HOBt/EDC.

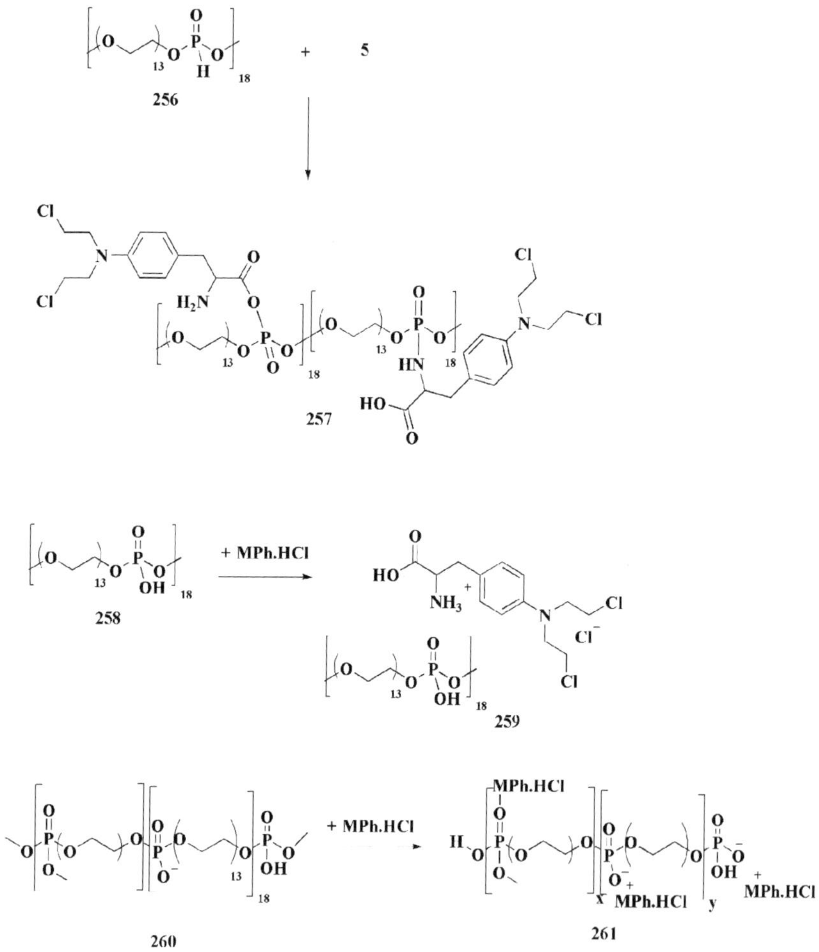

Scheme 8.41

Polymer-drug conjugates improve the selectivity of anticancer drugs and tumor cell targeting. Melphalan and quantum dots were linked to the backbone of hyaluronic acid to generate a polymer–drug conjugate. The conjugate was prepared (Scheme 8.45) and self-assembled into nanoparticles. Its receptor-mediated delivery system was exemplary and can be internalized into the human breast cancer cell. It possesses a poorer inhibition impact on normal breast cell and a better inhibition effect on human breast cancer cell than melphalan [246].

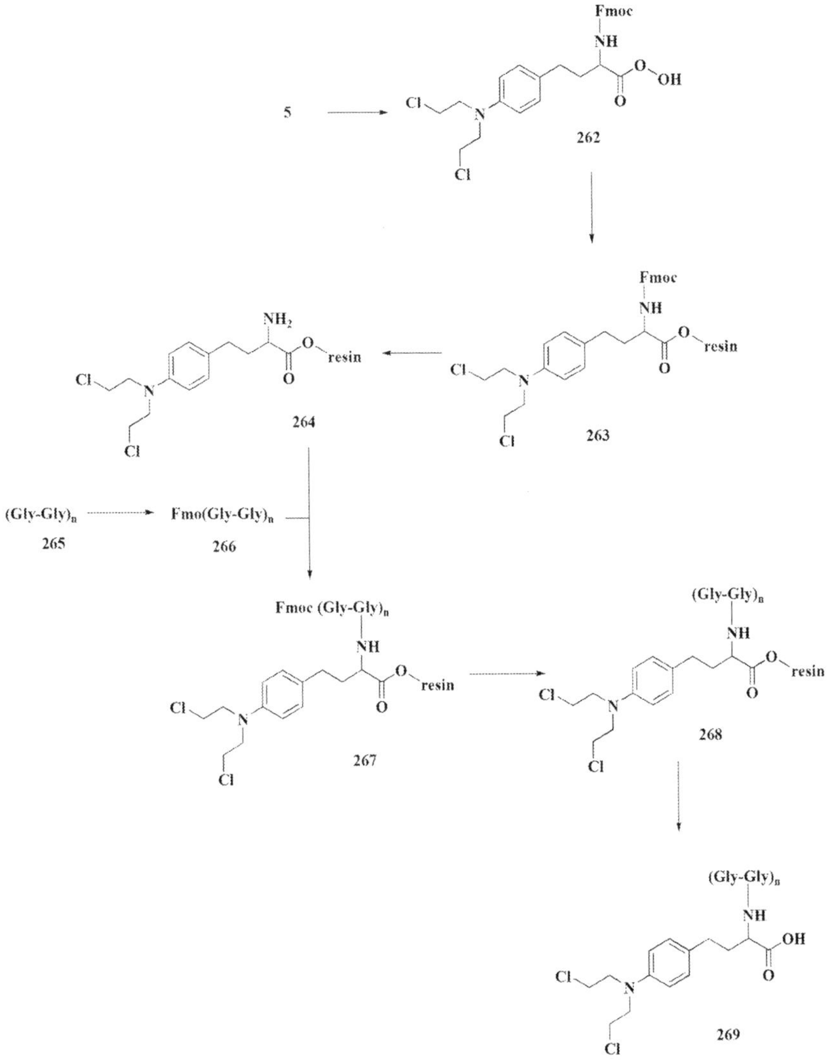

5 → 262 → 263 → 264; (Gly-Gly)$_n$ 265 → Fmo(Gly-Gly)$_n$ 266; 264 + 266 → 267 → 268 → 269

Scheme 8.42

12.3 Synthetic strategies of ifosfamide analogs

The bromo analog of ifosfamide, (*S*)-(7)-bromofosfamide (CBM-11), was a potent agent against different model tumors in mice [247]. Therapeutic indices of CBM-11 are suitable than those for ifosfamide. Following the steps shown in scheme, CBM-11 (**285**) was furnished upon utilizing the chiral α-methylbenzyl auxiliary attached to the nitrogen atom of

Scheme 8.43 Reagents and conditions: (i) EDC/NHS. (ii) TCA/IPA.

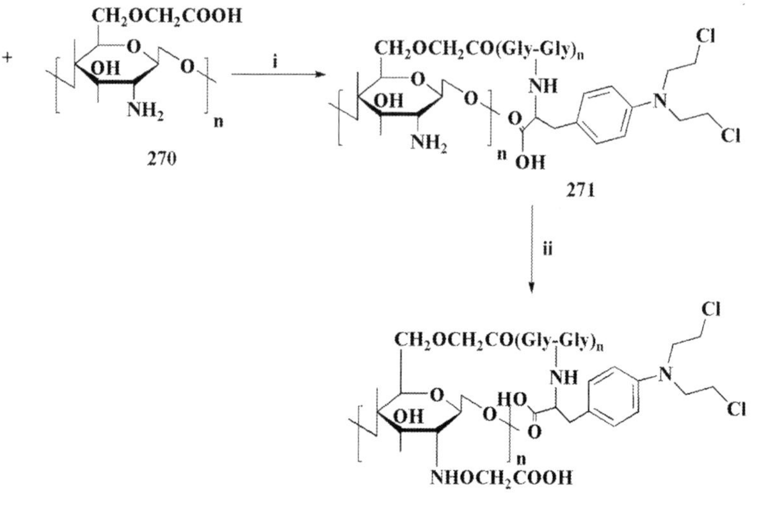

tetrahydro-2*H*-1,3,2-oxazaphosphorinane fragment, which enabled the formation of compound **282**. This procedure was utilized in the synthesis of enantiomers of ifosfamide, cyclophosphamide, and trofosfamide [248,249]. The development of its structure has been enrolled for the synthesis of CBM-11 (Scheme 8.46). So, replacing the carcinogenic aziridine, [248] 2-chloroethylamine hydrochloride was utilized for the synthesis of compound **283**. Compound **283** was acylated with bromoacetyl bromide. Compound **284** was generated in 78% yield. The formation of the hydrogen halides resulted from the condensation reactions trapped using triethylamine, but the latter cannot be used in catalyzing the reaction which furnish compound **284** as it is reactive toward the halogenoacetyl halides [250]. The base triethyl phosphate [251] could also be utilized. In compound **284**, further reduction of the carbonyl group was occurred by utilizing sodium tetrahydridoborate in the presence of boron trifluoride etherate [252].

In all three tumor models (B16 melanoma, L1210 leukemia, and Lewis lung carcinoma), CBM-11 was more potent than ifosfamide. The maximal therapeutic impact found after treatment using ifosfamide was remarkably less than that the case of administrating CBM-11 even at the doses close to MTD. There is a correspondence between the highly antineoplastic potency of CBM-11 and the rise in therapeutic index (TI) (MTD:CD50 ratio). TI of CBM-11 was remarkably increased as compared to IF in two

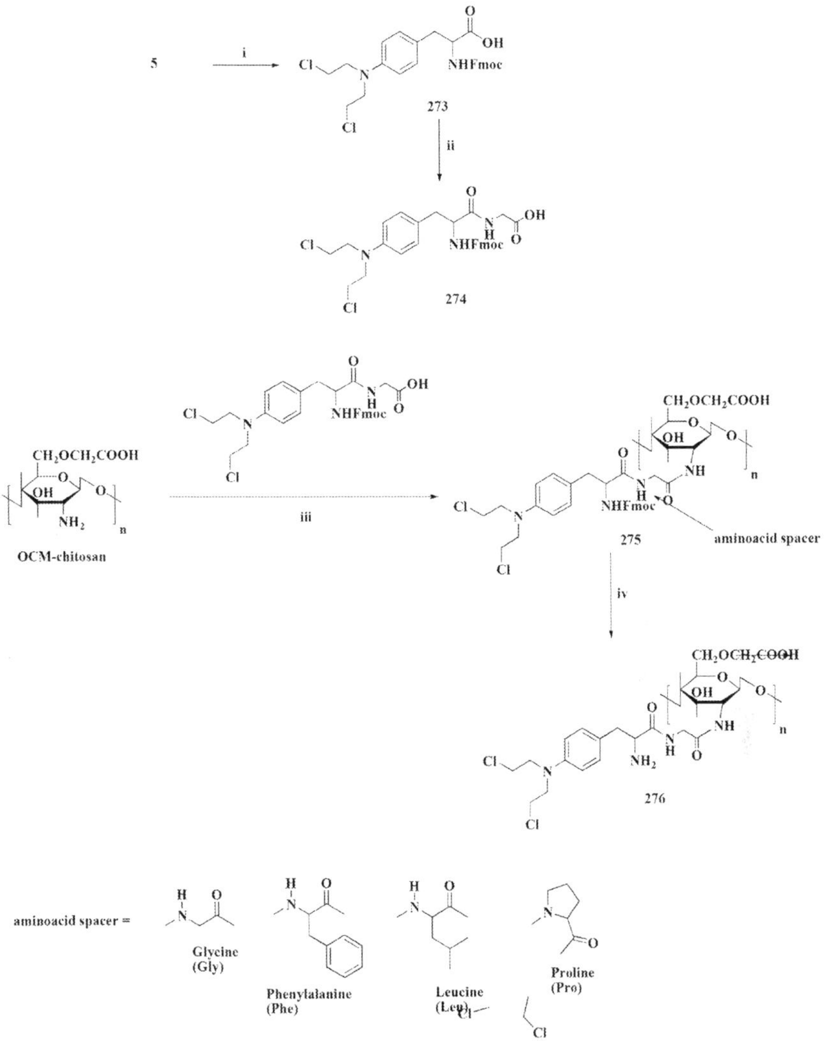

Scheme 8.44 Reagents and conditions: (i) Fmoc-Osu. (ii) NHS/DCC/Gly. (iii) NHS/EDC. (iv) Piperidine.

model systems, i.e., B16 melanoma and L1210 leukemia. The B16 melanoma is insensitive to the treatment with oxazaphosphorine drugs as ifosfamide or cyclophosphamide.

The influence of the various DNA damages caused by several isophosphoramide mustard derivatives on the 3-hydroxypropanal-assisted apoptotic antitumor potency of oxazaphosphorine cytostatics utilizing I-aldophosphamide-perhydrothiazine (IAP) and mesyl-I-aldophosphamide-

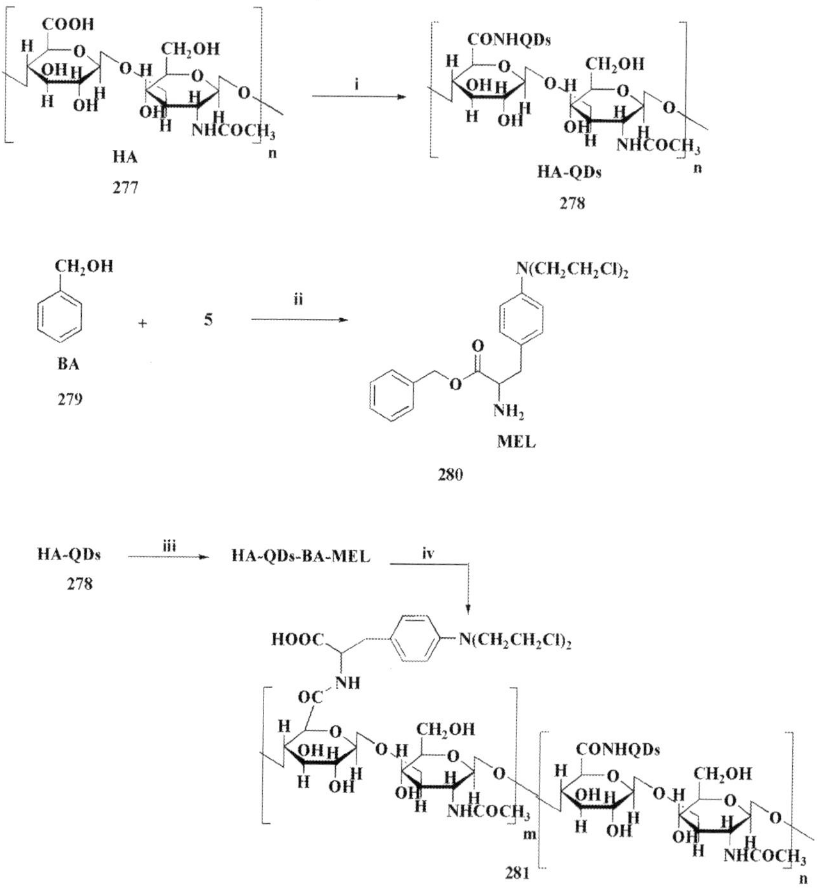

Scheme 8.45 Reagents and conditions: (i) EDC/QDs/NHS. (ii) $SOCl_2$. (iii) NHS/EDC/BA-MEL. (iv) Pd/C.

perhydrothiazine (SUM-IAP) for in vivo and in vitro experiments has been studied. Spontaneously, IAP and SUM-IAP hydrolyze to the corresponding I-aldophosphamide analogs. They are variant in the chemical structure of the alkylating moiety, while the IAP has two chlorethyl groups in the SUM-IAP molecule: one chlorethyl group is replaced by a mesylethyl group. Using the two substances, cytotoxicity investigations on P388 tumor cells in vitro and therapy experiments in mice bearing advanced growing P388 tumors were performed. IAP was remarkably more cytotoxic in vitro than SUM-IAP, but the antitumor potency of SUM-IAP was higher than that of IAP. Due to the enzymatic cleavage of several I-aldophosphamide analogs to the corresponding isophosphoramide

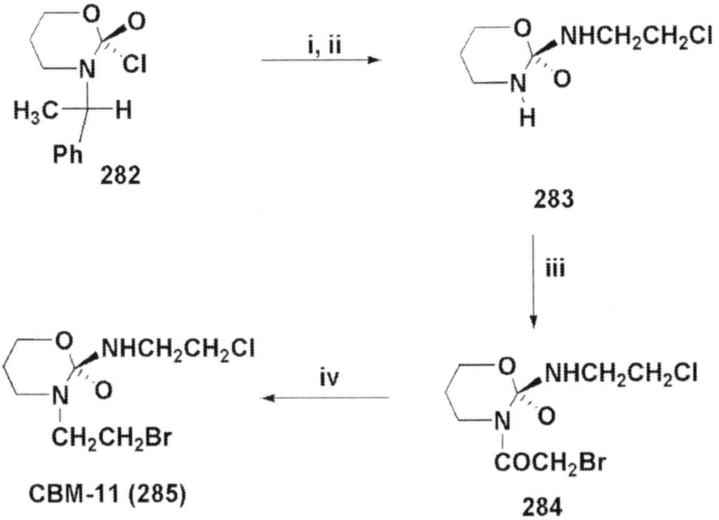

Scheme 8.46 Reagents and conditions: (i) $HCl.NH_2CH_2CH_2Cl$/TEA. (ii) Et_3NHBr/$BrCH_2COBr$. (iii) $BF_3xEt_2O/NaBH_4$. (iv) $BF_3xEt_2O/NaBH_4$.

mustards and 3-hydroxypropionaldehyde. These results support that antitumor potency of ifosfamide and analogs of ifosfamide can be enhanced through varying the alkylating moiety of the molecule, while the aldophosphamide structure is retained (Scheme 8.47) [253].

Deutero-substituted analogs of ifosfamide (IF-d4) were prepared as outlined in Schemes 8.48 and 8.49. The microsomal hydroxylation of IF-d4 is slower than for unlabeled compound, proposing that kinetic isotope impact operates during those conversions, which was indicated through the in vitro metabolic data. The negative role of side-chain hydroxylation metabolic pathways in the anticancer potency of ifosfamide and its analogs could be the reason that the deutero analogs are more potent against L1210 leukemia in mice than unlabeled compounds [254].

The prodrugs ifosfamide (IF) and cyclophosphamide (CP) each metabolize to an active alkylating agent via a cytochrome P450-mediated oxidation at the C-4 position. Competing with this activation pathway are enzymatic oxidations at the exocyclic α and α′ carbons, which lead to dechloroethylation of CP and IF [228].

[α,α,4,4,5,5-2H6]-2-Dechloroethylcyclophosphamide (equivalent to [α,α,4,4,5,5-2H6]-3-dechloroethylifosfamide), [4,4,5,5-2H4]-2-dechloroethylcyclophosphamide (equivalent to [4,4,5,5-2H4]-3-

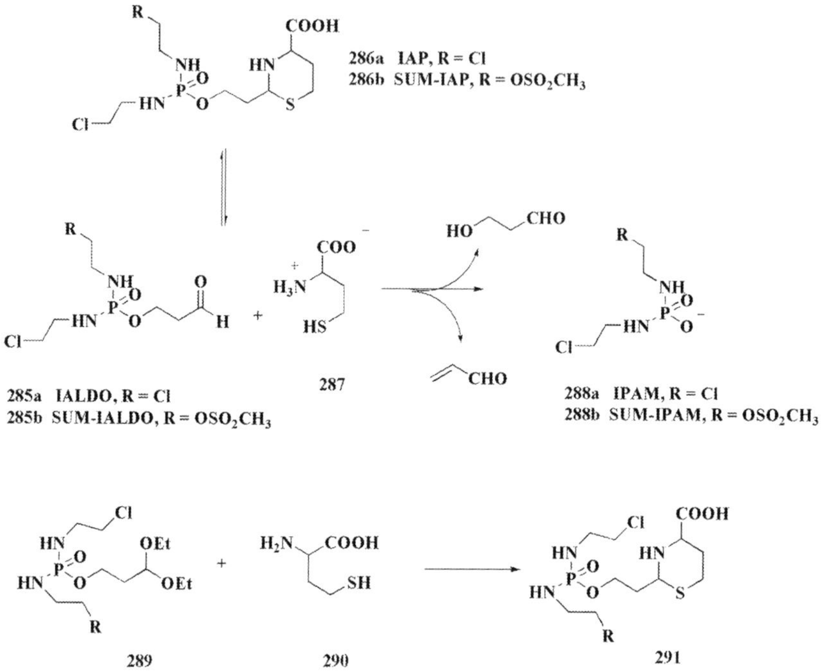

286a IAP, R = Cl
286b SUM-IAP, R = OSO_2CH_3

285a IALDO, R = Cl
285b SUM-IALDO, R = OSO_2CH_3

287

288a IPAM, R = Cl
288b SUM-IPAM, R = OSO_2CH_3

289 290 291

Scheme 8.47

dechloroethylifosfamide), and [α,α,4,4,5,5-2H6]-2-dechloroethylifosfamide were synthesized using the key compound [2,2,3,3-2H4]-3-aminopropanol. Reacting the latter compound with phosphorous oxy chloride and labeled or unlabeled 2-chloroethylamine hydrochloride afforded the d6- and d4-labeled 2-dechloroethylcyclophosphamides. Synthesis of 2-dechloroethylifosfamide from 1-aminopropan-1-ol required discreet steps (Schemes 8.50–8.52) [255].

Oxazaphosphorines are alkylating agents utilized in clinical practices for treating cancer [256]. They are antitumor prodrugs that need cytochrome P450 bio-activation, resulting in the formation of 4-hydroxy derivatives [257]. In the case of ifosfamide, the bio-activation yields two toxic metabolites; acrolein, an urotoxic compound, concomitantly furnished with the 2-chloroethanal and, isophosphoramide mustard, a nephrotoxic and neurotoxic compound, resulting from the oxidation of the side chains. To enhance the therapeutic index of IFO, preactivated IFO analogs were synthesized. Those derivatives with the covalent binding of different *S*- and *O*-alkyl scaffolds comprising poly-isoprenoid groups at the C-4 position

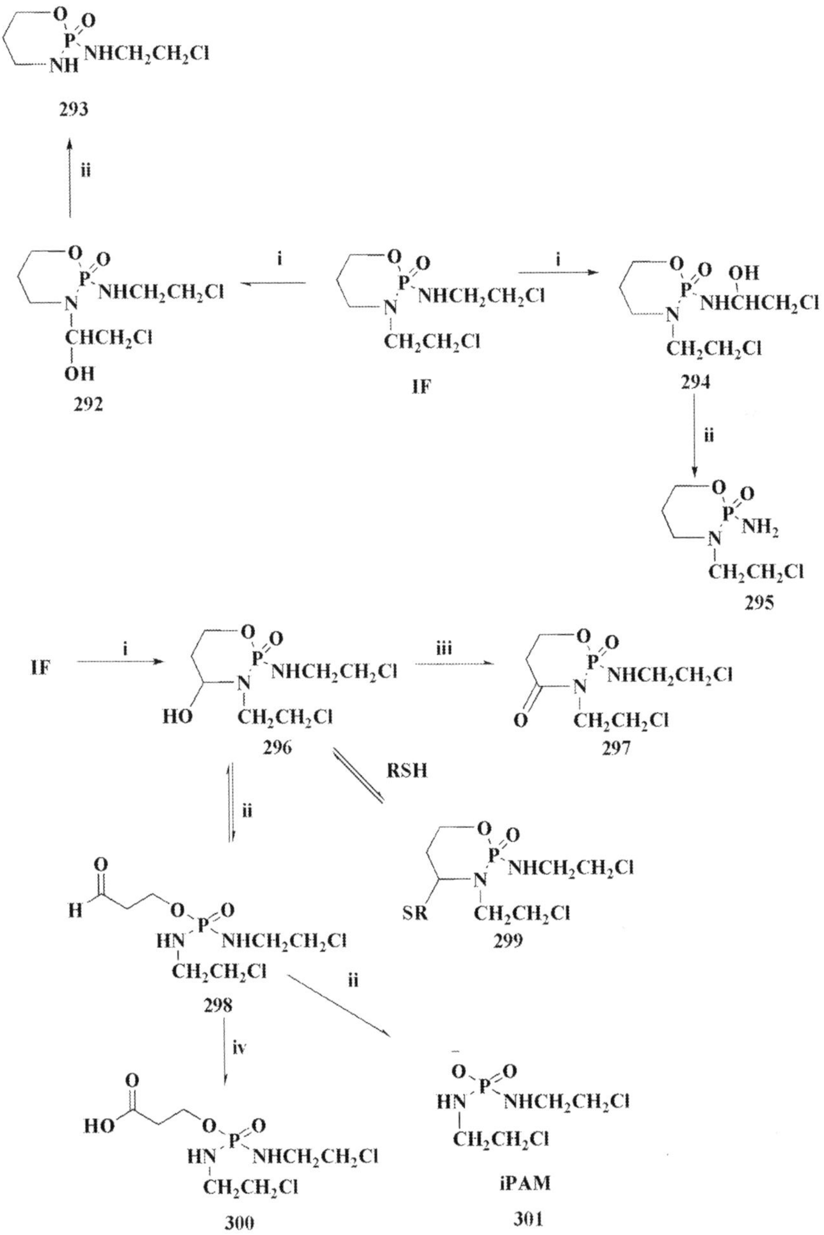

293

ii

292 i IF i 294

ii

295

IF i 296 iii 297

RSH ii

298 299

ii iv

300 iPAM 301

Scheme 8.48 Reagents and conditions: (i) Cytochrome P450. (ii) spontaneous. (iii) Mechanism unknown. (iv) Aldehyde dehydrogenase.

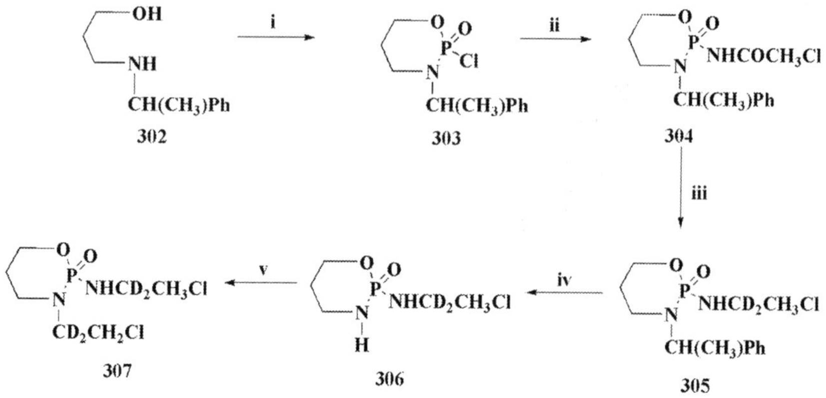

Scheme 8.49 Reagents and conditions: (i) $POCl_3$/TEA. (ii) $ClCH_2COCl/NH_3/Et_3NHCl$. (iii) $NaBD_4/CH_3OD/BF_3{\cdot}Et_2O/CH_3OH$. (iv) H_2/Pd/C. (v) $ClCH_2COCl/NaBD_4/EtNHCl/BF_3{\cdot}Et_2O$.

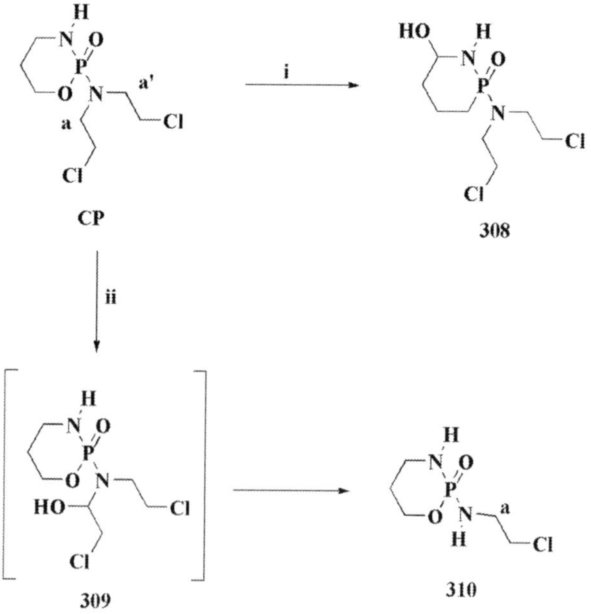

Scheme 8.50 Reagents and conditions: (i) C-4/[O]. (ii) a/a′[O].

of the oxazaphosphorine ring to prevent cytochrome bio-activation subscribing the release of the potent entity and restricting the chloroacetaldehyde release. Some of novel derivatives of the grafted terpene scaffolds indicated spontaneous self-assembling features into nanoassemblies when

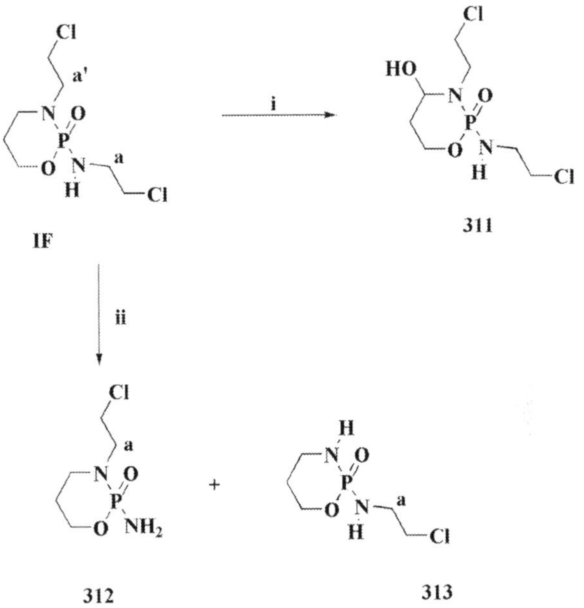

IF
311
312
313

Scheme 8.51 Reagents and conditions: (i) C-4/[O]. (ii) a/a′[O].

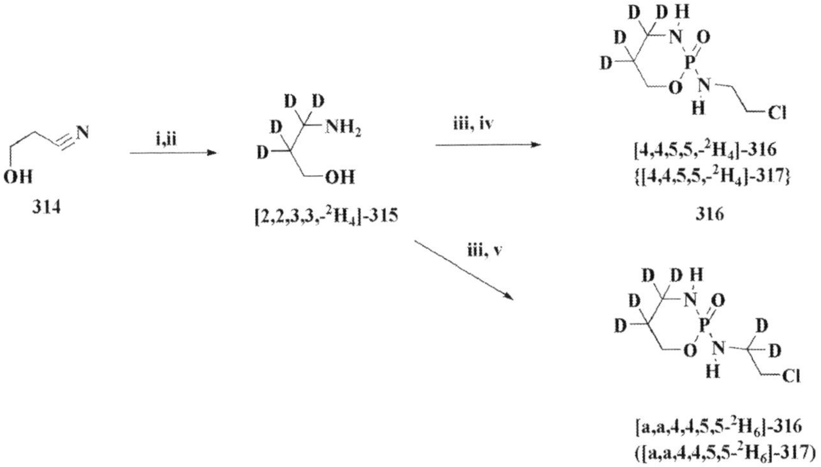

314
[2,2,3,3,-$^2H_4$]-315
[4,4,5,5,-$^2H_4$]-316
{[4,4,5,5,-$^2H_4$]-317}
316
[a,a,4,4,5,5-$^2H_6$]-316
([a,a,4,4,5,5-$^2H_6$]-317)

Scheme 8.52 Reagents and conditions: (i) D_2O/NaOD. (ii) $LiAlD_4$. (iv) $POCl_3$. (v) $H_2NCH_2CH_2Cl$. (vi) $H_2NCD_2CH_2Cl$.

dispersed in water. The cytotoxic potencies on a panel of human tumor cell lines of these new oxazaphosphorines, as bulk form or nano-assemblies, and the release of 4-hydroxy-IFO from these preactivated IFO derivatives in plasma are also studied [256] (Scheme 8.53).

The synthesis of preactivated conjugates **329a–f** is illustrated in Scheme 8.54 [256]. Following the previous reports, anodic oxidation of IFO in the presence of methanol yielded the mixture of isomers of the expected 4-methoxy-ifosfamide (MeO-IFO, **327)** as [258]. The alkyl chain replaced the methoxy group for producing the iminium salt **328** after treating compound **6** with the Lewis acid. $BF_3 \cdot Et_2O$ was a favorable Lewis acid reagent [256]. Compounds **329a–f** were furnished upon treating compound **327** using one equivalent of BF_3. Et_2O followed by the addition of a slight excess of nucleophiles **332** and **335**. The compounds generated as diastereomeric mixtures because of the presence of the C-4 asymmetric center and

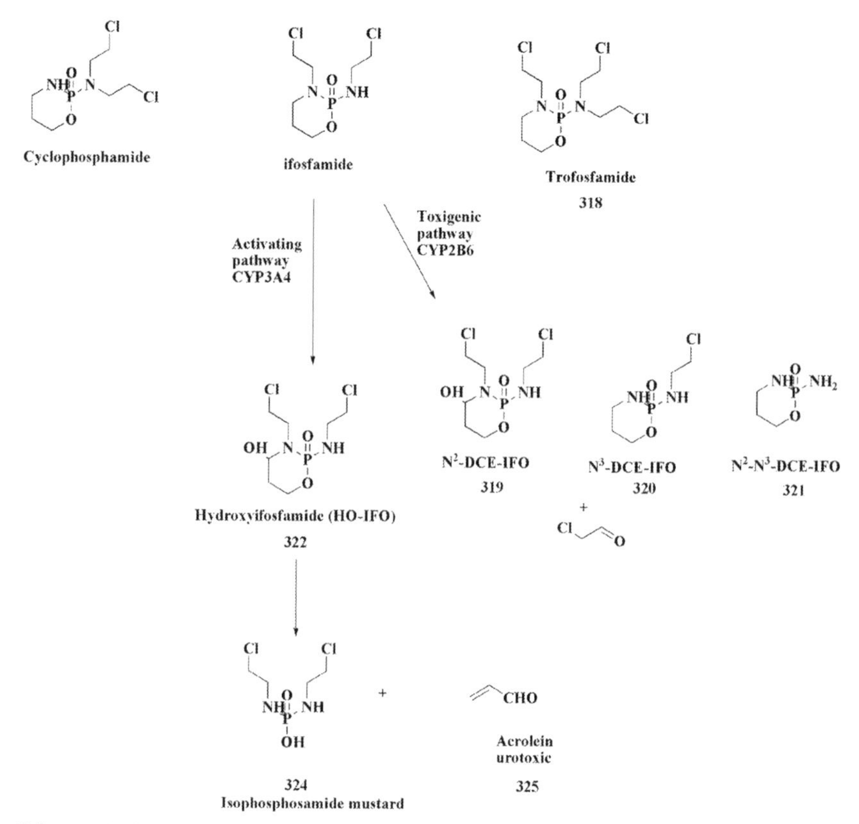

Scheme 8.53

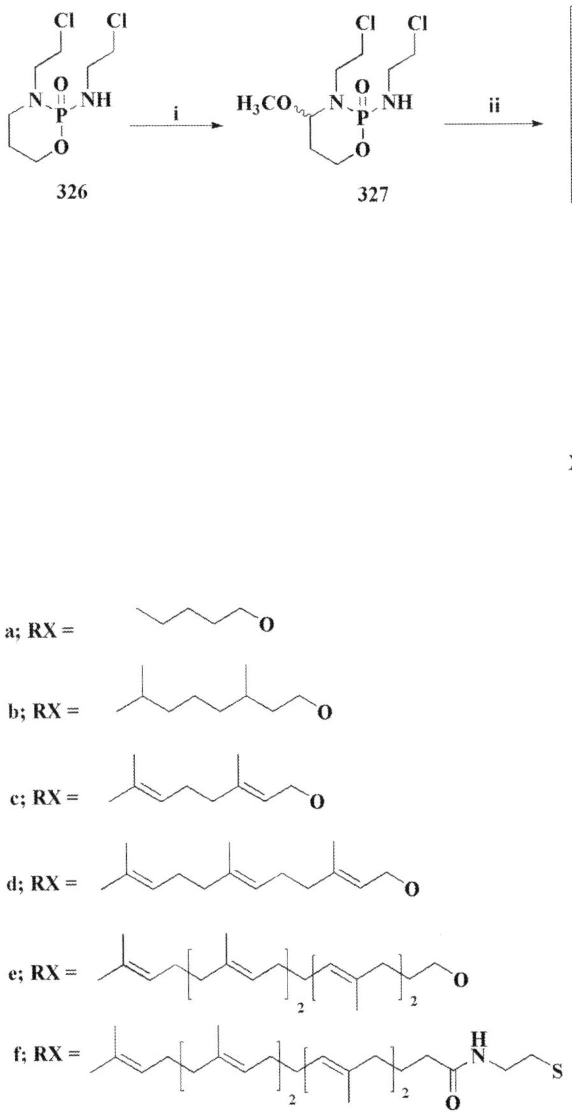

Scheme 8.54 Reagents and conditions: (i) Anodic oxidation/CH_3OH/Rdt. (ii) CH_2Cl_2/BF_3. OEt_2/−78°C. (iii) RXH.

the phosphorous stereogenic center. The squalene analog **335** was accessed from squalene as accomplished in Scheme 8.55. 1,1′,2-Trisnorsqualenol can also be generated from squalene using 1,1′,2-trisnorsqualenic aldehyde (**331)** [28,259]. 1,1′,2-Trisnorsqualenic acid (**333**) was furnished due to oxidation of compound **331** [28]. Acid activation by NHS utilizing DCC as coupling

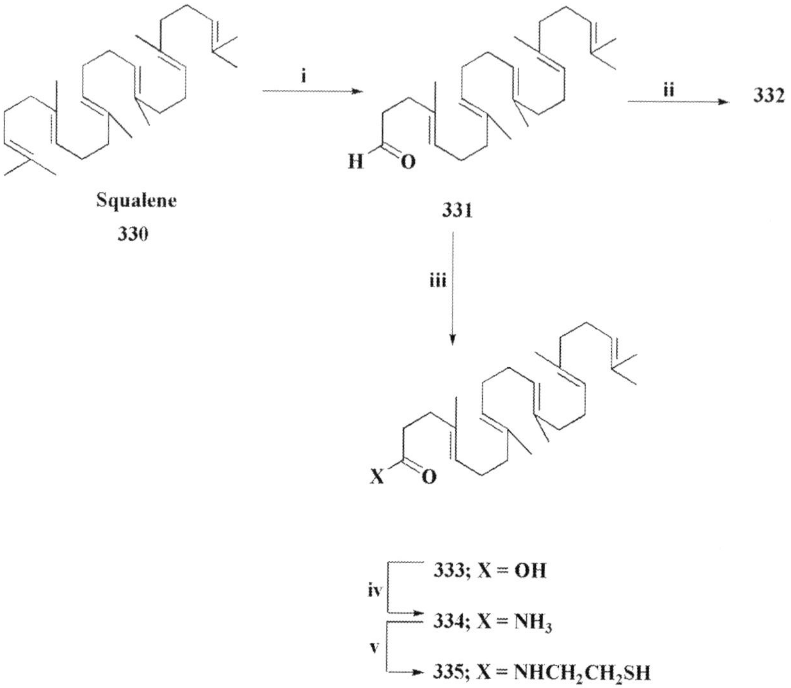

Squalene
330

i

H O

331

ii 332

iii

X O

iv 333; X = OH
334; X = NH_3
v 335; X = $NHCH_2CH_2SH$

Scheme 8.55 Reagents and conditions: (i) Ref. [28,29]. (ii) EtOH/$NaBH_4$/20°C/0.5 h. (iii) H_2SO_4/CrO_3/acetone/0°C. (iv) DCC/NHS/THF/24 h/20°C. (v) $HSCH_2CH_2NH_2$.HCl/TEA/CH_2Cl_2/24 h/20°C.

agent and then the subsequent simple cysteamine treating yielded compound **329f** [256].

12.4 Synthetic strategies of temozolomide analogs

Syntheses of many temozolomide analogs were furnished (Scheme 8.56) [260,261]. The preparation of functionalized tetrazines highly proper for the DARinv was reported [261]. The synthesis of tetrazine-based dienes in aqueous solution is accomplished. Cyanobenzoic acid **337** and cyanopyrimidine **336** reacted with hydrazine (aqueous) to produce the dihydro-1,-2,4,5-tetrazine **338**. Further oxidation to the corresponding 1, 4-diaryl-1,2,4,5-tetrazine subsequently occurred via the transformation to the acid chloride with sulfinyl chloride, which was then allowed to react with the Boc-mono-protected 1,3-propylenediamine to the propyleneamine-substituted acid amide **339**. After deprotection with TFA, the amino group was transferred with the acid chloride analog of TMZ **340** to TMZ-diaryl-tetrazine **341**, a diene compound poised for the DARinv [260].

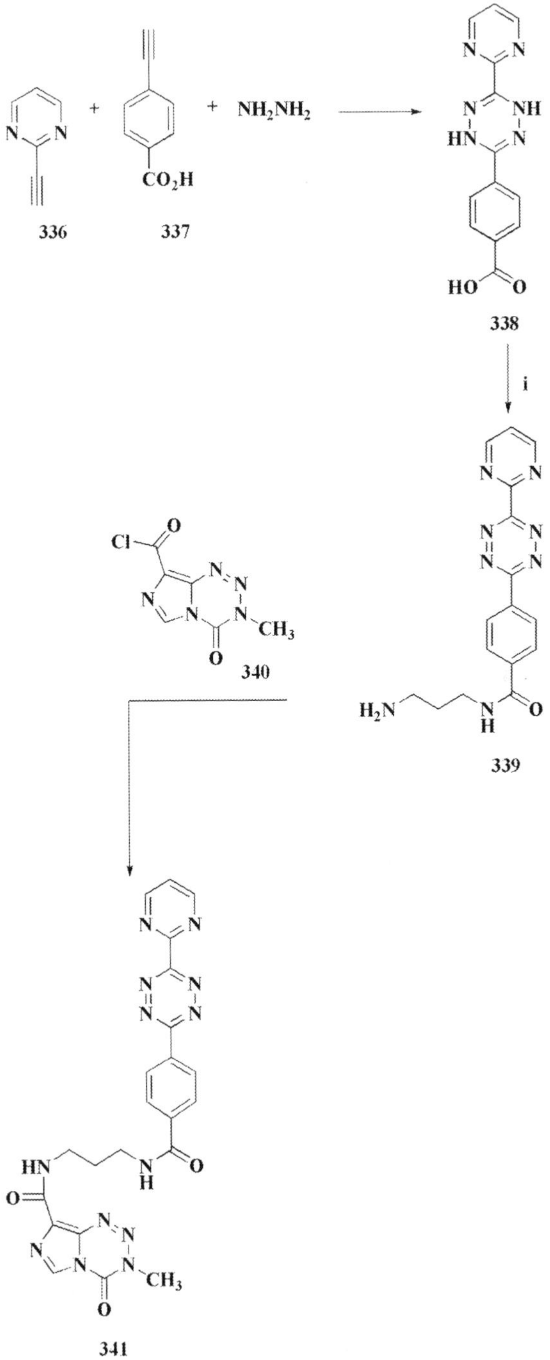

Scheme 8.56 Reagents and conditions: (i) Oxidation, $SOCl_2$, $NH_2(CH_2)_3NHBoc$.

One-pot radiosynthesis of [11C]TMZ was accessed in a radiochemical yield. [11C]TMZ is a useful radiopharmaceutical to monitor tumor cells through transfer of the 3-N-11C-methyl group to tumor cell DNA in vivo. The antitumor features of TMZ is the reason of its utility as a prognostic and diagnostic agent. Positron emission tomography (PET) imaging has been utilized with [11C]TMZ to affirm the drug's therapeutic action and estimate its metabolic potency, and other properties such as the biodistribution and pharmacokinetics. Challenges with reproducibly preparing [11C] TMZ, and the maintenance of its stability, have reduced its accessibility.

Aminoimidazole **347** was diazotized to yield diazoimidazole **345**. Similarly, further treatment of N-(tert-butoxycarbonyl) glycine **348** with ethylchloroformate and triethylamine afforded **347**. Subsequent treatment of compound **348′** upon isolation with **345** gave compound **349** starting from compound **345**. Compound **349** was deprotected to generate **350** (Schemes 8.57 and 8.58) [262].

A novel apprach to synthesize the 11C-labeled temozolomide and in particular pursuing alkylation with [11C]methyl iodide was developed. 3-(Hydroxymethyl)-4-oxo-3,4-dihydroimidazo[5,1-*d*][1,2,3,4]tetrazine-8-carboxamide (3-*N*-hydroxymethyl temozolomide) was accessed as depicted in Scheme 8.59. 2-(Trimethylsilyl)ethoxymethyl chloride (**351**) was transformed with silver isocyanate to (**352**), which then allowed to react with 5-diazoimidazole-4-carboxamide to generate 2-(trimethylsilyl)ethoxymethyl

$[^{11}C]CO_2$ (**342**) —i,ii→ $[^{11}C]CH_3I$ (**343**) —iii→ $[^{11}C]CH_3NCO$ (**344**)

iv, **345** → **346** ($[^{11}C]$TMZ)

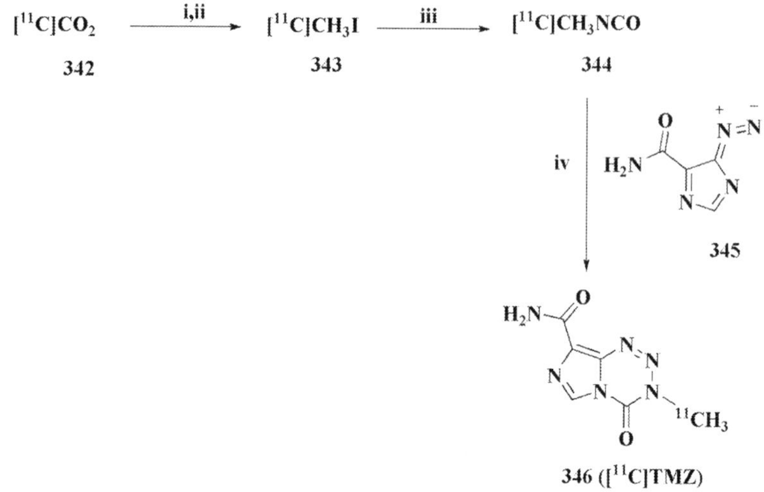

Scheme 8.57 Reagents and conditions: (i) Ni/H_2, 350°C. (ii) I2/740C. (iii) AgOCN/180°C. (iv) DMSO/100°C/10 min.

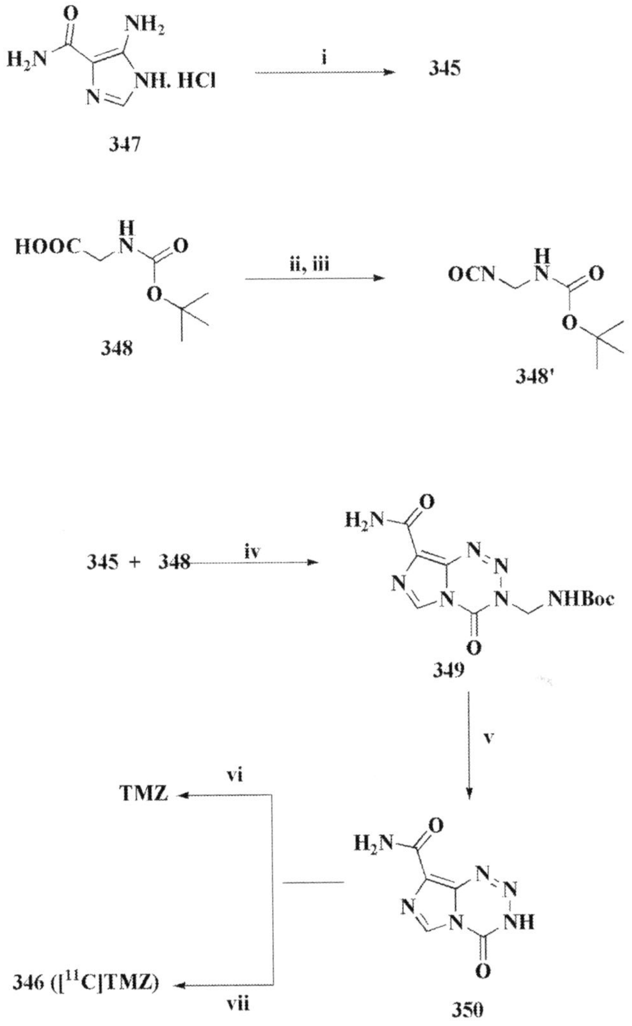

Scheme 8.58 Reagents and conditions: (i) $NaNO_2$, aqueous HCl, 0°C/5 min. (ii) $ClCO_2Et$, TEA/THF/0°C/45 min. (iii) NaN_3/H_2O C. (iv) DMSO/N_2 atmosphere/overnight/RT. (v) NaH/DMF/0°C/RT/5 h. (vi) [11C]CH3I/NaH/DMF/0°C/3 min./45°C/5 min.

analog of temozolomide (**353**). Further treating with Trifluoroborane afforded the starting material for the labeling reaction, 3-*N*-hydroxymethyl temozolomide (**354**) [263].

Rapid radiosynthetic methods for the synthesis of temozolomide labeled with the short-lived positron emitter were developed.

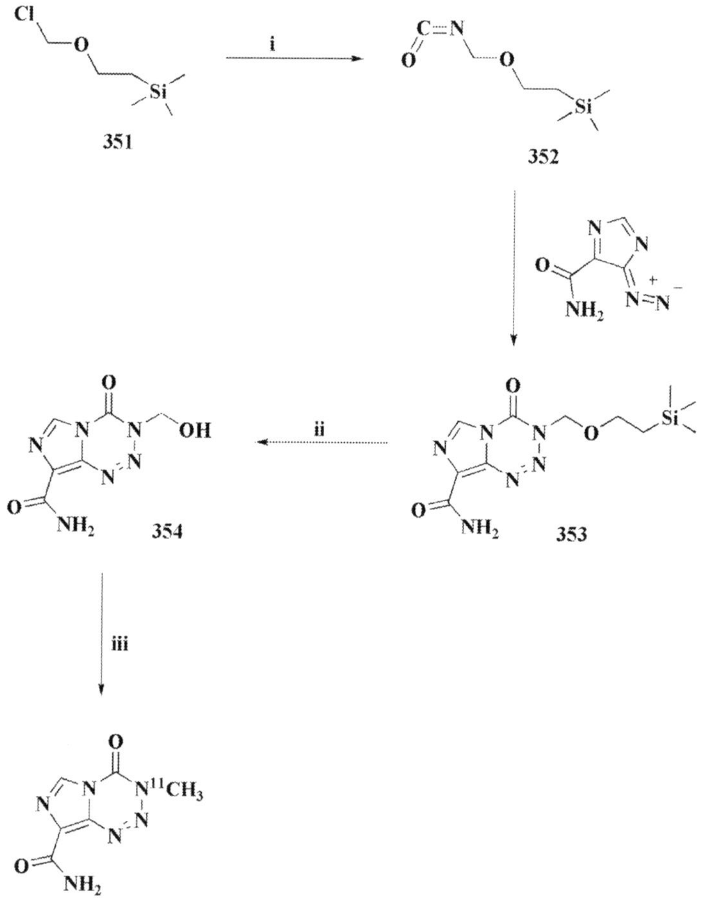

Scheme 8.59 Reagents and conditions: (i) AgNCO. (ii) BF_3. (iii) $[^{11}C]CH_3I$.

5-Diazoimidazole-4-carboxamide (**358**) reacted with the labeling agent [11C-*methyl*]methyl isocyanate afforded [3-*N*-11C-*methyl*]temozolomide (**357**) from [11C-*methyl*]methyl isocyanate. In a similar manner, the reaction of 5-diazoimidazole-4-carboxamide (**358**) and [11C-*carbonyl*]methyl isocyanate afforded [4-11C*carbonyl*]temozolomide (**362**). Apyrogenic samples of compound **357** and [4-11C*carbonyl*]temozolomide (**362**) have been synthesized and utilized in human PET studies (Schemes 8.60–8.62) [264].

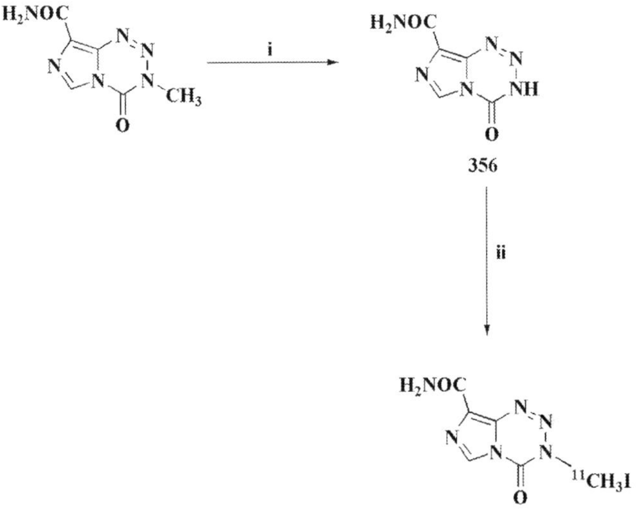

H2NOC N N N N N CH3 O → i → H2NOC N N N N NH O 356 → ii → H2NOC N N N N N 11CH3I O 357

Scheme 8.60 Reagents and conditions: (i) $ClCO_2CH(Cl)CH_3$. (ii) $^{11}CH_3I$

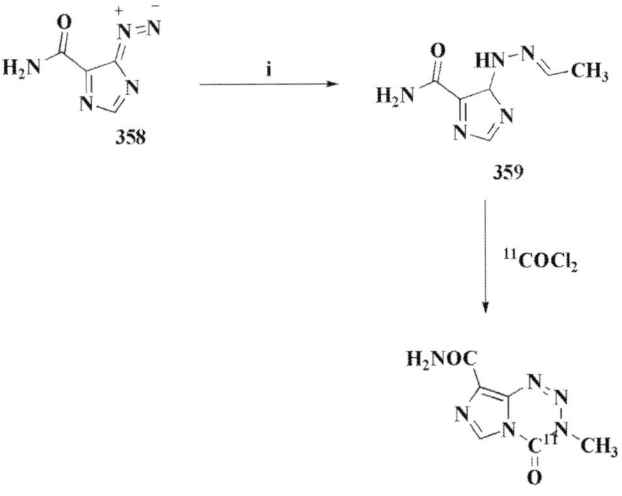

O H2N N=N N N 358 → i → O H2N HN N CH3 N N 359 → $^{11}COCl_2$ → H2NOC N N N N C11 N CH3 O 360

Scheme 8.61 Reagents and conditions: (i) CH_3NH_2

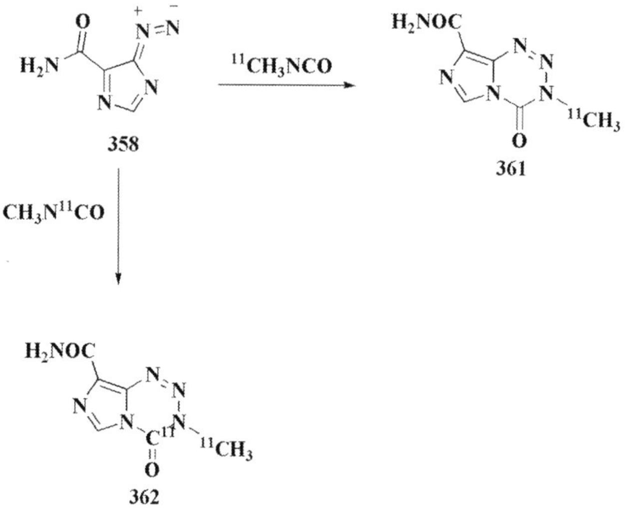

Scheme 8.62

12.5 Synthetic strategies of oxaliplatin analogs

Trans-S,S isomer of the oxaliplatin and the mixture of both enantiomers show less potency than oxaliplatin. The oxaliplatin derivatives (*SP*-4-3)-(4-ethyl-*trans*-cyclohexane-1,2-diamine)oxalatoplatinum(II) and (*SP*-4-3)-(4-methyl-*trans*-cyclohexane-1,2-diamine)oxalatoplatinum(II) have been furnished (Scheme 8.63), and their cytotoxicity has been examined in comparison with oxaliplatin, its corresponding *trans-S,S* isomer, and the mixture of both enantiomers. As compared to oxaliplatin, even the *trans-R,R*/*trans-S,S* mixture of the 4-ethyl- and 4-methyl-substituted oxaliplatin derivatives has indicated an equivalent cytotoxicity in ovarian cancer cells (CH1) and superior antiproliferative features in colon cancer cells (SW480) in the case of an equatorial position of the substituent at position 4 of the *trans*-cyclohexane-1,2-diamine ligand, while an axial substitution leads to a drop in the cytotoxic potency [265].

For the purpose of enhancing the pharmacological profile of oxaliplatin, the syntheses of novel mono- and dialkyl-substituted oxaliplatin derivatives have been performed (Scheme 8.64). The cytotoxicity was estimated in tumor cell lines, revealing that bulky substituents have a negative influence on the cytotoxic activity of the oxaliplatin analogs. With respect to the antiproliferative features, *cis*-4,5-dimethyl-, 4-methyl-, and especially 4,4-dimethyl-*trans*-cyclohexane-1,2-diamine(oxalato)platinum(II) complexes are the most promising candidates to be subsequent investigations. It has

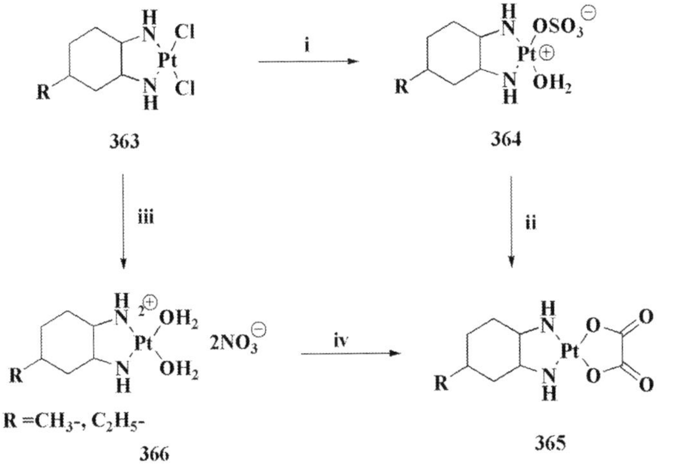

Scheme 8.63 Reagents and conditions: (i) Ag_2SO_4. (ii) $(COOH)_2{\cdot}2H_2O/2NaOH$. (iii) $2AgNO_3$. (iv) $2NaOH/(COOH)_2{\cdot}2H_2O$.

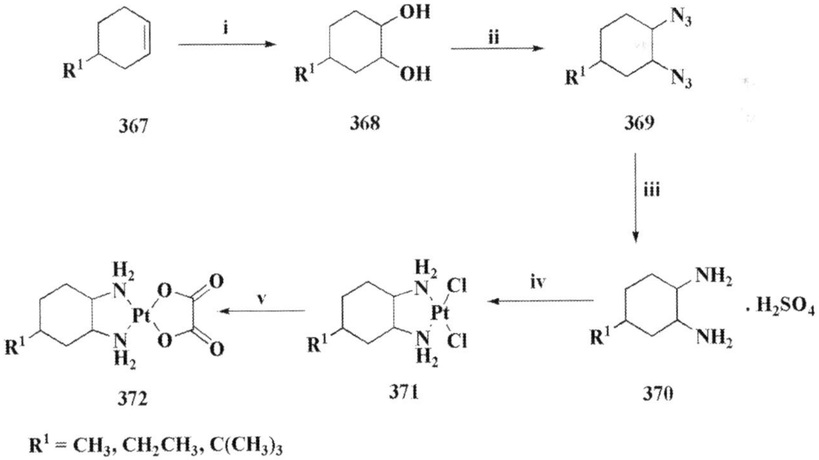

Scheme 8.64 Reagents and conditions: (i) $HCOOH/H_2O_2$. (ii) $NaN_3/MeCl/H_2/Pd/CaCO_3$. (iii) K_2PtCl_4. (iv) $(COONa)_2/AgNO_3$.

been shown that it is possible to improve the anticancer activity that has been strongly demonstrated by cytotoxicity comparison, and 4-alkyl-substituted oxaliplatin analogs have been prepared. Only the mixtures with substituents at C(4) of the cyclohexane ring with predominantly equatorial or axial position could be isolated, and this is considered as the main drawback of these compounds. Following another synthetic strategy, oxaliplatin derivatives were produced, and those derivatives have a stereochemistry defined at the carbon atoms 1, 2, 4, and 5, yielding the racemates and not in problematical mixtures of 4 isomers. This is not only the basis for reliable SAR but also important for enantiomer resolution. 4,4-Dimethyl, *cis*-4,5-dimethyl-, and 4-methyl-substituted analogs are the most promising oxaliplatin derivatives. Compound **372** ($R^1 = CH_3$) is a preferred structure, and this has been demonstrated by the cytotoxic profile as it exhibits only two chiral centers [266].

Estradiol-linked platinum(II) complex analogs to oxaliplatin E-OxaP and carboplatin (E-CarboP, **373a–379a**) were synthesized. E-OxaP and E-CarboP were synthesized based on the estradiol-linker moiety of E-CDDP analogs.

E-CarboP analogs (**373a–379a**) and E-OxaP derivatives were synthesized as shown in scheme. Compounds **373a–379b** were furnished during the treatment of compounds **373**, **378**, or **379** with silver oxalate or silver cyclobutane-1,1-dicarboxylate at room temperature for 1–15 days. The completion of the reaction took 1 day for compound **3** and from 7 to 15 days for compounds **373**, **378** [267].

The significance of the nature of platinum(II) salt on the potency of antiproliferation was assessed on MDA-MB-231 and MCF-7 human mammary carcinoma cell lines together with affinity for the ERa through substituting the dichloroplatinum(II) scaffold by an oxalateplatinum(II) or a cyclobutane-1,1-dicarboxylateplatinum(II) moiety. Excluding compound **379b**, which is inactive at the concentration examined, the antiproliferative potency of all compounds on both human mammary carcinoma cell lines are in μM range and are more potent than oxaliplatin and carboplatin alone but less potent than their ECDDP counterparts. E-CarboP derivatives **373a–379a** reveal very low affinity for Era, while E-OxaP **378b** indicate higher affinity for ERa than their parents E-CDDPs, proposing that the nature of the platinum(II) salt implicated in the vector complexes is essential to retain a remarkable antiproliferative potency and for the selectivity to the ERa and possibility to target estrogen-dependent tissues. E-OxaP **378b** are auspicious alternative vector complexes for estrogen-dependent tissues (Schemes 8.65 and 8.66) [267].

An oxaliplatin analog was furnished (Scheme 8.67); this analog has improved antitumor potency compared to oxaliplatin through the introduction of a small alkyl substituent on the cyclohexane ring in the equatorial position. Utilizing the L1210 murine leukemia model, the highest

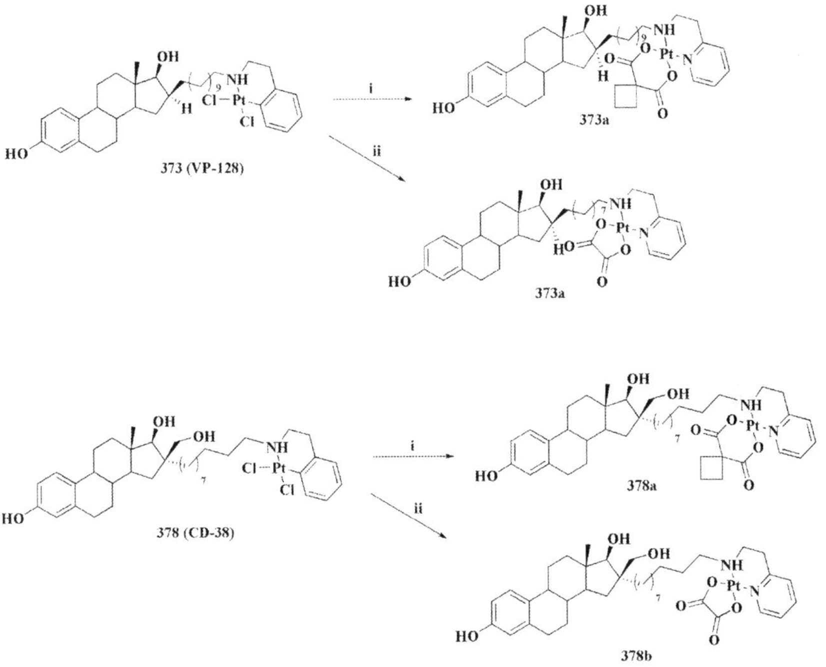

Scheme 8.65 (i) Silver cyclobutane-1,1-dicarboxylate/dimethyl ketone/H_2O. (ii) Silver ethanedioate/dimethyl ketone/H_2O.

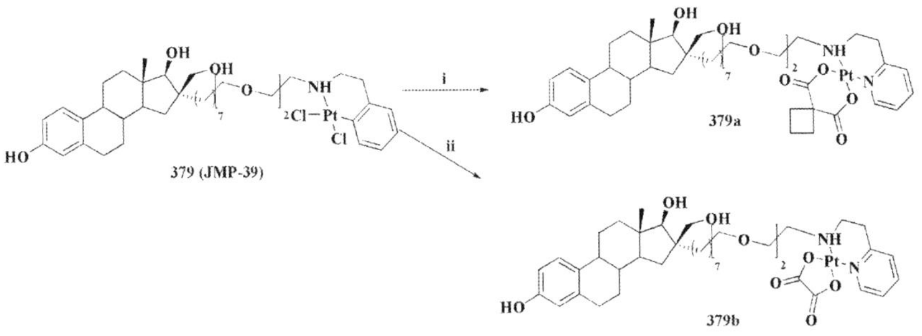

Scheme 8.66 Reagents and conditions: (i) Silver cyclobutane-1,1-dicarboxylate/acetone/H_2O (2:1). (ii) Silver oxalate/acetone/H_2O.

anticancer potency was affirmed for {(1*R*,2*R*,4*R*)-4-methyl-1,2-cyclohexanedi-amine}oxalatoplatinum(II). Although in vitro this compound is not more cytotoxic than oxaliplatin, the better tolerability enables the application of higher doses, leading to improved effectiveness in vivo [268].

The syntheses of zinc(II) phthalocyanine substituted with anoxaliplatin analog through a triethylene glycol linker have been accomplished as

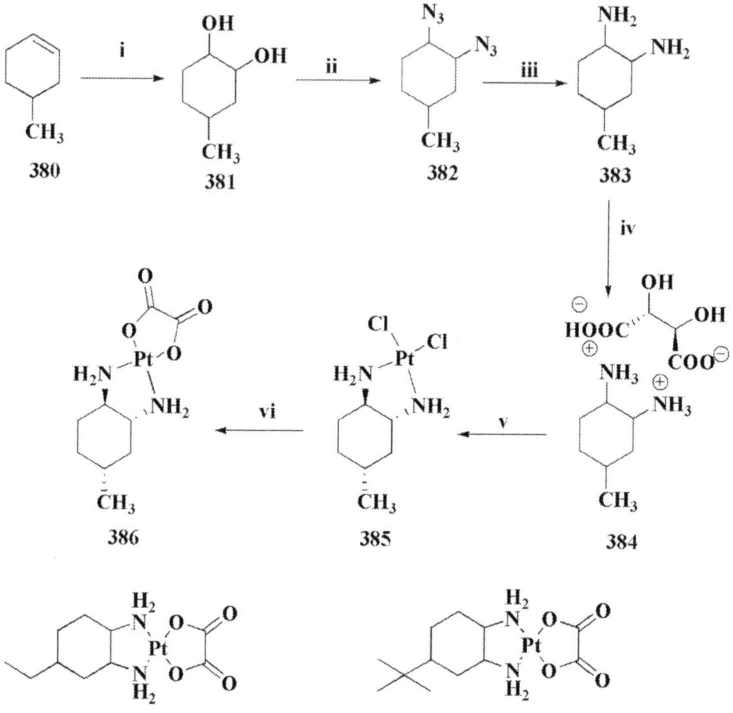

Scheme 8.67 Reagents and conditions: (i) $HCOOH/H_2O_2$. (ii) $MsCl/Py/NaN_3$, (iii) $CaCO_3/Pd/H_2$(3 bar). (iv) (2*R*,3*R*)-2,3-dihydroxysuccinic acid. (v) $NaOH/K_2PtCl_4$. (vi) $AgNO_3$

shown in Scheme 8.68. Its in vitro photodynamic potencies have been evaluated. The photo- and chemotherapeutic agents play a key role cooperatively in the antitumor action. The conjugate indicates a cytotoxic impact in the dark because of the cytostatic oxaliplatin scaffold and an improved cytotoxicity upon illumination due to the photosensitizing phthalocyanine unit against the HT29 human colon adenocarcinoma cells. The IC_{50} value of the conjugate is as low as 0.11 μM toward the HT29 cells, which is five times lower than that of the reference compound without the platinum complex. The high photodynamic potency of the conjugate is a result of its high cellular uptake and efficiency in producing intracellular reactive oxygen species. The conjugate also reveals a preferential localization in the lysosomes of the cells and induces cell death mainly via apoptosis. So this

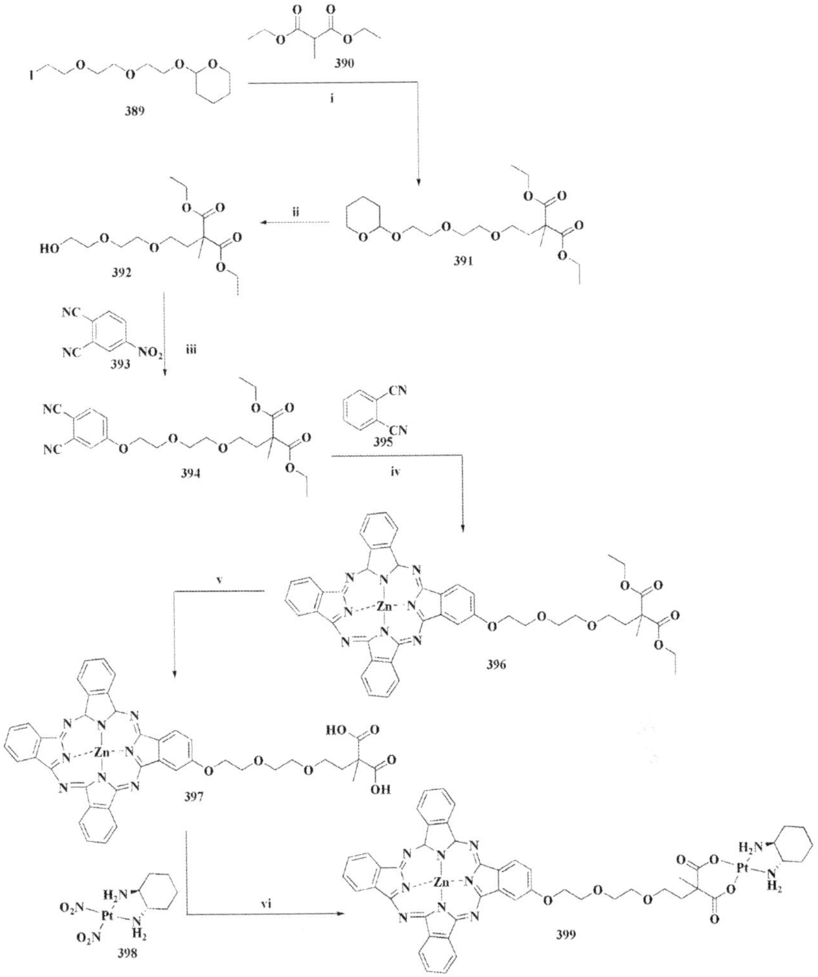

Scheme 8.68 Reagents and conditions: (i) DMF/NaH/60°C/overnight. (ii) EtOH/H_2SO_4/RT/3 h. (iii) DMF/K_2CO_3/110°C/24 h. (iv) $Zn(OAc)_2.2H_2O$/*N*-amyl alcohol/DBU/140–150°C/24 h. (v) acetone/NaOH/reflux/2 h/HCl. (vi) NaOH/H_2O/70°C/48 h.

conjugate is considered as a portentous antitumor agent for dual chemo- and photodynamic therapy [269].

Platinum(II) complexes, characteristic of chiral trans-bicyclo[2.2.2] octane-7,8-diamine as ligand having dicyclic steric hindrance, were prepared (Scheme 8.69). Biological estimation indicated that almost all complexes had a cytotoxic potency against the examined cancer cell lines,

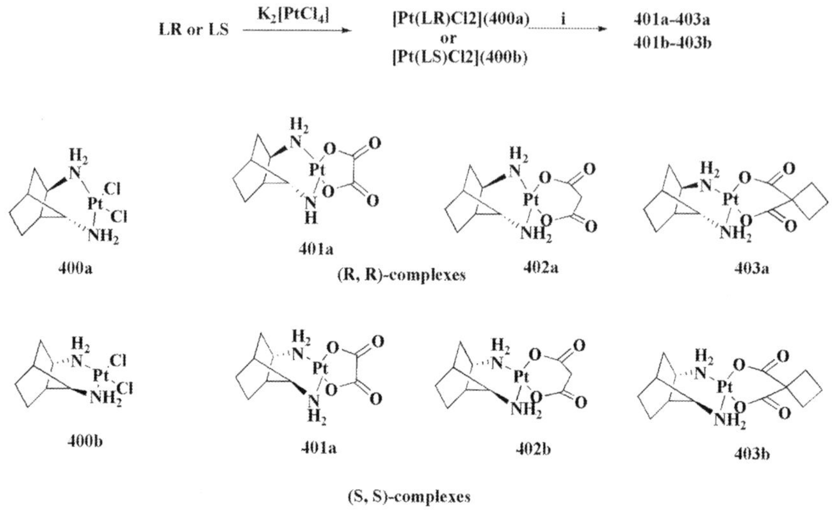

LR or LS $\xrightarrow{K_2[PtCl_4]}$ [Pt(LR)Cl2](400a) or [Pt(LS)Cl2](400b) $\xrightarrow{i}$ 401a-403a 401b-403b

400a 401a 402a 403a

(R, R)-complexes

400b 401a 402b 403b

(S, S)-complexes

Scheme 8.69 Reagents and conditions: (i) Silver dicarboxylate.

among which most of chiral (*R*,*R*)-enantiomeres had a stronger cytotoxicity than their (*S*,*S*)-counterparts, and compound **401a** is the most efficacious agent. Remarkably, its counterpart, **401b**, was much sensitive to cisplatin -resistant SGC7901/CDDP cancer cell line at a higher degree than **401a**. Agarose gel electrophoresis and docking study showed that the interaction of compound **401a** with DNA was similar to that of oxaliplatin. It was accomplished that compound **401a** could induce a better impact than cisplatin on the mitochondrial-dependent apoptosis pathway [270].

The synthesis of platinum(IV) complexes of the type [Pt(Am)-(R $(COO)_2)_2$] (whereas Am is two monodentate am(*m*)ine ligands or a chelating diamine and $R(COO)_2$ is a chelating dicarboxylato scaffold) was performed (Scheme 8.70). So the reaction between the corresponding tetrahydroxidoplatinum(IV)precursors and different dicarboxylic acids, such as malonic, oxalic, cyclobutanedicarboxylic acid, and 3-methylmalonic, was used. Using the MTT colorimetric assay, their in vitro cytotoxicity was identified in a panel of human tumor cell lines (SW480, A549, and CH1). The amine carrier ligands were responsible for their in vitro cytotoxicity, while the DACH-containing compounds affirmed to be the most active species, featuring IC_{50} values superior or comparable to carboplatin. The lipophilicity and redox features of the new complexes were estimated in order to study their pharmacological properties. Increasing the lipophilicity

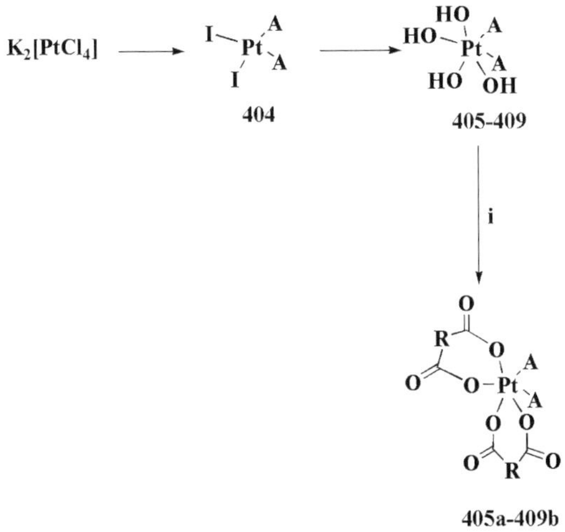

$K_2[PtCl_4]$ → 404 → 405-409 —(i)→ 405a-409b

$A = NH_3$, $EtNH_2$, or cha (cyclohexylamine) or A_2= en (ethane-1,2-diamine) or DACH ((1R,2R)-diaminocyclohexane); $R(COOH)_2$= oxalic, malonic, 3-methylmalonic,or 1,1-cyclobutanedicarboxylic acid.

Scheme 8.70 Reagents and conditions: (i) $R(COOH)_2$, H_2O.

through utilizing various carboxyls to ligands had only a minor impact on the cytotoxicity. Compound **408b** (Pt(DACH)$(mal)_2$) exhibited low in vivo toxicity in animal experiments but a profound anticancer potency against both the CT-26 colon carcinoma and L1210 leukemia models [271].

The oxaliplatin derivative, boldiplatin, has been prepared by the coordination of the boldine analog 3-carboxypredicentrinate to the corresponding platinum(II) scaffold. The complex was biologically estimated in vitro. As compared with oxaliplatin, boldiplatin indicated an equivalent potency over four human tumor cell lines (MCF-7, MDA-MB-231, HT-29, and PC-3) and a 10-fold decrease in toxicity over a nontumor cell line (DHF). This selectivity introduces boldiplatin as a candidate for subsequent investigations to access its prospect as antitumor drug. A boldine analog was used in the syntheses procedure as a ligand [272].

The natural product (**410**) is responsible for the antioxidant potency of **410**), which leads to a poor coordination ability to the platinum (II) scaffold. This is related to the presence of the two sterically hindered phenolic groups at each aromatic ring. To enhance this ability, compound **414**, an analog that contains two structural features that are needed in the preparation of the novel oxaliplatin derivative, was synthesized. The incorporation of a carboxylic group adjacent to the C-2 phenolic position of boldine furnished a salicylic acid-like structure, enhancing the capability of the natural product to coordinate to the platinum(II) unit. Additionally, the existence of this carboxylic group, directly bonded to the aporphine skeleton of boldine, averted the disruption of the functional groups responsible for its antioxidant potency.

3-Carboxypredicentrine (**414**) was synthesized (Scheme 8.71), after methylation of the C-9 phenolic group of boldine. Compound (**414**) and the novel Pt(II) derivative (**415**) were estimated in vitro toward nontumor and human tumor cell lines, and the data were compared with those of oxaliplatin. The preparation of boldiplatin (**415**) exemplifies the coordination of the cytotoxic species [Pt(1*R*,2*R*)-diaminocyclohexane)] + 2 with the cytoprotector ligand (3-carboxypredicentrinate) furnished from boldine. Boldiplatin (**415**) is as potent as oxaliplatin in the growth inhibition of some human tumor cell lines studied in vitro. Compound **415** displays a higher selectivity toward tumor cell lines than oxaliplatin (Scheme 8.72).

12.6 Synthetic strategies of quinazoline analogs

Compounds **423a–d** were synthesized (Scheme 8.73) to release an inhibitor of EGFR TK and a bifunctional alkylator in order to improve the activity of the "combi-molecules." When the basicity of the mustard scaffold increased the activity of the combi-molecules to block EGFR, TK is also increased. They selectively killed cells transfected with EGFR and were active against the DU145 prostate cancer cells. In an irreversible manner, the combi-molecule **423a** can block EGFR phosphorylation, induce the DNA-cross-links, and in mid-S, the cells are arrested [273].

The conjugates of the *N*-mustard-quinazoline were furnished as outlined in Schemes 8.74–8.76. Through a urea linker, the *N*-mustard scaffold was linked at C-6 of 4-anilinoquinazoline derivatives. Different substituents were inserted to the C-4 anilino scaffold. The preliminary antitumor data showed that these agents displayed a remarkable antitumor potency in preventing several human tumor cell growths in vitro. Compounds **436b**, **4361g**, and **436h** were elected for subsequent antitumor potency estimation against prostate PC-3 and human breast carcinoma MX-1 xenograft in an animal model. These agents revealed tumor suppression with 54%–75% and low toxicity. Through the alkaline agarose gel shift assay, the novel

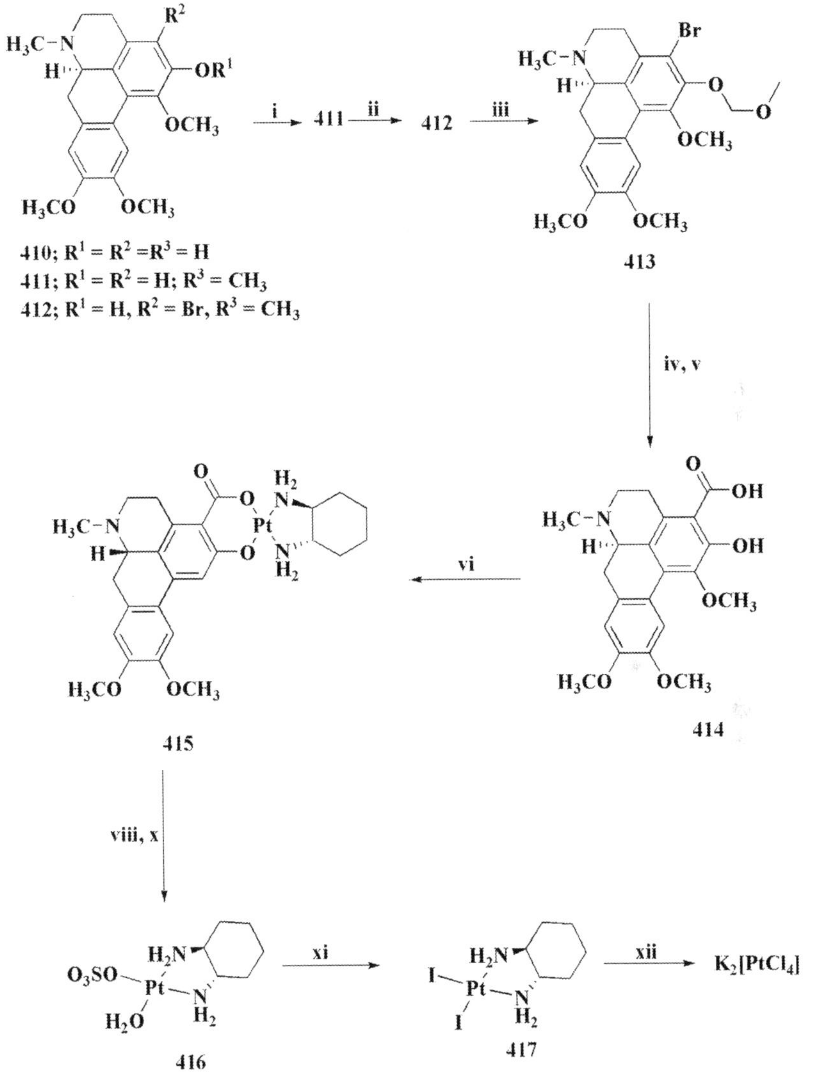

410; $R^1 = R^2 = R^3 = H$
411; $R^1 = R^2 = H$; $R^3 = CH_3$
412; $R^1 = H$, $R^2 = Br$, $R^3 = CH_3$

Scheme 8.71 Reagents and conditions: (i) Ether/CH_2N_2/24 h/RT. (ii) CF_3COOH/NBS/2 h RT. (iii) NaOH/CH_3OCH_2Cl/anhydrous Pyridine/anhydrous EtOH/2 h/RT. (iv) *n*-BuLi/anhydrous THF/−78°C -RT/CO_2/10 min. (v) HCl THF/*i*-PrOH/16 h/RT. (vi) H_2O/KI/5 min/RT. (vii) H_2O/ 1*R*,2*R*-DACH/30 min/RT. (viii) Ag_2SO_4/H_2O/24 h/RT/darkness. (x) H_2O/NaOH/RT.

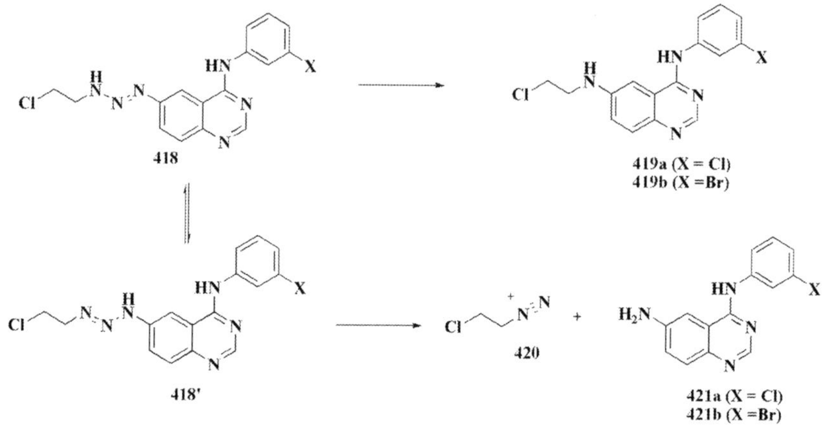

Scheme 8.72

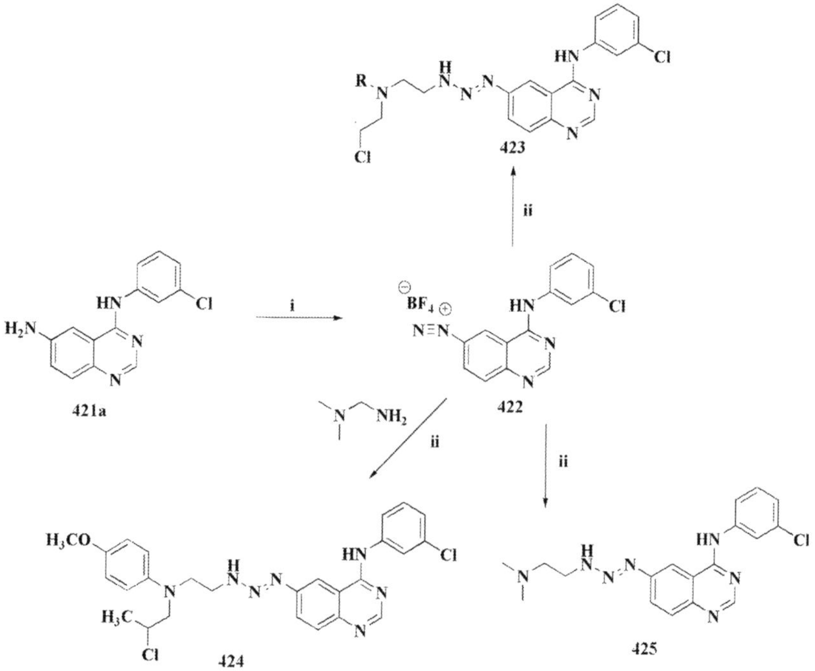

Scheme 8.73 Reagents and conditions: (i) NOBF4/0°C/CH_3CN. (ii) Et_3N/Et_2O/corresponding amine.

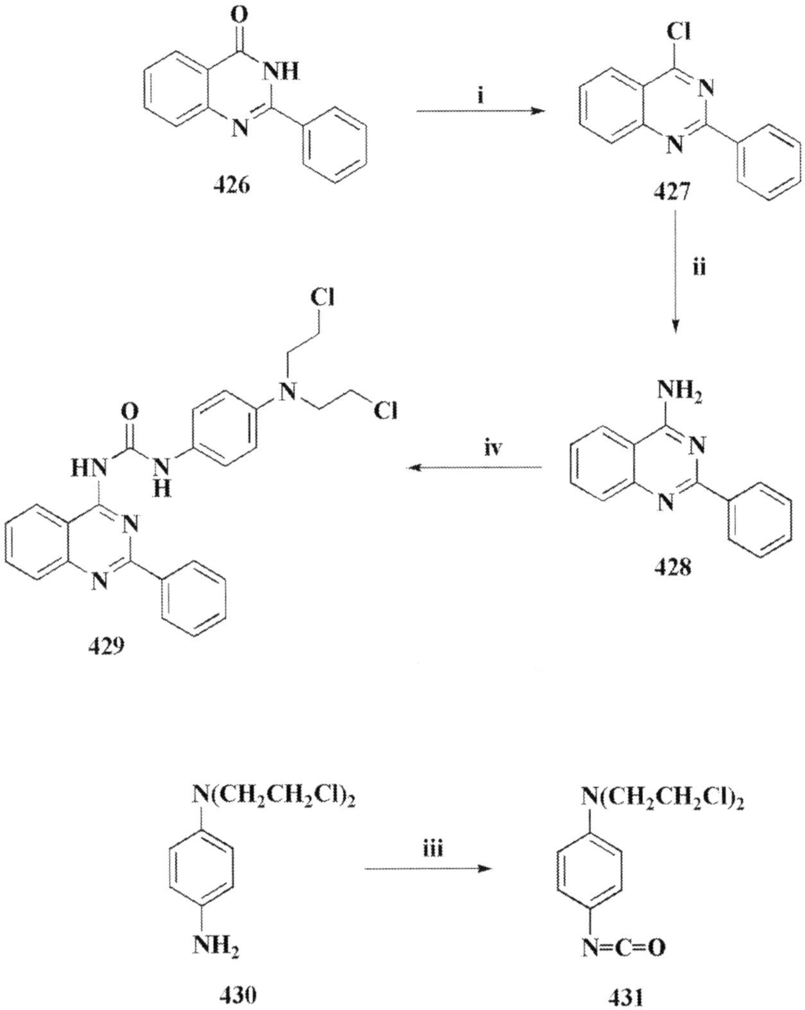

Scheme 8.74 Reagents and conditions: (i) $POCl_3$/reflux. (ii) NH_3(g)/phenol/170°C. (iii) Bis (trichloromethyl) carbonate/TEA/THF/$CHCl_3$/RT. (iv) TEA/$CHCl_3$/RT.

compounds tend to induce DNA cross-linking; also, the cell cycle arrest at G2/M phase was inhibited [274].

Compounds with phosphoramide mustard functionality linked to the quinazoline moiety of HER2/EGFR inhibitors were produced as multitarget-directed ligands against tumor cells. In vitro assays revealed that

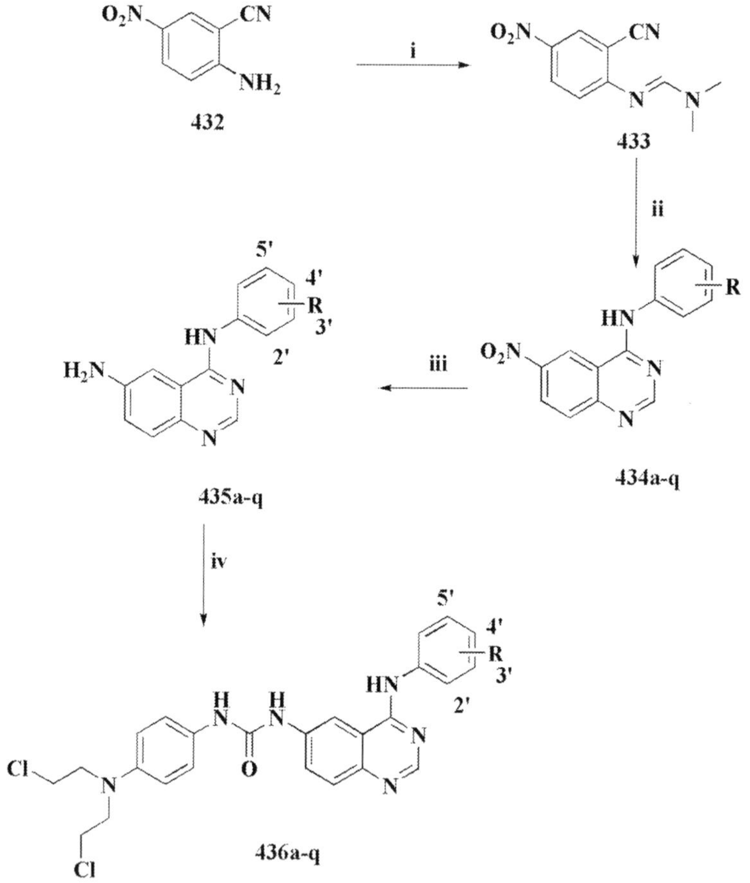

O_2N CN NH_2 432 → i → O_2N CN N N 433 → ii → HN R O_2N N N 434a-q → iii → 5' 4' R 3' HN 2' H_2N N N 435a-q → iv → 5' 4' R 3' HN 2' Cl N Cl H N H N O N N 436a-q

Scheme 8.75 Reagents and conditions: (i) DMF-DMA/reflux. (ii) AcOH/$ArNH_2$/reflux. (iii) AcOH/Fe/H_2O/EtOH/reflux. (iv) TEA/THF/RT.

tumor cell lines with high HER2 level were more sensitive to the compounds than tumor cells with low HER2 level. Compound **443** (EMB-3) was one of the most active inhibitors with IC_{50} of 82nM and 7.4nM against HER2 and EGFR. The mechanism investigations were also corroborated by the impact of **443**-induced DNA damage in MDA-MB-468 cells. In vivo efficacy data indicated that compound **443** could cause a significant inhibition of the H522 tumor xenograft model, and no remarkable body weight loss was found. No acute toxicity was indicated by compound **443** to mice at doses up to 900mg/kg (single dose) and this was revealed through the MTD data [275].

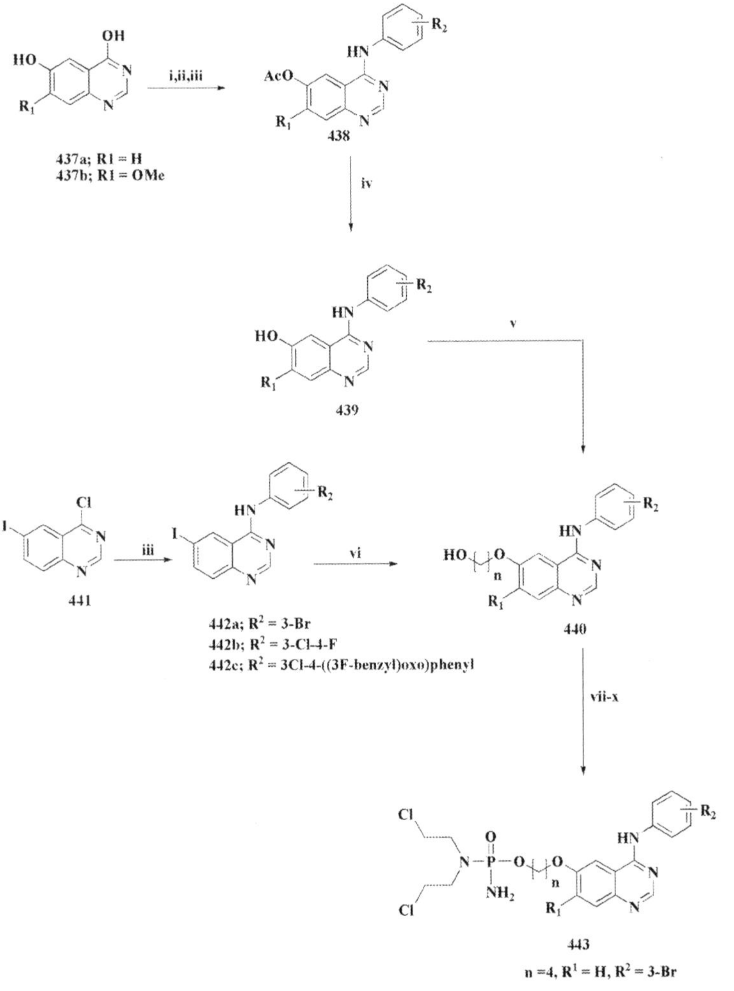

Scheme 8.76 Reagents and conditions: (i) Pyridine/Ac_2O/100°C. (ii) $SOCl_2$/DMF (cat.)/reflux. (iii) Substituted aniline/TEA/IPA isopropanol/reflux. (iv) Ammonia aqueous/CH_3OH/RT. (v) K_2CO_3/CH_3CN or DMF/reflux. (vi) alkyldiol/NaOBut/CuI/dry DMF/80°C. (g) *n*-BuLi/THF/0°C/N_2. (h) THF/$Cl_2P(O)N(CH_2CH_2Cl)_2$/0°C; (i) NH_3(g)/0°C.

Quinazoline nitrogen mustard analogs were accessed as depicted in Schemes 8.77 and 8.78. Their anticancer potencies were estimated in vitro and in vivo. Cytotoxicity assays were performed in cancer cell lines (DU145, SH-SY5Y, HepG2, A549, and MCF-7) and one normal human

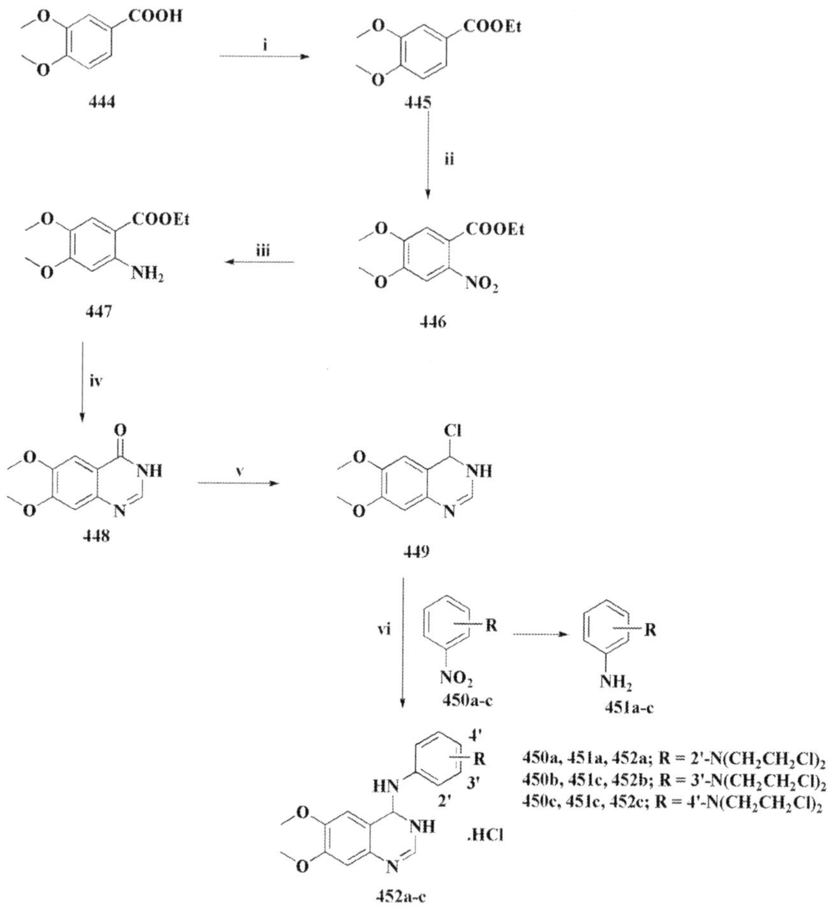

Scheme 8.77 Reagents and conditions: (i) $K_2CO_3/CH_3OCH_2CH_2Br/CH_3CN/Bu_4NBr$. (ii) HNO_3/AcOH/0°C. (iii) H_2/Pa/C/MeOH. (iv) carbamaldehyde/165–170°C. (v) Toluene/$POCl_3$. (vi) concentrated HCl/$SnCl_2$/reflux. (vii) Isopropyl alcohol/reflux.

cell line (GES-1), in which compound **458b** revealed very low IC_{50} to HepG2 (IC_{50}=3.06mM), which was lower than sorafenib. Inhibition of the cell cycle at S and G2/M phases could be happened in case of using compound **458b**, and cell apoptosis was induced. A remarkable cancer growth inhibition with low host toxicity in vivo was displayed by compound **458b** in the HepG2 xenograft model [276].

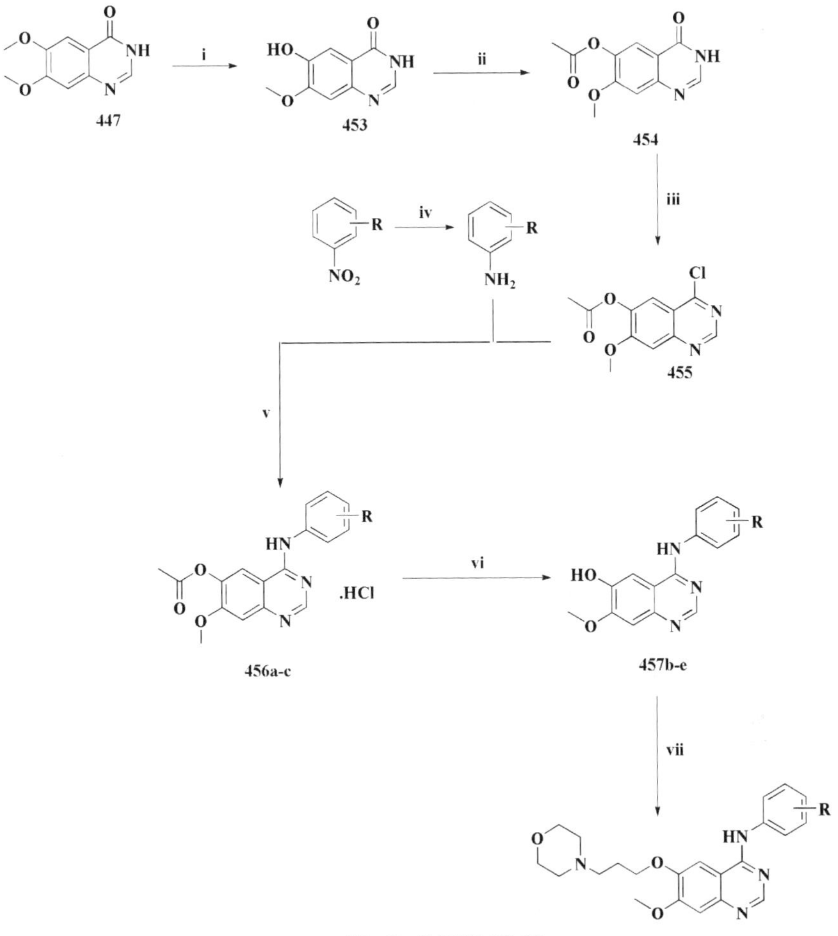

447 453 454 455 456a-c .HCl 457b-e 458b-c

456a; R = 2'-N(CH$_2$CH$_2$Cl)$_2$
456b, 457b, 458b; R = 3'-N(CH$_2$CH$_2$Cl)$_2$
456c, 457c, 458c; R = 4'-N(CH$_2$CH$_2$Cl)$_2$

Scheme 8.78 Reagents and conditions: (i) L-methionine/methansulfonic/120°C. (ii) DMAP/Ac_2O/pyridine. (iii) Toluene/$POCl_3$/reflux. (iv) Isopropyl alcohol/reflux. (v) Concentrated HCl/$SnCl_2$/reflux. (vi) CH_3OH/NH_3/H_2O. (vii) 3-Morpholinopropyl chloride/ DMF/K_2CO_3/N_2 protection.

References

[1] Nieto Y. DNA-binding agents. Cancer Chemother Biol Response Modif 2005; 22:163–203.
[2] Ben Abid F, Gazzah A, Ousbane A, Gutierrez M, Brain E. Alkylating agents. Oncologie 2007;9(11):751–7.
[3] Zanotto-Filho A, Masamsetti VP, Loranc E, Tonapi SS, Gorthi A, Bernard X, Goncalves RM, Moreira JCF, Chen Y, Bishop AJR. Alkylating agent-induced NRF2

blocks endoplasmic reticulum stress-mediated apoptosis via control of glutathione pools and protein thiol homeostasis. Mol Cancer Ther 2016;15(12):3000–14.
[4] Goeder A, Nagel G, Kraus A, Doeorsam B, Seiwert N, Kaina B, Fahrer J. Lipoic acid inhibits the DNA repair protein O6-methylguanine-DNA methyltransferase (MGMT) and triggers its depletion in colorectal cancer cells with concomitant autophagy induction. Carcinogenesis 2015;36(8):817–31.
[5] Calderon-Montano JM, Burgos-Moron E, Orta ML, Pastor N, Perez-Guerrero C, Austin CA, Mateos S, Lopez-Lazaro M. Guanidine-reactive agent phenylglyoxal induces DNA damage and cancer cell death. Pharmacol Rep 2012;64(6):1515–25.
[6] Iatsyshyna AP. Current approaches to improve the anticancer chemotherapy with alkylating agents: state of the problem in world and Ukraine. Biopolym Cell 2012; 28(2):83–92.
[7] Boldogh I, Bhakat KK, Bocangel D, Das GC, Mitra S, Panasci LC, et al. Regulation of DNA repair and apoptosis by p53 and its impact on alkylating drug resistance of tumor cells. In: DNA repair in cancer therapy. Springer; 2004. p. 73–108.
[8] Zheng C, Ji S, Zhang Y. DNA repair and alkylating chemotherapeutic agents. Bull Acad Mil Med Sci 2009;33(1):77–80.
[9] Ralhan R, Kaur J. Alkylating agents and cancer therapy. Expert Opin Ther Pat 2007;17(9):1061–75.
[10] Ma Y, Khojasteh SC, Hop CECA, Erickson HK, Polson A, Pillow TH, Yu S, Wang H, Dragovich PS, Zhang D. Antibody drug conjugates differentiate uptake and DNA alkylation of pyrrolobenzodiazepines in tumors from organs of xenograft mice From. Drug Metab Dispos 2016;44(12):1958–62.
[11] Kothandapani A, Patrick SM. Evidence for base excision repair processing of DNA interstrand crosslinks. Mutat Res Fundam Mol Mech Mutagen 2013;743–744:44–52.
[12] McHugh PJ, Sones WR, Hartley JA. Repair of intermediate structures produced at DNA interstrand cross-links in *Saccharomyces cerevisiae*. Mol Cell Biol 2000;20(10): 3425–33.
[13] Rodrigues-Ferreira C, Silva APP, Galina A. Effect of the antitumoral alkylating agent 3-bromopyruvate on mitochondrial respiration: role of mitochondrially bound hexokinase. J Bioenerg Biomembr 2012;44(1):39–49.
[14] Kaina B, Christmann M. DNA repair in resistance to alkylating anticancer drugs. Int J Clin Pharmacol Ther 2002;40(8):354–67.
[15] Marchini S, Broggini M, Sessa C, D'Incalci M. Development of distamycin-related DNA binding anticancer drugs. Expert Opin Investig Drugs 2001;10(9):1703–14.
[16] Izbicka E, Tolcher AW. Development of novel alkylating drugs as anticancer agents. Curr Opin Investig Drugs 2004;5(6):587–91.
[17] Li B, Gao J. Pharmaceutical composition for the treatment of tumor. Faming Zhuanli Shenqing; 2011. CN 102258526 A 20111130.
[18] Nieto Y. DNA-binding agents. Cancer Chemother Biol Response Modif 2003; 21:171–209.
[19] Roche VF. Cancer and chemotherapy. In: Lemke TL, Williams DA, editors. Foye's principles of medicinal chemistry. 6th ed; 2008. p. 1147–92.
[20] Boulikas T, Alevizopoulos N, Ladopoulou A, Belimezi M, Pantos A, Christofis P, et al. In: Missailidis S, editor. The Cancer Clock. West Sussex: John Wiley & Sons Ltd.; 2007. p. 173–218.
[21] Nieto Y, Jones RB. DNA-binding agents. Cancer Chemother Biol Response Modif 2002;20:197–225.
[22] Berte N, Piee-Staffa A, Piecha N, Wang M, Borgmann K, Kaina B, Nikolova T. Targeting homologous recombination by pharmacological inhibitors enhances the killing response of glioblastoma cells treated with alkylating drugs. Mol Cancer Ther 2016;15(11):2665–78.

[23] Roussel C, Witt KL, Shaw PB, Connor TH. Meta-analysis of chromosomal aberrations as a biomarker of exposure in healthcare workers occupationally exposed to antineoplastic drugs. Mutat Res Rev Mutat Res 2019;781:207–17.
[24] Hurley LH. DNA and its associated processes as targets for cancer therapy. Nat Rev Cancer 2002;2:188.
[25] Ali A, Bhattacharya S. For a review of DNA binders in clinical trials and chemotherapy. Bioorg Med Chem 2014;22:4506.
[26] Nelson SM, Ferguson LR, Denny WA. DNA and the chromosome—varied targets for chemotherapy. Cell Chromosome 2004;3:2.
[27] Blackburn GM, Gait MJ, Loakes D, Williams DM. Nucleic acids in chemistry and biology. 3rd ed. Washington, DC: RSC Publishing; 2006 [chapter 8].
[28] Van Tamelen EE, Curphey TJ. The selective *in vitro* oxidation of the terminal double bound in squalene. Tetrahedron Lett 1962;3:121–4.
[29] Maksimenko A, Mougin J, Mura S, Sliwinski E, Lepeltier E, Bourgaux C, Lepetre-Mouehli S, Zouhiri F, Desmaele D, Couvreur P. Poly-isoprenoyl gemcitabine conjugates self assemble as nano-assemblies, useful for cancer therapy. Cancer Lett 2013;334:346–53.
[30] Avendaño C, Menéndez JC. DNA alkylating agents. In: Medicinal chemistry of anticancer drugs. 2nd ed. Elsevier Science; 2015 [Chapter 5].
[31] Madkour LH. Clinical applications of nucleic acid (DNA) gene therapeutics delivery systems. In: Nucleic acids as gene anticancer drug delivery therapy. Academic Press; 2019 [Chapter 4].
[32] Adair FE, Bagg HJ. Experimental and clinical studies on the treatment of cancer by dichlorethylsulphide (Mustard gas). Ann Surg 1931;93:190.
[33] Busia A, Laffranchi A, Viviani S, Bonfante V, Villani F. Cardiopulmonary toxicity of different chemoradiotherapy combined regimens for Hodgkin's disease. Anticancer Res 2010;30:4381.
[34] Salmelin C, Vilpo J. Chlorambucil-induced high mutation rate and suicidal gene downregulation in a base excision repairdeficient *Escherichia coli* strain. Mutat Res Fundam Mol Mech Mutagen 2002;500(1,2):125–34.
[35] Mohamed D, Mowaka S, Thomale J, Linscheid MW. Chlorambucil-adducts in DNA analyzed at the oligonucleotide level using HPLC-ESI MS. Chem Res Toxicol 2009;22(8):1435–46.
[36] Zhang J, Ye Z, Lou Y. Metabolism of chlorambucil by rat liver microsomal glutathione S-transferase. Chem Biol Interact 2004;149(1):61–7.
[37] Roy G, Horton JK, Roy R, Denning T, Mitra S, Boldogh I. Acquired alkylating drug resistance of a human ovarian carcinoma cell line is unaffected by altered levels of pro- and anti-apoptotic proteins. Oncogene 2000;19(1):141–50.
[38] Omoomi FD, Siadat SD, Nourmohammadi Z, Tabasi MA, Pourhoseini S, Babaei RA, Saffari M, Ardestani MS. Molecular Chlorambucil-Methionine conjugate: novel anti-cancer agent against breast MCF-7 cell model. J Cancer Ther 2013;5(2):075–84.
[39] Amiot A, Levy M, Copie-Bergman C, Dupuis J, Szablewski V, Le Baleur Y, Baia M, Belhadj K, Sobhani I, Leroy K, et al. Rituximab, alkylating agents or combination therapy for gastric mucosa-associated lymphoid tissue lymphoma: a monocentric non-randomised observational study. Aliment Pharmacol Ther 2014;39(6):619–28.
[40] Salmelin C, Vilpo J. Induction of SOS response, cellular efflux and oxidative stress response genes by chlorambucil in DNA repair-deficient Escherichia coli cells (ada, ogt and mutS). Mutat Res Fundam Mol Mech Mutagen 2003;522(1,2):33–44.
[41] For reviews, see (a) Cheson BD, Rummel MJ. Bendamustine: rebirth of an old drug. J Clin Oncol 2009;27:1492–501. (b) Hoy SM. Bendamustine: a review of its use in the management of chronic lymphocytic leukaemia, rituximab-refractory indolent non-Hodgkin's lymphoma and multiple myeloma. Drugs 2012;72:1929.

[42] Liu B. Treatment of multiple myeloma using the combination of a class I histone deacetylase inhibitor and an antineoplastic alkylating agent. PCT Int. Appl.; 2013. WO 2013148919 A1 20131003.
[43] Werner W, Letsch G, Ihn W, Sohr R, Preiss R. Synthesis of a potential metabolite of the carcinostatic bendamustin (Cytostasen). Pharmazie 1991;46:113–4.
[44] Shmarina G, Pukhalsky A, Alioshkin V, Sabelnikov A. Melphalan reduces the severity of experimental colitis in mice by blocking tumor necrosis factor-α signaling pathway. Ann N Y Acad Sci 2007;1096(Signal Transduction Pathways, Part D):97–105.
[45] Bayraktar UD, Bashir Q, Qazilbash M, Champlin RE, Ciurea SO. Progress in haploidentical stem cell transplantation. Biol Blood Marrow Transplant 2012;18:372–80.
[46] Chauhan D, Ray A, Viktorsson K, Spira J, Paba-Prada C, Munshi N, Richardson P, Lewensohn R, Anderson KC. In vitro and in vivo antitumor activity of a novel alkylating agent, melphalan-flufenamide, against multiple myeloma cells. Clin Cancer Res 2013;19(11):3019–31.
[47] Wickstrom M, Johnsen JI, Ponthan F, Segestrom L, Sveinbjornsson B, Lovborg H, et al. The novel melphalan prodrug J1 inhibits neuroblastoma growth in vitro and in vivo. Mol Cancer Ther 2007;6:2409–17.
[48] Wickstrom M, Haglund C, Lindman H, Nygren P, Larsson R, Gullbo J. The novel alkylating prodrug J1: diagnosis directed activity profile ex vivo and combination analyses in vitro. Invest New Drugs 2008;26(3):195–204.
[49] Gullbo J, Wickstrom M, Tullberg M, Ehrsson H, Lewensohn R, Nygren P, et al. Activity of hydrolytic enzymes in tumour cells is a determinant for anti-tumour efficacy of the melphalan containing prodrug J1. J Drug Target 2003;11:355–63.
[50] Wickstrom M, Viktorsson K, Lundholm L, Aesoy R, Nygren H, Sooman L, et al. The alkylating prodrug J1 can be activated by aminopeptidase N, leading to a possible target directed release of melphalan. Biochem Pharmacol 2010;79:1281–90.
[51] Mikhael JR, Schuster SR, Jimenez-Zepeda VH, Bello N, Spong J, Reeder CB, et al. Cyclophosphamide-bortezomib-dexamethasone (CyBorD) produces rapid and complete hematologic response in patients with AL amyloidosis. Blood 2012;119:4391–4.
[52] Abed Alhassan SS, Baraaj AH. Histological and biochemical change in the liver of male albino rats treated with different doses of cyclophosphamide. Int Res J Pharm 2018;9 (6):64–70.
[53] Sakr SA, Shalaby SY, Beder RH. Ameliorative effect of fennel oil on cyclophosphamide induced hepatotoxicity in albino rats. Br J Pharm Res 2017;17(2). BJPR. 34197/1-BJPR.34197/12.
[54] Iqubal A, Iqubal MK, Sharma S, Ansari MA, Najmi A, Ali SM, Ali J, Haque SE. Molecular mechanism involved in cyclophosphamide-induced cardiotoxicity: old drug with a new vision. Life Sci 2019;218:112–31.
[55] O'Shaughnessy JA. Oral alkylating agents for breast cancer therapy. Drugs 1999;58 (Suppl. 3):1–9.
[56] Yasutake N, Urakawa M, Tokumitsu T, Gonda M, Ohyama W, Onoue M. Establishment of an in vivo genotoxicity test using bladder epithelial cells. 1. Detection of cell division and micronuclei induced by a mutagen and a carcinogen in isolated cells. Annual Report of Yakult Central Institute 2002;21:1–10.
[57] Bauer JA, Chiavaroli C. Combination anticancer therapy or OM-174 and pharmaceutical compositions therefor. PCT Int. Appl; 2006. WO 2006095270 A1 20060914.
[58] Zhang J, Tian Q, Zhou S. Clinical pharmacology of cyclophosphamide and ifosfamide. Curr Drug Ther 2006;1(1):55–84.
[59] Nejad MI, Johnson KM, Price NE, Gates KS. A new cross-link for an old cross-linking drug: the nitrogen mustard anticancer agent mechlorethamine generates cross-links derived from abasic sites in addition to the expected drug-bridged cross-links. Biochemistry 2016;55(50):7033–41.

[60] Kapuriya N, Kapuriya K, Dong H, Zhang X, Chou T, Chen Y, Lee T, Lee W, Tsai T, Naliapara Y, et al. Novel DNA-directed alkylating agents: design, synthesis and potent antitumor effect of phenyl N-mustard-9-anilinoacridine conjugates via a carbamate or carbonate linker. Bioorg Med Chem 2009;17(3):1264–75.
[61] Bartzatt R. Utilizing a D-amino acid as a drug carrier for antineoplastic nitrogen mustard groups. Drug Deliv 2005;12(3):141–7.
[62] Wilson HM. Primary carcinoma of the liver: report of 20 cases, including two treated with triethylene melamine. Ann Intern Med 1954;41(1):118–23.
[63] Eagan RT, Dinapoli RT, Cascino TL, Scheithauer B, O'Neill BP, O'Fallon JR. Comprehensive phase II evaluation of Aziridinylbenzoquinone (AZQ, Diaziquone) in recurrent human primary brain tumors. J Neuro-Oncol 1987;5:309–14.
[64] Obe G, Beek B. Trenimon: biochemical, physiological and genetic effects on cells and organisms. Mutat Res 1979;65:21–70.
[65] Phillips RM, Hendriks HR, Peters GJ. EORTC-Pharmacology and Molecular Mechanism Group. EO9 (Apaziquone): from the clinic to the laboratory and back again. Br J Pharmacol 2013;168(1):11–8.
[66] Begleiter A. Clinical applications of quinone-containing alkylating agents. Front Biosci 2000;5:E153–71.
[67] Danson S, Ward TH, Butler J, Ranson M. DT-diaphorase: a target for new anticancer drugs. DT-diaphorase: a target for new anticancer drugs. Cancer Treat Rev 2004;30:437–49.
[68] Gibson NW, Hartley JA, Butler J, Siegel D, Ross D. Relationship between DT-diaphorase-mediated metabolism of a series of aziridinylbenzoquinones and DNA damage and cytotoxicity. Mol Pharmacol 1992;42:531.
[69] Fourie J, Guziec Jr F, Guziec L, Monterrosa C, Fiterman DJ, Begleiter A. Structure-activity study with bioreductive benzoquinone alkylating agents: effects on DT-diaphorase-mediated DNA crosslink and strand break formation in relation to mechanisms of cytotoxicity. Cancer Chemother Pharmacol. Cancer Chemother Pharmacol 2004;53(3):191–203.
[70] Beall HD, Winski SL. Mechanisms of action of quinone-containing alkylating agents. I. NQO1-directed drug development. Front Biosci 2000;5:D639–48.
[71] Begleiter A. Clinical applications of quinone-containing alkylating agents. Front Biosci 2000;5:E153–71.
[72] Winski SL, Hargreaves RHJ, Butler J, Ross D. A new screening system for NAD(P)H: quinone oxidoreductase (NQO1)-directed antitumor quinones: identification of a new aziridinylbenzoquinone, RH1, as a NQO1-directed antitumor agent. Clin Cancer Res 1998;4:3083.
[73] Ward TH, Danson S, McGown AT, Ranson M, Coe NA, Jayson GC, et al. Preclinical evaluation of the pharmacodynamic properties of 2,5-diaziridinyl-3-hydroxymethyl-6-methyl-1,4-benzoquinone. Clin Cancer Res 2005;11:2695.
[74] Danson S, Ranson M, Denneny O, Cummings J, Ward TH. Validation of the comet-X assay as a pharmacodynamic assay for measuring DNA cross-linking produced by the novel anticancer agent RH1 during a phase I clinical trial. Cancer Chemother Pharmacol 2007;60:851–61.
[75] Danson SJ, Johnson P, Ward TH, Dawson M, Denneny O, Dickinson G, et al. Phase I pharmacokinetic and pharmacodynamic study of the bioreductive drug RH1. Ann Oncol 2011;22:1653–60.
[76] Hartley JA, O'Hare CC, Baumgart J. DNA alkylation and interstrand cross-linking by treosulfan. Br J Cancer 1999;79:264–6.
[77] Danylesko I, Shimoni A, Nagler A. Treosulfan-based conditioning before hematopoietic SCT: more than a BU look-alike. Bone Marrow Transplant 2012;47:5–14.

[78] Bernardo ME, Piras E, Vacca A, Giorgiani G, Zecca M, Bertaina A, et al. Allogeneic hematopoietic stem cell transplantation in thalassemia major: results of a reduced-toxicity conditioning regimen based on the use of treosulfan. Blood 2012;120:473–6.

[79] Szebeni J, Barna K, Uher F, Milosevits J, Paloczi K, Gaal D, et al. Comparison of the lymphoid toxicities of mitobronitol and busulphan in mice: reduced B cell toxicity and improved thymic recovery as possible contributors to the reduced risk for complications following BMT with mitobronitol preconditioning. Leukemia 1997; 11:1769–74.

[80] Coleman RS, Perez RJ, Burk CH, Navarro A. Studies on the mechanism of action of azinomycin B: definition of regioselectivity and sequence selectivity of dna cross-link formation and clarification of the role of the naphthoate. J Am Chem Soc 2002;124:13008.

[81] Foulke-Abel J, Agbo H, Zhang H, Mori S, Watanabe CMH. Mode of action and biosynthesis of the azabicycle-containing natural productsazinomycin and ficellomycin. Nat Prod Rep 2011;28:693–704.

[82] Ravdin PM, Havlin KA, Marshall MV, Brown TD, Koeller JM, Kuhn JG, et al. A phase I clinical and pharmacokinetic trial of hepsulfam. Cancer Res 1991;51(23 Pt 1):6268–72.

[83] Bartzatt R. Design of novel anticancer drugs utilizing busulfan for optimizing pharmacological properties and pattern recognition techniques for elucidation of clinical efficacy. In: Spencer P, Holt W, editors. Anticancer drugs. Nova Science Pub, Inc.; 2009. p. 195–210.

[84] Puppa A, Rossetto M, Ciccarino P, Denaro L, Rotilio A, d'Avella D, et al. Carmustine wafer implantation when surgical cavity is communicating with cerebral ventricles: technical considerations on a clinical series. World Neurosurg 2011;76(1–2):156–9.

[85] Glas M, Hundsberger T, Stuplich M, Wiewrodt D, Kurzwelly D, Nguyen-Huu B, et al. Nimustine (ACNU) plus teniposide (VM26) in recurrent glioblastoma. Oncology 2009;76:184–9.

[86] Samson MK, Baker LH, Cummings G, Talley RW, McDonald B, Bhathena DB. Clinical trial of chlorozotocin, DTIC, and dactinomycin in metastatic malignant melanoma. Cancer Treat Rep 1982;66(2):371–3.

[87] Ueda-Kawamitsu H, Lawson TA, Gwilt PR. In vitro pharmacokinetics and pharmacodynamics of 1,3-bis(2-chloroethyl)-1-nitrosourea (BCNU). Biochem Pharmacol 2002;63(7):1209–18.

[88] García del Muro X, López-Pousa A, Maurel J, Martín J, Martínez-Trufero J, Casado A, et al. Randomized phase II study comparing gemcitabine plus dacarbazine versus dacarbazine alone in patients with previously treated soft tissue sarcoma: a Spanish Group for Research on Sarcomas study. J Clin Oncol 2011;29:2528.

[89] Meer L, Janzer RC, Kleihues P, Kolar GF. In vivo metabolism and reaction with dna of the cytostatic agent, 5-(3,3-dimethyl-1-triazeno)imidazole-4-carboxamide (DTIC). Biochem Pharmacol 1986;35:3243–7.

[90] Stevens MFG. For a review of temozolomide development. In: Neidle S, editor. Cancer drug design and discovery. 2nd ed. New York: Elsevier; 2014 [chapter 5].

[91] Jiang G, Li LT, Xin Y, Zhang L, Liu YQ, Zheng JN. Strategies to improve the killing of tumors using temozolomide: targeting the DNA repair protein MGMT. Curr Med Chem 2012;19(23):3886–92.

[92] Chen X, Zhang K, Xin Y, Jiang G. Oncolytic adenovirus-expressed RNA interference of O6-methylguanine DNA methyltransferase activity may enhance the antitumor effects of temozolomide. Oncol Lett 2014;8(5):2201–2.

[93] Quiros S, Roos WP, Kaina B. Rad51 and BRCA2—new molecular targets for sensitizing glioma cells to alkylating anticancer drugs. PLoS One 2011;6(11), e27183.

[94] Shinoura N, Yamada R, Tabei Y, Saito K, Nakamura O, Takahashi M. Temozolomide (Temodal). Gan To Kagaku Ryoho 2008;35(3):543–7.
[95] Roos WP, Tsaalbi-Shtylik A, Tsaryk R, Guevercin F, de Wind N, Kaina B. The translesion polymerase Rev3L in the tolerance of alkylating anticancer drugs. Mol Pharmacol 2009;76(4):927–34.
[96] Beharry AA, Nagel ZD, Samson LD, Kool ET. Fluorogenic real-time reporters of DNA repair by MGMT, a clinical predictor of antitumor drug response. PLoS One 2016;11(4). e0152684/1-e0152684/15.
[97] Li X, Zhen Y, Zhang S, Ouyang Z, Shang Y. Lidamycin and temozolomide composition with synergetic effect for suppressing tumor cell proliferation. Faming Zhuanli Shenqing; 2009. CN 101554472 A 20091014.
[98] Djedid R, Kiss R, Lefranc F. Targeted therapy of glioblastomas: a 5-year view. Therapy 2009;6(3):351–70.
[99] D'Atri S, Graziani G, Lacal PM, Nistico V, Gilberti S, Faraoni I, Watson AJ, Bonmassar E, Margison GP. Attenuation of O6-methylguanine-DNA methyltransferase activity and mRNA levels by cisplatin and temozolomide in Jurkat cells. J Pharmacol Exp Ther 2000;294(2):664–71.
[100] Lefranc F, Sadeghi N, Camby I, Metens T, Dewitte O, Kiss R. Present and potential future issues in glioblastoma treatment. Expert Rev Anticancer Ther 2006;6 (5):719–32.
[101] Fukushima T, Takeshima H, Kataoka H. Anti-glioma therapy with temozolomide and status of the DNA-repair gene MGMT. Anticancer Res 2009;29:4845.
[102] Chan JK, Loizzi V, Manetta A, Berman ML. Oral altretamine used as salvage therapy in recurrent ovarian cancer. Gynecol Oncol 2004;92:368–71.
[103] Jackso C, Crabb TA, Gibson M, Godfrey R, Saunders R, Thurston DE. Studies on the stability of trimelamol, a carbinolamine-containing antitumor drug. J Pharm Sci 1991;80:245–51.
[104] Judson IR, Calvert AH, Gore ME, Balmanno K, Gumbrell LA, Perren T, et al. Phase II trial of trimelamol in refractory ovarian cancer. Br J Cancer 1991;63:311–3.
[105] Kehe K, Szinicz L. Medical aspects of sulphur mustard poisoning. Toxicology 2005;214:198–209.
[106] Monneret C. Platinum anticancer drugs. From serendipity to rational design [Les dérivés du platine en cancérologie. De la sérendipité à l'innovation rationnelle]. Ann Pharm Fr 2011;69:286–95.
[107] Wheate NJ, Walker S, Craig GE, Oun R. For a review of the status of platinum anticancer drugs in the clinic and in clinical trials. Dalton Trans 2010;39:8113.
[108] Boulikas T, Pantos A, Bellis E, Christofis P. Designing platinum compounds in cancer: structures and mechanisms. Cancer Ther 2007;5:537–83.
[109] Kelland L. The resurgence of platinum-based cancer chemotherapy. Nat Rev Cancer 2007;7:573.
[110] Ibrahim A, Hirschfeld S, Cohen MH, Griebel DJ, Williams GA, Pazdur R. FDA drug approval summaries: oxaliplatin. Oncologist 2004;9:8.
[111] Raymond E, Faivre S, Woynarowski JM, Chaney SG. Oxaliplatin: mechanism of action and antineoplastic activity. Semin Oncol 1998;25(2 Suppl 5):4–12.
[112] Graham J, Mushim M, Kirkpatrick P. Oxaliplatin. Nat Rev Drug Discov 2004;3:11.
[113] Wu Q, Qin S-K, Chen C-J, Wang R. Lobaplatin arrests cell cycle progression in human hepatocellular carcinoma cells. J Hematol Oncol 2010;3:43.
[114] Kim NK, Kim TY, Shin SG. A Phase I study of **cis** -malonato[(4 **R**,5 **R**)-4,5-bis(aminomethyl)-1,3-dioxolane] platinum(II) in patients with advanced malignancies. Cancer 2001;91:1549.
[115] Hoctin-Boes G, Cosaert J, Koehler M, Smith M. Safety profile of ZD0473 in phase II trials of patients with advanced cancers. Proc Am Soc Clin Oncol 2001;20:344a.

[116] Gelmon KA, Vandernberg TA, Panasci L, Norris B, Crump M, Douglas L, et al. A phase II study of ZD0473 given as a short infusion every 3 weeks to patients with advanced or metastatic breast cancer: a National Cancer Institute of Canada Clinical Trials Group trial, IND 129. Ann Oncol 2003;14:543–8.
[117] Kelland LR. An update on satraplatin: the first orally available platinum anticancer drug. Expert Opin Investig Drugs 2000;9:1373–82.
[118] Radulovic S, Tesic Z, Manic S. Trans-platinum complexes as anticancer drugs: recent developments and future prospects. Curr Med Chem 2002;9(17):1611–8.
[119] Perego P, Caserini C, Gatti L, Carenini N, Romanelli S, Supino R, Colangelo D, Viano I, Leone R, Spinelli S, Pezzoni G, Manzotti C, Farrell N, Zunino F. A novel trinuclear platinum complex overcomes cisplatin resistance in an osteosarcoma cell system. Mol Pharmacol 1999;55(3):528–34.
[120] Jodrell DI, Evans TRJ, Steward W, Cameron D, Prendiville J, Aschele C, et al. Phase II studies of BBR3464, a novel tri-nuclear platinum complex, in patients with gastric or gastro-oesophageal adenocarcinoma. Eur J Cancer 2004;40:1872–7.
[121] Mehrling T, Festuccia C. Pharmaceutical combination comprising dual-functional alkylating HDAC inhibitor fusion molecule EDOS101 and Poly(ADP)Ribose Polymerase inhibitiors for treating brain cancer. PCT Int. Appl; 2015. WO 2015180865 A1 20151203.
[122] Li D, Hua H, Li Z, Hu X, Shi H, Zhang Z, Li X. Preparation of evodiamine derivatives as antitumor agents. Faming Zhuanli Shenqing; 2018. CN 107602557 A 20180119.
[123] Ciavolella A, Montagnoli A, Pesenti E. Therapeutic combination comprising a PARP-1 inhibitor and an anti-neoplastic agent. PCT Int. Appl; 2012. WO 2012016876 A1 20120209.
[124] Moschel RC, Loktionova NA, Wei G, Pegg AE, Pauly GT, Moschel MK. Preparation of β-glucuronidase cleavable prodrugs of O6-alkylguanine-DNA alkyl-transferase inactivator. US Pat Appl Publ; 2007. US 20070213279 A1 20070913.
[125] Herzon S, Healy A, Crawford J, Vizcaino M, Nikolayevskiy H. DNA alkylation and cross-linking agents as compounds and payloads for targeted therapies. PCT Int. Appl; 2017. WO 2017132459 A1 20170803.
[126] Lewis J, Matteucci M, Chen T, Jiao H. Hypoxia activated drugs of nitrogen mustard alkylators as anticancer agents and their preparation. PCT Int. Appl; 2009. WO 2009140553 A2 20091119.
[127] Singh RK, Prasad DN, Bhardwaj TR. Hybrid pharmacophore-based drug design, synthesis, and antiproliferative activity of 1,4-dihydropyridineslinked alkylating anticancer agents. Med Chem Res 2015;24(4):1534–45.
[128] Croci VD, Valsasina B, Beria I, Ciavolella A, Ballinari D, Pesenti E, Moll J. Therapeutic combination comprising a PLK1 inhibitor and an antineoplastic agent for treatment of cancer. PCT Int. Appl; 2010. WO 2010136394 A1 20101202.
[129] Banerjee R, Rachid Z, McNamee J, Jean-Claude BJ. Synthesis of a prodrug designed to release multiple inhibitors of the epidermal growth factor receptor tyrosine kinase and an alkylating agent: a novel tumor targeting concept. J Med Chem 2003;46 (25):5546–51.
[130] Ono M, Koya K, Sun L, Wada Y, Wrona W, Dales N, Tao X, Holden S. Preparation of cytotoxic agents comprising an alkylating moiety linked to a bioreductive moiety. PCT Int. Appl; 1999. WO 9961409 A1 19991202.
[131] Su T, Chou T. Synthesis of 8H-3a-aza-cyclopenta[a]indenes and 5,10-dihydropyrrolo [1,2-b]isoquinolines derivatives and their use as DNA alkylator antitumor agents. U.S. Pat. Appl. Publ; 2009. US 20090117125 A1 20090507.
[132] Welford RWD, Hewitson KS, Mcneill LA, Schofield CJ, Schlemminger I. Anticancer agent combination using an AlkB-inhibiting oxoglutarate analog or benzopyran derivative and an alkylating agent. PCT Int. Appl; 2004. WO 2004058252 A2 20040715.

[133] Meyer C. Combination of chemotherapeutic compounds for treating cancer. PCT Int. Appl; 2007. WO 2007139556 A1 20071206.
[134] Sugiyama H, Iida H, Saito I, Saito T. Method of the solid phase synthesis of pyrrole-imidazole polyamide. PCT Int. Appl; 2003. WO 2003000683 A1 20030103.
[135] Vega-Perez JM, Candela JI, Blanco E, Iglesias-Guerra F. Potential anticancer drugs. Part 3. Alkylating agents from sugars. Stereoselective synthesis of 2,3-diaminoglucoses from 2-nitroalkenes, as intermediates in the synthesis of carriers of chlorambucil. Tetrahedron 1999;55(31):9641–50.
[136] Mao H, Zhan D, Zhu L. Medicinal composition of alkylating agent type antineoplastic compound, and preparation method thereof. Faming Zhuanli Shenqing; 2010. CN 101926809 A 20101229.
[137] Tian S. Compound antitumor sustained-release injection containing clofarabine. Faming Zhuanli Shenqing; 2007. CN 1887260 A 20070103.
[138] Kong Q, Sun J, Kong Q, Su H, Sun J. Sustained-release antitumor implant. Faming Zhuanli Shenqing; 2005. CN 1679948 A 20051012.
[139] Liu Y. Antitumor sustained-release injections or implants containing the combination of antitumor antibiotics and synergistic alkylating and/or antimetabolic agents. Faming Zhuanli Shenqing; 2007. CN 1923286 A 20070307.
[140] Sun J. New sustained-release microsphere formulations containing nitrosoureas and alkylating agents for cancer therapy. Faming Zhuanli Shenqing; 2007. CN 101011344 A 20070808.
[141] Sun J, Zou H, Liu E, He R. Composition of antitumor sustained-release injection containing interstitial hydrolytic agent. Faming Zhuanli Shenqing; 2007. CN 1969828 A 20070530.
[142] Kong Q, Liu E, He R. Sustained-release composition carrying both antitumor drugs and synergist. Faming Zhuanli Shenqing; 2007. CN 1969821 A 20070530.
[143] Kong Q, Yu J. Temperature-controlled sustained-release injection containing alkylating agent and preparation thereof. Faming Zhuanli Shenqing; 2008. CN 101273962 A 20081001.
[144] Kong Q, Sun J, Su H. Sustained-release injections containing combination of tetrazolium violet and synergistic antitumor agents. Faming Zhuanli Shenqing; 2007. CN 1927179 A 20070314.
[145] Sun J, Zou H, Liu E, Zhang J. Antitumor synergistic sustained-release composition containing antimetabolites and alkylating agents. Faming Zhuanli Shenqing; 2007. CN 101011340 A 20070808.
[146] Sun J, Yu J, Zhang H, Liu E. Antitumor sustained-release composition containing taxane derivatives and alkylating agents. Faming Zhuanli Shenqing; 2007. CN 101006978 A 20070801.
[147] Sun J, Liu Y, Kong Q. Antitumor sustained-release injection containing epothilone derivatives and synergistic compound. Faming Zhuanli Shenqing; 2007. CN 1969818 A 20070530.
[148] Kong Q. New sustained-release. microsphere formulations of antimetabolic drug and alkylating agent for cancer therapy. Faming Zhuanli Shenqing; 2007. CN 101040865 A 20070926.
[149] Parachalil DR, Commerford D, Bonnier F, Chourpa I, McIntyre J, Byrne HJ. Raman spectroscopy as a potential tool for label free therapeutic drug monitoring in human serum: the case of busulfan and methotrexate. Analyst (Cambridge, U K) 2019;144 (17):5207–14.
[150] Bantia S, Breitfeld P, Babu YS. Methods of treating hematologic cancers. PCT Int. Appl; 2009. WO 2009076455 A2 20090618.
[151] Schenk PW, Boersma AWM, Brok M, Burger H, Stoter G, Nooter K. Inactivation of the Saccharomyces cerevisiae SKY1 gene induces a specific modification of the yeast

anticancer drug sensitivity profile accompanied by a mutator phenotype. Mol Pharmacol 2002;61(3):659–66.
[152] Palom Y, Kumar GS, Tang L, Paz MM, Musser SM, Rockwell S, Tomasz M. Relative toxicities of DNA cross-links and monoadducts: new insights from studies of decarbamoyl mitomycin C and mitomycin C. Chem Res Toxicol 2002;15(11):1398–406.
[153] Mochida Y, Hayakawa H, Kuwano N, Onuma T, Takemura Y. Alkylating agents. Gan no Yakuzai Taisei to Sono Kokufuku 2001;233–41.
[154] Colvin ME, Quong JN. DNA-alkylating events associated with nitrogen mustard based anticancer drugs and the metabolic byproduct Acrolein. In: Advances in DNA sequence-specific agents, vol. 4. Elsevier; 2002. p. 29–46.
[155] Colvin ME, Sasaki JC, Tran NL. Chemical factors in the action of phosphoramidic mustard alkylating anticancer drugs: roles for computational chemistry. Curr Pharm Des 1999;5(8):645–63.
[156] Silber JR, Bobola MS, Blank A, Schoeler KD, Haroldson PD, Huynh MB, Kolstoe DD. The apurinic/apyrimidinic endonuclease activity of Ape1/Ref-1 contributes to human glioma cell resistance to alkylating agents and is elevated by oxidative stress. Clin Cancer Res 2002;8(9):3008–18.
[157] Mounetou E, Legault J, Lacroix J, C-Gaudreault R. Antimitotic antitumor agents: synthesis, structure-activity relationships, and biological characterization of N-aryl-N'-(2-chloroethyl)ureas as new selective alkylating agents. J Med Chem 2001;44 (5):694–702.
[158] Murray D. DNA repair in resistance to bifunctional alkylating and platinating agents. Cancer Treat Res 2002;112(Clinically Relevant Resistance in Cancer Chemotherapy):129–60.
[159] Masferrer JL. Method of using a COX-2 inhibitor and an alkylating-type antineoplastic agent as a combination therapy in the treatment of neoplasia. U.S. Pat. Appl. Publ; 2004. US 20040072889 A1 20040415.
[160] Dincer Y, Akcay T, Celebi N, Uslu I, Ozmen O, Hatemi H. Glutathione S-transferase and O6-methylguanine DNA methyl transferase activities in patients with thyroid papillary carcinoma. Cancer Invest 2002;20(7 & 8):965–71.
[161] Kong Q. Antitumor sustained-release composition containing nitrosourea drugs and alkylating agents. Faming Zhuanli Shenqing; 2007. CN 101053551 A 20071017.
[162] Capitan-Vallvey LF, Valencia Miron MC, Acosta Acosta R. Chemiluminescence determination of sodium 2-mercaptoethane sulfonate by flow injection analysis using cerium(IV) sensitized by quinine. Talanta 2000;51(6):1155–61.
[163] Carson DA, Leoni LM, Cottam HB. Indole compound-alkylating agent combination useful for the treatment of cancer. U.S. Pat. Appl. Publ; 2004. US 20040152672 A1 20040805.
[164] Kong Q, Zou H, Liu E, He R. Antitumor sustained-release composition containing glucocorticoid hormones, taxane derivatives, alkylating agents and/or plant alkaloids. Faming Zhuanli Shenqing; 2007. CN 101023921 A 20070829.
[165] Yamada T, Saeki K. Antitumor agents comprising alkylating agent and deoxyfluorothioarabinofuranosy
lcytosine, an antitumor effect potentiator, and antitumor kit. PCT Int. Appl; 2019. WO 2019176985 A1 20190919.
[166] Sun J, Zhang H, Yu J, Zou H. Antitumor sustained-release composition containing phosphoinositide 3-kinase inhibitors and alkylating agents. Faming Zhuanli Shenqing; 2007. CN 101019827 A 20070822.
[167] Moschel RC, Loktionova NA, Wei G, Pegg AE. Preparation of β-glucuronidase cleavable prodrugs of O6-alkylguanine-DNA alkyltransferase inactivator. PCT Int. Appl; 2006. WO 2006029065 A1 20060316.

[168] Fourie J. A structure-activity approach to drug action in pharmacology: (A) Structure-activity relationships involved in DT-diaphorase (NQO1) mediated reduction kinetics and mode of action of anticancer bioreductive benzoquinone alkylating agents. (B) Structural characteristics of novel NMDA receptor antagonists required for renal tubule organic cation secretion. No Corporate Source data available; 2004. 301 pp.
[169] Cai S, Xu Y, Cooper RJ, Ferkowicz MJ, Hartwell JR, Pollok KE, Kelley MR. Mitochondrial targeting of human O6-methylguanine DNA methyltransferase protects against cell killing by chemotherapeutic alkylating agents. Cancer Res 2005;65(8):3319–27.
[170] Petitclerc E, Deschesnes RG, Cote M, Marquis C, Janvier R, Lacroix J, Miot-Noirault E, Legault J, Mounetou E, Madelmont J, et al. Antiangiogenic and antitumoral activity of phenyl-3-(2-chloroethyl)ureas: a class of soft alkylating agents disrupting microtubules that are unaffected by cell adhesion-mediated drug resistance. Cancer Res 2004;64(13):4654–63.
[171] Bando T, Sugiyama H. Molecular design of effective sequence-specific DNA alkylating agents using pyrrole-imidazole polyamides. Seitai Zairyo Kogaku Kenkyusho Hokoku (Tokyo Ika Shika Daigaku) 2004;37:49–55.
[172] Legault J, Gaulin J, Mounetou E, Bolduc S, Lacroix J, Poyet P, Gaudreault RC. Microtubule disruption induced in vivo by alkylation of β-tubulin by 1-aryl-3-(2-chloroethyl)ureas, a novel class of soft alkylating agents. Cancer Res 2000;60(4):985–92.
[173] Kong Q, He R, Luan Y. Antitumor sustained-release injections containing vascular inhibitors and antimitotic drugs and alkylating agents. Faming Zhuanli Shenqing; 2006. CN 1883452 A 20061227.
[174] Kong Q, Zhang H, Zou H. Antitumor sustained-release injections containing platinum compounds in conjunction with taxanes and alkylating agents and plant alkaloids. Faming Zhuanli Shenqing; 2006. CN 1861051 A 20061115.
[175] Morales-Ramirez P, Vallarino-Kelly T, Cruz-Vallejo V. Kinetics of micronucleus induction and cytotoxicity caused by distinct antineoplastics and alkylating agents in vivo. Toxicol Lett 2014;224(3):319–25.
[176] Sun J. Antitumor sustained-release composition containing alkylating agents and hormones. Faming Zhuanli Shenqing; 2007. CN 101045058 A 20071003.
[177] Tercel M, Lee AE, Hogg A, Anderson RF, Lee HH, Siim BG, Denny WA, Wilson WR. Hypoxia-selective antitumor agents. 16. Nitroarylmethyl quaternary salts as bioreductive prodrugs of the alkylating agent mechlorethamine. J Med Chem 2001;44(21):3511–22.
[178] Larsson A, Shokeer A, Mannervik B. Molecular evolution of Theta-class glutathione transferase for enhanced activity with the anticancer drug 1,3-bis-(2-chloroethyl)-1-nitrosourea and other alkylating agents. Arch Biochem Biophys 2010;497(1-2):28–34.
[179] Bartzatt R. Design of novel anticancer drugs utilizing busulfan for optimizing pharmacological properties and pattern recognition techniques for elucidation of clinical efficacy. Horiz Cancer Res 2011;43:241–55.
[180] Mao H, Kong Q. Antitumor sustained-release gels for injection containing alkylating agents. Faming Zhuanli Shenqing; 2008. CN 101301266 A 20081112.
[181] Zolnierczyk JD, Komina O, Blonski JZ, Borowiak A, Cebula-Obrzut B, Smolewski P, Robak P, Kilianska ZM, Wesierska-Gadek J. Can ex vivo evaluation (testing) predict the sensitivity of CLL cells to therapy with purine analogs in conjunction with an alkylating agent? A comparison of in vivo and ex vivo responses to treatment. Med Oncol 2012;29(3):2111–26.
[182] Teicher BA. In vivo/ex vivo and in situ assays used in cancer research: a brief review. Toxicol Pathol 2009;37(1):114–22.

[183] Miot-Noirault E, Reux B, Debiton E, Madelmont J, Chezal J, Pascal C, Weber V. Preclinical investigation of tolerance and antitumour activity of new fluorodeoxyglucose-coupled chlorambucil alkylating agents. Invest New Drugs 2011; 29(3):424–33.

[184] Bielawski K, Bielawska A. Small-molecule based delivery systems for alkylating antineoplastic compounds. ChemMedChem 2008;3(4):536–42.

[185] Valero T, Steele S, Neumueller K, Bracher A, Niederleithner H, Pehamberger H, Petzelbauer P, Loewe R. Combination of dacarbazine and dimethylfumarate efficiently reduces melanoma lymph node metastasis. J Invest Dermatol 2010;130(4):1087–94.

[186] Kelner MJ, McMorris TC, Rojas RJ, Estes LA, Suthipinijtham P. Synergy of irofulven in combination with other DNA damaging agents: synergistic interaction with altretamine, alkylating, and platinum-derived agents in the MV522 lung tumor model. Cancer Chemother Pharmacol 2008;63(1):19–26.

[187] Vargiu AV, Ruggerone P, Magistrato A, Carloni P. Sliding of alkylating anticancer drugs along the minor groove of DNA: new insights on sequence selectivity. Biophys J 2008;94(2):550–61.

[188] Kawai H, Toyohara J, Kado H, Nakagawa T, Takamatsu S, Furukawa T, Yonekura Y, Kubota T, Fujibayashi Y. Acquisition of resistance to antitumor alkylating agent ACNU: a possible target of positron emission tomography monitoring. Nucl Med Biol 2006;33(1):29–35.

[189] Leyland-Jones B. Clinical resistance to alkylators: status and perspective. In: Panasci LC, Alaoui-Jamali MA, editors. DNA repair in cancer therapy. Springer; 2004. p. 1–7.

[190] Buschfort-Papewalis C, Moritz T, Liedert B, Thomale J. Down-regulation of DNA repair in human CD34+ progenitor cells corresponds to increased drug sensitivity and apoptotic response. Blood 2002;100(3):845–53.

[191] Mariani L, Citti L, Nevischi S, Eckstein F, Rainaldi G. Ribozyme and free alkylated base: a dual approach for sensitizing Mex+ cells to the alkylating antineoplastic drug. Cancer Gene Ther 2000;7(6):905–9.

[192] Lee P, Kakadiya R, Su T, Lee T. Combination of bifunctional alkylating agent and arsenic trioxide synergistically suppresses the growth of drug-resistant tumor cells. Neoplasia 2010;12(5):376–87.

[193] Ciavolella A, Montagnoli A, Pesenti E. Therapeutic combination comprising a parp-1 inhibitor and an anti-neoplastic agent. U.S. Pat. Appl. Publ; 2013. US 20130129841 A1 20130523.

[194] Doria F, Nadai M, Folini M, Di Antonio M, Germani L, Percivalle C, Sissi C, Zaffaroni N, Alcaro S, Artese A, et al. Hybrid ligand-alkylating agents targeting telomeric G-quadruplex structures. Org Biomol Chem 2012;10(14):2798–806.

[195] Heim MM, Eberhardt W, Seeber S, Muller MR. Differential modulation of chemosensitivity to alkylating agents and platinum compounds by DNA repair modulators in human lung cancer cell lines. J Cancer Res Clin Oncol 2000;126(4):198–204.

[196] Sun J. Antitumor compositions containing combinations of neovascularization inhibitors and alkylating agents. Faming Zhuanli Shenqing; 2007. CN 101041074 A 20070926.

[197] Sun J, Yu J, Zhang H, Liu E. Antitumor sustained-release composition containing phosphoinositide 3-kinase inhibitors and alkylating agents. Faming Zhuanli Shenqing; 2007. CN 101040838 A 20070926.

[198] Sun J, Zhang H, Yu J, Zou H. Antitumor sustained-release composition containing angiogenesis inhibitors and alkylating agents. Faming Zhuanli Shenqing; 2007. CN 101023930 A 20070829.

[199] Sun J, Zhang J, Zou H. Antitumor sustained-release composition containing tyrosine kinase inhibitors and alkylating agents. Faming Zhuanli Shenqing; 2007. CN 101073554 A 20071121.

[200] Yasunaga K, Kiyonari A, Nakagawa M, Yoshikawa K. Investigation into the ability of the Salmonella umu test to detect DNA damage using antitumor drugs. Toxicol In Vitro 2006;20(5):712–28.

[201] Sakumi K. Susceptibility to chemotherapeutic alkylating agents depends on the mgmt genotype. Saishin Igaku 2001;56(3):404–10.

[202] Ben-Efraim S. Immunomodulating anticancer alkylating drugs: targets and mechanisms of activity. Curr Drug Targets 2001;2(2):197–212.

[203] Shimizu T, Sasaki S, Minoshima M, Shinohara K, Bando T, Sugiyama H. Synthesis and evaluation of sequence-specific DNA alkylating agents: effect of alkylation subunits. Nucleic Acids Symp Ser 2006;50:155–6.

[204] Liu Y, Zhang H, Yu J. Antitumor sustained-release composition containing angiogenesis inhibitors and alkylating agents. Faming Zhuanli Shenqing; 2007. CN 101019829 A 20070822.

[205] Villarroel-Lecourt G, Carrasco-Carvajal J, Andrade-Villalobos F, Solis-Egana F, Merino- San Martin I, Robinson-Duggon J, Fuentealba D. Encapsulation of chemotherapeutic drug melphalan in cucurbit[7]uril: effects on its alkylating activity, hydrolysis, and cytotoxicity. acs. Omega 2018;3(7):8337–43.

[206] Tsaryk R, Fabian K, Thacker J, Kaina B. Xrcc2 deficiency sensitizes cells to apoptosis by MNNG and the alkylating anticancer drugs temozolomide, fotemustine and mafosfamide. Cancer Lett 2006;239(2):305–13.

[207] Gadhia PK, Gadhia M, Georje S, Vinod KR, Pithawala M. Induction of chromosomal aberrations in mitotic chromosomes of fish Boleophthalmus dussumieri after exposure in vivo to antineoplastics Bleomycin, Mitomycin-C and Doxorubicin. Indian. J Sci Technol 2008;1(7). No pp. given.

[208] Nagase H, Sugiyama H, Bando T. Novel alkylating agent for alkylating target with driver oncogene mutation, its manufacture, and use for pharmaceutical composition, kit, and anticancer agent. PCT Int. Appl; 2015. WO 2015053413 A1 20150416.

[209] Liu X. Antitumor pharmaceutical composition and its application for treating solid tumor and leukemia. Faming Zhuanli Shenqing; 2014. CN 103656646 A 20140326.

[210] Kang YH, Lee K, Yang Y, Kim S, Kim JH, Park SN, Paik S, Yoon D. The apoptotic effect of intercalating agents on HPV-negative cervical cancer C-33A cells. Amino Acids 2007;33(1):105–12.

[211] Horvath Z, Hoechtl T, Bauer W, Fritzer-Szekeres M, Elford HL, Szekeres T, Tihan T. Synergistic cytotoxicity of the ribonucleotide reductase inhibitor didox (3,4-dihydroxy-benzohydroxamic acid) and the alkylating agent carmustine (BCNU) in 9L rat gliosarcoma cells and DAOY human medulloblastoma cells. Cancer Chemother Pharmacol 2004;54(2):139–45.

[212] Mehta S, Umrethia M, Patel H, Mandal JK. Pharmaceutical composition of oral suspension of anti-neoplastic alkylating agents. PCT Int. Appl; 2018. WO 2018167627 A1 20180920.

[213] Sauter B, Gillingham D. Profiling the nucleobase and structure selectivity of anticancer drugs and other DNA alkylating agents by RNA sequencing. Chembiochem 2018;19 (15):1638–42.

[214] Nagase H, Hattori A, Watanabe T, Takatori A. Novel alkylating agent comprising immune checkpoint-related molecule gene-binding compound, pharmaceutical composition, and preparation thereof. PCT Int. Appl; 2018. WO 2018056361 A1 20180329.

[215] Bielawskaa A, Bielawskia K, Wołczyński S, Anchim T. Structure-activity studies of novel amidine analogues of chlorambucil: correlation of cytotoxic activity with dna-binding affinity and topoisomerase ii inhibition arch. Pharm Pharm Med Chem 2003;336:293–9.

[216] Bielawska A, Bielawski K, Muszyńska A. Synthesis and biological evaluation of new cyclic amidine analogs of chlorambucil. Il Farmaco 2004;59(59):111–7.
[217] Descôteaux C, Brasseur K, Leblanc V, Parent S, Asselin E, Bérubé G. SAR study of tyrosine–chlorambucil hybrid regioisomers; synthesis and biological evaluation against breast cancer cell lines. Amino Acids 2012;43:923–35.
[218] Descôteaux C, Leblanc V, Brasseur K, Gupta A, Asselin E, Bérubé G. Synthesis of D- and L-tyrosine-chlorambucil analogs active against breast cancer cell lines. Bioorg Med Chem Lett 2010;20:7388–92.
[219] Bartulewicz D, Bielawski K, Bielawska A, Różański A. Synthesis, molecular modelling, and antiproliferative and cytotoxic effects of carbocyclic derivatives of distamycin with chlorambucil moiety. Eur J Med Chem 2001;36:461–7.
[220] Clavel CM, Zava O, Schmitt F, Kenzaoui BH, Nazarov AA, Juillerat-Jeanneret L, Dyson PJ. Thermoresponsive chlorambucil derivatives for tumour targeting. Angew Chem Int Ed 2011;50:7124–7.
[221] Fousterisa MA, Koutsourea AI, Arsenoua ES, Papageorgiou A, Mourelatos D, Nikolaropoulos SS. Structure–anti-leukemic activity relationship study of B- and D-ring modified and non-modified steroidal esters of chlorambucil. Anticancer Drugs 2006;17(5).
[222] Pedersen PJ, Christensen MS, Ruysschaert. Synthesis and biophysical characterization of chlorambucil anticancer ether lipid prodrugs. J Med Chem 2009;52 (10):3408–15.
[223] Satpati D, Korde A, Venkatesh M, Banerjee S. Preparation and bioevaluation of a 99mTc-labeled chlorambucil analogas a tumor targeting agent. Appl Radiat Isot 2009;67:1644–9.
[224] Iglesias-guerra F, Candela JI, Blanco E, Alcudia F, Vega-perez JM. Alkylating agents from sugars: synthesis of chorambucil derivatives carried by chiral glycosyl glycerols derived from D-glucosamine. Chirality 2002;14:199–203.
[225] Iglesias-Guerra F, Candela JI, Bautista J, Alcudia F, Vega-Perez JM. Potential anticancer drugs, Part 2. Alkylating agents from sugars. Alkyl hexopyranoside derivatives as carrier systems for chlorambucil. Carbohydr Res 1999;316(1–4):71–84. https://doi.org/10.1016/S0008-6215(99)00030-0.
[226] Horton D. Organic Syntheses Collection. Vol. 5. New York: John Wiley and Sons; 1973. p. 1–5.
[227] Iglesias-Guerra F, Romero I, Alcudia F, Vega-Pérez JW. Alkylating agents from sugars. Cyclophosphamides derived from 2-amino-2-deoxy-d-allose. Carbohydr Res 1998;308:57–62.
[228] Charon D, Chaby R, Malinvaud A, Mondange M, Szabó L. Chemical synthesis and immunological activities of glycolipids structurally related to lipid A. Biochemistry 1985;24:2736–42.
[229] Gross PH, Jeanloz RW. Optically pure N-substituted derivatives of benzyl 2-amino-2-deoxy-.alpha.- and, beta.-D-glucopyranoside. J Org Chem 1967;32:2759–63.
[230] Iglesias-Guerra F., Vega-Perez J.W., unpublished results.
[231] Fujino M, Hatanaka C. A new procedure for the pentachlorophenylation of N-protected amino acids. Chem Pharm Bull 1968;16(5):929–32.
[232] Hongli ZHAO, Xianjiang MENG, Huihui YUAN, Minbo LAN. Novel melphalan and chlorambucil derivatives of 2,2,6,6-tetramethyl-1-piperidinyloxy radicals: synthesis, characterization, and biological evaluation *in vitro*. Chem Pharm Bull 2010;58 (3):332–5.
[233] Sachin M, Xueqin S, Balvinder SV, Christopher PL, Insook K, John MH, et al. Prolidase, a potential enzyme target for melanoma: design of proline-containing dipeptide-like prodrugs. Mol Pharm 2005;2(1):37–46.

[234] Bielawska A, Bielawski K, Anchim T. Amidine analogues of melphalan: synthesis, cytotoxic activity, and DNA binding properties. Arch Pharm Chem Life Sci 2007; 340:251–7.

[235] Ajazuddin AA, Basant A, Parijat K. Synthesis, characterization and *in vitro* studies of pegylated melphalan conjugates. Drug Dev Ind Pharm 2013;39(7):1053–62.

[236] Peyrode C, Weber V, David E, Vidal A, Philippe Auzeloux P, Communal Y, Dauplat MM, Besse S, Gouin F, Heymann D, Chezal JM, Rédini F, Miot-Noirault E. Quaternary ammonium-melphalan conjugate for anticancer therapy of chondrosarcoma: in vitro and in vivo preclinical studies. Invest New Drugs 2012;30:1782–90.

[237] (a) Giraud I, Rapp M, Maurizis JC, Madelmont JC. Synthesis and in vitro evaluation of quaternary ammonium derivatives of chlorambucil and melphalan, anticancer drugs designed for the chemotherapy of chondrosarcoma. J Med Chem 2002;45(10): 2116–9. (b) Wang J, Yang W, Xue J, Zhang Y, Liu Y. Synthesis, 99mTc-labeling, and preliminary biological evaluation of DTPA-melphalan conjugates. J Labelled Comp Radiopharm 2017;60(14):659–65.

[238] (a) Chezal JM, Papon J, Labarre P, Lartigue C, Galmier MJ, Decombat C, Chavignon O, Maublant J, Teulade JC, Madelmont JC, Moins N. Evaluation of radiolabeled (hetero)aromatic analogues of N-(2-diethylaminoethyl)-4-iodobenzamide for imaging and targeted radionuclide therapy of melanoma. J Med Chem 2008; 51:3133–44. (b) Scutaru AM, Wenzel M, Gust R. Bivalent bendamustine and melphalan derivatives as anticancer agents. Eur J Med Chem 2011;46:1604–15.

[239] Bielawski K, Bielawsk A. Small-molecule based delivery systems for alkylating antineoplastic compounds. Chem Med Chem 2008;3:536–42.

[240] Scutaru AM, Wenzel M, Scheffler H, Wolber G, Gust R. Optimization of the N-lost drugs melphalan and bendamustine: synthesis and cytotoxicity of a new set of dendrimer-drug conjugates as tumor therapeutic agents. Bioconjug Chem 2010; 21:1728–43.

[241] Kerr DE, Li Z, Siemers NO, Senter PD, Vrudhula VM. Development and activities of a new melphalan prodrug designed for tumor-selective activation. Bioconjug Chem 1998;9:255–9.

[242] Lidia K, Teruna JS, Anthony SB, Henry SF, Patricia EH, Di S, James MG. Modulation of melphalan resistance in glioma cells with a peripheral benzodiazepine receptor ligand-melphalan conjugate. J Med Chem 1997;40:1726–30.

[243] Bogomilova A, Höhn M, Günther M, Herrmann A, Troev K, Wagner E, Schreiner L. A polyphosphoester conjugate of melphalan as antitumoral agent. Eur J Pharm Sci 2013. https://doi.org/10.1016/j.ejps.2013.08.007.

[244] Lia D, Lua B, Huang Z, Xua P, Zhenga H, Yina Y, Xua H, Liu X, Chenc L, Loua Y, Zhanga X, Xionga F. A novel melphalan polymeric prodrug: preparation and property study. Carbohydr Polym 2014;111:928–35.

[245] Lu B, Huang D, Zheng H, Huang Z, Xu P, Xu H, Yin Y, Liu X, Li D, Zhang X. Preparation, characterization, and in vitro efficacy of O-carboxymethyl chitosan conjugate of melphalan. Carbohydr Polym 2013;98:36–42.

[246] Xua H, Heb J, Zhanga Y, Fanc L, Zhaoa Y, Xua T, Niea Z, Lia X, Huanga Z, Lua B, Xua P. Synthesis and in vitro evaluation of a hyaluronic acid–quantumdots–melphalan conjugate. Carbohydr Polym 2015;121:132–9.

[247] Misiura K, Kinas RW, KusÂ nierczyk H, Radzikowski C, Stec WJ. (S)-(7)-Bromofosfamide (CBM-11): synthesis and antitumor activity and toxicity in mice. Anticancer Drugs 2001;12:453–8.

[248] Pankiewicz K, Kinas R, Stec WJ, Foster AB, Jarman M, Van Maanen JMS. Synthesis and absolute configuration assignments of enantiomeric forms of ifosphamide, sulfosphamide, and trofosphamide. J Am Chem Soc 1979;101:7712±8.

[249] Sato T, Ueda H, Nakagawa K, Bodor N. Asymmetric synthesis of enantiomeric cyclophosphamides. J Org Chem 1983;48:98±101.
[250] March J. Advanced organic chemistry. 3rd ed. New York: Wiley; 1985. p. 916.
[251] Stec W, Michalski J. Volatile complexes of trialkyl phosphates and alkylphosphonates with protic acids. Z Naturforsch 1970;25b:554–5.
[252] Fieser LF, Fieser M. Reagents for organic synthesis. New York: Wiley; 1967. p. 1053–4.
[253] Voelcker G. Influence of the alkylating function of aldo-Ifosfamide on the anti-tumor activity. Anticancer Drugs 2018;29(1):75–9.
[254] Misiura K, Kinasa RW, Kusnierczyk H. Studies on the side-chain hydroxylation of ifosfamide and its bromo analogue. Bioorg Med Chem Lett 2002;12:427–31.
[255] Springer JB, Michael Colvin O, Ludeman SM. Labeled oxazaphosphorines for applications in mass spectrometry studies. 2. Synthesis of deuterium-labeled 2-dechloroethylcyclophosphamides
and 2- and 3-dechloroethylifosfamides. J Label Compd Radiopharm 2014;57:110–4.
[256] Skarbek C, Lesueur LL, Chapuis H, Deroussent A, Pioche-Durieu C, Daville A, Caron J, Rivard M, Martens T, Bertrand J, Le Cam E, Vassal G, Couvreur P, Desmaêle D, Paci A. Pre-activated oxazaphosphorines designed for isophosphoramide mustard delivery as bulk form or nano-assemblies: synthesis and proof of concept. J Med Chem 2015;58:705–17.
[257] Storme T, Deroussent A, Mercier L, Prost E, Re M, Munier F, Martens T, Bourget P, Vassal G, Royer J, Paci A. New ifosfamide analogs designed for lower associated neurotoxicity and nephrotoxicity with modified alkylating kinetics leading to enhanced in vitro anticancer activity. J Pharmacol Exp Ther 2009;328(2):598–609.
[258] Paci A, Martens T, Royer J. Anodic oxidation of ifosfamide and cyclophosphamide: a biomimetic metabolism model of the oxazaphosphorinane anticancer drugs. Bioorg Med Chem Lett 2001;11:1347–9.
[259] Sen SS, Prestwich GD. Trisnorsqualene alcohol, a potent inhibitor of vertebrate squaleneepoxidase. J Am Chem Soc 1989;111:1508–10.
[260] Wiessler M, Waldeck W, Kliem C, Pipkorn R, Braun K. The Diels-Alder-reaction with inverse-electron-demand, a very efficient versatile click-reaction concept for proper ligation of variable molecular partners. Int J Med Sci 2010;7(1):19–28.
[261] Pipkorn R, Waldeck W, Didinger B, Koch M, Mueller G, Wiessler M, Braun K. - Inverse-electron-demand Diels-Alder reaction as a highly efficient chemoselective ligation procedure: synthesis and function of a BioShuttle for temozolomide transport into prostate cancer cells. J Pept Sci 2009;15:235–41.
[262] Moseley CK, Carlin SM, Neelamegam R, Hooker JM. An efficient and practical radiosynthesis of [11C]temozolomide. Org Lett 2012;14(23):5872–5.
[263] Eriksson J, Van Kooij R, Schuit RC, Froklage FE, Reijneveld JC, Hendrikse NH, et al. Synthesis of [3-N-11C-methyl]temozolomide via in situ activation of 3-N-hydroxymethyl temozolomide and alkylation with [11C]methyl iodide. J Label Compd Radiopharm 2015;58(3):122–6.
[264] Brown GD, Luthra SK, Brock CS, Stevens MFG, Price PM, Brady F. Antitumor imidazotetrazines. 40.1 Radiosyntheses of [4-11C-Carbonyl]- and [3-N-11C-Methyl]-8-carbamoyl-3-methylimidazo[5,1-d]-1,2,3,5-tetrazin-4(3H)-one (Temozolomide) for Positron Emission Tomography (PET) Studies. J Med Chem 2002;45:5448–57.
[265] Galanski M, Yasemi A, Slaby S, Jakupe MA, Arion VB, Rausch M, Nazarov AA, Keppler BK. Synthesis, crystal structure and cytotoxicity of new oxaliplatin analogues indicating that improvement of anticancer activity is still possible. Eur J Med Chem 2004;39:707–14.

[266] Habala L, Galanski M, Yasemi A, Nazarov AA, Keyserlingk NG, Keppler BK. Synthesis and structure-activity relationships of monoand dialkyl-substituted oxaliplatin derivatives. Eur J Med Chem 2005;40:1149–55.
[267] Saha P, Descôteaux C, Brasseur K, Fortin S, Leblanc V, Parent S, Asselin E, Bérubé G. Synthesis, antiproliferative activity and estrogen receptor a affinity of novel estradiol-linked platinum(II) complex analogs to carboplatin and oxaliplatin. Potential vector complexes to target estrogen-dependent tissues. Eur J Med Chem 2012; 48:385–90.
[268] Abramkin SA, Jungwirth U, Valiahdi SM, Dworak C, Habala L, Meelich K, Berger W, Jakupec MA, Hartinger CG, Nazarov AA, Galanski M, Keppler BK. Oxalatoplatinum(II): A novel enantiomerically pure oxaliplatin derivative showing improved anticancer activity in vivo. J Med Chem 2010;53:7356–64.
[269] Lau JTF, Lo P, Fong W, DKP N. A Zinc(II) phthalocyanine conjugated with an oxaliplatin derivativefor dual chemo- and photodynamic therapy. J Med Chem 2012;55:5446–54.
[270] Liu F, Gou S, Chen F, Fang L, Zhao J. Study on antitumor platinum(II) complexes of chiraldiamines with dicyclic species as steric hindrance. J Med Chem 2015; 58:6368–77.
[271] Varbanov HP, Göschl S, Heffeter P, Theiner S, Roller A, Jensen F, et al. A novel class of bis- and tris-chelate diam(m)inebis(dicarboxylato)platinum(IV) complexes as potential anticancer prodrugs. J Med Chem 2014;57(15):6751–64.
[272] Thomet FA, Pinyol P, Villena GJ, Reveco PG. Towards a more selective analogue of oxaliplatin: synthesis of [Pt((1R,2R)-diaminocyclohexane)(3-carboxypredicentrinato)]. Inorg Chim Acta 2012;384:255–9.
[273] Rachid Z, Brahimi F, Qiu Q, Williams C, Hartley JM, Hartley JA, Jean-Claude B. Novel nitrogen mustard-armed combi-molecules for the selective targeting of epidermal growth factor receptor overexperessing solid tumors: discovery of an unusual structure-activity relationship. J Med Chem 2007;50:2605–8.
[274] Marvania B, Lee P, Chaniyara R, Dong H, Suman S, Kakadiya R, Chou T, Lee T, Shah A, Su T. Design, synthesis and antitumor evaluation of phenyl N-mustard-quinazoline conjugates. Bioorg Med Chem 2011;19:1987–98.
[275] Lin S, Li Y, Zheng Y, Luo L, Sun Q, Ge Z, Cheng T, Li R. Design, synthesis and biological evaluation of quinazoline–phosphoramidate mustard conjugates as anticancer drugs. Eur J Med Chem 2017;127:442–58.
[276] Li S, Wang X, He Y, Zhao M, Chen Y, Xu J, Feng M, Chang J, Ning H, Qi C. Design and synthesis of novel quinazoline nitrogen mustard derivatives as potential therapeutic agents for cancer. Eur J Med Chem 2013;67:293–301.

CHAPTER 9

Natural products in chemotherapy of cancers

1. Introduction

Natural products are considered as valuable structures for medicinal chemistry and drug discovery [1–5]. They are specialized metabolites, generated by plants, animals, and microorganisms, with diverse biological activities and chemical structures [6]. They are considered as essential original source of a variety of drugs, comprising anticancer drugs. Later research efforts for pursuing anticancer treatment from the natural products are mostly fixated on the compounds with cytotoxicity competency [7,8]. Plant-derived compounds have played a crucial role in developing several clinical useful anticancer agents [9]. Plants furnish a high diversity of secondary metabolites or natural products, which are essential for the communication of plants with other organisms [10]. The core bases of anticancer compounds are plants and microbes from the marine and terrestrial surroundings. The microbes serve as a main source of natural products with antitumor potency [11–18]. A sum of these products were first knowledgeable as antibiotics. Additional main role comes from podophyllotoxins, plant alkaloids, and taxoids. A huge collection of biological metabolites can be gained from the marine world, which can be utilized for an effective cancer treatment [16]. Mainly anticancer agents are generated from plants (vincristine, vinblastine, etoposide, camptothecin, irinotecan, paclitaxel, and topotecan), marine organisms (dolastatin 10, aplidine, and cytarabine), and microorganisms (doxorubicin, dactinomycin, and bleomycin); some of the vegetables and fruits also have anticancer potency [19]. Natural products and associated analogs, comprising their semi-synthetic or totally synthetic derivatives, have a crucial role in developing new therapeutic agents [20–22].

2. Podophyllotoxins

The podophyllotoxins are a significant category of anticancer agents comprising the compounds obtained from partial synthesis; etoposide and

https://doi.org/10.1016/B978-0-12-821783-2.00009-1

teniposide apply their cytotoxic potency through the inhibition of the topoisomerase II. Podophyllotoxin is a natural lignin isolated from podophyllin, a resin produced by rhizomes and roots of different plants of the genus, podophyllum (Berberidaceae family), and it is mainly abundant in American podophyllum, named *Podophyllum peltatum*, and Indian podophyllum, generally called *Podophyllum emodi* [23]. The podophyllotoxins have a lengthy therapeutic history as they have been utilized for over thousand years [24]. They possess the antitumor activity and also display remarkable antiviral features. It is used in treating condyloma acuminatum produced by human papilloma virus (HPV) and other veneral and perianal warts. The main active substance in podophyllin, which is the resin product accessed through the taking out of the rhizomes and dried roots with ethyl alchohol, was indicated to be the lignin lactone podophyllotoxin, although various lignan glycosides and lignans have also been isolated from podophyllin. Podophyllotoxins are potent antimitotic agents and inhibitors of microtubule assembly, which are essential for cell division. They fix to the main protein constituent of microtubules, the tubulin, and hinder its polymerization, which leads to inhibition for microtubule cytoskeletal structure construction and blocking of cell cycle in metaphase step. Since the discovery of the anticancer properties of podophyllotoxin, ongoing development of novel effective drugs with improved therapeutic index depending on structural reforms of the podophyllotoxin. Many podophyllotoxin analogs have been isolated from natural origins or considered as semi-synthetic compounds. Endeavors that were carried out to lessen its toxicity while holding antineoplastic potency lead to the development of two modified analogs, C-4 β-epipodophyllotoxin (EPPT) (VP16–213) or etoposide **1** (Fig. 9.1) and 4′-demethyl epipodophyllotoxin (DEPPT) or teniposide (VM26) **2** (Fig. 9.2).

These compounds revealed the greatest antitumor potency against L1210 leukemia and have been chosen for clinical trials and have showed effectiveness in treating various types of tumors [25–29]. The teniposide and etoposide are variant from podophyllotoxin in possessing a 4′-hydroxyl group rather than a 4′-methoxyl group, the *epi*-configuration at the 4-position, and a substituted glucose residue at the 4-position. They vary from one another in the nature of the substituent on the glucose ring: etoposide is the cyclic acetal synthesized from acetaldehyde and 4′-dimethylepipodophyllotoxin β-D-Glucopyranoside, while teniposide is the cyclic acetal synthesized when thiophene-2-carboxaldehyde replaces ethanal [30]. The SAR of etoposide have been depicted [31,32]. Teniposide and etoposide characterized by their poor water solubility lead to difficulty in their

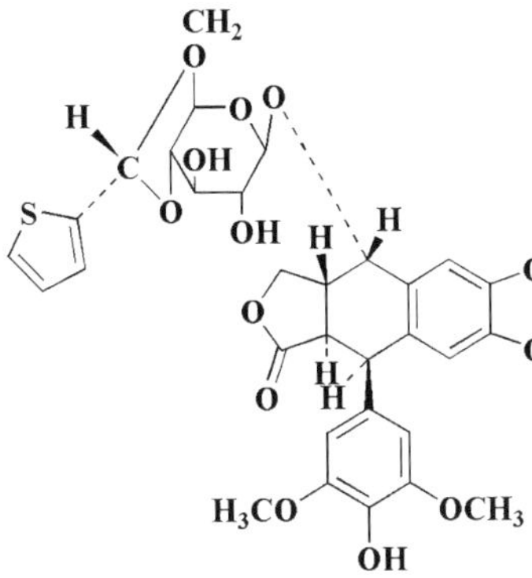

Fig. 9.1 Etoposide **1**.

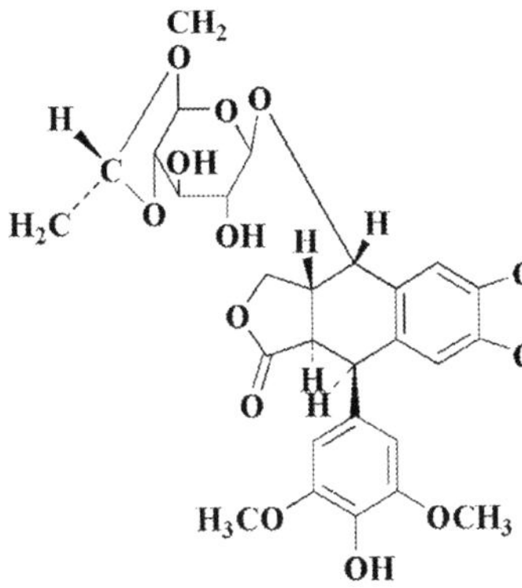

Fig. 9.2 Teniposide **2**.

administration, and this resulted in the development of the etoposide phosphate Etopophos **3** (Fig. 9.3), which has enhanced water solubility [33].

The podophyllotoxin analogs comprised four planar fused rings possessing a lactone ring, a dioxole ring, and an aromatic ring with α-configuration that minimizes its free rotation. The lactone ring is needed for inhibiting the polymerization of tubulin as it signifies the region that interacts with the protein. Introducing β-glycosidic moiety that arisen in the novel analogs (teniposide and etoposide) transforms these podophyllotoxins from inhibitors of polymerization of tubulin into effective irreversible inhibitors of DNA TOP II. The mechanism of action of teniposide and etoposide differs from that of the podophyllotoxin. Podophyllotoxin is a potent inhibitor of

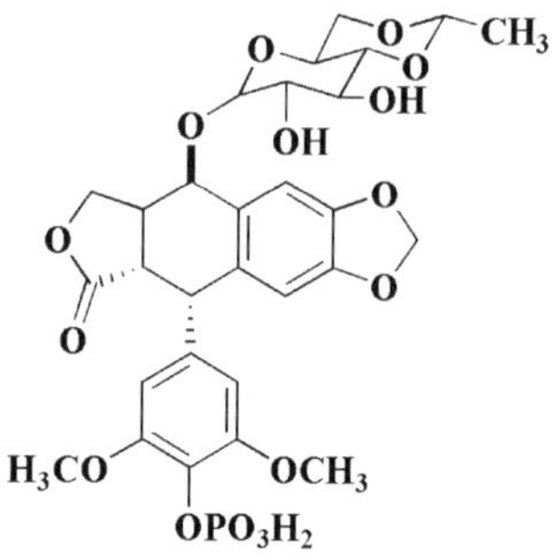

Fig. 9.3 Etopophos **3**.

microtubule assemblage in vitro, and it competitively prevents the binding of colchicines to microtubules [34,35]. So, it arrests cells at mitosis through the disruption of the equilibrium between tubulin dimer and tubulin polymer, in this manner destroying the cytoskeletal outline from chromosome separation and the arrest of the cell division at the mitotic stage of the cell cycle. In a different way, teniposide, etoposide, and Etopophos arrest cells in the late S and G2 phases of the cell cycle, and have no impact on tubulin assembly. As an alternative, they prompt single-strand breaks in DNA (etoposide) or in the DNA in L1210 cells in case of teniposide [36–38] in which these breaks are mainly double-stranded. These impacts returned to the capability of these compounds to inhibit the DNA topoisomerase II (topo II) [39]. DNA topoisomerase II mediates double-strand breaks through the formation of a complex called cleavable complex with the DNA. The later complex was stabilized via etoposide; it also causes the inhibition of the enzyme, resulting in the double-strand breaks and eventually cell death [40]. The antineoplastic potency for teniposide and etoposide in Hodgkin's disease, non-Hodgkin's lymphoma, acute myeloid leukemia, gastric cancer, lung cancer, ovarian cancer, and breast cancer was demonstrated. Etoposide is utilized in treating lymphomas and small cell lung cancer; it also has potency against hepatocellular carcinoma, monocytic or myelomonocytic leukemia, and non-Hodgkin's lymphomas. Teniposide has a role in treating neuroblastoma, childhood acute lymphoblastic leukemia, Hodgkin's disease, and non-Hodgkin's lymphomas. Development of novel derivatives of etoposide has been performed and reached clinical trials [29]. When teniposide and etoposide are used as a combined therapy, their toxicity is decreased and they act in a unique modes of action. Estimations of the various combinations have been performed in in vivo testing; few of them are clinically successful such as the combination of cytarabine (ara C) with teniposide in treating refractory pediatric acute lymphoblastic leukemia [27,41,42]. Etoposide is efficient in combination with cisplatin for treating non-small-cell lung cancer and small cell lung cancer. It is also utilized as part of multidrug treatment protocols for refractory childhood leukemia, non-Hodgkin's lymphomas, several lymphomas, and other different varieties of cancer [41,43]. Dose-limiting toxic effects of teniposide and etoposide are gastrointestinal disturbances, myelosuppression, peripheral neuropathy, alopecia, hypertension, bronchospasm, fever, and chills [43]. Novel derivatives of podophyllotoxin are developed and utilized in clinical trials. The analog NK-611 has a dimethylamino group instead of the 2-hydroxyl group of glucose in etoposide [44]. Clinical estimations of NK-611 **4** (Fig. 9.4) propose that it has enhanced bioavailability than etoposide, but indication for cross resistance between NK-611 and etoposide was observed [29]. The analog GL-331 **5** (Fig. 9.5) is a 4β-arylamino derivative of etoposide in which a *p*-nitro anilino group is placed instead of the sugar moiety.

GL-331 is more potent than etoposide in breaking the DNA double-strand and G2-phase arrest [45], and it is likewise more active against tumor cells both in vitro and in vivo. Preliminary consequences from phase I clinical evaluation in colon, non-small-cell, and small cell lung, head, and neck cancers were encouraging, with reduced side effects, while there is no objective response is indicated in phase II clinical trials in case of gastric cancer [46].

TOP-53 is an aminopodophyllotoxin analog [47]. It is considered as a more active inhibitor of topoisomerase II than etoposide and indicated high potency against a mutant yeast type II enzyme vastly resistant to etoposide.

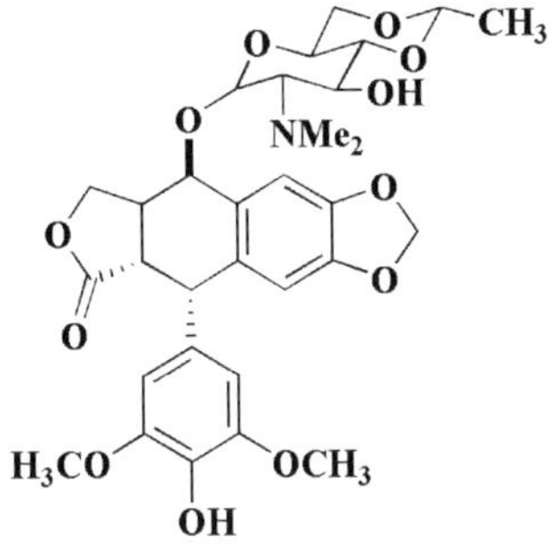

Fig. 9.4 NK-611 **4**.

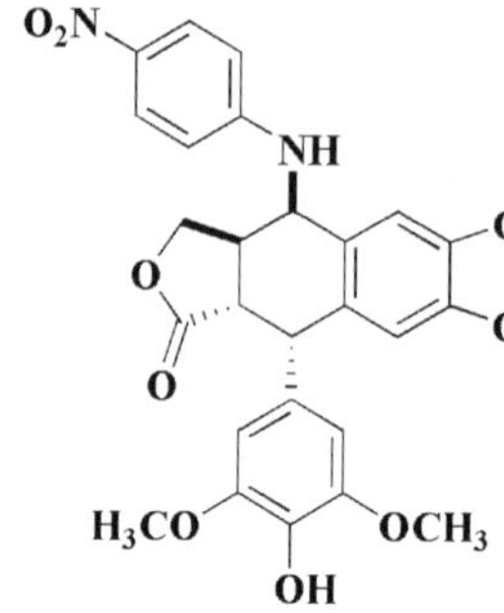

Fig. 9.5 GL-331 **5**.

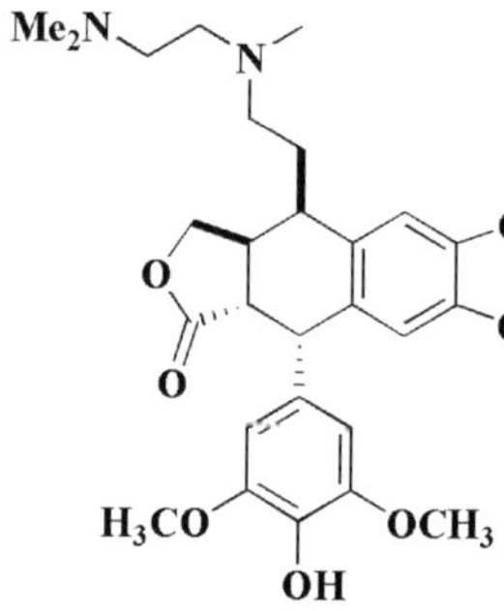

Fig. 9.6 TOP-53 **6**.

TOP-53 **6** (Fig. 9.6) displayed potency against non-small-cell lung cancer [48] and is currently in phase I clinical trials.

3. Vinca alkaloids

Vina alkaloids are a subset of drugs discovered by Robert noble and Charles Beer of Canada in the 1950s. While they have utilized in treating high blood pressure and diabetes or as disinfectants, their significance returned to their anticancer activity [23]. Vinca alkaloids are isolated from *Catharanthus roseus* "*Vinca rosea*" which are belong to the family apocynaceae and well-known as "Madagascar periwinkle." This *V. rosea* raises worldwide. Vinca alkaloids are characterized by their dimeric asymmetrical structure and the two multi-ringed subunits, catharanthine "indole nucleus" and vindoline "dihydroindole nucleus," are linked by a C—C bridge. Vindesine (VDS), vinblastine (Velban), and vincristine (Oncovin) are the first generation of vinca alkaloids with antitumor potency. In 1961 and 1963, vincristine and vinblastine were approved for clinical utility. The four main vinca alkaloids possessing medicinal assets are vincristine (VCR), vinblastine (VBL), vindesine (VDS), and vinorelbine (VRL). The naturally occurring vinca alkaloids (VCR and VBL) have been used in treating various malignancies, and hematological cancers as lymphoma and leukemia. The semi-synthetics (VDS and VRL) have displayed clinical potency against ovarian, lung, and breast malignancies [23].

The development of semi-synthetic agents, such as vinorelbine, vinflunine, and vindesine, has been accomplished. Vinca alkaloids have a well-established role in the treatment of various malignancies such as lymphatic neoplasms, hematological malignancy, and solid tumors as choriocarcinoma, non-small-cell lung cancer, and breast cancer. The usage of vinca alkaloids is restricted via side effects such as neurotoxicity and myelosuppression [49–51].

Vinblastine, vincristine **7** (Fig. 9.7), vindesine **8** (Fig. 9.8), and vinorelbine **9** (Fig. 9.9) are considered as vinca alkaloids utilized for treating cancer.

Vinca alkaloids's mechanism of action is revealed through stopping the cell cycle at M phase by interacting and disrupting of microtubule consisting of mitotic spindles; the endothelial proliferation and spreading of fibronectin were blocked by vinblastine and it also suppresses the malignant angiogenesis through binding to microtubule. Destabilization of the microtubules is performed due to the action of vincristine, which binds to the tubulin and blocks the polymerization leading to cell cycle arrest and apoptosis. The anticancer activity of the vinorelbine is recognized in patients suffering from small cell lung cancer and breast cancer and considered as better potency as

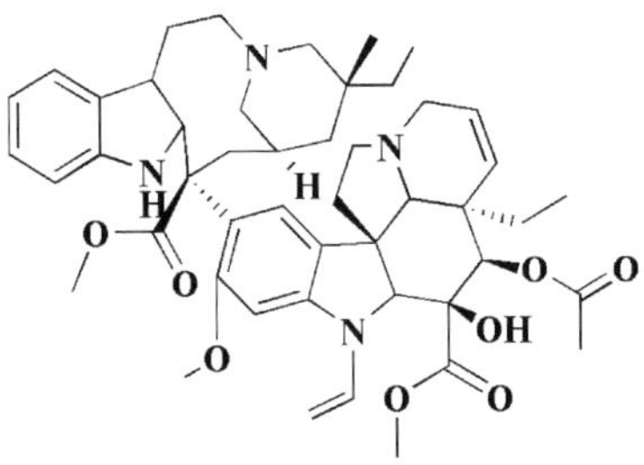

Fig. 9.7 Vincristine **7**.

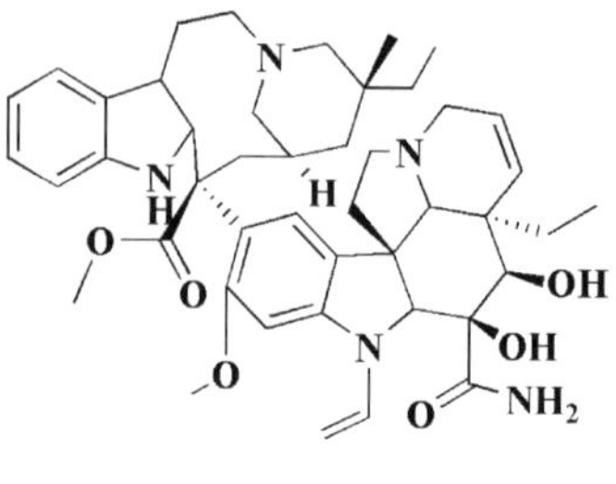

8

Fig. 9.8 Vindesine **8**.

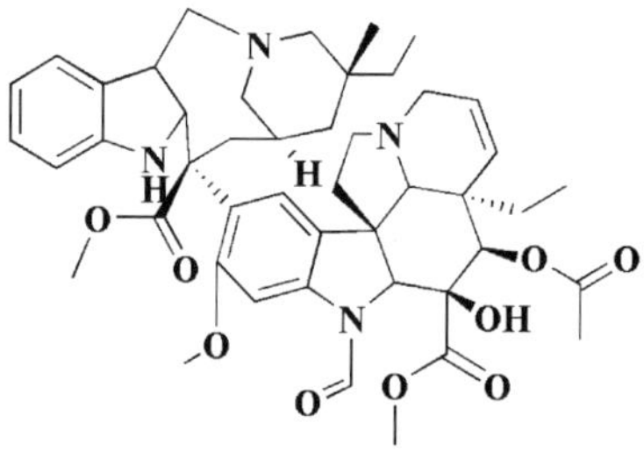

9

Fig. 9.9 Vinorelbine **9**.

compared to others. Vindesine and vinorelbine inhibit tubulin polymerization via disruption of the spindle formation and thereby arrest the cell cycle. Vinorelbine was approved for treating small cell lung carcinoma (SCLC) and breast cancer. Currently, vindesine and vinorelbine in combination with other drugs such as epirubicin, lapatinib, gefitinib, bevacizumab, gemcitabine, and cetuximab are under the clinical trial phase I, II, and III for treating lung, breast, skin, and colon cancers [52]. Patients are being enrolled for phase II clinical trials for the treatment non-small-cell lung cancer (NSCLC) and metastatic breast cancer utilizing vinorelbine. Vincristine has been affirmed for treating SCLC, leukemia, and Hodgkin's disease. Vinblastine is utilized in treating number of cancers such as bladder, brain, lung cancer, and Hodgkin's lymphoma. Vinblastine acts by mechanisms at low concentrations, and it suppresses the microtubule action, however, at higher concentrations, it minimizes the microtubule mass, which is in the enrollment stage of phase II trials for treating low-grade glioma. Formerly, vinblastine and vincristine with other compounds as alisertib, sirolimus, bryostatin

I, nilotinib, volasertib, cytarabine, doxorubicin, and bevacizumab are clinically under the trial of phase I/II/III for treating lymphoma, acute lymphoblastic leukemia (ALL), and sarcoma [52].

The first vinca alkaloid fluorinated agent, vinflunine (Javlor) **10** (Fig. 9.10), is currently still in a phase III clinical trial [53,54]. It is a novel synthetic vinca alkaloid, which has been permitted in Europe for the second-line treatment of advanced urothelial and metastatic cancer after miscarriage of platin-containing therapy [23]. It is furnished through the semi-synthesis via the superacid chemistry to introduce two fluorine atoms at the 2′ position of the catharanthine scaffold selectively. The fluorine atoms have an essential contribution to the antitumor potency. Like other vinca alkaloids, vinflunine inhibits the dynamics of microtubule through interacting with the vinca alkaloid binding site on tubulin. It has the weakest affinity for tubulin than the other vinca alkaloids, and the binding of vinflunine to tubulin is more easily reversible. These variations might result in specific effects of vinflunine, the low toxicity of vinflunine, and cell killing. Drug resistance is induced via vinflunine which was mediated by Pgp but the activity of vinflunine to prompt drug resistance is far weaker than that of vinorelbine. Additionally, vinflunine has effectiveness on human tumor cell lines with atypical multidrug-resistant phenotype. It has also antiangiogenic and vascular-disrupting potencies both in vitro and in vivo [53,54]. The major side effects induced, via vinflunine, constipation and myelosuppression in clinical trials, and they are more manageable compared to those of other vinca alkaloids [55–58].

The vincristine analog **11** (Fig. 9.11) was reacted with compound **12** (Fig. 9.12) to give a novel compound. The latter inventive compound has good inhibitory tumor cell proliferation, especially the proliferation of drug-resistant tumor cells, and has little side effect on normal cells. For

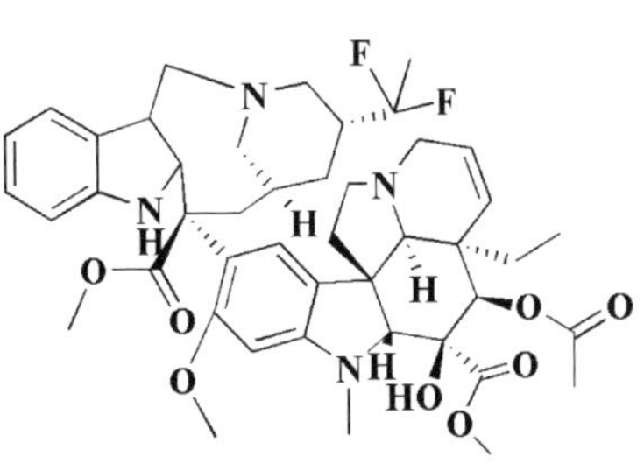

10

Fig. 9.10 Vinflunine **10**.

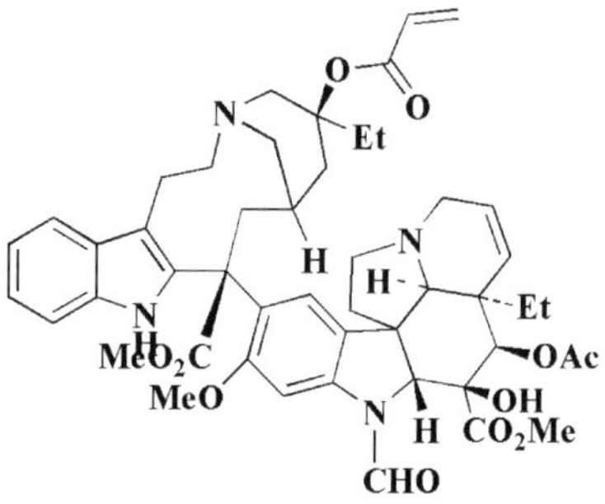

Fig. 9.11 Vincristine analog **11**.

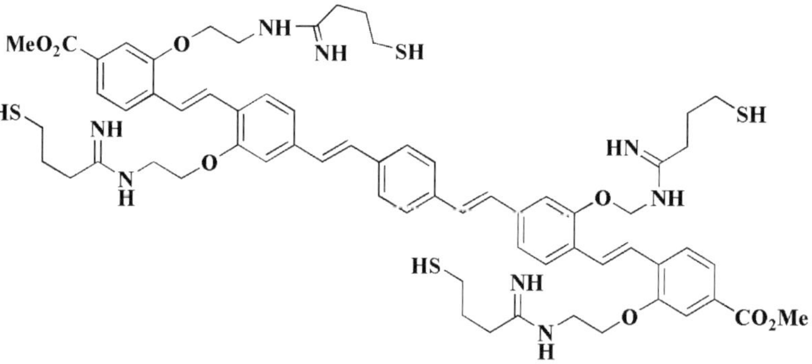

Fig. 9.12 Compound **12**.

example, the final compound containing vincristine indicated in vitro antitumor activities against cancer cell line HCT-8/V ($IC_{50}=0.27\,\mu M$), which is 5.5 times more efficient than vincristine [59].

4. Camptothecin (CPT)

A naturally occurring alkaloid, camptothecin **13** (Fig. 9.13), was accessed upon its isolation from the bark of Camptotheca acuminate, a tree native to central china. The pentacyclic alkaloid structure of camptothecin was determined using X-ray crystallography. Camptothecin received a great attention because of its conspicuous potency in in vivo L1210 mouse life prolongation assays and the in vitro 9KB cytotoxicity tests. Clinical investigations were performed with the water-soluble sodium salt **14** (Fig. 9.14),

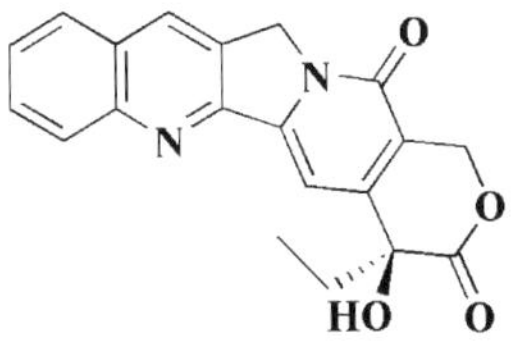

Fig. 9.13 CPT **13**.

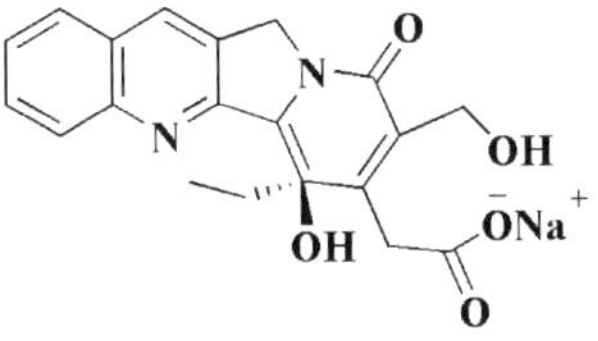

Fig. 9.14 CPT's water-soluble sodium salt **14**.

which furnished modest results and many side effects. Topoisomerase I was reported the cellular target of camptothecin [60].

Many camptothecin analogs have been synthesized (Fig. 9.15) [61]. Studies of compounds modified on the quinolone ring system have indicated that the substitutions at C-12 and C-11 lead to reduced potency, while substitutions at C-10, C-9, and C-7 can result in improved potency [61]. The E-ring lactone is essential for the compound potency, and entirely all modifications to this ring have resulted in less potent compounds; the homocamptothecins represent an exclusion. While the camptothecins are considered the only topo 1 inhibitors used clinically, various categories such

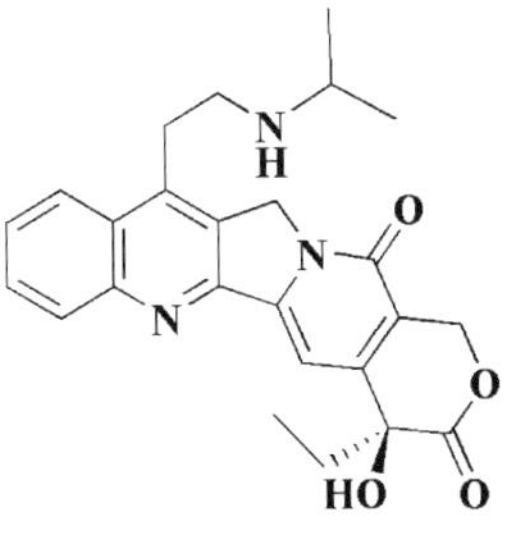

Fig. 9.15 Belotecan **15**.

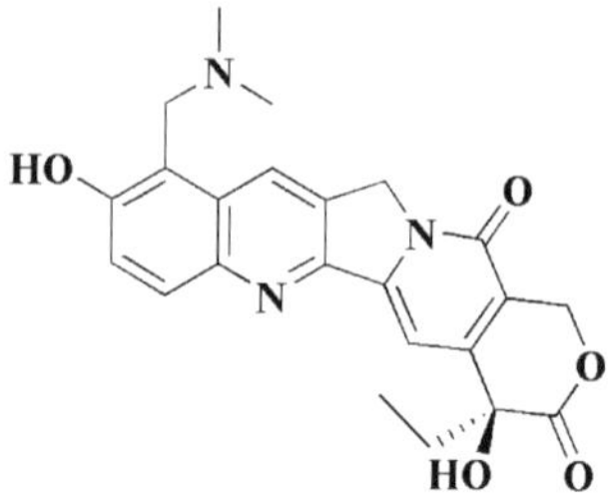

16

Fig. 9.16 Topotecan **16**.

as the indenoisoquinolines, indolocarbazoles, benzimidazoles, and benzacridines are under estimations as sources of alternative inhibitors [62].

Topotecan (Hycamtin) **16** (Fig. 9.16) has a *N,N*-dimethylaminomethyl substituent at C-9; this essential functional group afforded enhanced water solubility. Hycamtin is exploited as second-line treatment for advanced ovarian cancer in patients who did not respond to therapy that comprise paclitaxel or platinum. Recent data from a phase III study, however, have indicated that longstanding survival of patients with advanced epithelial ovarian cancer was comparable for those on paclitaxel and those on topotecan [63]. Recurrent small-cell lung cancer is also an affirmed indication; it has been revealed to rise the time to disease progression for patients who are formerly treated with etoposide in combination with cisplatin [64]. Administration is usually performed through an intravenous (IV) infusion. As Hycamtin binds poorly to human plasma proteins, its half-life is much shorter than that of other CPT analogs, and thus, drug accumulation is not obvious. The neutropenia and thrombocytopenia are the causes of the dose-limiting toxicity. Hycamtin also has potency against hematological malignancies, small cell lung and ovarian cancer. Combination therapies with various different agents such as cisplatin and paclitaxel are under development [65].

Irinotecan (Camptosar) **17** (Fig. 9.17) is a water-soluble prodrug of the hydroxycamptothecin analog SN-38. Carboxylesterase cleavage of the bispiperidine group at the 10-position furnished the potent compound SN-38, which is three significant degrees more active than irinotecan as an inhibitor of topoisomerase I in vitro. Irinotecan has been affirmed for therapy of the advanced colorectal cancer. It is affirmed both as first-line treatment (plus 5-fluorouracil (5-FU)) and as salvage therapy for 5-FU-resistant tumors. It is most frequently administered as an intravenous infusion [64]. One preferred position of irinotecan over topotecan is that the biological half-life of

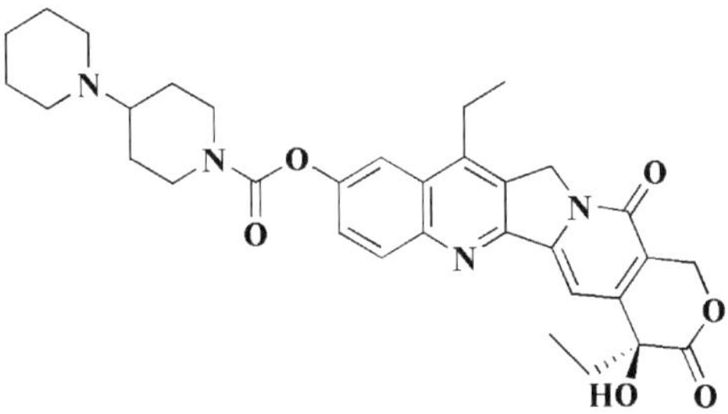

17

Fig. 9.17 Irinotecan **17**.

the lactone form of SN-38 surpasses that of Hycamtin. As the lactone form binds advantageously to serum albumin, this leads to the perseverance in the plasma after drug administration of great percentage of the integral lactone form of both SN-38 and irinotecan. Glucuronidation and biliary excretion are the main pathway mechanisms of elimination of SN-38. The delayed diarrhea, with or without neutropenia, is the principal cause of the dose-limiting toxicity, and it has been proposed that the diarrhea's risk is in reverse linked to the level of glucuronidation [66]. Auspicious antitumor potency against small cell and non-small-lung cancer, cervical cancer, and ovarian cancer has also been indicated, and also clinical trials upon patients with malignant gliomas were performed. Approaches to investigate irinotecan drug combined with alkylating agents, *Vinca* alkaloids, anthracyclines, or taxanes, are in advancement.

5. Paclitaxel

Taxanes are a category of natural products (NP) that include the well-known paclitaxel as an anticancer compound [67] [68]. Paclitaxel (trademark "Taxol") **18** (Fig. 9.18) is a natural diterpene isolated from the bark of *Taxus*

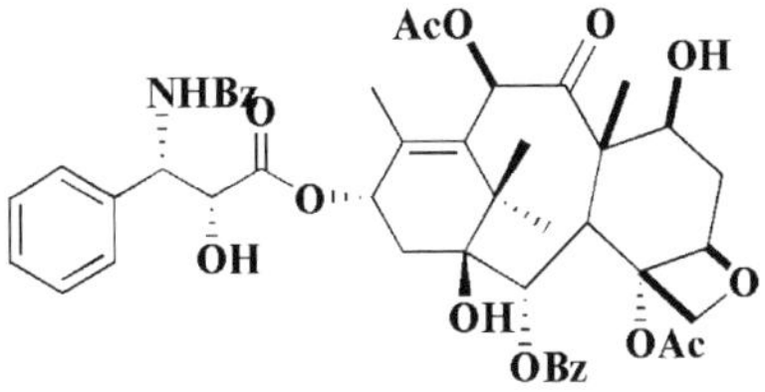

18

Fig. 9.18 Paclitaxel (Taxol) **18**.

brevifolia [69] [70]. Paclitaxel was initially isolated from the Pacific yew tree in 1967 and prompted an extreme engineering and scientific endeavor to furnish the compound consistently to cancer patients [71]. It has been heralded as the antitumor agent of the 1990s because of its promising clinical potency against several human solid tumors such as lung, ovary, head and neck, breast, and melanoma. Taxol seems to provide good antitumor potency [72–75]. Interest in various aspects of this drug has mushroomed due to the clinical importance of its antitumor profile in the treatment of common human cancers. Paclitaxel was used as therapeutic anticancer drug in clinics [76]. Paclitaxel binds specifically with β-tubulin and causes the prevention of the depolymerization during the process of cell division in cancer cells [77]. The structurally complex natural product paclitaxel is among the most significant antitumor drug present in the clinic.

Paclitaxel is widely used in chemotherapy for cancer treatment and is utilized in clinical therapy of breast cancer and advanced ovarian cancer [78] Docetaxel and paclitaxel are among the most potent chemotherapeutic agents for non-small-cell lung cancer patients [79]; the antitumor agent paclitaxel, including initial clinical outcomes, clinical toxicity, clinical pharmacokinetic, pharmacologic, pharmacodynamic correlations, and combination chemotherapy [80], have drawn great attention in drug discovery [81]. Currently, paclitaxel is approved as an cancer agent by the FDA in the United States for second-line treatment against metastatic breast cancer and *cis*-platinum refractory ovarian cancer. It has also been affirmed on the basis of broader ongoing clinical trials, becoming a first-line chemotherapeutic agent in the future [82–84].

Paclitaxel stands as one of the most economically and medically significant natural products [85,86]. The biosynthesis of paclitaxel has requested many efforts of many collaborative research groups to hold the synthesis and labeling of supposed biosynthetic intermediates; in concert with the identification, cloning, and functional expression of the biosynthetic genes necessary for constructing this complex natural product [87] after the discovery of its distinctive pharmacological features, the clinical supply with material of such a vastly sophisticated molecular structure suited a thoughtful concern [88,89]. Its poor water solubility stands as a marked disadvantage in its clinical utility, and new analogs with enhanced pharmacologic properties are needed [69]. Paclitaxel (PTX) has some serious side effects [90]. So its clinical application to brain tumors is restricted by drug resistance, and by these side effects, besides its lower brain penetration [91–94]. Paclitaxel affects the immune system, and the toxic side effects on the immune system are more reversible [95,96]. As the generation of paclitaxel from plants is an expensive process, biotechnological substitutes to yield it more economically become essential [10]. The lack of knowledge and the chemical complexity of the complete biosynthetic pathway of paclitaxel impede many of these options; plant cell culture technology is an striking substitute for supply [14].

Paclitaxel is considered as the prototype of the taxane family of antitumor drugs, the first of a novel category of microtubule-stabilizing antitumor agents [97]. Paclitaxel is a unique tubulin-interacting agent and has unique mechanism of action. Unlike other clinical antimitotic agents such as the vinca alkaloids, which inhibit the microtubule assemblage process, paclitaxel promotes tubulin polymerization and stabilizes the resulting microtubules toward depolymerization. This change in the normal dynamics in the cellular tubulin-microtubule system by paclitaxel is presently recognized as its mode of action for cell cytotoxicity. Taxanes block cell cycle progression via centrosomal impairment, induction of abnormal spindles, and suppression of spindle microtubule dynamics. Triggering of apoptosis by aberrant mitosis, or by consequent multinucleated G1-like state associated with mitotic slippage, relies on cell type and drug schedule. The development of fluorescent analogs of paclitaxel resulted in the set of the location of the spindle pole microtubules and centrosomes as key subcellular targets of cytotoxic taxoids in living cells [98]. It is a microtubule-stabilizing agent, which has been indicated to block dissimilar cells in the G2/M phase of the cell cycle and then modify their radioresponsiveness. Paclitaxel exhibits anticancer effects by inducing the paraptosis-associated cell death [99]. Paclitaxel is also being clinically tested in combination with other chemotherapeutic drugs to treat a variety of different malignancies [100]. Combination effects of paclitaxel and other agents have been studied such as its combination with carboplatin [101]; the natural marine product, ET-743 (Yondelis, trabectedin) [102] or *Ocimum basilicum* polysaccharide [103,104], was demonstrated to generate synergistic and additive inhibitory effects.

The mechanism of action of paclitaxel is known, and the biological active conformation has been much debated. The bioactive conformation of paclitaxel in tubulin/microtubules has been studied. The bioactive conformation of paclitaxel is significant as it could provide critical information that could lead to the development of new analogs with simpler structures and/or higher anticancer activity [105]. Potent and novel synthetic analogs were synthesized [75,106–117] and some of them are listed below:

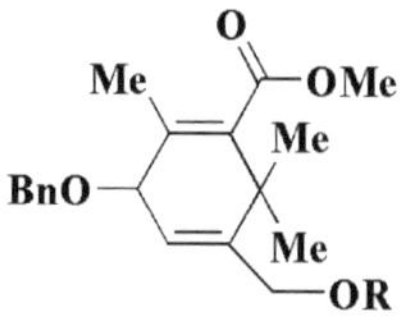

19

Fig. 9.19 Substituted cyclohexene derivatives **19**.

Highly substituted cyclohexenes [**19** (Fig. 9.19); R = H or $C_6H_3(NO_2)$ 2-3,5] corresponding to the A-ring of the anticancer diterpene natural product paclitaxel are prepared via a Diels-Alder reaction and decarboxylative elimination as the key steps [116].

Novel compounds and formulations of turmeric oil, fish oil, aspirin, and anticancer drugs (paclitaxel) were invented possessing antiinflammatory, analgesic, and/or anticancer activity. E.g., ar-turmerone, α-turmerone, and β-turmerone were isolated from turmeric oil. Also, curcumin conjugates were synthesized as well as aspirin analogs (Fig. 9.20) [117].

A convergent synthesis of the ABC ring **21** (Fig. 9.21) of antitumor natural product paclitaxel is reported. SmI2-mediated reductive cyclization of an allylic benzoate having an aldehyde function, prepared from 1,3-cyclohexanedione and tri-O-acetyl-D-glucal, smoothly yielded the 6-8-6 tricarbocyclic structure which is highly strained (66% yield) [75].

The ongoing efforts have been occurred to optimize the utility of paclitaxel in the treatment for curative cancer therapy. The drug has presented problems during its intravenous administration to patients. A close analog of paclitaxel, namely, docetaxel (Taxotere) **22** (Fig. 9.22), which is pending FDA approval for the clinical utility, is devoid of such problems. Docetaxel possesses a slightly better aqueous solubility than paclitaxel [82].

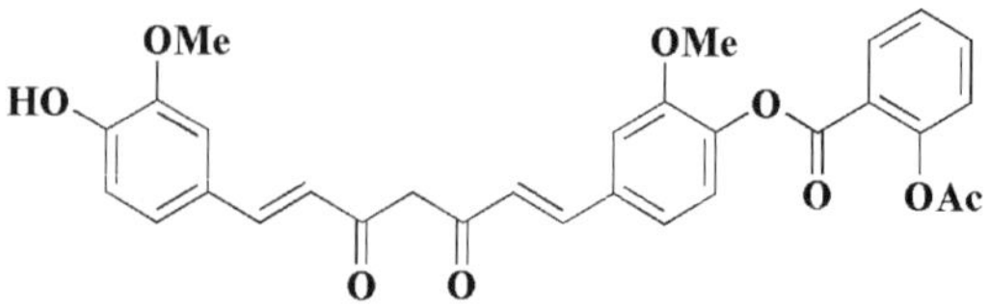

20

Fig. 9.20 Compound **20**.

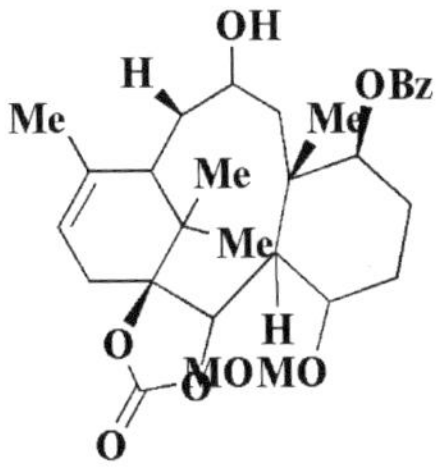

Fig. 9.21 Compound **21**.

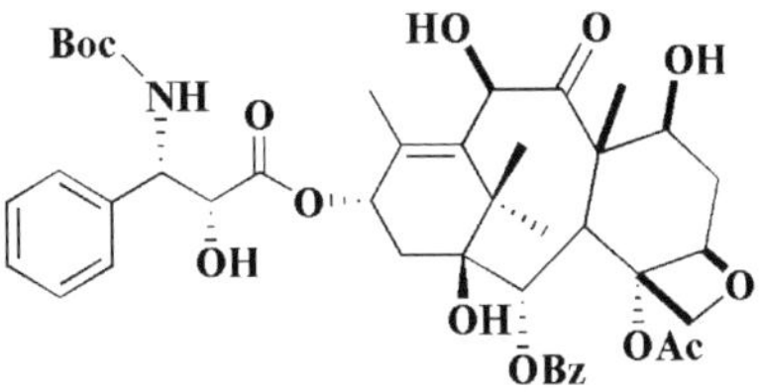

Fig. 9.22 Docetaxel (Taxotere) **22**.

6. Epothilones

Natural products exhibit taxol-like antitumor potency, such as epothilones A and B, eleutherobin, sarcodictyins and B, laulimalide, isolaulimalide, and discodermolide. These natural products exert their cytotoxic effect through the destabilization of microtubule structure and promotion of disassembly of microtubules into tubulin [118]. These significance inhibitors can act as antitumor agents [119]. The epothilones, a family of macrolactone natural products, are a novel category of antimicrotubule agents that were isolated as cytotoxic metabolites of the *Myxobacterium sorangium cellulosum* [120,121]. They are active microtubule-stabilizing agents having antiproliferative potency, which have functioned as significant lead compounds for the discovery of various clinical candidates for treating cancer. In 2007, epothilone-type agent is FDA-approved for clinical utility for treating breast cancer [122–124].

Two main fermentation products were reported by Reichenbach's group, epothilone A and epothilone B, which vary only by the presence or absence of the methyl group at the trisubstituted epoxide scaffold. Epothilones were also indicated to inhibit the growth of cells overexpressing the

P-glycoprotein efflux pump, thus showing that these compounds might finally be beneficial for treating multidrug-resistant tumors [125]. Epothilones cannot be generated in huge amounts due to their lower output in fermentation and complicated methods in chemical synthesis. The generation of epothilones in heterologous hosts furnishes various advantages over the use of the natural host, including easy genetic manipulation and quick replication of heterologous hosts. Production of epothilones in heterologous hosts has become a hot spot [126]. They were originally described as antifungal macrolides; they can stabilized microtubules at submicromolar concentrations [121,127]. The epothilone synthetase is a decamodular megasynthase necessary for the biosynthesis of a category of polypetide natural products with therapeutically promising antitumor activity [128]; the epothilones are from the main categories of tubulin polymerization promoters presented. Epothilone D is a member of a category of potent antineoplastic natural products [129].

Natural products are mixed polyketide/nonribosomal peptide molecules, comprising epothilones and bleomycin; they are in clinical utility as anticancer agents [130–133]. Epothilones are microtubule depolymerization inhibitors; they act via a microtubule stabilization mechanism [134]; like to taxol, the epothilones provide a main prospective therapeutic benefit in that they preserve their potency against multidrug-resistant cell lines [135–137]. Like taxanes, they induce microtubule bundling, generation of multipolar spindles, and mitotic arrest [127]. Epothilones like paclitaxel in that they compete with paclitaxel for binding to the microtubule and suppress the potency of these structures [138,139]. They are cytotoxic polyketides that mimic the effects of taxol on cancer cells (i.e., microtubule stabilization). Currently, there are several epothilones or their chemical modified analogs undergoing clinical evaluation for cancer treatments in humans [140,141]. Cell lines that are resistant to epothilones have a mutation in the β-tubulin–*binding* site which is located neighboring the binding site for taxanes. This has led to the hypothesis that microtubule binding has a common pharmacophore. Nettles et al. utilizing electron crystallography, nuclear magnetic resonance spectroscopy, and molecular modeling derived the conformation of epothilone A in complex with β-tubulin subunits in zinc-stabilized tubulin sheets [142]. This research showed that although paclitaxel and epothilones overlap in their occupation of a large common binding site, the ligands exploit the pocket in a specific and qualitatively distinct manner. A latest study established variances in binding between these two classes of compounds. Bode et al. discovered that epothilones stabilize microtubules in

Saccharomyces cerevisiae but paclitaxel did not, implying that the two compounds' binding interactions are different [143].

The hypothesis that the two groups of compounds have identical but unique binding sites is confirmed by these studies. Various epothilone derivatives have been prepared, and many are presently in clinical trials, such as patupilone (epothilone B, EPO906), BMS-310705 (a water-soluble semi-synthetic analog of epothilone B), ixabepilone (azaepothilone B; BMS-247550), ZK-EPO, and KOS-862 (epothilone D) [144]. Epothilones (A and B) have a remarkable antitumor potency even against taxane-resistant cell lines, however, they have limited success in clinic due to their toxicity. This has resulted in developing many semi-synthetic derivatives. The compounds have been modified near C12–13 to generate the greatest impact on microtubule-stabilizing potency [145]. The addition of —CH_3 group at this position gives epothilone B, which is twice as active as epothilone A in induction of the microtubule polymerization [146]. Epothilones have indicated a broad-spectrum potency against various human malignances in cell culture and xenograft models [144]. Epothilone (Epo) D, an antitumor agent currently in clinical trials, is a hybrid natural product produced by the united action of polyketide synthases (PKS) and nonribosomal peptide synthetases (NRPS) [147,148].

In cell culture studies the epothilones show more cytotoxicity than paclitaxel in which the typical IC_{50} values in these investigations is in the nano- or subnanomolar concentration range [149–151]. These same preclinical studies have demonstrated that there are significant variations in refer to the mechanism of drug resistance between epothilones and paclitaxel. Overexpression of *p*-glycoprotein has the least impact on the cytotoxicity of desoxyepothilone, epothilone B, and aza-epothilone B in cell culture models [149–151]. The differences between the IC_{50} values for epothilones in the sensitive and *p*-glycoprotein expressing resistant cells are very small as compared to the variances found in cells which are resistant to paclitaxel. So epothilones are proposed to be more potent in *p*-glycoprotein-expressing cancers. Aza-epothilone B is vastly potent in colon, breast, and ovarian cancer xenograft models and has induced treatments in the ovarian xenograft model Par-7 that is resistant to paclitaxel. Aza-epothilone B also shows potency when administered orally. This presumably refers to the circumstance that p-glycoprotein is expressed in the GI mucosa, so the inhibition of the absorption of paclitaxel is provided [152]. New details about biological active conformation of epothilones could help guide efforts to design enhanced derivatives of these natural products; various alternates of which are in clinical investigations as anticancer agents. Epothilones implement two families of conformations in the bioactive epoxide region of each molecule [153].

The framework structure of epothilone might introduce a remarkable template for developing developed antitumor agents. Many epothilone analogs with altered conformational profiles have been designed and synthesized [154–157]; for example, the compounds which introduce a novel structural scaffold for microtubule stabilization and are promising lead compounds for anticancer drug discovery were synthesized such as azathilones **23** (Fig. 9.23) (R=Bu-t, Et); those derivatives were estimated for tubulin polymerization and anticancer potency. They are active inducers of tubulin polymerization and inhibit human cancer cell growth in vitro [158].

Various lactam derivatives of the epothilones were also synthesized utilizing a semi-synthetic approach starting with the unprotected natural products. The intact three-step sequence was stream-lined into a "one-pot" process for the epothilone B-lactam, BMS-247550 (**24**) (Fig. 9.24), which is currently undergoing phase I clinical trials [159].

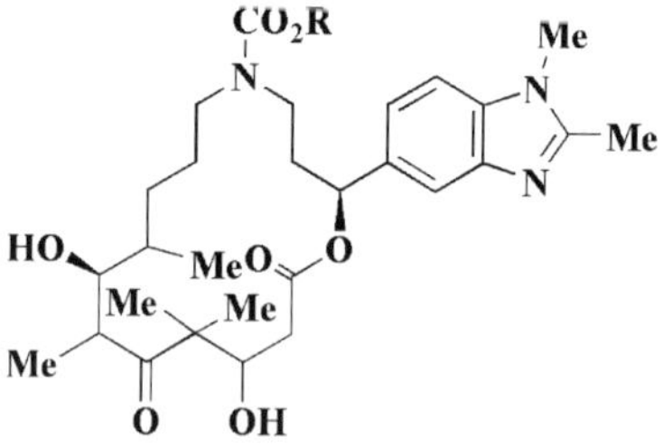

Fig. 9.23 Azathilone derivative **23**.

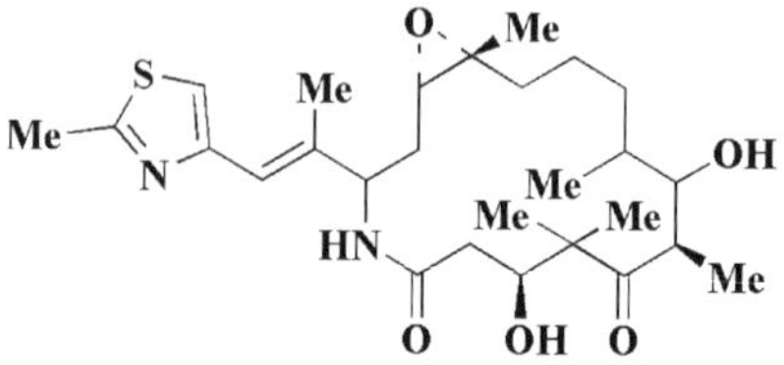

Fig. 9.24 The epothilone B-lactam, BMS-247550 **24**.

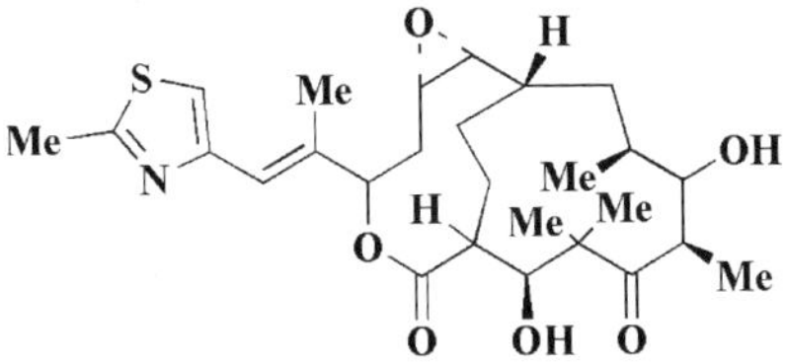

25

Fig. 9.25 An epothilone analog **25**.

The effective synthesis of both right- and left-hand halves of a constrained derivative **25** (Fig. 9.25) of the anticancer natural product epothilone is reported. The 11-membered rings common to both compounds are synthesized through olefin metathesis [160].

7. Dolastatin

Dolastatin (Dol) **26** (Fig. 9.26) is marine natural product that is cytotoxic for cancer cells [161,162]. Dolastatins are pentapeptides originally isolated in 1970 from the Indian Ocean from the sea hare *Dolabella auricularia* [163,164] and then from cyanobacterium (genus *Symploca*), which has been called *Caldora penicillata*. Dolastatin 10 is the most potent compound of this category, with its ED50 being 4.6.10–5 μg/mL against murine PS leukemia cells. Dolastatin 10 is an active antimitotic agent in which its mechanism of action is similar to that of vinca alkaloids and taxenes. Many studies propose that it binds to β-subunit of tubulin; forms tubulin bundles; then the inhibition of the formation of microtubules occurred, which arrest the cells in G2/M phase; and leads to Bcl2 mediated apoptotic cell death. Additionally, it acts synergistically with bryostatin and vinca alkaloids against various

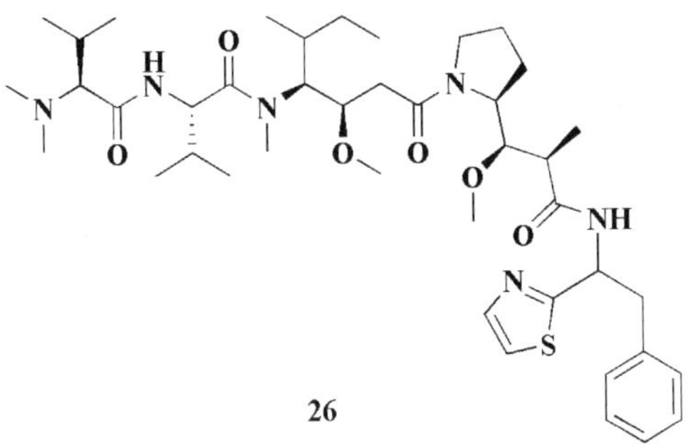

26

Fig. 9.26 Dolastatin 10 **26**.

cancers. Dolastatin 10 was investigated in many clinical trials (phases I and II) for treating solid tumors of kidney and pancreas, besides a multi-institutional phase II clinical trial for treating metastatic soft tissue sarcoma. However, because of its side effects such as peripheral neuropathy and minimal responses, it was withdrawn from the subsequent clinical trials. Nevertheless, due to its antimitotic activity against tumors, developing its synthetic analogs was prompted. Auristatin and its synthetic analogs of naturally occurring dolastatin 10, which is a simple linear peptide [165] (an antineoplastic agent [166,167], are highly active antimitotic agents due to their cytotoxic potency when targeting tumor cells in the form of antibody-drug conjugates (ADCs) [168]. Auristatins are ultrapotent cytotoxic microtubule inhibitors that are clinically utilized as payloads in ADCs [169,170]. Some marine blue-green algae are evolving as a significant source of anticancer drugs. The mainstream of these active biomolecules, comprising the curacin A, dolastatins, hectochlorin, the apratoxins, and the lyngbyabellins, are from the mixed polyketide/polypeptide structural category [171]. Multistep total synthesis of dolastatin 10 was developed due to its limited availability from natural origin [172]. Many analogs have been designed and synthesized [173,174]. The focus of modification of dolastatin 10 core moiety is mainly on the P1, N-terminus, and P5, C-terminus, with the least consideration to the P2 subunit [175].

8. Eribulin (E7389)

Eribulin mesylate (Halaven) (E7389) **27** (Fig. 9.27), a macrocyclic ketone [176] and halichondrin B synthetic analog [177–181], binds to tubulin and microtubules and possesses broad anticancer activity [182] E7389

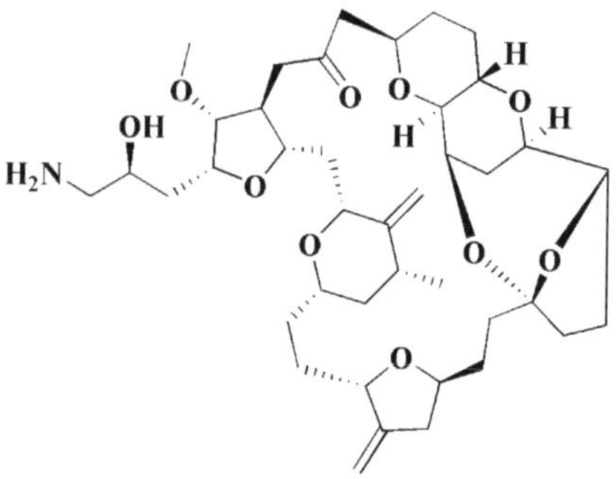

27

Fig. 9.27 Eribulin **27**.

(eribulin mesylate, anticancer agent) [183–191], available in market, is also developed wherein halichondrin B is considered as a lead chemical compound [192–195]. Eribulin (E7389) shows a mechanistically unique microtubule dynamics inhibitory activity [196–202]. It acts as a nontaxane microtubule dynamics inhibitor [203,204]. It is considered as a global inhibitor of tubulin polymer formation, causing the disruption of the tubulin/tubulin contacts at the interdimer interface [205–212]; it acts by inhibiting microtubule dynamics through mechanisms distinct from those of other conventional tubulin-targeted agents [213] Eribulin targets microtubules, suppressing dynamic instability at microtubule plus ends via an inhibition of microtubule growth with little or no effect on shortening [214]. Eribulin binds to the vinca domain of tubulin; then the polymerization of tubulin is inhibited and the assembly of microtubules, leading to the inhibition of mitotic spindle assemblage, induction of cell cycle arrest at G2/M phase, and tumor regression [215].

Eribulin is a promising novel cytotoxic chemotherapy agent owed to its importance in treating cancers that are refractory or resistant to other drugs besides its convenient toxicity profile [186,216,217]. The microtubule growth is inhibited by the Eribulin mesylate, without effects on shortening, and causes the promotion of the nonproductive tubulin aggregate formation. Eribulin was well-tolerated in children with recurrent or refractory solid tumors with neutropenia recognized as the principal DLT [218]. Eribulin was potent and tolerated, and represents a potential treatmenting for non-small-cell lung cancer (NSCLC) [219,220], and it shows efficacy in some cases of patients who are previously treated with a taxane [220]. Additionally, eribulin mesylate (E7389) exhibited effectiveness in patients with different solid tumors, particularly, those with heavily pretreated metastatic breast cancer [221]; it is currently used in clinical trials [222–226]. It is considered as a novel active agent for MBC and has a crucial role in the management of breast disease [227–229]. Up to the present time Eribulin is considered as the only single agent that has been indicated to persist survival in patients with heavily pretreated MBC [223,230]. Additionally, it has been approved for treating acute myelogenous leukemia [231] and some other types of cancer [224,232] Eribulin was approved for treating metastatic liposarcoma for patients who have progressed with anthracycline treatment [233–237]. Additionally, it has combination potency with multiple agents from various mechanistic classes in many human cancer models, including melanoma, ovarian, and breast [238]. Eribulin mesylate indicates potency superior to paclitaxel in certain human epithelial ovarian cancer xenograft models [199].

Additionally, the synthesis of Halaven (INN eribulin mesylate, E7389) from a medicinal chemical method to the implementation of the final route on pilot scales is studied [239]. The synthesis of potent antitumor compound E7389 [240–242] analogs was reported [243–246].

A 12-step kilogram-scale synthesis of the C1–C13 fragment **28** (Fig. 9.28), common to halichondrin B and the entirely synthetic analog Halaven (INN eribulin mesylate, E7389), is reported [240].

Cr-Mediated coupling reactions are usually performed with a minor excess of an assumed nucleophile. For the development of a profitable usage of this procedure, two various tactics have been studied. The first tactic relies on two successive catalytic asymmetric Cr-mediated couplings, with the utility of the coupling partners intentionally being of unbalanced complexity and molecular size. The second tactic rests on the achievement in recognizing the nucleophile. The C23—O bond is stereospecifically assembled through reductive cyclization of the oxonium ion, or oxy-Michael cyclization. Both syntheses have a great efficacy: E7389 C14–C35 and halichondrin C14–C38 building blocks, **29** (Fig. 9.29) ($R^2=CMe_2$, $R^2=COCMe_3$; $R=COPh$, $R^2=SiPh_2CMe_3$) and compound **30** (Fig. 9.30) ($R^1=COPh$, $R^2=SiPh_2CMe_3$; $R^1-SiMe_2CMe_3$, $R^2=COCMe_3$), respectively, have been furnished from the corresponding C27–C35 and C27–C38 aldehydes, respectively, with an excellent stereoselectivity in high yields [247].

Tert-butyldimethylsilyl polyethers are deprotected utilizing using simple nonaqueous method; the procedure is utilized for deprotections in the preparation of the core structure of halichondrin B and of the polycyclic core of E7389. Desilylation of the tert-butyldimethylsilyl ethers with tetrabutylammonium fluoride (TBAF) is followed by the addition of calcium carbonate, a sulfonic acid resin such as DOWEX 50WX8-400, and methanol, stirring, filtration, and removal of solvent; the desired alcohols can be isolated free

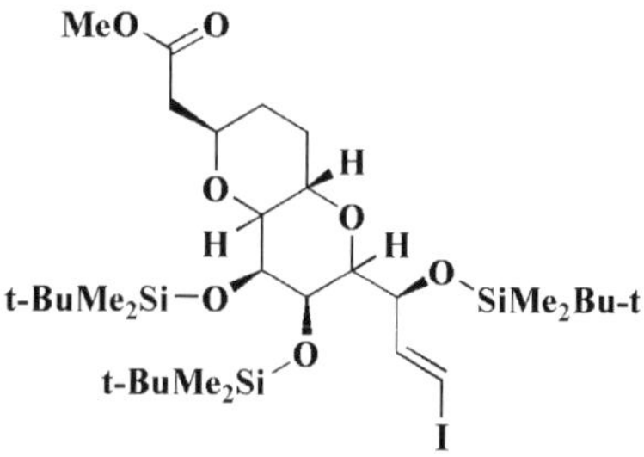

Fig. 9.28 Compound **28**.

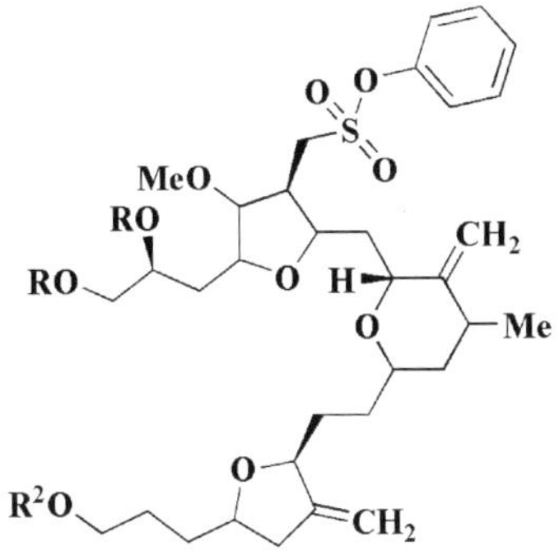

Fig. 9.29 Compound **29**.

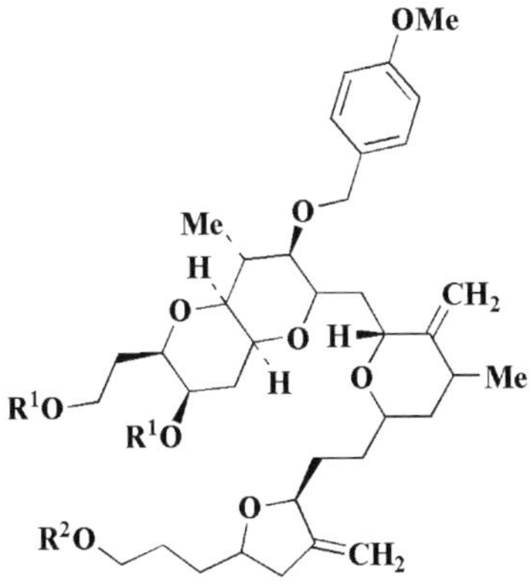

Fig. 9.30 Compound **30**.

of tetrabutylammonium fluoride. The method is utilized for deprotection of three TBDMS-protected diols and for the deprotection of three monosaccharide tetra-TBDMS ethers. Compound **31** (Fig. 9.31) is synthesized from a macrocyclic precursor by acetic acid and TBAF-mediated desilylation and Michael addition followed by ketalization utilizing resin-bound reagents; the polycyclic core of E7389 is synthesized using the identical method [248].

With consecutive usage of catalytic asymmetric Cr-mediated coupling reactions, E7389 C14–C35 and halichondrin C14–C38 building blocks, **32** (Fig. 9.32) and **33** (Fig. 9.33), have been stereoselectively prepared. The C19-C20 bond is first formed through the catalytic asymmetric Ni/Cr-

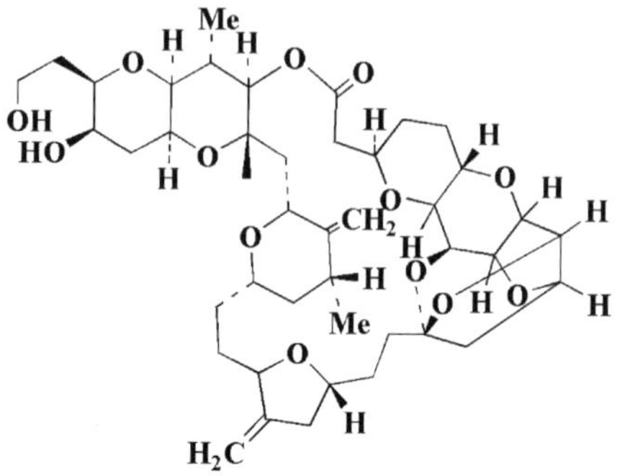

31

Fig. 9.31 Compound **31**.

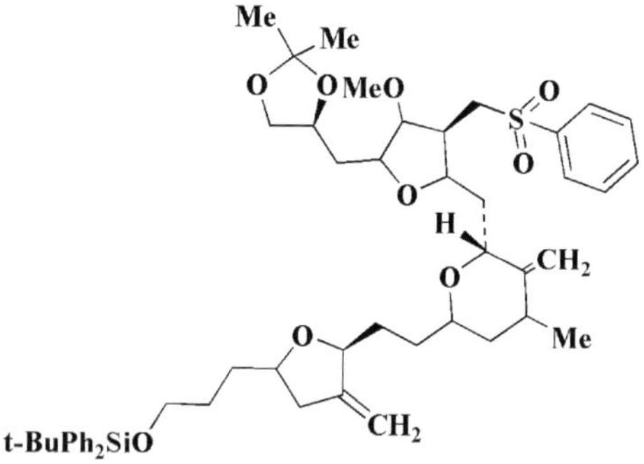

32

Fig. 9.32 Compound **32**.

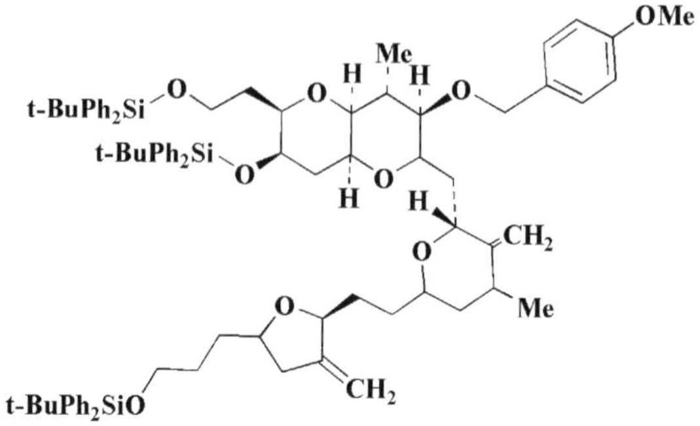

33

Fig. 9.33 Compound **33**.

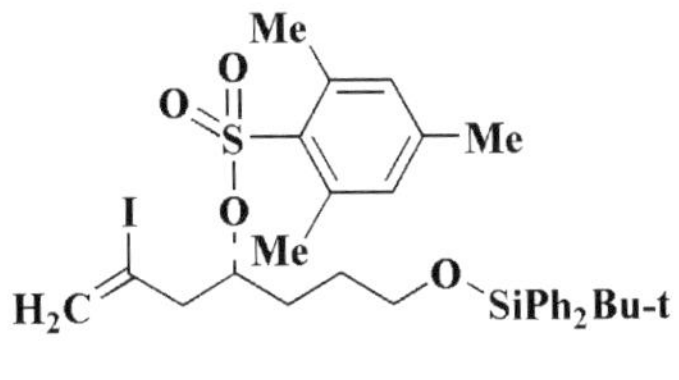

Fig. 9.34 Compound **34**.

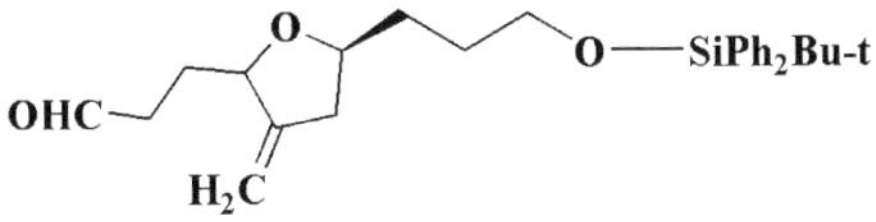

Fig. 9.35 Compound **35**.

mediated coupling, i.e., iodoalkene **34** (Fig. 9.34) + $PhCO(CH_2)_3CHO$, in which iodoalkene **34** is utilized as the restraining substrate. Then the formation of the C23–C24 bond is performed through the catalytic asymmetric Co/Cr-mediated coupling, i.e., (Fig. 9.35) + (*R*)-$H_2C{:}CICHMeCH_2I$, in which the alkyl-iodide bond is selectively activated over the vinyl-iodide bond. The catalytic asymmetric Ni/Cr-mediated reaction is employed to couple C14–C26 segment with E7389 C27–C35 segment. In this synthesis, the C23—O bond is stereoselectively assembled through a double-inversion process to yield the E7389 C14–C35 building block **32**. Then the halichondrin C14–C38 building block **33** is synthesized using the matching synthetic sequence [249].

References

[1] Rodrigues T, Reker D, Schneider P, et al. Counting on natural products for drug design. Nat Chem 2016;8:531–41.

[2] Koehn F, Carter G. The evolving role of natural products in drug discovery. Nat Rev Drug Discov 2005;4:206–20.

[3] Cragg GM. Paclitaxel (Taxol): a success story with valuable lessons for natural product drug discovery and development. Med Res Rev 1998;18(5):315–31.

[4] Tan LT. Marine cyanobacteria: a prolific source of bioactive natural products as drug leads. Mar Microbiol 2013;59–81.

[5] Mukhopadyay A. Natural products in cancer therapy and prevention. Indian J Nutr Diet 2010;47(8):364–77.

[6] Smanski M, Zhou H, Claesen J, et al. Synthetic biology to access and expand nature's chemical diversity. Nat Rev Microbiol 2016;14:135–49.

[7] Hua F, Shang S, Hu Z. Seeking new anti-cancer agents from autophagy-regulating natural products. J Asian Nat Prod Res 2017;19(4):305–13.
[8] Hanauske AR. The development of new chemotherapeutic agents. Anticancer Drugs 1996;7(Suppl):229–32.
[9] Zishan M, Saidurrahman A, Azeemuddin AZ, Hussain W. Natural products used as anti-cancer agents. J Drug Deliv Ther 2017;7(3):11–8.
[10] Schaefer H, Wink M. Medicinally important secondary metabolites in recombinant microorganisms or plants: progress in alkaloid biosynthesis. Biotechnol J 2009;4 (12):1684–703.
[11] Mann J. Natural products in cancer chemotherapy: past, present and future. Nat Rev Cancer 2002;2:143–8.
[12] Kingston DGI, Newman DJ. Natural products as anticancer agents. In: Begley TP, editor. Wiley encyclopedia of chemical biology, vol. 3. Wiley; 2009. p. 249–60.
[13] Liu EH, Qi L, Wu Q, Peng Y, Li P. Anticancer agents derived from natural products. Mini-Rev Med Chem 2009;9(13):1547–55.
[14] Kolewe ME, Gaurav V, Roberts SC. Pharmaceutically active natural product synthesis and supply via plant cell culture technology. Mol Pharm 2008;5(2):243–56.
[15] Altmann K, Gertsch J. Anticancer drugs from nature-natural products as a unique source of new microtubule-stabilizing agents. Nat Prod Rep 2007;24(2):327–57.
[16] Demain AL, Vaishnav P. Natural products for cancer chemotherapy. Microb Biotechnol 2011;4(6):687–99.
[17] Huang D, Wu J, Gu J. Overview about the research of apratoxins family-antitumor marine natural products. Guowai Yiyao Kangshengsu Fence 2012;33(5):193–8.
[18] Wood HBJ. Development of natural products as antitumor drugs. In: Simkins MA, editor. Med Chem, Proc Int Symp, 6th; 1979. p. 265–80.
[19] Srinivasan R, Lakshmana G, Anjinayalu B, Kumar DA, Raju KS. Naturevolution effective against cancer therapy-review (treating cancer diseases using plant products). Int J Pharm Bio Sci 2014;5(2):357–65.
[20] Wang X, Itokawa H, Lee K. Structure-activity relationships of toxoids. Med Aromat Plants- -Ind Profiles 2003;32(Taxus):298–386.
[21] Abdul Fattah T, Saeed A. Applications of Keck allylation in the synthesis of natural products. New J Chem 2017;41(24):14804–21.
[22] Van Lanen SG, Shen B. Combinatorial biosynthesis of anticancer natural products. In: Cragg GM, Kingston DGI, Newman DJ, editors. Anticancer agents from natural products. 2nd ed; 2012. p. 671–98. 1 plate.
[23] Akhtar MS, Swamy MK. 1st. Anticancer plants: properties and application, vol. 1. Springer Nature Singapore Pte Ltd.; 2018.
[24] Kelly MG, Hartwell JL. The biological effects and the chemical composition of podophyllin. A review. J Natl Cancer Inst 1954;14:967–1010.
[25] Bohlin L, Rosén B. Podophyllotoxin derivatives: drug discovery and development. Drug Discov Today 1996;8:343–51.
[26] Damayanthi Y, Lown JW. Podophyllotoxins: current status and recent developments. Curr Med Chem 1998;5:205–52.
[27] Gordaliza M, Castro MA, Miguel Del Corral JM, San Feliciano A. Antitumor properties of podophyllotoxin and related compounds. Curr Pharm Des 2000;6:1811–39.
[28] Podophyllotoxins JI. In: Cassady JM, Douros JD, editors. Anticancer agents based on natural product models. New York: Academic; 1980. p. 319–51.
[29] Lee K-H, Xiao Z. The podophyllotoxins and analogs. In: Cragg GM, Kingston DGI, Newman DJ, editors. Anticancer agents from natural products. New York: CRC Press; 2005.
[30] Keller-Juslén C, Kuhn M, Von Warburg A, Stahelin H. Synthesis and antimitotic activity of glycosidic lignan derivatives related to podophyllin. J Med Chem 1971;14:936–40.

[31] Doyle TW. The chemistry of etoposide. In: Issell BF, Muggia FM, Carter SK, editors. Etoposide (VP16). Current status and new developments. New York: Academic Press; 1984. p. 15–32.
[32] Meresse P, Dechaux E, Monneret C, Bertounesque E. Etoposide: discovery and medicinal chemistry. Curr Med Chem 2004;11:2443–66.
[33] Saulnier MG, Langley DR, Kadow JF, Senter PD, Knipe JO, Jay O, et al. Synthesis of etoposide phosphate, BMY-40481: a water-soluble clinically active prodrug of etoposide. Bioorg Med Chem Lett 1994;4:2567–72.
[34] Cortese F, Bhattacharyya B, Wolff J. Podophyllotoxin as a probe for the colchicine binding site of tubulin. J Biol Chem 1977;252:1134–40.
[35] Loike JD. VP16–213 and podophyllotoxin. A study on the relationship between chemical structure and biological activity. Cancer Chemother Pharmacol 1982;7:103–11.
[36] Wozniak AJ, Ross WE. DNA damage as a basis for 4-demethylepipodophyllotoxicity. Cancer Res 1983;43:120–4.
[37] Loike JB, Horwitz SB, Grollman AP. Effect of podophyllotoxin and VP-16 on microtubule assembly in vitro and nucleoside transport in HeLa cells. Biochemistry 1976;15:5435–42.
[38] Roberts D, Hilliard S, Peck C. Sedimentation of DNA from L1210 cells after treatment with 4′-demethylepipodophyllotoxin-9-(4,6-O-2-thenylidene-β-D-glucopyramoside) of 1-β-D-arabinofuranosylcytosineor both drugs. Cancer Res 1980;40:4225–31.
[39] Ross W, Rowe T, Glisson B, Yalowich J, Liu L. Role of topoisomerase II in mediating epipodophyllotoxin-induced DNA cleavage. Cancer Res 1984;44:5857–60.
[40] Berger JM, Wang JC. Recent developments in DNA topoisomerase II structure and mechanism. Curr Opin Struct Biol 1996;6:84–90.
[41] Ayres DC, Loike JD. Lignans chemical, biological and clinical properties. Cambridge: University Press; 1990. p. 113.
[42] Rivera G, Dahl GV, Bowman WP, Avery TL, Wood A, Aur RJ. VM26 and cytosine arabinoside combination chemotherapy for initial induction failures in childhood lymphocytic leukemia. Cancer 1980;46:1727–30.
[43] Issell BF. The podophyllotoxin derivatives VP16–213 and VM26. Cancer Chemother Pharmacol 1982;7:73–80.
[44] Saito H, Yoshikawa H, Nishimura Y, Kondo S, Takeuchi T, Umezawa H. Studies on lignan lactone antitumor agents. II. Synthesis of N-alkylamino- and 2,6-dideoxy-2-aminoglycosidic lignan variants related to podophyllotoxin. Chem Pharm Bull 1986;34:3741–6.
[45] Wang ZQ, Kuo YH, Schnur D, Bowen JP, Liu SY, Han FS. Antitumor agents 113. New 4β-arylamino derivatives of 4'-O-demethylepipodophyllotoxin and related compounds as potent inhibitors of human DNA topoisomerase II. J Med Chem 1990;33:2660–6.
[46] Liu JM, Chen LT, Chao YL, Anna FY, Wu CW, Liu TS, Shiah HS, Chang JY, Chen JD, Wu HW, Lin WC, Lan C, Whang-Peng J. Phase II and pharmacokinetic study of GL331 in previously treated Chinese gastric cancer patients. Cancer Chemother Pharmacol 2002;49:425–8.
[47] Terada T, Fujimoto K, Nomura M, Yamashita J, Wierzba K, Yamazaki R. Antitumor agents 3. Synthesis and biological activity of 4β-alkyl derivatives containing hydroxyl, amino, and amido groups of 4'-O-demethyl-4-deoxypodophyllotoxin as antitumor agents. J Med Chem 1993;36:1689–99.
[48] Byl JAW, Cline SD, Utsugi T, Kounai T, Yamada Y, Osheroff N. DNA topoisomerase II as the target for the anticancer drug TOP-53: mechanistic basis for drug action. Biochemistry 2001;40:712–8.

[49] Gupta S, Bhattacharyya B. Antimicrotubular drugs binding toVincado-main of tubulin. Mol Cell Biochem 2003;253:41–7.
[50] Johnson IS, Armostrong JG, Gorman M, Burnett Jr JP. TheVincaalkaloids: a new class of oncolytic agents. Cancer Res 1967;23:1390–427.
[51] Jordan MA, Thrower D, Wilson L. Mechanism of inhibition of cell prolif-eration byVincaalkaloids. Cancer Res 1991;51:2212–22.
[52] Akhtar MS, Swamy MK. 1st. Anticancer plants: clinical trials and nanotechnology, vol. 3. Springer; 2017.
[53] Kruczynski A, Poli M, Dossi R, Chazottes E, Berrichon G, Ricome C, Giavaz-zi R, Hill BT, Taraboletti G. Anti-angiogenic, vascular-disrupting and antimetastatic activities of vinflunine, the latestVincaalkaloid in clin-ical development. Eur J Cancer 2006;42:2821–32.
[54] Holwell SE, Hill BT, Bibby MC. Anti-vascular effects of vinflunine in theMac 15A transplantable adenocarcinoma model. Br J Cancer 2001;84:290–5.
[55] Johnson P, Geldart T, Fumoleau P, Pinel MC, Nguyen L, Judson I. Phase I study of vinflunine administered as a 10-minute infusion on days 1and 8 every 3 weeks. Invest Drugs 2006;24:223–31.
[56] Yun-San Yip A, Yuen-Yuan Ong E, Chow LW. Vinflunine: clinical perspec-tives of an emerging anticancer agent. Expert Opin Invest Drugs 2008;17:583–91.
[57] Bennouna J, Delord JP, Campone M, Nguyen L. Vinflunine: a new micro-tubule inhibitor agent. Clin Cancer Res 2008;14:1625–32.
[58] Kruczynski A, Hill BT. Vinflunine, the latest Vincaalkaloid in clinical de-velopment. A review of its preclinical anticancer properties. Crit Rev Oncol Hematol 2001;40:159–73.
[59] Wang S, Zhou L, Liu L, Lv F. Preparation of natural products containing oligomeric phenylene-ethylene compounds as antitumor agents. Faming Zhuanli Shenqing; 2017. CN 107382786 A 20171124.
[60] Rahier NJ, Thomas CJ, Hecht SM. Camptothecin and its analogs. In: Cragg GM, Kingston DGI, Newman DJ, editors. Anticancer agents from natural products. 2nd. CRC Press; 2012.
[61] Thomas CJ, Rahier NJ, Hecht SM. Camptothecin: current perspectives. Bioorg Med Chem 2004;12:1585–604.
[62] Meng L, Liao Z, Pommier Y. Non-camptothecin DNA topoisomerase I inhibitors in cancer therapy. Curr Top Med Chem 2003;3:305–20.
[63] Ten Bokkel HW, Lane SR, Ross GA. Longterm survival in a phase III, randomized study of topotecan versus paclitaxel in advanced epithelial ovarian carcinoma. Ann Oncol 2004;15:100–3.
[64] Garcia-Carbonero R, Supko JG. Current perspectives on the clinical experience, pharmacology, and continued development of the camptothecins. Clin Cancer Res 2002;8:641–61.
[65] Ulukan H, Swaan PW. Camptothecins. A review of their chemotherapeutic potential. Drugs 2002;62:2039–57.
[66] Abigerges D, Chabot GG, Armand JP, Herait P, Gouyette A, Gandia G. Phase I and pharmacologic studies of the camptothecin analog irinotecan administered every 3 weeks in cancer patients. J Clin Oncol 1995;13:210–21.
[67] MacEachern-Keith GJ, Butterfield LJW, Mattina MJI. Paclitaxel stability in solution. Anal Chem 1997;69(1):72–7.
[68] Kingston DGI, Newman DJ. Natural products as anticancer agents. In: Civjan N, editor. Natural products in chemical biology. Wiley; 2012. p. 325–49.
[69] Jiang S, Zu Y, Fu Y, Zhang Y, Efferth T. Activation of the mitochondria-driven pathway of apoptosis in human PC-3 prostate cancer cells by a novel hydrophilic paclitaxel derivative, 7-xylosyl-10-deacetylpaclitaxel. Int J Oncol 2008;33(1):103–11.

[70] Pradier O, Rave-Frank M, Schmidberger H, Bomecke M, Lehmann J, Meden H, Hess CF. Effects of paclitaxel in combination with radiation on human head and neck cancer cells (ZMK-1), cervical squamous cell carcinoma (CaSki), and breast adenocarcinoma cells (MCF-7). J Cancer Res Clin Oncol 1999;125(1):20–7.
[71] Hanauske AR. The development of new chemotherapeutic agents. Anti-Cancer Drugs 1996;7(Suppl. 2, Management of Advanced Breast Cancer: Patient Needs, Challenges and New Treatment Options):29–32.
[72] Riondel J, Jacrot M, Picot F, Beriel H, Mouriquand C, Potier P. Therapeutic response to taxol of six human tumors xenografted into nude mice. Cancer Chemother Pharmacol 1986;17(2):137–42.
[73] Teicher BA, Holden SA, Ara G, Dupuis NP, Goff D. Restoration of tumor oxygenation after cytotoxic therapy by a perflubron emulsion/carbogen breathing. Cancer J Sci Am 1995;1(1):43–8.
[74] Cowan KH, Moscow JA, Huang H, Zujewski JA, O'Shaughnessy J, Sorrentino B, Hines K, Carter C, Schneider E, Cusack G, et al. Paclitaxel chemotherapy after autologous stem-cell transplantation and engraftment of hematopoietic cells transduced with a retrovirus containing the multidrug resistance complementary DNA (MDR1) in metastatic breast cancer patients. Clin Cancer Res 1999;5(7):1619–28.
[75] Fukaya K, Tanaka Y, Sato AC, Kodama K, Yamazaki H, Ishimoto T, Nozaki Y, Iwaki YM, Yuki Y, Umei K. Synthesis of paclitaxel. 1. Synthesis of the ABC ring of paclitaxel by SmI2-mediated cyclization. Org Lett 2015;17(11):2570–3.
[76] Dhiman K, Agarwal SM. NPred: QSAR classification model for identifying plant based naturally occurring anti-cancerous inhibitors. RSC Adv 2016;6(55):49395–400.
[77] Kani BC, Kumar SS, Pandi M. Screening and characterization of fungal taxol from leaf spot fungi. Am J Biosci Bioeng 2017;5(6):113–20.
[78] Fu Z, Yuan J, Huang X, Huang B. Advances in study on anticancer paclitaxel. Xiandai Zhongyao Yanjiu Yu Shijian 2006;20(3):58–61.
[79] Francis PA, Kris MG, Rigas JR, Grant SC, Miller VA. Paclitaxel (Taxol) and docetaxel (Taxotere): active chemotherapeutic agents in lung cancer. Lung cancer (Amsterdam, Netherlands) 1995;12(Suppl 1):S163–72.
[80] De Furia MD, Paclitaxel (Taxol). A new natural product with major anticancer activity. Phytomedicine 1997;4(3):273–82.
[81] Kim J, Pramanick S, Lee J, Lee YM, Im S, Kim WJ. Andrographolide-loaded polymerized phenylboronic acid nanoconstruct for stimuli-responsive chemotherapy. J Control Release 2017;259203–11.
[82] The chemistry and pharmacology of taxol and its derivatives edited by Vittorio F. Vol 22.
[83] Zheng L, Wen G, Yao Y, Li X, Gao F. Design, synthesis, and anticancer activity of natural product hybrids with paclitaxel side chain inducing apoptosis in human colon cancer cells. Nat Prod Commun 2020;15(4), 1934578X20917298.
[84] Li B, Kuang Y, Zhang M, He J, Xu L, Leung C, Ma D, Lo J, Qiao X, Ye M. Cytotoxic triterpenoids from *Antrodia camphorata* as sensitizers of paclitaxel. Org Chem Front 2020;7(5):768–79.
[85] Edgar S, Zhou K, Qiao K, King JR, Simpson JH, Stephanopoulos G. Mechanistic insights into Taxadiene epoxidation by Taxadiene-5α-hydroxylase. ACS Chem Biol 2016;11(2):460–9.
[86] Gao J, Wang LW, Zheng HC, Damirin A, Ma CM. Cytotoxic constituents of *Lasiosphaera fenzlii* on different cell lines and the synergistic effects with paclitaxel. Nat Prod Res 2016;30(16):1862–5.
[87] Guerra-Bubb J, Croteau R, Williams RM. The early stages of taxol biosynthesis: an interim report on the synthesis and identification of early pathway metabolites. Nat Prod Rep 2012;29(6):683–96.

[88] Schaefer B. Hope against cancer: taxol. Chem Unserer Zeit 2011;45(1):32–46.
[89] Ma Y, Feng C. Progress in the studies on antitumoral natural products from endophytic fungi. Youji Huaxue 2008;28(10):1697–706.
[90] Yuan H, Sun B, Gao F, Lan M. Synergistic anticancer effects of andrographolide and paclitaxel against A549 NSCLC cells. Pharm Biol (Abingdon, United Kingdom) 2016;54(11):2629–35.
[91] Deborah F, Molonia MS, Romina B, Claudia M, Guido F, Gregorio C, Antonella S, Francesco C, Antonio S. Curcumin potentiates the antitumor activity of paclitaxel in rat glioma C6 cells. Phytomedicine 2019;55:23–30.
[92] Jiang Q, Yang M, Qu Z, Zhou J, Zhang Q. Resveratrol enhances anticancer effects of paclitaxel in HepG2 human liver cancer cells. BMC Complement Altern Med 2017;17. 477/1-477/12.
[93] Xu Y. Paclitaxel detoxification compound extract with antitumor effect and preparation method thereof. Faming Zhuanli Shenqing; 2017. CN 106421553 A 20170222.
[94] Kim J, Lee J, Lee YM, Pramanick S, Im S, Kim WJ. Andrographolide-loaded polymerized phenylboronic acid nanoconstruct for stimuli-responsive chemotherapy. J Control Release 2017;259:203–11.
[95] Senthilnathan P, Padmavathi R, Banu SM, Sakthisekaran D. Enhancement of antitumor effect of paclitaxel in combination with immunomodulatory *Withania somnifera* on benzo(a)pyrene induced experimental lung cancer. Chem Biol Interact 2006;159(3):180–5.
[96] Edler MC, Buey RM, Gussio R, Marcus AI, Vanderwal CD, Sorensen EJ, Diaz JF, Giannakakou P, Hamel E. Cyclostreptin (FR182877), an antitumor tubulin-polymerizing agent deficient in enhancing tubulin assembly despite its high affinity for the taxoid site. Biochemistry 2005;44(34):11525–38.
[97] De Furia MD. Paclitaxel (Taxol®): a new natural product with major anticancer activity. Phytomed: Int J Phytother phytopharmacol 1997;4(3):273–82.
[98] Abal M, Andreu JM, Barasoain I. Taxanes: microtubule and centrosome targets, and cell cycle dependent mechanisms of action. Curr Cancer Drug Targets 2003;3 (3):193–203.
[99] Lee D, Kim IY, Saha S, Choi KS. Paraptosis in the anti-cancer arsenal of natural products. Pharmacol Ther 2016;162120–33.
[100] Asif M, Rizwani GH, Zahid H, Khan Z, Qasim R. Pharmacognostic studies on *Taxus baccata* L.: a brilliant source of anti-cancer agents. Pak J Pharm Sci 2016;29(1):105–9.
[101] Uesato S, Yamashita H, Maeda R, Hirata Y, Yamamoto M, Matsue S, Nagaoka Y, Shibano M, Taniguchi M, Baba K, et al. Synergistic antitumor effect of a combination of paclitaxel and carboplatin with nobiletin from citrus depressa on non-small-cell lung cancer cell lines. Planta Med 2014;80(6):452–7.
[102] D'Incalci M, Jimeno J. Preclinical and clinical results with the natural marine product ET-743. Expert Opin Investig Drugs 2003;12(11):1843–53.
[103] Zhu Q, Zhang D, Xiao L, Zhou H, Yu H, Li L. Application of *Ocimum basilicum* polysaccharide as paclitaxel synergist in preparation of antitumor agent. Faming Zhuanli Shenqing; 2015. CN 104382927 A 20150304.
[104] Altmann K, Gertsch J. Anticancer drugs from nature- natural products as a unique source of new microtubule-stabilizing agents [erratum to document cited in CA146:474606]. Nat Prod Rep 2012;29(12):1481.
[105] Sun L, Simmerling C, Ojima I. Recent advances in the study of the bioactive conformation of taxol. ChemMedChem 2009;4(5):719–31.
[106] Xie B, Lu Y, Luo Z, Qu Z, Zheng C, Huang X, Zhou H, Hu Y, Shen X. Tenacigenin B ester derivatives from Marsdenia tenacissima actively inhibited CYP3A4 and enhanced in vivo antitumor activity of paclitaxel. J Ethnopharmacol 2019;235:309–19.

[107] Yahyaei B, Pourali P. One step conjugation of some chemotherapeutic drugs to the biologically produced gold nanoparticles and assessment of their anticancer effects. Sci Rep 2019;9(1):1–15.

[108] Zhao W, Qi Z, Ying W, Hou J. Preparation, characterization and in vitro antitumor effect of cholesterol succinyl Bletillastriata polysaccharide-loaded paclitaxel nanoparticles. Biomed Res (Aligarh, India) 2017;28(21):9638–46.

[109] Lin S. Paclitaxel anticancer patch. Faming Zhuanli Shenqing; 2012. CN 102824543 A 20121219.

[110] Lin S. Preparation of oral anticancer agent containing *Ornithogalum caudatum* saponin OSW-I and paclitaxel. Faming Zhuanli Shenqing; 2012. CN 102580065 A 20120718.

[111] Chen J, Qiu X, Wang R, Duan L, Chen S, Luo J, et al. Iinhibition of human gastric carcinoma cell growth in vitro and in vivo by cladosporol isolated from the paclitaxel-producing strain *Alternaria alternata* var. monosporus. Biol Pharm Bull 2009;32 (12):2072–4.

[112] Li Y, Zhang G, Pfeifer BA. Current and emerging options for Taxol production. Adv Biochem Eng Biotechnol 2015;148(Biotechnology of Isoprenoids):405–25.

[113] Li S, Yao H, Xu J, Jiang S. Synthetic routes and biological evaluation of largazole and its analogues as potent histone deacetylase inhibitors. Molecules (Basel, Switzerland) 2011;16(6):4681–94.

[114] Braga SF, Galvao DS. A structure-activity study of taxol, taxotere, and derivatives using the electronic indices methodology (EIM). J Chem Inf Comput Sci 2003;43 (2):699–706.

[115] Li WW, Takahashi N, Jhanwar S, Cordon-Cardo C, Elisseyeff Y, Jimeno J, Faircloth G, Bertino JR. Sensitivity of soft tissue sarcoma cell lines to chemotherapeutic agents: identification of ecteinascidin-743 as a potent cytotoxic agent. Clin Cancer Res 2001;7(9):2908–11.

[116] Carballares S, Craig D, Lane CAL, Mitchell WP, MacKenzie AR, Wood A. Paclitaxel synthetic studies. A Diels-Alder approach to the A-ring. Chem Commun (Camb) 2000;18:1767–8.

[117] Jacob JN. Formulations from natural products, turmeric, paclitaxel, and aspirin. U.S. Pat. Appl. Publ.; 2013. US 20130029922 A1 20130131.

[118] Ceccarelli S, Bell AA, Gennari C. Natural products with Taxol-like anti-tumour activity. In: Seminars in Organic Synthesis, Summer School "A. Corbella", 25th, Gargnano, Italy, June 12–16, 2000; 2000. p. 91–115.

[119] Osada H, Nakazawa J, Usui T. Microtubule proteins and chemical inhibitors. Baiosaiensu to Indasutori 2000;58(11):789–92.

[120] Alexandre J, et al. Novel action of paclitaxel against cancer cells: bystander effect mediated by reactive oxygen species. Cancer Res 2007;67(8):3512–7.

[121] Gerth K, et al. Epothilons A and B: antifungal and cytotoxic compounds from *Sorangium cellulosum* (Myxobacteria). Production, physico-chemical and biological properties. J Antibiot (Tokyo) 1996;49(6):560–3.

[122] Altmann K, Cachoux F, Feyen F, Gertsch J, Kuzniewski CN, Wartmann M. Natural products as leads for anticancer drug discovery: discovery of new chemotypes of microtubule stabilizers through reengineering of the epothilone scaffold. Chimia 2010;64(1–2):8–13.

[123] Tang L, Qiu R, Li Y, Katz L. Generation of novel epothilone analogs with cytotoxic activity by biotransformation. J Antibiot 2003;56(1):16–23.

[124] Wessjohann LA, Scheid G. Synthetic access to epothilones - natural products with extraordinary anticancer activity. In: Schmalz H, editor. Organic synthesis highlights IV; 2000. p. 251–67.

[125] Altmann KH, Wartmann M, O'Reilly T. Epothilones and related structures - a new class of microtubule inhibitors with potent in vivo antitumor activity. Biochim Biophys Acta Rev Cancer 2000;1470(3):M79–91.

[126] Liu S, Wang Y, Zhang Q, Wu C. Advance in research on heterologous production of epothilones. Junshi Yixue Kexueyuan Yuankan 2009;33(1):84–7.
[127] Bollag DM, et al. Epothilones, a new class of microtubule-stabilizing agents with a taxollike mechanism of action. Cancer Res 1995;55(11):2325–33.
[128] Lovato TM, Edward WR, Chaitan K. Substrate tolerance of module 6 of the epothilone synthetase. Biochemistry 2007;46(11):3385–93.
[129] Frykman SA, Tsuruta H, Licari PJ. Assessment of fed-batch, semicontinuous, and continuous epothilone D production processes. Biotechnol Prog 2005;21(4):1102–8.
[130] Wenzel SC, Mueller R. Host organisms: myxobacterium. Adv Biotechnol 2017;3B (Industrial Biotechnology):453–85.
[131] Altmann K. The merger of natural product synthesis and medicinal chemistry: on the chemistry and chemical biology of epothilones. Org Biomol Chem 2004;2 (15):2137–52.
[132] Altmann K, Floersheimer A, O'Reilly T, Wartmann M. The natural products epothilones A and B as lead structures for anticancer drug discovery: chemistry, biology, and SAR studies. Prog Med Chem 2004;42:171–205.
[133] Miller DA, Walsh CT, Luo L. C-methyltransferase and cyclization domain activity at the intraprotein PK/NRP switch point of Yersiniabactin synthetase. J Am Chem Soc 2001;123(34):8434–5.
[134] Nora de Souza MV. New natural products able to act on the stabilization of microtubules, an important target against cancer. Quimica Nova 2004;27(2):308–12.
[135] Ting-Chao C, Xiuguo Z, Zi-Yang Z, Yong L, Li F, Sara E, David RM, Robert Jr J, Nian W, Ye Ingrid Y, et al. Therapeutic effect against human xenograft tumors in nude mice by the third generation microtubule stabilizing epothilones. Proc Natl Acad Sci U S A 2008;105(35):13157–62.
[136] Altmann K, Floersheimer A, Bold G, Caravatti G, Wartmann M. Natural product-based drug discovery—epothilones as lead structures for the discovery of new anticancer agents. Chimia 2004;58(10):686–90.
[137] Starks CM, Zhou Y, Liu F, Licari PJ. Isolation and characterization of new epothilone analogues from recombinant *Myxococcus xanthus* fermentations. J Nat Prod 2003;66 (10):1313–7.
[138] Kowalski RJ, Giannakakou P, Hamel E. Activities of the microtubule-stabilizing agents epothilones A and B with purified tubulin and in cells resistant to paclitaxel (Taxol(R)). J Biol Chem 1997;272(4):2534–41.
[139] Kamath K, Jordan MA. Suppression of microtubule dynamics by epothilone B is associated with mitotic arrest. Cancer Res 2003;63(18):6026–31.
[140] Gong G, Jia L, Huang F, Xu Z. Current status of the biosynthesis of the anti-tumor agents, epothilones. Zhongguo Xinyao Zazhi 2009;18(16):1515–20.
[141] Taylor RE, Zajicek J. Conformational properties of epothilone. J Org Chem 1999;64 (19):7224–8.
[142] Nettles JH, et al. The binding mode of epothilone A on alpha,beta-tubulin by electron crystallography. Science 2004;305(5685):866–9.
[143] Bode CJ, et al. Epothilone and paclitaxel: unexpected differences in promoting the assembly and stabilization of yeast microtubules. Biochemistry 2002;41(12):3870–4.
[144] Cortes J, Baselga J. Targeting the microtubules in breast cancer beyond taxanes: the epothilones. Oncologist 2007;12(3):271–80.
[145] Wartmann M, Altmann KH. The biology and medicinal chemistry of epothilones. Curr Med Chem Anticancer Agents 2002;2(1):123–48.
[146] Nicolaou KC, et al. Synthesis of epothilones a and B in solid and solution phase. Nature 1997;387(6630):268–72.
[147] Liu F, Garneau S, Walsh CT. Hybrid nonribosomal peptide-polyketide interfaces in epothilone biosynthesis: minimal requirements at N and C termini of EpoB for elongation. Chem Biol 2004;11(11):1533–42.

[148] Kelly WL, Hillson NJ, Walsh CT. Excision of the epothilone synthetase B cyclization domain and demonstration of in transcondensation/cyclodehydration activity. Biochemistry 2005;44(40):13385–93.
[149] Lee FY, et al. BMS-247550: a novel epothilone analog with a mode of action similar to paclitaxel but possessing superior antitumor efficacy. Clin Cancer Res 2001;7(5):1429–37.
[150] Chou TC, et al. Desoxyepothilone B: an efficacious microtubule-targeted antitumor agent with a promising in vivo profile relative to epothilone B. Proc Natl Acad Sci U S A 1998;95(16):9642–7.
[151] Newman RA, et al. Antitumor efficacy of 26-fluoroepothilone B against human prostate cancer xenografts. Cancer Chemother Pharmacol 2001;48(4):319–26.
[152] Schinkel AH. The physiological function of drug-transporting P-glycoproteins. Semin Cancer Biol 1997;8(3):161–70.
[153] Anon. Concentrates: conformational dependence of epothilone activity. Chem Eng News 2002;80(49):29.
[154] Taylor RE, Chen Y, Galvin GM, Pabba PK. Conformation-activity relationships in polyketide natural products. Towards the biologically active conformation of epothilone. Org Biomol Chem 2004;2(1):127–32.
[155] Storer RI, Takemoto T, Jackson PS, Brown DS, Baxendale IR, Ley SV. Multi-step application of immobilized reagents and scavengers: a total synthesis of epothilone C. Chem A Eur J 2004;10(10):2529–47.
[156] Zhu B, Panek JS. Methodology based on chiral silanes in the synthesis of polypropionate-derived natural products - total synthesis of epothilone a. Euro J Org Chem 2001;9:1701–14.
[157] Zhu B, Panek JS. Total synthesis of epothilone a. Org Lett 2000;2(17):2575–8.
[158] Feyen F, Gertsch J, Wartmann M, Altmann K. Design and synthesis of 12-aza- epothilone (Azathilones)-"non- natural" natural products with potent anticancer activity. Angew Chem Int Ed 2006;45(35):5880–5.
[159] Borzilleri RM, Zheng X, Schmidt RJ, Johnson JA, Kim S, DiMarco JD, Fairchild CR, Gougoutas JZ, Lee FYF, Long BH, et al. A novel application of a Pd(0)-catalyzed nucleophilic substitution reaction to the Regio- and Stereoselective synthesis of lactam analogues of the epothilone natural products. J Am Chem Soc 2000;122(37):8890–7.
[160] Winkler JD, Holland JM, Kasparec J, Axelsen PH. Design and synthesis of constrained epothilone analogs: the efficient synthesis of eleven-membered rings by olefin metathesis. Tetrahedron 1999;55(27):8199–214.
[161] Akhtar MS, Swamy MK. Anticancer plants: clinical trials and nanotechnology, vol. 3. Springer; 2017. p. 86.
[162] Moody TW, Pradhan T, Mantey SA, Jensen RT, Dyba M, Moody D, Tarasova NI, Michejda CJ. Bombesin marine toxin conjugates inhibit the growth of lung cancer cells. Life Sci 2008;82(15–16):855–61.
[163] Bates RB, Brusoe KG, Burns JJ, Caldera S, Cui W, Gangwar S, Gramme MR, McClure KJ, Rouen GP, Schadow H, et al. Dolastatins 26. Synthesis and stereochemistry of Dolastatin 11. J Am Chem Soc 1997;119(9):2111–3.
[164] Casalme LO, Yamauchi A, Sato A, Petitbois JG, Nogata Y, Yoshimura E, Okino T, Umezawa T, Matsuda F. Total synthesis and biological activity of Dolastatin 16. Org Biomol Chem 2017;15(5):1140–50.
[165] Simmons TL, Andrianasolo E, McPhail K, Flatt P, Gerwick WH. Marine natural products as anticancer drugs. Mol Cancer Ther 2005;4(2):333–42.
[166] Zhou W, Nie X, Zhang Y, Si C, Zhou Z, Sun X, Wei B. A practical approach to asymmetric synthesis of dolastatin 10. Org Biomol Chem 2017;15(29):6119–31.
[167] Roberson RW, Tucker B, Pettit GR. Microtubule depolymerization in Uromyces appendiculatus by three new antineoplastic drugs: combretastatin A-4, dolastatin 10 and halichondrin B. Mycol Res 1998;102(3):378–82.

[168] Abdollahpour-Alitappeh M, Habibi-Anbouhi M, Balalaie S, Golmohammadi F, Lotfinia M, Abolhassani M. A new and simple non-chromatographic method for isolation of drug/linker constructs: vc-MMAE evaluation. J HerbMed Pharmacol 2017;6 (4):153–9.

[169] Maderna A, Doroski M, Subramanyam C, Porte A, Leverett CA, Vetelino BC, Chen Z, Risley H, Parris K, Pandit J, et al. Discovery of cytotoxic dolastatin 10 analogues with N-terminal modifications. J Med Chem 2014;57(24):10527–43.

[170] Akaiwa M, Martin T, Mendelsohn BA. Synthesis and evaluation of linear and macrocyclic dolastatin 10 analogues containing pyrrolidine ring modifications. ACS Omega 2018;3(5):5212–21.

[171] Tan LT. Filamentous tropical marine cyanobacteria: a rich source of natural products for anticancer drug discovery. J Appl Phycol 2010;22(5):659–76.

[172] Subramanian PK, Tseng CC, Vishnuvajjala R. Synthesis and cytotoxicity of androstanyl ureas as dolastatin 10 model candidates. In: Book of Abstracts, 213th ACS National Meeting, San Francisco, April 13–17 1997; MEDI-210; 1997.

[173] Wang X, Dong S, Feng D, Chen Y, Ma M, Hu W. Synthesis and biological activity evaluation of dolastatin 10 analogs with N-terminal modifications. Tetrahedron 2017;73(16):2255–66.

[174] Paterson I, Findlay AD. Recent advances in the total synthesis of polyketide natural products as promising anticancer agents. Aust J Chem 2009;62(7):624–38.

[175] Dugal-Tessier J, Barnscher SD, Kanai A, Mendelsohn BA. Synthesis and evaluation of Dolastatin 10 analogs containing heteroatoms on the amino acid side chains. J Nat Prod 2017;80(9):2484–91.

[176] Kuznetsov G, Towle MJ, Cheng H, Kawamura T, TenDyke K, Liu D, Kishi Y, Yu MJ, Littlefield BA. Induction of morphological and biochemical apoptosis following prolonged mitotic blockage by halichondrin B macrocyclic ketone analog E7389. Cancer Res 2004;64(16):5760–6.

[177] Morgan RJ, Synold TW, Longmate JA, Quinn DI, Gandara D, Lenz H, Ruel C, Xi B, Lewis MD, Colevas AD, et al. Pharmacodynamics (PD) and pharmacokinetics (PK) of E7389 (eribulin, halichondrin B analog) during a phase I trial in patients with advanced solid tumors: a California cancer consortium trial. Cancer Chemother Pharmacol 2015;76(5):897–907.

[178] Yu MJ, Kishi Y, Littlefield BA. Discovery of E7389, a fully synthetic macrocyclic ketone analog of halichondrin B. In: Cragg GM, Kingston DGI, Newman DJ, editors. Anticancer agents from natural products. 2nd ed. CRC Press; 2012. p. 317–45.

[179] Yoshimatsu K. Creation of eribulin, a new drug for breast cancer. Farumashia 2013;49 (6):534–8.

[180] Okouneva T, Azarenko O, Wilson L, Littlefield BA, Jordan MA. Inhibition of centromere dynamics by eribulin (E7389) during mitotic metaphase. Mol Cancer Ther 2008;7(7):2003–11.

[181] Dabydeen DA, Burnett JC, Bai R, Verdier-Pinard P, Hickford SJH, Pettit GR, Blunt JW, Munro Murray HG, Gussio R, Hamel E. Comparison of the activities of the truncated halichondrin B analog NSC 707389 (E7389) with those of the parent compound and a proposed binding site on tubulin. Mol Pharmacol 2006;70(6):1866–75.

[182] Arnold SM, Moon J, Williamson SK, Atkins JN, Sai-HI O, LeBlanc M, Urba SG. Phase II evaluation of eribulin mesylate (E7389, NSC 707389) in patients with metastatic or recurrent squamous cell carcinoma of the head and neck: southwest oncology group trial S0618. Invest New Drugs 2011;29(2):352–9.

[183] Edited TK, By: Kita Y. Development of industrial production process of a new anticancer agent E7389 (eribulin mesylate). Tennenbutsu Zengosei no Saishin Doko 2009;293–308.

[184] Cigler T, Vahdat LT. Eribulin mesylate for the treatment of breast cancer. Expert Opin Pharmacother 2010;11(9):1587–93.
[185] Wozniak KM, Wu Y, Farah MH, Littlefield BA, Nomoto K, Slusher BS. Neuropathy-inducing effects of eribulin mesylate versus paclitaxel in mice with pre-existing neuropathy. Neurotox Res 2013;24(3):338–44.
[186] Nastrucci C, Cesario A, Russo P. Anticancer drug discovery from the marine environment. Recent Pat Anticancer Drug Discov 2012;7(2):218–32.
[187] Goel S, Mita AC, Mita M, Rowinsky EK, Chu QS, Wong N, Desjardins C, Fang F, Jansen M, Shuster DE, et al. A phase I study of eribulin mesylate (E7389), a mechanistically novel inhibitor of microtubule dynamics, in patients with advanced solid malignancies. Clin Cancer Res 2009;15(12):4207–12.
[188] Jimeno A. Eribulin: rediscovering tubulin as an anticancer target. Clin Cancer Res: Off J Am Assoc Cancer Res 2009;15(12):3903–5.
[189] Yang Q. A new antimetastatic breast cancer agent eribulin mesylate. Zhongguo Xinyao Zazhi 2012;21(1):3–5. 25.
[190] Choi H, Demeke D, Kang F, Kishi Y, Nakajima K, Nowak P, Wan Z, Xie C. Synthetic studies on the marine natural product halichondrins. Pure Appl Chem 2003;75(1):1–17.
[191] Nomoto K, Wu J. Use of eribulin and lenvatinib as combination therapy for treatment of cancer. PCT Int. Appl.; 2014. WO 2014208774 A1 20141231.
[192] Kawazoe Y. Uemura D eribulin (halaven): development of anticancer drug from marine natural products. Saibo Kogaku 2013;32(6):675–81.
[193] Chiba H, Tagami K. Research and development of HALAVEN (eribulin Mesylate). Yuki Gosei Kagaku Kyokaishi 2011;69(5):600–10.
[194] Bauer A. Story of eribulin Mesylate: development of the longest drug synthesis. In: Topics in heterocyclic chemistry. Synthesis of heterocycles in contemporary medicinal chemistry, vol. 44; 2016. p. 209–70.
[195] Swami U, Shah U, Goel S. Eribulin in non-small cell lung cancer: challenges and potential strategies. Expert Opin Investig Drugs 2017;26(4):495–508.
[196] Towle MJ, Salvato KA, Wels BF, Aalfs KK, Zheng W, Seletsky BM, Xiaojie Z, Lewis BM, Kishi Y, Yu MJ, et al. Eribulin induces irreversible mitotic blockade: implications of cell-based pharmacodynamics for in vivo efficacy under intermittent dosing conditions. Cancer Res 2011;71(2):496–505.
[197] Umang S, Umang S, Sanjay G. Eribulin in cancer treatment. Mar Drugs 2015;13 (8):5016–58.
[198] Toru M, Shunji N, Hirofumi M, Masayuki N, Hironobu M. Eribulin mesylate in patients with refractory cancers: a phase I study. Invest New Drugs 2012;30 (5):1926–33.
[199] Hensley ML, Kravetz S, Jia X, Iasonos A, Tew W, Pereira L, Sabbatini P, Whalen C, Aghajanian CA, Zarwan C, et al. Eribulin mesylate (halichondrin B analog E7389) in platinum-resistant and platinum-sensitive ovarian cancer a 2-cohort, phase 2 study. Cancer (Hoboken, NJ, United States) 2012;118(9):2403–10.
[200] Wach J, Gademann K. Reduce to the maximum: truncated natural products as powerful modulators of biological processes. Synlett 2012;23(2):163–70.
[201] Yeung BKS. Natural product drug discovery: the successful optimization of ISP-1 and halichondrin B. Curr Opin Chem Biol 2011;15(4):523–8.
[202] Kingston DGI. Tubulin-interactive natural products as anticancer agents. [erratum to document cited in CA150:274732]. J Nat Prod 2011;74(5):1352.
[203] Cortes J, Vahdat L, Blum JL, Twelves C, Campone M, Roche H, Bachelot T, Awada A, Paridaens R, Goncalves A, et al. Phase II study of the halichondrin B analog eribulin mesylate in patients with locally advanced or metastatic breast cancer previously treated with an anthracycline, a taxane, and capecitabine. J Clin Oncol: Off J Am Soc Clin Oncol 2010;28(25):3922–8.

[204] Scarpace SL. Eribulin Mesylate (E7389): review of efficacy and tolerability in breast, pancreatic, head and neck, and non-small cell lung cancer. Clin Ther 2012;34 (7):1467–73.
[205] Vahdat LT, Pruitt B, Fabian CJ, Rivera RR, Smith DA, Tan-Chiu E, Wright J, Tan AR, DaCosta NA, Chuang E, et al. Phase II study of eribulin mesylate, a halichondrin B analog, in patients with metastatic breast cancer previously treated with an anthracycline and a taxane. J Clin Oncol 2009;27(18):2954–61.
[206] Tan AR, Rubin EH, Walton DC, Shuster DE, Wong YN, Fang F, Ashworth S, Rosen LS. Phase I study of eribulin mesylate administered once every 21 days in patients with advanced solid tumors. Clin Cancer Res 2009;15(12):4213–9.
[207] Folmer F, Schumacher M, Diederich M, Jaspars M. Finding NEMO (inhibitors) the search for marine pharmacophores targeting the nuclear factor-κB. Chim Oggi 2008;26(4):40–2. 44–46.
[208] Kingston DGI. Tubulin-interactive natural product as anticancer agents. J Nat Prod 2009;72(3):507–15.
[209] DesJardins C, Saxton P, Lu SX, Li X, Rowbottom C, Wong YN. A high-performance liquid chromatography-tandem mass spectrometry method for the clinical combination study of carboplatin and anti-tumor agent eribulin mesylate (E7389) in human plasma. J Chromatogr B Analyt Technol Biomed Life Sci 2008;875(2):373–82.
[210] de Bono JS, Molife LR, Sonpavde G, Maroto JP, Calvo E, Cartwright TH, Loesch DM, Feit K, Das A, Zang EA, et al. Phase II study of eribulin mesylate (E7389) in patients with metastatic castration-resistant prostate cancer stratified by prior taxane therapy. Anna Oncol: Off J Euro Soc Med Oncol 2012;23(5):1241–9.
[211] Smith JA, Wilson L, Azarenko O, Zhu X, Lewis BM, Littlefield BA, Jordan MA. Eribulin binds at microtubule ends to a single site on tubulin to suppress dynamic instability. Biochemistry 2010;49(6):1331–7.
[212] Alday PH, Correia JJ. Macromolecular interaction of halichondrin B analogues eribulin (E7389) and ER-076349 with tubulin by analytical ultracentrifugation. Biochemistry 2009;48(33):7927–38.
[213] Zhang ZY, King BM, Pelletier RD, Wong YN. Delineation of the interactions between the chemotherapeutic agent eribulin mesylate (E7389) and human CYP3A4. Cancer Chemother Pharmacol 2008;62(4):707–16.
[214] Liu X, Henderson JA, Sasaki T, Kishi Y. Dramatic improvement in catalyst loadings and molar ratios of coupling partners for Ni/Cr-mediated coupling reactions: heterobimetallic catalysts. J Am Chem Soc 2009;131(46):16678–80.
[215] Wu L, Peng J. A study on mechanistically-unique microtubule inhibitor- eribulin. Zhongliu Yaoxue 2011;1(4):327–9.
[216] Dieras V, Pivot X, Brain E, Roche H, Extra J, Monneur A, Provansal M, Tarpin C, Bertucci F, Viens P, et al. Safety results and analysis of eribulin efficacy according to previous microtubules- inhibitors sensitivity in the french prospective expanded access program for heavily pre-treated metastatic breast cancer. Cancer Res Treat 2018;50 (4):1226–37.
[217] Preston JN, Trivedi MV. Eribulin: a novel cytotoxic chemotherapy agent. Anna Pharmacother 2012;46(6):802–11. 10 pp.
[218] Schafer ES, Rau RE, Berg S, Minard CG, Blaney SM, Liu X, D'Adamo D, Reyderman L, Martinez G, Scott R, et al. A phase 1 study of eribulin mesylate (E7389), a novel microtubule-targeting chemotherapeutic agent, in children with refractory or recurrent solid tumors: A Children's Oncology Group Phase 1 Consortium study (ADVL1314). Pediatr Blood Cancer 2018;65(8), e27066.
[219] Spira AI, Iannotti NO, Savin MA, Neubauer M, Gabrail NY, Yanagihara RH, Zang EA, Cole PE, Shuster D, Das A. A phase II study of eribulin mesylate (E7389) in

patients with advanced, previously treated non-small-cell lung cancer. Clin Lung Cancer 2012;13(1):31–8.
[220] Gitlitz Barbara J, Tsao-Wei DD, Groshen S, Davies A, Koczywas M, Belani CP, Argiris A, Ramalingam S, Vokes EE, Edelman M, et al. A phase II study of halichondrin B analog eribulin mesylate (E7389) in patients with advanced non-small cell lung cancer previously treated with a taxane: a California cancer consortium trial. J Thorac Oncol 2012;7(3):574–8.
[221] Cortes J, Lorca R. Eribulin mesylate: a promising new antineoplastic agent for locally advanced or metastatic breast cancer. Future Oncol (Lond, Engl) 2011;7(3):355–64.
[222] Taur J, DesJardins CS, Schuck EL, Wong YN. Interactions between the chemotherapeutic agent eribulin mesylate (E7389) and P-glycoprotein in CF-1 abcb1a-deficient mice and Caco-2 cells. Xenobiotica 2011;41(4):320–6.
[223] Gradishar WJ. The place for eribulin in the treatment of metastatic breast cancer. Curr Oncol Rep 2011;13(1):11–6.
[224] Twelves C, Cortes J, Vahdat LT, Wanders J, Akerele C, Kaufman PA. Phase III trials of eribulin mesylate (E7389) in extensively pretreated patients with locally recurrent or metastatic breast cancer. Clin Breast Cancer 2010;10(2):160–3.
[225] Wozniak KM, Nomoto K, Lapidus RG, Wu Y, Carozzi V, Cavaletti G, Hayakawa K, Hosokawa S, Towle MJ, Littlefield BA, et al. Comparison of neuropathy-inducing effects of eribulin mesylate, paclitaxel, and ixabepilone in mice. Cancer Res 2011;71(11):3952–62.
[226] O'Shaughnessy J, Cortes J, Twelves C, et al. Efficacy of eribulin for metastatic breast cancer based on localization of specific secondary metastases: a post hoc analysis. Sci Rep 2020;10:11203.
[227] Munoz-Couselo E, Perez-Garcia J, Cortes J. Eribulin mesylate as a microtubule inhibitor for treatment of patients with metastatic breast cancer. OncoTargets Ther 2011;4185–92.
[228] Dubbelman AC, Rosing H, Thijssen B, Lucas L, Copalu W, Wanders J, Schellens JHM, Beijnen JH. Validation of high-performance liquid chromatography-tandem mass spectrometry assays for the quantification of Eribulin (E7389) in various biological matrices. J Chromatogr B Anal Technol Biomed Life Sci 2011;879 (15–16):1149–55.
[229] Cortes J, Montero AJ, Glueck S. Eribulin mesylate, a novel microtubule inhibitor in the treatment of breast cancer. Cancer Treat Rev 2012;38(2):143–51.
[230] Dalby SM, Paterson I. Synthesis of polyketide natural products and analogs as promising anticancer agents. Curr Opin Drug Discov Devel 2010;13(6):777–94.
[231] Cao J, Yang F, Yuan H, Huang Z. Clinical trail research of antitumor components from sponge. Zhongnan Yaoxue 2013;11(6):447–50.
[232] Singh R, Sharma M, Joshi P, Rawat DS. Clinical status of anti-cancer agents derived from marine sources. Anticancer Agents Med Chem 2008;8(6):603–17.
[233] Seetharam M, Kolla KR, Chawla SP. Eribulin therapy for the treatment of patients with advanced soft tissue sarcoma. Future Oncol (Lond, Engl) 2018;14(16):1531–45.
[234] Aftimos P, Polastro L, Ameye L, Jungels C, Vakili J, Paesmans M, van den Eerenbeemt J, Buttice A, Gombos A, de Valeriola D, et al. Results of the Belgian expanded access program of eribulin in the treatment of metastatic breast cancer closely mirror those of the pivotal phase III trial. Eur J Cancer 2016;60:117–24.
[235] Sharon WJ. Use of Eribulin and poly(ADP)ribose polymerase (PARP) inhibitors as combination therapy for the treatment of cancer. PCT Int. Appl.; 2015. WO 2015184145 A1 20151203.
[236] Kinghorn AD. Review of anticancer agents from natural products. J Nat Prod 2015;78 (9):2315.

[237] Littlefield BA, Funahashi Y, Uenaka T. Use of eribulin and mTOR inhibitors as combination therapy for the treatment of cancer. PCT Int. Appl.; 2015. WO 2015134399 A1 20150911.

[238] Asano M, Matsui J, Towle MJ, Littlefield BA, Wu J, McGonigle S, De Boisferon MH, Uenaka T, Nomoto K. Broad-spectrum preclinical antitumor activity of eribulin (Halaven®): combination with anticancer agents of differing mechanisms. Anticancer Res 2018;38(6):3375–85.

[239] Austad BC, Calkins TL, Chase CE, Fang FG, Horstmann TE, Hu Y, Lewis BM, Niu X, Noland TA, Orr JD, et al. Commercial manufacture of Halaven: chemoselective transformations En route to structurally complex macrocyclic ketones. Synlett 2013;24 (3):333–7.

[240] Kaburagi Y, Kishi Y. Effective procedure for selective ammonolysis of monosubstituted oxiranes: application to E7389 synthesis. Tetrahedron Lett 2007;48 (51):8967–71.

[241] Aftimos P, Polastro L, Jungels C, Vakili J, van den Eerenbeemt J, Buttice A, Gombos A, de Valeriola D, Gil T, Piccart-Gebhart M, et al. Results of the Belgian expanded access program of eribulin in the treatment of metastatic breast cancer closely mirror those of the pivotal phase III trial. Eur J Cancer (Oxf, Engl: 1990) 2016;60117–24.

[242] Yu MJ, Kishi Y, Littlefield BA. Discovery of E7389, a fully synthetic macrocyclic ketone analog of halichondrin B. In: Anticancer agents from natural products; 2005. p. 241–65. 2 plates.

[243] Yang Y, Kim D, Kishi Y. Second generation synthesis of C27-C35 building block of E7389, a synthetic halichondrin analogue. Org Lett 2009;11(20):4516–9.

[244] Liu S, Kim JT, Dong C, Kishi Y. Catalytic enantioselective Cr-mediated propargylation: application to Halichondrin synthesis. Org Lett 2009;11(20):4520–3.

[245] Chase CE, Fang FG, Lewis BM, Wilkie GD, Schnaderbeck MJ, Zhu X. Process development of Halaven: synthesis of the C1-C13 fragment from D-(−)-gulono-1, 4-lactone. Synlett 2013;24(3):323–6.

[246] Chavan LN, Chegondi R, Chandrasekhar S. Tandem organocatalytic approach to C28-C35 fragment of eribulin mesylate. Tetrahedron Lett 2015;56(29):4286–8.

[247] Dong C, Henderson JA, Kaburagi Y, Sasaki T, Kim D, Kim JT, Urabe D, Guo H, Kishi Y. New syntheses of E7389C14-C35 and halichondrin C14-C38 building blocks: reductive cyclization and oxy-Michael cyclization approaches. J Am Chem Soc 2009;131(43):15642–6.

[248] Kaburagi Y, Kishi Y. Operationally simple and efficient workup procedure for TBAF-mediated desilylation: application to Halichondrin synthesis. Org Lett 2007;9 (4):723–6.

[249] Kim D, Dong C, Kim JT, Guo H, Huang J, Tiseni PS, Kishi Y. New syntheses of E7389 C14-C35 and halichondrin C14-C38 building blocks: double-inversion approach. J Am Chem Soc 2009;131(43):15636–41.

CHAPTER 10

Synthetic strategies for antimetabolite analogs in our laboratory

1. Introduction

Antimetabolites are drugs that are structurally similar to essential metabolites; they may inhibit the synthesis of the essential metabolites or their critical intracellular reactions. The new analogs can interfere with their formation or utilization, thus inhibiting essential metabolic routes. The marketed antimetabolites are categorized as antifolates, pyrimidine analogs, and purine analogs. New antimetabolites have recently become a focus for anticancer medication development. Antimetabolites play an important role in the treatment of a variety of both malignant and nonmalignant diseases, such as rheumatoid arthritis and antiviral infections.

During the last decade, one of our key research projects at our laboratory has focused on the development of new chemical strategies in the syntheses of antimetabolites. Thus, the design and synthesis of diverse derivatives and analogs of the naturally occurring metabolites is a focus of research in our group [1–13].

Our recent studies of the synthetic routes of the new class of agents were demonstrated. This chapter will describe the most recent synthetic approaches of antimetabolites. New trends and several examples of analogs of the main antimetabolite are discussed, such as antifolate analogs, mercaptopurine antimetabolite analogs, pyrimidines, and other heterocyclic thioglycosides. The objective of this chapter would allow the development of drug design of the new active structures.

2. Synthesis of antifolate analogs

The antifolates, or folate antagonists, are compounds that inhibit the synthesis of folate coenzymes. The clinically available antifolates are structural analogs of folic acid. The antifolate drug in general use is methotrexate. It is a potent

New Strategies Targeting Cancer Metabolism
https://doi.org/10.1016/B978-0-12-821783-2.00008-X

inhibitor of the enzyme folate reductase, which catalyzes the formation of dihydrofolic acid and tetrahydrofolic acid from folic acid. The latter compound is the precursor of $N^{5,10}$-methylene-tetrahydrofolic acid, which is an essential cofactor involved in the conversion of deoxyuridylic to thymidylic acid, which is required for DNA synthesis. This appears to be the major site of action of the drug, although there is some evidence to suggest that a direct inhibition of thymidylate synthetase may be the primary site of action. Formylation reactions of various purine and amino acid precursors are also inhibited, with consequent reduction in the synthesis of RNA and protein.

The extreme toxicity of methotrexate, coupled with its inactivity toward most forms of human cancers, continues to stimulate an intensive research for less toxic and more selective agents for cancer chemotherapy based upon the inhibition of dihydrofolate reductase and thymidylate synthetase. Among the most promising modified folate derivatives are different deaza analogs of the pteridine ring system. 8-Deazafolic acid, for example, indicates a significant potency against mouse L1210 leukemia and has better transport features than folic acid itself. 10-Deaza-aminopterin is an extremely promising compound that displays greater potency than methotrexate or aminopterin against various solid tumors. As a result of these findings, we have started a synthetic initiative focused on developing novel synthetic approaches to deazapteridines and synthesizing the missing deaza analogs of folic acid, aminopterin, and methotrexate. Recent research from our laboratory has demonstrated a new strategy for the synthesis of a variety of 5-deaza classical antifolate antimetabolites. The results obtained have demonstrated the effectiveness of 5-deaza antifolates as antineoplastic agents in a number of experimental murine tumor systems (Scheme 10.1) [14].

A synthesis of 5-deazaaminopterines by utilizing our previously reported pyridine-(1*H*)-thiones **3** is depicted in Scheme 10.1. Compounds **3** were synthesized via reacting alkylidene cyanothioacetamide **1** and ethyl acetoacetate **2**. Compounds **3** were reacted with methyl iodide in methylene chloride-sodium hydroxide to furnish the corresponding methylthio derivatives **4**. After the reduction of compounds **4** with LAH to give the corresponding 5-(hydroxymethyl)pyridine derivatives **5**, the hydroxymethyl function was protected by methoxymethylation to give compounds **6**. When compounds **6** were subjected to selective oxidation, the corresponding sulfone derivatives **7** were obtained. Compound **7** was treated with guanidine to afford the 5-deazapteridines **8**. The latter on deprotection with HCl afforded the corresponding 6-hydroxymethylpyrido[2,3-*d*]pyrimidines **9**, which were converted into the corresponding bromides **10**. Condensation of **10** with *N*-(4-aminobenzoyl)-L-glutamic acid diethyl ester followed by saponification of the resulting diethyl ester resulted in L-5-deazaaminopterines **11b**.

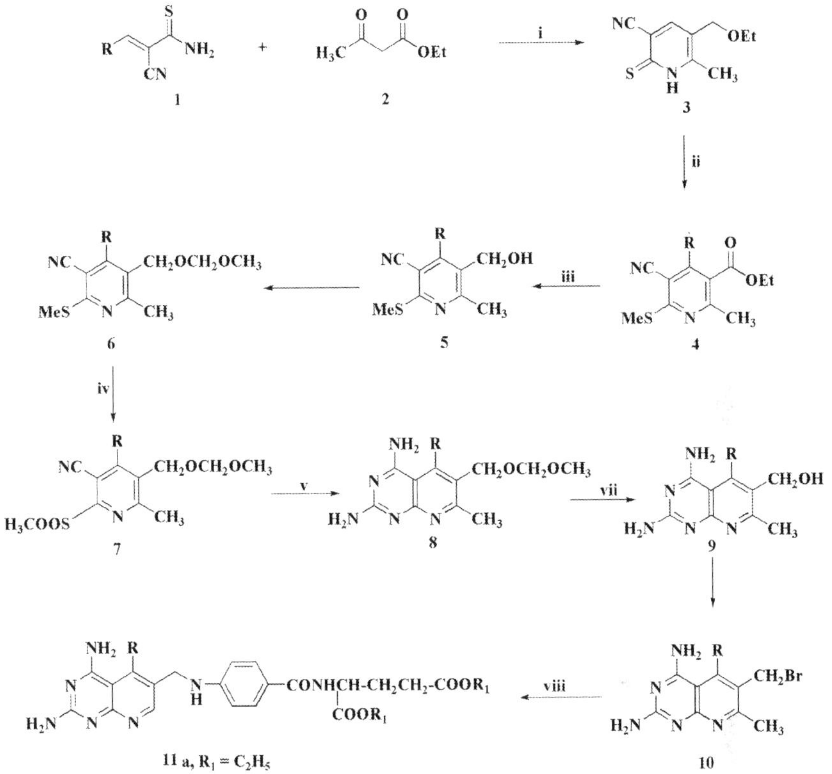

Scheme 10.1 Reagents and conditions: (i) EtOH/pip/heat; (ii) methyl iodide in methylene chloride-sodium hydroxide; (iii) LAH; (iv) oxidation; (v) guanidine; (vii) HCl; (viii) diethyl(*p*-aminobenzoyl)-L-glutamate.

Recently, classical folates have the disadvantage of requiring a transport mechanism into the cell. Cells that lack this transport mechanism are not susceptible to the action of classical antifolates. As a result, various nonclassical antifolates have been developed that lack the L-glutamate component. These nonclassical analogs have a wide range of activity and are very lipophilic. They are transported into cells via passive diffusion.

Total synthesis of several biologically active 5-deaza nonclassical antifolates is being pursued in our laboratory (Scheme 10.2) [8].

3. Synthesis of mercaptopurine antimetabolite analogs

Many analogs of purine bases have been synthesized and investigated. Two of these have clinical importance. Thioguanine (6-TG) and mercaptopurine

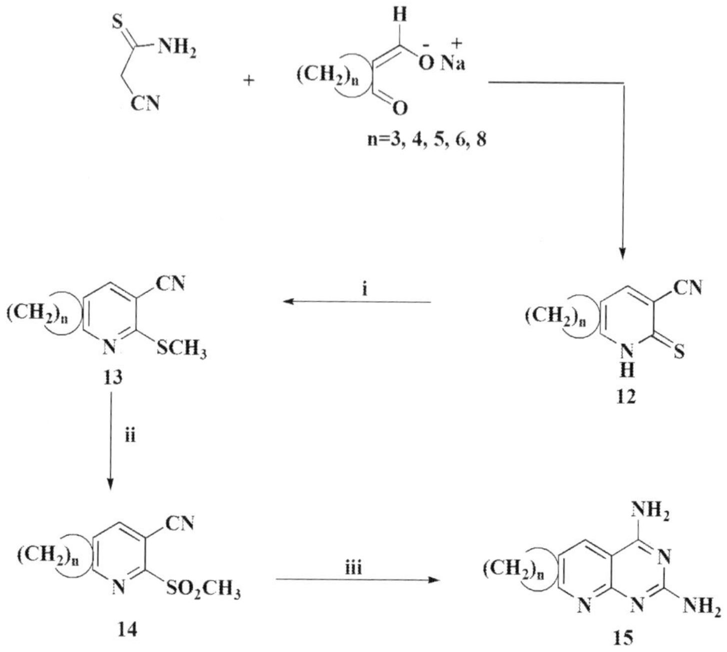

Scheme 10.2 Reagents and conditions: (i) $CH_3I/NaOH/CH_2Cl_2$; (ii) [O]; (iii) guanidine hydrochloride.

(Purinethol, 6-MP) are valuable antileukemic agents. Both 6-MP and 6-TG must be activated by intracellular conversion to ribonucleotides in order to exert cytotoxic effects. These nucleotides are potent inhibitors of the de novo synthesis of purines and limit the synthesis of the purine precursor ribosylamine-5-phosphate. They also inhibit the synthesis of purine nucleotides from preformed purines. As a consequence, nucleic acid synthesis is diminished. Several other purine analogs are also now commercially available. Recently, different successful methodologies for the syntheses of methylsulfanylpyrazolopyrimidines, methylsulfanylpyrazolopyridines, and methylsulfanylpyrazolotriazines were reported [15–25]. Purine analogs are noteworthy because they can act as antimetabolites in purine biochemical reactions. The synthesis of 7-methylthiopyrazolo[1,5-*a*]pyrimidines through reacting [bis(methylthio)methylene]malononitrile and ethyl 2-cyano-3,3-bis(methylthio)acrylate with 5-aminopyrazoles is demonstrated in Scheme 10.3 [26].

Novel 4-methylthiopyrazolo[1,5-*a*]-1,3,5-triazines were synthesized through reacting dimethyl *N*-cyanodithioiminocarbonate with 5-aminopyrazoles as shown in Scheme 10.4 [20].

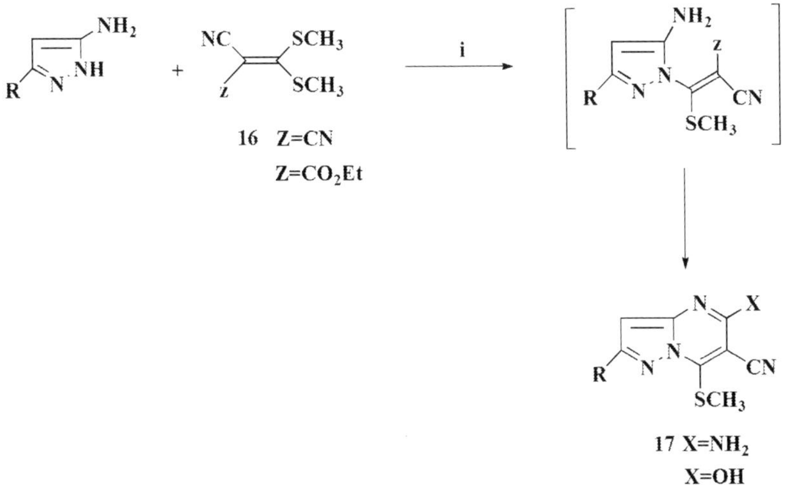

NH_2 NH R N + NC SCH_3 Z SCH_3 i NH_2 Z R N N CN SCH_3

16 Z=CN

Z=CO_2Et

N X R N N CN SCH_3

17 X=NH_2

X=OH

Scheme 10.3 Reagents and conditions: (i) EtOH/pip./reflux.

NH_2 NH Ar N + N NC SCH_3 SCH_3 i

18

N NH_2 Ar N N N SCH_3 H_2 N: C≡N Ar N N N SCH_3

19

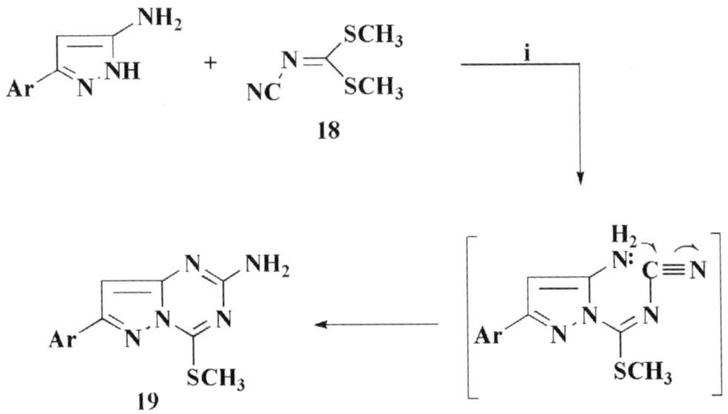

Scheme 10.4 Reagents and conditions: (i) EtOH/pip./reflux/3 h.

A novel regiospecific synthesis of the nonclassical thioguanine and sulfanylpurine derivatives and other antimetabolites through reacting the heterocyclic ketone dithioacetals with active methylene compounds and hydrazine derivatives was achieved (Scheme 10.5) [18].

The substituted pyrazolo [4,3-b]pyridin-3(2*H*)-ones were furnished through reacting the substituted pyrazolin-5-one with ketene dithioacetals and ethyl ethoxymethylene-cyanoacetate or ethoxymethylenemalononitrile. Intramolecular cyclization of the products has been accomplished (Scheme 10.6) [24].

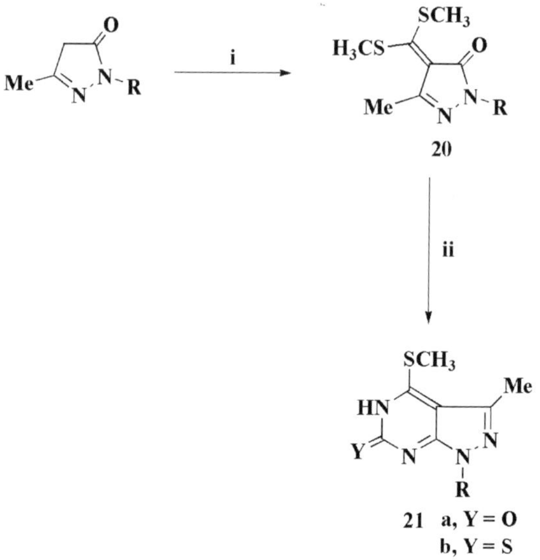

Scheme 10.5 Reagents and conditions: (i) $CS_2/CH_3I/C_2H_5ONa/EtOH$; (ii) urea or thiourea/$C_2H_5ONa/EtOH$.

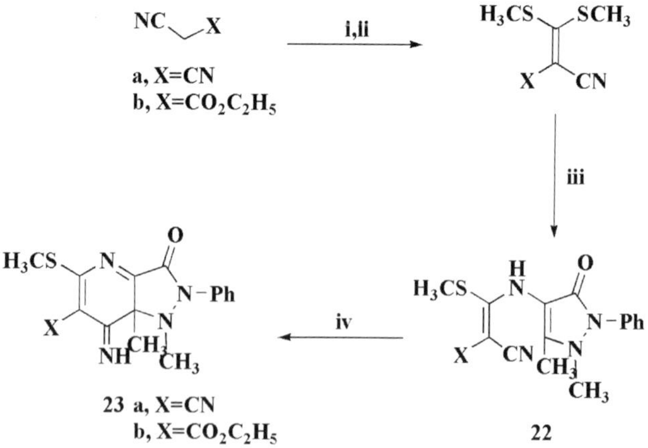

Scheme 10.6 Reagents and conditions: (i) CS_2/C_2H_5ONa; (ii) CH_3I/CH_3OH; (iii) 4-amino-1,2-dihydro-1,5-dimethyl-2-phenylpyrazol-3-one/EtOH/reflux; (iv) HCl/EtOH/reflux.

The synthesis of new pyrazolopyrimidine and pyrazolopyridine thioglycosides was performed. These compounds were furnished via reacting sodium cyanocarbonimidodithioate and sodium 2,2-dicyanoethene-1,1-bis(thiolate) with 4-aminoantipyrine. Ammonolysis of the thioglycosides yielded the corresponding free thioglycosides (Schemes 10.7 and 10.8) [27].

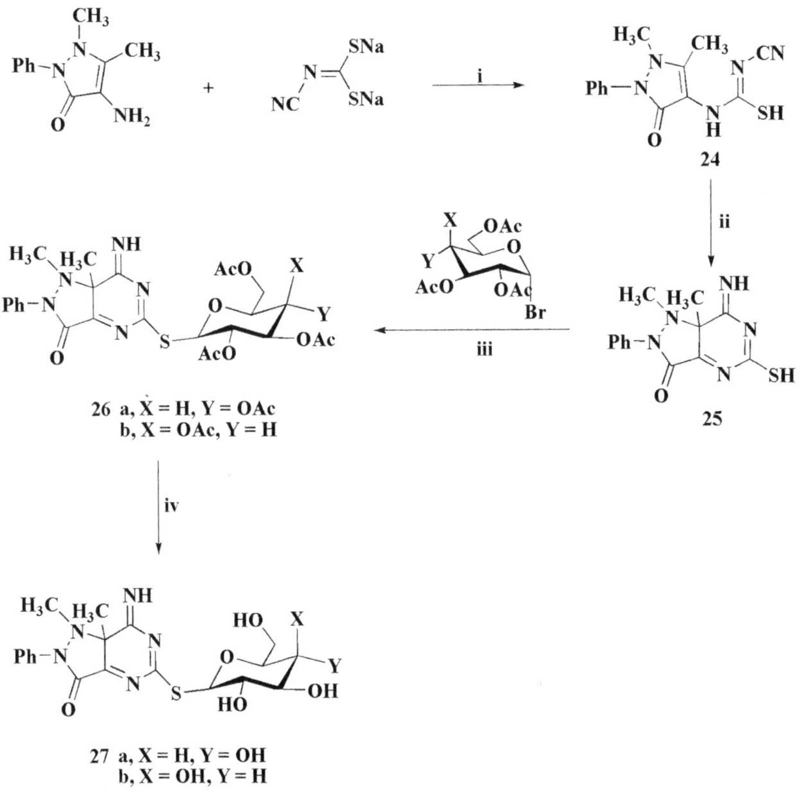

26 a, X = H, Y = OAc
b, X = OAc, Y = H

27 a, X = H, Y = OH
b, X = OH, Y = H

Scheme 10.7 Reagents and conditions: (i) AcOH/reflux/5 min. (ii) EtOH/reflux/HCl. (iii) KOH/acetone/RT. (iv) NH_3/CH_3OH/RT/10 min.

A novel variety of substituted methylsulfanylazoloazines was synthesized through reacting dimethyl *N*-cyanodithioiminocarbonate with diazoles comprising amino and active methylene function. Thus, the substituted pyrimidino [1,6-*a*]benzimidazole **31** was furnished through the reaction of dimethyl *N*-cyanodithioiminocarbonate **18** with 2-cyanomethylbenzimidazole **30**. The formation of compound **33** from reacting compounds **32** and **18** may proceed via the initial Michael addition of the active methylene group of compound **32** to the double bond of compound **18**. The Michael adduct that was generated was then cyclized through eliminating methanethiol and addition to the cyano group to afford compound **33** (Scheme 10.9) [5,21,28].

The synthesis of pyrazolo[1,5-*a*]pyrimidine derivatives was done via reacting unsaturated keto compounds with 5-aminopyrazoles. The sodium salts of 2-(hydroxymethylene)-1-cycloalkanones **34** reacted with 5-aminopyrazoles

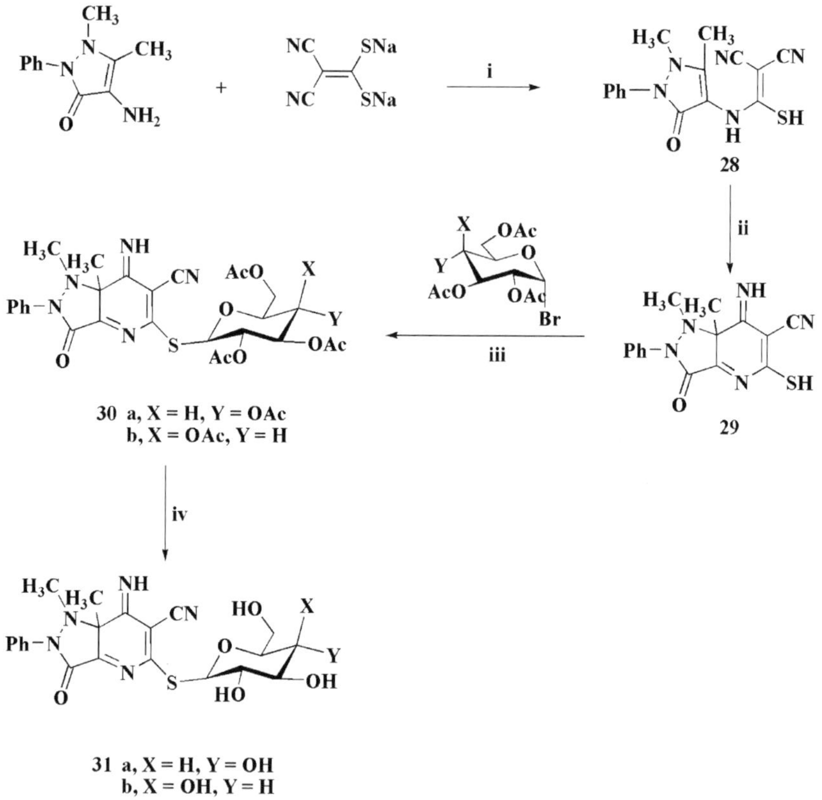

Scheme 10.8 Reagents and conditions: (i) AcOH/reflux/5 min. (ii) EtOH/reflux/HCl. (iii) KOH/acetone/RT. (iv) NH_3/CH_3OH/RT/10 min.

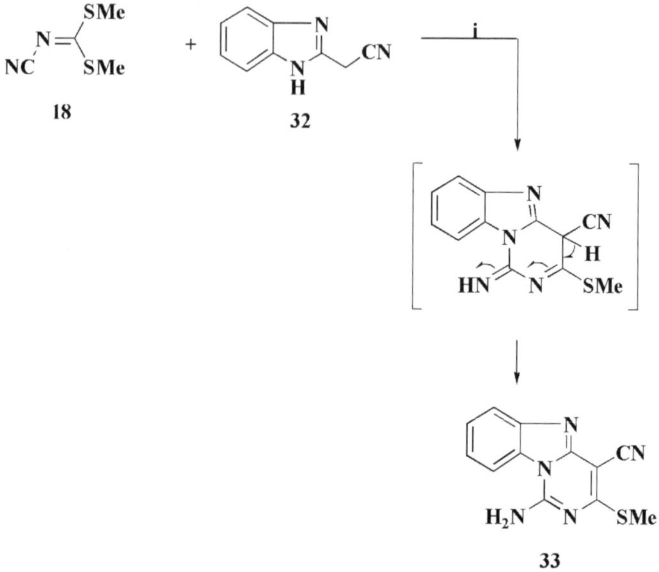

Scheme 10.9 Reagents and conditions: (i) dioxane/KOH/RT/24 h.

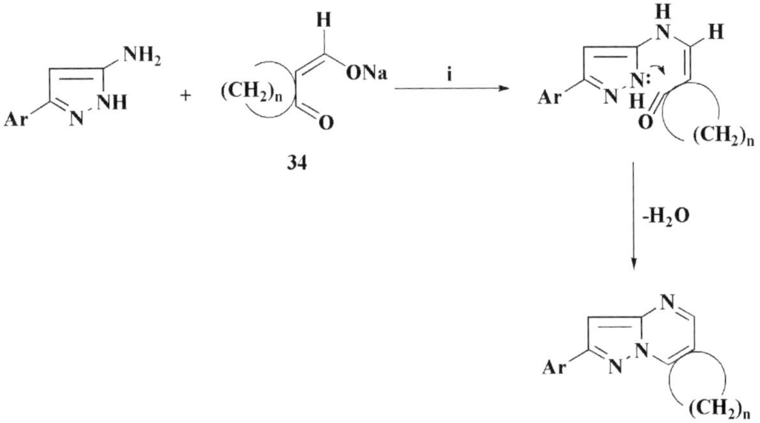

Scheme 10.10 Reagents and conditions: (i) pip. acetate/AcOH/reflux/3 h.

to furnish an adduct for which two isomeric structures are possible. Initial nucleophilic attack by the exocyclic amino group at the formyl group was followed by cyclization and elimination of water to afford angular tricyclic compounds **35** (Scheme 10.10) [29].

4. Pyrimidine and heterocyclic thioglycosides

The two major types of pyrimidine analogs in clinical use are the fluorinated pyrimidines and cytarabine. Fluorouracil, a clinically useful anticancer drug, is converted in vivo into fluorodeoxyuridylate (F-dUMP). This analog of dUMP irreversibly inhibits thymidylate synthase after acting as a normal substrate through part of the catalytic cycle. 5-Fluorouracil is very effective in treating some forms of cancer of the digestive tract and breast. There is a great interest in the synthesis of nucleoside derivatives and their incorporation into DNA sequences for the study of ligand-DNA and protein-DNA interactions. The deaza pyrimidine nucleosides constitute a category of analogs with biological potency. In recent reports from our laboratory, we demonstrated the synthesis of a new class of pyridine thioglycosides, which revealed an antagonistic activity [30]. The structure and stereochemistry of the target molecules are determined by NMR spectroscopy and X-ray crystallography [31–34]. Also, we have developed a new synthetic methodology based on the coupling of piperidinium salts of dihydropyridine thiolates with α-halogeno sugars to furnish highly functionalized dihydropyridine thioglycosides, which have been studied extensively [35]. This is the first known coupling reaction of this type for glycoside production [36]. Recently, we have described a series of dihydropyridine thioglycosides as the first example to be used as a substrates or inhibitors in the protein glycosylation process [37]. Many of these nucleosides have been scrutinized by many biological evaluation committees by virtue of their activities against human tumors.

These common features prompted us to devise a new, simple method for synthesizing heterocyclic thioglycosides. Several antiviral heterocyclic thioglycosides have been synthesized, which provided an active cytotoxicity, such as purines [27,38], pyrimidine derivatives [39–50], pyridine thioglycosides [51,52], quinolone thioglycosides [53], triazole thioglycosides [54], thiazole thioglycosides [55], oxadiazole thioglycosides [56], imidazole thioglycosides [57], pyrazole thioglycosides [58], and thienopyrazole thioglycosides [59], and in recent work, we described the synthesis of pyrimidine and pyridine phosphoramidates [60].

4.1 Synthesis of pyrimidine thioglycosides

Many pyrimidines were synthesized using ketene dithioacetals, which are important reagents in synthesizing many heterocycles [61–68].

The dihydropyrimidine-4-thiolate derivatives **37** were afforded as a result of reaction with cyanoacetohydrazides **36** with the sodium cyanocarbonimidodithioate salt. Compounds **37** reacted with halosugars to afford the corresponding *S*-glycosides **38**. The latter glycosides **38** were treated with methanolic ammonia, and the deprotected analogs **39** were accessed (Scheme 10.11) [69].

2-Cyano-*N*-arylacetamides reacted with sodium cyanocarbonimidodithioate salt to give the corresponding dihydropyrimidine-4-thiolates **41**. The latter compound reacted with halosugars to give the corresponding pyrimidine *S*-glycosides **43**. The glycoside derivatives **43** were treated with MeOH ammonia to afford deprotected analogs **44** (Scheme 10.12) [70].

Compounds **47** were synthesized via the condensation of thiourea and aromatic aldehydes with cyclopentanone in acidic condition. Compounds **47** could also be furnished by the reaction of thiourea and the diarylmethylenecyclopentanones. The *S*-alkyl derivatives were obtained through the methylation of compounds **49**. The reaction of the latter compounds with

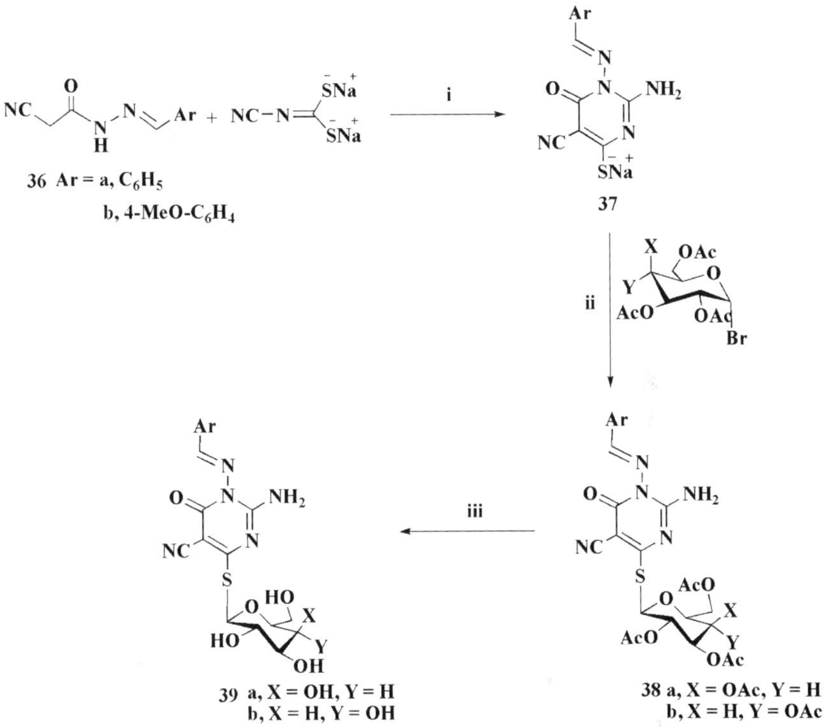

Scheme 10.11 Reagents and conditions: (i) KOH/dioxane/RT/24 h; (ii) DMF/stirring/RT; (iii) CH_3OH/NH_3/RT/10 min.

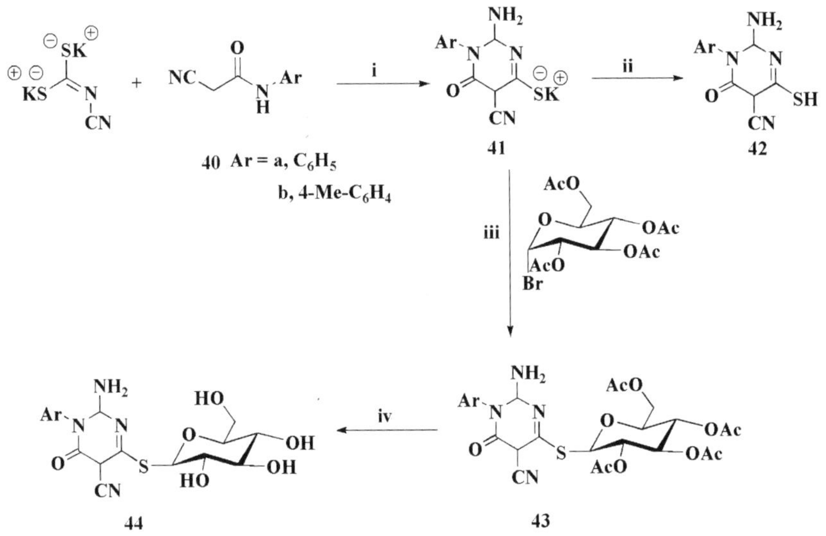

Scheme 10.12 Reagents and conditions: (i) C_2H_5ONa/EtOH/reflux/3 h; (ii) HCl.; (iii) DMF/RT; (iv) CH_3OH/NH_3.

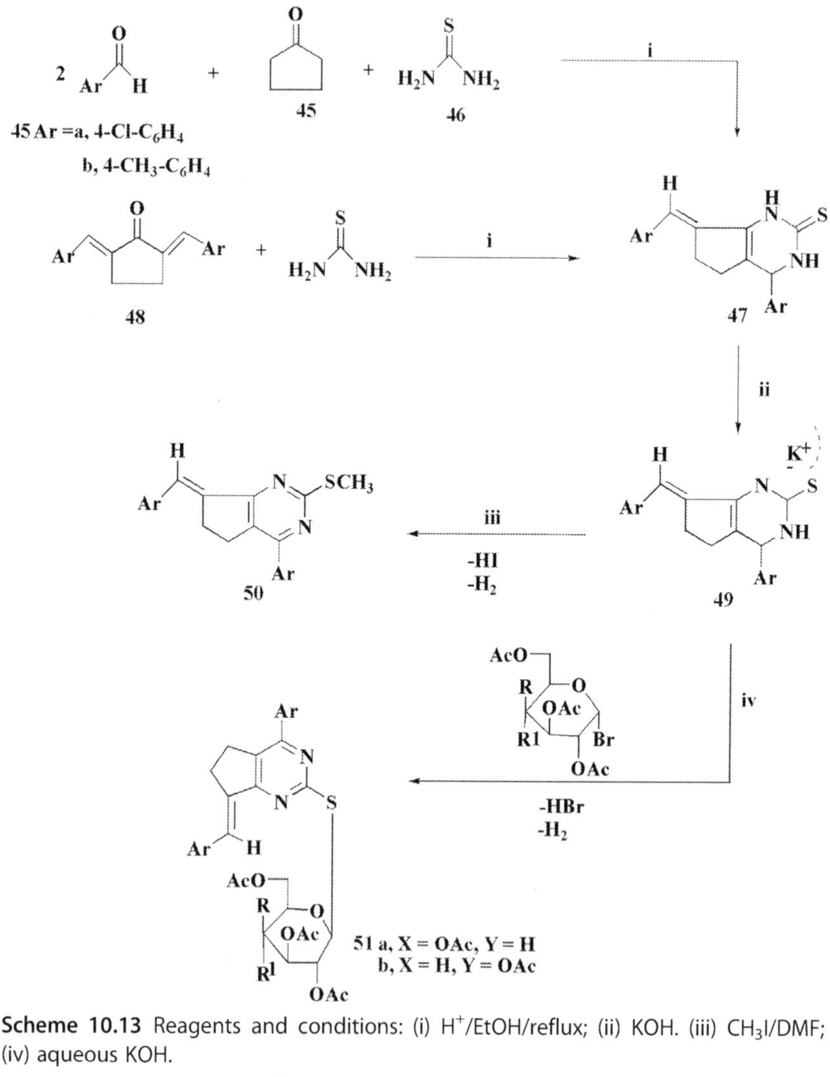

Scheme 10.13 Reagents and conditions: (i) H^+/EtOH/reflux; (ii) KOH. (iii) CH_3I/DMF; (iv) aqueous KOH.

gluco- or galactopyranosyl bromides gave the *S*-glycosylated pyrimidines (Scheme 10.13) [43].

The pyrimidine 2-thioglycosides were prepared by utilizing pyrimidinethiones **54**, which were synthesized through the reaction of thiourea and the sodium salts of 2-(hydroxymethylene)-1-cycloalkanones **52** in piperidine acetate. The reaction of compounds **54** with tetra-*O*-acetyl-α-D-gluco- or galactopyranosyl bromides furnished the *S*-glycosylated pyrimidines **55** as shown in Scheme 10.14 [71].

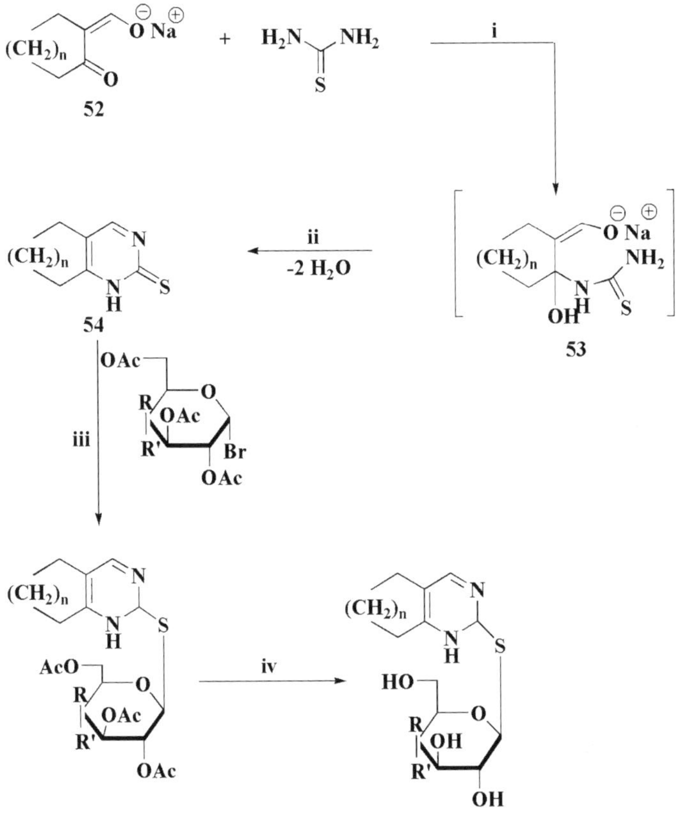

55	n	R	R′
a,	1	H	OAc
b,	2	H	OAc
c,	3	H	OAc
d,	1	OAc	H
e,	2	OAc	H
f,	3	OAc	H

56	n	R	R′
a,	1	H	OH
b,	2	H	OH
c,	3	H	OH
d,	1	OH	H
e,	2	OH	H
f,	3	OH	H

Scheme 10.14 Reagents and conditions: (i) pip. Acetate/reflux/10 min.; (ii) HCl; (iii) KOH/acetone/RT; (iv) $NH_3/C_2H_5OH/0°C$.

Amino pyrimidine thioglycosides were obtained from the reaction between guanidine hydrochloride and sodium 2-cyano-3-(arylamino)prop-1-ene-1,1-bis(thiolate) **57a–d** to give the corresponding sodium 2,6-diamino-5-aryl-1,2-dihydropyrimidine-4-thiolate, which in coupling with peracylated α-D-gluco- and galactopyranosyl bromides in dimethylformamide yielded the corresponding pyrimidine thioglycosides **60a–h**. Treatment of 2,6-diamino-5-aryl-1,2-dihydropyrimidine-4-thiolate salts with hydrochloric acid yielded the pyrimidine-4-thioles **59a–d**. The latter reacted with peracetylated halo sugars α-D-gluco- and galactopyranosyl bromides to furnish the pyrimidine thioglycosides **60a–h**. The pyrimidine thioglycosides were deacetylated to generate the corresponding free pyrimidine thioglycosides **61a–h** (Schemes 10.15–10.17) [72,73].

Pyrimidine and pyrimidine thioglycoside analogs were accessed by utilizing urea and sodium 2-cyano-3-(arylamino)prop-1-ene-1,1-bis(thiolates) to afford the corresponding sodium 6-amino-5-aryl-2-oxo-1,2-dihydropyrimidine-4-thiolates. The latter reacted with peracetylated α-D-gluco- and galactopyranosyl bromides to generate the pyrimidine thioglycosides. 6-Amino-5-aryl-2-oxo-1,2-dihydropyrimidine-4-thiolate salts were treated with hydrochloric acid to produce the corresponding pyrimidine-4-thioles, upon stirring the latter with α-D-gluco- or galactopyranosyl bromides in NaH and dimethylformamide the corresponding pyrimidine thioglycosides were yielded. The protected pyrimidine thioglycosides were deacetylated to afford the free pyrimidine thioglycosides (Schemes 10.18 and 10.19) [74].

Dimethyl *N*-cyanodithioiminocarbonate [75,76] reacted with cyanothioacetamide to yield compound **74**. Treatment of the latter compound with hydrochloric acid yielded compound **76**. Another way to access **75** was accomplished by reacting compound **76** with halosugars (Scheme 10.20) [70].

4.2 Synthesis of pyridine thioglycosides

Recently, the syntheses of several pyridinethiones [77–118] and other various pyridine thioglycosides were reported [119–127].

In Scheme 10.21, the synthesis of dihydropyridine thioglycosides is accomplished. The reaction of 1,3-diketones with arylmethylene cyanothioacetamide afforded the corresponding piperidinium salts of 1,4-dihydropyridine-2-thiones **77**. The reaction of the latter compound with gluco- or galactopyranosyl bromides generated the corresponding *S*-galactosides or *S*-glucosides [128].

The reflux of 3-oxo-*N*-aryl butanamide **79** and prop-2-enethioamide **80** was accomplished to generate pyridine-2(*1H*)-thiones **81**, which were then transformed to pyridine thioglycosides **82** (Scheme 10.22) [128].

A novel substituted pyridine thioglycoside **84** was accessed through the reaction of compound **83** with glycopyranosyl bromide. Compound **84** was then deprotected to give free sugar pyridine-2-thioglycoside **85** (Scheme 10.23) [32].

The arylmethylidenecyanothioacetamides reacted with acetylacetone to yield compound **86**. The conversion of the latter compound to compound

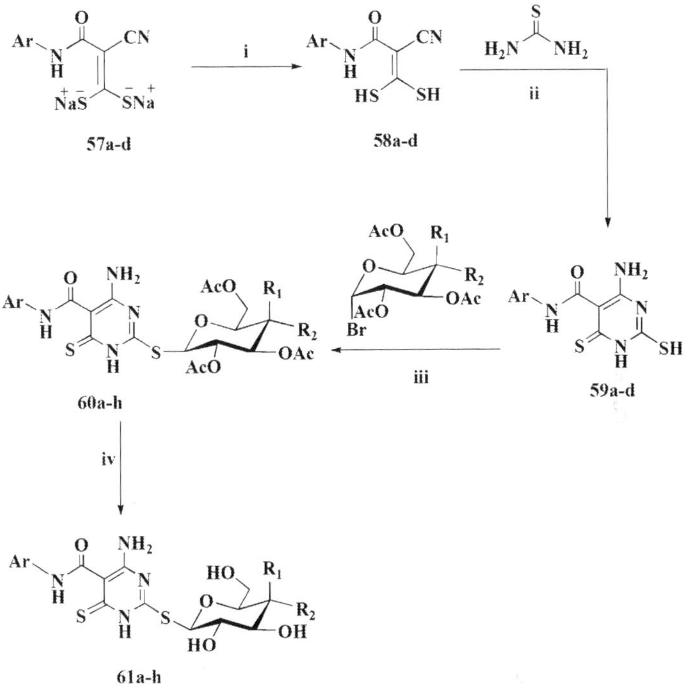

57, 58, 59	Ar
a	C_6H_5
b	$4\text{-}Cl\text{-}C_6H_4$
c	$4\text{-}CH_3\text{-}C_6H_4$
d	$4\text{-}CH_3O\text{-}C_6H_4$

60	Ar	R_1	R_2	60	Ar	R_1	R_2
a	C_6H_5	H	OAc	e	C_6H_5	OAc	H
b	$4\text{-}Cl\text{-}C_6H_4$	H	OAc	f	$4\text{-}Cl\text{-}C_6H_4$	OAc	H
c	$4\text{-}CH_3\text{-}C_6H_4$	H	OAc	g	$4\text{-}CH_3\text{-}C_6H_4$	OAc	H
d	$4\text{-}CH_3O\text{-}C_6H_4$	H	OAc	h	$4\text{-}CH_3O\text{-}C_6H_4$	OAc	H

61	Ar	R_1	R_2	6	Ar	R_1	R_2
a	C_6H_5	H	OH	e	C_6H_5	OH	H
b	$4\text{-}Cl\text{-}C_6H_4$	H	OH	f	$4\text{-}Cl\text{-}C_6H_4$	OH	H
c	$4\text{-}CH_3\text{-}C_6H_4$	H	OH	g	$4\text{-}CH_3\text{-}C_6H_4$	OH	H
d	$4\text{-}CH_3O\text{-}C_6H_4$	H	OH	h	$4\text{-}CH_3O\text{-}C_6H_4$	OH	H

Scheme 10.15 Reagents and conditions: (i) HCl; (ii) EtOH/pip./reflux/2 h; (iii) NaH/DMF/RT/8 h; (iv) NH_3/CH_3OH/RT/10 min.

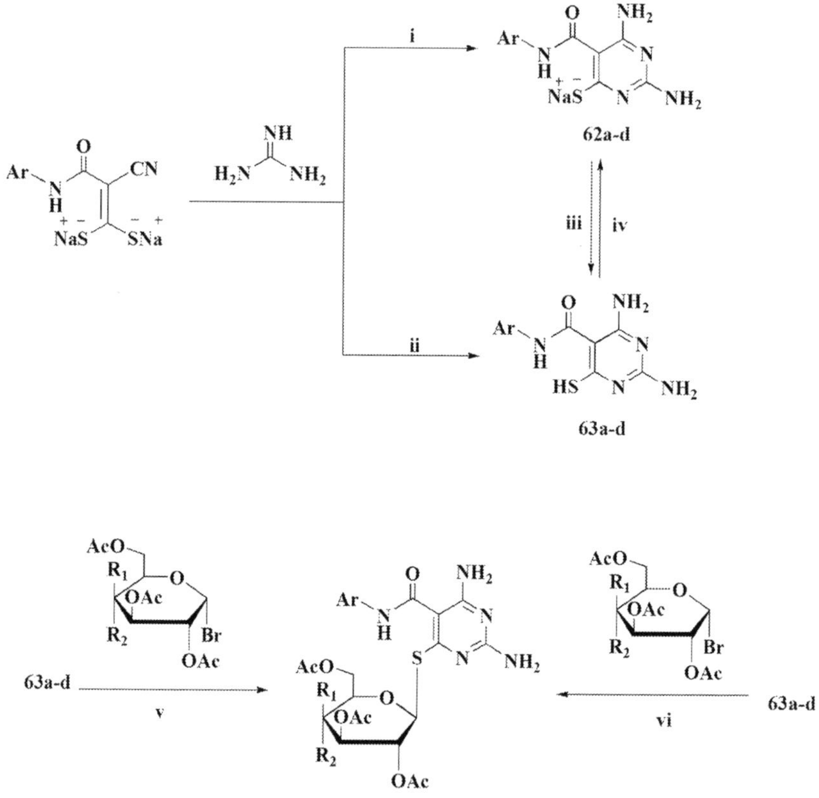

Scheme 10.16 Reagents and conditions: (i) EtOH/TEA; (ii) HCl; (iii) HCl; (iv) NaOH/EtOH; (v) DMF/RT; (vi) NaH/DMF/RT.

87 took place by utilizing iodomethane. Additionally, the thioribosides **88** were obtained through the reaction of compound **86** with 1-*O*-acetyl-2,3,5-tri-*O*-benzoyl-D-ribofuranose. Compound **88** is then converted to free riboside **89** (Scheme 10.24) [79].

Arylbutanamides were allowed to react with arylmethylidene-cyanothioacetamides to generate the corresponding piperidinium salts of 1,4-dihydropyridine-2-thiones. By reacting the latter compounds with 2,3,4-tri-*O*-acetyl-β-D-arabino- and α-D-xylopyranosyl bromides, pyridine thioarabinosides and thioxylosides were furnished (Schemes 10.25 and 10.26) [124].

Arylacetamides reacted with sodium 2,2-dicyanoethene-1,1-bis(thiolate) to furnish the substituted dihydropyridine-4-thiolates **95**. Compounds **95** reacted with 2,3,4,6-tetra-*O*-acetyl-α-D-gluco- and galactopyranosyl

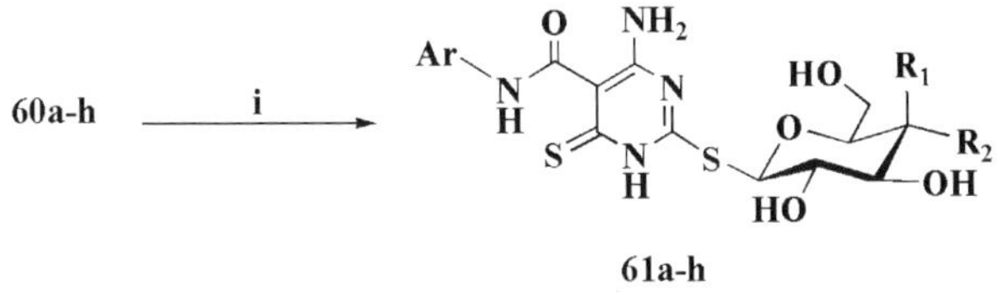

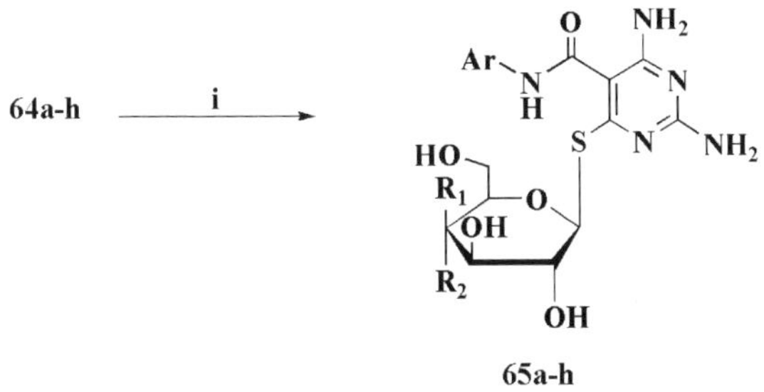

60a-h —i→ 61a-h

64a-h —i→ 65a-h

61	Ar	R_1	R_2	61	Ar	R_1	R_2
a	C_6H_5	H	OH	e	C_6H_5	OH	H
b	4-Cl-C_6H_4	H	OH	f	4-Cl-C_6H_4	OH	H
c	4-CH_3-C_6H_4	H	OH	g	4-CH_3-C_6H_4	OH	H
d	4-CH_3O-C_6H_4	H	OH	h	4-CH_3O-C_6H_4	OH	H

64	Ar	R_1	R_2	64	Ar	R_1	R_2
a	C_6H_5	H	OAc	e	C_6H_5	OAc	H
b	4-Cl-C_6H_4	H	OAc	f	4-Cl-C_6H_4	OAc	H
c	4-CH_3-C_6H_4	H	OAc	g	4-CH_3-C_6H_4	OAc	H
d	4-CH_3O-C_6H_4	H	OAc	h	4-CH_3O-C_6H_4	OAc	H

65	Ar	R_1	R_2	65	Ar	R_1	R_2
a	C_6H_5	H	OH	e	C_6H_5	OH	H
b	4-Cl-C_6H_4	H	OH	f	4-Cl-C_6H_4	OH	H
c	4-CH_3-C_6H_4	H	OH	g	4-CH_3-C_6H_4	OH	H
d	4-CH_3O-C_6H_4	H	OH	h	4-CH_3O-C_6H_4	OH	H

Scheme 10.17 Reagents and conditions: (i) NH_3/CH_3OH/RT/10 min.

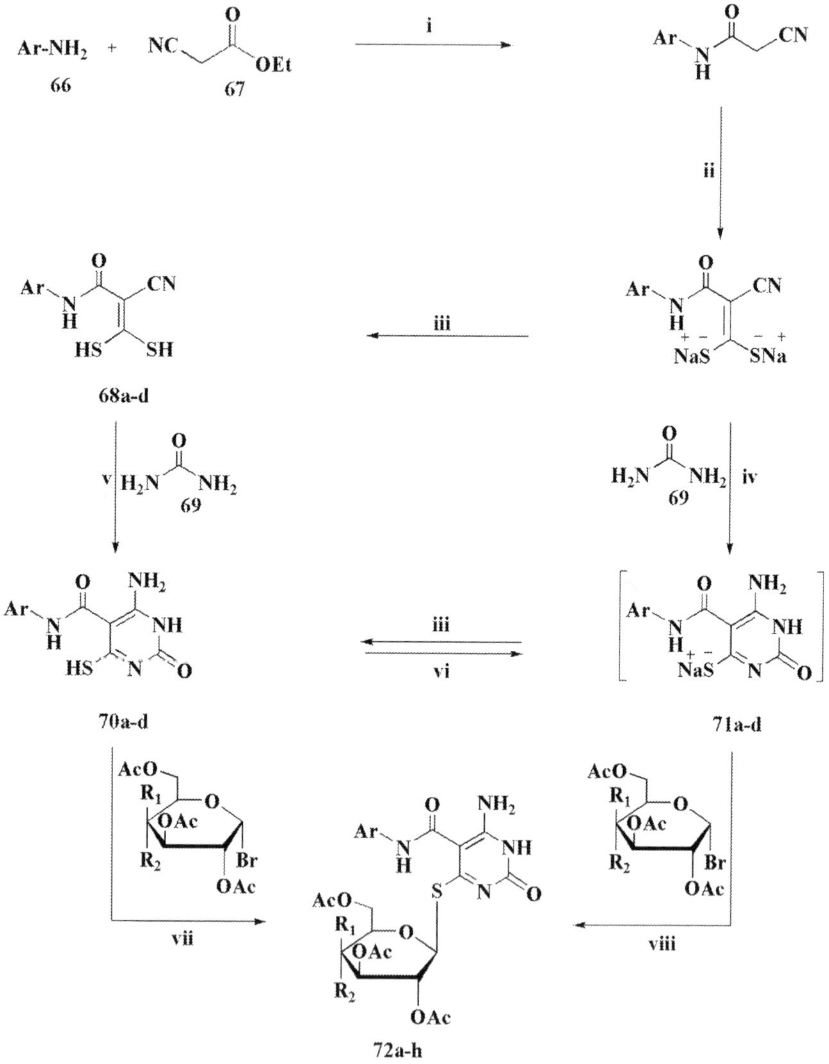

Scheme 10.18 Reagents and conditions: (i) EtOH/pip.; (ii) $CS_2/C_2H_5ONa/EtOH$; (iii) HCl; (iv) EtOH/reflux; (v) NaOH/EtOH/reflux; (vi) NaOH/EtOH; (vii) NaH/DMF/RT; (viii) DMF/RT.

bromides to give compound **96**. The free glycoside analogs **97** were generated from the glycosides **96**, as indicated in Scheme 10.27 [51].

Compounds **100** reacted with gluco- or galactopyranosyl bromides to access the corresponding *S*-glycosides or *S*-galactosides. Compounds **99** reacted with hexamethyldisilazane in the presence of sulfuric acid

72a-h →(i) 73a-h

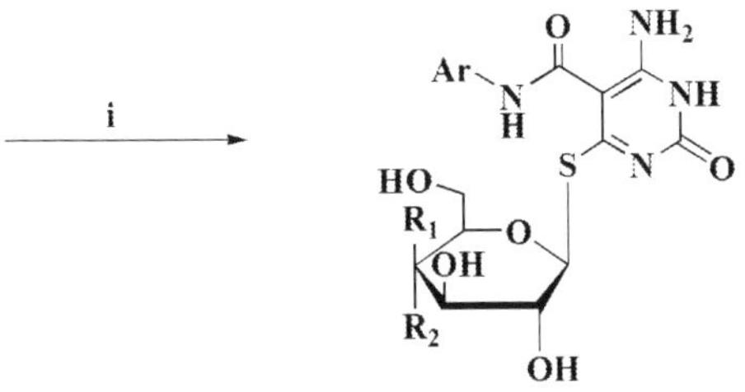

72	Ar	R_1	R_2	72	Ar	R_1	R_2
a	C_6H_5	H	OAc	e	C_6H_5	OAc	H
b	4-Cl-C_6H_4	H	OAc	f	4-Cl-C_6H_4	OAc	H
c	4-CH_3-C_6H_4	H	OAc	g	4-CH_3-C_6H_4	OAc	H
d	4-CH_3O-C_6H_4	H	OAc	h	4-CH_3O-C_6H_4	OAc	H

73	Ar	R_1	R_2	73	Ar	R_1	R_2
a	C_6H_5	H	OH	e	C_6H_5	OH	H
b	4-Cl-C_6H_4	H	OH	f	4-Cl-C_6H_4	OH	H
c	4-CH_3-C_6H_4	H	OH	g	4-CH_3-C_6H_4	OH	H
d	4-CH_3O-C_6H_4	H	OH	h	4-CH_3O-C_6H_4	OH	H

Scheme 10.19 Reagents and conditions: (i) NH_3/CH_3OH/RT/10 min.

diammonium salt to yield the corresponding 2-trimethylsilylthiopyridines **100**, which were then treated with peracetylated sugars to give *S*-glycosyl compounds **101** (Scheme 10.28) [129].

A novel category of biheterocyclic thioglycosides was furnished through the reaction of 2,3,4,6-tetra-*O*-acetyl-α-D gluco- and galactopyranosyl bromides with pyrimidine thiones. The deprotected forms **105** were yielded by utilizing methanolic ammonia (Scheme 10.29) [130].

The reaction of glucopyranosyl bromide with pyridine-2(1*H*)-thione derivative **106** yielded pyridine thioglycoside. The formation of the product was affirmed using single-crystal X-ray diffraction technique. It was proved that the product was in the form of *S*-glycoside **107** and not *N*-glycoside **108** (Scheme 10.30) [31,131].

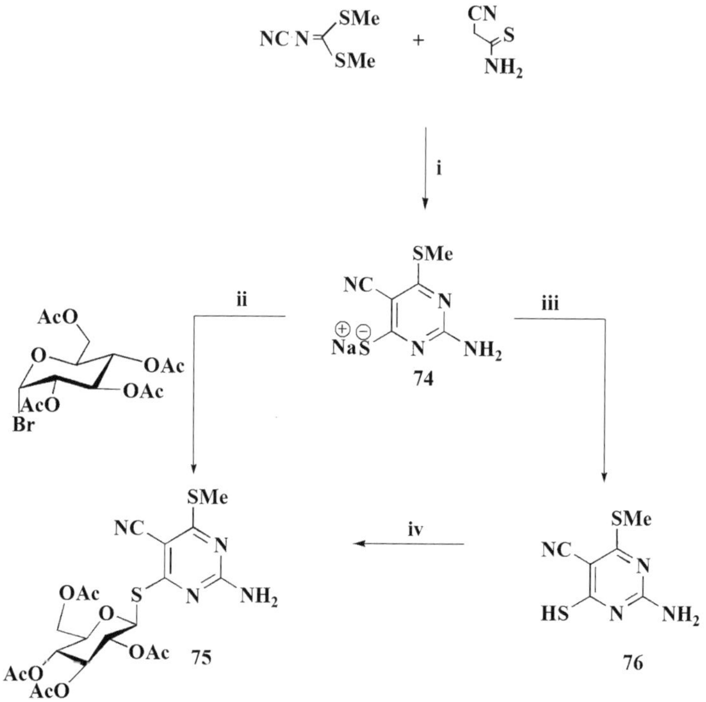

NC N=C(SMe)SMe + CN–CH2–C(=S)NH2

i

SMe
NC
N
⊕⊖ NaS
N
NH_2
74

ii

AcO
O
OAc
OAc
AcO
Br

iii

SMe
NC
N
OAc
S
N
NH_2
O
OAc 75
AcO
AcO

iv

SMe
NC
N
HS
N
NH_2
76

Scheme 10.20 Reagents and conditions: (i) C_2H_5ONa/EtOH/reflux/1 h; (ii) DMF/RT/24 h; (iii) HCl/RT; (iv) DMF/KOH/RT/24 h.

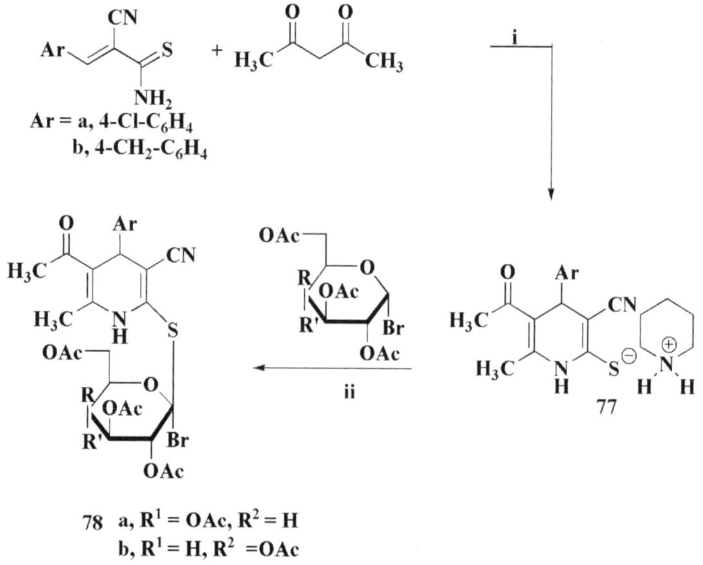

CN
Ar
S
+ H_3C
O O
CH_3
NH_2

Ar = a, 4-Cl-C_6H_4
b, 4-CH_2-C_6H_4

i

O Ar
H_3C
CN
H_3C
N H
S
OAc
R
O
OAc
R'
Br
OAc

OAc
R
O
OAc
R'
Br
OAc

ii

O Ar
H_3C
CN
H_3C
N H
S⊖
⊕N
H H
77

78 a, R^1 = OAc, R^2 = H
b, R^1 = H, R^2 = OAc

Scheme 10.21 Reagents and conditions: (i) EtOH/pip./0°C; (ii) acetone/0°C.

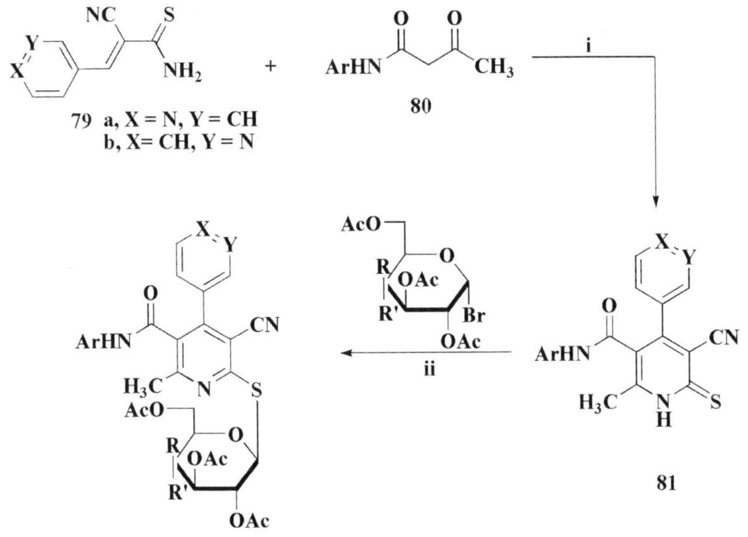

82 a, R^1 = H, R^2 = OAc
b, R^1 = OAc, R^2 = H

Scheme 10.22 Reagents and conditions: (i) pip/EtOH/reflux/3 min.; (ii) KOH/acetone/30°C.

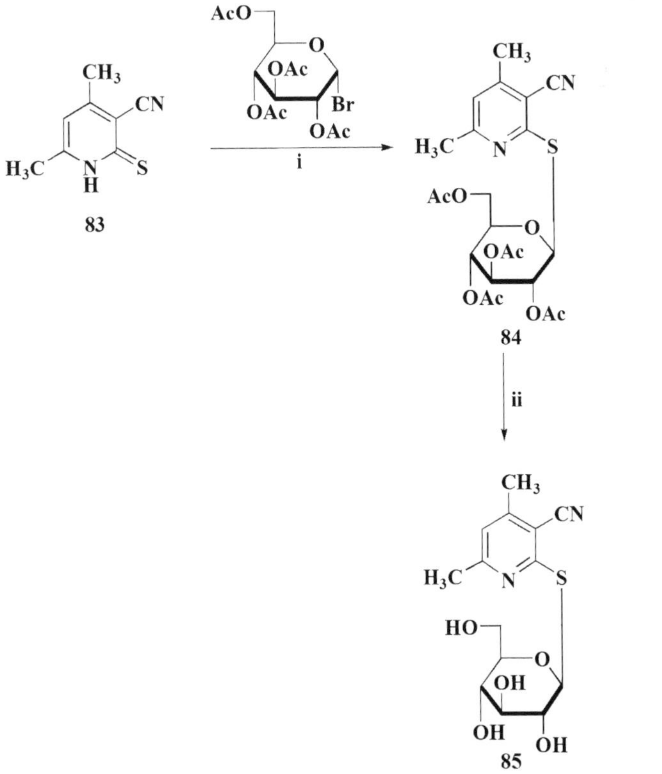

Scheme 10.23 Reagents and conditions: (i) KOH/acetone/RT; (ii) NH_3/CH_3OH/0°C.

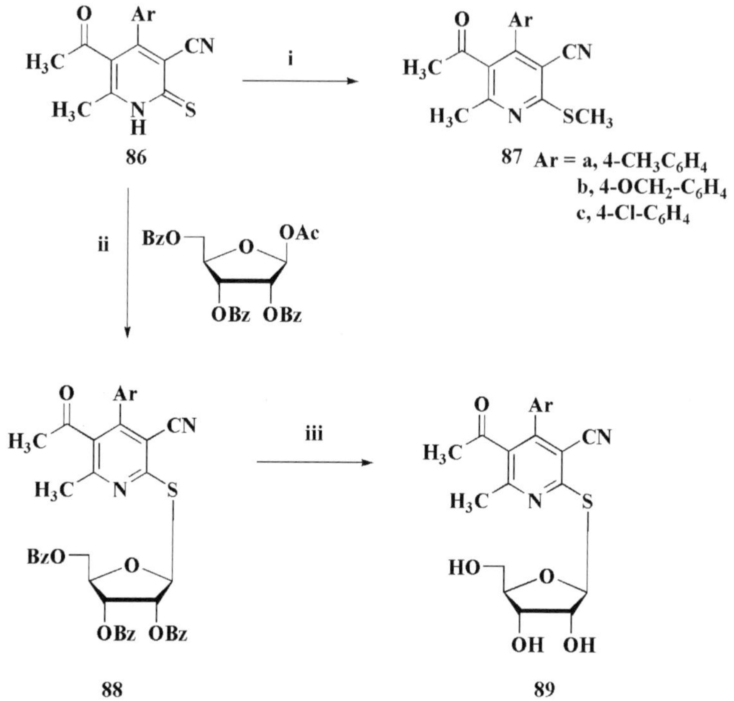

Scheme 10.24 Reagents and conditions: (i) aq. KOH/CH_3I; (ii) $HMDS/(NH_4)_2SO_4/CH_2Cl_2$; (iii) $CH_3ONa/CH_3OH/CO_2$.

4.3 Synthesis of thiophene thioglycosides

Thiophene and its substituted derivatives are a significant category of heterocyclic compounds, which reveals an interesting implementation in the field of medicinal chemistry [132–135].

A novel synthesis of thiophene thioglycosides is obtained through one-pot reaction of α-halogeno sugars with sodium thiophenethiolates. By utilizing sodium cyanoethylene thiolate salts, the sodium thiophenethiolate salts were synthesized (Schemes 10.31 and 10.32) [136].

4.4 Synthesis of Indenopyridine thioglycosides

Some indenopyridine thioglycosides were synthesized through reacting (E)-2-cyano-3-(furan/or thiophene-2-yl)prop-2-enethioamide **120a,b** with 1-indanone **121** to afford 2-thiooxo-1H-indeno[1,2-*b*]pyridine-3-carbonitriles **122a,b**. Treating the latter compounds with peracetylated

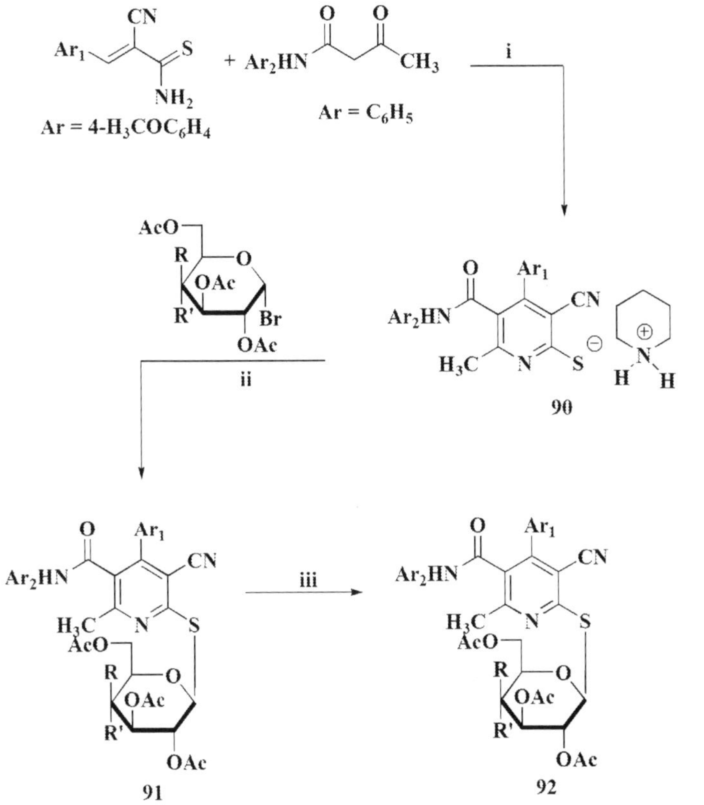

CN
Ar_1 S
+ Ar_2HN O O CH_3
i
NH_2
Ar = 4-$H_3COC_6H_4$
Ar = C_6H_5
AcO
R O
OAc
R' Br
OAc
ii
O Ar_1
Ar_2HN CN
H_3C N S
H H
90
O Ar_1
Ar_2HN CN
H_3C N S
AcO
R O
OAc
R'
OAc
91
iii
O Ar_1
Ar_2HN CN
H_3C N S
AcO
R O
OAc
R'
OAc
92

Scheme 10.25 Reagents and conditions: (i) EtOH/pip/0°C; (ii) acetone/0°C; (iii) NH_3/CH_3OH.

sugar bromides in the presence of potassium hydroxide in acetone gave the corresponding indenopyridine thioglycosides **124a–h**. The corresponding free indenopyridine thioglycosides **125a–h** were furnished via ammonolysis of the protected indenopyridine thioglycosides **124a–h** (Scheme 10.33) [137].

4.5 Synthesis of thiazole thioglycosides

New thiazole thioglycosides were synthesized via reacting potassium cyanocarbonimidodithioate with 3-oxo-3-phenylpropanenitrile and bromoacetic acid ethyl ester to yield potassium 4-amino-5-substituted-thiazole-2-thiolates **125a,b**. The latter compounds were treated with peracetylated sugar bromides to afford thiazole thioglycosides **128a–d**. Thiazole salts

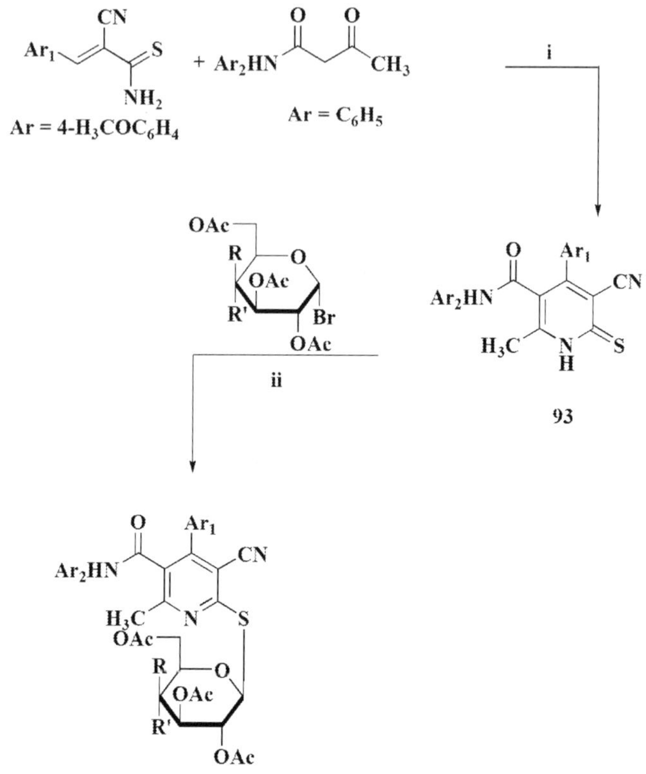

Ar = 4-$H_3COC_6H_4$

Ar = C_6H_5

93

94 a,R = OAc, R' = H
b,R = H,R' =OAc

Scheme 10.26 Reagents and conditions: (i) EtOH/pip./reflux; (ii) KOH/acetone/30°C.

125a,b were treated with hydrochloric acid to give 3-mercaptothiazole analogs **127a,b**. The reaction of the latter compounds with peracetylated sugars in the presence of hydridosodium in dimethylformamide produced S-glycosyl compounds. Ammonolysis of the protected thiazole thioglycosides **128a–h** furnished free thiazole thioglycosides **129a–d** (Scheme 10.34) [55].

4.6 Synthesis of pyrazole thioglycosides

Pyrazoles are significant compounds showing various biological potencies, and many derivatives of pyrazole were synthesized [59,138–158]. The synthesis of thiopyrazoles and their corresponding thioglycosides was furnished. These compounds were synthesized via reacting the hydrazine derivatives

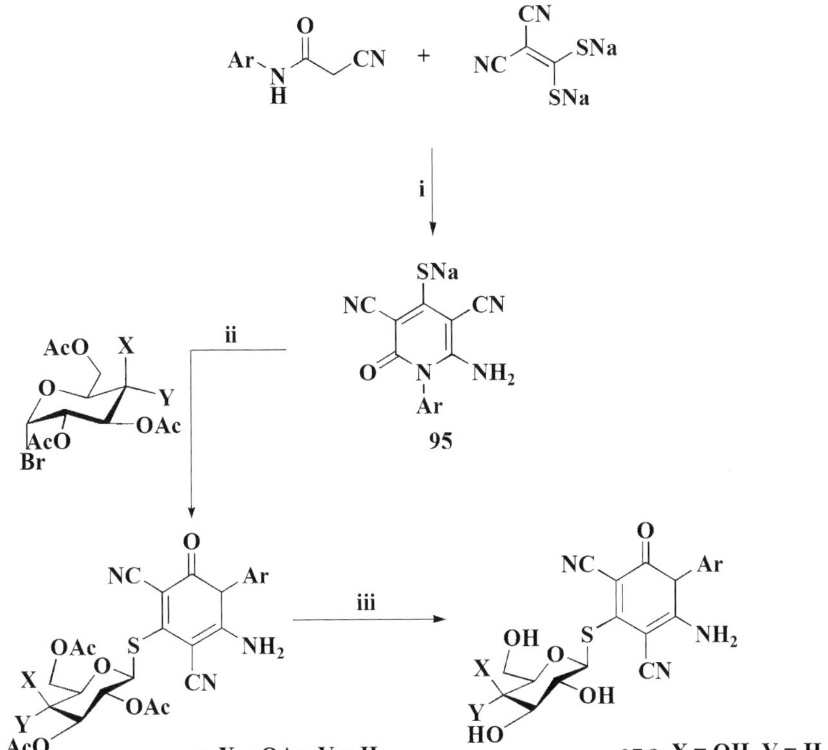

96 a, X = OAc, Y = H
b, X = H, Y = OAc

97 a, X = OH, Y = H
b, X = H, Y = OH

Scheme 10.27 Reagents and conditions: (i) C_2H_5ONa/EtOH/reflux/4 h; (ii) DMF/RT/6 h; (iii) NH_3/CH_3OH/RT/10 min.

with sodium dithiolate salt in ethanol to yield the corresponding sodium 5-amino-4-cyano-1*H*-pyrazole-3-thiolates **132a–d**. The latter compounds were treated with α-acetobromo*glucos*e and α-acetobromogalactose in dimethylformamide to afford the corresponding pyrazole *S*-glycosides **133a–h**. Ammonolysis of the pyrazole thioglycosides **133a–h** generated free thioglycosides **134a–h** (Scheme 10.35) [58].

The sodium dithiolate salts were readily monoalkylated to obtain the stable sodium salts of monoalkylthio analogs. Thus, one equivalent of 2-bromo-1-phenylethanone or iodomethane afforded the corresponding sodium salts of monoalkylated products **136**. Compound **136** reacted with 2,3,4,6-tetra-*O*-acetyl-α-D-gluco- or galactopyranosyl bromides to yield S-glucosides **137** (Scheme 10.36) [159].

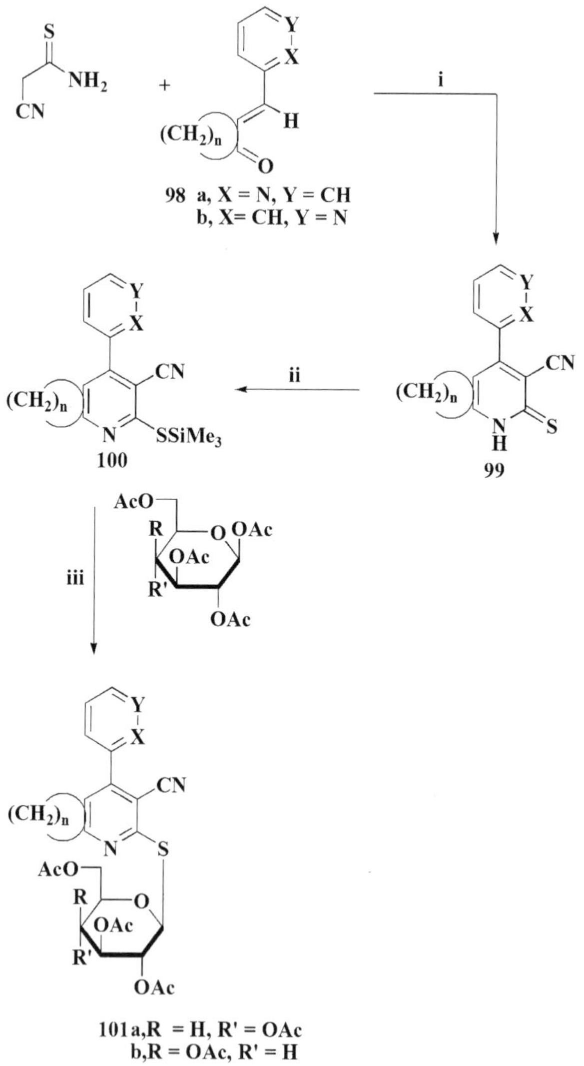

Scheme 10.28 Reagents and conditions: (i) EtOH/CH_3COONH_4/reflux; (ii) HMDS; (iii) $SnCl_4$/CH_3CN.

4.7 Synthesis of thienopyrazole thioglycosides

The pathways accomplished for the synthesis of thienopyrazole thioglycosides are demonstrated in Scheme 10.37. While refluxing, sodium ethylate compound **138** cyclized to afford sodium thieno[3,4-*c*]pyrazolethiolate **139** that upon alkylation with halosugars furnished the corresponding 2-(glycopyranosylthio)thienopyrazole derivatives **140**. When compound

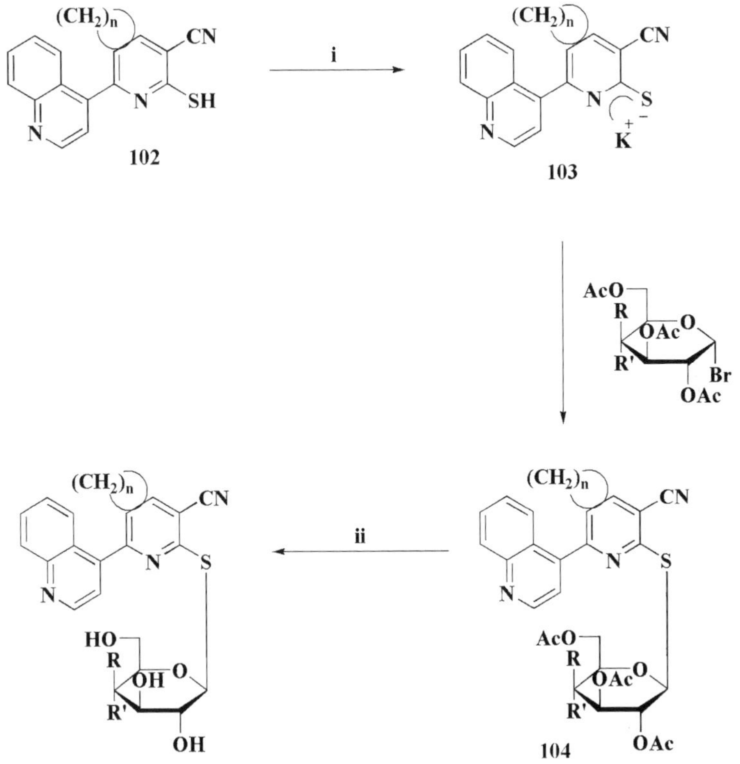

105	n	R	R`
a	3	OH	H
b	4	OH	H
c	5	OH	H
d	6	OH	H
e	3	H	OH
f	4	H	OH
g	5	H	OH
h	6	H	OH

Scheme 10.29 Reagents and conditions: (i) aq. KOH; (ii) NH_3/CH_3OH.

138 cyclized during refluxing sodium ethylate followed by acidification, it yielded thienopyrazole-4-thiol **139**, which was then coupled with halosugars in the presence of potassium hydroxide in acetone to afford also compounds **140** [59].

4.8 Synthesis of imidazole thioglycosides

The hydantoin **142** was allowed to react with phenylisothiocyanate to generatepotassium-2-thioxoimidazolidin-4-one-5-(phenylamino)-

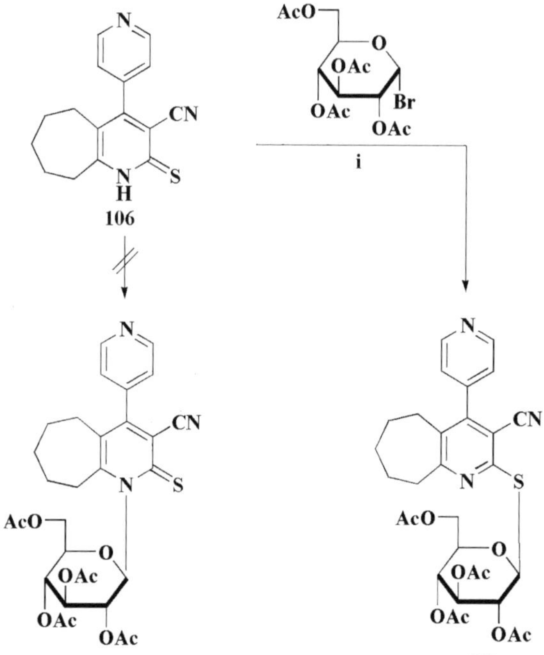

Scheme 10.30 Reagents & conditions: (i) KOH/acetone.

methylenethiolate salt **143**. The reaction of the latter with gluco- and galactopyranosyl bromides afforded the corresponding S-glucoside and S-galactoside (Scheme 10.38) [57].

Thus, one equivalent of iodomethane or 2-bromo-4′-chloroacetophenone afforded sodium-2-thioxoimidazolidin-4-one-5-(methylthio)[2-oxo-2-(4-chlorophenylethyl)thio]-methylenethiolate salts, which were treated with tetra-acetylated gluco/galactopyranosyl bromides to generate the corresponding S-galactoside and S-glucosides (Schemes 10.39 and 10.40) [22].

4.9 Synthesis of thienoimidazole thioglycosides

2-Thiohydantoin reacted with carbon disulfide and sodium ethylate to obtain the corresponding stable sodium 2-thiohydantoin-5-methylenedithiolate. The latter reacted either with 2 mol 2-bromo-4′-chloroacetophenone or with 1 mol 2-bromo-4′-chloroacetophenone and 1 mol chloromethane and then reacted with 2,3,4,6-tetra-*O*-acetyl-*α*-D-gluco- and galactopyranosyl bromides to afford the corresponding S-glucosides or S-galactosides.

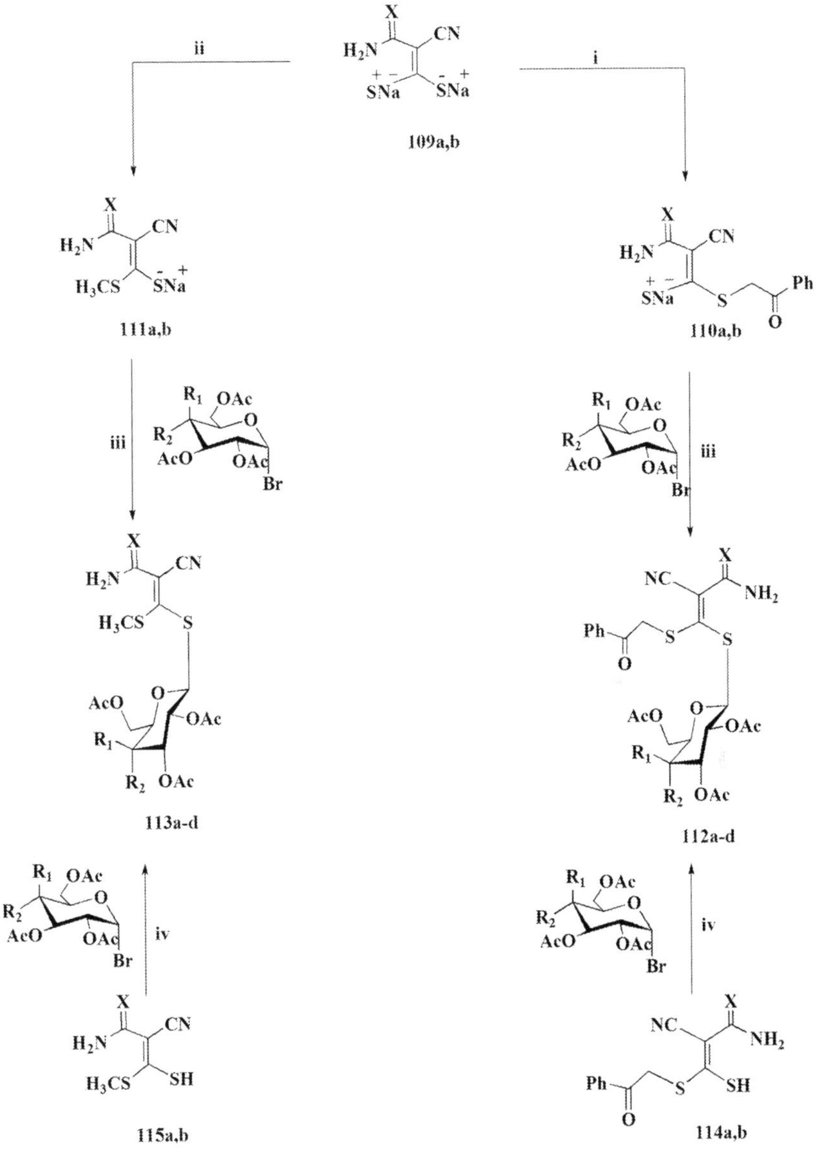

112, 113	X	R_1	R_2
a,	O	H	OAc
b,	S	H	OAc
c,	O	OAc	H
d,	S	OAc	H

Scheme 10.31 Reagents and conditions: (i) 2-bromo-1-phenylethanone/EtOH/RT; (ii) CH_3I/EtOH/RT; (iii) EtOH/RT/3 h. (iv) KOH/EtOH/RT/3 h.

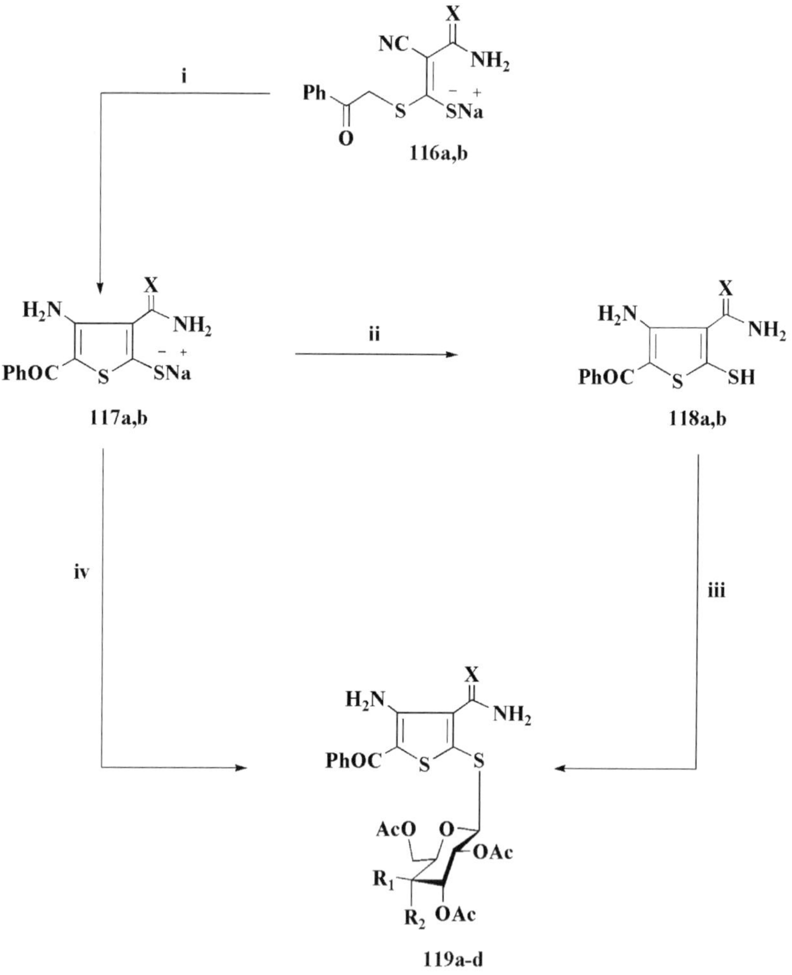

119	X	R_1	R_2
a,	O	H	OAc
b,	S	H	OAc
c,	O	OAc	H
d,	S	OAc	H

Scheme 10.32 Reagents and conditions: (i) C_2H_5ONa/EtOH/reflux; (ii) HCl; (iii) KOH/EtOH/RT/3 h; (iv) EtOH/RT/3 h.

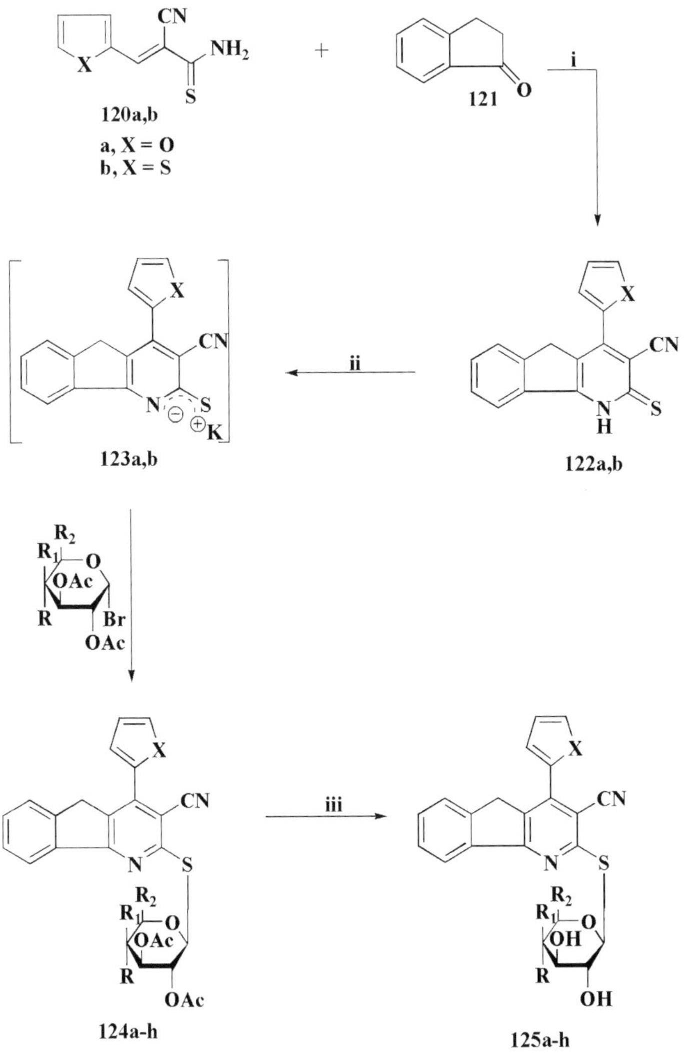

Scheme 10.33 Reagents and conditions: (i) CH_3COONH_4/EtOH/heat/3 h; (ii) KOH/acetone/RT; (iii) NH_3/CH_3OH/0°C.

When compound sodium 2-thiohydantoin-5-methylenedithiolate di-alkylated with two *p*-chloro phenacyl bromide, it afforded (*5Z*)-5-bis-[(2-oxo-2-(4-chlorophenyl ethyl) thio) methylene]-2-thiohydanyoin and/or *N*-phenyl-2-thiohydantoin, which then reacted with α-halosugars in acetone and aqueous potassium hydroxide to afford the corresponding S-galactosides or S-glucosides.

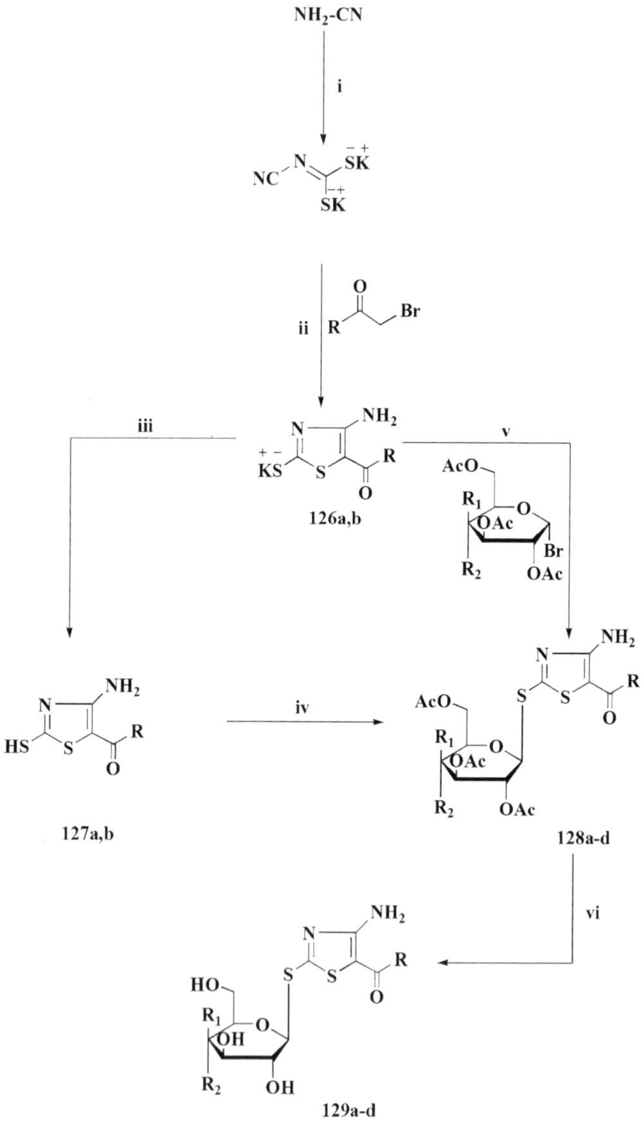

126, 127	R
a,	C_6H_5
b,	OC_2H_5

128	R	R^1	R^2
a,	C_6H_5	H	OAc
b,	OC_2H_5	H	OAc
c,	C_6H_5	OAc	H
d,	OC_2H_5	OAc	H

129	R	R^1	R^2
a,	C_6H_5	H	OH
b,	OC_2H_5	H	OH
c,	C_6H_5	OH	H
d,	OC_2H_5	OH	H

Scheme 10.34 Reagents and conditions: (i) CS_2/KOH/EtOH/RT; (ii) KOH/EtOH/RT; (iii) HCl/RT; (iv) NaH/DMF/RT; (v) DMF/RT/8 h; (vi) CH_3OH/NH_3/RT/10 min.

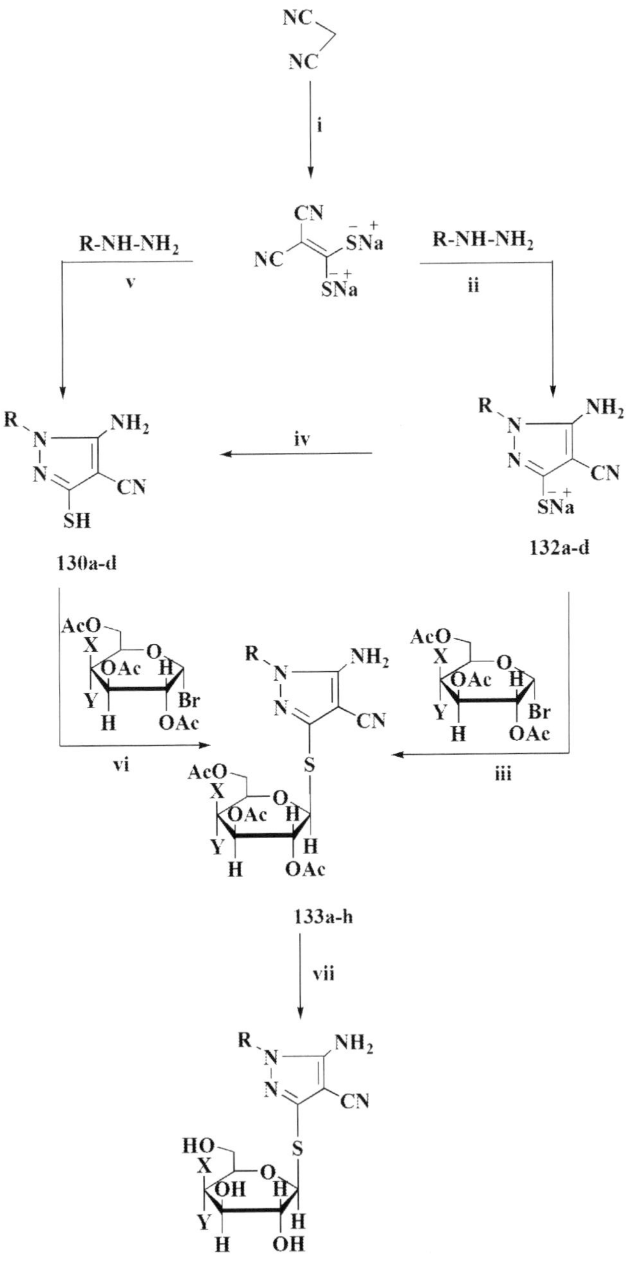

Scheme 10.35 Reagents and conditions: (i) CS_2/C_2H_5ONa/EtOH/RT; (ii) EtOH/RT/24 h; (iii) DMF/RT/8 h; (iv) HCl; (v) AcOH/EtOH/RT/10 h; (vi) NaH/DMF/RT/3-8 h; (vii) NH_3/CH_3OH/0°C.

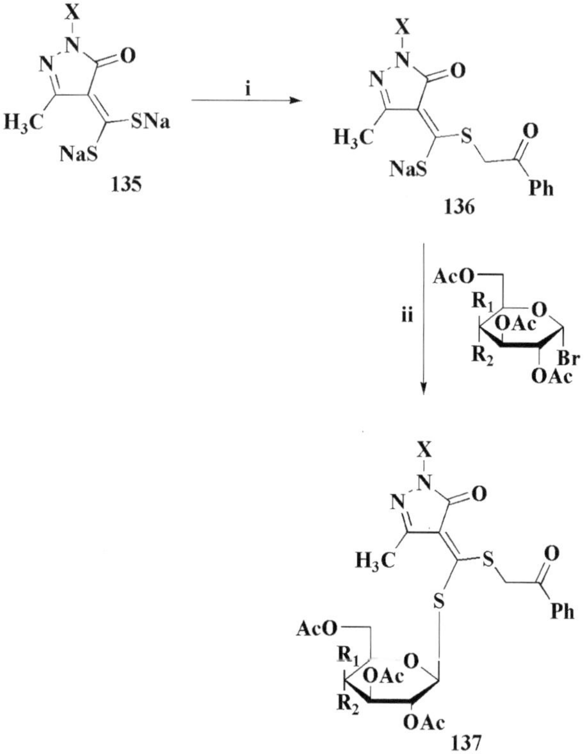

Scheme 10.36 Reagents and conditions: (i) $PhCOCH_2Br$/EtOH/RT; (ii) EtOH/RT.

The thieno-imidazole thioglycosides **152** and **154** were produced as a further modification of the aglycon molecule. Compound sodium 2-thiohydantoin-5-methylenedithiolate refluxed with one iodomethane and one 2-bromo-4′-chloroacetophenone to yield 4-[methylthio]-6-[(4-chlorophenyl)-oxo-methyl]-2-thio-2,3 dihydro-1-H and/or phenyl -thieno [3,4-*d*] imidazole. The latter reacted with 2,3,4,6-tetra-*O*-acetyl-α-D-gluco- and galactopyranosyl bromides to furnish the corresponding S-glucosides or S-galactosides **154** (Schemes 10.41 and 10.42) [160].

4.10 Synthesis of triazole thioglycosides

A novel approach for the syntheses of new category of triazole thioglycosides is reported. The compounds were synthesized via reacting potassium cyanocarbonimidodithioate with hydrazine derivatives to afford the

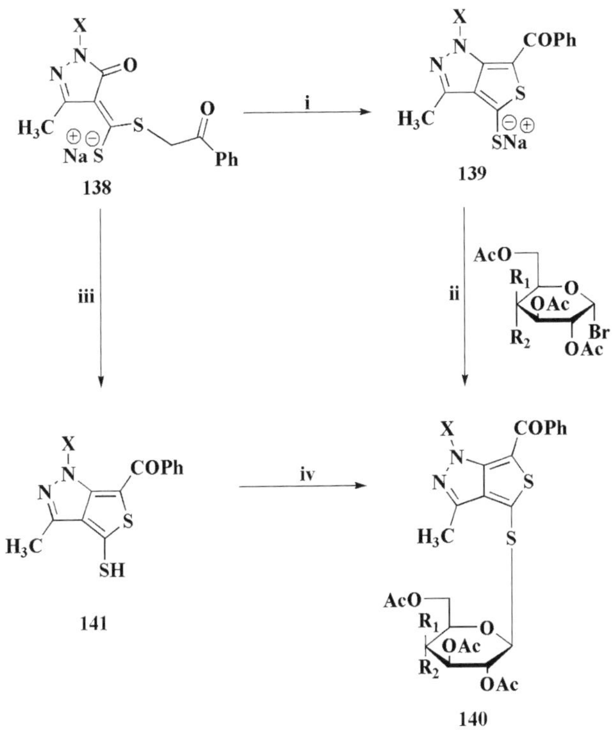

Scheme 10.37 Reagents and conditions: (i) C_2H_5ONa/reflux/3 h; (ii) EtOH/RT; (iii) EtOH/reflux/3 h/H^+/RT; (iv) KOH/acetone/RT.

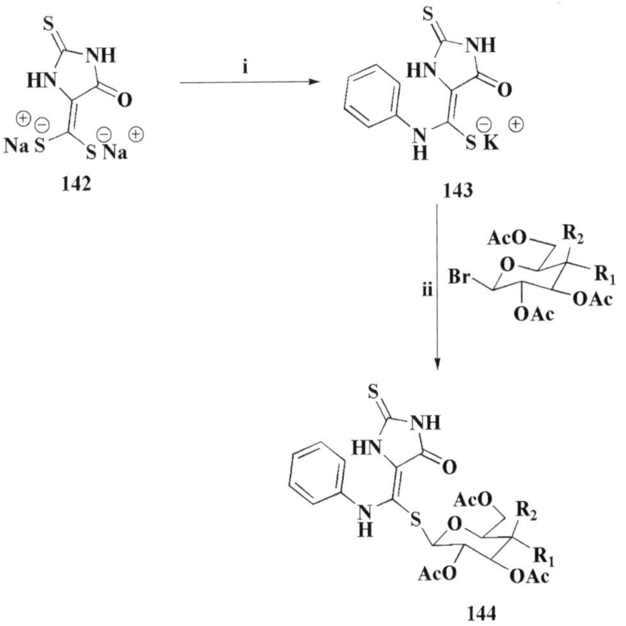

Scheme 10.38 Reagents and conditions: (i) PhNCS/KOH/reflux; (ii) EtOH/RT.

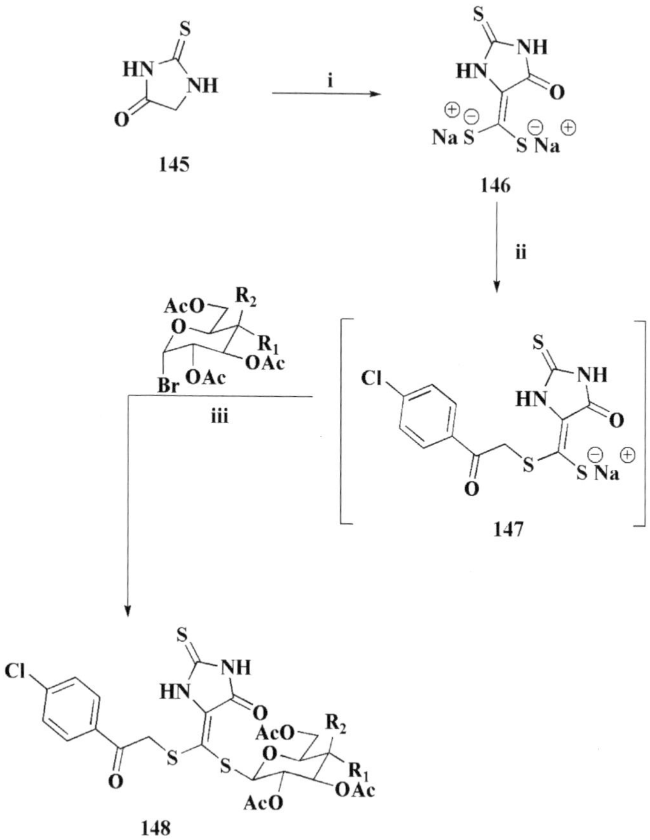

Scheme 10.39 Reagents and conditions: (i) CS_2/C_2H_5ONa; (ii) 2-bromo-1-(4-chlorophenyl)ethanone/EtOH/RT; (iii) EtOH/RT.

corresponding potassium 5-amino-1*H*-1,2,4-triazole-3-thiolates **156a–d**. Treatment of the latter compounds with tetra-*O*-acetyl-*α*-D-galactopyranosyl bromide and tetra-*O*-acetyl-*α*-D-glucopyranosyl bromide was done to furnish the corresponding triazole thioglycosides **157a–h**. The triazole salts **156a–d** were treated with hydrochloric acid to yield 3-mercaptotriazoles **155a–d**. Compounds **155a–d** then reacted with bromoperacetylated sugars in NaH-DMF to generate thioglycosyl compounds **157a–h**. Ammonolysis of triazole thioglycosides **157a–h** generated free thioglycosides **158a–h** (Scheme 10.43) [54].

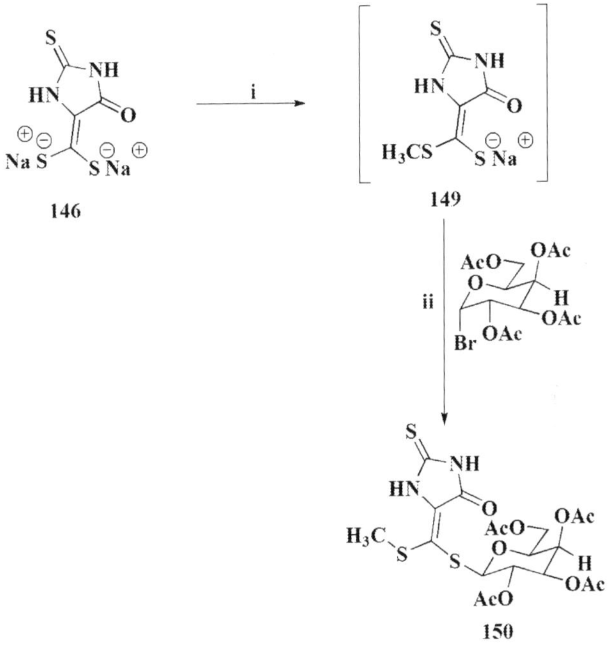

Scheme 10.40 Reagents and conditions: (i) CH_3I/EtOH/RT; (ii) EOH/RT.

4.11 Synthesis of oxadiazole thioglycosides

An interesting 1,3,4-oxadiazoline-2-thiones **163a–c** were prepared via the straight forward reaction pathway, as indicated in Scheme 10.20, and reacted with activated acetobromoglucose in the presence of sodium hydride as a base in dimethylformamide to afford 1,3,4-oxadiazole thioglycosides **164a–c** in good yields (Scheme 10.44) [56].

4.12 Synthesis of benzothiazole thioglycosides

Benzothiazole moiety is of great importance as it is found in a large variety of naturally occurring compounds and possesses a broad spectrum of medicinal and pharmacological properties. Recently, many benzothiazole derivatives have been extensively reported [161–166].

The benzothiazole acetonitrile **165** reacted with isothiocyanatobenzene to give potassium 2-[(2*E*)-2-(phenylamino)-1-(cyano) vinyl] 1,3-benzothiazole-2-thiolates salt **193**. The latter reacted with 2,3,4,6-tetra-*O*-acetyl-*α*-D-gluco- and galactopyranosyl bromides to furnish the corresponding S-glucosides or S-galactosides **167** (Scheme 10.45) [57].

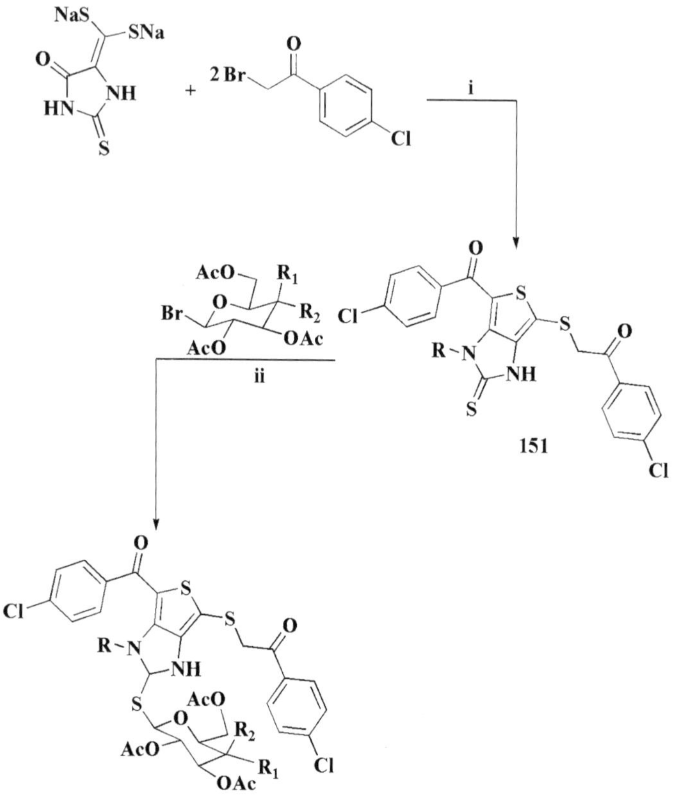

152

Scheme 10.41 Reagents and conditions: (i) EtOH/reflux; (ii) aqueous KOH.

4.13 Synthesis of pyridine phosphoramidates

A series of pyridine-based thioglycoside phosphoramidates are synthesized through coupling α-bromo per-acetylated sugars with mercapto-derivatized heterocyclic bases. Under basic conditions, the acetate esters were hydrolyzed that were consequently conjugated with the phosphoramidating reagent to generate the targeted thioglycoside (Scheme 10.46) [60].

4.14 Synthesis of pyrimidine phosphoramidates

Pyrimidine-based thioglycoside phosphoramidates were synthesized as depicted in Scheme 10.47.

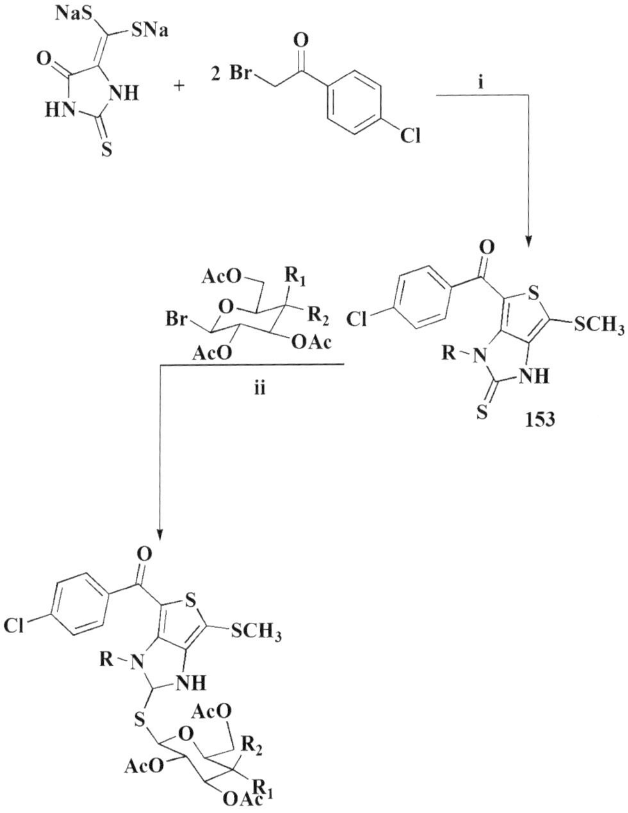

154

Scheme 10.42 Reagents and conditions: (i) CH_3I/EtOH/reflux; (ii) aqueous KOH.

4.15 Synthesis of azoloazine thioglycosides

A new pyrimidobenzothiazole derivative **170** was furnished through the reaction of dimethyl *N*-cyanodithioiminocarbonate with 2-cyanomethyl-benzo[1,2-*b*]thiazole **169**. Sodium cyanocarbonimidodithioate salt reacted with compound **169** to give compound **171**. And further reaction of compound **171** with halosugars yielded compound **172**. Treatment of the latter with hydrochloric acid gave compound 3-mercapto-1-imino-pyrido[6,1-*a*] benzothiazol-4-carbonitrile **173**. Compound **172** was accessed by starting with compound **177** and then reacting it with halosugar to give the corresponding S-glycosides (Scheme 10.48) [167].

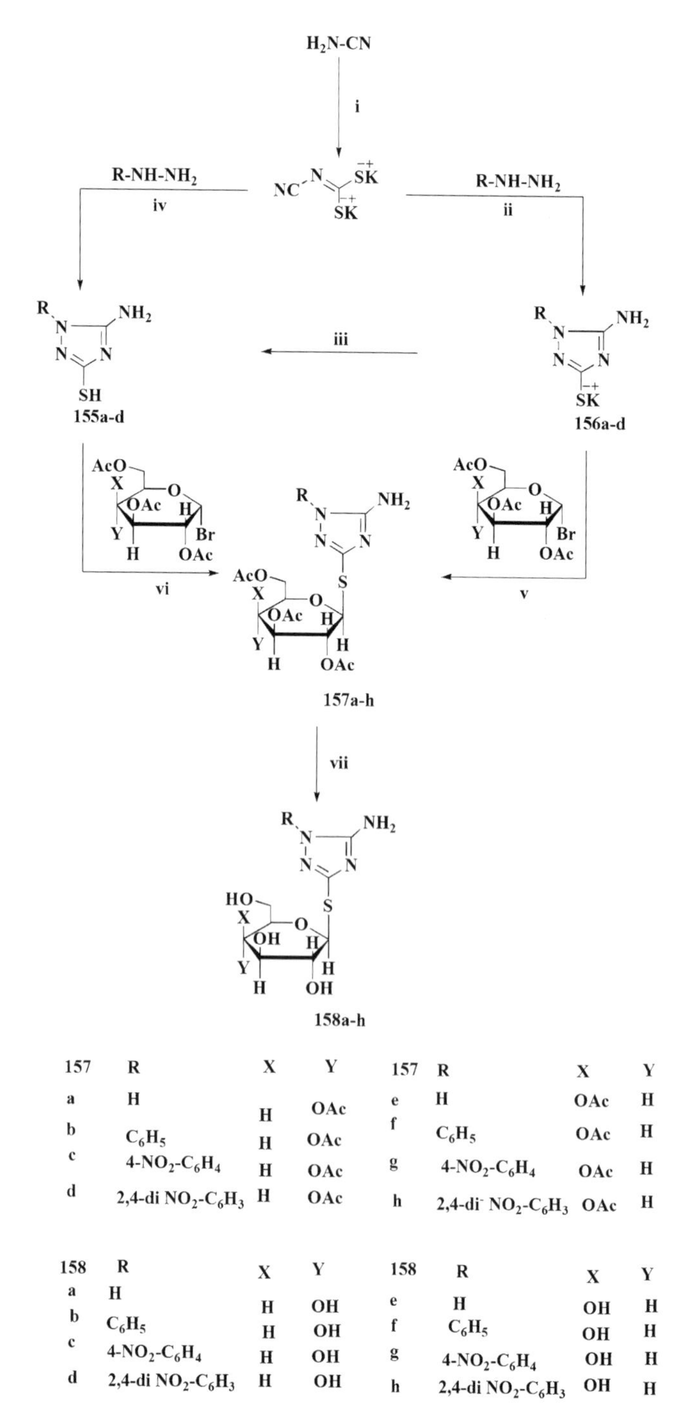

157	R	X	Y
a	H	H	OAc
b	C_6H_5	H	OAc
c	4-NO_2-C_6H_4	H	OAc
d	2,4-di NO_2-C_6H_3	H	OAc
e	H	OAc	H
f	C_6H_5	OAc	H
g	4-NO_2-C_6H_4	OAc	H
h	2,4-di NO_2-C_6H_3	OAc	H

158	R	X	Y
a	H	H	OH
b	C_6H_5	H	OH
c	4-NO_2-C_6H_4	H	OH
d	2,4-di NO_2-C_6H_3	H	OH
e	H	OH	H
f	C_6H_5	OH	H
g	4-NO_2-C_6H_4	OH	H
h	2,4-di NO_2-C_6H_3	OH	H

Scheme 10.43 Reagents and conditions: (i) CS_2/EtOH/KOH/RT; (ii) EtOH/RT/24h; (iii) HCl/RT; (iv) AcOH/EtOH/RT/8h; (v) NaH/DMF/RT/3–10h; (vi) DMF/RT/8h; (vii) NH_3/CH_3OH/RT/10min.

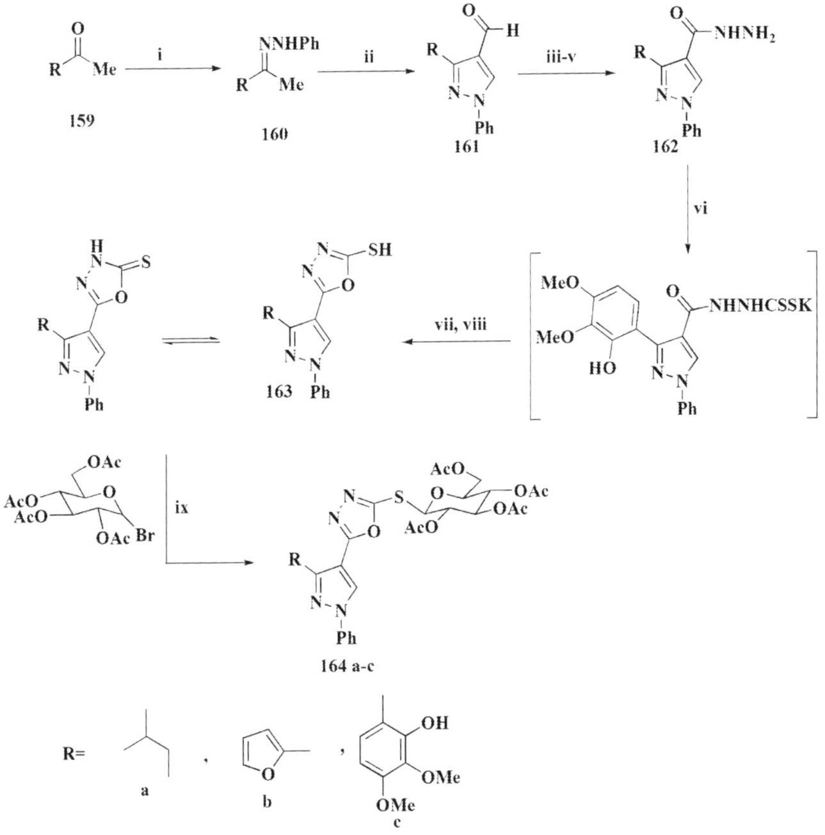

Scheme 10.44 Reagents and conditions: (i) $PhNHNH_2/AcOH$; (ii) $POCl_3/DMF$; (iii) $K_2Cr_2O_7/H_2SO_4$; (iv) $EtOH/H_2SO_4$; (v) $NH_2{\cdot}NH_2.\ H_2O$; (vi) CS_2/KOH; (vii) EtOH/reflux; (viii) HCl; (ix) NaH/DMF.

5. Pyrimidine and pyridine *N*-non-nucleoside analogs

The syntheses of *N*-aryl-6-methylsulfanyl-4-oxopyrimidine-5-carbonitriles were accomplished through reacting dimethyl *N*-cyanodithioiminocarbonate with substituted cyanoacetanilides. (3) *N*-Aryl-2,5-diamino-7*H*-pyrazolo[3,4-*d*]-pyrimidin-4(3*H*)-one has also been synthesized by reacting hydrazines with 6-sulfanylthio-*N*-aryl-4-pyrimidinones (Scheme 10.49) [25,167–171].

A new variety of methylsulfanylpyrimidines were synthesized by reacting the substituted hydrazides with dimethyl *N*-cyanodithioiminocarbonate (Scheme 10.50) [168].

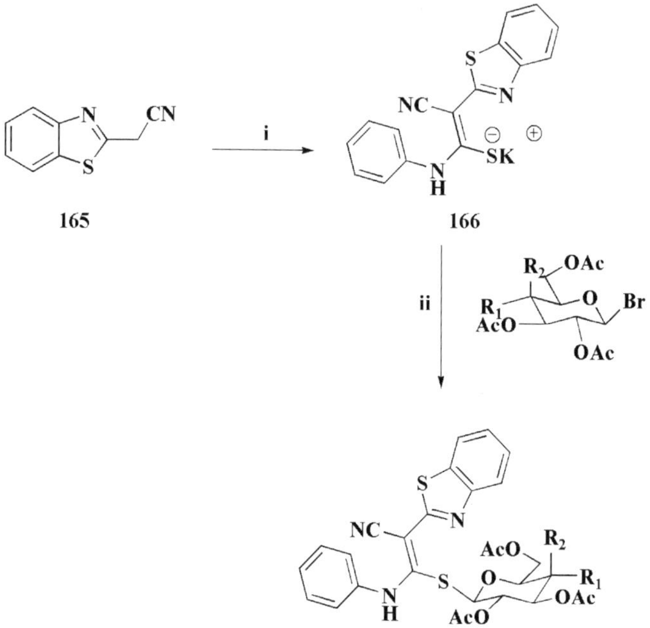

165 166 167

Scheme 10.45 Reagents and conditions: (i) PhNCS/KOH/EtOH/RT; (ii) EtOH/RT.

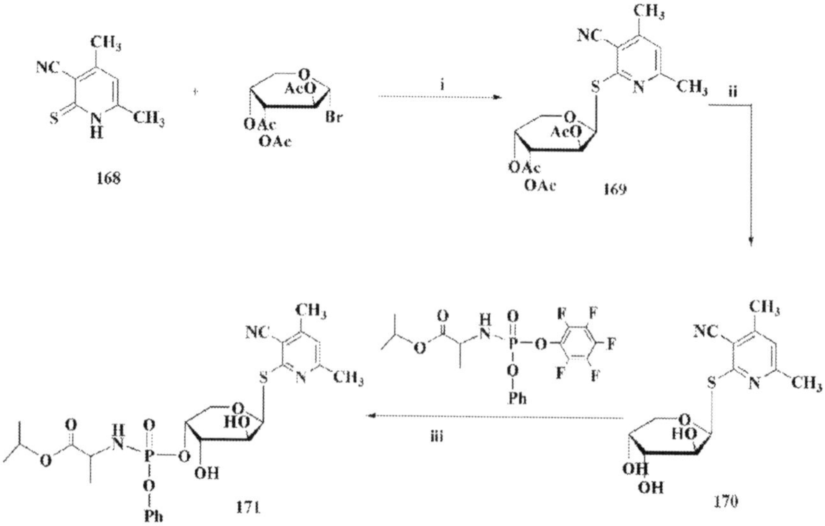

168 169 170 171

Scheme 10.46 Reagents and conditions: (i) NaH/DMF; (ii) NH_3/CH_3OH; (iii) iPrMgCl/THF/0–5°C.

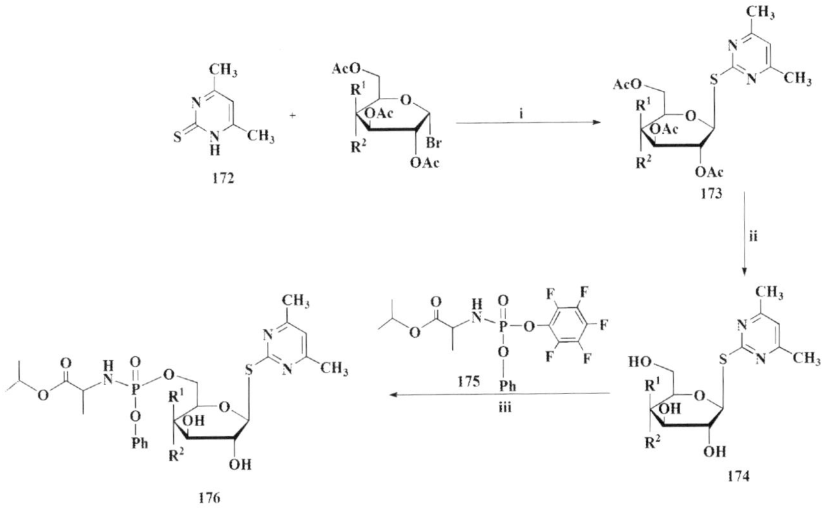

173	R^1	R^2	174, 175	R^1	R^2
a	H	OAc	a	H	OH
b	OAc	H	b	OH	H

Scheme 10.47 Reagents and conditions: (i) NaH/DMF/RT/6h; (ii) NH_3/CH_3OH/RT/10min; (iii) iPrMgCl/THF/0–5°C.

The reactions of 2-cyano-*N*′-(1-phenylethylidene)acetohydrazide (**192a**) or 2-cyano-*N*′-(1-(pyridin-2-yl)ethylidene)acetohydrazide (**192b**) with dimethyl *N*-cyanodithioiminocarbonate in KOH/EtOH were studied and identified by X-ray crystallography as compounds **193a,b** (Scheme 10.51) [172].

Novel *N*-substituted derivatives of 4-ethylsulfanyl-2-pyridones and triazolopyridines were designed and synthesized by reacting 2,2-dicyanoethene-1,1-bis(ethylthiolate) with **N-**cyanoacetohydrazide, *N*′-[(aryl)-methylene]-2-cyanoacetohydrazides, cyanoaceto-*N*-phenylsulfonylhydrazide, or cyanoacetanilides (Schemes 10.52–10.54) [114,115,118, 173–175].

A novel strategy for designing and assembling a new category of functionalized pyridine-based benzothiazole by incorporating sulfonamide scaffolds was synthesized. The synthesis was performed via reacting *N*-cyanoacetoarylsulfonylhydrazide with different electrophiles such as 2-(benzo[*d*]imidazol-2-yl)-3,3-bis(methylthio)-acrylonitriles and 2-(benzo[*d*]thiazol-2-yl)-3,3-bis(alkylthio)acrylonitriles as well as 2-ethoxyl acrylonitrile derivatives (Schemes 10.55–10.57) [162,176].

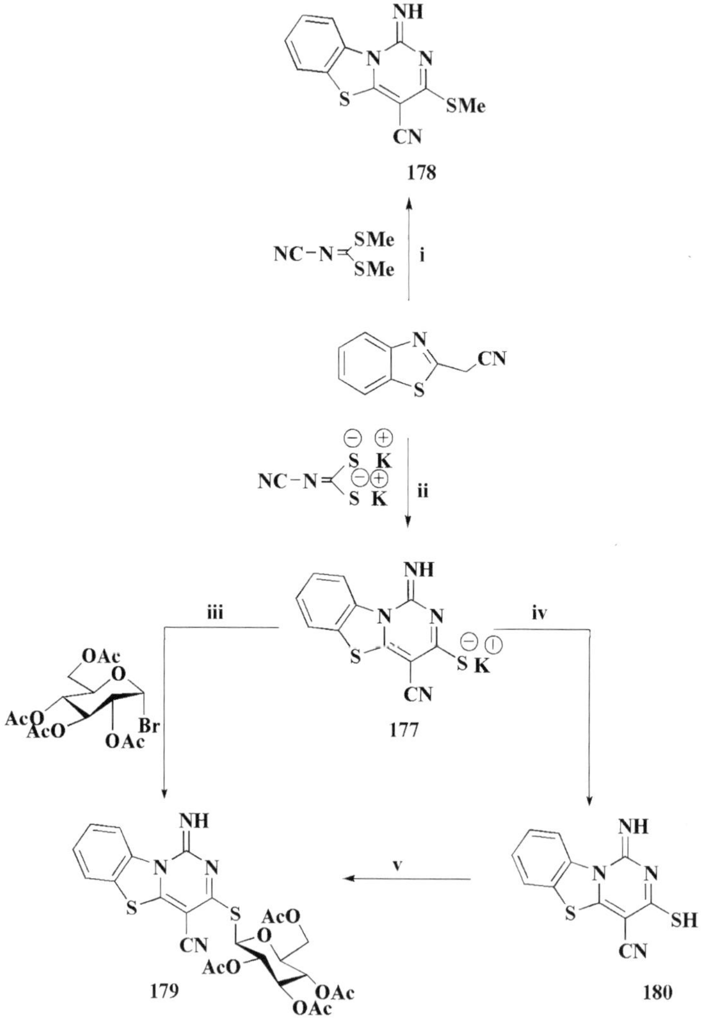

Scheme 10.48 Reagents and conditions: (i) dioxane/KOH/RT/1 h; (ii) C_2H_5ONa/EtOH/reflux/3 h; (iii) DMF/RT/24 h; (iv) HCl/RT; (v) DMF/KOH/RT/24 h.

The synthesis of benzothiazole-bearing *N*-sulfonamide 2-pyridone derivatives was made through reacting benzothiazole sulfonylhydrazide with sodium salts of both (hydroxymethylene) cycloalkanones and unsaturated ketones, as well as ethoxymethylene derivatives (Schemes 10.58 and 10.59) [177].

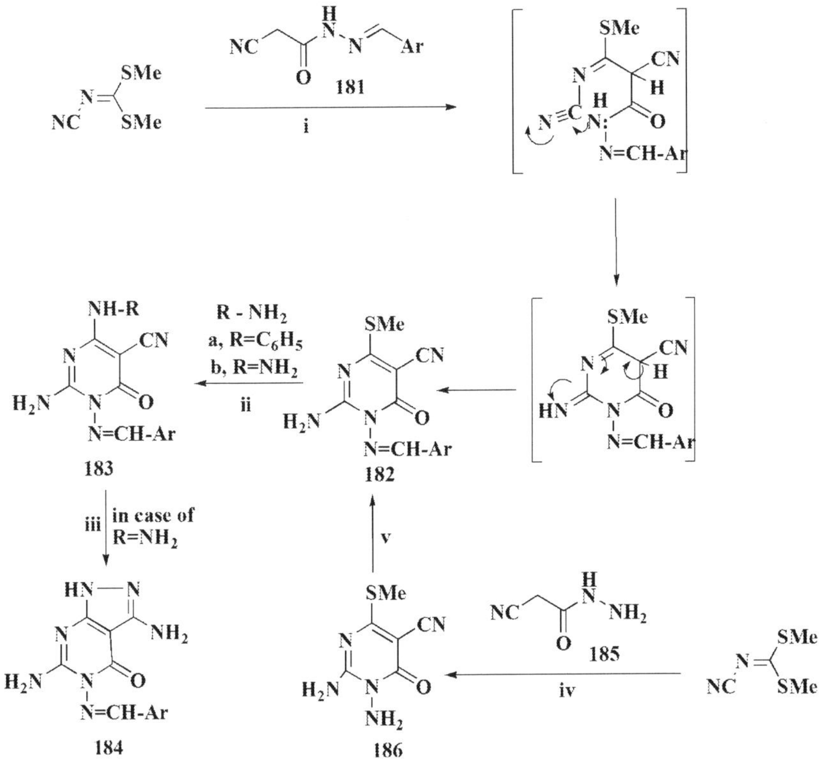

181 - 184, 186	Ar		
a	C_6H_5	d	$4\text{-}H_3CO\text{-}C_6H_4$
b	$4\text{-}H_3C\text{-}C_6H_4$	e	$2\text{-}Cl\text{-}C_6H_4$
c	$2\text{-}HO\text{-}C_6H_4$	f-	$4\text{-}Cl\text{-}C_6H_4$

Scheme 10.49 Reagents and conditions: (i) dioxane/KOH/RT/24 h; (ii) reflux; (iii) EtOH/pip./reflux; (iv) dioxane/KOH/RT/24 h; (v) EtOH/pip./reflux.

Sulfonamides and trimethoprim (TMP) drugs are normally utilized to inhibit the action of dihydrofolate reductase (DHFR) and dihydropteroate synthase (DHPS) enzymes, respectively. A novel series of *N*-sulfonamide 2-pyridone derivatives were synthesized. The synthesis was performed through reacting benzothiazole sulfonyl hydrazide with ketene dithioacetal analogs (Schemes 10.60 and 10.61) [178].

The synthesis of the pyrido[2,1-*b*]benzothiazole **232** was performed through reacting *N*-aryl-2-cyano-3,3-bis(methylthio)acrylamide with benzothiazol-2-ylacetonitrile, while *N*-substituted 2-pyridylbenzothiazole

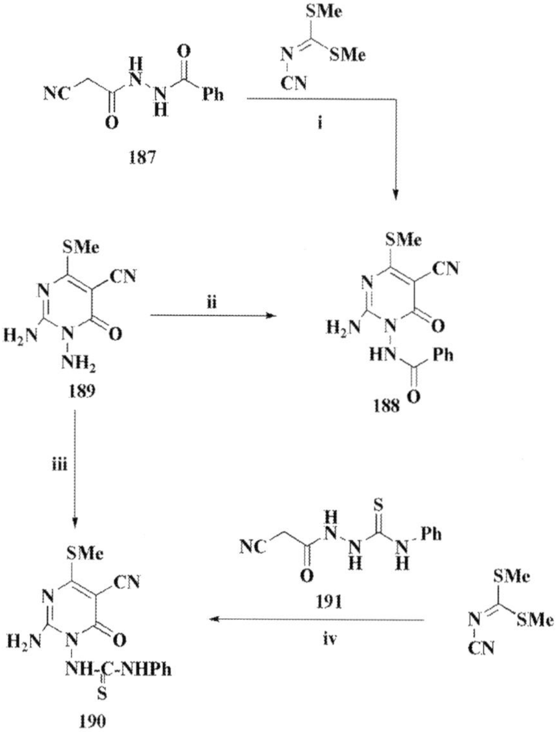

Scheme 10.50 Reagents and conditions: (i) dioxane/KOH/RT/24 h; (ii) PhCOCl/EtOH/pip./reflux; (iii) PhNCS/EtOH/pip./reflux; (iv) dioxane/KOH/RT/24 h.

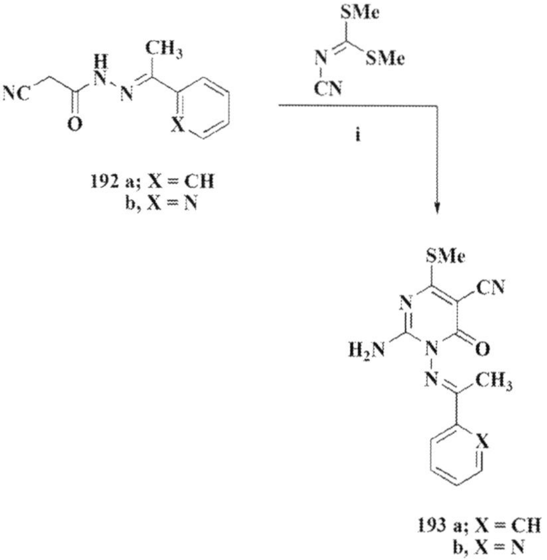

Scheme 10.51 Reagents and conditions: (i) EtOH/KOH/reflux.

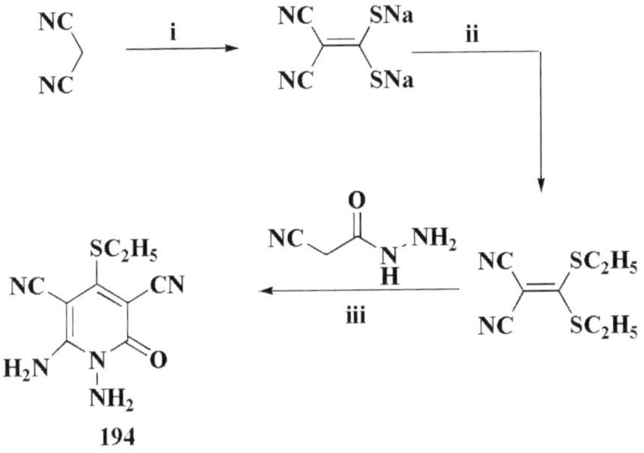

Scheme 10.52 Reagents and conditions: (i) C_2H_5ONa/CS_2/RT; (ii) C_2H_5I/EtOH/reflux/1 h; (iii) dioxane/KOH/RT/12 h.

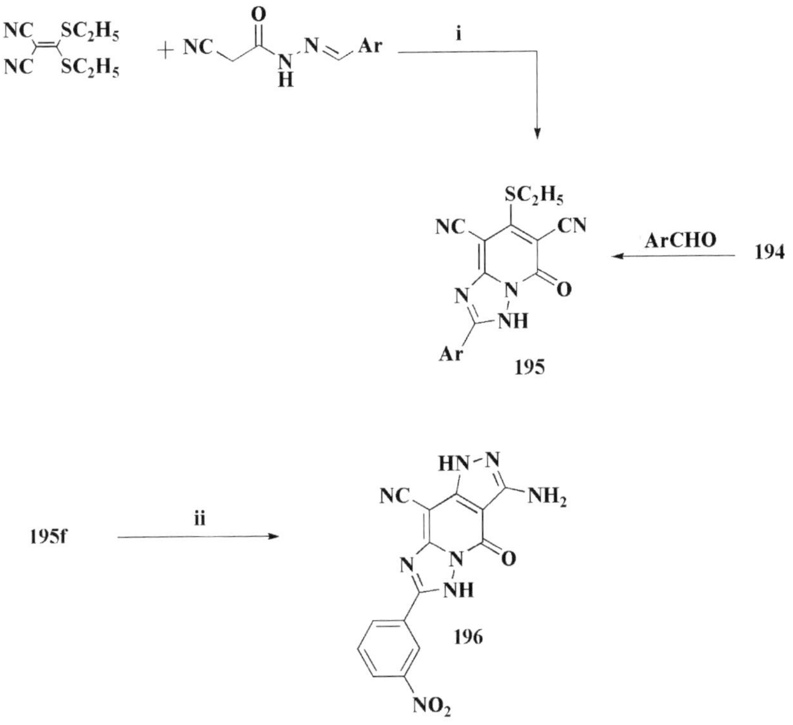

195	Ar		
a	C_6H_5	d	4-CH_3-C_6H_4
b	4-Cl-C_6H_4	e	4-F-C_6H_4
c	4-OCH_3-C_6H_4	f	3-NO_2-C_6H_4

Scheme 10.53 Reagents and conditions: (i) dioxane/KOH/RT/12 h; (ii) hydrazine hydrate/pip./EtOH/reflux/1 h.

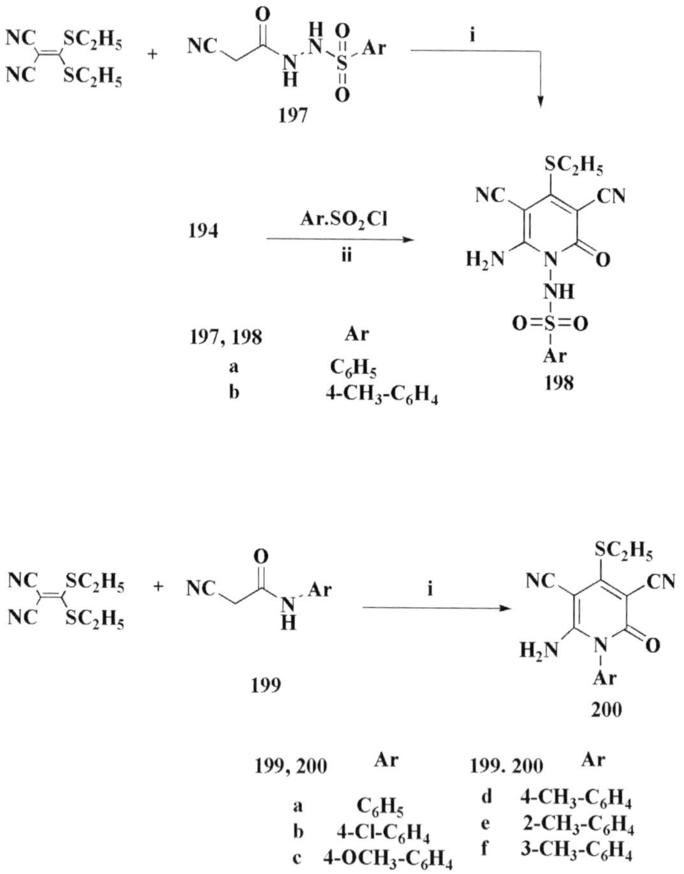

197, 198	Ar
a	C_6H_5
b	$4\text{-}CH_3\text{-}C_6H_4$

199, 200	Ar	199. 200	Ar
a	C_6H_5	d	$4\text{-}CH_3\text{-}C_6H_4$
b	$4\text{-}Cl\text{-}C_6H_4$	e	$2\text{-}CH_3\text{-}C_6H_4$
c	$4\text{-}OCH_3\text{-}C_6H_4$	f	$3\text{-}CH_3\text{-}C_6H_4$

Scheme 10.54 Reagents and conditions: (i) dioxane/KOH/RT/12 h; (ii) EtOH/reflux/1 h.

derivatives were furnished via the reaction of 2-(benzo[*d*]thiazol-2-yl)-3-(dimethylamino)acrylonitrile with either 2-cyano-*N*′-(4-substituted benzylidene)acetohydrazide, cyanoacetamide, or aryl cyanoacetamides. The synthesized *N*-substituted 2-pyridylbenzothiazole derivatives displayed significant fluorescence properties based on the steady-state fluorescence measurements (Scheme 10.62) [179].

The synthesis of N-arylsulfonylamino derivatives of 2-pyridone was accomplished. N-substituted cyanoacetyl sulfonylhydrazide reacted with compound **233** in piperidine acetate, to furnish compound **235** (Scheme 10.63) [180,181].

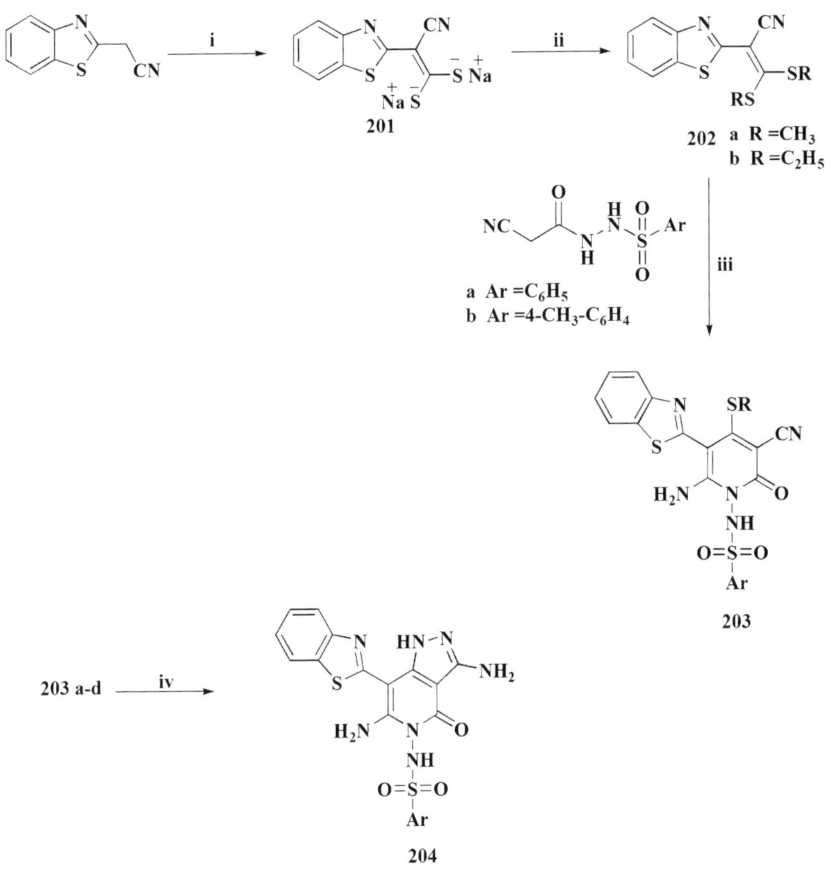

202 a R $=CH_3$
b R $=C_2H_5$

a Ar $=C_6H_5$
b Ar $=4\text{-}CH_3\text{-}C_6H_4$

203	Ar	R
a	C_6H_5	CH_3
b	$4\text{-}CH_3\text{-}C_6H_4$	CH_3
c	C_6H_5	C_2H_5
d	$4\text{-}CH_3\text{-}C_6H_4$	C_2H_5

204	Ar
a	C_6H_5
b	$4\text{-}CH_3\text{-}C_6H_4$

Scheme 10.55 Reagents and conditions: (i) CS_2/C_2H_5ONa/EtOH/reflux/20 min; (ii) RI/EtOH/RT/24 h; (iii) KOH/DMF/reflux/3 h; (iv) hydrazine hydrate/DMF/pip./reflux/2 h.

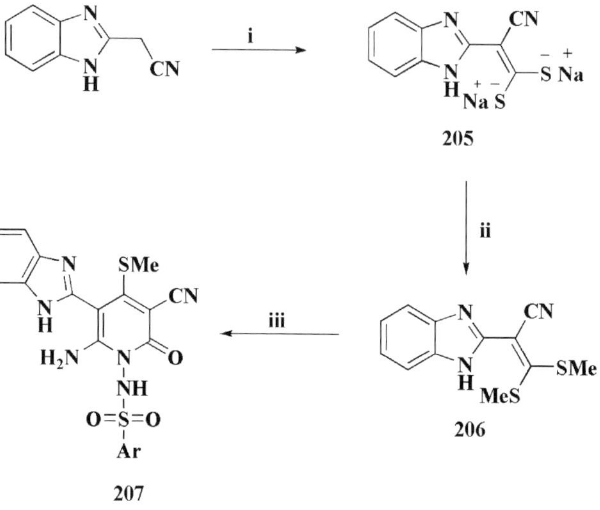

207	Ar
a	C_6H_5
b	4-CH_3-C_6H_4

Scheme 10.56 Reagents and conditions: (i) CS_2/C_2H_5ONa/EtOH/reflux/20 min; (ii) CH_3I/RT/24 h; (iii) KOH/dioxane/RT/24 h.

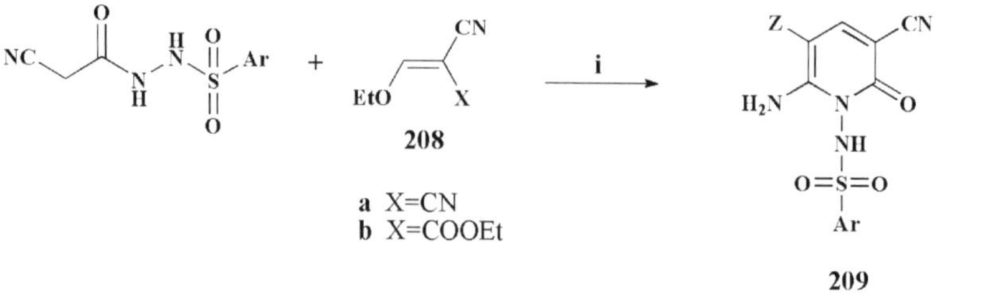

208

a X=CN
b X=COOEt

209	Ar	Z
a	C_6H_5	CN
b	4-CH_3-C_6H_4	CN
c	C_6H_5	COOEt
d	4-CH_3-C_6H_4	COOEt

Scheme 10.57 Reagents and conditions: (i) C_2H_5ONa/EtOH/reflux/45 min.

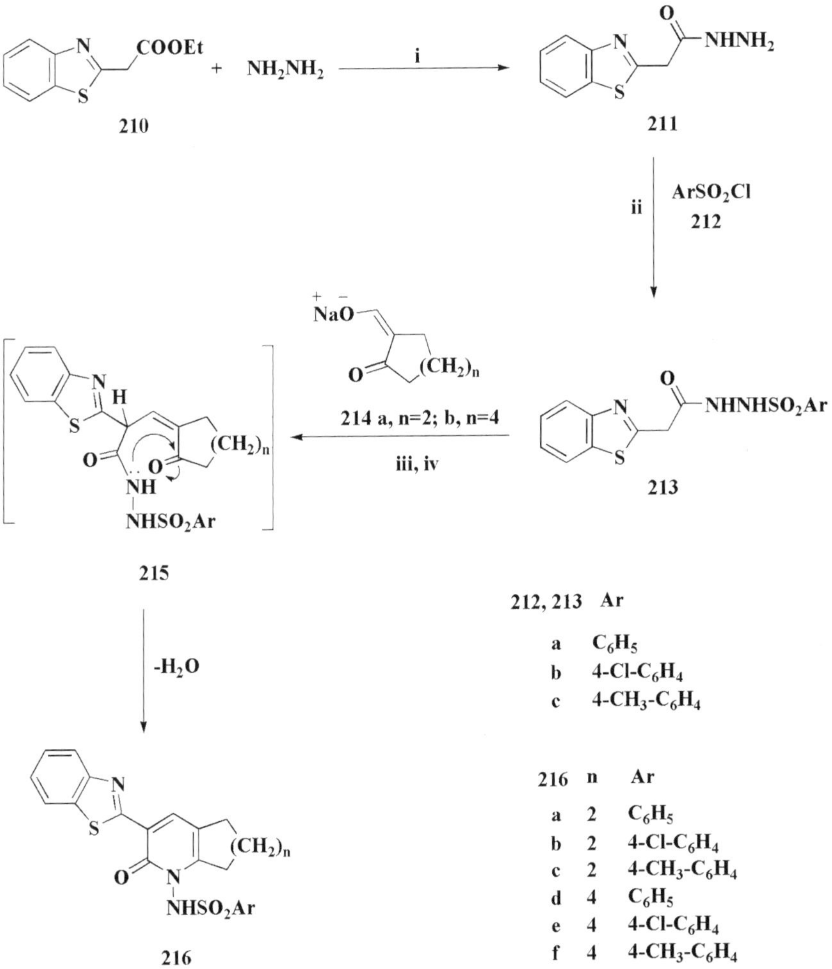

212, 213	Ar
a	C_6H_5
b	$4\text{-}Cl\text{-}C_6H_4$
c	$4\text{-}CH_3\text{-}C_6H_4$

216	n	Ar
a	2	C_6H_5
b	2	$4\text{-}Cl\text{-}C_6H_4$
c	2	$4\text{-}CH_3\text{-}C_6H_4$
d	4	C_6H_5
e	4	$4\text{-}Cl\text{-}C_6H_4$
f	4	$4\text{-}CH_3\text{-}C_6H_4$

Scheme 10.58 Reagents and conditions: (i) EtOH/RT/12 h; (ii) pyridine/RT/3 h; (iii) EtOH/pip. Acetate/reflux/10 min.; (iv) AcOH/reflux/12 h.

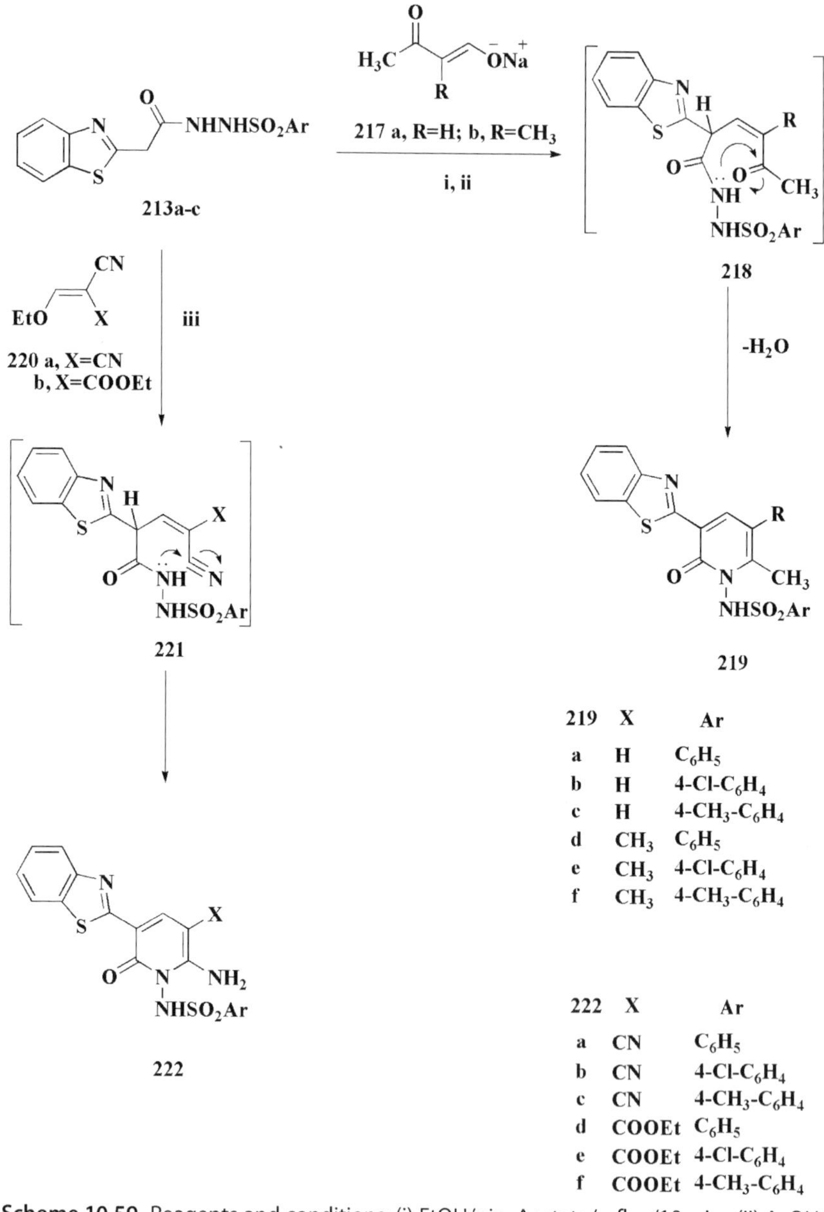

219	X	Ar
a	H	C_6H_5
b	H	4-Cl-C_6H_4
c	H	4-CH_3-C_6H_4
d	CH_3	C_6H_5
e	CH_3	4-Cl-C_6H_4
f	CH_3	4-CH_3-C_6H_4

222	X	Ar
a	CN	C_6H_5
b	CN	4-Cl-C_6H_4
c	CN	4-CH_3-C_6H_4
d	COOEt	C_6H_5
e	COOEt	4-Cl-C_6H_4
f	COOEt	4-CH_3-C_6H_4

Scheme 10.59 Reagents and conditions: (i) EtOH/pip. Acetate/reflux/10 min.; (ii) AcOH/reflux/12 h; (iii) EtOH/C_2H_5ONa/reflux/2 h.

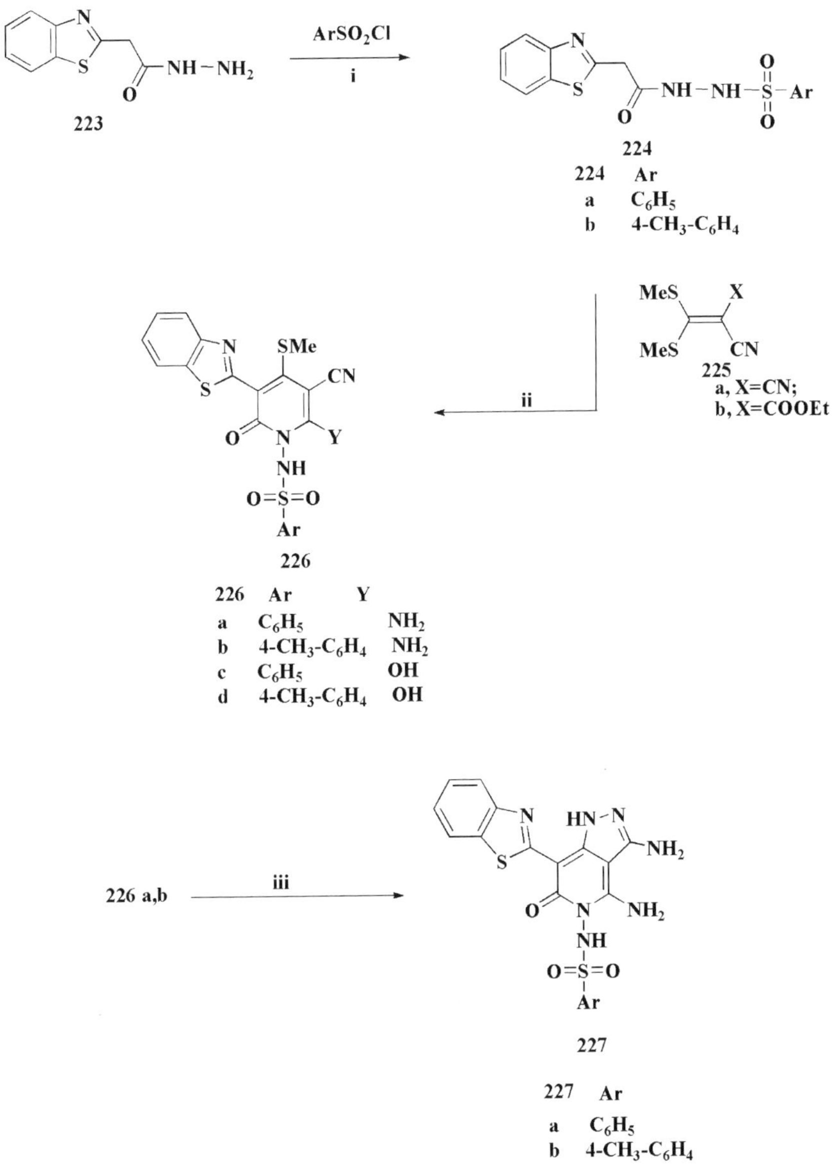

223 —($ArSO_2Cl$, i)→ 224

224	Ar
a	C_6H_5
b	$4\text{-}CH_3\text{-}C_6H_4$

225: a, X=CN; b, X=COOEt

224 + 225 —(ii)→ 226

226	Ar	Y
a	C_6H_5	NH_2
b	$4\text{-}CH_3\text{-}C_6H_4$	NH_2
c	C_6H_5	OH
d	$4\text{-}CH_3\text{-}C_6H_4$	OH

226 a,b —(iii)→ 227

227	Ar
a	C_6H_5
b	$4\text{-}CH_3\text{-}C_6H_4$

Scheme 10.60 Reagents and conditions: (i) pyridine/RT/3 h; (ii) KOH/DMF/reflux/2 h; (iii) hydrazine hydrate/DMF/pip./reflux/2 h.

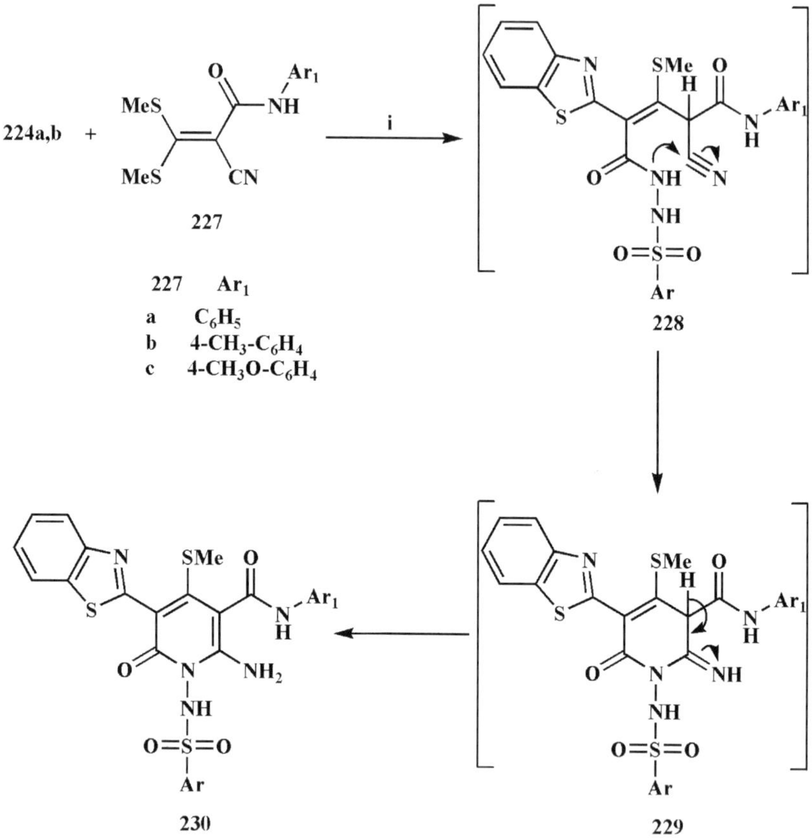

224a,b + 227 →(i) [228] → [229] → 230

227	Ar_1
a	C_6H_5
b	$4\text{-}CH_3\text{-}C_6H_4$
c	$4\text{-}CH_3O\text{-}C_6H_4$

228-230	Ar	Ar_1
a	C_6H_5	C_6H_5
b	$4\text{-}CH_3\text{-}C_6H_4$	C_6H_5
c	C_6H_5	$4\text{-}CH_3\text{-}C_6H_4$
d	$4\text{-}CH_3\text{-}C_6H_4$	$4\text{-}CH_3\text{-}C_6H_4$
e	C_6H_5	$4\text{-}CH_3O\text{-}C_6H_4$
f	$4\text{-}CH_3\text{-}C_6H_4$	$4\text{-}CH_3O\text{-}C_6H_4$

Scheme 10.61 Reagents and conditions: (i) KOH/dioxane/reflux/4–8 h.

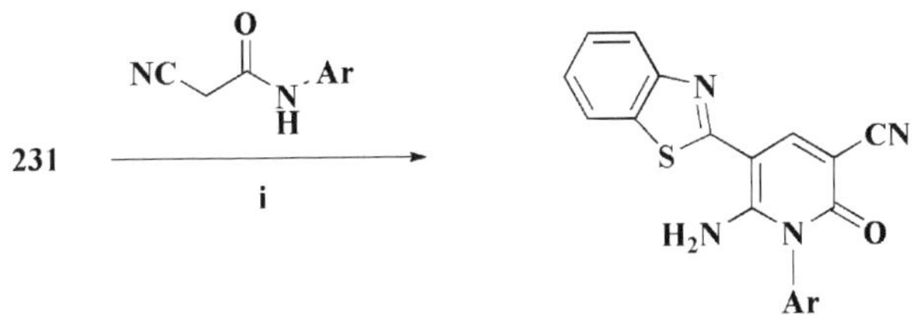

231 → (i) NC-CH2-C(=O)-NH-Ar → 232

232	Ar
a	C_6H_5
b	4-Cl-C_6H_4
c	4-CH_3-C_6H_4
d	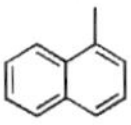 1-naphthyl

Scheme 10.62 Reagents and conditions: (i) KOH/dioxane/reflux/2 h.

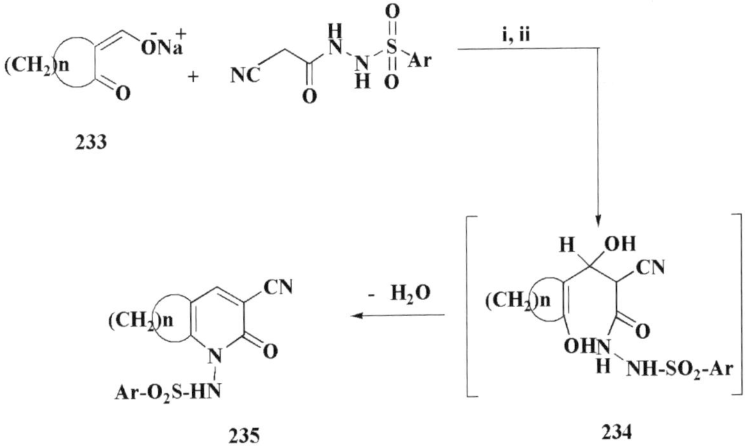

233 + NC-CH2-C(=O)-NH-NH-SO_2-Ar → (i, ii) → [234] → $-H_2O$ → 235

235	n	Ar	235	n	Ar
a	3	C_6H_5	d	4	4-H_3C-C_6H_4
b	3	4 -H_3C-C_6H_4	e	5	C_6H_5
c	4	C_6H_5	f	5	4 -H_3C-C_6H_4

Scheme 10.63 Reagents and conditions: (i) pip. acetate; (ii) HCl.

References

[1] Abdallah A, Elgemeie G, Ahmed E. Synthesis, docking and antimicrobials evaluation of novel pyrazolotriazines as RNA polymerase inhibitors. Curr Org Chem 2021;25 (14):1715–30.

[2] Elgemeie GH. Thioguanine, mercaptopurine, their analogs, and nucleosides as antimetabolites. Curr Pharm Des 2003;9(31):2627–42.

[3] Elgemeie GH, Mohamed RA. New anticancer pyrimidine nucleoside analogs. Germany: Lap Lampert Academic Publishing; 2016. p. 1–300. ISBN:978-3330-00001-8.

[4] Elgemeie GH, Mohamed RA. New antimetabolites: strategies for discovery. Germany: Lap Lampert Academic Publishing; 2017. p. 1–681. ISBN: 978-3330-03593-5.

[5] Elgemeie GH, Elkaradawy SY. Novel mercaptopurine and thioguanine analogues: the reaction of dimethyl N-cyanodithioiminocarbonate with oxo- and amino diazoles. Synth Commun 2004;34:805.

[6] Elgemeie GH, Ali HA, Mansour AK. Antimetabolites: a convenient synthesis of mercaptopurine and thioguanine analogues. Phosphorus Sulfur Silicon 1994;90:143.

[7] Attia AM, Elgemeie GH. Nucleic acid related compounds: a convenient synthesis of 3 deazauridine analogues. Nucleosides Nucleotides 1995;14:1211.

[8] Elgemeie GH, Hussain BA. A convenient synthesis of 5-deaza nonclassical antifolates: reaction of cyanothioacetamide with sodium salts of 2-(hydroxymethylene)-1-cycloalkanones. Tetrahedron 1994;50:199.

[9] Elgemeie GH, Fathy NM, Hopf H, Jones PG. 1,2,3,4-Tetrahydrobenzimidazo[2,1-b] quinazoline. Acta Crystallogr 1998;C54:1109.

[10] Elgemeie GH, Mansour OA, Metwally NH. Synthesis and anti-HIV activity of different novel nonclassical nucleosides. Nucleosides Nucleotides 1999;18:113.

[11] Sharaf MF, Abdel-Aal FA, Elgemeie GH, El-Damaty A. Reactions with diethyl acetonedicarboxylate: novel synthesis of pyrazolo[3,4-d]pyridazine derivatives. Arch Pharm (Weinheim) 1991;324:585.

[12] Elgemeie GH, Saber NM. Synthesis and in vitro anti-tumor activity of a new class of acyclic thioglycosides. Nucleosides Nucleotides 2015;34:463–74.

[13] Elgemeie GH, Abu-Zaied MA, Hebishy A, Abbas N, Hamed M. A first microwave-assisted synthesis of a new class of purine and guanine thioglycoside analogs. Nucleosides Nucleotides 2016;35:459–78.

[14] Elgemeie GH. Unpublished data, grant no. 90051, "USA-Egyptian coordination unit"; 1991.

[15] Elgemeie GH, El-Ezbawy SE, Ali HA, Mansour AK. Novel synthesis of mercaptopurine and pentaaza-as-indacene analogues: reaction of [bis(methylthio)methylene]malononitrile and ethyl 2-cyano-3,3-bis(methylthio)acrylate with 5-aminopyrazoles. Bull Chem Soc Jpn 1994;67:738.

[16] Elgemeie GH, Ali HA, Elzanate AM. The reaction of cyanoketene dithioacetals with imidazolidinones: a synthesis of methylsulfanylpyrrolo[1,2-c]imidazolones. J Chem Res 1996;340.

[17] Elgemeie GH, Elghandour AH, Elzanate AM, Hussain AA. Synthetic strategies to novel condensed methylsulfanylazoles: reaction of ketene dithioacetals with amino- and oxo-azoles. J Chem Res 1997;256.

[18] Elgemeie GH, Elghandour AH, Elzanate AM, Ahmed SA. Novel synthesis of thioguanine and sulfanylpurine analogues: reaction of heterocyclic ketene dithio-acetals with nucleophiles. J Chem Res 1998;162–3.

[19] Elgemeie GH, Metwally NH. Synthesis of structurally related purines: benzimidazo [1,2-a]pyridines, Benzimidazo[1,2-c]pyrimidines, and pyrazolo[1,5-a]pyrimidines. Monatsh Chem 2000;131:779.

[20] Elgemeie GH, El-Ezbawy SR, El-Aziz HA. The design and synthesis of structurally related mercaptopurine analogues: reaction of dimethyl N-cyano-dithioiminocarbonate with 5-aminopyrazoles. Synth Commun 2001;31:3453.

[21] Elgemeie GH, Sood SA. The reaction of dimethyl N-cynothioiminocarbonate with amino- and oxo-azoles: a new general synthesis of methylsulfanylazoloazines. J Chem Res 2001;439.

[22] Elgemeie GH, El-Aziz HA. Potential purine analogue antagonists: synthesis of novel cycloalkane ring-fused pyrazolo[1,5-a]pyrimidines. Synth Commun 2002; 32:253.

[23] Elgemeie GH, Elghandour AH, Ali HA, Hussein AM. Novel 2-thioxohydantoin ketene dithioacetals: versatile intermediates for synthesis of methylsulfanylimidazo [4,5-c]pyrazoles and methylsulfanylpyrrolo[1,2-c]imidazoles. Synth Commun 2002;32:2245.

[24] Elgemeie GH, Elghandour AH, Elzanate AM, Ahmed SA. Novel intramolecular cyclization of pyrazolone ketene S,N-acetals for the construction of methylsulfanyl-pyrazolo[4,3-b]pyridines. Synth Commun 2002;32:3509.

[25] Elgemeie GH, Sood SA. A novel synthesis of N-aryl-6-methylsulfanyl-4-pyrimidinones and purine analogues: the reaction of dimethyl N-cyanothioiminocarbonate with cyanoacetanilide. Synth Commun 2003;33:2095.

[26] Azzam RA, Elgemeie GH, Ramadan R, Jones PG. Crystal structure of 2-cyano-3, 3-bis-(ethyl-sulfan-yl)-*N*-*O*-tolyl-acryl-amide. Acta Crystallogr Sect E Crystallogr Commun 2017;73:752–4.

[27] Elgemeie GH, Abu-Zaied MA, Loutfy SA. 4-Aminoantipyrine in carbohydrate research: design, synthesis and anticancer activity of a novel class of derivatives of 4-aminoantipyrine thioglycosides and their corresponding pyrazolopyrimidine and pyrazolopyridine thioglycosides. Tetrahedron 2017;73:5853–61.

[28] Elgemeie GH, El-Ezbawy SR, Ali HA. Reactions of chlorocarbonyl isocyanate with 5-aminopyrazoles and active methylene nitriles: a novel synthesis of pyrazolo[1,5-a]-1,3,5-triazines and barbiturates. Synth Commun 2001;31(22):3459–67.

[29] Elgemeie GH, Ali HA. Potential purine analogue antagonists: synthesis of novel cycoalkane ring-fused pyrazolo[1,5-a]pyrimidines. Synth Commun 2002;32(2): 253–64.

[30] Elgemeie GH, Mahdy EM, Elgawish MA, Ahmed MM, Shousha WG, Eldin ME. A new class of antimetabolites: pyridine thioglycosides as potential anticancer agents. Z Naturforsch C J Biosci 2010;65(9–10):577–87.

[31] Elgemeie GH, Hussein MM, Jones PG. 2-(2′,3′,4′,6′-tetra-O-acetyl-β-D-glucopyranosylthio)-4-pyridin-4-yl 6,7,8,9-tetrahydro-5H-cyclohepta[b] pyridin-3-carbonitrile. Acta Crystallogr 2002;E58:1244.

[32] Masoud DM, Hammad SF, Elgemeie GH, Jones PG. Crystal structure of 4,6-dimethyl-2-(3,4,5-trihydroxy-6 (hydroxymethyl)tetrahydro-2H-pyran-2-ylthio) nicotinonitrile. Acta Crystallogr 2017;E73:1751–4.

[33] Hammad SF, Masoud DM, Elgemeie GH, Jones PG. Crystal structure of racemic 2-[(β-arabinopyranosyl)sulfanyl]-4,6-diphenylpyridine-3-carbonitrile. Acta Crystallogr 2018;E74:853–6.

[34] Abu-Zaied MA, Elgemeie GH, Jones PG. Crystal structure of 4,6-dimethyl-2-[(2,3,4,6-tetra-O-acetyl-β-D-galactopyranosyl)sulfanyl]pyrimidine. Acta Crystallogr 2019;E75:1820–3.

[35] Elgemeie GH, Attia AM. A new class of dihydropyridine thioglycosides via piperdinium salts. Synth Commun 2003;33:2243.

[36] Attia AM, Elgemeie GH. First gylcoside synthesis via piperidinium salts of heterocyclic nitrogen bases: the synthesis of a new class of dihydropyridine thioglycosides. J Carbohydr Chem 2002;21:325–39.

[37] Scala S, Akhmed N, Rao US, Paull K, Lan LB, Dickstein B, Lee JS, Elgemeie GH, Stein WD, Bates SE. P-glycoprotein substrates and antagonists cluster into two distinct groups. Mol Pharmacol 1997;51(6):1024–33.
[38] Abu-Zaied MA, Loutfy SA, Hassan AE, Elgemeie GH. Novel purine thioglycoside analogs: synthesis, nanoformulation and biological evaluation in in-vitro human liver and breast cancer models. Drug Des Dev Ther 2019;13:2437–57.
[39] Elnagdi MH, Elfahham HA, Ghozlan SA, Elgemeie GH. Activated nitriles in heterocyclic synthesis: a new procedure for the synthesis of pyrimidine derivatives. J Chem Soc, Perkin Trans 1 1982;2667.
[40] Elnagdi MH, Elgemeie GH, Elmoghayer MR. Chemistry of pyrazolopyrimidines. In: Katritzky AA, editor. Advances in heterocyclic chemistry, vol. 41. USA: Academic Press; 1987. p. 319–76.
[41] Elgemeie GH, Fathy NM, Faddah LM, Ebeid MY. Reactions with 3,5-diaminopyrazoles: new routes to pyrazolo[1,5-a]-pyrimidines. Arch Pharm (Weinheim) 1991;324:149.
[42] Elgemeie GH, Kamal EA. Pyrimidinethione nucleosides and their deaza analogues. Nucleosides Nucleotides 2002;21:287.
[43] Elgemeie GH, Attia AM, Al-Kabai SS. Nucleic acid components and their analogues: new synthesis of bicyclic thiopyrimidine nucleosides. Nucleosides Nucleotides 2000;19:723.
[44] Elgemeie GH, Gohar AM, Regiala HA, Elfahham HA. Activated nitriles in heterocyclic synthesis: novel syntheses of pyrano[2,3-b]pyridines and pyrano[2,3-d] pyrimidines. Arch Pharm (Weinheim) 1988;321(131).
[45] Hammad MA, Nawwar GA, Elgemeie GH, Elnagdi MH. Nitriles in heterocyclic synthesis: novel synthesis of benzo[g]-imidazo[1,2-c]pyrimidines and benzo[g]imidazo [1,2-a]pyridine derivatives. Heterocycles 1985;23:2177.
[46] Gaber HM, Elgemeie GH, Ouf SA, Sherif SM. Heterocyclic synthesis with 4-hydrazinopyridothienopyrimidines: synthesis of pyridothienotriazolopyrimidines and heterocyclic pyridothienopyrimidines with biological interest. Heteroat Chem 2005;16:298.
[47] Abu-Zaied MA, Elgemeie GH. A facile synthesis of novel pyrazolopyrimidine thioglycosides as purine thioglycoside analogues. Nucleosides Nucleotides 2018;37: 67–77.
[48] Abdallah AEM, Elgemeie GH. Design, synthesis, docking, and antimicrobial evaluation of some novel pyrazolo[1,5-a]pyrimidines and their corresponding cycloalkane ring-fused derivatives as purine analogs. Drug Des Devel Ther 2018;12:1785–98.
[49] Elgemeie GH, Farag AB. Design, synthesis, and in vitro anti-hepatocellular carcinoma of novel thymine thioglycoside analogs as new antimetabolic agents. Nucleosides Nucleotides Nucleic Acids 2017;36:355–77.
[50] Elgemeie GH, Alkhursani SA, Mohamed RA. New synthetic strategies for acyclic and cyclic pyrimidinethione nucleosides and their analogues. Nucleosides Nucleotides 2019;38:12–87.
[51] Elgemeie GH, Abu-Zaied M, Alsaid S, Hebishy A, Essa H. Novel nucleoside analogues: first synthesis of pyridine-4-thioglycosides and their cytotoxic evaluation. Nucleosides Nucleotides Nucleic Acids 2015;34:659–73.
[52] Elgemeie GH, Fathy NM, Farag AB, Kursani SA. Novel synthesis of dihydropyridine thioglycosides and their cyctotoxic activity. Nucleosides Nucleotides 2017;36: 198–212.
[53] Elgemeie GH, El-Enany MM, Ismail MM, Ahmed EK. Nucleic acid components and their analogues: a novel and efficient method for the synthesis of a new class of bipyridyl and biheterocyclic-nitrogen thioglycosides from pyridine-2(1H)-thiones. Nucleosides Nucleotides Nucleic Acids 2002;21:477–93.

[54] Elgemeie GH, Abu-Zaied MA, Nawwar GA. First novel synthesis of triazole thioglycosides as ribavirin analogues. Nucleosides Nucleotides Nucleic Acids 2018;37:112–23.
[55] Abu-Zaied MA, Elgemeie GH. Thiazoles in glycosylation reactions: novel synthesis of thiazole thioglycosides. Heteroat Chem 2017;28, e21404.
[56] Abu-Zaied MA, El-Telbani EM, Elgemeie GH, Nawwar GA. Synthesis and in vitro anti-tumor activity of new oxadiazole thioglycosides. Eur J Med Chem 2011;46:229–35.
[57] Elgemeie GH, Farag AB, Amin KM, El-Badry OM, Hassan GS. Design, synthesis and cytotoxic evaluation of novel heterocyclic thioglycosides. J Med Chem 2014;4:814–20.
[58] Abu-Zaied MA, Elgemeie GH. Synthesis of the first novel pyrazole thioglycosides as deaza ribavirin analogues. Nucleosides Nucleotides Nucleic Acids 2017;36:713–25.
[59] Elgemeie GH, Zaghary WA, Amin KM, Nasr TM. First synthesis of thienopyrazole thioglycosides. J Carbohydr Chem 2008;27:345.
[60] Abu-Zaied MA, Hammad SF, Halaweish FT, Elgemeie GH. Sofosbuvir thio-analogues: synthesis and antiviral evaluation of the first novel pyridine- and pyrimidine-based thioglycoside phosphoramidates. Am Chem Soc (Omega) 2020;5:14645–55.
[61] Elgemeie GH, Elghandour AH, Elzanate AM, Ahmed SA. Synthesis of some novel α-cyanoketene S,S-acetals and their use in heterocyclic synthesis. J Chem Soc, Perkin Trans 1 1997;3285.
[62] Elgemeie GH, Abd-Elaziz GW. Recent trends in synthesis of heterocycles using ketene dithioacetals with electron attracting groups. Targets Heterocycl Syst 2002;6:325–68.
[63] Elgemeie GH, Elghandour AH, Abd-Elaziz GW. Novel synthesis of heterocyclic ketene N,N-, N,O-, and N,S-acetals using cyanoketene dithioacetals. Synth Commun 2003;33:1659.
[64] Elgemeie GH, Elghandour AH, Elzanaty AM, Ahmed SA. Novel 1,3-dithiolanes using sodium α-cyano-ketene dithiolates. Synth Commun 2006;36:755.
[65] Helal MH, Elgemeie GH, Masoud DM. Preparation and characterization of novel methylsulfanyl-pyrazolopyrimidines as heterocyclic dyes from ketene dithioacetals. Pigm Resin Technol 2007;36:306.
[66] Elgemeie GH, Ali HA, Elghandour AH, Hussein AM. Synthesis of benzimidazole ketene N,S-acetals and their reactions with nucleophiles. Synth Commun 2003;33:555.
[67] Elgemeie GH, Ahmed KA, Ahmed EA, Helal MH, Masoud DM. Synthesis of novel tetrasubstituted thiophenes based dye using sodium α-cyanoketene dithiolates as starting materials. Pigm Resin Technol 2015;44:339–46.
[68] Elgemeie GH, Ahmed SH. Synthesis and chemistry of dithioles. Synthesis 1747; 2001.
[69] Elgemeie GH, Abou-Zeid M, Azzam R. Antimetabolites: a first synthesis of a new class of cytosine thioglycoside analogs. Nucleosides Nucleotides 2016;35:211–22.
[70] Elgemeie GH, Salah AM, Abbas NS, Hussein HA, Mohamed RA. Nucleic acid components and their analogs: design and synthesis of novel cytosine thioglycoside analogs. Nucleosides Nucleotides 2017;36:139–50.
[71] Elgemeie GH, Farag AB. Design, synthesis and in vitro anti-hepatocellular carcinoma of novel thymine thioglycoside analogs as new antimetabolic agents. Nucleosides Nucleotides 2017;36:328–42.
[72] Abu-Zaied MA, Elgemeie GH, Mahmoud NM. Synthesis of novel pyrimidine thioglycosides as structural analogs of favipiravir (avigan) and their antibird flu virus activity. Nucleosides Nucleotides Nucleic Acids 2021;40(3):336–56.

[73] Abu-Zaied MA, Elgemeie GH, Mahmoud NM. Anti-covid-19 drug analogues: synthesis of novel pyrimidine thioglycosides as antiviral agents against SARS-COV-2 and avian influenza H5N1 viruses. ACS Omega 2021;6(26):16890–904.
[74] Abu-Zaied MA, Mahmoud NM, Elgemeie GH. Toward developing therapies against corona virus: synthesis and anti-avian influenza virus activity of novel cytosine thioglycoside analogues. Am Chem Soc (Omega) 2020;5:20042–50.
[75] Elgemeie GH, Mohamed RA. Recent trends in synthesis of five- and six-membered heterocycles using dimethyl N-cyanodithioiminocarbonate. Heterocycl Commun 2014;20:257.
[76] Elgemeie GH, Mohamed RA. Application of dimethyl *N*-cyanodithioiminocarbonate in synthesis of fused heterocycles and in biological chemistry. Heterocycl Commun 2014;20:313.
[77] Elgemeie GH, Mohamed RA. Microwave chemistry: synthesis of purine and pyrimidine nucleosides using microwave radiation. J Carbohydr Chem 2019; 38:20–66.
[78] Elgemeie GH, Ahmed MM. A new class of biheterocyclic thioglycosides from pyridine-2(1H) thiones. Nucleosides Nucleotides 2002;21:837.
[79] Elgemeie GH, Heikel AZ, Ahmed MA. A direct route to 2-(β-D-ribofuranosylthio) pyridine glycosides. Nucleosides Nucleotides 2002;21:411.
[80] Elgemeie GH, Elfahham HA, Nabey HA. α,β-Unsaturated nitriles in heterocyclic synthesis: novel synthesis of pyridines and thieno[2,3-b]pyridines derivatives. Bull Chem Soc Jpn 1988;61:4431.
[81] Nawwar GA, Elbayouki KM, Elgemeie GH, Elnagdi MH. Nitriles in heterocyclic synthesis: new routes for synthesis of pyran, pyridine and pyrrole derivatives. Heterocycles 1985;23:2983.
[82] Elgemeie GH, El-Enany MM, Ahmed EK. Nucleic acid components and their analogues: a novel and efficient method for the synthesis of a new class of bipyridyl and biheterocyclic-nitrogen thioglycosides from pyridine-2(1H)-thiones. Nucleosides Nucleotides 2002;21:477.
[83] Elgemeie GH, Helal MH, Abbas EM, Abdel Mowla EA. Novel pyridine-2(1H)-thione and thieno-[2,3-b]pyridine derivatives containing arylazo moiety: synthesis, characterization and dyeing properties. Pigm Resin Technol 2002;31:365.
[84] Elgemeie GH, Ahmed SH. A novel synthesis of thiazoles and thiazolopyridines using N-cyanoacetoarylsulfonylhydrazid
es. Synth Commun 2003;33:535.
[85] Elgemeie GH, Jones PG. 6-Amino-4-(methylsulfanyl)-2-oxo-1-tolyl-1,2-dihydropyridine-3,5-dicarbonitrile. Acta Crystallogr 2004;E60:02107.
[86] Elgemeie GH, Abu-Zaied M, Alsaid S, Hebishy A, Essa H. Novel nucleoside analogues: first synthesis of pyridine 4-thioglycosides and their cytotoxic evaluation. Nucleosides Nucleotides 2015;34:659.
[87] Elgemeie GH, Jones PG. Crystal structure of 1-amino-2-oxo-2,5,6,7,8,9-hexahydro-1H-cyclohepta[b]pyridine-3-carbonitrile. Acta Crystallogr 2016;E72: 1239.
[88] Elgemeie GH, Fathy NM, Farag AB, Al-Kursani SS. Design, synthesis, molecular docking and anti-hepatocellular carcinoma evaluation of novel acyclic pyridine thioglycosides. Nucleosides Nucleotides 2018;37:186–98.
[89] Elgemeie GH, Elghandour AH. Activated nitriles in heterocyclic synthesis: novel synthesis of 5-imino-5H-[1]benzopyrano[3,4-c]pyridine-4(3H)-thiones and their oxo analogues. Bull Chem Soc Jpn 1990;63:1230.
[90] Elgemeie GH, Sherif SM, Elaal FA, Elnagdi MH. Nitriles in heterocyclic synthesis: novel synthesis of 4H-thiopyran and of 2-hydroxy-6-pyridine-thione derivatives. Z. Naturforsch 1986;41b:781.

[91] Elgemeie GH, Regaila HA, Shehata N. Unexpected products of the reaction of cycloalkylidene(cyano)-thio-acetamides with arylmethylenemalononitriles: a different novel routes to condensed pyridine-2(1H)-thiones and condensed carbocyclic nitriles. J Chem Soc, Perkin Trans 1990;1:1267.
[92] Elgemeie GH, Zohdi HF, Sherif SM. Activated nitriles in heterocyclic synthesis: a novel synthetic route to furyl- and theinyl-substituted pyridine derivatives. Phosphorus Sulfur Silicon 1990;54:215.
[93] Elgemeie GH, El-Ezbawy SR, Ramiz MM, Mansour OA. Novel synthesis of pyridine-2(1H)-thiones, N-amino-2-pyridones and pyridazine derivatives. Org Prep Proced Int 1991;23:645.
[94] Elgemeie GH, Alnaimi IS, Alarab HF. Synthesis of pyridine-2(1H)-thione and thieno [2,3-b]pyridine derivatives. Heterocycles 1992;34:1721.
[95] Elgemeie GH, Elzanate AM, Mansour AK. Novel synthesis of indans, tetralones, condensed benzocarbocyclic nitriles and condensed pyridine-2(1H)-thiones from the reactions of cycloalkylidenemalononitriles and arylmethylene(cyano)-acetamide and -thioacetamide. J Chem Soc, Perkin Trans 1992;1:1073.
[96] Al-Kaabi SS, Elgemeie GH. Studies on fused 2(1H)-pyridinethiones: new routes for the synthesis of fused 1H-pyrazolo[3,4-b]pyridines and fused thieno-[2,3-b]pyridines. Bull Chem Soc Jpn 1992;65:2241.
[97] Elgemeie GH, El-Zanate AM, Mansour AK. Reaction of (cyano)thioacetamide with arylhydrazones of β-diketones: novel synthesis of 2(1H)-pyridinethiones, thieno[2,3-b]pyridines and pyrazolo[3,4-b]pyridines. Bull Chem Soc Jpn 1993;66:555.
[98] Elgemeie GH, Attia AM. Novel synthesis of condensed pyridinethione carbocyclic nucleosides. Phosphorus Sulfur Silicon 1994;92:95.
[99] Elgemeie GH, Fathy NM. Rearrangment studies on 1-tetralylidenecyanothioacetamide: a different novel synthetic routes to strong fluorescent phenanthridine and phenanthrene analogues. Tetrahedron 1995;51:3345.
[100] Elgemeie GH, Attia AM. Synthesis of some N-hexopyranosyl-2-pyridones and 2-pyridinethione. Carbohydr Res 1995;268:295.
[101] Ibraheim ES, Elgemeie GH, Abbasi MM, Attia AM. Synthesis of N-glycosylated pyridines as new antiviral agents. Nucleosides Nucleotides 1995;14:1415.
[102] Elgemeie GH, Hanafi NM. Reactions of sodium salts of 3-(hydroxymethylene) alkan-2-ones with enamines: synthesis of polysubstituted pyridines. J Chem Res 1999;208–9.
[103] Elgemeie GH, Attia AM, Hussain BA. Synthesis of N-glycosylated pyridines as new antimetabolite agents. Nucleosides Nucleotides 1999;18:2335.
[104] Elgemeie GH, Attia AM, Fathy NM. Novel synthesis of a new class of strongly fluorescent phenanthridine analogues. J Chem Res 1997;112.
[105] Elgemeie GH, Elghandour AH, Ali HA, Hussain AA. Novel synthesis of pyridine-2 (1H)-thiones: reaction of imino esters with cyanothioacetamide. J Chem Res 1997;260.
[106] Elgemeie GH, Attia AM, Hussain BA. A synthetic strategy to a new class of cycloalkane ring-fused pyridine nucleosides as potential anti-HIV agents. Nucleosides Nucleotides 1998;17:855.
[107] Elgemeie GH, Hanfy N, Hopf H, Jones PG. 2-Dicyanomethylene-5,6-dimethyl-1,2-dihydropyridine-3-carbonitrile. Acta Crystallogr 1998;C54:820–2.
[108] Elgemeie GH. Novel synthetic route to pyridine-2(1H)-thiones: unexpected products of the reaction of β-phenathylidenemalononitriles with arylmethylenecyanothioacetamides. Heterocycles 1990;31:123.
[109] Elgemeie GH, Hafez EA, Nawwar GA, Elnagdi MH. Activated nitriles in heterocyclic synthesis: a new synthesis of 3-furan-2-ylidene- and 3-thiophen-2-ylidene-3,6-dihydropyridine derivatives. Heterocycles 1984;22:2829.

[110] Elgemeie GH, Elghandour AH, Elzanate AM, Sayed MM. Reaction of thioglycolic acid with N-cyanoacetoarylsulphonyl-hydrazides: novel synthesis of 2-(N-acetoarylsulfonylhydrazid)-2-thiazolin-4-ones and their corresponding thiazolo[2,3-a]pyridines. Heterocycl Commun 2002;8:573.
[111] Elgemeie GH, El-Ezbawy SR, Ali HA, Mansour AK. Synthesis of several N-substituted amino-2-pyridone derivatives. Org Prep Proced Int 1994;26:465.
[112] Elgemeie GH, Elghandour AH, Elzanate AM, Masoud WA. Novel N-substituted amino-4-methylsulfanyl-2-pyridones and deazapurine analogues from ketene dithioacetals. J Chem Res 1998;164.
[113] Elgemeie GH, Elghandour AH, Ali HA, Abdel-Azzez HM. A novel and efficient method for the synthesis of N-arylsulfonylamino-2-pyridones. J Chem Res 1999;6.
[114] Elgemeie GH, Elghandour AH, Elzanate AM, Masoud WA. Design and synthesis of a new class of N-arylsulfonylaminated pyridones. Phosphorus Sulfur Silicon 2000;163:91.
[115] Elgemeie GH, Elghandour AH, Ali HA, Abdel-Azzez HA. A new general method for substituted 4-alkylthio-N-arylsulfonylamino-2-pyridones: reaction of ketene-SS-acetals with arylsulfonyl-hydrazides. Phosphorus Sulfur Silicon 2001;170:171.
[116] Elgemeie GH, Elzanate AM. Reaction of oxime derivatives of β-diketones and β-ketoesters with substituted hydrazides: novel synthesis of nitroso-N-sulfonyl- and nitroso-N-substituted amino pyridones. Synth Commun 2003;33:2087.
[117] Elgemeie GH, Elghandour AH, Elzanate AM, Abdel-Azzez GA. Novel synthesis of N-aroylaminated pyridones via reaction of ketene dithioacetals with cynoaceto-N-aroylhydrazides. Synth Commun 2003;33:253.
[118] Elgemeie GH, Altalbawy F, Alfaidi M, Azab R, Hassan A. Synthesis, characterization and antimicrobial evaluation of novel 5-benzoyl-N-substituted amino- and 5-benzoyl-N-sulfonylamino-4-alkylsulfanyl-2-pyridones. Drug Des Devel Ther 2017; 11:3389–99.
[119] Elgemeie GH, Fathy NM, Farag AB, Al-Kursani SS. Antimetabolites: design, synthesis, and cytotoxic evaluation of novel dihydropyridine thioglycosides and pyridine thioglycosides. Nucleosides Nucleotides 2017;36:355–77.
[120] Elgemeie GH, Elnaggar DH. Novel dihydropyridine thioglycosides and their corresponding dehydrogenated forms as potent anti-hepatocellular carcinoma agents. Nucleosides Nucleotides 2018;37:199–216.
[121] Elgemeie GH, Attia AM, Elzanate AM, Mansour AK. Convenient synthesis of 2(1H)-pyridinethione glycosides. Bull Chem Soc Jpn 1994;67:1627.
[122] Elgemeie GH, Attia AM, Farag DS, Sherif SM. Nucleotides and nucleosides: direct route to condensed pyridinethione carbocyclic nucleosides related to 3-deazauridine. J Chem Soc, Perkin Trans 1994;1:1285.
[123] Elgemeie GH, Attia AM, Fathy NM. Glycosides of heterocycles: a direct route to 1-(β-D-glycopyranosyl)-pyridinethione nucleosides. Liebigs Ann Chem 1994;955.
[124] Elgemeie GH, Eltamny EH, Elgawad II, Mahmoud NM. Direct route to novel 2-(β-D-xylo- and arabinopyranosylthio)dihydropyridine glycosides and their corresponding dehydrogenated forms. Synth Commun 2009;39:443.
[125] Attia M, Elgemeie GH, Shehada L. Synthesis of some novel condensed pyridine-2 (1H)-thiones and related glycosides. Tetrahedron 1997;53:17441.
[126] Attia M, Elgemeie GH, Alnaimi IS. Synthesis of 1-(β-D-glycopyranosyl)-3-deazapyrimidines from 2 hydroxy and 2-mercaptopyridines. Nucleosides Nucleotides 1998;17:1355.
[127] Elgemeie GH, Attia AM, Fathy NM. Novel synthesis of a new class of polynuclear pyridinethione nucleosides. Nucleosides Nucleotides 1997;16:485.
[128] Elgemeie GH, Fathy NM, Farag AB, Alkursani SA. Antimetabolites: design, synthesis, and cytotoxic evaluation of novel dihydropyridine thioglycosides and pyridine thioglycosides. Nucleosides Nucleotides Nucleic Acids 2017;36:355–77.

[129] Elgemeie GH, Hussein MM, Al-Khursani SA. A total synthesis of a new class of biazinethioglycosides. J Carbohydr Chem 2004;23:465–81.
[130] Elgemeie GH, Ahmed MA. A new class of biheterocyclicthioglycosides from pyridine-2-(1H)-thiones. Nucleosides Nucleotides Nucleic Acids 2002;21:837–47.
[131] Ding Y, Hofstadler SA, Swayze EE, Risen L, Griffey RH. Design and synthesis of paromomycin-related heterocycle-substituted aminoglycoside mimetics based on a mass spectrometry RNA binding assay. Angew Chem Int Ed 2003;42:3409–12.
[132] Elgemeie GH, Zaghary WA, Amin KM, Nasr TM. First synthesis of thiophene thioglycosides. J Carbohydr Chem 2009;28:161.
[133] Elgemeie GH, Elsayed SH, Hassan AS. Design and synthesis of the first thiophene thioglycosides. Synth Commun 2009;39:1781.
[134] Elgemeie GH, Mohamed MA. α-Cyanodithioic acids and their corresponding mono and dithiolate salts as building blocks for the synthesis of novel mercaptothiophenes. Synth Commun 2006;36:1025.
[135] Elgemeie GH, Elzanaty AM, Elghandour AH, Ahmed SA. Novel alkylsulfanylisothiazoles and alkylsulfanylthiophenes using sodium α-cyanoketene dithiolates as starting materials. Synth Commun 2006;36:825.
[136] Elgemeie GH, Abu-Zaied MA. Heterocyclic thioglycosides in carbohydrate research: synthesis of thiophene thioglycosides. Nucleosides Nucleotides Nucleic Acids 2017;36 (8):511–9.
[137] Elgemeie GH, Fathy NM, Farag AB, Bin Yahab AM. Design and synthesis of a new class of indeno[1,2-b]pyridine thioglycosides. Nucleosides Nucleotides Nucleic Acids 2020;39(8):1134–49.
[138] Elgemeie GH, Elghandour AH, Abd Elaziz GW. Potassium 2-cyanoethylene-1-thiolate: a new preparative route to 2-cyanoketene S,N-acetals and pyrazole derivatives. Synth Commun 2004;34:3281.
[139] Elgemeie GH, Kurz T, Widyan K. Novel synthesis of fluorinated cyanoketene N, S-Acetals and their conversions to fluorinated pyrazole derivatives. Phosphorus Sulfur Silicon 2006;181:299.
[140] Hebishy AMS, Elgemeie GH, Elwahy AHM. Novel bis(2-cyanoketene-S,S-/S, N-acetals): versatile precursors for novel bis(aminopyrazole) derivatives. J Heterocyclic Chem 2019;56:1581–7.
[141] Elfahham HA, Elgemeie GH, Ibraheim YR, Elnagdi MH. Studies on 3,5-diaminopyrazoles: new routes for the synthesis of new pyrazoloazines and pyrazoloazoles. Liebigs Ann Chem 1988;819.
[142] Elgemeie GH, Hanfy N. Novel synthesis of 5-amino-1-arylsulfonyl-4-pyrazolin-3-one as a new class of N-sulfonylated pyrazoles. J Chem Res 1999;385.
[143] Elgemeie GH, Ali HA, Elghandour AH, Hussein AM. A convenient synthesis of pyrazolo[3,4-c]pyrazoles using some novel α-cyanoketene dithioacetals. Heterocycl Commun 2002;8:443.
[144] Elgemeie GH, Abu-Zaied MA, Mossa AH. Novel synthesis and biological evaluation of the first pyrazole thioglycosides as pyrazofurin analogues. Nucleosides Nucleotides 2019;38:183–202.
[145] Abu-Zaied MA, Elgemeie GH. Novel synthesis of new pyrazole thioglycosides as pyrazomycin analogues. Nucleosides Nucleotides 2019;38:374–89.
[146] Elgemeie GH, Sayed SH, Jones PG. (E)-3-Amino-4-(2-phenylhydrazinylidene)-1H-pyrazol-5(4H)-one. Acta Crystallogr 2013;E69:0187.
[147] Elgemeie GH, Jones PG. 5-Amino-3-anilino-N-(chlorophenyl)-1H-pyrazole-4-carboxamide ethanol solvent. Acta Crystallogr 2004;E60, 01616.
[148] Elgemeie GH, Zaghary WA, Amin KM, Nasr TM. New trends in synthesis of pyrazole nucleosides as new antimetabolites. Nucleosides Nucleotides 2005; 24:1227–47.

[149] Elgemeie GH, Elsayed SH, Hassan AS. Direct route to a new class of acrylamide thioglycosides and their conversions to pyrazole derivatives. Synth Commun 2008;38:2700.
[150] Elnagdi MH, Elfahham HA, Elmoghayer MR, Elgemeie GH. Reactions with heterocyclic diazonium salts: novel synthesis of pyrazolo[4,3-c]pyridazines and pyrazolo[4,3-c]pyrazoles. J Chem Soc, Perkin Trans 1 1982;989.
[151] Elfahham HA, Sadek KU, Elgemeie GH, Elnagdi MH. Novel synthesis of pyrazolo [5,1-c]-1,2,4-triazoles, imidazo[1,2-b]pyrazoles and [1,2,4]-triazolo[4,3-a]benzimidazoles. Reaction of nitrile imines with amino- and oxo-substituted diazoles. J Chem Soc, Perkin Trans 1 1982;2663.
[152] Elgemeie GH, Elfahham HA, Ghozlan SA, Elnagdi MH. Synthesis of several new pyrazolo[5,1-c]-1,2,4-triazoles, imidazo-[1,2-b]pyrazoles and pyrazolo[3,4-b] pyrazines: reaction of nitrile imines with amino- and oxo-substituted azoles. Bull Chem Soc Jpn 1984;57:1650.
[153] Elgemeie GH, Elfahham HA, Elgamal S, Elnagdi MH. Activated nitriles in heterocyclic synthesis: novel synthesis of pyridazines, pyridines, pyrazoles and polyfunctionally substituted benzene derivatives. Heterocycles 1985;23:1999.
[154] Elgemeie GH, Hanfy N, Hopf H, Jones PG. 5-Amino-1-phenylsulfonyl-4-pyrazolin-3-one. Acta Crystallogr 1998;C54:136.
[155] Elnagdi MH, Elgemeie GH, Elaal FE. Recent developments in the synthesis of pyrazole derivatives. Heterocycles 1985;23:3121.
[156] Elgemeie GH, Sayed SH, Jones PG. True symmetry or pseudosymmetry: 5-amino-1-(4-methylphenylsulfonyl)-4- pyrazolin-3-one and a comparison with its 1-phenylsulfonyl analogue. Acta Crystallogr 2013;E69:90.
[157] Elgemeie GH, Abu-Zaied M, Jones PG. Crystal structure of 4–1,5-dimethyl-2-phenyl-2,3-dihydro-1H-pyrazol-3-one. Acta Crystallogr 2015;E71:104.
[158] Elgemeie GH, Riad BY, Nawwar GA, Elgamal S. Nitriles in heterocyclic synthesis: synthesis of new pyrazolo[1,5-a]pyrimidines, pyrano[2,3-c]pyrazoles and pyrano[3,4-c]pyrazoles. Arch Pharm (Weinheim) 1987;320:223.
[159] Elgemeie GH, Zaghary WA, Amin KM, Nasr TM. A direct route to a new class of acrylamide thioglycosides. J Carbohydr Chem 2008;27:373.
[160] Elgemeie GH, Amin KM, El-Badry OM, Hassan GS, Farag AB, Velazequez C, El-kadi AO. Synthesis and in vitro anti-tumor activity of a new imidazole and thienoimidazole thioglycosides. J Am Sci 2012;8:1071.
[161] Azzam RA, Elgemeie GH, Elsayed RE, Jones PG. Crystal structure of N'-[2-(benzo [d]thiazol-2-yl)acetyl]-4-methylbenzenesulfonohydrazide. Acta Crystallogr 2017; E73:1041–3.
[162] Azzam RA, Elgemeie GH, Elsayed RE, Jones PG. Crystal structure of N-[6-amino-5-(benzo[d]thiazol-2-yl)-3-cyano-4-methylsulfanyl-2-oxo-1,2-dihydropyridin-1-yl]-4-methylbenzenesulfonamide dimethylformamide monosolvate. Acta Crystallogr 2017;E73:1820–2.
[163] Elgemeie GH, Shams HZ, Elkholy YM, Abbaas N. Novel synthesis of N-amino-2-pyridones and cycloalkane ring-fused pyridines containing benzothiazole moiety. Heterocycl Commun 2000;6:363.
[164] Elgemeie GH, Shams HZ, Elkholy YM, Abbaas N. Novel synthesis of pyrido[2,1-b] benzothiazoles and 1,3-benzothiazole derivatives. Phosphurs Sulfur Silicon 2000;165:265.
[165] Elgemeie GH, Azzam RA, Osman RR. Recent advances in synthesis, metal complexes and biological evaluation of 2-aryl, 2-pyridyl and 2-pyrimidylbenzothiazoles as potential chemotherapeutics. Inorg Chim Acta 2020;502.
[166] Fathy NM, Motti FM, Elgemeie GH. Nitriles in heterocyclic synthesis: novel synthesis of pyrido[2,1-b]-benzothiazoles and 1,3-benzothiazole derivatives. Arch Pharm (Weinheim) 1988;321:509.

[167] Elgemeie GH, Salah AM, Abbas NS, Hussein HA, Mohamed RA. Pyrimidine non-nucleoside analogs: a direct synthesis of a novel class of N-substituted amino and N-sulfonamide derivatives of pyrimidines. Nucleosides Nucleotides 2017;36:213–23.

[168] Elgemeie GH, Salah AM, Mohamed RA, Jones PG. Crystal structure of (E)-2-amino-4-methylsulfanyl-6-oxo-1–1,6-dihydropyrimidine-5-carbonitrile. Acta Crystallogr 2015;E71:1319.

[169] Elgemeie GH, Mohamed RA, Hussein HA, Jones PG. Crystal structure of N-(2-amino-5-cyano-4-methylsulfanyl-6-oxo-1,6-dihydropyrimidin-1-yl)-4-bromoben zenesulfonamide dimethylformamide monosolvate. Acta Crystallogr 2015;E71: 1322.

[170] Azzam RA, Elgemeie GH, Osman RR, Jones PG. Crystal structure of potassium [4-amino-5-(benzo[d]thiazol-2-yl)-6-(methylsulfanyl)pyrimidin-2-yl](phenylsulfonyl) azanide dimethylformamide monosolvate hemihydrate. Acta Crystallogr 2019; E75:367–71.

[171] Azzam RA, Osman RR, Elgemeie GH. Efficient synthesis and docking studies of novel benzothiazole-based pyrimidinesulfonamide scaffolds as new antiviral agents and Hsp90α inhibitors. Am Chem Soc (Omega) 2020;5:1640–55.

[172] Mohamed-Ezzat RA, Elgemeie GH, Jones PG. Crystal structures of (E)-2-amino-4-methylsulfanyl-6-oxo-1-(1-phenylethylideneamino)-1,6-dihydropyrimidine-5-carbonitrile and (E)-2-amino-4-methylsulfanyl-6-oxo-1-[1-(pyridin-2-yl) ethylideneamino]-1,6-dihydropyrimidine-5-carbonitrile. Acta Crystallogr 2021; E77:547–50.

[173] Azzam RA, Elgemeie GH. Synthesis and antimicrobial evaluation of novel N-substituted 4-ethylsulfanyl-2-pyridones and triazolopyridines. Med Chem Res 2019;28:62–70.

[174] Elgemeie GH, Azzam RA, Elsayed RE. Sulfa drug analogs: new classes of N-sulfonyl aminated azines and their biological and preclinical importance in medicinal chemistry (2000–2018). Med Chem Res 2019;28:1099–131.

[175] Elgemeie GH, Ali HA, Elghandour AH, Abd-Elaziz GW. Synthesis of novel derivatives of 4-methylthio-N-aryl-2-pyridone and deazapurine analogues: the reaction of ketene dithioacetals with substituted acetanilides. Phosphorus Sulfur Silicon 2000; 164:189.

[176] Azzam RA, Elsayed RE, Elgemeie GH. Design and synthesis of a new class of pyridine-based N-sulfonamides exhibiting antiviral, antimicrobial, and enzyme inhibition characteristics. Am Chem Soc (Omega) 2020;5:26182–94.

[177] Azzam RA, Elboshi HA, Elgemeie GH. Novel synthesis and antiviral evaluation of new benzothiazole-bearing N-sulfonamide 2-pyridone derivatives as USP7 enzyme inhibitors. Am Chem Soc (Omega) 2020;5:30023–36.

[178] Azzam RA, Elsayed RE, Elgemeie GH. Design, synthesis, and antimicrobial evaluation of a new series of N-sulfonamide 2-pyridones as dual inhibitors of DHPS and DHFR enzymes. Am Chem Soc (Omega) 2020;5:10401–14.

[179] Azzam RA, Elgemeie GH, Osman RR. Synthesis of novel pyrido[2,1-b] benzothiazole and N-substituted 2-pyridylbenzothiazole derivatives showing remarkable fluorescence and biological activities. J Mol Struct 2020;1173:707–42.

[180] Elgemeie GH, Jones PG. N-[3-Cyano-2-oxo-5,6,7,8-tetrahydroquinoline-1(2H)-yl]-4- methylbenzenesulfonamide. Acta Crystallogr 2002;E58:1250.

[181] Elgemeie GH, Mahmoud MA, Jones PG. N-(3-Cyano-2-oxo-2,5,6,7,8,9-hexahydro-1H-cyclohepta[b]pyridin-1-yl)-4-methylbenzenesulfonamide. Acta Crystallogr 2002;E58:1293.

Index

Note: Page numbers followed by *f* indicate figures, *t* indicate tables, and *s* indicate schemes.

B

Q

R

S

Printed in the United States
by Baker & Taylor Publisher Services